D0204867

ELECTROMAGNETIC WAVES
and
RADIATING SYSTEMS

PRENTICE-HALL ELECTRICAL ENGINEERING SERIES

WILLIAM L. EVERITT, editor

PRENTICE-HALL INTERNATIONAL, INC., *London*
PRENTICE-HALL OF AUSTRALIA, PTY. LTD., *Sydney*
PRENTICE-HALL OF CANADA, LTD., *Toronto*
PRENTICE-HALL OF INDIA PRIVATE LTD., *New Delhi*
PRENTICE-HALL OF JAPAN, INC., *Tokyo*

ELECTROMAGNETIC WAVES and RADIATING SYSTEMS

SECOND EDITION

EDWARD C. JORDAN

Professor of Electrical Engineering
College of Engineering
University of Illinois

KEITH G. BALMAIN

Associate Professor
Department of Electrical Engineering
University of Toronto

PRENTICE-HALL, INC.

Englewood Cliffs, New Jersey

Current printing (last digit):
15 14

Library of Congress Catalog Card No. 68-16319
Printed in the United States of America

PREFACE

The enthusiastic reception of the first edition, and the sustained demand for it over a period of nearly two decades, have encouraged the preparation of this new edition. Since the first appearance of *Electromagnetic Waves and Radiating Systems*, the field has seen many significant advances and new developments which naturally have influenced the choice of material to be included in the revision. Although these advances are reflected in changes throughout the book, the authors (remembering that the student's time is not unlimited) have resisted the strong temptation to include a detailed account of all the new developments. Fortunately fundamentals do not change, so that except for some necessary updating of symbols and terminology, the early chapters are little affected. The significant changes are noted below:

1. Because the Dirac-delta and Green's-function methods of problem solution are now commonplace in more advanced texts, these approaches have been introduced, but in such a manner that they may be utilized or not, as circumstances dictate. The availability of these alternatives may facilitate using the book for both honors and regular sections.

2. A chapter has been added on the interaction of fields and matter. Increasingly, the electrical engineer is finding that advances in his technology are directly related to development of new materials, and a basic understanding of why materials behave as they do is now essential for the engineer.

3. The first part of the book is designed for undergraduates. By bringing together in chapters 10 and 11 most of the practical information on radiation and antennas, the undergraduate is provided with sufficient working knowledge to solve some real problems of the physical world.

4. New developments in satellite communications, radio astronomy and holography are reflected in additions and changes of treatment in chapters 11, 12 and 13. Unquestionably the major recent advance in the antenna field has been the discovery of frequency-independent and log-periodic structures; this topic is covered in a new chapter 15.

5. The increasingly close interweaving of science and engineering requires of the engineer a deeper understanding of the physics of his problems. The relationship between Maxwell's electromagnetic theory and special relativity theory deserves particular attention. There may come a time when electromagnetic theory is introduced to the electrical engineer using Coulomb's law and special relativity as a starting point. Meanwhile an awareness of the

interrelationships between special relativity and electromagnetic theory leads to an enhanced understanding of both. This is the subject of a new chapter 18.

It is a pleasure to acknowledge the helpful comments of professors and former students who have used the first edition. The authors are indebted to many colleagues for suggestions and advice. Particular thanks are due to W. G. Albright, A. D. Bailey, G. A. Deschamps, J. D. Dyson, P. W. Klock, Y. T. Lo, P. E. Mayes, C. F. Sechrist, and J. T. Verdeyen at the University of Illinois, and S. Dmitrevsky, G. Sinclair and J. L. Yen at the University of Toronto.

Grateful acknowledgment is expressed to Dean W. L. Everitt for helpful counsel and guidance. As editor and friend he contributed many ideas to the first edition, and to this revision.

EDWARD C. JORDAN
KEITH G. BALMAIN

CONTENTS

Chapter 10 RADIATION 311

Chapter 11 ANTENNA FUNDAMENTALS 345

Chapter 12 ANTENNA ARRAYS 422

Chapter 1

FUNDAMENTALS OF ELECTROMAGNETIC ANALYSIS

1.01 Circuits and Fields. The rapid advances that have been made in electrical engineering during the past few decades have been due largely to the ability of the engineer to predict with accuracy the performance of complicated electrical networks. The secret of this ability lies chiefly in the use of a simple but powerful tool called *circuit theory*. The power of the circuit approach depends upon its simplicity, and this simplicity is due to the fact that circuit theory is a simplified approximation of a more exact field theory. In chap. 14 familiar circuit relations are derived directly from the more general field relations, and in the process the assumptions and approximations involved in the use of circuit theory are made apparent.

Despite the power and usefulness of the circuit approach the communications engineer concerned with microwaves or with radio transmission problems quickly becomes aware of its limitations. In the overall design of a radio communication system the engineer can use circuit theory to design the terminal equipment, but between the output terminals of the transmitter and the input terminals of the receiver, circuit theory fails to give him answers, and he must turn to field theory. Electromagnetic field theory deals directly with the field vectors **E** and **H**, whereas circuit theory deals with voltages and currents that are the integrated effects of electric and magnetic fields. Of course voltages and currents are the end results in which the engineer is interested, but the intermediate step, the electromagnetic field, is now a necessary one. It is the purpose of this book to familiarize the student and the engineer with the fundamental relations of the electromagnetic field, and to demonstrate how such relations are used in the solution of engineering problems.

Field theory is more difficult than circuit theory only because of the larger number of variables involved. When current is constant around a circuit, the voltages and currents are functions of one variable—time. In uniform transmission-line theory, the distance along the line is an added variable, but the engineer has learned to treat this distributed-constants circuit by means of an extension of ordinary circuit theory, which he calls transmission-line theory. In the most general electromagnetic field problems there are three space variables involved, and the solutions tend to become correspondingly complex. The additional complexity that results from having to deal with vector quantities in three dimensions can largely be overcome by use of vector analysis. The small amount of effort required to become familiar with vector analysis is soon amply repaid by the simplification that results from its use. For this reason the first topic to be treated will be vector analysis.

1.02 Vector Analysis. The use of vector analysis in the study of electromagnetic field theory results in a real economy of time and thought. Even more important, the vector form helps to give a clearer understanding of the physical laws that mathematics describes. To express these essentially simple physical relations in the longhand scalar form is like trying to sing a song note-by-note, or like sending a code message dot-by-dash, instead of in letter or word groups. The more concise vector form states each relation as a whole, rather than in its component parts. The brief introduction to vector analysis included here is for the benefit of those readers not already familiar with this useful tool. This treatment is adequate for present purposes, but it is expected that the student may later find it desirable to refer to some standard vector analysis text for a more thorough presentation.

Scalar. A quantity that is characterized only by magnitude and algebraic sign is called a scalar. Examples of physical quantities that are scalars are mass, time, temperature, and work. They are represented by italic letters, such as A, B, C, a, b, and c.

Vector. A quantity that has direction as well as magnitude is called a vector. Force, velocity, displacement, and acceleration are examples of vector quantities. They are represented by letters in boldface roman type, such as **A**, **B**, **C**, **a**, **b**, and **c**. A vector can be represented geometrically by an arrow whose direction is appropriately chosen and whose length is proportional to the magnitude of the vector.

Field. If at each point of a region there is a corresponding value of some physical function, the region is called a field. Fields may be classified as either scalar or vector, depending upon the type of function involved.

If the value of the physical function at each point is a scalar quantity, then the field is a *scalar field*. The temperature of the atmosphere, the height of the surface of the earth above sea level, and the density of a nonhomogeneous body are examples of scalar fields.

When the value of the function at each point is a vector quantity, the field is a *vector field*. The wind velocity of the atmosphere, the force of gravity on a mass in space, and the force on a charged body placed in an electric field, are examples of vector fields.

Sum and Difference of Two Vectors. The sum of any two vectors **A** and **B** is illustrated in Fig. 1-1(*a*). It is apparent that it makes no difference whether **B** is added to **A** or **A** is added to **B**. Hence

$$\mathbf{A} + \mathbf{B} = \mathbf{B} + \mathbf{A} \tag{1-1}$$

When the order of the operation may be reversed with no effect on the result, the operation is said to obey the *commutative law*.

Figure 1-1(*b*) illustrates the difference of any two vectors **A** and **B**.

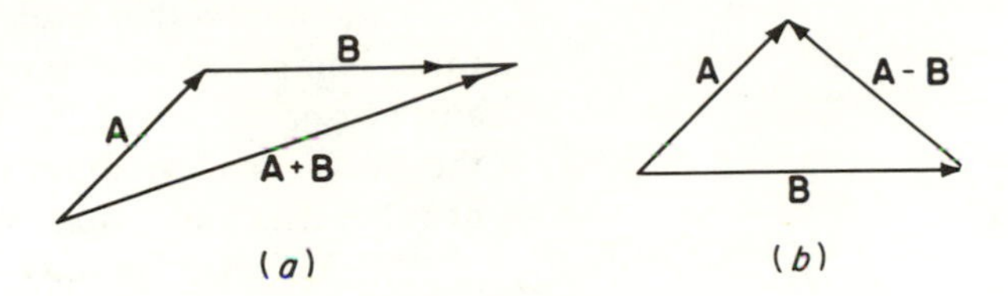

Figure 1-1.

It is to be remembered that the negative of a vector is a vector of the same magnitude, but with a reversed direction.

Multiplication of a Scalar and a Vector. When a vector is multiplied by a scalar, a new vector is produced whose direction is the same as the original vector and whose magnitude is the product of the magnitudes of the vector and scalar. Thus

$$\mathbf{C} = a\mathbf{B} \tag{1-2}$$

Note the absence of any multiplication symbols between a and **B**. The symbols $\cdot$ and $\times$ are reserved for special types of multiplication, which will be discussed later.

The mathematician finds in vector analysis a tool by which relationships can be expressed without reference to a co-ordinate system. The engineer, however, generally needs a reference set of co-ordinates to solve problems. The text will use rectangular or Cartesian co-ordinates, except in those cases where other co-ordinate systems reduce the complexity of the problems. It will be assumed that all vectors and fields are three-dimensional.

A three-dimensional vector is completely described by its projections on the x, y, and z axes. Therefore it can be said that a three-dimensional vector specifies three scalars (the scalar magnitudes of the three mutually orthogonal vector components). Also, a vector field specifies three scalar fields (the scalar magnitudes of the three component vector fields). This idea of component vectors can be represented by

$$\mathbf{A} = A_x\hat{\mathbf{x}} + A_y\hat{\mathbf{y}} + A_z\hat{\mathbf{z}} \tag{1-3}$$

where A_x, A_y, and A_z are the magnitudes of the projections of the vector on the x, y, and z axes respectively, and $\hat{\mathbf{x}}$, $\hat{\mathbf{y}}$, and $\hat{\mathbf{z}}$ are unit vectors in the direction of the axes, (Fig. 1-2).

If any two vectors **A** and **B** are added, there results

$$\mathbf{A} + \mathbf{B} = A_x\hat{\mathbf{x}} + A_y\hat{\mathbf{y}} + A_z\hat{\mathbf{z}} + B_x\hat{\mathbf{x}} + B_y\hat{\mathbf{y}} + B_z\hat{\mathbf{z}} \tag{1-4}$$

which can be grouped as

$$\mathbf{A} + \mathbf{B} = (A_x + B_x)\hat{\mathbf{x}} + (A_y + B_y)\hat{\mathbf{y}} + (A_z + B_z)\hat{\mathbf{z}} \tag{1-5}$$

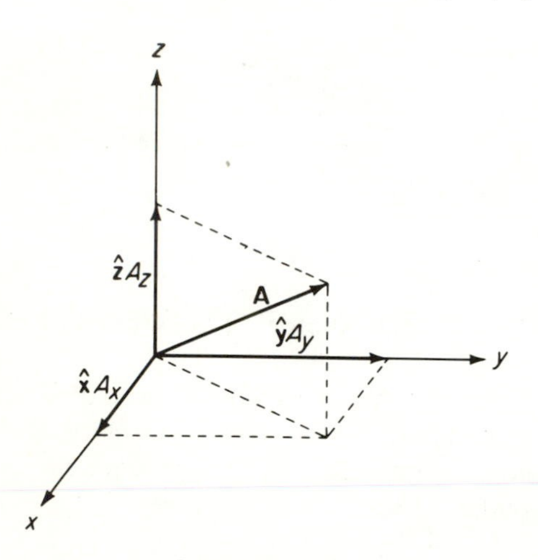

Figure 1-2.

This shows that each of the three components of the resultant vector is found by adding the two corresponding components of the individual vectors.

Furthermore, in any vector *equation*, the sum of the $\hat{\mathbf{x}}$ components on the left-hand side is equal to the sum of the $\hat{\mathbf{x}}$ components on the right-hand side. The same is true also of the $\hat{\mathbf{y}}$ and $\hat{\mathbf{z}}$ components. *Therefore, a vector equation can be written as three separate and distinct equations.* For example, the equation

$$\mathbf{A} + \mathbf{B} = \mathbf{C} + \mathbf{D} + \mathbf{E} \tag{1-6}$$

could be written as the three equations

$$A_x + B_x = C_x + D_x + E_x \tag{1-6a}$$

$$A_y + B_y = C_y + D_y + E_y \tag{1-6b}$$

$$A_z + B_z = C_z + D_z + E_z \tag{1-6c}$$

The ease with which three component equations can be written as one vector equation makes vector analysis particularly useful in field theory.

Scalar Multiplication. It was just shown that a vector could be multiplied by a scalar. It is also possible to multiply a vector by a vector, but first the meaning of such multiplication must be defined and suitable rules formulated. Two types of vector multiplication have been defined, namely "scalar product" and "vector product." The meaning of such multiplications and the necessary rules are briefly discussed in the following. The *scalar product* of two vectors is a *scalar* quantity whose magnitude is equal to the product of the magnitudes of the two vectors and the cosine of the angle between them. This type of multiplication is often called the *dot product* and is indicated by a · (dot) placed between the two vectors to be multiplied. Hence in Fig. 1-3,

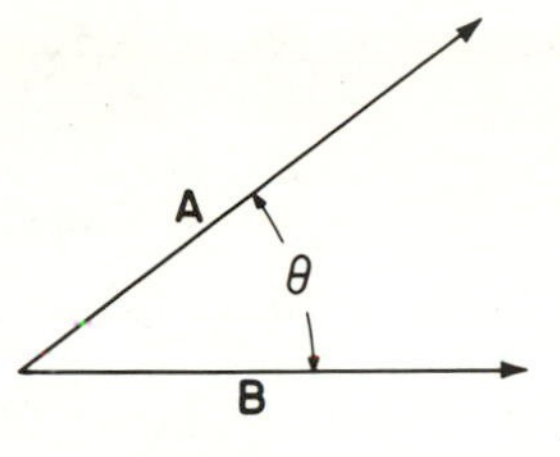

Figure 1-3.

$$\mathbf{A}\cdot\mathbf{B} = AB\cos\theta \tag{1-7}$$

It is seen that the dot product obeys the commutative law, that is,

$$\mathbf{A}\cdot\mathbf{B} = \mathbf{B}\cdot\mathbf{A} \tag{1-8}$$

A physical example of the dot product can be found in the relationship between force and distance. If **F** represents a force that acts through the distance **D** (Fig. 1-4), then the work done would be given by the equation

$$\text{Work} = \mathbf{F}\cdot\mathbf{D} \tag{1-9}$$

Again notice that the dot product, which in this case is work, is a scalar quantity.

The dot product of two vectors can be found by using ordinary algebraic rules.

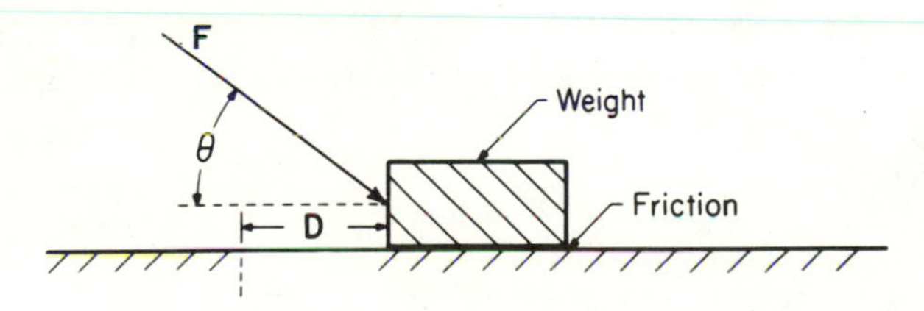

Figure 1-4.

Let

$$\mathbf{A} = A_x\hat{\mathbf{x}} + A_y\hat{\mathbf{y}} + A_z\hat{\mathbf{z}}$$
$$\mathbf{B} = B_x\hat{\mathbf{x}} + B_y\hat{\mathbf{y}} + B_z\hat{\mathbf{z}}$$

Therefore

$$\begin{aligned}\mathbf{A}\cdot\mathbf{B} = {} & A_xB_x(\hat{\mathbf{x}}\cdot\hat{\mathbf{x}}) + A_xB_y(\hat{\mathbf{x}}\cdot\hat{\mathbf{y}}) + A_xB_z(\hat{\mathbf{x}}\cdot\hat{\mathbf{z}}) \\ & + A_yB_x(\hat{\mathbf{y}}\cdot\hat{\mathbf{x}}) + A_yB_y(\hat{\mathbf{y}}\cdot\hat{\mathbf{y}}) + A_yB_z(\hat{\mathbf{y}}\cdot\hat{\mathbf{z}}) \\ & + A_zB_x(\hat{\mathbf{z}}\cdot\hat{\mathbf{x}}) + A_zB_y(\hat{\mathbf{z}}\cdot\hat{\mathbf{y}}) + A_zB_z(\hat{\mathbf{z}}\cdot\hat{\mathbf{z}})\end{aligned} \tag{1-10}$$

But it can be seen from eq. (7) that

$$\hat{\mathbf{x}}\cdot\hat{\mathbf{x}} = \hat{\mathbf{y}}\cdot\hat{\mathbf{y}} = \hat{\mathbf{z}}\cdot\hat{\mathbf{z}} = 1 \tag{1-11a}$$

$$\hat{\mathbf{x}}\cdot\hat{\mathbf{y}} = \hat{\mathbf{y}}\cdot\hat{\mathbf{z}} = \hat{\mathbf{z}}\cdot\hat{\mathbf{x}} = \hat{\mathbf{y}}\cdot\hat{\mathbf{x}} = \hat{\mathbf{z}}\cdot\hat{\mathbf{y}} = \hat{\mathbf{x}}\cdot\hat{\mathbf{z}} = 0 \tag{1-11b}$$

Therefore eq. (10) reduces to

$$\mathbf{A}\cdot\mathbf{B} = A_xB_x + A_yB_y + A_zB_z \tag{1-12}$$

Vector Multiplication. The *vector product* of two vectors is defined as a *vector* whose magnitude is the product of the magnitudes of the two vectors and the sine of the angle between them, and whose direction is perpendicular to the plane containing the two vectors. If a right-handed screw is rotated from the first vector to the second (through the smaller included angle), it moves in the positive direction of the resultant vector. This type of multiplication is often called the *cross product* and is indicated by a × (cross) placed between the two vectors to be multiplied. Hence, in Fig. (1-3)

$$|\mathbf{A} \times \mathbf{B}| = AB \sin \theta \tag{1-13}$$

where the bars | | indicate "magnitude of."

The direction of the vector $\mathbf{A} \times \mathbf{B}$ would be into the paper away from the reader. The vector $\mathbf{B} \times \mathbf{A}$ would have the same magnitude but the opposite direction, that is, toward the reader. Therefore

$$\mathbf{A} \times \mathbf{B} = -\mathbf{B} \times \mathbf{A} \tag{1-14}$$

and the commutative law does *not* apply.

A physical example of vector multiplication can be found in the lifting force of a screw jack. If friction is neglected and a force $\mathbf{f}$ is applied at the end of a lever arm of length $\mathbf{l}$, then the lifting force $\mathbf{F}$ produced by the jack will be

$$\frac{p}{2\pi}\mathbf{F} = \mathbf{f} \times \mathbf{l}$$

where the constant p is the pitch of the screw.

The vector product may also be obtained by straightforward algebraic multiplication and a result similar to that of eq. (10) obtained. Thus

$$\begin{aligned}\mathbf{A} \times \mathbf{B} &= A_xB_x(\hat{\mathbf{x}} \times \hat{\mathbf{x}}) + A_xB_y(\hat{\mathbf{x}} \times \hat{\mathbf{y}}) + A_xB_z(\hat{\mathbf{x}} \times \hat{\mathbf{z}}) \\ &\quad + A_yB_x(\hat{\mathbf{y}} \times \hat{\mathbf{x}}) + A_yB_y(\hat{\mathbf{y}} \times \hat{\mathbf{y}}) + A_yB_z(\hat{\mathbf{y}} \times \hat{\mathbf{z}}) \\ &\quad + A_zB_x(\hat{\mathbf{z}} \times \hat{\mathbf{x}}) + A_zB_y(\hat{\mathbf{z}} \times \hat{\mathbf{y}}) + A_zB_z(\hat{\mathbf{z}} \times \hat{\mathbf{z}})\end{aligned} \tag{1-15}$$

By using eqs. (13) and (14) and a right-handed system of co-ordinates (Fig. 1-5) it is found that

$$\hat{\mathbf{x}} \times \hat{\mathbf{y}} = \hat{\mathbf{z}} = -\hat{\mathbf{y}} \times \hat{\mathbf{x}} \tag{1-16a}$$

$$\hat{\mathbf{y}} \times \hat{\mathbf{z}} = \hat{\mathbf{x}} = -\hat{\mathbf{z}} \times \hat{\mathbf{y}} \tag{1-16b}$$

$$\hat{\mathbf{z}} \times \hat{\mathbf{x}} = \hat{\mathbf{y}} = -\hat{\mathbf{x}} \times \hat{\mathbf{z}} \tag{1-16c}$$

$$\hat{\mathbf{x}} \times \hat{\mathbf{x}} = \hat{\mathbf{y}} \times \hat{\mathbf{y}} = \hat{\mathbf{z}} \times \hat{\mathbf{z}} = 0 \tag{1-16d}$$

Therefore eq. (15) reduces to

$$\begin{aligned}\mathbf{A} \times \mathbf{B} = {} & (A_yB_z - A_zB_y)\hat{\mathbf{x}} \\ & + (A_zB_x - A_xB_z)\hat{\mathbf{y}} \\ & + (A_xB_y - A_yB_x)\hat{\mathbf{z}}\end{aligned} \tag{1-17}$$

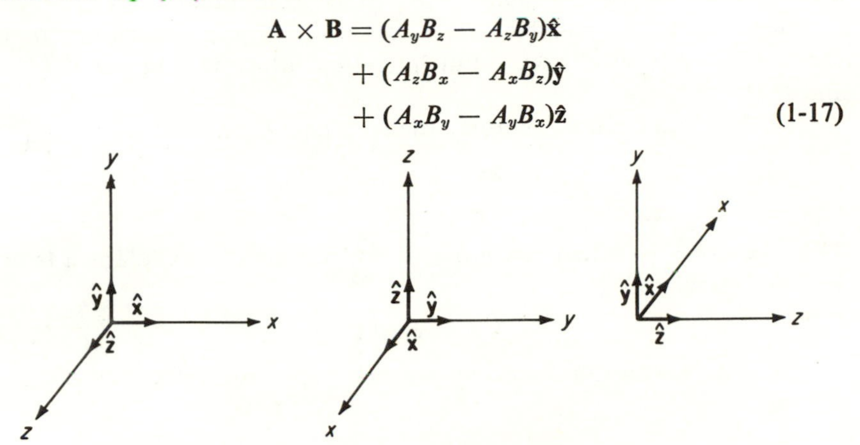

Figure 1-5. Right-handed co-ordinate system.

This result may be remembered easily by noting that the subscripts of the first (positive) part of each term are cyclic with an *x-y-z* rotation when combined with the axis direction of the associated unit vector. For example, in the first part of the first term, the subscript order is *y-z-x* ($\hat{\mathbf{x}}$ is in the *x* direction). The subscript order of the positive part of the second term is *z-x-y*, and for the positive part of the third term it is *x-y-z*. The second or negative part of each term is obtained by reversing the subscripts of the first part of the term. The correct order also may be found from the determinant

$$\mathbf{A} \times \mathbf{B} = \begin{vmatrix} A_x & A_y & A_z \\ B_x & B_y & B_z \\ \hat{\mathbf{x}} & \hat{\mathbf{y}} & \hat{\mathbf{z}} \end{vmatrix} \quad \text{or} \quad \mathbf{A} \times \mathbf{B} = \begin{vmatrix} \hat{\mathbf{x}} & \hat{\mathbf{y}} & \hat{\mathbf{z}} \\ A_x & A_y & A_z \\ B_x & B_y & B_z \end{vmatrix}$$

Differentiation—The ∇ *Operator.* The differential vector operator ∇, called *del* or *nabla*, has many important applications in physical problems. It is defined as*

*The ∇ operator is defined here only in rectangular co-ordinates because in that co-ordinate system the dot and cross products give the correct expressions for the divergence and curl, respectively. In all other co-ordinate systems the divergence and curl cannot be derived in this manner (see sec. 1.04 for cylindrical and spherical co-ordinates) and thus in those systems the notations $\nabla\times$ and $\nabla\cdot$ are symbolic and are not to be taken literally.

$$\nabla = \frac{\partial}{\partial x}\hat{\mathbf{x}} + \frac{\partial}{\partial y}\hat{\mathbf{y}} + \frac{\partial}{\partial z}\hat{\mathbf{z}} \tag{1-18}$$

A differential operator can be treated in much the same way as any ordinary quantity. For example, with the operator $D = \partial/\partial x$, the operation Dy means the quantity $\partial y/\partial x$ is to be obtained.

There are three possible operations with ∇ corresponding to the three possible types of vector multiplication, illustrated in eqs. (2), (12), and (17).

(1) If V is a scalar function, then by eqs. (2) and (18)

$$\nabla V = \frac{\partial V}{\partial x}\hat{\mathbf{x}} + \frac{\partial V}{\partial y}\hat{\mathbf{y}} + \frac{\partial V}{\partial z}\hat{\mathbf{z}} \tag{1-19}$$

This operation is called the *gradient* (for reasons to be explained later), and is abbreviated

$$\nabla V = \text{grad } V \tag{1-20}$$

(2) If $\mathbf{A}$ is a vector function, we can apply eqs. (10), (12) and (18) and get

$$\nabla \cdot \mathbf{A} = \frac{\partial A_x}{\partial x} + \frac{\partial A_y}{\partial y} + \frac{\partial A_z}{\partial z} \tag{1-21}$$

This operation is called the *divergence* and is abbreviated

$$\nabla \cdot \mathbf{A} = \text{div } \mathbf{A} \tag{1-22}$$

(3) If $\mathbf{A}$ is a vector function, we can use eqs. (15), (17), and (18) to show that

$$\nabla \times \mathbf{A} = \left(\frac{\partial A_z}{\partial y} - \frac{\partial A_y}{\partial z}\right)\hat{\mathbf{x}} + \left(\frac{\partial A_x}{\partial z} - \frac{\partial A_z}{\partial x}\right)\hat{\mathbf{y}} + \left(\frac{\partial A_y}{\partial x} - \frac{\partial A_x}{\partial y}\right)\hat{\mathbf{z}} \tag{1-23}$$

$$\nabla \times \mathbf{A} = \begin{vmatrix} \dfrac{\partial}{\partial x} & \dfrac{\partial}{\partial y} & \dfrac{\partial}{\partial z} \\ A_x & A_y & A_z \\ \hat{\mathbf{x}} & \hat{\mathbf{y}} & \hat{\mathbf{z}} \end{vmatrix}$$

This operation is called the *curl* and can be written as

$$\nabla \times \mathbf{A} = \text{curl } \mathbf{A} \tag{1-24}$$

Identities. The identities that follow are useful in deriving field equations. The student can verify them by direct expansions.

$$\text{div curl } \mathbf{A} = \nabla \cdot (\nabla \times \mathbf{A}) = 0 \tag{1-25}$$

$$\text{curl grad } V = \nabla \times (\nabla V) = 0 \tag{1-26}$$

$$\text{div grad } V = \nabla \cdot (\nabla V) = \nabla^2 V$$

where ∇^2 is defined (in Cartesian co-ordinates) as the operation*

$$\nabla^2 = \frac{\partial^2}{\partial x^2} + \frac{\partial^2}{\partial y^2} + \frac{\partial^2}{\partial z^2} \tag{1-27}$$

$$\nabla \times \nabla \times \mathbf{A} = \nabla(\nabla \cdot \mathbf{A}) - \nabla^2 \mathbf{A} \tag{1-28}$$

$$\nabla \cdot \mathbf{A} \times \mathbf{B} = \mathbf{B} \cdot \nabla \times \mathbf{A} - \mathbf{A} \cdot \nabla \times \mathbf{B} \tag{1-29}$$

$$\nabla(ab) = a\nabla b + b\nabla a \tag{1-30}$$

$$\nabla \cdot (a\mathbf{B}) = \mathbf{B} \cdot \nabla a + a\nabla \cdot \mathbf{B} \tag{1-31}$$

$$\nabla \cdot (a\nabla b) = \nabla a \cdot \nabla b + a\nabla^2 b \tag{1-32}$$

$$\nabla \times (a\mathbf{B}) = \nabla a \times \mathbf{B} + a\nabla \times \mathbf{B} \tag{1-33}$$

$$\nabla \times (\mathbf{A} \times \mathbf{B}) = \mathbf{A}\nabla \cdot \mathbf{B} - \mathbf{B}\nabla \cdot \mathbf{A} + (\mathbf{B} \cdot \nabla)\mathbf{A} - (\mathbf{A} \cdot \nabla)\mathbf{B} \tag{1-34}$$

$$\nabla(\mathbf{A} \cdot \mathbf{B}) = (\mathbf{A} \cdot \nabla)\mathbf{B} + (\mathbf{B} \cdot \nabla)\mathbf{A} + \mathbf{A} \times (\nabla \times \mathbf{B}) + \mathbf{B} \times (\nabla \times \mathbf{A}) \tag{1-35}$$

Direction Cosines. The *component* of a vector in a given direction is the projection of the vector on a line in that direction. Thus A_x, the x component of **A**, is equal to $A \cos \alpha$, where α is the angle between **A** and the x axis. Then

$$A_x = \mathbf{A} \cdot \hat{\mathbf{x}}$$

That is, the component of a vector in a given direction is equal to the dot product of the vector and a unit vector in that direction.

If a vector makes angles α, β, γ, with the co-ordinate axes, then

$$l = \cos \alpha, \qquad m = \cos \beta, \qquad n = \cos \gamma$$

are known as the *direction cosines* of the vector.

Problem 1. The scalar product of two vectors may be written in terms of the sum of the products of their direction components.

$$\mathbf{A} \cdot \mathbf{B} = A_x B_x + A_y B_y + A_z B_z$$

Show that the cosine of the angle ψ between the vectors is given by the sum of the products of their direction cosines:

$$\cos \psi = \cos \alpha_A \cos \alpha_B + \cos \beta_A \cos \beta_B + \cos \gamma_A \cos \gamma_B$$
$$= l_A l_B + m_A m_B + n_A n_B$$

*The operator ∇^2 (del squared) is called the *Laplacian*. The Laplacian of a scalar V is given by eq. (27). The Laplacian of a vector **A** is defined as the vector whose Cartesian components are the Laplacians of the Cartesian components of **A**, that is,

$$\nabla^2 \mathbf{A} = \hat{\mathbf{x}}\nabla^2 A_x + \hat{\mathbf{y}}\nabla^2 A_y + \hat{\mathbf{z}}\nabla^2 A_z \qquad \text{(9 terms)}$$

1.03 Physical Interpretation of Gradient, Divergence, and Curl. The three operations which can be performed with the operator del have important physical significance in scalar and vector fields. They will be considered in turn.

Gradient. The gradient of any scalar function is the maximum space rate of change of that function. If the scalar function V represents temperature, then $\nabla V = \text{grad}\, V$ is a temperature gradient, or rate of change of temperature with distance. It is evident that although the temperature V is a scalar quantity—having magnitude but no direction—the temperature gradient ∇V is a vector quantity, its direction being that in which the temperature changes most rapidly. This vector quantity may be expressed in terms of its components in the x, y, and z direction. These are respectively $\partial V/\partial x, \partial V/\partial y$, and $\partial V/\partial z$. The resultant temperature gradient is the vector sum of these three components:

$$\nabla V = \frac{\partial V}{\partial x}\hat{\mathbf{x}} + \frac{\partial V}{\partial y}\hat{\mathbf{y}} + \frac{\partial V}{\partial z}\hat{\mathbf{z}}$$

If the scalar V represents electric potential in volts, ∇V represents potential gradient or electric field strength in volts per meter (MKS).

Divergence. As a mathematical tool, vector analysis finds great usefulness in simplifying the expressions of the relations that exist in three-dimensional fields. A consideration of fluid motion gives a direct interpretation of divergence and curl.

Consider first the flow of an incompressible fluid. (Water is an example of a fluid that is almost incompressible.) In Fig. 1-6 the

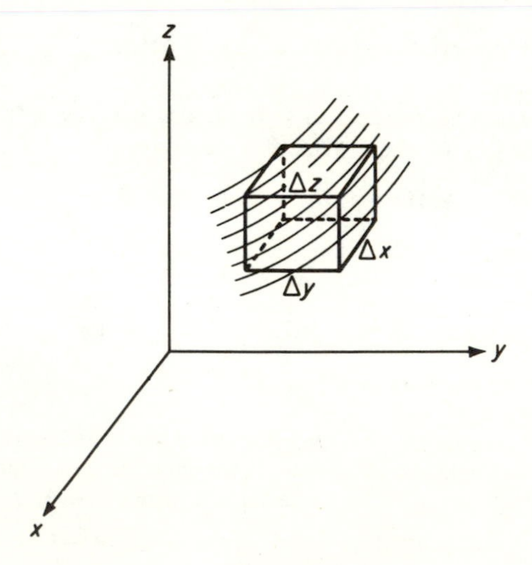

Figure 1-6.

rectangular parallelepiped $\Delta x, \Delta y, \Delta z$, is an infinitesimal volume element within the fluid. If ρ_m is the mass density of the fluid, the flow into the volume through the left-hand face is $\rho_m v_y \, \Delta x \, \Delta z$ where v_y is the average of the y component of fluid velocity through the left-hand face. The corresponding velocity through the right-hand face will be $[v_y + (\partial v_y/\partial y)\Delta y]$ so that the flow through this face is

$$\left[\rho_m v_y + \frac{\partial(\rho_m v_y)}{\partial y} \Delta y\right] \Delta x \, \Delta z$$

The net outward flow in the y direction is therefore

$$\frac{\partial(\rho_m v_y)}{\partial y} \Delta x \, \Delta y \, \Delta z$$

Similarly the net outward flow in the z direction is

$$\frac{\partial(\rho_m v_z)}{\partial z} \Delta x \, \Delta y \, \Delta z$$

and in the x direction it is

$$\frac{\partial(\rho_m v_x)}{\partial x} \Delta x \, \Delta y \, \Delta z$$

The total net outward flow, considering all three directions, is then

$$\left[\frac{\partial(\rho_m v_x)}{\partial x} + \frac{\partial(\rho_m v_y)}{\partial y} + \frac{\partial(\rho_m v_z)}{\partial z}\right] \Delta x \, \Delta y \, \Delta z$$

The net outward flow per unit volume is

$$\frac{\partial(\rho_m v_x)}{\partial x} + \frac{\partial(\rho_m v_y)}{\partial y} + \frac{\partial(\rho_m v_z)}{\partial z} = \operatorname{div}(\rho_m \mathbf{v}) = \nabla \cdot (\rho_m \mathbf{v})$$

This is the *divergence* of the fluid at the point x, y, z. Evidently, for an *incompressible* fluid $\nabla \cdot (\rho_m \mathbf{v})$ always equals zero. An incompressible fluid cannot diverge from, nor converge toward, a point.

The case of a compressible fluid or gas such as steam is different. When the valve on a steam boiler is opened, there is a value for the divergence at each point within the boiler. There is a net outward flow of steam for each elemental volume. In this case the divergence has a positive value. On the other hand, when an evacuated light bulb is broken, there is momentarily a negative value for divergence in the space that was formerly the interior of the bulb.

Curl. The concept of curl or rotation of a vector quantity is clearly illustrated in the stream flow problems. Figure 1-7 shows a stream on the surface of which floats a leaf (in the x-y plane).

If the velocity at the surface is entirely in the y direction and is uniform over the surface, there will be no rotational motion of the leaf but only a translational motion downstream. However, if there are

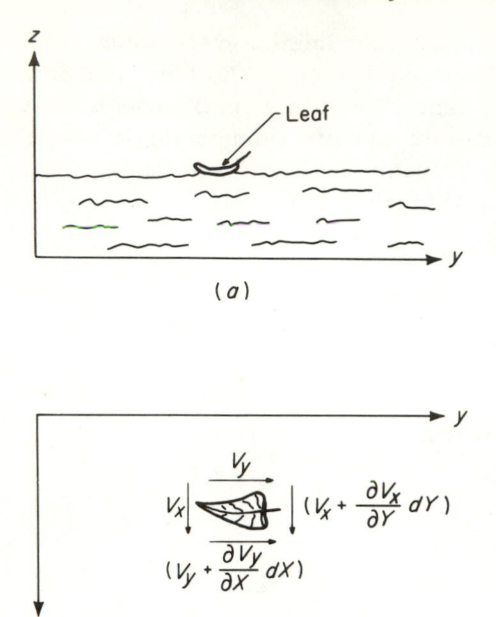

Figure 1-7. Rotation of a floating leaf.

eddies or vortices in the stream flow, there will in general be a rotational as well as translational motion. The rate of rotation or angular velocity at any point is a measure of the *curl* of the velocity of the water at that point. In this case, where the rotation is about the z axis, the curl of $\mathbf{v}$ is in the z direction and is designated by $(\nabla \times \mathbf{v})_z$. A positive value of $(\nabla \times \mathbf{v})_z$ denotes a rotation from x to y, that is, a counterclockwise rotation. From Fig. 1-7(b) it is seen that a positive value for $\partial v_y/\partial x$ will tend to rotate the leaf in a counterclockwise direction, whereas a positive value for $\partial v_x/\partial y$ will tend to produce a clockwise rotation. The rate of rotation about the z axis is therefore proportional to the *difference* between these two quantities. By definition of the curl in rectangular co-ordinates,

$$(\nabla \times \mathbf{v})_z = \frac{\partial v_y}{\partial x} - \frac{\partial v_x}{\partial y}$$

More generally, considering any point within the fluid, there may be rotations about the x and y axes as well. The corresponding components of the curl are given by

$$(\nabla \times \mathbf{v})_x = \frac{\partial v_z}{\partial y} - \frac{\partial v_y}{\partial z}$$

$$(\nabla \times \mathbf{v})_y = \frac{\partial v_x}{\partial z} - \frac{\partial v_z}{\partial x}$$

A rotation about any axis can always be expressed as the sum of the component rotations about the x, y, and z axes. Since the rotations have direction as well as magnitude this will be a vector sum and the resultant rate of rotation or angular velocity will be proportional to

$$\nabla \times \mathbf{v} = \left(\frac{\partial v_z}{\partial y} - \frac{\partial v_y}{\partial z}\right)\hat{\mathbf{x}} + \left(\frac{\partial v_x}{\partial z} - \frac{\partial v_z}{\partial x}\right)\hat{\mathbf{y}} + \left(\frac{\partial v_y}{\partial x} - \frac{\partial v_x}{\partial y}\right)\hat{\mathbf{z}}$$

The direction of the resultant curl is the axis of rotation.

It should be observed that it is not necessary to have circular motion or eddies in order to have a value for curl. In the example of Fig. 1-7, if v_x were everywhere zero but v_y were greater in midstream than near the bank (that is, v_y varies in the x direction), the leaf would tend to rotate and there would be a value for curl given by

$$(\nabla \times \mathbf{v})_z = \frac{\partial v_y}{\partial x}$$

1.04 Vector Relations in Other Co-ordinate Systems. In order to simplify the application of the boundary conditions in particular problems, it is often desirable to express the various vector relations in co-ordinate systems other than the rectangular or Cartesian system. Two other systems are of great importance. They are cylindrical and spherical polar systems. The expressions for gradient, divergence, curl, and so on, in these co-ordinate systems can be obtained directly by setting up a mathematical statement for the particular physical operation to be carried out.

Cylindrical Co-ordinates. The *gradient* of a scalar quantity is the space rate of change of that quantity. In cylindrical co-ordinates the elements of length along the three co-ordinate axes* are $d\rho$, $\rho\, d\phi$, and dz (Fig. 1-8). The respective components of the gradient of a scalar V are therefore

$$(\nabla V)_\rho = \frac{\partial V}{\partial \rho}, \qquad (\nabla V)_\phi = \frac{\partial V}{\rho\, \partial \phi}, \qquad (\nabla V)_z = \frac{\partial V}{\partial z} \tag{1-36}$$

If the unit vectors are designated by $\hat{\boldsymbol{\rho}}$, $\hat{\boldsymbol{\phi}}$, and $\hat{\mathbf{z}}$, the gradient may be written in cylindrical co-ordinates as

$$\nabla V = \text{grad } V = \frac{\partial V}{\partial \rho}\hat{\boldsymbol{\rho}} + \frac{\partial V}{\rho\, \partial \phi}\hat{\boldsymbol{\phi}} + \frac{\partial V}{\partial z}\hat{\mathbf{z}} \tag{1-37}$$

The *divergence* was found to represent the net outward flow per unit volume. The expression for it can be obtained as before by determining the flow through the six surfaces of an elemental volume. Considering an incompressible fluid, the mass density ρ_m will be a constant and so this factor can be dropped from the expressions. Then

*The symbol ρ is used for radial distance in cylindrical co-ordinates ($\rho = \sqrt{x^2 + y^2}$) in order to distinguish it from r, the radial distance in spherical co-ordinates ($r = \sqrt{x^2 + y^2 + z^2}$). This is necessary because these co-ordinate systems are often used together in problems. No confusion with ρ_m, used for mass density, or ρ, used for volume charge density, is anticipated. If it should ever happen that volume charge density and radial distance in cylindrical co-ordinates appear in the same equation the symbol ρ_v can be used for volume charge density. This is consistent with the notation ρ_s for *surface* charge density, which is used later. When no confusion results, volume charge density is represented by the symbol ρ (without subscript).

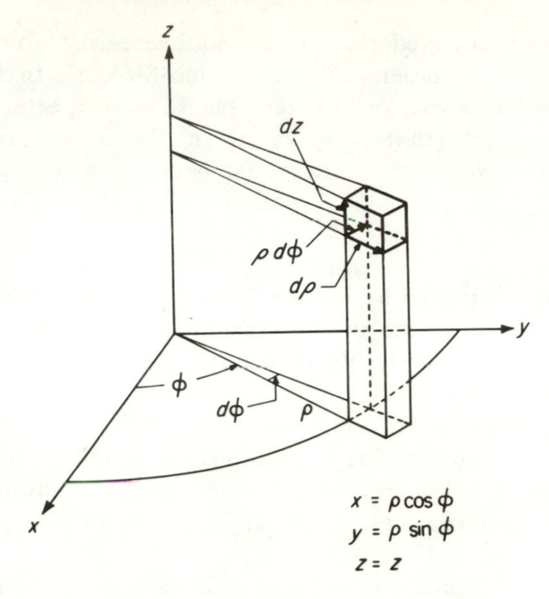

Figure 1-8. A cylindrical co-ordinate system.

in the ρ direction the flow in through the left-hand face is proportional to

$$v_\rho \rho \, d\phi \, dz$$

The flow out of the right-hand face is proportional to

$$\left(v_\rho + \frac{\partial v_\rho}{\partial \rho} d\rho\right)(\rho + d\rho) \, d\phi \, dz$$

The difference between these two quantities (neglecting the second-order differential) is

$$\frac{\partial v_\rho}{\partial \rho} \rho \, d\rho \, d\phi \, dz + \frac{v_\rho}{\rho} \rho \, d\rho \, d\phi \, dz = \frac{1}{\rho} \frac{\partial(\rho v_\rho)}{\partial \rho} d\rho \, \rho \, d\phi \, dz$$

In the ϕ direction the difference is $(\partial v_\phi / \rho \, \partial \phi) \, d\rho \, \rho \, d\phi \, dz$, and in the z direction it is $(\partial v_z / \partial z) \, d\rho \, \rho \, d\phi \, dz$. The net flow out is therefore proportional to

$$\left(\frac{1}{\rho} \frac{\partial(\rho v_\rho)}{\partial \rho} + \frac{\partial v_\phi}{\rho \, \partial \phi} + \frac{\partial v_z}{\partial z}\right) d\rho \, \rho \, d\phi \, dz$$

The net flow out per unit volume is proportional to

$$\nabla \cdot \mathbf{v} = \text{div } \mathbf{v} = \frac{1}{\rho} \frac{\partial(\rho v_\rho)}{\partial \rho} + \frac{\partial v_\phi}{\rho \, \partial \phi} + \frac{\partial v_z}{\partial z}$$

In terms of any vector **A**, the divergence in cylindrical co-ordinates is

$$\nabla \cdot \mathbf{A} = \operatorname{div} \mathbf{A} = \frac{1}{\rho}\frac{\partial(\rho A_\rho)}{\partial \rho} + \frac{\partial A_\phi}{\rho\, \partial \phi} + \frac{\partial A_z}{\partial z} \tag{1-38}$$

Curl. The three cylindrical components of curl are:

$$(\nabla \times \mathbf{A})_\rho = \frac{\partial A_z}{\rho\, \partial \phi} - \frac{\partial A_\phi}{\partial z} \tag{1-39a}$$

$$(\nabla \times \mathbf{A})_\phi = \frac{\partial A_\rho}{\partial z} - \frac{\partial A_z}{\partial \rho} \tag{1-39b}$$

$$(\nabla \times \mathbf{A})_z = \frac{1}{\rho}\left[\frac{\partial}{\partial \rho}(\rho A_\phi) - \frac{\partial A_\rho}{\partial \phi}\right] \tag{1-39c}$$

In chap. 3, the expression for curl in rectangular co-ordinates will be developed in connection with Ampere's law. The expression for curl in cylindrical co-ordinates can be derived in exactly the same manner.

The Laplacian Operator. The operator $\nabla^2 = \nabla \cdot \nabla$ is the divergence of the gradient of the (scalar) quantity upon which ∇^2 operates. Carrying out this operation, it will be found that in cylindrical co-ordinates,

$$\nabla^2 V = \frac{1}{\rho}\frac{\partial}{\partial \rho}\left(\rho \frac{\partial V}{\partial \rho}\right) + \frac{1}{\rho^2}\frac{\partial^2 V}{\partial \phi^2} + \frac{\partial^2 V}{\partial z^2} \tag{1-40}$$

For a *vector*, the symbol $\nabla \cdot \nabla \mathbf{A}$ so far has no meaning except in Cartesian co-ordinates where it has been defined [see footnote following eq. (27)]. The definition for the Laplacian of a vector can be generalized for other orthogonal co-ordinate systems by writing,

$$\nabla^2 \mathbf{A} = \nabla \cdot (\nabla \mathbf{A}) \tag{1-41}$$

where $\nabla \mathbf{A}$ is defined to mean* (in cylindrical co-ordinates)

$$\nabla \mathbf{A} = \hat{\boldsymbol{\rho}} \frac{\partial \mathbf{A}}{\partial \rho} + \hat{\boldsymbol{\phi}} \frac{\partial \mathbf{A}}{\rho\, \partial \phi} + \hat{\mathbf{z}} \frac{\partial \mathbf{A}}{\partial z} \tag{1-42}$$

Spherical Polar Co-ordinates. In the spherical polar co-ordinate system the elements of length along the three co-ordinates are dr, $r\, d\theta$, and $r \sin\theta\, d\phi$ (Fig. 1-9).

Gradient. The three components of the gradient in spherical co-ordinates are

$$(\nabla V)_r = \frac{\partial V}{\partial r} \qquad (\nabla V)_\theta = \frac{\partial V}{r\, \partial \theta} \qquad (\nabla V)_\phi = \frac{1}{r \sin\theta}\frac{\partial V}{\partial \phi} \tag{1-43}$$

Divergence. The expression for divergence in spherical polar co-ordinates is

*The definitions of the symbol $\nabla \mathbf{A}$, given by eqs. (42) and (48), have significance only when $\nabla \mathbf{A}$ is associated with the divergence operation as in eqs. (41) and (47).

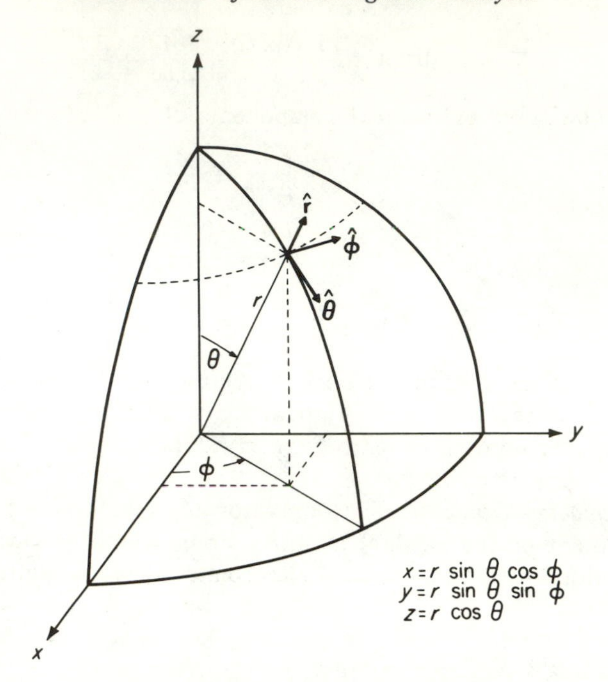

Figure 1-9. A spherical co-ordinate system.

$$\nabla\cdot\mathbf{A} = \frac{1}{r^2}\frac{\partial}{\partial r}(r^2 A_r) + \frac{1}{r\sin\theta}\frac{\partial}{\partial\theta}(\sin\theta A_\theta) + \frac{1}{r\sin\theta}\frac{\partial A_\phi}{\partial\phi} \tag{1-44}$$

Curl. The three spherical polar components for the curl of a vector are

$$(\nabla\times\mathbf{A})_r = \frac{1}{r\sin\theta}\left[\frac{\partial}{\partial\theta}(\sin\theta A_\phi) - \frac{\partial A_\theta}{\partial\phi}\right] \tag{1-45a}$$

$$(\nabla\times\mathbf{A})_\theta = \left[\frac{1}{r\sin\theta}\frac{\partial A_r}{\partial\phi} - \frac{1}{r}\frac{\partial(rA_\phi)}{\partial r}\right] \tag{1-45b}$$

$$(\nabla\times\mathbf{A})_\phi = \frac{1}{r}\left[\frac{\partial}{\partial r}(rA_\theta) - \frac{\partial A_r}{\partial\theta}\right] \tag{1-45c}$$

The Laplacian in Spherical Polar Co-ordinates. For a scalar V,

$$\nabla^2 V = \frac{1}{r^2}\frac{\partial}{\partial r}\left(r^2\frac{\partial V}{\partial r}\right) + \frac{1}{r^2\sin\theta}\frac{\partial}{\partial\theta}\left(\sin\theta\frac{\partial V}{\partial\theta}\right) + \frac{1}{r^2\sin^2\theta}\frac{\partial^2 V}{\partial\phi^2} \tag{1-46}$$

The Laplacian of a vector quantity is defined by

$$\nabla^2\mathbf{A} = \nabla\cdot(\nabla\mathbf{A}) \tag{1-47}$$

where in spherical co-ordinates, $\nabla\mathbf{A}$ is defined as

$$\nabla\mathbf{A} = \hat{\mathbf{r}}\frac{\partial\mathbf{A}}{\partial r} + \hat{\boldsymbol{\theta}}\frac{\partial\mathbf{A}}{r\,\partial\theta} + \hat{\boldsymbol{\phi}}\frac{1}{r\sin\theta}\frac{\partial\mathbf{A}}{\partial\phi} \tag{1-48}$$

Problem 2. In the illustration of the leaf floating on the surface of the water (the x-y plane) show that for a very small circular leaf, $(\nabla \times \mathbf{v})$ is equal to twice the angular velocity of rotation of the leaf, that is, that

$$\left(\frac{\partial v_y}{\partial x} - \frac{\partial v_x}{\partial y}\right) = 2\frac{d\theta}{dt}$$

(Suggestion: Assume that the tangential force on the leaf per unit area at any point is a constant times the relative velocity between leaf and water at that point. The sum of all the torques on the leaf must be zero.)

Problem 3. For a two-dimensional system in which $r = \sqrt{x^2 + y^2}$ determine $\nabla^2 V$ (use rectangular co-ordinates and then check in cylindrical co-ordinates) (a) when $V = 1/r$, (b) when $V = \ln 1/r$.

Problem 4. Repeat problem 3 for a three-dimensional system in which $r = \sqrt{x^2 + y^2 + z^2}$ (use rectangular co-ordinates, and check with spherical co-ordinates).

1.05 Integral Theorems. The mathematical review up to this point has included vector analysis and differential vector calculus. There are in addition a number of integral relations which are exceedingly useful in field theory and which are summarized in this section.

The Divergence Theorem. If S is a *closed* surface surrounding the volume V, the divergence theorem for any vector $\mathbf{A}$ is expressed as

$$\oint_S \mathbf{A} \cdot d\mathbf{a} = \int_V \nabla \cdot \mathbf{A} \, dV \tag{1-49}$$

in which $d\mathbf{a} = \hat{\mathbf{n}}\, da$ with da an element of area on S and $\hat{\mathbf{n}}$ the unit outward normal to S.

Stokes' Theorem. If C is the closed contour around the edge of the open surface S, then Stokes' theorem may be expressed as

$$\oint_C \mathbf{A} \cdot d\mathbf{s} = \int_S \nabla \times \mathbf{A} \cdot d\mathbf{a} \tag{1-50}$$

in which $d\mathbf{s} = \hat{\boldsymbol{\tau}}\, ds$ and $d\mathbf{a} = \hat{\mathbf{n}}\, da$. The relationship between $\hat{\boldsymbol{\tau}}$ and $\hat{\mathbf{n}}$ follows the right-hand rule as illustrated in Fig. 1-10.

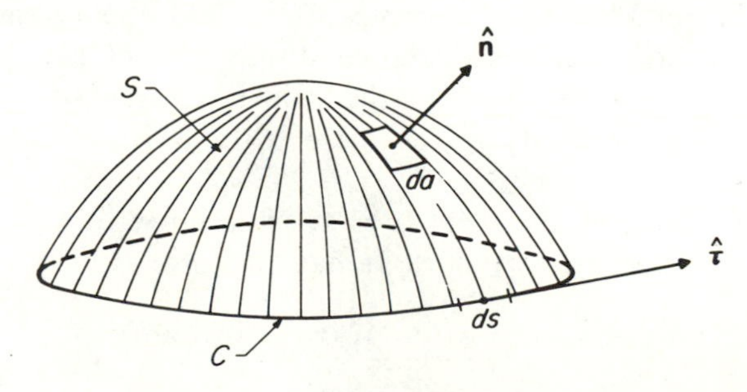

Figure 1-10.

The vector identity (32) leads directly to *Green's first identity* (*Green's theorem*)

$$\oint_S (b\nabla a - a\nabla b)\cdot d\mathbf{a} = \int_V (b\nabla^2 a - a\nabla^2 b)\, dV \tag{1-51}$$

and to *Green's second identity*

$$\oint_S a\nabla b\cdot d\mathbf{a} = \int_V (a\nabla^2 b + \nabla a\cdot\nabla b)\, dV \tag{1-52}$$

Other useful integral relations are

$$\oint_S b\, d\mathbf{a} = \int_V \nabla b\, dV \tag{1-53}$$

$$\oint_S \hat{\mathbf{n}} \times \mathbf{B}\, da = \int_V \nabla \times \mathbf{B}\, dV \tag{1-54}$$

$$\oint_C b\, d\mathbf{s} = \int_S \hat{\mathbf{n}} \times \nabla b\, da \tag{1-55}$$

Differentiation of an integral with variable limits is carried out as follows:

$$\frac{d}{dt}\int_{a(t)}^{b(t)} f(x, t)\, dx = \int_a^b \frac{\partial f}{\partial t} dx + f(b, t)b'(t) - f(a, t)a'(t) \tag{1-56}$$

1.06 The Dirac Delta. In field theory one deals for the most part with well-behaved functions, that is, functions which are continuous and have continuous derivatives. Under these conditions discrete sources such as point charges, filamentary currents, and current shells, constitute singularities of the field, and require special treatment. One common method is to treat these discrete sources as limiting cases of volume distributions for which one or more dimensions are allowed to become vanishingly small. An alternative technique which also proves very convenient is use of the *Dirac delta.* Under the name of "unit impulse" the Dirac delta has been widely used in circuit theory to represent a very short pulse of high amplitude. Because of its compactness it is now finding increasing use in field theory for the representation of discrete sources. Detailed discussions of the properties of the Dirac delta may be found in the books referenced at the end of the chapter. The summary given here will be sufficient for present purposes.

The Dirac delta at the point $x = x_0$ is designated by $\delta(x - x_0)$ and at $x = 0$ it is designated by $\delta(x)$. It has the property

$$\begin{aligned}\int_a^b \delta(x - x_0)\, dx &= 1 \qquad \text{if } x_0 \text{ is in } (a, b)\\ &= 0 \qquad \text{if } x_0 \text{ is not in } (a, b)\end{aligned} \tag{1-57}$$

Thus the delta behaves as if it were a very sharply peaked function of unit area; in fact, the delta may be represented rigorously as a vanishingly thin Gaussian function of unit area as suggested by the sketch in Fig. 1-11(*a*).

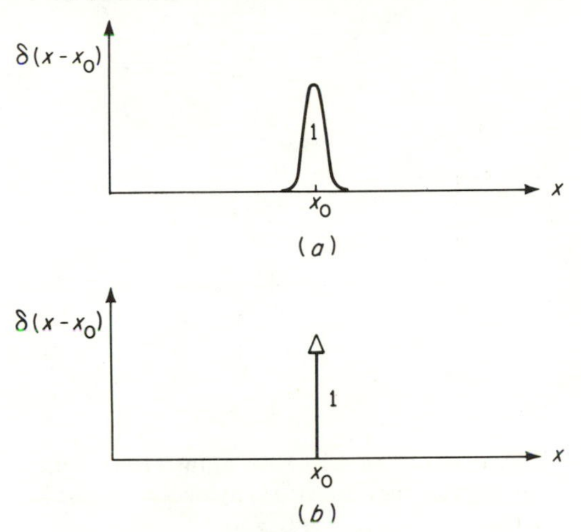

Figure 1-11. The Dirac delta: (*a*) Approximate form; (*b*) Symbolic representation.

Another fundamental property may be stated as follows:

$$\int_a^b f(x)\,\delta(x - x_0)\,dx = f(x_0) \qquad \text{if } x_0 \text{ is in } (a, b)$$
$$= 0 \qquad \text{if } x_0 \text{ is not in } (a, b) \tag{1-58}$$

Under the integral operation, the delta has the property of selecting the value of the function $f(x)$ at the point x_0; thus the delta is a kind of mathematical "sampling" device or gating operation.

The derivative of the Dirac delta (sketched in Fig. 1-12) is useful in representing charge dipoles and electric double layers. Its most important property may be displayed by carrying out the following integration by parts:

$$\int_a^b f(x)\,\delta'(x - x_0)\,dx = [f(x)\,\delta(x - x_0)]_a^b - \int_a^b f'(x)\,\delta(x - x_0)\,dx$$
$$= -f'(x_0) \qquad \text{if } x_0 \text{ is in } (a, b) \tag{1-59}$$
$$= 0 \qquad \text{if } x_0 \text{ is not in } (a, b)$$

For a derivative of order n, one may obtain

$$\int_a^b f(x)\,\delta^{(n)}(x - x_0)\,dx = (-1)^n f^{(n)}(x_0) \qquad \text{if } x_0 \text{ is in } (a, b)$$
$$= 0 \qquad \text{if } x_0 \text{ is not in } (a, b) \tag{1-60}$$

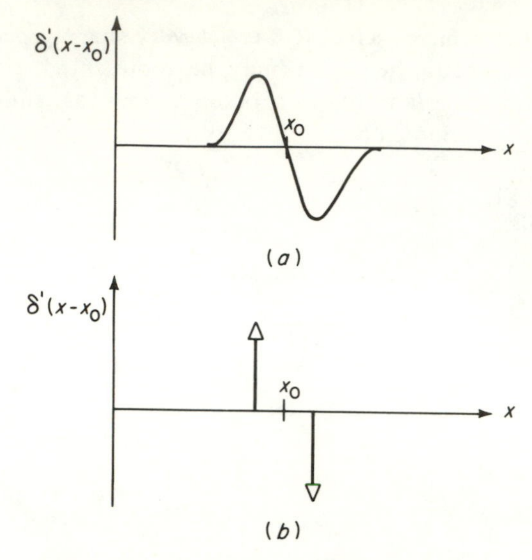

Figure 1-12. The derivative of the Dirac delta: (*a*) Approximate form; (*b*) Symbolic representation.

The representation of a point charge or a current element requires the use of a three-dimensional Dirac delta. If $\mathbf{r}$ denotes the point (x, y, z) then the three-dimensional delta at the point $\mathbf{r} = \mathbf{r}_0$ is designated by $\delta(\mathbf{r} - \mathbf{r}_0)$ and at the origin it is designated by $\delta(\mathbf{r})$. It has the property

$$\begin{aligned} \int_V \delta(\mathbf{r} - \mathbf{r}_0)\, dV &= 1 \qquad \text{if } \mathbf{r}_0 \text{ is in } V \\ &= 0 \qquad \text{if } \mathbf{r}_0 \text{ is not in } V \end{aligned} \tag{1-61}$$

Analogous to (58) there is also the property

$$\begin{aligned} \int_V f(\mathbf{r})\, \delta(\mathbf{r} - \mathbf{r}_0)\, dV &= f(\mathbf{r}_0) \qquad \text{if } \mathbf{r}_0 \text{ is in } V \\ &= 0 \qquad \text{if } \mathbf{r}_0 \text{ is not in } V \end{aligned} \tag{1-62}$$

In rectangular coordinates the three-dimensional delta may be expressed readily in terms of one-dimensional deltas as

$$\delta(\mathbf{r} - \mathbf{r}_0) = \delta(x - x_0)\, \delta(y - y_0)\, \delta(z - z_0). \tag{1-63}$$

This is not quite so easy in other coordinate systems because of the necessity of satisfying eq. (61). In cylindrical co-ordinates,

$$\delta(\mathbf{r} - \mathbf{r}_0) = \frac{\delta(\rho - \rho_0)\, \delta(\phi - \phi_0)\, \delta(z - z_0)}{\rho_0} \tag{1-64}$$

and in spherical co-ordinates,

$$\delta(\mathbf{r} - \mathbf{r}_0) = \frac{\delta(r - r_0)\,\delta(\theta - \theta_0)\,\delta(\phi - \phi_0)}{r_0^2 \sin\theta_0} \tag{1-65}$$

The reader should verify the last two representations by substituting them into (61).

1.07 Matrices. The matrix A is a two-dimensional arrangement of elements set in rows and columns, a typical element being expressed as a_{ij}, meaning the element in the ith row and jth column. For example, a typical square matrix has the form

$$A = \begin{bmatrix} a_{11} & a_{12} & \cdots & a_{1n} \\ a_{21} & a_{22} & \cdots & a_{2n} \\ \vdots & \vdots & & \vdots \\ a_{n1} & a_{n2} & \cdots & a_{nn} \end{bmatrix} \tag{1-66}$$

and typical column matrices have the form

$$X = \begin{bmatrix} x_1 \\ x_2 \\ \vdots \\ x_n \end{bmatrix} \qquad B = \begin{bmatrix} b_1 \\ b_2 \\ \vdots \\ b_n \end{bmatrix} \tag{1-67}$$

Matrices are used frequently in the solution of sets of linear equations such as

$$\begin{aligned} a_{11}x_1 + a_{12}x_2 + \cdots + a_{1n}x_n &= b_1 \\ a_{21}x_1 + a_{22}x_2 + \cdots + a_{2n}x_n &= b_2 \\ \vdots \qquad \vdots \qquad\qquad \vdots \quad &\quad \vdots \\ a_{n1}x_1 + a_{n2}x_2 + \cdots + a_{nn}x_n &= b_n \end{aligned} \tag{1-68}$$

which may be expressed in matrix form as

$$AX = B \tag{1-69}$$

If A and B are known, then the solution of the problem involves finding the *inverse* of the coefficient matrix, designated A^{-1}, such that

$$X = A^{-1}B \tag{1-70}$$

Matrix theory shows that the inverse is given by

$$A^{-1} = \left[\frac{A_{ij}}{\det A}\right]^T \tag{1-71}$$

in which $[A_{ij}]^T$ is called the *adjoint matrix*, det A means the *determinant* of A, and A_{ij} is the *cofactor* of a_{ij}. The cofactor of a_{ij} is $(-1)^{i+j}$ times the determinant of the submatrix formed by deleting the ith row and jth column from A. The T superscript in $[A_{ij}]^T$ indicates taking the transpose of the matrix A, that is, interchanging its rows and columns.

1.08 Units and Dimensions. Although several systems of units are used in electromagnetic theory, most engineers now use some form of the practical meter-kilogram-second (MKS) system. It is to be expected that the marked advantages of this system will prompt its universal adoption.

The existence of the large number of systems of electric and magnetic units requires some explanation. The units used to describe electric and magnetic phenomena can be quite arbitrary, and a complete system of units can be built up from any of a large number of starting points. It is necessary only to define the units of length, mass, time, and one electrical quantity (such as charge, current, permeability or resistance) in order to have the *basic units* from which all other required units can be derived. Unfortunately, in the orginal CGS (centimeter-gram-second) systems the defined units of length and mass were so small that the derived electrical units were unsuitable for practical use. It was found necessary to set up the so-called practical system with units that were related to the corresponding CGS units by some power of 10 (volt, ampere, ohm, and so on). In 1901 Professor Giorgi showed that this practical series could be made part of a complete system, based upon the meter, kilogram, and second, provided that μ_v, the magnetic permeability of a vacuum or free space, is given the value 10^{-7} instead of unity as in the CGS system. The resulting (MKS) system has the advantage that it utilizes units already in use in electrical engineering. In addition, it is a complete and self-consistent system.

The problem of selecting a suitable system of electric and magnetic units has been further complicated by the question of *rationalization.* As was pointed out by Heaviside, the CGS system is unrationalized in that the factor 4π occurs in the wrong places, that is, where logically it is not expected. It would be expected that 4π would occur in problems having spherical symmetry, 2π in problems having circular or cylindrical symmetry and no π in problems involving rectangular shapes. In the ordinary CGS system that is not the case, and Heaviside proposed to *rationalize* the system. However his proposal involved changing the values of the volt, ampere, ohm, and so on, by nonintegral values and so was not considered feasible for practical reasons. It was pointed out later that, if the permeability μ_v of a vacuum or free space were changed from 1 to 4π in the CGS system, rationalization could be effected without changing the magnitude of the practical units. In the rationalized MKS system of units this requires that μ_v have the value of $4\pi \times 10^{-7}$. In any system of units the product $1/\sqrt{\mu_v \epsilon_v}$ must be equal to c, the velocity of light. This requires that in the rationalized MKS system

$$\epsilon_v = 8.854 \times 10^{-12} \approx \frac{1}{36\pi \times 10^9}$$

In the rationalized system the factor 4π occurs explicitly in Coulomb's law and in Ampere's law for the current element, but it does not occur in Maxwell's field equations. It is for this latter reason that the rationalized system is favored in electromagnetic theory.

In 1935 the Giorgi (MKS) system was adopted as the international standard, with the question of rationalization left unsettled. In 1960 the International System of Units (designated SI, for Système International d'Unités) was defined and given official status at a general conference on weights and measures.* The International System uses the MKS rationalized system as a base, but expands it by adding units of temperature and luminous intensity to the defined or elemental units. In this book the expanded rationalized MKS or International System of Units will be used; the common mechanical and electrical quantities as they appear in this system are listed below.†

International System of Units

Elemental Units

Length. The unit of length is the *meter* (m).
Mass. The unit of mass is the *kilogram* (kg).
Time t. The unit of time is the *second* (s).
Current I. The unit of electric current is the *ampere* (A).
Temperature. The unit of temperature is the *degree Kelvin* (°K).
Luminous Intensity. The unit of luminous intensity is the *candela* (cd).

Derived Units

Frequency f. The unit of frequency is the *hertz* (Hz). (1 hertz = 1 cycle per sec)

Force F. The unit of force is the *newton.* It is the force required to accelerate 1 kg at the rate of 1 meter/sec^2 (1 newton = 10^5 dynes).

Energy. The unit of electrical energy is the same as the unit of mechanical energy. It is the *joule.* A joule is the work done when a force of 1 newton is exerted through a distance of 1 meter (1 joule = 10^7 ergs).

Power. The unit of power is the *watt.* It represents a rate of energy expenditure of 1 joule/sec.

Charge Q or q. The unit of charge is the *coulomb.* One ampere of current flowing for 1 sec transports 1 coulomb of charge.

*"Actions of the 11th General Conference on Weights and Measures," *NBS Tech. News Bulletin* 44, 199 (1960).

†A complete listing, including the definitions of the most important SI Units, is given in a National Aeronautics and Space Administration publication: *International System of Units: Physical Constants and Conversion Factors,* by E. A. Mechtly, NASA SP-7012, 1964. See also, "IEEE Recommended Practice for Units in Published Scientific and Technical Work," *IEEE Spectrum,* **3,** No. 3, March, 1966. p. 169.

Resistance R. The unit of resistance is the *ohm.* If 1 watt of power is dissipated in a resistance when 1 amp of current flows through it, the value of the resistance is 1 ohm.

Conductance G. Conductance is the reciprocal of resistance. The reciprocal ohm is known as the *mho* (or *siemens*).

Resistivity. The resistivity of a medium is the resistance measured between two parallel faces of a unit cube. The unit of resistivity is the *ohm-meter.*

Conductivity σ. The conductivity of a medium is the reciprocal of resistivity. The unit of conductivity is the *mho/meter.*

Electromotive Force V. The unit of electromotive force (emf) or voltage is the *volt,* which is defined as 1 watt/amp. It is also equal to 1 joule/coulomb and so has the dimensions of work per unit charge. (It is not a force.)

Electric Field Strength **E**. Electric field strength is measured in *volts/meter.* The electric field strength at any point in a medium is the electric force per unit positive charge at the point. It has the dimension newton/coulomb.

Current Density **J**. The unit of current density is the *ampere/square meter.*

Electric Displacement Ψ. The electric displacement through a closed surface is equal to the charge enclosed by the surface. The unit of electric displacement is the *coulomb.*

Displacement Density **D**. The unit of electric displacement density (usually called just displacement density) is the *coulomb/square meter.*

Magnetic Flux Φ. The voltage V between the terminals of a loop of wire due to a changing magnetic field is related to the magnetic flux through any surface enclosed by the loop by $V = -d\Phi/dt$. The unit of magnetic flux is defined by this relation and is called the *weber.* A weber is 1 *volt-sec.*

Magnetic Flux Density **B**. The unit of magnetic flux density is the *tesla* or *weber/square meter.* (1 weber/sq m = 10^4 gauss)

Magnetic Field Strength **H**. The magnetic field strength between two parallel plane sheets carrying equal and oppositely directed currents is equal to the current per meter width (amperes per meter) flowing in the sheets. The unit of magnetic field strength is the *ampere/meter.*

Magnetomotive Force $\mathscr{F}$. The magnetomotive force between two points a and b is defined as the line integral $\int_a^b \mathbf{H} \cdot d\mathbf{s}$. The unit of magnetomotive force is the *ampere.* The magnetomotive force around a *closed* path is equal to the current enclosed by the path.

Capacitance C. A conducting body has a capacitance of 1 *farad* if it requires a charge of 1 coulomb to raise its potential by 1 volt. A farad is equal to *1 coulomb/volt.*

Inductance L. A circuit has an inductance of 1 *henry* if a changing current of 1 amp/sec induces in the circuit a "back-voltage" of 1 volt. The dimensions of the henry are

$$\frac{\text{volt-seconds}}{\text{ampere}} = \text{ohm-seconds}$$

Permittivity ϵ. In a homogeneous medium the electrical quantities **D** and **E** are related by the equation $\mathbf{D} = \epsilon\mathbf{E}$, where ϵ is the permittivity or dielectric constant of the medium. It has the dimensions *farad/meter*. The permittivity of evacuated (free) space is

$$\epsilon_v = 8.854 \times 10^{-12} \approx \frac{1}{36\pi \times 10^9} \text{ farads/m}$$

The permittivity of a medium may be written as $\epsilon = \epsilon_r \epsilon_v$ where ϵ_r is a dimensionless constant known as the *relative permittivity* of the medium.

Permeability μ. The magnetic flux density and magnetic field strength in a homogeneous medium are related by $\mathbf{B} = \mu\mathbf{H}$ where μ is the magnetic permeability of the medium. It has the dimensions henry/meter. The permeability of a medium may be written as $\mu = \mu_r \mu_v$ where μ_r is the *relative* permeability of the medium and μ_v is the absolute permeability of free space.

Permeability of Free Space μ_v. The (absolute) permeability of free space has the value of $4\pi \times 10^7$ henry/meter.

By definition the *ampere* is the constant current which, if maintained in two straight conductors of infinite length, of negligible circular sections, and placed one meter apart in a vacuum, will produce between these conductors a force equal to 2×10^{-7} newton per meter of length. From the experimental law of force (Ampere's law) for this case, viz.,

$$F = \frac{\mu_v I_1 I_2 L}{2\pi d}$$

it is evident that the above definition for the unit of current is equivalent to defining the permeability of a vacuum, as $\mu_v = 4\pi \times 10^{-7}$. Hence the defined electrical quantity may be considered to be either current or the permeability of free space.

Table 1-1 gives the dimensions of the units of the MKS or International System. In this table the dimensions of all of the units have been expressed in terms of mass M, length L, time T, and charge Q. By expressing the *dimensions* in terms of charge Q, rather than μ_v, fractional exponents in the dimensional equations are avoided.

A table that can be used for converting from the MKS practical system to the CGS systems or vice versa is shown inside the back cover.

When the size of a unit is inconveniently small or large, multiples or submultiples of the units may be employed. The names of the multiples or submultiples are formed with the prefixes shown in the table at the top of page 27.

Table 1-1

Dimensions of Units in the MKS System

Quantity	*Symbol*	*MKS Unit*	*(Abbrev.)*	*Dimensional Equivalent*	*Dimensions*
Length	l	meter	m		L
Mass	m	kilogram	kg		M
Time	t	second	s		T
Charge	q	coulomb	C		Q
Force	$\mathbf{F}$	newton	N	joule per meter	MLT^{-2}
Energy	U	joule	J	volt-coulomb	ML^2T^{-2}
Power	W	watt	W	joule per second	ML^2T^{-3}
Current	I	ampere	A	coulomb per second	$T^{-1}Q$
Current density	$\mathbf{J}$	ampere/square meter	A/m^2		$L^{-2}T^{-1}Q$
Charge density (volume)	ρ(or ρ_v)	coulomb/cubic meter	C/m^3		$L^{-3}Q$
Charge density (surface)	ρ_s	coulomb/square meter	C/m^2		$L^{-2}Q$
Resistance	R	ohm		volt per ampere	$ML^2T^{-1}Q^{-2}$
Conductivity	σ	mho/meter			$M^{-1}L^{-3}TQ^2$
Electromotive force or voltage	V	volt	V	joule per coulomb	$ML^2T^{-2}Q^{-1}$
Electric field strength	$\mathbf{E}$	volt/meter	V/m	newton per coulomb	$MLT^{-2}Q^{-1}$
Capacitance	C	farad	F	coulomb per volt	$M^{-1}L^{-2}T^2Q^2$
Permittivity	ϵ	farad/meter			$M^{-1}L^{-3}T^2Q^2$
Electric displacement	Ψ	coulomb			Q
Electric displacement density	$\mathbf{D}$	coulomb/square meter			$L^{-2}Q$
Magnetic flux	Φ	weber	Wb	volt-second	$ML^2T^{-1}Q^{-1}$
Magnetic flux density	$\mathbf{B}$	tesla	T		$MT^{-1}Q^{-1}$
Magnetomotive force	$\mathscr{F}$ (or M)	ampere (turn)	A		$T^{-1}Q$
Magnetic field strength	$\mathbf{H}$	ampere (turn)/meter	A/m		$L^{-1}T^{-1}Q$
Inductance	L	henry	H	ohm-second	ML^2Q^{-2}
Permeability	μ	henry/meter			MLQ^{-2}
Frequency	f	hertz	Hz		T^{-1}

Factor by Which Unit Is Multiplied	*Prefix*	*Symbol*
10^{12}	tera	T
10^{9}	giga	G
10^{6}	mega	M
10^{3}	kilo	k
10^{2}	hecto	h
10	deka	da
10^{-1}	deci	d
10^{-2}	centi	c
10^{-3}	milli	m
10^{-6}	micro	μ
10^{-9}	nano	n
10^{-12}	pico	p
10^{-15}	femto	f
10^{-18}	atto	a

1.09 Order of Magnitude of the Units. A concept of the order of magnitude of the units of the MKS practical system can be obtained from a few examples. A meter is equal to 3.281 ft, and roughly 3 meters equal 10 ft. A kilogram is slightly more than 2 lb (1 kg = 2.205 lb). A newton is approximately the force required to lift 1/4 lb. (more accurately 0.225 lb). A joule is the work done in lifting this 1/4 lb weight 1 meter. To raise the weight through 1 meter in 1 sec requires the expenditure of 1 watt of power. Whereas the watt is usually thought of as a rather small unit of power (the smallest lamp in general household use requires 15 watts, and it takes 2 or 3 watts to run an electric clock), it represents a considerable amount of mechanical power. A man can do work for a 12-hour day at the rate of about 40 watts, which is less than the power required to run his wife's electric washing machine. The coulomb, which is about the amount of charge passing through a 100-watt lamp in one second would charge a sphere the size of the earth to about 1400 volts. If it were possible to place a coulomb of charge on each of two small spheres placed 1 meter apart, the force between them would be 9×10^9 newtons, or about the force required to lift a million tons. The farad is a large unit of capacitance, and the terms microfarad (10^{-6} F) and picofarad (10^{-12} F) are in common use. The filter capacitors on a radio set are usually 8 or 16 μF (microfarads). The capacitance of a sphere 1 cm in radius is approximately 1 pF (picofarad). The inductance of the primary winding of an iron-core audio transformer may be the order of 50 henrys, whereas the inductance of the radio frequency "tuning-coils" for the broadcast band is about 300 μH (microhenrys).

A tesla or weber per square meter is about one-half the saturation flux density of iron used in transformer cores.

BIBLIOGRAPHY

Barrow, Bruce B., "IEEE Takes a Stand on Units," *IEEE Spectrum,* 3, No. 3, March, 1966, p 164.

Lighthill, M. J., *Fourier Analysis and Generalized Functions,* Cambridge University Press, London, 1960.

Papoulis, A., *The Fourier Integral and its Applications,* McGraw-Hill Book Company, New York, 1962.

Chapter 2

ELECTROSTATICS

2.01 Introduction. The sources of electromagnetic fields are electric charges, and the strength of a field at any point depends upon the magnitude, position, velocity, and acceleration of the charges involved. An *electrostatic* field can be considered as a special case of an electromagnetic field in which the sources are stationary,* so that only the magnitude and position of the charges need be considered. The study of this relatively simple case lays the foundations for solving problems of the more general time-varying electromagnetic field. In what follows it is assumed that the reader has had an elementary course covering the subject of electrostatics and has some general knowledge of the experimental facts and their theoretical interpretation. The purpose of this chapter is to review the subject briefly, not as a study in itself but as an introduction to the electromagnetic field.

2.02 Fundamental Relations of the Electrostatic Field. *Coulomb's Law.* It is found experimentally that between two charged bodies there exists a force that tends to push them apart or pull them together, depending on whether the charges on the bodies are of like or opposite sign. If the two bodies are spheres whose radii are very small compared with their distance apart, and if the spheres are sufficiently remote from conducting surfaces and from other dielectric media (more technically if the spheres are immersed in an infinite homogeneous insulating medium), the magnitude of the force between them due to their charges obeys an inverse square law. That is

$$F = \frac{q_1 q_2}{kr^2} \tag{2-1}$$

where q_1 is the net charge on one sphere, and q_2 the net charge on the other. This is Coulomb's law of force. In the CGS electrostatic

*Individual charges (e.g., electrons) are of course never stationary, having random velocities, which depend among other things upon the temperature. This statement regarding stationary sources simply means that when any elemental *macroscopic* volume is considered, the *net* movement of charge through any face of the volume is zero.

system of units the constant k is arbitrarily put equal to unity for a vacuum and relation (1) is used to define the unit of charge for the electrostatic system of units. However, in the MKS system the unit of charge has already been determined from other considerations, and since units of length and force have also been defined, the constant k can be determined from experiment. In order to *rationalize* the units and so leave Maxwell's field equations free from the factor 4π, it is convenient to show a factor 4π explicitly in the constant k and write

$$k = 4\pi\epsilon$$

The "constant" ϵ depends upon the medium or dielectric in which the charges are immersed. It is called the *dielectric constant* or *permittivity* of the medium. For free space, that is for a vacuum, but also very closely for air, the value of ϵ is

$$\epsilon_v = 8.854 \times 10^{-12} \qquad \text{F/m} \quad (2\text{-}2)$$

To a very good approximation (the same approximation involved in writing the velocity of light as $c \approx 3 \times 10^8$ meter/sec) the value of ϵ_v is given by

$$\epsilon_v \approx \frac{1}{36\pi \times 10^9} \quad (2\text{-}3)$$

The subscript, v, indicates that this is the dielectric constant of a vacuum or free space. For other media the value of ϵ will be different. Then Coulomb's law in MKS units is

$$F = \frac{q_1 q_2}{4\pi\epsilon r^2} \qquad \text{newtons} \quad (2\text{-}4)$$

The direction of the force is along the line joining the two charges.

Electric Field Strength **E**. If a small probe charge Δq is located at any point near a second fixed charge q, the probe charge experiences a force, the magnitude and direction of which will depend upon its location with respect to the charge q. About the charge q there is said to be an electric field of strength **E**, and the magnitude of **E** at any point is measured simply as the *force per unit charge* at that point. The direction of **E** is the direction of the force on a positive probe charge, and is along the outward radial from the (positive) charge q.

From eq. (4) the magnitude of the force on Δq will be

$$\Delta F = \frac{q\,\Delta q}{4\pi\epsilon r^2} \quad (2\text{-}5)$$

and the magnitude of the electric field strength is

$$E = \frac{q}{4\pi\epsilon r^2} \quad (2\text{-}6)$$

The force on the probe charge is dependent upon the strength of the probe charge, but the electric field strength is not. If the charge on the probe is allowed to approach zero, then the force acting on it does also, but the force *per unit charge* remains constant; that is, the electric field due to the charge q is considered to exist, whether or not there is a probe charge to detect its presence.

The direction, as well as the magnitude, of the electric field about a point charge is indicated by writing the vector relation

$$\mathbf{E} = \frac{q}{4\pi\epsilon r^2}\hat{\mathbf{r}} \tag{2-7}$$

where $\hat{\mathbf{r}}$ is a unit vector along the outward radial from the charge q.

Electric Displacement Ψ *and Displacement Density* **D**. It is seen from eq. (7) that at any particular point the electric field strength **E** depends not only upon the magnitude and position of the charge q, but also upon the dielectric constant of the medium (air, oil and others) in which the field is measured. It is desirable to associate with the charge q a second electrical quantity that will be *independent* of the medium involved. This second quantity is called *electric displacement* or *electric flux* and is designated by the symbol Ψ. An understanding of what is meant by electric displacement can be gained by recalling Faraday's experiments with concentric spheres. A sphere with charge Q was placed within, but not touching, a larger hollow sphere. The outer sphere was "earthed" momentarily, and then the inner sphere was removed. The charge remaining on the outer sphere was then measured. This charge was found to be equal (and of opposite sign) to the charge on the inner sphere *for all sizes of the spheres* and *for all types of dielectric media* between the spheres. Thus it could be considered that there was an *electric displacement* from the charge on the inner sphere through the medium to the outer sphere, the amount of this displacement depending only upon the magnitude of the charge Q. In MKS units the displacement Ψ is equal in magnitude to the charge that produces it, that is

$$\Psi = Q \qquad \text{coulombs} \tag{2-8}$$

For the case of an isolated point charge q remote from other bodies the outer sphere is assumed to have infinite radius. The electric displacement per unit area or *electric displacement density* **D** at any point on a spherical surface of radius r centered at the isolated charge q will be

$$D = \frac{\Psi}{4\pi r^2} = \frac{q}{4\pi r^2} \qquad \text{coulomb/sq m} \tag{2-9}$$

The displacement per unit area at any point depends upon the *direction* of the area. Displacement density **D** is therefore a vector quantity, its direction being taken as that direction of the normal to the surface element which makes the displacement through the element of area a maximum. For the case of displacement from an isolated charge this direction is along the radial from the charge and is the same as the direction of **E**. Therefore the vector relation corresponding to (9) is

$$\mathbf{D} = \frac{q}{4\pi r^2}\hat{\mathbf{r}} \tag{2-10}$$

Comparing eqs. (7) and (10) shows that **D** and **E** are related by the vector relation

$$\mathbf{D} = \epsilon\mathbf{E} \tag{2-11}$$

Equation (11) is true in general for all *isotropic* media. For certain crystalline media, the dielectric constant ϵ is different for different directions of the electric field, and for these media **D** and **E** will generally have different directions. Such substances are said to be *anisotropic.** In the first few chapters of this book only *homogeneous isotropic* media will be considered. For these ϵ is constant, that is, independent of position (homogeneous) and independent of the magnitude and direction of the electric field (isotropic).

It is possible to measure the displacement density at a point by the following experimental procedure. Two small thin metallic disks are put in contact and placed together at the point at which **D** is to be determined. They are then separated and removed from the field, and the charge upon them is measured. The charge per unit area is a direct measure of the component of **D** in the direction of the normal to the disks. If the experiment is performed for all possible orientations of the disks at the point in question, the direction (of the normal to the disks) that results in maximum charge on the disks is the direction of **D** at that point, and this maximum value of charge per unit area is the magnitude of **D**.

*In an anisotropic dielectric, the relation between **D** and **E** may be expressed as

$$\begin{aligned} D_x &= \epsilon_{11}E_x + \epsilon_{12}E_y + \epsilon_{13}E_z \\ D_y &= \epsilon_{21}E_x + \epsilon_{22}E_y + \epsilon_{23}E_z \\ D_z &= \epsilon_{31}E_x + \epsilon_{32}E_y + \epsilon_{33}E_z \end{aligned}$$

or in matrix form as

$$\begin{bmatrix} D_x \\ D_y \\ D_z \end{bmatrix} = \begin{bmatrix} \epsilon_{11} & \epsilon_{12} & \epsilon_{13} \\ \epsilon_{21} & \epsilon_{22} & \epsilon_{23} \\ \epsilon_{31} & \epsilon_{32} & \epsilon_{33} \end{bmatrix} \begin{bmatrix} E_x \\ E_y \\ E_z \end{bmatrix}$$

Anisotropic media will be discussed later in connection with radio wave propagation in the ionosphere.

Lines of Force and Lines of Flux. In an electric field *a line of electric force* is a curve drawn so that at every point it has the direction of the electric field. The number of lines per unit area is made proportional to the magnitude of the electric field strength, E. A line of *electric flux* is a curve drawn so that at every point it has the direction of the electric flux density or displacement density. The number of flux lines per unit area is used to indicate the magnitude of the displacement density, D. In homogeneous isotropic media lines of force and lines of flux always have the same direction.

2.03 Gauss's Law. Gauss's law states that *the total displacement or electric flux through any closed surface surrounding charges is equal to the amount of charge enclosed.* This may be regarded as a generalization of a fundamental experimental law (recall Faraday's experiments) or it may be deduced from Coulomb's inverse-square law, and the relation $\mathbf{D} = \epsilon\mathbf{E}$ (now used to define $\mathbf{D}$).

Consider a point charge q located in a homogeneous isotropic medium whose dielectric constant is ϵ. The electric field strength at any point a distance r from the charge q will be

$$\mathbf{E} = \frac{q}{4\pi\epsilon r^2}\hat{\mathbf{r}}$$

and the displacement density or electric flux density at the same point will be

$$\mathbf{D} = \epsilon\mathbf{E} = \frac{q}{4\pi r^2}\hat{\mathbf{r}}$$

Now consider the displacement through some surface enclosing the charge (Fig. 2-1). The displacement or electric flux through the element of surface da is

$$d\Psi = D\, da \cos\theta \tag{2-12}$$

where θ is the angle between $\mathbf{D}$ and the normal to da. From the figure it is seen that $da \cos\theta$ is the projection of da normal to the radius vector. Therefore, by definition of a solid angle,

$$da \cos\theta = r^2\, d\Omega \tag{2-13}$$

where $d\Omega$ is the solid angle subtended at q by the element of area da.

The total displacement through the surface is obtained by integrating eq. (12) over the entire surface.

$$\Psi = \oint D\, da \cos\theta \tag{2-14}$$

(The circle on the integral sign indicates that the surface of integration is a closed surface.) Using eq. (13) the displacement is given by

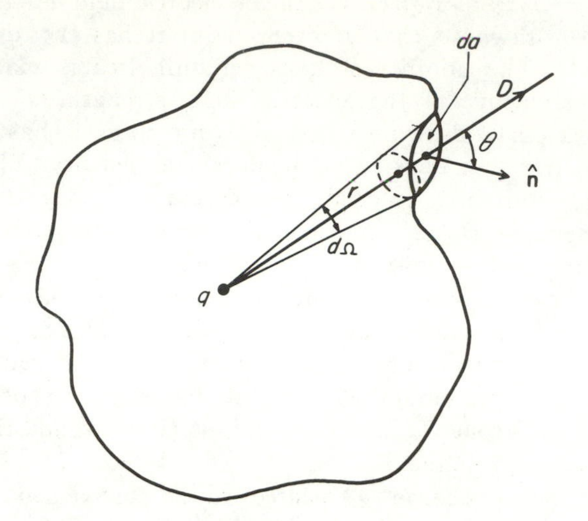

Figure 2-1. Displacement through a surface enclosing a charge.

$$\Psi = \oint D r^2 \, d\Omega$$

and substituting for D from (9)

$$\Psi = \frac{q}{4\pi} \oint d\Omega \tag{2-15}$$

But the total solid angle subtended at q by the *closed surface* is

$$\Omega = \oint d\Omega = 4\pi \text{ solid radians}$$

Therefore from (15) the total displacement through the *closed surface* will be

$$\Psi = q \tag{2-16}$$

If there are a number of charges within the volume enclosed by the surface the total displacement through the surface will be equal to the sum of all the charges. If the charge is continuously* distributed

*Actual charge distributions consist of aggregations of discrete particles or corpuscles. However, since there will always be an enormous number of these microscopic particles in any *macroscopic* element of volume ΔV, is is permissible to speak of the charge density ρ where $\rho = \Delta q/\Delta V$ is the charge per unit volume in elemental volume ΔV. Thus by "charge density at a point" is really meant the charge per unit volume in the elemental volume ΔV containing the point. Although ΔV may be made very small, it is always kept large enough to contain many charges.

throughout the volume with a charge density ρ (coulombs per cubic meter), the total displacement through the surface is

$$\Psi = \int_V \rho \, dV \tag{2-17}$$

where the right-hand side represents the total charge contained within the volume V.

It is often desirable to state the above relations in vector form. By definition of the dot product, the expression $D\,da \cos\theta$ in eq. (12) can be written as $\mathbf{D}\cdot d\mathbf{a}$. In this case the element of area $d\mathbf{a}$ is considered to be a vector quantity having the magnitude da and the direction of the normal to its surface. Then eq. (14) would be written

$$\Psi = \oint \mathbf{D}\cdot d\mathbf{a} \tag{2-18}$$

When da is a part of a *closed* surface as it is here the direction of the *outward* normal usually is taken to be positive. The right-hand side of eq. (18) is the integration over a closed surface of the normal component of the displacement density, that is, it is the total (outward) electric displacement or electric flux through the surface.

Combining eqs. (17) and (18) the vector statement of Gauss's law is

$$\oint_S \mathbf{D}\cdot d\mathbf{a} = \int_V \rho \, dV \tag{2-19}$$

In words, *the net outward displacement through a closed surface is equal to the charge contained in the volume enclosed by the surface.*

2.04 The Potential Function. An electric field is a field of force, and a force field can be described in an alternative manner to that given above. If a body being acted upon by a force is moved from one point to another, work will be done on or by the body. If there is no mechanism by which the energy represented by this work can be dissipated, then the field is said to be *conservative*, and the energy must be stored in either potential or kinetic form. If a charge is moved in a static electric field or a mass is moved in a gravitational field and no friction is present in the region, then no energy is dissipated. Hence these are examples of conservative fields. If some point is taken as a reference or zero point the field of force can be described by the work that must be done in moving the body from the reference point up to any point in the field. A reference point that is commonly used is a point at infinity. For example, if a small body has a charge q and a second body with a small test charge Δq is moved from infinity along a radius line to a point p at a distance R from the charge q, then the work done *on* the system in moving the test charge against the force $\mathbf{F}$ will be

$$\text{work} = -\int_{\infty}^{R} F_r \, dr$$

and since $\Delta F = q \, \Delta q / 4\pi\epsilon r^2$

$$\text{work on test charge} = -\frac{q \, \Delta q}{4\pi\epsilon} \int_{\infty}^{R} \frac{1}{r^2} \, dr$$

$$= \frac{q \, \Delta q}{4\pi\epsilon R}$$

The work done on the test charge *per unit charge* is

$$V = \frac{q}{4\pi\epsilon R} \tag{2-20}$$

V is called the *potential* at the point p due to the charge q. Because it is a scalar quantity, having only magnitude and no direction, it is often called the *scalar potential.*

In a *conservative* field the work done in moving from one point to another is independent of the path. This is easily proven. If it were not independent of the path and a charge were moved from point P_1 to point P_2 over one path, and then from point P_2 back to point P_1 over a second path, the work done *on* the body on one path could be different from the work done *by* the body on the second path. If this were true, a net (positive or negative) amount of work would be done when the body returned to its original position P_1. In a conservative field there is no mechanism for dissipating energy corresponding to positive work done and no source from which energy could be absorbed if the work were negative. Hence, it is apparent that the assumption that the work done is different over two paths is untenable, and so the work must be independent of the path. Thus for every point in the static electric field there corresponds one and only one scalar value of the work done in bringing the charge from infinity up to the point in question by any possible path. This scalar value at any point is called the *potential* of that point. The potential* is measured in volts where 1 volt = 1 joule per coulomb.

If two points are separated by an infinitesimal distance ds, the work done by an external force in moving a unit positive charge from one point to the other will be

$$dW = dV = -\mathbf{E} \cdot d\mathbf{s}$$

*In electrostatics the terms potential or potential difference and voltage are used interchangeably. For time-varying electromagnetic fields, potential, as defined here, has no meaning. However, the voltage between two points a and b, defined by

$$V_a - V_b = V_{ab} = \int_a^b \mathbf{E} \cdot d\mathbf{s}$$

continues to have meaning as long as the path is specified.

Since V is a function of x, y, z, the above relation may be expressed in the following forms:

$$\frac{\partial V}{\partial x}dx + \frac{\partial V}{\partial y}dy + \frac{\partial V}{\partial z}dz = -\mathbf{E}\cdot d\mathbf{s} \tag{2-21}$$

$$\left(\hat{\mathbf{x}}\frac{\partial V}{\partial x} + \hat{\mathbf{y}}\frac{\partial V}{\partial y} + \hat{\mathbf{z}}\frac{\partial V}{\partial z}\right)\cdot(\hat{\mathbf{x}}\,dx + \hat{\mathbf{y}}\,dy + \hat{\mathbf{z}}\,dz) = -\mathbf{E}\cdot d\mathbf{s} \tag{2-22}$$

$$\nabla V\cdot d\mathbf{s} = -\mathbf{E}\cdot d\mathbf{s} \tag{2-23}$$

from which it follows that

$$\mathbf{E} = -\nabla V \tag{2-24}$$

Thus the electric field strength at any point is just the negative of the potential gradient at that point. The direction of the electric field is the direction in which the gradient is greatest or in which the potential changes most rapidly.

When the system of charges is specified and the problem is that of determining the resultant electric field due to the charges, it is often simpler to find first the potential field and then determine $\mathbf{E}$ as the potential gradient according to eq. (24). This is so because the electric field strength is a vector quantity, and when the electric field produced by several charges is found directly by adding the field strengths caused by the individual charges, the addition of fields is a vector addition. This relatively complicated operation is carried out by resolving each vector quantity into (generally) three components, adding these components separately, and then combining the total values of the components to obtain the resultant field. On the other hand the potential field is a scalar field and the total potential at any point is found simply as the algebraic sum of the potentials due to each charge. If the potential is known, the electric field can be found from eq. (24).

In a simple problem there may be little advantage, if any, to using the potential method, but in more complex problems it will be found that the use of the scalar potential results in a real simplification.

Example 1: *Field of an Infinitesimal Electric Dipole.* The concept of the *electric dipole* is extremely useful in electromagnetic field theory. Two equal and opposite charges of magnitude q separated by an infinitesimal distance l are said to constitute an electric dipole or *electric doublet*. The electric field due to such an arrangement can be found readily by first finding the potential V at the point P. In Fig. 2-2

$$V = \frac{1}{4\pi\epsilon}\left\{\frac{q}{r_1} + \frac{-q}{r_2}\right\}.$$

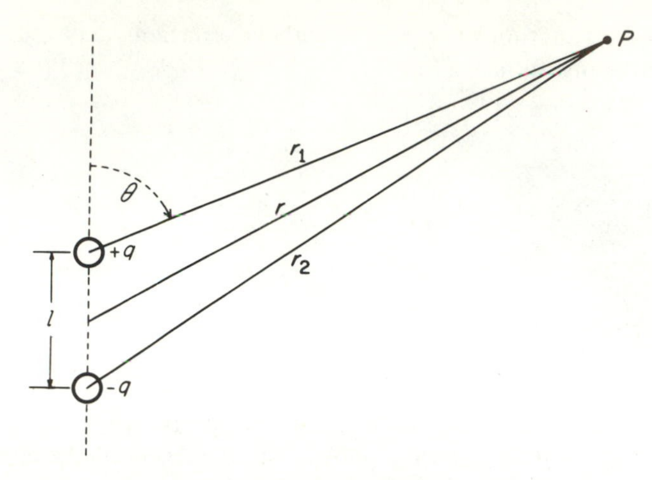

Figure 2-2. An infinitesimal electric dipole.

Because l is infinitesimally small

$$r_1 \approx r - \frac{l}{2}\cos\theta$$

$$r_2 \approx r + \frac{l}{2}\cos\theta$$

$$4\pi\epsilon V = \frac{q}{r - \frac{l}{2}\cos\theta} - \frac{q}{r + \frac{l}{2}\cos\theta} = \frac{ql\cos\theta}{r^2 - \frac{l^2}{4}\cos^2\theta} \approx \frac{ql\cos\theta}{r^2} \qquad (\text{for } l^2 \ll r^2)$$

The electric field is found from $\mathbf{E} = -\nabla V$. The three components in spherical co-ordinates are

$$E_r = -\frac{\partial V}{\partial r} = \frac{2ql\cos\theta}{4\pi\epsilon r^3}$$

$$E_\theta = -\frac{\partial V}{r\partial\theta} = \frac{ql\sin\theta}{4\pi\epsilon r^3}$$

$$E_\phi = -\frac{1}{r\sin\theta}\frac{\partial V}{\partial\phi} = 0$$

2.05 Field Due to a Continuous Distribution of Charge. The potential at a point p due to a number of charges is obtained as a simple algebraic addition or *superposition* of the potentials produced at the point by each of the charges acting alone. If $q_1, q_2, q_3, \ldots, q_n$ are charges located at distances $R_1, R_2, R_3, \ldots, R_n$, respectively, from the point p, the potential at p is given by

$$V = \frac{1}{4\pi\epsilon}\left(\frac{q_1}{R_1} + \frac{q_2}{R_2} + \cdots + \frac{q_n}{R_n}\right) = \frac{1}{4\pi\epsilon}\sum_{i=1}^{i=n}\frac{q_i}{R_i}$$

If the charge is distributed continuously throughout a region, rather than being located at a discrete number of points, the region can be divided into elements of volume ΔV each containing a charge $\rho\Delta V$, where ρ is the *charge density* in the volume element. The potential at a point p will then be given as before by

$$V = \frac{1}{4\pi\epsilon}\sum_{i=1}^{i=n}\frac{\rho_i\,\Delta V_i}{R_i}$$

where R_i is the distance to p from the ith volume element. As the size of volume element chosen is allowed to become very small, the summation becomes an integration, that is

$$V = \frac{1}{4\pi\epsilon}\int_V \frac{\rho\,dV}{R} \tag{2-26}$$

The integration is performed throughout the volume where ρ has value. However it must be noted that (26) is not valid for charge distributions which extend to infinity.

Equation (26) is often written in the form

$$V = \int_V \rho G\,dV$$

in which $G = 1/4\pi\epsilon R$. The function G is the potential of a unit point charge and is often referred to as the *electrostatic Green's function for an unbounded homogeneous region.*

Example 2: *Potential Distribution about Long Parallel Wires.* Determine the potential distribution about a long parallel pair of wires of negligible cross section when the wires have equal and opposite charges distributed along their length.

Assume that a linear charge density ρ_L coulombs per meter is distributed along wire a and $-\rho_L$ coulombs per meter along wire b (Fig. 2-3). Then $\rho\,dV$ becomes $\rho_L\,dz$ so that the expression for potential at the point p will be

$$V = \frac{1}{4\pi\epsilon}\int_{-H}^{+H}\left(\frac{\rho_L}{r_1} - \frac{\rho_L}{r_2}\right)dz = \frac{\rho_L}{2\pi\epsilon}\int_0^H\left(\frac{1}{r_1} - \frac{1}{r_2}\right)dz$$

Substituting $r_1 = \sqrt{r_a^2 + z^2}$ and $r^2 = \sqrt{r_b^2 + z^2}$,

$$V = \frac{\rho_L}{2\pi\epsilon}\int_0^H\left(\frac{1}{\sqrt{r_a^2+z^2}} - \frac{1}{\sqrt{r_b^2+z^2}}\right)dz$$

$$= \frac{\rho_L}{2\pi\epsilon}[\ln(z+\sqrt{r_a^2+z^2}) - \ln(z+\sqrt{r_b^2+z^2})]_0^H$$

$$= \frac{\rho_L}{2\pi\epsilon}\left[\ln\frac{z+\sqrt{r_a^2+z^2}}{z+\sqrt{r_b^2+z^2}}\right]_0^H$$

As z approaches infinity the fraction $(z+\sqrt{r_a^2+z^2})/(z+\sqrt{r_b^2+z^2})$ approaches unity. Therefore, if $H \gg r_a$ and $H \gg r_b$, the expression for potential at the

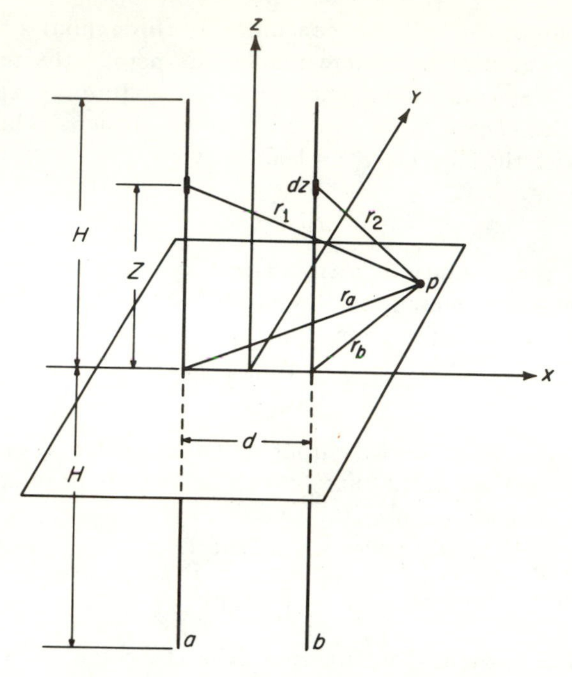

Figure 2-3. A pair of parallel line charges.

point p becomes

$$V = -\frac{\rho_L}{2\pi\epsilon}\ln\frac{r_a}{r_b}$$

$$= \frac{\rho_L}{2\pi\epsilon}\ln\frac{r_b}{r_a} \tag{2-27}$$

It will be observed that in the plane of symmetry between the wires ($r_a = r_b$) the potential is zero.

2.06 Equipotential Surfaces. The solutions to many problems involving electric fields are simplified by making use of equipotential surfaces. An *equipotential surface* is a surface on which the potential is everywhere the same. The movement of charge over such a surface would require no work. Since any two points on the surface have the same potential, there is zero potential difference and therefore zero electric field everywhere along (tangential to) the surface. This means that the electric field must always be perpendicular to an equipotential surface.

A very simple example of equipotential surfaces exists in the case of a point charge. Since $V = q/4\pi r\epsilon$, a surface with a fixed r would have a constant potential. The constant potential surfaces therefore are concentric spherical shells.

In the problem of the parallel line charges the equipotential surfaces can be determined with little difficulty. The locus of a constant potential is obtained by setting the potential of eq. (27) equal to a constant, that is

$$k_1 = \frac{\rho_L}{2\pi\epsilon} \ln \frac{r_b}{r_a}$$

This requires that

$$\left(\frac{r_b}{r_a}\right)^2 = k^2$$

where k^2 is another constant. From Fig. 2-3

$$r_a^2 = \left(x + \frac{d}{2}\right)^2 + y^2$$

$$r_b^2 = \left(x - \frac{d}{2}\right)^2 + y^2$$

Therefore

$$\frac{\left(x - \frac{d}{2}\right)^2 + y^2}{\left(x + \frac{d}{2}\right)^2 + y^2} = k^2$$

$$x^2(1 - k^2) + y^2(1 - k^2) - xd(1 + k^2) + \frac{d^2}{4}(1 - k^2) = 0$$

$$x^2 + y^2 - xd\frac{1 + k^2}{1 - k^2} + \frac{d^2}{4} = 0$$

$$x^2 + y^2 - xd\frac{1 + k^2}{1 - k^2} + \frac{d^2}{4}\frac{(1 + k^2)^2}{(1 - k^2)^2} = \frac{d^2}{4}\left[\frac{(1 + k^2)^2}{(1 - k^2)^2} - 1\right]$$

$$\left[x + \frac{d}{2}\left(\frac{k^2 + 1}{k^2 - 1}\right)\right]^2 + y^2 = \frac{k^2 d^2}{(k^2 - 1)^2}$$

This is the equation of a family of circles with radius $kd/(k^2 - 1)$ and center at $(d/2)[(k^2 + 1)/(1 - k^2)], 0$. Because of invariance in the z direction the equipotential surfaces will be cylinders. The cylinders are *not* concentric because k will depend on the potential selected.

Figure 2-4 shows a plot of the equipotential surfaces about the parallel line charges. It is seen that for *small* values of radius the equipotential cylinders about each line are nearly concentric, with the line charges as the center.

Conductors. A conducting medium is one in which an electric field or difference of potential is always accompanied by a movement of charges. The theory explaining this phenomenon is that a conductor contains *free electrons* or *conduction electrons* that are relatively free to move through the ionic crystal lattice of the conducting medium. It follows that in a conductor there can be no *static* electric field, because any electric field originally present causes the charges to redistribute themselves until the electric field is zero. The electric field being zero

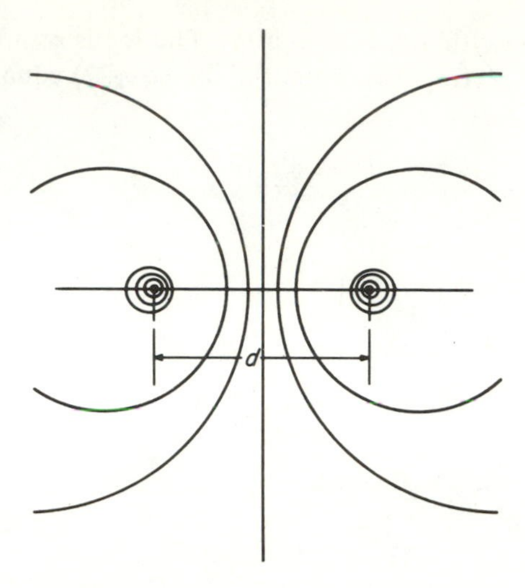

Figure 2-4. Equipotential surfaces about parallel line charges.

within a conductor means that there is no difference of potential between any two points on the conductor. For static electric fields, therefore, a conductor surface is always an *equipotential* surface.

It also follows that within a conductor there can be no *net* charge (excess positive or negative charge). If there were a net charge anywhere within the conductor, then by Gauss's law there would be a displacement away from this charge and therefore a displacement density **D** in the conductor. Since $\mathbf{E} = \mathbf{D}/\epsilon$ this requires (for any finite value of dielectric constant ϵ) that there be an electric field **E** in the conductor. But the possibility of this has already been ruled out for the electrostatic case. Therefore, the (net) charge density ρ must be zero within the conductor. There can, however, be a distribution of charge on the surface of the conductor, and this gives rise to a normal component of electric field in the dielectric medium outside the conductor. The strength of this normal component of electric field strength in terms of the surface charge is obtained directly from Gauss's law.

Electric Field Due to Surface Charge. Let the charge per unit area or *surface charge density* on the surface of a conductor be ρ_s coulombs per square meter. Enclose an element of the surface in a volume of "pillbox" shape with its flat surfaces parallel to the conductor surface as in Fig. 2-5. Then, if the depth d of the pillbox is made extremely small compared with its diameter, the electic displacement through its edge surface will be negligible compared with any displacement through its flat surfaces. There can be no displacement through the left-hand surface submerged in the conductor (because no **E** exists in the conductor) so all the electric flux must emerge through the right-hand surface. Applying Gauss's law to this case gives

$$D_n da = \rho_s da$$

where da is the area of one face of the pillbox and D_n is the displacement density normal to the surface. Therefore,

$$D_n = \rho_s \qquad \text{and} \qquad E_n = \frac{\rho_s}{\epsilon}$$

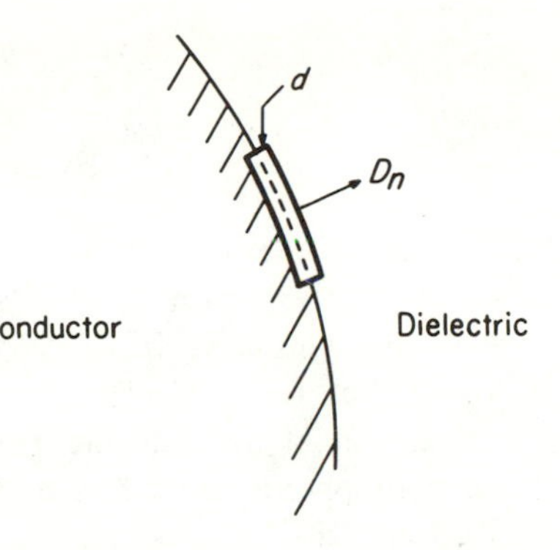

Figure 2-5. Boundary surface between a conductor and a dielectric.

The electric displacement density at the surface of a conductor is normal to the surface and equal in magnitude to the surface charge density. The electric field strength is also normal to the surface and is equal to the surface charge density divided by the dielectric constant.

2.07 Divergence Theorem. The divergence theorem (also called Gauss's theorem) relates an integration throughout a volume to an integration over the surface surrounding the volume.

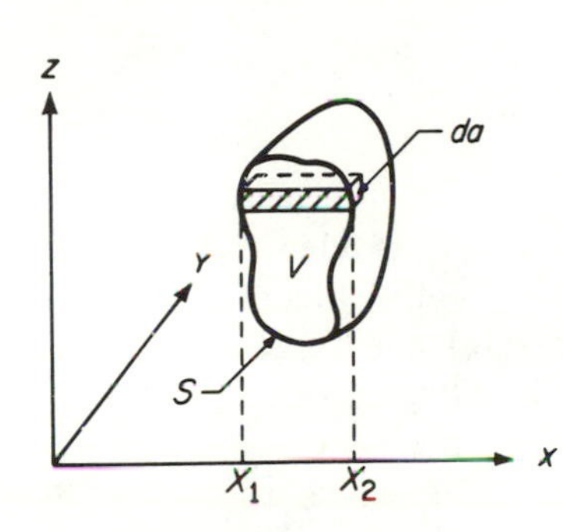

Figure 2-6. Section of a volume V.

Figure 2-6 shows a closed surface S enclosing a volume V that contains charges (or a charge density) that produce an electric flux density $\mathbf{D}$.

By the definition of divergence,

$$\nabla \cdot \mathbf{D} = \frac{\partial D_x}{\partial x} + \frac{\partial D_y}{\partial y} + \frac{\partial D_z}{\partial z}$$

so that

$$\int_V \nabla \cdot \mathbf{D}\, dV = \iiint \left(\frac{\partial D_x}{\partial x} + \frac{\partial D_y}{\partial y} + \frac{\partial D_z}{\partial z}\right) dx\, dy\, dz \tag{2-28}$$

where

$$dV = dx\, dy\, dz$$

Consider now the elemental rectangular volume shown shaded, which has dimensions dy and dz in the y and z directions respectively. Let D_{x_1} and D_{x_2} respectively be the x component of the electric flux entering the left-hand side and leaving the right-hand side of the rectangular volume. The total flux emerging is the algebraic difference of these two. But

$$D_{x_2} - D_{x_1} = \int_{x_1}^{x_2} \frac{\partial D_x}{\partial x}\, dx$$

$$\therefore \quad \iiint \frac{\partial D_x}{\partial x}\, dx\, dy\, dz = \iint (D_{x_2} - D_{x_1})\, dy\, dz \tag{2-29}$$

Now $dy\, dz$ is the x component of the surface element $d\mathbf{a}$, and so (29) is just the integration of the product of D_x times the x component of $d\mathbf{a}$ over the whole surface. (Note that for the right face $D_x\, da_x = D_{x_2}\, dy\, dz$, but for the left face $D_x\, da_x = -D_{x_1}\, dy\, dz$. This is because the direction of $d\mathbf{a}$ is along the *outward* normal and for the left face the x component of $d\mathbf{a}$ has a direction opposite to that of D_{x_1}.)

By definition of a scalar product

$$\mathbf{D} \cdot d\mathbf{a} = D_x\, da_x + D_y\, da_y + D_z\, da_z$$

where da_x indicates the x component of $d\mathbf{a}$, and so on. Then, making use of (29), eq. (28) may be written

$$\int_V \nabla \cdot \mathbf{D}\, dV = \oint_S \mathbf{D} \cdot d\mathbf{a} \tag{2-30}$$

This is the *divergence theorem.*

Although derived here for the particular case of electric displacement density $\mathbf{D}$ it is a quite general and very useful theorem of vector analysis. For *any* vector, it relates the integral over a closed surface of the normal component of the vector to the integral over the volume (enclosed by the surface) of the divergence of the vector.

Integral Definition of Divergence. The divergence theorem (30) provides a definition of divergence of a vector in the integral form which is easy to put into words. The expression on the right-hand side of (30) is the net outward electric flux through the closed surface S. The expression on the left represents the average divergence of $\mathbf{D}$ multiplied by the volume V that is enclosed by S. Thus the *average divergence* of a vector is the net outward flux of the vector through a closed surface S divided by the volume V enclosed. The limit of the average divergence as S is allowed to shrink to zero about a point is the divergence of the vector at that point; that is,

$$\nabla \cdot \mathbf{D} = \lim_{S \to 0} \frac{\oint_S \mathbf{D} \cdot d\mathbf{a}}{V}$$

In words, the divergence of the vector $\mathbf{D}$ is the net outward flux of $\mathbf{D}$ per unit volume.

Alternative Statement of Gauss's Law. Making use of Gauss's law which states

$$\oint_S \mathbf{D} \cdot d\mathbf{a} = \int_V \rho\, dV \tag{2-31}$$

and applying the divergence theorem (30), gives

$$\int_V \nabla \cdot \mathbf{D}\, dV = \int_V \rho\, dV$$

This holds for any volume whatsoever. As the volume considered is reduced to an elemental volume, this becomes the point relation,

$$\nabla \cdot \mathbf{D} = \rho \tag{2-32}$$

This is the alternative statement of Gauss's law. It states that at every point in a medium the *divergence* of electric displacement density is equal to the charge density. Recalling the physical interpretation of the term divergence, eq. (32) might be stated as follows: The net outward flux of electric displacement per unit volume is equal to the charge per unit volume. Equation (32) will often be found to be a more useful form for mathematical manipulation than the corresponding integral statement (31).

2.08 Poisson's Equation and Laplace's Equation. Equation (32) is a relation between the electric displacement density and the charge density in a medium. If the medium is homogeneous and isotropic so that ϵ is constant and a scalar quantity, eq. (32) can be written as

$$\nabla \cdot \epsilon\mathbf{E} = \epsilon\nabla \cdot \mathbf{E} = \rho$$

or

$$\nabla \cdot \mathbf{E} = \frac{\rho}{\epsilon} \tag{2-33}$$

Recall that **E** is related to the potential V by

$$\mathbf{E} = -\nabla V$$

Substituting this into (33)

$$\nabla \cdot \nabla V = -\frac{\rho}{\epsilon} \tag{2-34a}$$

or

$$\nabla^2 V = -\frac{\rho}{\epsilon} \tag{2-34b}$$

Equation (34) is known as *Poisson's equation.* In free space, that is, in a region in which there are no charges ($\rho = 0$), it becomes

$$\nabla^2 V = 0 \tag{2-35}$$

This special case for source-free regions is *Laplace's equation.*

Laplace's Equation. Laplace's equation is a relation of prime importance in electromagnetic field theory. Expanded in rectangular coordinates it becomes

$$\nabla^2 V = \frac{\partial^2 V}{\partial x^2} + \frac{\partial^2 V}{\partial y^2} + \frac{\partial^2 V}{\partial z^2} = 0 \tag{2-36}$$

This is a second-order partial differential equation relating the rate of change of potential in the three component directions. In any charge-free region the potential distribution must be such that this relation is satisfied. An alternative form of (36) in terms of electric field strength is

$$\nabla \cdot \mathbf{E} = 0 \tag{2-37}$$

In this form the statement is that in a homogeneous charge-free region the number of lines of electric field strength emerging from a unit volume is zero, or (in such a region) lines of electric field strength are continuous.

The Problem of Electrostatics. In a homogeneous charge-free region the potential distribution, whatever it may be, must be a solution of the Laplace equation. The problem is to find a potential distribution that will satisfy (35) as well as the boundary conditions of the particular problem. When the charges are given the potential can be found directly from

$$V = \frac{1}{4\pi\epsilon}\int_V \frac{\rho}{R}\,dV$$

This is a simple problem and the solution is straightforward. On the other hand, if the potential distribution is given for a certain configuration of conductors, the charge distribution on the conductors can be found from

$$\rho_s = D_n = \epsilon E_n$$

In the general problem as it exists, however, neither the potential distribution nor the charge distribution is known. These are the quantities to be found. A certain configuration of conductors is specified and the voltages or potential differences between conductors are given (or the total charge on each conductor may be given). The charges on the conductors will then distribute themselves to make the conductors equipotential surfaces and at the same time produce a potential distribution between conductors which will satisfy Laplace's equation.

Thus the problem is that of finding a solution to a second-order differential equation (Laplace's equation) that will fit the boundary conditions. The problem is one of integration and therefore straightforward methods of solution are not generally available. In fact, only in a relatively small number of cases, where symmetry or some other consideration makes it possible to specify the charge distribution, can an exact solution in closed form be found. Of course, an approximate solution can always be obtained (facilitated by digital computers), and the degree of approximation can usually be improved to any desired extent by a systematic method of successive approximations.

A similar situation exists in the more general electromagnetic field problem where the fields and charge distributions are varying in time. Although it is this more general problem that is of primary concern in electromagnetic wave theory, it is helpful to consider some of the special methods and solutions that exist for the electrostatic case. It will be found that some of these special methods can be extended to the general case. Moreover, a knowledge of the actual electrostatic solutions for certain simple configurations is required for later use.

Solutions for Some Simple Cases. It is instructive first to obtain the solutions for the simplest possible cases in which, because of symmetry, the field is constant along two axes of the co-ordinate system and variations occur in one direction only.

EXAMPLE 3: *In Rectangular Co-ordinates—Two Parallel Planes.* Two parallel planes of infinite extent in the x and y directions and separated by a distance d in the z direction have a potential difference applied between them (Fig. 2-7). It is required to find the potential distribution and electric field strength in the region between the planes.

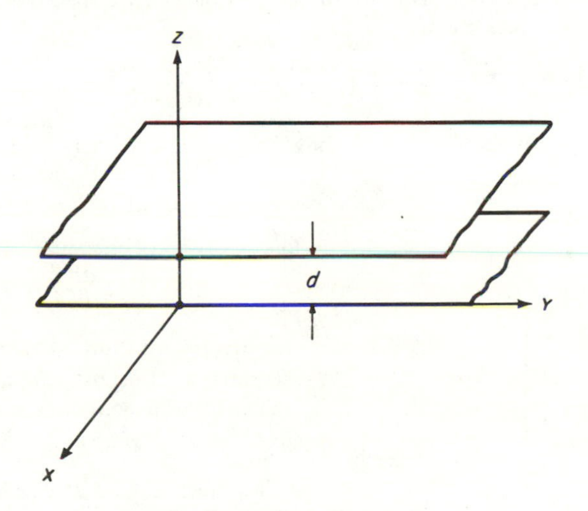

Figure 2-7. Two parallel planes.

In rectangular co-ordinates Laplace's equation is

$$\nabla^2 V = \frac{\partial^2 V}{\partial x^2} + \frac{\partial^2 V}{\partial y^2} + \frac{\partial^2 V}{\partial z^2} = 0$$

From symmetry it is evident that there is no variation of V with x or y, but only with z. For this simple case Laplace's equation reduces to

$$\nabla^2 V = \frac{\partial^2 V}{\partial z^2} = 0$$

which has a solution

$$V = k_1 z + k_2$$

where k_1 and k_2 are arbitrary constants. Substituting the boundary conditions

$$V = V_0 \quad \text{at } z = 0, \qquad V = V_1 \quad \text{at } z = d$$

gives

$$k_2 = V_0 \qquad \text{and} \qquad k_1 = \frac{V_1 - V_0}{d}$$

so that

$$V = \frac{V_1 - V_0}{d} z + V_0$$

The electric field strength is obtained from the relation

$$\mathbf{E} = -\nabla V = -\frac{\partial V}{\partial z}\hat{\mathbf{z}} = -\frac{V_1 - V_0}{d}\hat{\mathbf{z}}$$

The electric field strength is constant in the region between the plates. It is directed along the z axis and toward the plate of lower potential.

EXAMPLE 4: *In Cylindrical Co-ordinates—Concentric Cylinders.* In cylindrical co-ordinates Laplace's equation is

$$\nabla^2 V = \frac{1}{\rho}\frac{\partial}{\partial \rho}\left(\rho \frac{\partial V}{\partial \rho}\right) + \frac{1}{\rho^2}\frac{\partial^2 V}{\partial \phi^2} + \frac{\partial^2 V}{\partial z^2} = 0 \tag{2-38}$$

For the space between two very long concentric cylinders (Fig. 2-8), in which case there will be no variations with respect to either ϕ or z, but only in the ρ direction, eq. (38) becomes

$$\frac{1}{\rho}\frac{\partial}{\partial \rho}\left(\rho \frac{\partial V}{\partial \rho}\right) = 0 \tag{2-39}$$

A trivial solution to this equation is V equals a constant. A useful solution that fits the boundary condition is

$$V = k_1 \ln \rho + k_2$$

The electric field strength in the region between the cylinders will be

$$\mathbf{E} = -\nabla V$$

$$= -\frac{\partial V}{\partial \rho}\hat{\rho}$$

$$= -\frac{k_1}{\rho}\hat{\rho}$$

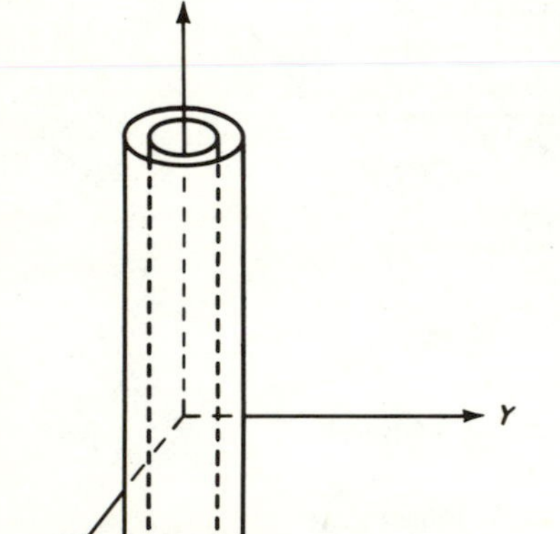

Figure 2-8. Two concentric cylinders.

EXAMPLE 5: *In Spherical Co-ordinates—Concentric Spheres.* In spherical co-ordinates with no variations in θ or ϕ directions, Laplace's equation is

$$\nabla^2 V = \frac{1}{r^2}\frac{\partial}{\partial r}\left(r^2 \frac{\partial V}{\partial r}\right) = 0$$

A solution is

$$V = \frac{k_1}{r} + k_2$$

The electric field strength between the spheres is

$$\mathbf{E} = -\nabla V = -\frac{\partial V}{\partial r}\,\hat{\mathbf{r}} = \frac{k_1}{r^2}\,\hat{\mathbf{r}}$$

In the three examples just solved, the simplicity of the boundary conditions (due to symmetry) made it possible to guess the solution and write it down from inspection. Only in rare cases is it possible to do this. However, there is an important group of problems that can be solved almost by inspection because their boundary conditions are similar to those of problems which have already been solved. These make use of the principle of the electrical image.

Solution by Means of the Electrical Image. As a simple example of this method of solution consider the problem of a line charge ρ_L coulombs per meter parallel to and at a distance $d/2$ from a perfectly conducting plane of infinite extent. It is required to determine the resulting potential distribution and the electric field. The boundary condition in this case is that the conducting plane must be an equipotential surface. Also if the potential at infinity is considered to be zero, the potential of the conducting plane must be zero since it extends to infinity.

The lines of electric flux, which start on the positive line charge, must terminate on negative charges on the plate and at infinity. These negative charges on the conducting surface are required to distribute themselves so that there is no tangential component of electric field along the surface of the conductor; i.e., so that the conducting plane is an equipotential surface (Fig. 2-9).

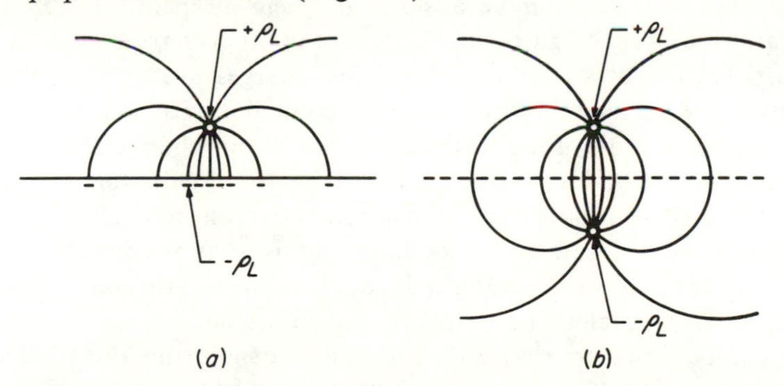

Figure 2-9. (*a*) Line charge near a conducting surface. (*b*) Charges on the conducting surface have been replaced by an appropriately located "image" charge.

If this distribution can be found, the potential at any point can then be determined from the relation $V = \int_V \rho/4\pi\epsilon r \, dV$. As it stands, this is not a simple problem. However now recall the problem of two equal and opposite line charges for which the solution has already been obtained (sec. 2.05). It will be remembered that the plane of symmetry between the wires is an equipotential plane of zero potential. Hence a conducting surface could be placed at the location of this equipotential plane without affecting the potential distribution in any manner whatsoever. If this were done, the negative line charge then would have no effect on the field on the opposite side of the conductor and, so as far as *that* field is concerned, could be removed. The problem is now just the one for which a solution is required. The solution can be set down directly. The field due to a line charge at a distance $d/2$ from an infinite conducting plane is exactly the same as the field (on one side of the zero potential plane) produced by that line charge and an equal and opposite line charge located parallel to it and a distance d away. This second (hypothetical) line charge is called the electrical image of the other, from the analogy with optical images.

Thus in any problem involving charges and conductors, if an additional distribution of charges can be found which will make the surfaces to be occupied by the conductors equipotential surfaces having the correct potential, the conductors can be removed and the field in the volume, originally outside the conductors, will not be changed. Then this field can be computed by methods already developed. The problem is now simply one of finding the potential and electric field strength for a distribution of charges without conductors.

As a second example of this method let it be required to find the potential distribution and electric field about a pair of parallel cylindrical conductors which have applied to them a specified voltage or potential difference. Again referring to the line charges in sec. 2.05, the equipotential surfaces about these line charges are cylinders. If the conductors are located in this field to coincide with an appropriate pair of equipotentials, their introduction will not change the field configuration outside the volume occupied by the conductors. Thus the potential distribution outside the conductors is just the same as it was before the conductors were introduced and is that produced by a pair of line charges of proper strength located along appropriate axes. The solution to this problem has already been obtained.

It must be observed that this process of computing the effects of a conductor is an inverse one, i.e., a solution must be found by experience, and there is no straightforward method of finding an analytical solution in every case. This is analogous to the problem of differentiation and

integration. In differentiation a straightforward method is available for finding derivatives, but the determination of integrals depends on the experience of the operator, or the recorded experience of those who have gone before him. In the case of electric fields, an analytical expression for the charge distribution that can replace a given conductor is not always known, just as the integral of every function is not known. On the other hand, with a given configuration, there are approximate methods available for determining the effect of a conductor in a field just as there are approximate methods for the integration of any curve that can be graphed.

2.09 Capacitance. The capacitance between two conductors is defined by the relation

$$C = \frac{Q}{V}$$

where V is the voltage or potential difference between the conductors due to equal and opposite charges on them of magnitude Q. Gauss's law may be employed to demonstrate that the total charge on one of the conductors is proportional to the potential difference; by definition, C is the constant of proportionality. When the capacitance of a single conductor is referred to, it is implied that the other conductor is a spherical shell of infinite radius.

EXAMPLE 6: *Parallel-Plate Capacitor.* Consider the parallel-plate capacitor having plates of area A and separation d (Fig. 2-10). (d is assumed to be very

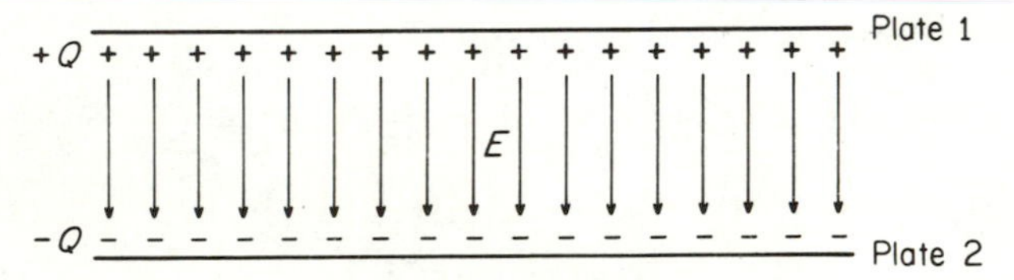

Figure 2-10. Parallel-plate capacitor.

small compared with the length and width of the plates so that the effect of flux fringing may be neglected.) If the plates have a charge of magnitude Q, the surface charge density will be

$$\rho_s = \frac{Q}{A}$$

The electric field strength **E** between the plates is uniform and of magnitude

$$E = \frac{\rho_s}{\epsilon}$$

where ϵ is the dielectric constant of the medium between the plates.

The voltage between the plates will be

$$V = \int_1^2 \mathbf{E} \cdot d\mathbf{s} = Ed$$

The capacitance is
$$C = \frac{Q}{V} = \frac{Q}{Ed} = \frac{\epsilon A}{d} \qquad \text{farads}$$

Because of their usefulness in later work the capacitances for two other simple cases will be found.

EXAMPLE 7: *Concentric Conductors.* It is required to determine the capacitance per unit length between two infinitely long concentric conducting cylinders (Fig. 2-11). The outside radius of the inner conductor is a and the inside radius of the outer conductor is b.

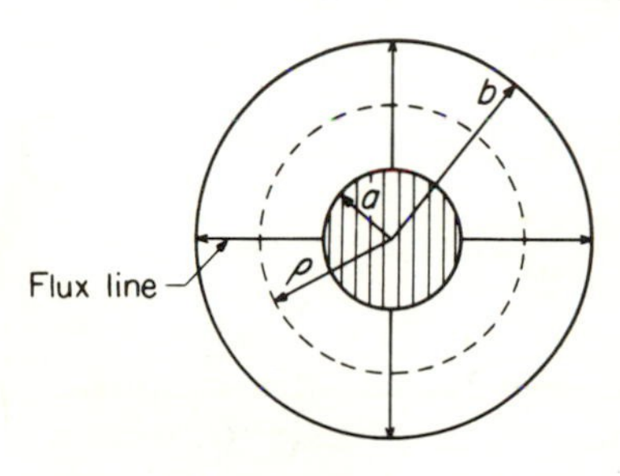

Figure 2-11. Concentric conductors.

Assume a charge distribution ρ_L coulombs per meter on the inner conductor and an equal and opposite charge on the outer conductor. Because of symmetry the lines of electric flux will be radial and the displacement through any cylindrical shell will be ρ_L coulombs per unit length. The magnitude of the displacement density will be

$$D = \frac{\rho_L}{2\pi\rho}$$

and the magnitude of the electric field strength will be

$$E = \frac{\rho_L}{2\pi\rho\epsilon}$$

The voltage between the conductors is

$$V = \int_a^b E\,d\rho = \int_a^b \frac{\rho_L}{2\pi\rho\epsilon}\,d\rho = \frac{\rho_L}{2\pi\epsilon}\ln\rho\Big]_a^b = \frac{\rho_L}{2\pi\epsilon}\ln\frac{b}{a}$$

The capacitance per meter will be

$$C = \frac{\rho_L}{V} = \frac{2\pi\epsilon}{\ln b/a} \qquad \text{F/m} \quad (2\text{-}40)$$

For the air dielectric for which $\epsilon = 1/(36\pi \times 10^9)$

$$C = \frac{10^{-9}}{18\ln b/a} \qquad \text{F/m} \quad (2\text{-}41)$$

EXAMPLE 8: *Parallel Cylindrical Conductors.* The method for determining the electric field for this case has already been considered. A pair of line charges, appropriately located, would make the surfaces occupied by the conductors equipotentials (Fig. 2-12). If the radius of the cylinders is a and the separation between their axes is b, then, in terms of the notation used in connection with Fig. 2-3,

$$b = d\frac{k^2 + 1}{k^2 - 1}$$

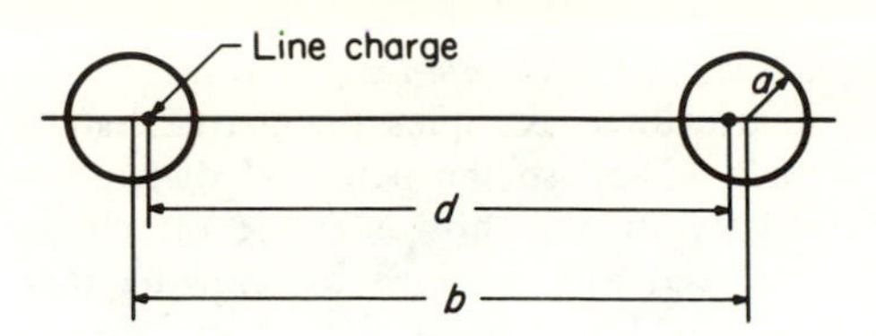

Figure 2-12. Parallel cylindrical conductors.

$$a = \frac{kd}{k^2 - 1}$$

$$\frac{b}{a} = \frac{k^2 + 1}{k}$$

$$ak^2 - bk + a = 0$$

$$k = \frac{b + \sqrt{b^2 - 4a^2}}{2a}$$

But
$$k = \frac{r_b}{r_a}$$

$$\therefore \frac{r_b}{r_a} = \frac{b + \sqrt{b^2 - 4a^2}}{2a}$$

The potential at the surface of one conductor is given by eq. (27)

$$V_1 = \frac{\rho_L}{2\pi\epsilon} \ln \frac{b + \sqrt{b^2 - 4a^2}}{2a}$$

where ρ_L is the charge per unit length.

When the separation is large compared with the radius, that is when $b \gg a$, this becomes

$$V_1 = \frac{\rho_L}{2\pi\epsilon} \ln \frac{b}{a}$$

The potential at the other conductor will be equal and opposite. Hence

$$V = V_1 - V_2 = \frac{\rho_L}{\pi\epsilon} \ln \frac{b + \sqrt{b^2 - 4a^2}}{2a}$$

The capacitance per unit length is

$$C = \frac{\rho_L}{V} = \frac{\pi\epsilon}{\ln \dfrac{b + \sqrt{b^2 - 4a^2}}{2a}} \qquad \text{F/m}$$

If $b \gg a$ the capacitance is given very closely by

$$C \approx \frac{\pi\epsilon}{\ln b/a} \qquad \text{F/m} \quad (2\text{-}42)$$

For an air dielectric between the conductors

$$C \approx \frac{10^{-9}}{36 \ln b/a} \qquad \text{F/m} \quad (2\text{-}43)$$

EXAMPLE 9: *Capacitance of a (Finite-length) Wire or Cylindrical Rod.* In the first two of the above three examples the charge distribution was uniform over the conductor surfaces and so the potential distribution could be determined directly and exactly. In the third example the charge distribution was not known but the potential was obtained by showing that the problem was similar to that of a pair of line charges for which the exact solution was known. In practice there are very few problems that can be solved so simply. In most actual problems the charge distribution is unknown and there are no methods available for obtaining an exact solution. It is then necessary to set about finding an approximate solution.

In the present problem it is required to determine the capacitance of a straight horizontal wire or conducting rod elevated at a height h above the earth. The rod has a length $L = 1$ meter and a radius $a = 0.5$ cm, and is elevated at a height $h = 10$ meters (Fig. 2-13).

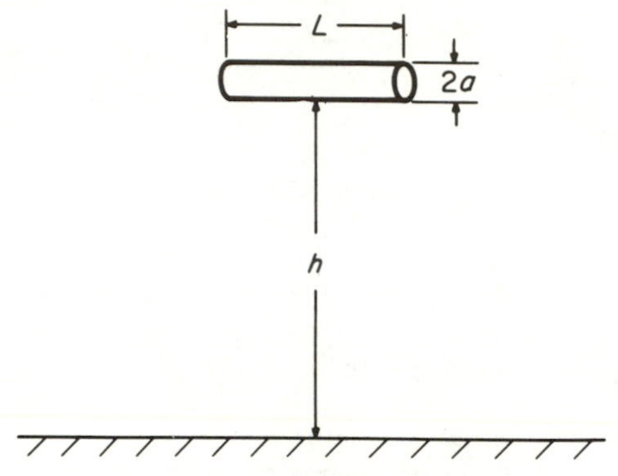

Figure 2-13. An elevated wire or rod.

For a first attack on the problem it will be assumed that the height above the earth is very great so that the problem is that of determining the capacitance of a cylindrical rod remote from the earth. The boundary condition is that the surface of the rod be an equipotential surface. Obviously the charge distribution cannot be uniform along the length of the rod because such a distribution produces a potential that varies along the length of the wire. Moreover there is apparently no straightforward method available for finding the correct charge distribution, which will make the surface an equipotential. This is a *typical practical* problem.

This particular problem was solved many years ago by G.W.O. Howe, using a method of attack that is now used very frequently in electrostatic and electromagnetic problems. It is first assumed that the charge distribution is uniform (even though such an assumption is known to be incorrect). The potential along the wire due to this uniform charge distribution is calculated. It is then assumed that the true potential, which actually exists along the surface of the wire, is equal to the average value of this calculated potential. Knowing the potential for a given total charge the capacitance of the wire is obtained from $C = Q/V_{\text{avg}}$.

Solution: Figure 2-14 shows the rod, which is assumed to have a uniform charge distribution on its surface of amount ρ_L coulombs per meter of length.

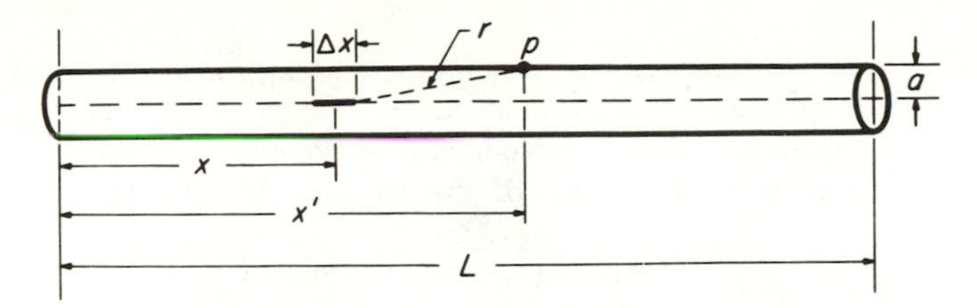

Figure 2-14. Surface charge is replaced by a line charge along the axis for the purpose of computing potential.

The surface charge density is then $\rho_L/2\pi a$ coulombs per square meter. The charge on each element of area contributes to the potential at a point p on the surface and the total potential at p can be obtained by integrating these contributions over the surface of the wire. It is possible to simplify this part of the problem in the following manner. It is known that the equipotential surfaces about a line charge of infinite length are cylinders whose axes coincide with the line charge. If a conducting cylinder is made to coincide with one of these equipotential surfaces and is given a charge per unit length equal to that of the line charge, the electric field in the region about the cylinder will be exactly the same as that produced originally by the line charge.

Thus, as far as the potential outside of it (and on its surface) is concerned, a long charged cylinder may be replaced by a line charge situated along its axis and having the same charge per unit length as the cylinder. Applying this principle in Fig. 2-14, the contribution to the potential at a point x' on the surface due to the charge on an element of length Δx located at point x along the axis will be

$$\Delta V = \frac{\rho_L \Delta x}{4\pi\epsilon\sqrt{(x'-x)^2 + a^2}} \tag{2-44}$$

where ρ_L is the charge per unit length. The total potential at x' due to the assumed charge distribution along the axis is

$$\begin{aligned} V_{x'} &= \frac{\rho_L}{4\pi\epsilon}\int_0^L \frac{dx}{\sqrt{(x'-x)^2 + a^2}} \\ &= \frac{\rho_L}{4\pi\epsilon}\left[-\sinh^{-1}\left(\frac{x'-x}{a}\right)\right]_0^L \\ &= \frac{\rho_L}{4\pi\epsilon}\left[-\sinh^{-1}\left(\frac{x'-L}{a}\right) + \sinh^{-1}\left(\frac{x'}{a}\right)\right] \end{aligned} \tag{2-45}$$

From eq. (45) the potential at the middle of the rod will be

$$\begin{aligned} V_{x'=L/2} &= \frac{\rho_L}{4\pi\epsilon}\left[-\sinh^{-1}\left(-\frac{L}{2a}\right) + \sinh^{-1}\left(\frac{L}{2a}\right)\right] \\ &= \frac{2\rho_L}{4\pi\epsilon}\sinh^{-1} 100 = \frac{10.6}{4\pi\epsilon}\rho_L \end{aligned}$$

and the potential at each end is

$$V_{x'=0} = V_{x'=L} = \frac{5.99\,\rho_L}{4\pi\epsilon}$$

The potential can be calculated at other points along the length to obtain the resulting distribution, shown by Fig. 2-15.

The *average* potential along the rod may be found by integrating eq. (45) (with respect to x') over the length of the rod and dividing by L.

$$V_{\text{avg}} = \frac{\rho_L}{4\pi\epsilon L}\int_0^L \left[-\sinh^{-1}\left(\frac{x'-L}{a}\right) + \sinh^{-1}\left(\frac{x'}{a}\right)\right] dx'$$

$$= \frac{\rho_L}{4\pi\epsilon L}\Bigg| -(x'-L)\sinh^{-1}\left(\frac{x'-L}{a}\right) + \sqrt{(x'-L)^2 + a^2}$$

$$+ x'\sinh^{-1}\left(\frac{x}{a}\right) - \sqrt{x^2 + a^2}\Bigg|_0^L$$

$$= \frac{\rho_L}{2\pi\epsilon}\left[\frac{a}{L} + \sinh^{-1}\left(\frac{L}{a}\right) - \sqrt{1 + \frac{a^2}{L^2}}\right]$$

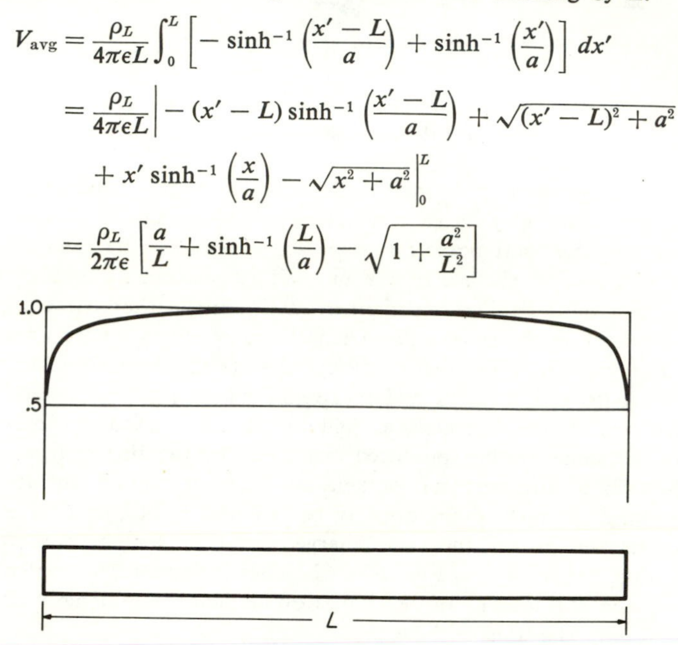

Figure 2-15. Potential distribution along the rod calculated from assumed uniform charge distribution.

Substituting numerical values

$$V_{\text{avg}} = \frac{\rho_L}{2\pi\epsilon}\,[0.005 + \sinh^{-1} 200 - (1 + 0.00001)]$$

$$= \frac{5.00\,\rho_L}{2\pi\epsilon}$$

The capacitance of the rod (remote from the earth) will be (approximately)

$$C = \frac{\rho_L L}{V_{\text{avg}}} = 11.11 \text{ pF}$$

The effect of the proximity of the earth can be accounted for by means of the image principle. A negative charge $-\rho_L L$ located at the position of the image will decrease the average potential of the rod slightly. With negligible error this negative charge can be considered as being located at a point at the center of the image a distance $2h$ from the rod and the potential at the rod due to this negative charge will be

$$V_{\text{image}} = \frac{-\rho_L L}{4\pi\epsilon \times 2h} = -\frac{0.05\,\rho_L L}{4\pi\epsilon}$$

The average potential of the rod including the contribution from the image charge is

$$V_{\text{avg}} = \frac{\rho_L}{4\pi\epsilon}(10.0 - 0.05) = \frac{9.95\,\rho_L}{4\pi\epsilon}$$

The capacitance of the rod including the effect of the presence of the earth will be

$$C = \frac{4\pi\epsilon}{9.95} = 11.16 \text{ pF}$$

The proximity of the earth has increased the capacitance by about $\frac{1}{2}$ of 1 per cent. It will be observed that in this case a 50 per cent error in computing the contribution from the image would affect the final answer a negligible amount. Therefore there is usually no justification for seeking a more accurate solution for this part of the problem.

The method outlined above gives an approximate answer for the capacitance of the rod. The degree of approximation can be improved by assuming a second and different charge distribution, which will produce a more nearly uniform potential distribution. (This is easy to do once the potential distribution due to a uniform charge distribution has been found.) It will be found (for this case) that the answers obtained with more nearly correct charge distributions do not differ appreciably from that obtained above. The *correct* value for capacitance will always be a little larger than that calculated from any assumed charge distribution. This is because the actual charge distribution is always such as to make the potential energy of the system, and therefore the potential of the rod, a minimum.

2.10 Electrostatic Energy. When a capacitor is charged so that there exists a voltage V between its plates, there is a storage of energy, which can be converted into heat by discharging the capacitor through a resistance. The amount of energy stored can be found by calculating the work done in charging the capacitor. Since potential was defined in terms of work per unit charge, the work done in moving a small charge dq against a potential difference V is $V\,dq$. But the voltage V can be expressed in terms of the capacitance C and the charge q by

$$V = \frac{q}{C}$$

Therefore the work done in increasing the charge on a capacitor by an amount dq is

$$\frac{q}{C}\,dq$$

The total work done in charging a capacitor to Q coulombs is

$$\text{Total work} = \int_0^Q \frac{q}{C}\,dq = \frac{1}{2}\frac{Q^2}{C}$$

Therefore the energy stored by a charged capacitor is

$$\text{Stored energy} = \frac{1}{2}\frac{Q^2}{C} = \frac{1}{2}VQ = \frac{1}{2}V^2C \tag{2-46}$$

Electrostatic energy also may be looked upon as the energy necessary to establish a given charge distribution in space. Suppose that all of space is initially field-free and that N point charges are brought in from infinity and located at specific points. The energy expended in locating the ith charge at the point $\mathbf{r}_i$ (a vector indicating the point x_i, y_i, z_i, where the potential is V_i) is given by

$$W_i = q_iV_i = \frac{q_i}{4\pi\epsilon}\sum_{j=1}^{i-1}\frac{q_j}{R_{ij}} \tag{2-47}$$

in which $R_{ij} = |\mathbf{r}_i - \mathbf{r}_j|$.

No energy is used up in locating the first charge and thus the total energy is

$$W = \sum_{i=2}^{N} W_i = \frac{1}{4\pi\epsilon}\sum_{i=2}^{N}\sum_{j=1}^{i-1}\frac{q_iq_j}{R_{ij}} \tag{2-48}$$

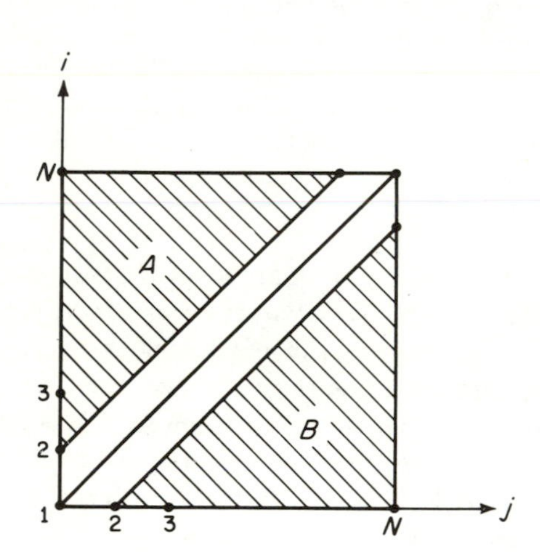

Figure 2-16.

Note that the energy necessary to form each point charge has been ignored in the above derivation. The numbers over which the summation in (48) is carried out may be depicted as points in the *i-j* plane as shown in Fig. 2-16. The summation in (48) is clearly over the triangular region marked *A*. Since the quantity being summed is symmetric in i and j, the same energy W would be obtained by a summation over triangle *B*. Summation over both triangles must give $2W$, a result which may be stated as follows:

$$W = \frac{1}{8\pi\epsilon}\sum_{i=1}^{N}\sum_{j=1}^{N}\frac{q_iq_j}{R_{ij}}, \qquad i \neq j \tag{2-49}$$

A continuous distribution of charge may be dealt with similarly by writing $\rho(\mathbf{r})\,dV$ in place of q_i and $\rho(\mathbf{r}')\,dV'$ in place of q_j. The summations now may be replaced by integrations over volumes V and V' which must be large enough to contain all the charge present. Note that the stipulation $i \neq j$ in (49) now may be waived because point charges are not present (with their associated infinite self-energies). Thus the energy necessary to form a continuous charge distribution is

$$W = \frac{1}{8\pi\epsilon}\int_V \int_{V'} \frac{\rho(\mathbf{r})\rho(\mathbf{r}')}{R}\,dV'\,dV \tag{2-50}$$

in which $\quad R = |\mathbf{r} - \mathbf{r}'|$.

Previous work on the field of a charge distribution shows that

$$V(\mathbf{r}) = \frac{1}{4\pi\epsilon}\int_{V'} \frac{\rho(\mathbf{r}')}{R}\,dV' \tag{2-51}$$

Substitution of (51) into (50) gives the very important formula

$$W = \tfrac{1}{2}\int_V \rho(\mathbf{r})V(\mathbf{r})\,dV \tag{2-52}$$

The above formula for W may be put into another form if use is made of the identity

$$\nabla\cdot(V\mathbf{D}) = V\nabla\cdot\mathbf{D} + \mathbf{D}\cdot\nabla V \tag{2-53}$$

Equation (52) may be transformed as follows:

$$\begin{aligned} W &= \tfrac{1}{2}\int_V \rho V\,dV = \tfrac{1}{2}\int_V V\nabla\cdot\mathbf{D}\,dV \\ &= \tfrac{1}{2}\int_V [\nabla\cdot(V\mathbf{D}) - \mathbf{D}\cdot\nabla V]\,dV \\ &= \tfrac{1}{2}\int_S V\mathbf{D}\cdot d\mathbf{a} + \tfrac{1}{2}\int_V \mathbf{D}\cdot\mathbf{E}\,dV \end{aligned} \tag{2-54}$$

If the surface S is allowed to approach infinity, the integrand in the surface integral must drop off at least as fast as r^{-3} and thus the surface integral vanishes. For this case, (54) becomes

$$W = \tfrac{1}{2}\int_{\text{all space}} \epsilon E^2\,dV \tag{2-55}$$

Equation (55) is often *interpreted* as an assertion that everywhere in space there exists an energy density w given by

$$w = \tfrac{1}{2}\epsilon E^2 \tag{2-56}$$

This interpretation is frequently very useful but one must be cautious in applying it—only the energy *integral* (55) may be used without question. Note that energy may be computed either from the charge distribution (50) or from the electric field strength (55). Thus the energy is said to be "associated with the electric charge" or alternatively "associated with the electric field."*

Charge on a conducting surface. If charge is assembled on the surface of a conductor, that surface must be an equipotential. Under such circumstances (52) becomes

$$W = \tfrac{1}{2} V \int_S \rho_s \, ds$$

$$= \tfrac{1}{2} QV \tag{2-57}$$

in which V is the potential of the surface and Q is the total charge on it. Note that (57) is identical to (46).

Force on a charged conductor. The force on a charged conductor may be calculated using the expression for energy density (56). If an elemental area ΔS on a charged conductor is depressed a distance Δl, the increase in stored energy is

$$\Delta W = w \, \Delta S \, \Delta l \tag{2-58}$$

Such a depression must be carried out against a force F and thus

$$\Delta W = F \, \Delta l = f \, \Delta S \, \Delta l \tag{2-59}$$

in which f is the force per unit area (pressure). Comparison of (59) with (58) shows that

$$f = w \tag{2-60}$$

If the surface charge density is ρ_s, then (56) and (60) give

$$f = \tfrac{1}{2} \epsilon E^2 = \frac{1}{2\epsilon} D^2 = \frac{1}{2\epsilon} \rho_s^2 \tag{2-61}$$

If the concept of energy density is not used, derivation of (61) can be carried out using the concept of an electric field exerting a force on a charge distribution. In this case the electric field *acting on* the charge does not include the field *due to* the charge. A simple example is the parallel-plate capacitor. A "model" of the capacitor may be constructed using two charge sheets in space; such a model produces the same fields as shown in Fig. 2-17. The electric field strength at the positive sheet *due to the negative sheet* is given by $\rho_s/2\epsilon$ and con-

*These statements represent two different points of view or two interpretations of a single set of experimental facts. The question of just where the energy "resides" in this case is similar to the question of where the potential energy is stored when a weight has been raised. The question seems to be one of philosophy or interpretation and as such is unanswerable on the basis of any physical measurements that can be made by the engineer.

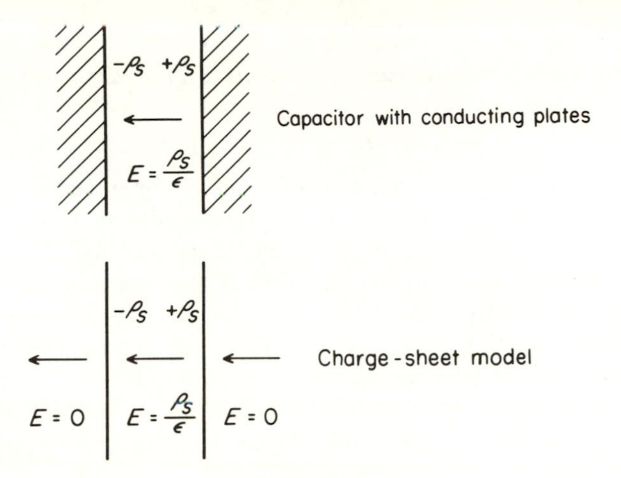

Figure 2-17.

sequently

$$f = \frac{1}{2\epsilon}\rho_s^2 \tag{2-62}$$

as already derived using the energy density concept.

2.11 Conditions at a Boundary between Dielectrics. Consider conditions at the interface between two dielectrics in an electric field. The dielectric constants of the media are ϵ_1 and ϵ_2 respectively, and it is assumed that there are no free charges on the boundary surface.

Apply Gauss's law to the shallow pillbox volume that encloses a portion of the boundary (Fig. 2-18). Since there are no charges within the volume the net outward displacement through the surface of the box is zero. As the depth of the box is allowed to approach zero, always keeping the boundary surface between its two flat faces, the displacement through the curved-edge surface becomes negligible. Gauss's law then requires that the displacement through the upper face be

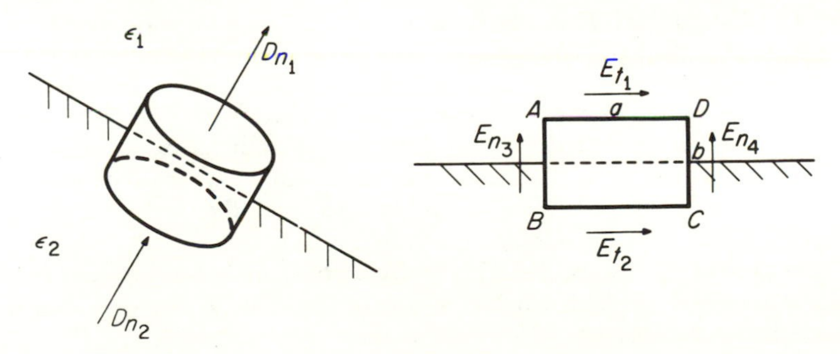

Figure 2-18. Boundary surface between two dielectric media.

equal to the displacement through the lower face. Because the areas of the faces are equal, the *normal components of the displacement densities must be equal*, that is,

$$D_{n_1} = D_{n_2}$$

Thus there are the same number of lines of displacement flux entering one face as are leaving the other face and the lines of electric displacement are continuous across a boundary surface.

Whereas the normal component of **D** is the same on both sides of a boundary, it is easily shown that the *tangential* component of the electric field strength **E** must be continuous across the boundary. Referring to Fig. 2-18(*b*) it is supposed that there are electric field strengths $\mathbf{E}_1$ and $\mathbf{E}_2$ respectively in medium (1) and medium (2). In the electrostatic field the voltage around any closed path must be zero, that is,

$$V_{\text{closed path}} \equiv \oint \mathbf{E} \cdot d\mathbf{s} \equiv 0$$

Apply this to the rectangular path *ABCD*, in which *AD* is *just* inside medium (1) and *BC just* inside medium (2). The length of the rectangle is *a*, and its width is *b*.

$$\oint \mathbf{E} \cdot d\mathbf{s} = -E_{n_3} b + E_{t_2} a + E_{n_4} b - E_{t_1} a \tag{2-63}$$

where E_{t_1} and E_{t_2} are the average tangential components of **E** along paths *AD* and *BC* and E_{n_3} and E_{n_4} are the average normal components of **E** along the paths *BA* and *CD*. As the sides *AD* and *BC* are brought closer together, always keeping the boundary between them, the lengths *AB* and *CD* approach zero and the first and third terms in eq. (63) become zero (assuming that the electric field never becomes infinite). Therefore

$$-E_{t_1} a + E_{t_2} a = 0 \qquad \text{and} \qquad E_{t_1} = E_{t_2}$$

The tangential component of **E** *is continuous at the boundary.*

The two conditions: (a) Normal *D* is continuous at the boundary, and (b) Tangential *E* is continuous at the boundary are used to solve problems involving dielectrics.

EXAMPLE 10: *Refraction.* Consider the problem of Fig. 2-19 where an infinite slab of dielectric whose dielectric constant is ϵ_2, is immersed in a medium of ϵ_1. Let θ_1 be the angle that the normal to the boundary makes with the lines of electric force in medium (1). Then the lines of **E** and **D** will be refracted in passing through the slab.

Let $\mathbf{D}_1$ and $\mathbf{D}_2$ be the electric displacement density outside and inside the slab respectively, and $\mathbf{E}_1$ and $\mathbf{E}_2$ be the electric field strength outside and inside the slab. Then

$$\mathbf{D}_1 = \epsilon_1 \mathbf{E}_1$$

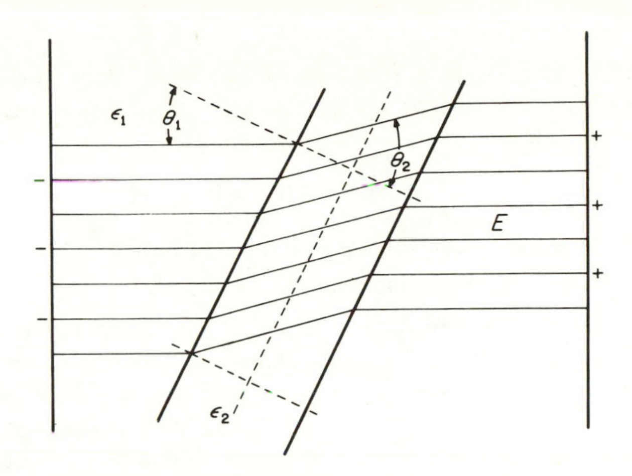

Figure 2-19. Refraction of an electric field.

$$\mathbf{D}_2 = \epsilon_2 \mathbf{E}_2$$

By the two fundamental principles stated above

$$D_1 \cos \theta_1 = D_2 \cos \theta_2$$

$$E_1 \sin \theta_1 = E_2 \sin \theta_2$$

$$\frac{D_1}{E_1} \cot \theta_1 = \frac{D_2}{E_2} \cot \theta_2$$

Therefore

$$\epsilon_1 \cot \theta_1 = \epsilon_2 \cot \theta_2$$

$$\frac{\tan \theta_1}{\tan \theta_2} = \frac{\epsilon_1}{\epsilon_2} \tag{2-64}$$

Equation (64) gives the relation between the tangents of the angle of incidence θ_1, and the angle of refraction θ_2 in terms of the dielectric constants of the media involved.

2.12 Cylindrical and Spherical Harmonics. It was pointed out in earlier sections of this chapter that, except for a few special cases, solution of Laplace's equation, subject to the appropriate boundary conditions, was in general a quite difficult problem. There is a group of problems having a certain symmetry that may be solved approximately by use of cylindrical or spherical harmonics. Because these functions are also required for later use in electromagnetic problems, they will be considered briefly here.

For those problems which can be set up in cylindrical co-ordinates, and for which there is no variation of the field in the z direction, Laplace's equation (38) may be written as

$$\frac{1}{\rho} \frac{\partial}{\partial \rho} \left(\rho \frac{\partial V}{\partial \rho} \right) + \frac{1}{\rho^2} \frac{\partial^2 V}{\partial \phi^2} = 0 \tag{2-65}$$

If a solution of the form $V = \rho^n Q_n$

is assumed (where Q_n is a function of ϕ alone), then substitution of this solution back into (65) shows that Q_n must satisfy the following differential equation:

$$\frac{\partial^2 Q_n}{\partial \phi^2} + n^2 Q_n = 0$$

The solution of this equation is well known and has the form

$$Q_n = A_n \cos n\phi + B_n \sin n\phi$$

where A_n and B_n are arbitrary constants. It will be noted that when $-n$ is substituted for $+n$, the same differential equation for Q results, so that Q_{-n} can be put equal to Q_n. Then, if $\rho^n Q_n$ is a solution of (65), $r^{-n} Q_{-n} = Q_n/r^n$ is also a solution. By inspection it is seen that $V = \ln \rho$ is a solution of (65). Now if a function is a solution of Laplace's equation, each of its partial derivatives with respect to any of the *rectangular* co-ordinates x, y, or z, (but not in general with respect to cylindrical or spherical co-ordinates) is also a solution. That this is so, may be verified by differentiating Laplace's equation partially in rectangular co-ordinates. Differentiating the solution $V = \ln \rho$ with respect to x yields $(\cos \phi)/\rho$ as another solution, while differentiation with respect to y yields $(\sin \phi)/\rho$. Successive differentiation leads to the following set of possible solutions of (65):

$$\ln \rho; \quad \frac{\cos \phi}{\rho}; \quad \frac{\sin \phi}{\rho}; \quad \frac{\cos 2\phi}{\rho^2}; \quad \frac{\sin 2\phi}{\rho^2}; \quad \frac{\cos 3\phi}{\rho^3}; \quad \frac{\sin 3\phi}{\rho^3}; \ldots$$

Replacing ρ^{-n} by ρ^n gives a second set, viz.:

$$\rho \cos \phi; \quad \rho \sin \phi; \quad \rho^2 \cos 2\phi; \quad \rho^2 \sin 2\phi; \quad \rho^3 \cos 3\phi; \quad \rho^3 \sin 3\phi$$

These solutions of Laplace's equation (65) are known as *circular harmonics* or *cylindrical harmonics*. These harmonic functions may be used to solve problems in which there is no variation of the field in the z direction.

EXAMPLE 11: *Conducting Cylinder in an Electric Field.* A long conducting cylinder is placed in, and perpendicular to, a uniform electric field E_x with the axis of the cylinder coincident with the z axis (Fig. 2-20). Determine the field distribution in the region about the cylinder.

Although the field in the neighborhood of the cylinder will be disturbed by its presence, the distant field will be unaffected and will be just E_x. Therefore, if the potential of the cylinder is taken as zero potential, the potential at a great distance ρ will be $-E_x \rho \cos \phi$. Also the surface of the cylinder, $\rho = a$, is an equipotential surface, which has arbitrarily been set at zero potential. The problem can be solved by finding that combination of the given cylindrical harmonic solutions that will also satisfy these two boundary conditions. The answer in this case happens to be quite simple, for it is

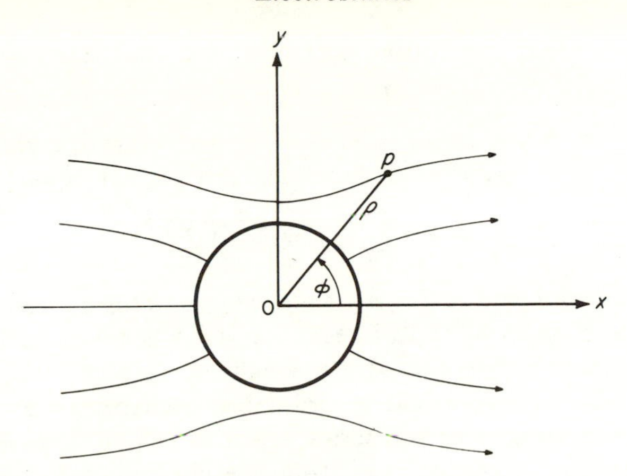

Figure 2-20. Conducting cylinder in a uniform field.

evident that the following combination of cylindrical harmonics, selected from those listed in the table, can be made to satisfy the boundary conditions:

$$V = A\rho \cos \phi + \frac{B \cos \phi}{\rho}$$

For ρ very large ($\rho \to \infty$)

$$V = A\rho \cos \phi = -E_x \rho \cos \phi$$

Therefore $$A = -E_x$$

For $\rho = a$, $$V = Aa \cos \phi + \frac{B \cos \phi}{a} = 0$$

Therefore $$B = -Aa^2 = a^2 E_x$$

Then $$V = \left(\frac{a^2}{\rho} - \rho\right) E_x \cos \phi$$

The components of electric field strength in the region outside the cylinder are given by

$$E_\rho = -\frac{\partial V}{\partial \rho} = \left(\frac{a^2}{\rho^2} + 1\right) E_x \cos \phi$$

$$E_\phi = -\frac{1}{\rho} \frac{\partial V}{\partial \phi} = \left(\frac{a^2}{\rho^2} - 1\right) E_x \sin \phi \qquad (2\text{-}66)$$

Spherical Harmonics. For problems that can be set up in spherical co-ordinates and for which there is no variation in the ϕ direction, Laplace's equation is

$$\nabla^2 V = \frac{1}{r^2} \frac{\partial}{\partial r} \left(r^2 \frac{\partial V}{\partial r}\right) + \frac{1}{r^2 \sin \theta} \frac{\partial}{\partial \theta} \left(\sin \theta \frac{\partial V}{\partial \theta}\right) = 0 \qquad (2\text{-}67)$$

Letting $u = \cos \theta$, so that $du = -\sin \theta \, d\theta$, eq. (67) becomes

$$\frac{\partial}{\partial r} \left(r^2 \frac{\partial V}{\partial r}\right) + \frac{\partial}{\partial u} \left[(1 - u^2) \frac{\partial V}{\partial u}\right] = 0 \qquad (2\text{-}68)$$

Again assuming that a solution may be found that has the form

$$V = r^n P_n$$

(where P_n is a function of $u = \cos\theta$ alone), substitution into (68) shows that P_n must satisfy the following differential equation:

$$\frac{d}{du}\left[(1 - u^2)\frac{dP_n}{du}\right] + n(n + 1)P_n = 0 \tag{2-69}$$

Equation (69) is known as *Legendre's equation.* This is an important equation in field theory for it is encountered whenever solutions (involving variations with r and θ) are sought to Laplace's equation or the wave equation in spherical co-ordinates. Solution to eq. (69) may be found by assuming a power series solution, which is inserted back into the differential equation. Equating the coefficients of corresponding powers, relations among these coefficients are found. The result is the following set of solutions for (69)

$$\left.\begin{aligned} P_0 &= 1 \\ P_1 &= u = \cos\theta \\ P_2 &= \tfrac{1}{2}(3u^2 - 1) = \tfrac{1}{2}(3\cos^2\theta - 1) = \tfrac{1}{4}(3\cos 2\theta + 1) \\ P_3 &= \tfrac{1}{2}(5u^3 - 3u) = \tfrac{1}{2}(5\cos^3\theta - 3\cos\theta) \\ &= \tfrac{1}{8}(5\cos 3\theta + 3\cos\theta) \end{aligned}\right\} \tag{2-70}$$

and so on.

The function P_n is called a *Legendre function* of the order n. Substitution of $-(n + 1)$ for n in (69) results in the same equation, showing that $P_{-(n+1)} = P_n$.

Solutions to Laplace's equation (67) can now be found by trial, using these Legendre functions. Alternatively, the solutions may be found as in the cylindrical harmonic case by a process of partial differentiation. By trial it is found that

$$V = \frac{1}{r}$$

is a solution of (67). Then differentiating partially with respect to z the following solutions are obtained:

$$\left.\begin{aligned} &\frac{1}{r} && 1 \\ &\frac{1}{r^2}\cos\theta && r\cos\theta \\ &\frac{1}{r^3}(3\cos^2\theta - 1) && r^2(3\cos^2\theta - 1) \\ &\frac{1}{r^4}(5\cos^3\theta - 3\cos\theta) && r^3(5\cos^3\theta - 3\cos\theta) \end{aligned}\right\} \tag{2-71}$$

The second set has been obtained from the first set by replacing $r^{-(n+1)}$ by r^n. The solutions to Laplace's equation in spherical co-ordinates are called *spherical harmonics*. The particular sets (71), obtained for no variation with ϕ are known more specifically as *zonal harmonics* because the potential is constant in each zone of latitude.

Zonal harmonics can be used to obtain solutions to problems in spherical co-ordinates for which there is no variation in the ϕ direction. A simple example would be that of a conducting sphere placed in a uniform field which is parallel to the z axis. The solution to this problem follows in a manner similar to that of the conducting cylinder and is left as an exercise for the student.

It is important to realize that only certain very special problems yield to an exact solution such as was obtained in the two examples above. In general, an *infinite* number of harmonic solutions would be required to satisfy the boundary conditions. However, just as any periodic function (satisfying certain conditions) may be approximated by a finite number of terms of a Fourier series, so any problem having a geometry suitable for the application of these harmonic functions may be solved approximately by an appropriate combination of a finite number of them.

The methods of this last section are also applicable in the solution of certain electromagnetic problems. Examples of such problems will be encountered in chaps. 13 and 14.

2.13 The Electrostatic Uniqueness Theorem. Up to this point, boundary conditions have been specified without the assurance that they are sufficient to produce a unique solution. The adequacy of the boundary conditions may be determined by postulating two different solutions and looking for the conditions that will make those solutions identical. Consider a problem in which a given charge density ρ is contained in a volume V surrounded by a surface S. Suppose that V_1 and V_2 are two potentials satisfying Poisson's equation:

$$\nabla^2 V_1 = -\frac{\rho}{\epsilon}, \qquad \nabla^2 V_2 = -\frac{\rho}{\epsilon} \tag{2-72}$$

Suppose also that the potentials satisfy boundary conditions (as yet unspecified) on the surface S.

Equations (72) suggest the formation of a difference potential $V_0 = V_1 - V_2$ which satisfies Laplace's equation,

$$\nabla^2 V_0 = 0 \tag{2-73}$$

It is convenient now to introduce the identity

$$\nabla \cdot (V_0 \nabla V_0) = V_0 \nabla^2 V_0 + \nabla V_0 \cdot \nabla V_0 \tag{2-74}$$

Equation (74) may be simplified using (73) and integrated over the volume V. Application of the divergence theorem gives

$$\int_S V_0 \nabla V_0 \cdot d\mathbf{a} = \int_V |\nabla V_0|^2 \, dV \tag{2-75}$$

The component of the gradient normal to S may be expressed as $\partial/\partial n$ so that (75) may be expressed as

$$\int_S V_0 \frac{\partial V_0}{\partial n} \, da = \int_V |\nabla V_0|^2 \, dV \tag{2-76}$$

Uniqueness of the electric field strength means that $\nabla V_1 = \nabla V_2$ or $\nabla V_0 = 0$ everywhere in the volume V. Equation (76) shows that such uniqueness is guaranteed if either $V_0 = 0$ or $\partial V_0/\partial n = 0$ at all points on S. The condition $V_0 = 0$ means that $V_1 = V_2 = V_S$ on S; specification of the potential function on S is called the *Dirichlet boundary condition.* The condition $\partial V_0/\partial n = 0$ means that

$-\dfrac{\partial V_1}{\partial n} = -\dfrac{\partial V_2}{\partial n} = E_S$ on S; specification of the normal derivative of the potential on S is called the *Neumann boundary condition* and it corresponds to a specification of normal electric field strength or charge density. If both the Dirichlet and the Neumann conditions are given over part of S then the problem is said to be *overspecified* or *improperly posed.*

An interesting corollary concerns the uniqueness of the potential. The boundary conditions established above state that

$$\nabla V_1 = \nabla V_2$$

or that

$$V_1 = V_2 + \text{a constant}$$

throughout the volume V, the presence of the arbitrary constant indicating that the potential is not unique. However, if the potential is specified at any point on S, then $V_1 = V_2$ at that point and the constant must be zero. In other words, the Dirichlet boundary condition specified on any portion of S guarantees the uniqueness of the potential.

The uniqueness theorem may be applied easily to example 11. The boundaries are the cylinder of radius a at the origin and what may be regarded as a rectangular "box" at infinity, the sides of the box being given by $x = \pm\infty$ and $y = \pm\infty$. On the cylinder $V = 0$ and thus the Dirichlet condition has been used. At $x = \pm\infty$, $-\partial V/\partial x = E_x$ while at $y = \pm\infty$, $\partial V/\partial y = 0$ and thus at infinity the Neumann condition has been used. As a result of these conditions both the electric field strength and the potential must be unique.

2.14 Far Field of a Charge Distribution. In the derivation of the electric dipole field it was found that the field expressions became relatively simple at observation points far from the dipole (far compared to the length of the dipole). Far-field approximations are frequently used and thus it is worth our while to derive a general far-field expansion valid for any charge distribution.

Consider a charge distribution ρ to exist within the volume V_0 shown in Fig. 2-21. The potential V anywhere in space is given by (26) which may be re-stated as

$$V(\mathbf{r}) = \frac{1}{4\pi\epsilon}\int_{V_0} \frac{\rho(\mathbf{r}')}{R}\,dV' \tag{2-77}$$

in which $dV' = dx'\,dy'\,dz'$.

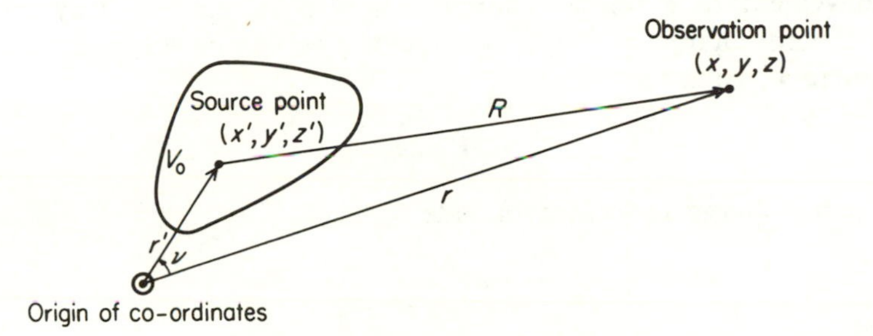

Figure 2-21. Far-field co-ordinates.

The problem is one of finding an expansion for

$$\begin{aligned} R^{-1} &= [(x - x')^2 + (y - y')^2 + (z - z')^2]^{-(1/2)} \\ &= [r^2 - 2rr'\cos\nu + r'^2]^{-(1/2)} \\ &= r^{-1}\left[1 - 2\left(\frac{r'}{r}\right)\cos\nu + \left(\frac{r'}{r}\right)^2\right]^{-(1/2)} \end{aligned} \tag{2-78}$$

The binomial theorem may be used to expand (78) in an infinite series of the form

$$R^{-1} = r^{-1}\left[1 + \cos\nu\left(\frac{r'}{r}\right) + \tfrac{1}{2}(3\cos^2\nu - 1)\left(\frac{r'}{r}\right)^2 + \cdots\right] \tag{2-79}$$

which is uniformly convergent for $r'/r < \sqrt{2} - 1$. Comparison of (79) with (70) shows that the coefficients are in the form of Legendre functions. Thus (79) may be expressed as

$$R^{-1} = r^{-1}\sum_{n=0}^{\infty} P_n(\cos\nu)\left(\frac{r'}{r}\right)^n \tag{2-80}$$

An especially convenient form of (79) results if $r'\cos\nu$ is written as $\mathbf{r}'\cdot\hat{\mathbf{r}}$:

$$\frac{1}{R} = \frac{1}{r}\left[1 + \frac{\mathbf{r}'\cdot\hat{\mathbf{r}}}{r} + \frac{1}{2r^2}\{3(\mathbf{r}'\cdot\hat{\mathbf{r}})^2 - r'^2\} + \ldots\right] \tag{2-81}$$

Substitution of (81) into (77) and term-by-term integration gives the complete far-field expansion

$$4\pi\epsilon V(r) = \frac{1}{r}\int_{V_0} \rho(\mathbf{r}')\, dV' + \frac{1}{r^2}\int_{V_0} \rho(\mathbf{r}')(\mathbf{r}'\cdot\hat{\mathbf{r}})\, dV'$$
$$+ \frac{1}{r^3}\int_{V_0} \rho(\mathbf{r}')\left[\frac{3}{2}(\mathbf{r}'\cdot\hat{\mathbf{r}})^2 - \frac{r'^2}{2}\right] dV + \ldots \quad (2\text{-}82)$$

The far field is always given by the first nonzero term in this expansion. Of course this far-field expansion is not valid for time-varying charges which produce radiation fields (to be discussed in chap. 10).

2.15 Dirac Delta Representation for a Point Charge. Frequently it is convenient to write the charge density ρ in such a way as to denote a point charge. If a point charge of q coulombs is located within the volume V, then

$$\int_V \rho\, dV = q$$

If the point charge is located outside V, then

$$\int_V \rho\, dV = 0$$

The above conditions are satisfied if ρ is set equal to q multiplied by the three-dimensional Dirac delta discussed in chap. 1. Thus the charge density of a point charge of q coulombs located at $\mathbf{r} = \mathbf{r}_0$ may be written as

$$\rho(\mathbf{r}) = q\,\delta(\mathbf{r} - \mathbf{r}_0) \quad (2\text{-}83)$$

or in terms of one-dimensional deltas as

$$\rho(\mathbf{r}) = q\,\delta(x - x_0)\,\delta(y - y_0)\,\delta(z - z_0) \quad (2\text{-}84)$$

For a point charge at the origin, (84) becomes

$$\rho(\mathbf{r}) = q\,\delta(x)\,\delta(y)\,\delta(z) \quad (2\text{-}85)$$

Line and sheet distributions of charge also may be represented using the Dirac delta. For instance, a line charge of ρ_L coulombs per meter along the z axis has a volume charge density given by

$$\rho(\mathbf{r}) = \rho_L\,\delta(x)\,\delta(y) \quad (2\text{-}86)$$

and a charge sheet of ρ_s coulombs per square meter lying in the x-y plane has a volume charge density given by

$$\rho(\mathbf{r}) = \rho_s\,\delta(z) \quad (2\text{-}87)$$

The Dirac delta representation is used chiefly when writing Poisson's equation. For a point charge of q coulombs, Poisson's equation takes the form

$$\nabla^2 V = -\frac{q}{\epsilon}\,\delta \quad (2\text{-}88)$$

In sec. 2.05 the Green's function G was introduced as the potential of a unit point charge (valid for an unbounded, homogeneous region). If the point charge is located at $\mathbf{r}_0$, the function G must satisfy Poisson's equation in the form

$$\nabla^2 G(\mathbf{r}, \mathbf{r}_0) = -\frac{1}{\epsilon}\,\delta(\mathbf{r} - \mathbf{r}_0) \tag{2-89}$$

EXAMPLE 12: *Field of an Infinitesimal Dipole (Far-Field Expansion).* The charge dipole of example 1 (Fig. 2-2, p. 38) may be represented by the volume charge density

$$\rho(\mathbf{r}) = q\,\delta(x)\,\delta(y)\left[\delta\left(z - \frac{l}{2}\right) - \delta\left(z + \frac{l}{2}\right)\right]$$

The potential may be obtained using the far-field expansion (82). The first term in (82) must be zero because the total charge is zero. The second term may be evaluated as follows:

$$\begin{aligned} 4\pi\epsilon V(\mathbf{r}) &= \frac{1}{r^2}\int_{V_0} \rho(\mathbf{r}')(\mathbf{r}' \cdot \hat{\mathbf{r}})\,dV' \\ &= \frac{1}{r^2}\iiint q\,\delta(x')\,\delta(y')\left[\delta\left(z' - \frac{l}{2}\right) - \delta\left(z' + \frac{l}{2}\right)\right] z' \cos\theta\,dx'\,dy'\,dz' \\ &= \frac{ql\cos\theta}{r^2} \end{aligned}$$

2.16 Dirac Delta Representation for an Infinitesimal Dipole. In examples 1 and 12, the potential due to an infinitesimal dipole was found by evaluating the far field of a finite dipole. Alternatively, one might look for a suitable representation for an infinitesimal dipole so that far-field approximations would be unnecessary. Such a distribution may be derived by writing an expression for the charge density with finite spacing and then letting the spacing approach zero. With reference to example 12,

$$\begin{aligned} \rho(\mathbf{r}) &= \lim_{l\to 0}\left\{q\,\delta(x)\,\delta(y)\left[\delta\left(z - \frac{l}{2}\right) - \delta\left(z + \frac{l}{2}\right)\right]\right\} \\ &= \lim_{l\to 0}\left\{-ql\,\delta(x)\,\delta(y)\left[\frac{\delta\left(z + \frac{l}{2}\right) - \delta\left(z - \frac{l}{2}\right)}{l}\right]\right\} \end{aligned}$$

If it is noted that the quantity in square brackets is the derivative of the Dirac delta and if the *dipole moment* p is defined as

$$p = \lim_{l\to 0}\,(ql) \tag{2-90}$$

then the charge density of an infinitesimal dipole becomes

$$\rho(\mathbf{r}) = -p\,\delta(x)\,\delta(y)\,\delta'(z) \tag{2-91}$$

Frequently the dipole moment p is expressed as a vector $\mathbf{p}$ by assigning to it the direction of a line drawn from the negative point charge to the positive point charge. Thus for the charge distribution in eq. (91),

$$\mathbf{p} = \hat{\mathbf{z}}p \tag{2-92}$$

EXAMPLE 13: *Potential of an Infinitesimal Dipole (δ' Charge Distribution).* The charge density given in (91) may be substituted into (77) in order to calculate the potential V.

$$4\pi\epsilon V(\mathbf{r}) = -p \iiint_{-\infty}^{\infty} \frac{\delta(x')\,\delta(y')\,\delta'(z')}{\sqrt{(x-x')^2 + (y-y')^2 + (z-z')^2}}\, dx'\, dy'\, dz'$$

$$= -p \int_{-\infty}^{\infty} \frac{\delta'(z')}{\sqrt{x^2 + y^2 + (z-z')^2}}\, dz'$$

As shown in chap. 1, integration over a δ' amounts to an evaluation of minus the derivative at the point where the δ is located. Thus

$$4\pi\epsilon V(\mathbf{r}) = p \left\{ \frac{\partial}{\partial z'} [x^2 + y^2 + (z-z')^2]^{-(1/2)} \right\}_{z'=0}$$

$$= p \frac{z}{(x^2 + y^2 + z^2)^{3/2}}$$

$$= \frac{p \cos\theta}{r^2}$$

in which $\quad r^2 = x^2 + y^2 + z^2 \quad$ and $\quad r\cos\theta = z.$

PROBLEMS

1. If a flat conducting surface could have placed on it a surface charge density $\rho_s = 1$ coulomb per square meter, what would be the value of the electric field strength E at its surface?

2. A point charge q is located a distance h above an infinite conducting plane. Using the method of images find the displacement density normal to the plane and hence show that the surface charge density on the plane is

$$\rho_s = -\frac{qh}{2\pi r^3}$$

where r is the distance from the charge q to the point on the plane. Integrate this expression over the plane to show that the total charge on its surface is $-q$.

3. Show that the capacitance of an isolated sphere of radius R is

$$4\pi\epsilon_0 R \text{ farads}$$

4. Verify that the capacitance between two spheres, whose separation d is very much larger than their radii R, is given approximately by

$$C \approx \frac{4\pi\epsilon_0 Rd}{2(d-R)} \approx 2\pi\epsilon_0 R$$

Hence show that the capacitance of a sphere above an infinite ground plane is independent of the height h above the plane when $h \gg R$.

5. Verify that the expression for the potential due to an electric dipole satisfies the Laplace equation.

6. Verify that the expression obtained for the potential due to two parallel oppositely charged wires, viz.,

$$V = \frac{q'}{2\pi\epsilon} \ln \frac{r_b}{r_a} = \frac{q'}{2\pi\epsilon} \ln \frac{\sqrt{\left(x - \frac{d}{2}\right)^2 + y^2}}{\sqrt{\left(x + \frac{d}{2}\right)^2 + y^2}}$$

is a solution of the Laplace equation.

7. A *very long* cylindrical conductor of radius a has a charge q coulombs per meter distributed along its length. Find the electric field strength E in air normal to the surface of the conductor (a) by applying Gauss's law; (b) by finding the potential V and deriving the electric field strength from $E = -\nabla V$.

8. (a) Verify that $V = \ln \cot \theta/2$ is a solution of $\nabla^2 V = 0$. (b) Hence show that the capacitance per unit length between two infinitely long coaxial cones (Fig. 2-22), placed tip to tip with an infinitesimal gap between them, is

$$C = \frac{\pi\epsilon}{\ln \cot \theta_1/2} \approx \frac{\pi\epsilon}{\ln 2/\theta_1}$$

for small angles of θ_1.

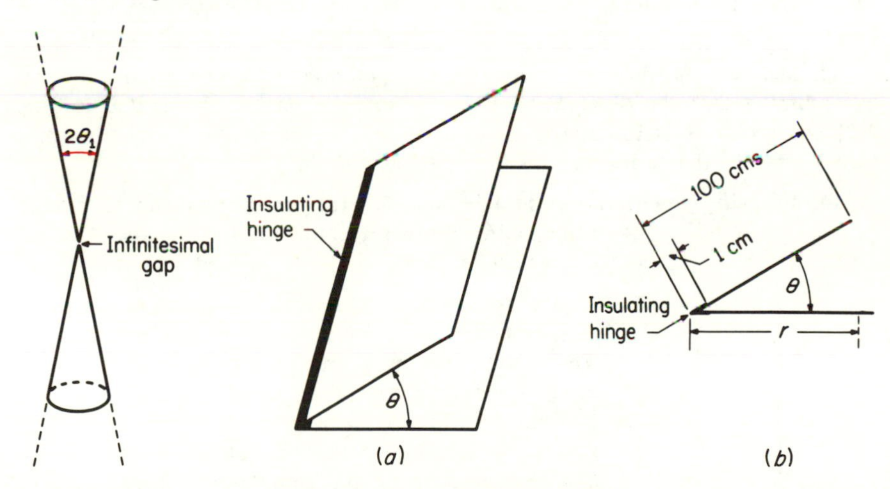

Figure 2-22. Figure 2-23.

9. (a) Find the electric field distribution between the hinged plates [Fig. 2-23(*a*)] and the charge distribution on the plates in a region not too close to the edges (that is, neglect fringing). The plates are insulated at the hinge.

(b) If the plates are 1 meter wide and very long [Fig. 2-23(*b*)], estimate roughly the capacitance between them per meter length when $\theta = 10$ degrees; when $\theta = 180$ degrees. The insulating hinge extends from $r = 0$ to $r = 1$ cm. (Don't forget the outside surfaces).

10. The general definition for the voltage between two points in an electromagnetic field is

$$V_{ab} = \int_a^b \mathbf{E} \cdot d\mathbf{s}$$

By taking the point b to infinity, show that in an electrostatic field due to a charge q the voltage at a (with respect to the voltage at infinity) is the same as the potential at a as defined on page 36. That is, show that

$$V_{a-\infty} = \frac{q}{4\pi\epsilon R}$$

where R is the distance of a from the charge q.

11. By the methods of sec. 2.12 derive a set of solutions to Laplace's equation (a) in cylindrical co-ordinates, starting with

$$V = k\phi$$

(b) in spherical co-ordinates starting with

$$V = k \ln \tan \frac{\theta}{2}$$

12. In example 9, the capacitance of a thin, isolated wire was calculated approximately using the average potential (at the wire surface) of a uniform line charge. Re-formulate the solution to the same order of accuracy using the concept of the energy required to assemble a charge distribution in space.

13. Derive a far-field expansion for the potential of a charge distribution by assuming that the observation point is so far from the charge that in Fig. 2-21 the vector $\mathbf{R}$ is essentially parallel to the vector $\mathbf{r}$. Work out the first three terms in the series and compare them with (82).

14. Find the potential in the far field for the linear quadrupole having three point charges located on the z axis. Assume charges $2Q$ at $z = 0$, $-Q$ at $z = a$ and $-Q$ at $z = -a$.

15. A grounded spherical conductor of radius a is surrounded by a charge cloud of uniform density. The cloud contains a total charge Q and is spherically symmetric, extending from radius a to radius b. Find $\mathbf{E}$ and V everywhere in space ($V = 0$ at $r = \infty$). Also find both the total induced charge and the pressure on the conducting sphere.

16. A distributed dipole charge distribution has a charge density given by (see Fig. 2-24)

$$\rho(\mathbf{r}) = \frac{Q}{L} \delta(x)\, \delta(y)\, T(z)$$

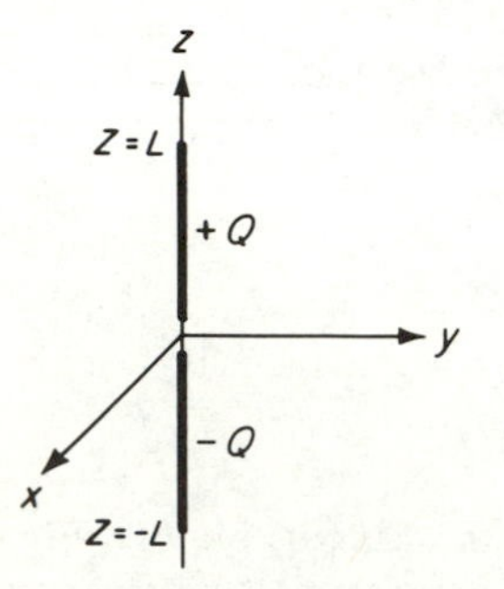

Figure 2-24.

$$
\begin{aligned}
T(z) &= +1 \quad \text{for} \quad 0 < z < L \\
&= -1 \quad \text{for} \quad -L < z < 0 \\
&= 0 \quad \text{elsewhere}
\end{aligned}
$$

(a) Find the potential V in the far field.

(b) Find a closed-form expression for E_z valid everywhere in space.

BIBLIOGRAPHY

Durand, E., *Electrostatique et Magnetostatique*, Masson, Paris, 1953.

Howe, G. W. O., "On the Capacity of Radio-Telegraphic Antennae," *The Electrician*, p. 829, Aug. 28, 1914; also LXXV, p. 870.

Jackson, J. D., *Classical Electrodynamics*, John Wiley & Sons, Inc., New York, 1962.

Jeans, J. H., *The Mathematical Theory of Electricity and Magnetism*, Cambridge University Press, London, 1946.

Page, L., and N. I. Adams, Jr., *Principles of Electricity*, D. Van Nostrand Co., Inc., Princeton, N. J., 1931.

Panofsky, W. K. H., and M. Phillips, Classical Electricity and Magnetism (2nd ed.), Addison-Wesley Publishing Co., Inc., Reading, Mass., 1962.

Phillips, H. B., *Vector Analysis*, John Wiley & Sons, Inc., New York, 1933.

Plonsey, R., and R. E. Collin, *Principles and Applications of Electromagnetic Fields*, McGraw-Hill Book Company, New York, 1961.

Smythe, W. R., *Static and Dynamic Electricity*, McGraw-Hill Book Company, New York, 1939.

Weber, Ernst, *Electromagnetic Fields: Theory and Applications*, Vol. I, *Mapping of Fields*, John Wiley & Sons, Inc., New York, 1950.

Chapter 3

THE STEADY MAGNETIC FIELD

Electric charges at rest produce an electric field—the electrostatic field. Electric charges in motion, that is, electric currents, produce a *magnetic field.* This is evidenced by the fact that in the region about a wire carrying a current, each end of a magnetic compass needle experiences a force dependent upon the magnitude of the current. There is said to be a magnetic field about the wire, and the direction of the magnetic field is taken to be that in which the north-seeking pole of the compass needle is urged. The strength H of the magnetic field was originally defined by Coulomb in a manner similar to that for the electric field strength E. A unit magnetic pole was first defined in terms of the force between two similar poles, and then the magnetic field strength was defined in terms of the force per unit pole. In electromagnetic wave theory, magnetic fields due to electric currents are of chief concern and the effects of permanent magnets are of little importance. Therefore the above approach will be discarded for one that leads more directly to a solution of the type of problems encountered in electromagnetic engineering.

3.01 Theories of the Magnetic Field. It is possible to develop a quantitative theory of the magnetic field from any of several different starting points. Rowland's experiments showed that moving charges produce magnetic effects. Therefore a theory based upon the magnetic forces between individual moving charges would be logical. In this theory permanent-magnet effects are ascribed to the motion of external electrons about the atomic nuclei. This theory is used in modern physics, and can be developed to answer most of the questions that arise in connection with magnetism. Some such fundamental approach is required whenever it is necessary to deal with individual charges, but in most engineering problems, where only macroscopic effects are considered, such a procedure involves an unnecessary complexity. The motion of a single electron in a wire is an erratic and highly unpredictable affair, subject to forces that vary greatly in the small length of interatomic distances. Yet, the intelligent sophomore experiences little difficulty in predicting with fair accuracy the statistical *average* motion of millions of electrons by a simple application of Ohm's law. For most engineering problems it will be the magnetic effect of *currents* rather than the motion of individual charges that will be of importance, and it would seem reasonable to use the forces between currents as a starting point. Ampere's experiments on the force between current-carrying conductors form a logical starting point for this development and lead to quite satisfactory engineering definitions, especially when the end result desired is in terms of mechanical forces; for completeness, this approach is outlined briefly in sec. 3.10. In electromagnetic wave theory, however, primary interest is in the relations between electric and magnetic fields, and a different starting point proves to be convenient. This starting point is Faraday's induction law, which relates the magnetic flux through a closed path to the voltage induced around the path. This relation, which *defines* magnetic flux in terms of a measurable electric voltage, is the starting point that will be used in the present discussion of magnetic fields.

Still another attack that is often used in electromagnetic theory is to *postulate* a vector potential due to the currents, and then obtain a magnetic field in terms of this potential. This vector-potential method has the marked advantage that it can be readily extended to the general case where the currents vary with time—the electromagnetic field—and in this latter case it will also yield directly the electric field produced by changing currents. In general, use of the vector-potential method simplifies the mathematical analyses and facilitates the solution of electromagnetic problems. Therefore it will be developed and used. However, instead of starting with a postulated potential and deducing from it the electric and magnetic fields, the reverse procedure will be used. The electric and magnetic vectors will be defined in terms of relations derived from experiments, often performed under restricted conditions. These definitions will then be generalized for use in the electromagnetic field, and in the process a potential will be found such that the space and time derivatives of this potential will give the magnetic and electric fields. The generalizations may be considered valid as long as conclusions derived from them agree with subsequent experiment.

Finally, a most logical approach to a theory of the magnetic field is through an application of special relativity theory to an "electrostatic" field, the source charges of which are in motion relative to the observer. By this means, it can be demonstrated that "magnetic" effects follow directly from Coulomb's law and the Lorentz contraction applied to moving charge densities. Because of its generality, and the deeper understanding of electromagnetic theory that it brings, this method will be treated in some detail (in chap. 18). However, it is believed that better appreciation of the power of this approach is gained after the student has attained some familiarity with Maxwell's equations and basic electromagnetic theory. Therefore for the present the magnetic field will be introduced through Faraday's law.

3.02 Magnetic Induction and Faraday's Law. In the experimental setup indicated in Fig. 3-1, a ballistic galvanometer is connected to a loop placed near a long straight wire, carrying a current I. Probing with a magnetic compass needle shows that there is a magnetic field in the region about the wire. At the position of the loop shown, the direction of the field is out of the plane of the paper for an upward flow of the current I. If now the current I is reduced to zero, the galvanometer is deflected, the amount of the deflection being independent of the rate at which the current is reduced to zero, so long as the time required is short compared with the period of the galvanometer. The current I_g through the galvanometer flows as a result of a voltage V "induced" in the loop and is given by

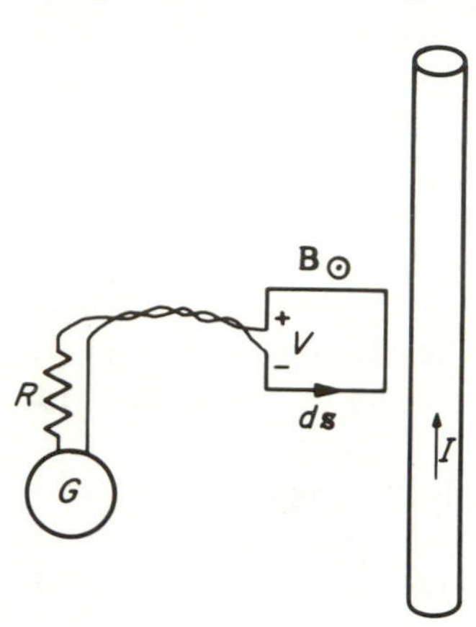

Figure 3-1. Measurement of magnetic flux.

$$I_g = \frac{V}{R}$$

where R is the total resistance in the galvanometer circuit (R is a very large resistance). The galvanometer deflection is a measure of the charge Q or the time integral of the current through it, so that

$$Q = \int_0^t I_g \, dt = \frac{1}{R}\int_0^t V \, dt$$

is an experimentally determinable quantity. *Magnetic flux* Φ through the loop is then defined as the time integral of voltage induced in the loop throughout the interval during which the magnetic field is being established; or having been established, as the time integral of voltage throughout the interval in which the field is being reduced to zero. (These quantities are equal but of opposite sign.) That is,

$$\Phi = \pm \int_0^t V \, dt \tag{3-1}$$

where the time interval 0 to t is that required to establish the field or reduce it to zero. Differentiating with respect to time gives

$$V = -\frac{d\Phi}{dt} \tag{3-2}$$

which is Faraday's induction law. Consistent with a right-hand coordinate system, the negative sign has been used to indicate that when the flux is increasing in the positive direction (out of the paper through the loop in Fig. 3-1), the induced voltage is negative. It is evident from eq. (1) or (2) that the unit of magnetic flux is the volt-second; this unit has been named the *weber*.

3.03 Magnetic Flux Density B. The magnetic flux per unit area through a loop of small area is called the *magnetic flux density* **B** at the location of the loop. Because the flux through the loop depends upon the orientation of the loop as well as upon its area, magnetic flux density is a *vector* quantity. The direction of **B** is taken as the normal to the plane of the loop when oriented to enclose maximum flux. The positive sense of **B** is the direction of the magnetic field at the point in question. The unit of magnetic-flux density is the *tesla* or *weber per square meter*. The magnetic flux through any surface is the surface integral of the normal component of **B**, that is

$$\Phi = \int_S B_n \, da = \int_S \mathbf{B} \cdot d\mathbf{a}$$

3.04 Magnetic Field Strength H and Magnetomotive Force $\mathscr{F}$. Using a small probe loop and galvanometer as in Fig. 3-1, it is possible to determine **B** at all points in a region about a long current-carrying wire. Experiment shows that for a homogeneous medium, B is related to the current I through

$$B \propto \frac{\mu I}{r} \tag{3-3}$$

where r is the distance from the wire and μ is a constant that depends upon the medium. The constant μ, called the *permeability* of the medium, may be written as

$$\mu = \mu_r \, \mu_v$$

where μ_v is the absolute permeability of a vacuum, and μ_r is the *relative permeability* (relative to a vacuum). μ_v is the basic *defined* electrical unit, which has been assigned the value

$$\mu_v = 4\pi \times 10^{-7} \qquad \text{henry/meter}$$

in the rationalized MKS system of units. Using this value of μ_v, and probing the field about the wire in a vacuum for which $\mu_r = 1$ (or in air for which $\mu_r \approx 1$), the proportionality factor in (3) is found to be $1/2\pi$, so that the relation becomes

$$B = \frac{\mu I}{2\pi r} = \mu H \qquad \text{(3-4)}$$

where

$$H = \frac{I}{2\pi r} \qquad \text{A/m}$$

The *magnetic field strength* H is thus defined by this relation in terms of the current which produces it and the geometry of the system. Magnetic field strength is a vector quantity, having the same direction as the magnetic flux density (in isotropic media), so the equality expressed by (4) can be stated as the vector relation

$$\mathbf{B} = \mu \mathbf{H} \qquad \text{(3-4a)}$$

Although no longer defined in terms of unit poles, the relative value of H at any point may be indicated by the force on one end of a magnetized compass needle.

Under the conditions of the above (long-wire) experiment, H, the magnitude of the magnetic field strength is independent of the permeability of the medium, depending only on the current and distance from it, while B is dependent on the permeability of the medium. In this sense H may be pictured as a magnetic "stress" that drives a "resultant" flux density through the medium. This particular viewpoint, and indeed the names selected for the vectors E, D, B and H, suggest that H is analogous to E and B is analogous to D. However, as is discussed in sec. 3.14, an equally valid analogy relates E to B and D to H.

The line integral

$$\mathscr{F} = \int_a^b \mathbf{H} \cdot d\mathbf{s}$$

is defined as the *magnetomotive force* between the points a and b. For a circular path about the wire, with the wire at the center, $\mathbf{H}$ has the constant value $I/2\pi r$ and is directed along the path, so that

$$\mathscr{F} = \oint \mathbf{H} \cdot d\mathbf{s} = I \qquad \text{(3-5)}$$

It is easily demonstrated that this same result (5) will be obtained for *any* closed path about the current. Equation (5) is *Ampere's work law* or *Ampere's circuital law.* The positive directions (or senses) of magnetomotive force and current are related by the familiar "right-hand rule."

Ampere's work law makes it easy to compute H in certain problems. For example, consider the toroidal coil of Fig. 3-2, consisting of

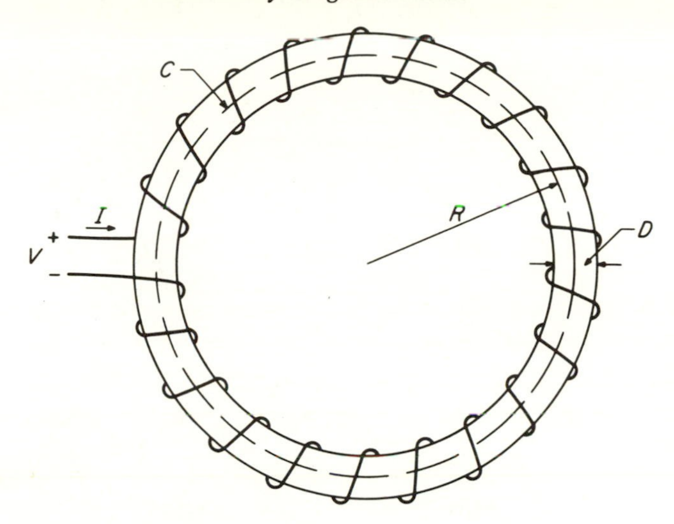

Figure 3-2. Toroidal coil.

a large number of closely spaced turns on a tubular core. For any closed path C taken around the core inside the winding, the magnetomotive force will be

$$\mathscr{F} = nI$$

where n is the number of turns and, therefore, the number of times the path links with the current I. If D, the thickness of the core, is small compared with R, the radius of the ring, the radii of all circular paths through the core are approximately equal to R, so that at any point within the core

$$H \approx \frac{\mathscr{F}}{2\pi R} = \frac{nI}{2\pi R} = \frac{nI}{l} \quad \text{ampere turns/m}$$

with $l = 2\pi R$ denoting the length of the coil. The magnetic field strength is nearly uniform throughout the cross section of the core and is equal to the ampere turns per unit length.

Another example, in which **H** is simply related to the current that produces it, is the case of two very large closely spaced parallel planes carrying equal and oppositely directed currents (Fig. 3-3). The magnetic field is confined to the region between the planes and is found to be uniform (except near the edges) and independent of the distance apart of the planes as long as this distance is small compared with the other dimensions. In Fig. 3-3 the current is assumed to be flowing in the positive x direction (outward) in the upper plate.

Then if J_{sx} represents the current *per meter width* flowing in this plate, Ampere's work law states that

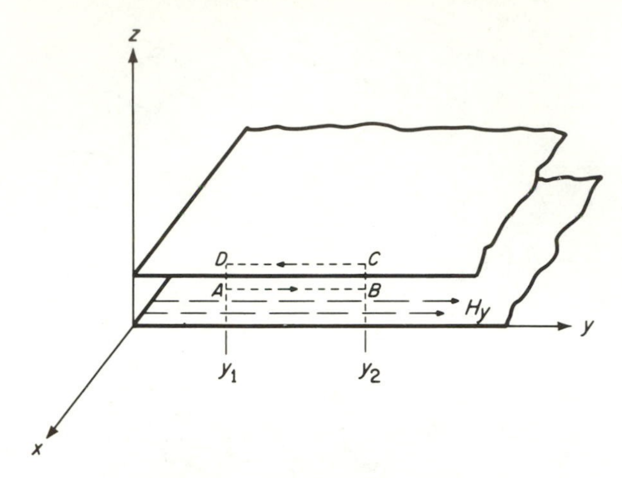

Figure 3-3. Parallel-plane conductors.

$$\oint_{ABCDA} \mathbf{H} \cdot d\mathbf{s} = H_y(y_2 - y_1) = J_{sx}(y_2 - y_1)$$

from which

$$H_y = J_{sx} \qquad \text{A/m}$$

The magnetic field strength is equal in magnitude to the linear current density (amperes per meter width) flowing in each of the planes. It is parallel to the planes, but perpendicular to the direction of current flow.

3.05 Ampere's Work Law in the Differential Vector Form. Ampere's work law states that the magnetomotive force around a closed path is equal to the current enclosed by the path. That is

$$\oint \mathbf{H} \cdot d\mathbf{s} = I \qquad \text{amperes} \tag{3-5}$$

This law may be put into an alternative form as follows: Consider a conducting region in which there is a current density **J** (Fig. 3-4). Let $ABCD$ be an element of area parallel to the x-y plane, and let the co-ordinates of the point A be (x, y, z). The magnetomotive force around the closed path $ABCDA$ can be obtained by summing the magnetomotive forces along the four sides of the rectangle. If the average value of H_x over the path AB is represented by $\bar{H}_x$ and the average value of H_y over the path AD is represented by $\bar{H}_y$, then the following relations will hold:

$$\text{mmf from } A \text{ to } B = \bar{H}_x \, \Delta x$$

$$\text{mmf from } B \text{ to } C = \left(H_y + \frac{\partial \bar{H}_y}{\partial x} \Delta x\right) \Delta y$$

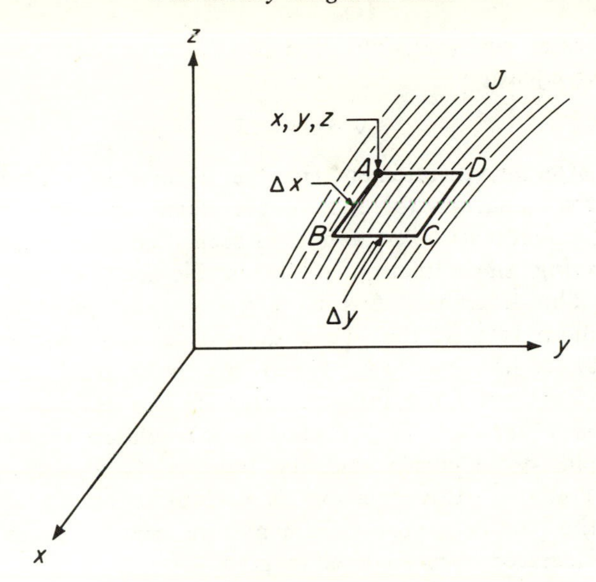

Figure 3-4.

$$\text{mmf from } C \text{ to } D = -\left(\bar{H}_x + \frac{\partial \bar{H}_x}{\partial y}\Delta y\right)\Delta x$$

$$\text{mmf from } D \text{ to } A = -\bar{H}_y\,\Delta y$$

Adding on both sides,

$$\text{mmf around closed path} = \left(\frac{\partial \bar{H}_y}{\partial x} - \frac{\partial \bar{H}_x}{\partial y}\right)\Delta x\,\Delta y$$

The current flowing through this rectangle is

$$dI = J_z\,\Delta x\,\Delta y$$

Therefore by Ampere's law,

$$\left(\frac{\partial \bar{H}_y}{\partial x} - \frac{\partial \bar{H}_x}{\partial y}\right)\Delta x\,\Delta y = J_z\,\Delta x\,\Delta y$$

As Δx and Δy are allowed to approach zero, $\bar{H}_x$ becomes H_x, and $\bar{H}_y$ becomes H_y, so that in the limit

$$\frac{\partial H_y}{\partial x} - \frac{\partial H_x}{\partial y} = J_z \tag{3-6a}$$

Next if the element of area is taken parallel to the y-z plane, and then parallel to the z-x plane, the following relations are obtained:

$$\frac{\partial H_z}{\partial y} - \frac{\partial H_y}{\partial z} = J_x \tag{3-6b}$$

$$\frac{\partial H_x}{\partial z} - \frac{\partial H_z}{\partial x} = J_y \tag{3-6c}$$

The three scalar eqs. (6a), (6b), and (6c) can be combined into the single vector equation

$$\nabla \times \mathbf{H} = \mathbf{J} \tag{3-6}$$

This is an alternative statement (in the differential vector form) of Ampere's law. Equations (5) and (6) are stated correctly for a right-hand set of co-ordinate axes, and it is seen that the right-hand rule for determining the direction of **H** is included in both of these statements. The differential forms (6) of the equation require a *homogeneous* medium because the space derivative has no meaning at a discontinuity of the medium. When the path under consideration crosses a discontinuity the integral form (5) is suitable.

Interpretation of $\nabla \times \mathbf{H}$. Equation (6) relates the curl of the magnetic field to the current density that exists at any point in a region. A study of this relation is helpful in obtaining a physical picture of the curl of a vector. The picture can be made clearer if eq. (6) is integrated over an area to give

$$\int_S \nabla \times \mathbf{H} \cdot d\mathbf{a} = \int_S \mathbf{J} \cdot d\mathbf{a} \tag{3-7}$$

The right-hand side of (7), being the current density integrated over a surface S, is just the total current I flowing through the surface. Recalling the original form of the statement of Ampere's law in eq. (5) shows that the following relation must be true:

$$\int_S \nabla \times \mathbf{H} \cdot d\mathbf{a} = \oint \mathbf{H} \cdot d\mathbf{s} \tag{3-8}$$

This relates the integral of $\nabla \times \mathbf{H}$ over a surface to the line integral of **H**, or magnetomotive force, around the closed path bounding the surface. If the surface is reduced to an element of area da, the left-hand side becomes $(\nabla \times \mathbf{H}) \cdot d\mathbf{a}$. Dividing through by da, the result is $\nabla \times \mathbf{H} \cdot \hat{\mathbf{n}} = (\oint \mathbf{H} \cdot d\mathbf{s})/da$, which may be interpreted as: $\nabla \times \mathbf{H}$ equals the magnetomotive force per unit area. The *direction* of $\nabla \times \mathbf{H}$ is that direction $\hat{\mathbf{n}}$ of the area $d\mathbf{a}$ that results in a maximum magnetomotive force around its edge.

Relation (8), derived here for the magnetic vector **H**, is simply a statement of the general relation (1-50) known as Stokes' theorem. Stokes' theorem, relating a line integral to a surface integral, provides a convenient means for interchanging an integral formulation and a differential vector formulation. For example, it is evident that in electrostatics the relation

$$\nabla \times \mathbf{E} \equiv 0$$

follows directly from the statement that the voltage around any closed path is zero.

3.06 Permeability μ. In the examples of sec. 3.04 the magnetic field strength H has been related directly to the current that produces it. Using the toroidal coil example the magnetic flux Φ and therefore the magnetic flux density $\mathbf{B}$ within the core can be measured; the magnetomotive force $\oint \mathbf{H} \cdot d\mathbf{s}$, and therefore H, is known in terms of the current; therefore μ, and hence μ_r, the relative permeability of the medium composing the core, can be determined from the relations

$$\mu = \frac{B}{H}; \qquad \mu = \mu_r \mu_v; \qquad \mu_v = 4\pi \times 10^{-7} \qquad \text{henry/meter} \quad (3\text{-}9)$$

For air and most materials the relative permeability is very nearly unity. For *paramagnetic* substances μ_r is very slightly greater than unity; thus for air it is 1.00000038 and for aluminum it is 1.000023. For *diamagnetic* substances μ_r is slightly less than unity; for copper $(1 - 8.8 \times 10^{-6})$; for water $(1 - 9.0 \times 10^{-6})$. However, for that exceptional class of materials known as *ferromagnetic* materials (iron and certain alloys) the relative permeability may have a value of several hundred or even several thousand. In general, the permeability of these materials is not constant but depends upon the strength of the magnetic field and upon their past magnetic history. However, for most applications of interest in electromagnetic wave theory, the range of flux densities involved is small enough that μ may be considered constant.

3.07 Energy Stored in a Magnetic Field. It is found experimentally that a certain amount of work is required to establish a current in a circuit. This work is done in establishing the current against the electromotive force induced in the circuit by the increasing magnetic flux, and the energy thus transferred to the circuit is said to be stored in the magnetic field. The amount of the energy so stored can be determined in terms of the extent and strength of the magnetic field by considering the elementary example of current flow in a toroidal coil as shown in Fig. 3-13. In this case, when the turns are closely spaced, the magnetic field is confined to the core of the toroid, and the magnetic field strength is given by

$$H = \frac{nI}{l}$$

where n is the number of turns and l is the mean length of path through the core of the coil. The back voltage induced in the coil is, by definition of magnetic flux Φ,

$$V = -n\frac{d\Phi}{dt}$$

$$= -nA\frac{dB}{dt}$$

where B is the magnetic flux density, and A is the cross-sectional area of the core. The work done in establishing the current I in the coil is

$$W = -\int_0^{t_1} VI\,dt$$

$$= \int_0^{t_1} lAH\frac{dB}{dt}\,dt$$

$$= \int_0^{H_1} \mu lAH\,dH$$

$$= lA\left[\frac{\mu H_1^2}{2}\right] \tag{3-10}$$

This is the total energy stored in the field, and since lA is the volume of the region in which the magnetic field exists, it is inferred that the quantity

$$\frac{\mu H_1^2}{2}$$

represents the energy density of the magnetic field. Whether or not it is considered desirable to ascribe a certain energy density to each small volume of space and so "locate" the energy, it is nevertheless true in general that the quantity $\mu H^2/2$, when integrated over the whole volume (in which H has value), does give the correct value for the total stored magnetic energy of the system.

3.08 Ampere's Law for a Current Element. When a current I flows in a closed circuit the magnetic field strength $\mathbf{H}$ at any point is a result of this flow in the complete circuit. For computational purposes it is convenient to consider the total magnetic field strength at any point as the sum of contributions from elemental lengths ds of the circuit, each carrying the current I. The quantity $I\,ds$ is called a *current element*. It is a vector quantity having the direction of the current, or what amounts to the same thing, the direction of the element ds in which the current flows. This may be indicated by writing $\mathbf{I}\,ds$ or alternatively $I\,d\mathbf{s}$. The two notations are used interchangeably and to suit convenience in the particular problem.

The magnitude of the contribution to $\mathbf{H}$ from each current element $I\,d\mathbf{s}$ cannot be measured directly, but is inferred from experimental results to be

$$dH = \frac{I\,ds \sin \psi}{4\pi R^2} \tag{3-11}$$

R is the distance measured outward from the current element $I\,ds$ to the point p at which H is being evaluated (Fig. 3-5). ψ is the angle between the direction of $I\,d\mathbf{s}$ and the direction of $\mathbf{R}$. The direction of $\mathbf{H}$ is perpendicular to the plane containing $I\,d\mathbf{s}$ and $\mathbf{R}$, in the direction in which a right-hand screw would progress in turning from $I\,d\mathbf{s}$ to $\mathbf{R}$. This complete statement can be written in vector notation simply as

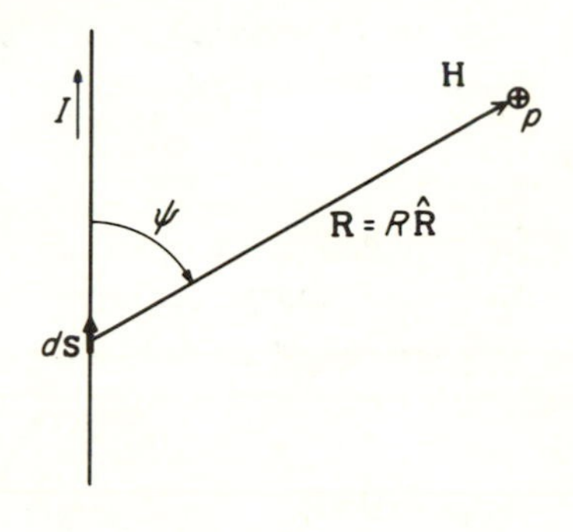

Figure 3-5.

$$d\mathbf{H} = \frac{I\,d\mathbf{s} \times \hat{\mathbf{R}}}{4\pi R^2} \tag{3-12}$$

where $\hat{\mathbf{R}}$ is a unit vector in the $\mathbf{R}$ direction. Equation (12) is known as *Ampere's law for a current element* (or sometimes as the *Biot-Savart law*).

The total magnetic field strength $\mathbf{H}$ at a point p will be the sum or integration of the contributions from all the current elements of the circuit and will be

$$\mathbf{H} = \oint \frac{I\,d\mathbf{s} \times \hat{\mathbf{R}}}{4\pi R^2} \tag{3-13}$$

Examples of the use of (13) are given in sec. 3.10.

3.09 Volume Distribution of Current and the Dirac Delta. Since currents are not necessarily confined to filamentary paths, it is useful to generalize (13) to include a current density $\mathbf{J}$ flowing within a volume V, as shown in Fig. 3-6. Under these conditions, the magnetic field strength may be expressed as

$$\mathbf{H}(\mathbf{r}) = \frac{1}{4\pi} \int_V \frac{\mathbf{J}(\mathbf{r}') \times \hat{\mathbf{R}}}{R^2}\, dV' \tag{3-14}$$

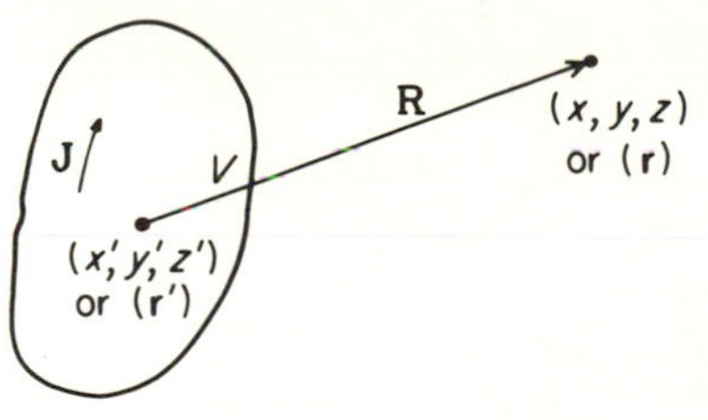

Figure 3-6. A volume distribution of current.

where the prime is used to denote the co-ordinates of the source region. The current density may be used to represent filamentary currents by making use of the properties of the Dirac delta (discussed in chaps. 1 and 2). If the current is confined to a straight filament along the z axis and if the total current is I amperes, then

$$\mathbf{J} = \hat{\mathbf{z}} I\, \delta(x)\, \delta(y)$$

Similarly if the current I is confined to a filamentary loop of radius a, then the current density is given by

$$\mathbf{J} = \hat{\boldsymbol{\phi}}\, I\, \delta(\rho - a)\, \delta(z)$$

The above representations satisfy the relation between current and current density

$$I = \int_S \mathbf{J} \cdot d\mathbf{a} \tag{3-15}$$

in which S is an *open* surface. If S is a closed surface, then conservation of charge requires that the net, steady current passing through the surface be zero. Thus

$$\oint_S \mathbf{J} \cdot d\mathbf{a} = 0$$

Application of the divergence theorem to the above yields

$$\int_V \nabla \cdot \mathbf{J}\, dV = 0$$

in which V is the volume enclosed by S. Since conservation of charge must hold for any arbitrary volume V, it follows that

$$\nabla \cdot \mathbf{J} = 0 \tag{3-16}$$

This is frequently called the *equation of continuity for steady currents.*

3.10 Ampere's Force Law. Ampere's experiments on the forces between current-carrying loops of thin wire led him to a formulation of the law obeyed by these forces. Suppose that two loops designated C_1 and C_2 carry currents I_1 and I_2 as shown in Fig. 3-7. The quanti-

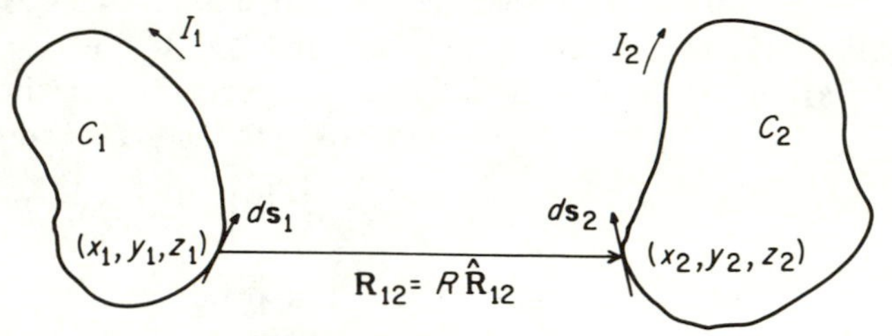

Figure 3-7. Current loops.

ties $d\mathbf{s}_1$ and $d\mathbf{s}_2$ are directed elements of length and the directed distance between them is $R\hat{\mathbf{R}}_{12}$ in which $\hat{\mathbf{R}}_{12}$ is the unit vector drawn from point (1) to point (2). In its most general form, Ampere's force law may be expressed as

$$\mathbf{F}_{21} = \frac{\mu\, I_1\, I_2}{4\pi} \oint_{C_1} \oint_{C_2} \frac{d\mathbf{s}_2 \times (d\mathbf{s}_1 \times \hat{\mathbf{R}}_{12})}{R^2} \tag{3-17}$$

in which $\mathbf{F}_{21}$ is the force on circuit C_2.

It is frequently convenient to regard the force as acting at a distance on circuit C_2 through the intermediary of a magnetic field produced by current flowing in C_1. This can be accomplished by separating (17) as

$$\mathbf{F}_{21} = \mu I_2 \oint_{C_2} d\mathbf{s}_2 \times \mathbf{H}_{21} \tag{3-18}$$

where [corresponding to eq. (13)]

$$\mathbf{H}_{21} = \frac{I_1}{4\pi} \oint_{C_1} \frac{d\mathbf{s}_1 \times \hat{\mathbf{R}}_{12}}{R^2} \tag{3-19}$$

is the magnetic field strength due to the current I_1 in C_1. Expression (19) was first deduced by Biot and Savart and is usually referred to as the *Biot-Savart* law.

It is clear that the separation of force and "field" in (17) could have been effected differently by including μ in (19) and deleting it in (18) to give

$$\mathbf{F}_{21} = I_2 \oint_{C_2} d\mathbf{s}_2 \times \mathbf{B}_{21} \tag{3-20}$$

where

$$\mathbf{B}_{21} = \mu \mathbf{H}_{21} = \frac{\mu I_1}{4\pi} \oint_{C_1} \frac{d\mathbf{s}_1 \times \hat{\mathbf{R}}_{21}}{R^2} \tag{3-21}$$

Example 1: *The field of a long, straight wire.* A long, straight wire may be regarded as part of a loop which is closed at infinity. The magnetic field

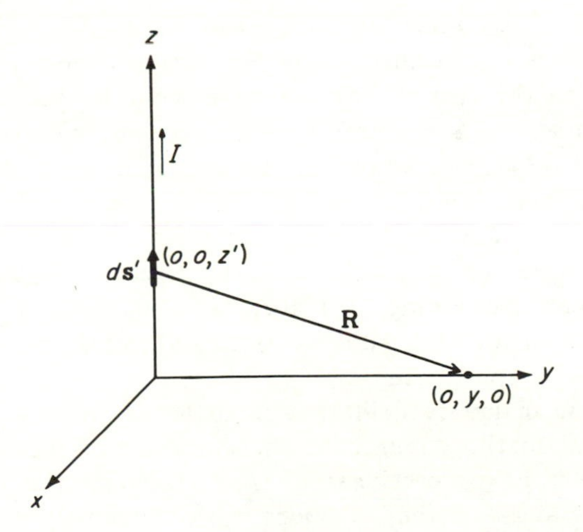

Figure 3-8.

strength may be calculated as follows:

$$\begin{aligned}
\mathbf{H}(o, y, o) &= \frac{I}{4\pi}\int_{-\infty}^{\infty} \frac{d\mathbf{s}' \times \hat{\mathbf{R}}}{R^2} \\
&= -\hat{\mathbf{x}}\frac{I}{4\pi}\int_{-\infty}^{\infty} \frac{y}{\sqrt{z'^2 + y^2}}\frac{dz'}{(z'^2 + y^2)} \\
&= -\hat{\mathbf{x}}\frac{I}{4\pi}\frac{1}{y}\frac{z'}{\sqrt{z'^2 + y^2}}\bigg|_{-\infty}^{\infty} \\
&= -\hat{\mathbf{x}}\frac{I}{2\pi y}
\end{aligned}$$

This result usually is expressed in cylindrical co-ordinates as

$$\mathbf{H} = \hat{\phi}\frac{I}{2\pi\rho}$$

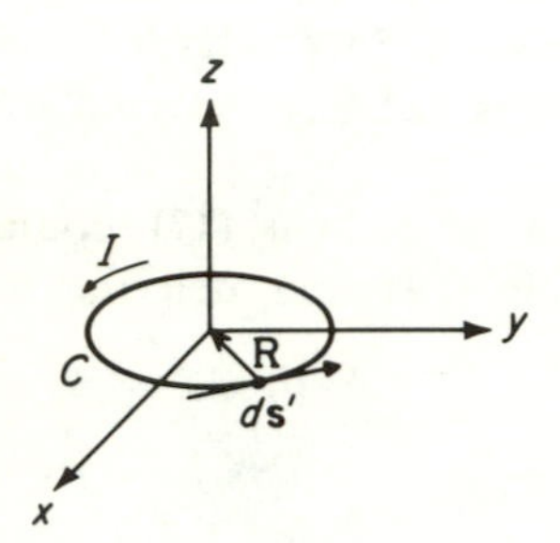

Figure 3-9.

EXAMPLE 2: *The Field at the Center of a Circular Loop.*

$$\begin{aligned}
\mathbf{H}(0, 0, 0) &= \frac{I}{4\pi}\oint_C \frac{d\mathbf{s}' \times \hat{\mathbf{R}}}{R^2} \\
&= \hat{\mathbf{z}}\frac{I}{4\pi}\int_0^{2\pi} \frac{R\, d\phi}{R^2} \\
&= \hat{\mathbf{z}}\frac{I}{2R}
\end{aligned}$$

3.11 Magnetic Vector Potential. In the electric field it was found desirable to introduce the concept of potential. In that case the electric potential was a space function that depended upon the magnitude and location of the charges, the charges being the *sources* of the electric field. The electric field was obtained from the potential V by taking the gradient or space derivative of V. This procedure was often found to be much simpler than that of trying to obtain $\mathbf{E}$ directly in terms of the magnitude and location of the charges.

Similarly in the case of the magnetic field it would be desirable to be able to set up a *magnetic potential*, the space derivative of which would give $\mathbf{B}$ or $\mathbf{H}$. Corresponding to the individual charges in the electric field case, the sources of the magnetic field would be the current elements $I\,ds$ of the circuits that produce the field. The magnetic potential being sought would therefore depend upon these current elements. Assuming that a suitable magnetic potential can be found, the properties that such a potential must possess are easily determined by simple reasoning.

Because the magnetic field that is to be derived from the potential is proportional to the strength of the current element $I\,ds$, the potential itself must be proportional to $I\,ds$. Because the magnetic field strength due to a current element varies inversely as the *square* of the distance R from the element (Ampere's law), the magnetic potential due to the current elements must vary inversely as the *first* power of the distance because the magnetic field is to be obtained by taking the space derivative of the potential. This is equivalent to dividing by R, as far as dimensions are concerned. The electrostatic potential due to charges was a scalar quantity. This was adequate in that case, because the charges themselves were scalars having magnitudes only. In the present case, the current elements have directions as well as magnitudes, and it is necessary that this additional information on the *direction* of the source be contained in the potential due to the source. Therefore the potential in this case must be a *vector* quantity, the direction of which will somehow be related to the direction of the current-element source. If this *vector magnetic potential* is designated by the vector $\mathbf{A}$, then it should be possible to obtain $\mathbf{B}$ or $\mathbf{H}$ as the space derivative of $\mathbf{A}$. There are two possible space-derivative operations on a vector quantity, namely the divergence and the curl. The divergence operation yields a scalar quantity, whereas the curl operation yields a vector quantity. Inasmuch as the resulting magnetic field is a vector quantity, the curl is the only space-derivative operation which can be used. Therefore, if there is a suitable vector magnetic potential $\mathbf{A}$, the vectors $\mathbf{H}$ or $\mathbf{B}$ should be derivable from it through the relations

$$\mathbf{H} = \nabla \times \mathbf{A} \qquad \text{or} \qquad \mathbf{B} = \nabla \times \mathbf{A}$$

Both of these relations have been used in the past, but in recent years the latter has achieved wider acceptance; accordingly

$$\mathbf{B} = \nabla \times \mathbf{A} \tag{3-22}$$

will be adopted for this book.* For magnetically isotropic and homogeneous media (with which this book is solely concerned) the permeability μ is a simple constant, so the two forms are equivalent except for a constant multiplying factor. As indicated above, the relation between the magnetic vector potential and the current element source must be of the form

$$d\mathbf{A} = k\left(\frac{I\,d\mathbf{s}}{R}\right) \tag{3-23}$$

where the constant k is still to be determined. With one eye on eq. (12) a reasonable guess for the expression for the vector potential of a current element $I\,d\mathbf{s}$ would appear to be

$$d\mathbf{A} = \frac{\mu I\,d\mathbf{s}}{4\pi R} \tag{3-24}$$

*The first edition used $\mathbf{H} = \nabla \times \mathbf{A}$. The authors have taken considerable pains to make all the necessary changes, but the alert student may discover omissions.

The vector magnetic potential (usually called just vector potential) due to current flow in a complete circuit is obtained as a summation or integration of vector potentials caused by all the current elements that comprise the circuit. That is

$$\mathbf{A} = \int \frac{\mu I\, d\mathbf{s}}{4\pi R} \tag{3-25}$$

where the integration extends over the complete circuit in which I flows. As mentioned previously, the *direction* of a current element can be indicated by making either ds or I the vector quantity. In the latter case the expression for **A** would be

$$\mathbf{A} = \int \frac{\mu \mathbf{I}\, ds}{4\pi R}$$

This expression can be written in a more general form by replacing the current I by a current density **J** and then integrating over the *volume* in which this current density exists. Then the expression for the vector potential **A** is

$$\mathbf{A} = \int_V \frac{\mu \mathbf{J}\, dV}{4\pi R}$$

This reduces to the previous expression when the current flows in a filamentary circuit.

Problem 1. Verify that if the proposed expression (24) for vector potential of a current element is used with (22), the correct expression for magnetic field (11 or 12) is obtained.

3.12 The Vector Potential (Alternative derivation). The integrand in (14) may be modified by making use of the ∇ operator operating on the unprimed co-ordinates.

$$\begin{aligned} \frac{\mathbf{J}(\mathbf{r}') \times \hat{\mathbf{R}}}{R^2} &= -\mathbf{J}(\mathbf{r}') \times \nabla \frac{1}{R} \\ &= \nabla \times \frac{\mathbf{J}(\mathbf{r}')}{R} \end{aligned} \tag{3-26}$$

This modification makes use of the vector identity $\nabla \times (\phi \mathbf{A}) = \phi \nabla \times \mathbf{A} + \nabla \phi \times \mathbf{A}$ and of the fact that **J** is a constant with respect to the unprimed co-ordinates. When (26) is substituted into (14), the curl operation may be taken outside the integral. Use of eq. (4a) permits **B** to be expressed as

$$\mathbf{B}(\mathbf{r}) = \nabla \times \frac{1}{4\pi} \int_V \frac{\mu \mathbf{J}(\mathbf{r}')}{R}\, dV' \tag{3-27}$$

The above formula suggests immediately a definition for the *vector potential* **A**

$$\mathbf{A}(\mathbf{r}) = \frac{1}{4\pi}\int_V \frac{\mu\mathbf{J}(\mathbf{r}')}{R}\,dV' \tag{3-28}$$

Thus the magnetic flux density may be obtained by taking the curl of **A**:

$$\mathbf{B} = \nabla \times \mathbf{A} \tag{3-29}$$

Because div curl of any vector is zero, it follows directly that the divergence of **B** is zero.

The use of the vector potential is a great convenience in finding the magnetic field due to a given current. The reason for this is the simplicity of the integral in (28) compared to the integral in (14). This simplicity is largely due to the fact that **A** is parallel to **J** (or, more precisely, to each volume element of current density there corresponds a parallel element of vector potential). Evaluation of (28) is most easily accomplished by evaluating separately its three rectangular components.

EXAMPLE 3: *Magnetic Field about a Long Straight Wire.* Using the vector potential, let it be required to find the magnetic field strength about a long straight wire carrying a current I.

The general expression for vector potential is

$$\mathbf{A} = \int_V \frac{\mu\mathbf{J}\,dV'}{4\pi R}$$

For this problem the current density **J**, integrated over the cross section of the wire, gives the total current I. Also the current is entirely in the z direction (Fig. 3-10) so that **A** has only one component, A_z.

Then, in a homogeneous medium,

$$A_z = \frac{\mu}{4\pi}\int_{-L}^{+L}\frac{I\,dz}{R}$$

If the point P is taken in the y-z plane, $R = \sqrt{z'^2 + y^2}$ and

$$A_z = \frac{\mu}{2\pi}\int_0^L \frac{I}{\sqrt{y^2 + z'^2}}\,dz'$$

$$= \frac{\mu I}{2\pi}[\ln(z' + \sqrt{y^2 + z'^2})]_0^L$$

$$= \frac{\mu I}{2\pi}[\ln(L + \sqrt{y^2 + L^2}) - \ln y]$$

For $L \gg y$, the vector potential is given approximately by

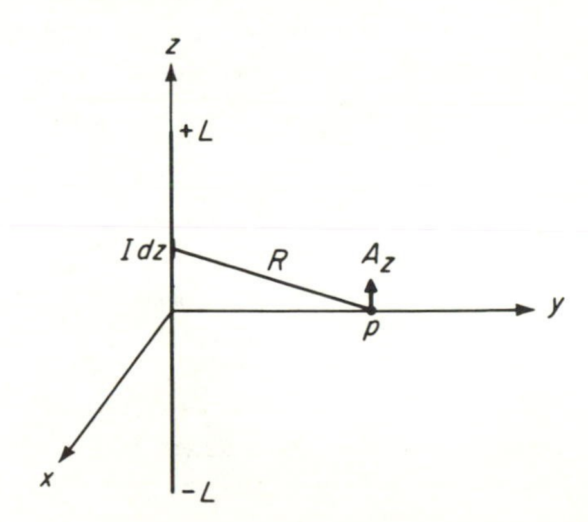

Figure 3-10. Vector potential about a long, straight wire.

$$A_z \approx \frac{\mu I}{2\pi}(\ln 2L - \ln y)$$

Then for a point in the y-z plane

$$H_x = \frac{1}{\mu}(\nabla \times \mathbf{A})_x = \frac{1}{\mu}\frac{\partial A_z}{\partial y} = -\frac{I}{2\pi y}$$

The lines of magnetic field strength will be circles about the wire, that is in the ϕ direction. For any arbitrary point P, not necessarily in the y-z plane,

$$H_\phi = \frac{I}{2\pi r}$$

where $r = \sqrt{x^2 + y^2}$ is the distance of the point P from the wire.

Problem 2. Let it be required to derive the expressions for the magnetic field about two long straight parallel wires, carrying equal and oppositely directed currents. Start with $A_z \approx \mu(I/2\pi)(\ln 2L - \ln R)$ for a single wire.

Answer:

$$H_x = \frac{1}{\mu}(\nabla \times \mathbf{A})_x = \frac{1}{\mu}\frac{\partial A_z}{\partial y}$$
$$= \frac{I}{2\pi}\left[\frac{y + (d/2)}{r_2^2} - \frac{y - (d/2)}{r_1^2}\right]$$
$$H_y = \frac{1}{\mu}(\nabla \times \mathbf{A})_y = -\frac{1}{\mu}\frac{\partial A_z}{\partial x}$$
$$= -\frac{I}{2\pi}\left(\frac{x}{r_2^2} - \frac{x}{r_1^2}\right)$$

where r_1, r_2 are the distances from each of the two wires to the observation point.

3.13 The Far Field of a Current Distribution.

The vector potential integral (28) is valid for any observation point. If the observation point is far from the current distribution (that is, far compared to the largest dimension of the current distribution) it is possible to approximate (28) using the R^{-1} expansion already discussed in chap. 2.

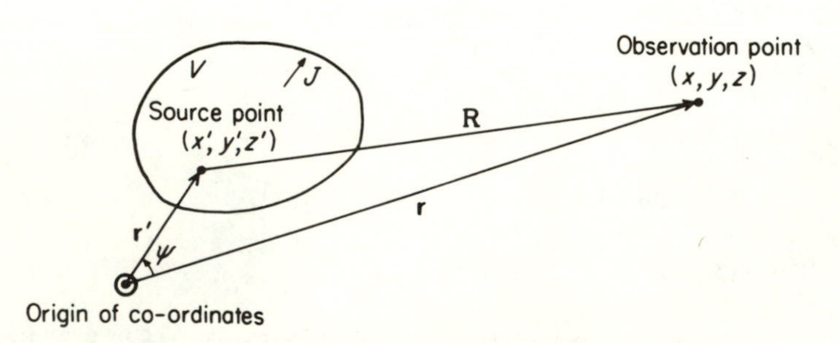

Figure 3-11. Far-field co-ordinates.

With reference to Fig. 3-11, the expansion is given by

$$\frac{1}{R} = \frac{1}{r\sqrt{1 - 2\cos\psi\,\dfrac{r'}{r} + \left(\dfrac{r'}{r}\right)^2}} \qquad r \gg r'$$

$$= \frac{1}{r}\left[1 + \frac{r'}{r}\cos\psi + \frac{1}{2}(3\cos^2\psi - 1)\left(\frac{r'}{r}\right)^2 + \cdots\right]$$

$$= \frac{1}{r}\sum_{n=0}^{\infty} P_n(\cos\psi)\left(\frac{r'}{r}\right)^n$$

$$= \frac{1}{r}\left[1 + \frac{\mathbf{r}'\cdot\hat{\mathbf{r}}}{r} + \frac{1}{2}\left\{\frac{3(\mathbf{r}'\cdot\hat{\mathbf{r}})^2}{r^2} - \frac{r'^2}{r^2}\right\} + \cdots\right\} \tag{3-30}$$

Substitution into (28) gives the required far-field expansion.

$$\mathbf{A}(\mathbf{r}) = \frac{\mu}{4\pi r}\int_V \mathbf{J}(\mathbf{r}')dV' + \frac{\mu}{4\pi r^2}\int_V \mathbf{J}(\mathbf{r}')(\mathbf{r}'\cdot\hat{\mathbf{r}})dV'$$

$$+ \frac{\mu}{8\pi r^3}\int_V \mathbf{J}(\mathbf{r}')[3(\mathbf{r}'\cdot\hat{\mathbf{r}})^2 - r'^2]dV' + \cdots \tag{3-31}$$

Since the current must flow in closed loops, the first term must be zero. The first nonzero term predominates for arbitrarily large values of r.

EXAMPLE 4: *The Magnetic Dipole*. The circular loop of radius a shown in Fig. 3-12 carries a current I and it is required to find the magnetic field far

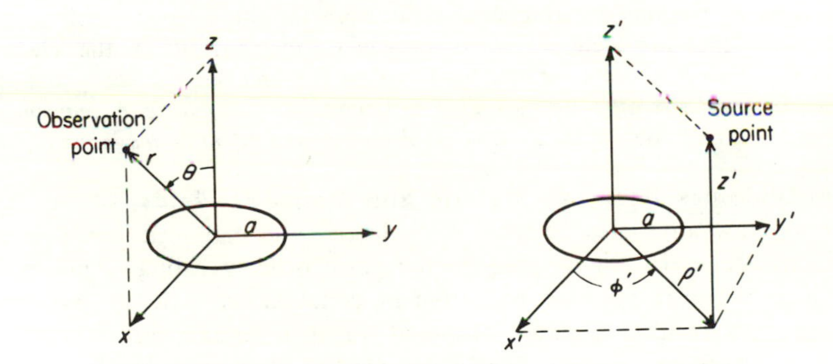

Figure 3-12. Observation-point and source-point co-ordinates.

from the loop. Without loss of generality, the observation point may be assumed to be in the x-z plane. The current distribution is given in cylindrical coordinates by

$$\mathbf{J}(\mathbf{r}') = \hat{\boldsymbol{\phi}} I\,\delta(\rho' - a)\,\delta(z')$$

At the observation point the vector potential is entirely in the y direction since the x components from the current elements cancel when integrated around the loop. Thus $\mathbf{A} = \hat{\boldsymbol{\phi}}A_\phi = \hat{\mathbf{y}}A_y$ and in the far field (31) shows that

$$A_\phi = \frac{\mu}{4\pi r^2}\int J_y(\mathbf{r}')\,\mathbf{r}'\cdot\hat{\mathbf{r}}\,dV'$$

$$J_y(\mathbf{r}') = I\cos\phi'\,\delta(\rho' - a)\,\delta(z')$$

$$\begin{aligned}\mathbf{r}'\cdot\hat{\mathbf{r}} &= [\hat{\mathbf{x}}x' + \hat{\mathbf{y}}y' + \hat{\mathbf{z}}z']\cdot\left[\frac{1}{r}(\hat{\mathbf{x}}x + \hat{\mathbf{z}}z)\right]\\ &= [\hat{\mathbf{x}}\rho'\cos\phi' + \hat{\mathbf{y}}\rho'\sin\phi' + \hat{\mathbf{z}}z']\cdot[\hat{\mathbf{x}}\sin\theta + \hat{\mathbf{z}}\cos\theta]\\ &= \sin\theta\,\rho'\cos\phi' + \cos\theta\,z'\end{aligned}$$

$$\begin{aligned}A_\phi &= \frac{\mu}{4\pi r^2}\int_{-\infty}^{\infty}\int_0^{2\pi}\int_0^{\infty} I\cos\phi'\delta(\rho' - a)\,\delta(z')[\sin\theta\,\rho'\cos\phi' + \cos\theta\,z']\\ &\qquad\qquad\qquad\qquad\qquad\qquad\qquad\qquad\qquad\qquad \rho'\,d\rho'\,d\phi'\,dz'\\ &= \frac{\mu}{4\pi r^2}\int_0^{2\pi} Ia^2\sin\theta\cos^2\phi'\,d\phi'\\ &= \frac{\mu I a^2\sin\theta}{4r^2}\end{aligned}$$

There will be two components of **H** at the observation point. Expanding $\mathbf{H} = 1/\mu\,(\nabla\times\mathbf{A})$ in spherical co-ordinates gives

$$H_\theta = -\frac{1}{\mu r}\frac{\partial}{\partial r}(rA_\phi) = \frac{I(\pi a^2)\sin\theta}{4\pi r^3}$$

$$H_r = \frac{1}{\mu r\sin\theta}\frac{\partial}{\partial\theta}(\sin\theta\,A_\phi) = \frac{I(\pi a^2)\cos\theta}{2\pi r^3}$$

If these expressions are compared with those for the electric dipole it will be seen that they are identical when the electric moment ql of the electric dipole is replaced by $\pi a^2 I$ for the loop. πa^2 is the area of the loop, and the product of this area and the current I is known as the *magnetic moment* of the loop. A small loop such as this is often referred to as a *magnetic dipole.*

3.14 Analogies between Electric and Magnetic Fields. It is natural to draw analogies between the electric and magnetic fields. Such analogies are useful in helping to maintain orderly thought processes and often make it possible to arrive at conclusions quickly by comparison with results already obtained in a diflerent but analogous problem. There are several possible analogies that can be drawn between electric and magnetic fields, but two of these are particularly applicable to later work in the (time-varying) electromagnetic field. The first analogy considers **D** and **H** as analogous quantities and **E** and **B** as analogous quantities. This is based on consideration of the fact that displacement density **D** is related directly to its source, the charge, and is independent of the characteristics of the (homogeneous) medium in which the charge is immersed. Similarly the magnetic vector **H** can be related directly to its source, the current, and is independent of the (homogeneous) medium in which the magnetic field exists. The vectors **E** and **B** are also related to their respective sources, charge and current, but show a dependence on the characteristics of the medium, that is on the dielectric constant, and magnetic permeability respectively. This analogy is correct in the sense that it is self-consistent and can be made to give useful interpretations. The second analogy, which is equally valid, considers **E** and **H** as analogous and **D** and **B** as analogous. It is no more "correct" than the first analogy, but has the advantage in electromagnetic field theory that it gives a symmetry to Maxwell's equations that otherwise would be lacking. Inasmuch as these equations form the starting point for every problem of the electromagnetic field, this is a very useful result. Two simple experiments serve to point up this analogy. In Fig. 3-13(*a*) voltage V produces an electric field E in the space

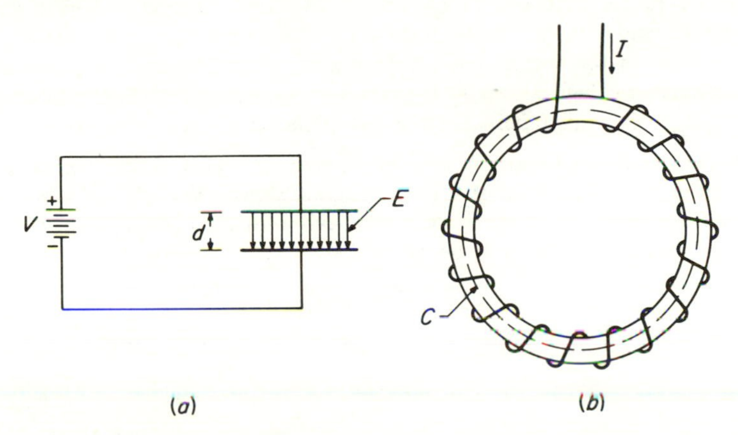

Figure 3-13. Circuits illustrating analogies between electric and magnetic fields.

between the capacitor plates. E is equal to V/d and is independent of the dielectric constant ϵ of the dielectric. However the displacement density depends upon ϵ (for a constant applied voltage and therefore constant E) and is given by $D = \epsilon E$. In Fig. 3-13(*b*) the current I results in a magnetomotive force nI around the closed path l. The magnetic field strength H within the core is equal to nI/l. The magnetic flux density B depends upon the permeability μ (for a constant applied mmf) and is given by $B = \mu H$. In this analogy E and H are sometimes pictured as electric and magnetic field strengths or forces that result in electric and magnetic flux densities, D and B respectively.

In the above experiments, if the charge Q (instead of the voltage) is held constant in Fig. 3-13(*a*) and the current is held constant as before in Fig. 3-13(*b*), then the first analogy results. That is, D and H are the analogous quantities that remain unchanged for different dielectric and core materials.

ADDITIONAL PROBLEMS

3. Verify that *within* a conductor carrying a current I the magnetic field strength at a distance r from the center of the wire is given by

$$H = \frac{Ir}{2\pi R^2}$$

where R is the radius of the wire. The current density is constant across the cross section of the conductor.

4. Verify that expressions (3-22) and (3-24) combine to give the same result for magnetic field strength as is given by (3-12); that is, verify that the curl of the expression for vector potential due to a current element does indeed yield the magnetic field strength as given by Ampere's law (Suggestion: Solve for the special case of a point in the y-z plane, and then generalize).

5. A very long thin sheet of copper having a width b meters carries a direct current I in the direction of its length. Show that if the sheet is assumed to lie in the x-z plane with the z axis along its center line, the magnetic field about the strip will be given by

$$H_x = -\frac{I}{2\pi b}\left(\tan^{-1}\frac{\frac{b}{2}+x}{y} + \tan^{-1}\frac{\frac{b}{2}-x}{y}\right)$$

$$H_y = \frac{I}{4\pi b}\ln\left[\frac{\left(\frac{b}{2}+x\right)^2 + y^2}{\left(\frac{b}{2}-x\right)^2 + y^2}\right]$$

(NOTE: Solve by first setting up the vector potential due to long narrow strips.)

6. By setting up the statement of Ampere's work law for elemental areas in cylindrical co-ordinates derive the expansion for $\nabla \times \mathbf{H}$ in these co-ordinates.

7. Show that the answers to problem 5 agree with the answers to example 1 for (a) a point on the y axis when $y \gg b$; (b) a point on the x axis when $x \gg b$.

8. The familiar statement of Ohm's law is $I = V/R$, where the direction of current flow is in the direction of the voltage drop. Show that for an *elemental volume* this law may be written as the vector point relation $\mathbf{J} = \sigma\mathbf{E}$. (Recall that the resistance R of a conductor of length l and cross-sectional area A is given by $R = l/\sigma A$ where σ is the conductivity of the material.)

BIBLIOGRAPHY

See bibliography of chap. 2.

Chapter 4

MAXWELL'S EQUATIONS

Up to the present, the fields considered have been mainly the static electric field due to charges at rest and the static magnetic field due to steady currents. The equations governing these fields may be summarized as follows:

$$\nabla \times \mathbf{E} = 0 \qquad \oint \mathbf{E} \cdot d\mathbf{s} = 0 \tag{4-1}$$

$$\nabla \cdot \mathbf{D} = \rho \qquad \oint \mathbf{D} \cdot d\mathbf{a} = \int \rho \, dV \tag{4-2}$$

$$\nabla \times \mathbf{H} = \mathbf{J} \qquad \oint \mathbf{H} \cdot d\mathbf{s} = \int \mathbf{J} \cdot d\mathbf{a} \tag{4-3}$$

$$\nabla \cdot \mathbf{B} = 0 \qquad \oint \mathbf{B} \cdot d\mathbf{a} = 0 \tag{4-4}$$

Contained in the above is the equation of continuity for steady currents,

$$\nabla \cdot \mathbf{J} = 0 \qquad \oint \mathbf{J} \cdot d\mathbf{a} = 0 \tag{4-5}$$

In order to discuss magnetic induction and energy it was found necessary to include time-varying fields, but only to the extent of introducing Faraday's law, which states that the electromotive force around a closed path is equal to the negative of the time rate of change of magnetic flux enclosed by the path:

$$\oint \mathbf{E} \cdot d\mathbf{s} = -\frac{d\Phi}{dt} = -\frac{d}{dt}\int_s \mathbf{B} \cdot d\mathbf{a}$$

Faraday's law states that the voltage around the closed path can be generated by a time-changing magnetic flux through a fixed path (transformer action) or by a time-varying path in a steady magnetic field (electric generator action). In *electromagnetics*, interest centers on the relations between time-changing electric and magnetic fields and the path of integration can be considered fixed. In this case Faraday's law reduces to

$$\oint \mathbf{E} \cdot d\mathbf{s} = -\int_S \frac{\partial \mathbf{B}}{\partial t} \cdot d\mathbf{a}$$

where the partial derivative with respect to time is used to indicate that only variations of magnetic flux with time through a *fixed* closed path or a fixed region in space are being considered.*

Hence for time-varying fields, instead of eq. (1) we have

$$\nabla \times \mathbf{E} = -\frac{\partial \mathbf{B}}{\partial t} \qquad \oint \mathbf{E} \cdot d\mathbf{s} = -\int_S \frac{\partial \mathbf{B}}{\partial t} \cdot d\mathbf{a} \tag{4-6}$$

4.01 The Equation of Continuity for Time-Varying Fields. Since current is simply charge in motion, the total current flowing out of some volume must be equal to the rate of decrease of charge within the volume, assuming that charge cannot be created or destroyed. This concept is essential in understanding why current flows in the leads to a capacitor during charge or discharge when no current flows between the capacitor plates. The explanation is simply that the current flow is accompanied by a charge buildup on the plates. In mathematical terms, this *conservation of charge* concept can be stated as

$$\oint \mathbf{J} \cdot d\mathbf{a} = -\frac{d}{dt}\int \rho \, dV \tag{4-7a}$$

If the region of integration is stationary, the above relation becomes

$$\oint \mathbf{J} \cdot d\mathbf{a} = -\int \frac{\partial \rho}{\partial t} dV \tag{4-7b}$$

The divergence theorem now may be applied in order to change the surface integral into a volume integral:

$$\int \nabla \cdot \mathbf{J} \, dV = -\int \frac{\partial \rho}{\partial t} \, dV$$

If the above relation is to hold for any arbitrary volume, then it must be true that

$$\nabla \cdot \mathbf{J} = -\frac{\partial \rho}{\partial t} \tag{4-8}$$

This is the time-varying form of the *equation of continuity* and thus it replaces (5).

4.02 Inconsistency of Ampere's Law. Taking the divergence of Ampere's law (3) yields the equation of continuity for *steady* currents (5). Thus Ampere's law is not consistent with the time-varying equa-

*For a discussion of induced-emf under other conditions, refer to any text on electricity and magnetism. A thorough treatment is given in E. G. Cullwick, *The Fundamentals of Electromagnetism* (2nd ed.), The Macmillan Co., Ltd., Cambridge, England, 1949.

tion of continuity. This difficulty may have been one of the factors which led James Clerk Maxwell in the mid-1860's to seek a modification to Ampere's law. A suggestion as to the correct modification may be found by substituting Gauss's law (2) into the equation of continuity (8), giving

$$\nabla \cdot \mathbf{J} = -\frac{\partial}{\partial t} \nabla \cdot \mathbf{D}$$

Interchanging the differentiations with respect to space and time gives

$$\nabla \cdot \left(\frac{\partial \mathbf{D}}{\partial t} + \mathbf{J} \right) = 0 \tag{4-9}$$

This may be put into integral form by integrating over a volume and then applying the divergence theorem:

$$\oint \left(\frac{\partial \mathbf{D}}{\partial t} + \mathbf{J} \right) \cdot d\mathbf{a} = 0 \tag{4-10}$$

The two preceding equations suggest that $(\partial \mathbf{D}/\partial t + \mathbf{J})$ may be regarded as the *total* current density for time-varying fields. Since $\mathbf{D}$ is the displacement density, $\partial \mathbf{D}/\partial t$ is known as the *displacement current density*. When applied as shown in Fig. 4-1 to a surface enclosing one plate

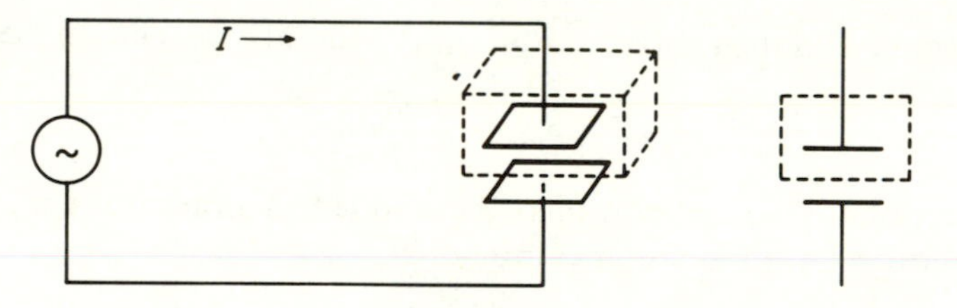

Figure 4-1.

of a two-plate capacitor, (10) shows that, during charge or discharge, the conduction current in the wire attached to the plate is equal to the displacement current passing between the plates.

Maxwell reasoned that the total current density should replace $\mathbf{J}$ in Ampere's law with the result that

$$\nabla \times \mathbf{H} = \frac{\partial \mathbf{D}}{\partial t} + \mathbf{J} \tag{4-11}$$

Taking the divergence of (11) gives (9) and thus the inconsistency has been removed. It should be made perfectly clear that (11) was not *derived* from the preceding equations but rather *suggested* by them. Therefore when it was obtained by Maxwell, (11) was a postulate whose validity had to be established finally by experiment.

Maxwell's assumption that a changing displacement density (that is, a changing electric field) was equivalent to an electric current density, and as such would produce a magnetic field, has had most far-reaching effects. Combined with Faraday's law which indicates that a changing magnetic field will produce an electric field it leads directly to the "wave equations." This result enabled Maxwell to predict electromagnetic wave propagation some thirty years before Hertz's brilliant researches gave experimental verification. This is an interesting example of one of those rather rare cases where the mathematical reasoning has preceded and pointed the way for experiment.

Integration of (11) over a surface and application of Stokes' theorem leads to the integral form

$$\oint \mathbf{H} \cdot d\mathbf{s} = \int \left(\frac{\partial \mathbf{D}}{\partial t} + \mathbf{J} \right) \cdot d\mathbf{a} \tag{4-12}$$

Equation (12) states that the magnetomotive force around a closed path is equal to the *total* current enclosed by the path. Thus equations (11) and (12) replace the static form of Ampere's law (3).

4.03 Maxwell's Equations. The electromagnetic equations are known as *Maxwell's equations* since Maxwell contributed to their development and established them as a self-consistent set. Each differential equation has its integral counterpart; one form may be derived from the other with the help of Stokes' theorem or the divergence theorem. In summary the equations are as follows, with the dot superscripts indicating partial derivatives with respect to time.

$$\nabla \times \mathbf{H} = \dot{\mathbf{D}} + \mathbf{J} \qquad \oint \mathbf{H} \cdot d\mathbf{s} = \int (\dot{\mathbf{D}} + \mathbf{J}) \cdot d\mathbf{a} \tag{I}$$

$$\nabla \times \mathbf{E} = -\dot{\mathbf{B}} \qquad \oint \mathbf{E} \cdot d\mathbf{s} = -\int \dot{\mathbf{B}} \cdot d\mathbf{a} \tag{II}$$

$$\nabla \cdot \mathbf{D} = \rho \qquad \oint \mathbf{D} \cdot d\mathbf{a} = \int \rho \, dV \tag{III}$$

$$\nabla \cdot \mathbf{B} = 0 \qquad \oint \mathbf{B} \cdot d\mathbf{a} = 0 \tag{IV}$$

Contained in the above is the equation of continuity,

$$\nabla \cdot \mathbf{J} = -\dot{\rho} \qquad \oint \mathbf{J} \cdot d\mathbf{a} = -\int \dot{\rho} \, dV$$

In all the above cases, the regions of integration are assumed to be stationary.

Word Statement of the Field Equations. A word statement of the significance of the field equations is readily obtained from their mathematical statement in the integral form. It would be somewhat as follows:

(I) The magnetomotive force around a closed path is equal to the conduction current plus the time derivative of the electric displacement through any surface bounded by the path.

(II) The electromotive force around a closed path is equal to the time derivative of the magnetic displacement through any surface bounded by the path.

(III) The total electric displacement through the surface enclosing a volume is equal to the total charge within the volume.

(IV) The net magnetic flux emerging through any closed surface is zero.

As indicated previously the time derivative of electric displacement is called displacement current. The term electric current is then generalized in meaning to include both conduction currents and displacement currents.* Furthermore, if the time derivative of electric displacement is called an electric current, the time derivative of magnetic displacement can be considered as being a magnetic current. Finally, electromotive force is called electric voltage, so that magnetomotive force may be called magnetic voltage.

The first two Maxwell equations can then be stated:

(I) The magnetic voltage around a closed path is equal to the electric current through the path.

(II) The electric voltage around a closed path is equal to the magnetic current through the path.

Interpretation of the field equations. Maxwell's equations contain *curl* operations which can be interpreted in terms of measurable quantities. For example, consider $\nabla \times \mathbf{E}$ which occurs in the differential form of (II). Stokes' theorem states that

$$\int \nabla \times \mathbf{E} \cdot d\mathbf{a} = \oint \mathbf{E} \cdot d\mathbf{s}$$

If the surface integral is reduced to an integral over the element of area da and if $\hat{\mathbf{n}}$ is the unit normal in the direction of $d\mathbf{a}$, then the above may be expressed as

$$\nabla \times \mathbf{E} \cdot \hat{\mathbf{n}} = \frac{\oint \mathbf{E} \cdot d\mathbf{s}}{da}$$

Suppose that $\hat{\mathbf{n}}$ is oriented in such a way that $\oint \mathbf{E} \cdot d\mathbf{s}$ is a maximum. Under these conditions the direction of $\nabla \times \mathbf{E}$ is given by $\hat{\mathbf{n}}$ and the magnitude of $\nabla \times \mathbf{E}$ is given by $\oint \mathbf{E} \cdot d\mathbf{s}/da$.

*Also *convection* currents (e.g., electron beam currents). Conduction currents obey Ohm's law, $\mathbf{J} = \sigma \mathbf{E}$; convection currents do not.

The above discussion may be applied to a very small, single-turn loop of wire with area *da*. If the loop of wire is connected to a voltmeter, the voltage measured is $\oint \mathbf{E} \cdot d\mathbf{s}$ according to Faraday's law. Suppose that the orientation of the loop is changed until the voltmeter reading indicates a maximum. Under these conditions the magnitude of $\nabla \times \mathbf{E}$ is given by the voltmeter reading divided by the loop area and the direction of $\nabla \times \mathbf{E}$ is given by the direction of the loop axis. Faraday's law indicates that this measurement of $\nabla \times \mathbf{E}$ is also a measurement of $-\partial \mathbf{B}/\partial t$, the rate of decrease of the magnetic flux density. In a region in which there is no time-changing magnetic flux, the voltage around the loop would be zero and thus $\nabla \times \mathbf{E} = 0$; such an electric field occurs in electrostatics and is said to be *irrotational.*

Another equation worthy of notice is

$$\nabla \cdot \mathbf{B} = 0$$

This states that there are no isolated magnetic poles or "magnetic charges" on which the lines of magnetic flux can terminate. To put it another way, the number of lines of flux entering any region must be equal to the number of lines leaving the region or, as is often said, the lines of magnetic flux are *continuous.*

4.04 Conditions at a Boundary Surface. Maxwell's equations in the differential form express the relationship that must exist between the four field vectors **E**, **D**, **H**, and **B** at any point within a continuous medium. In this form, because they involve space derivatives, they cannot be expected to yield information at points of discontinuity in the medium. However, the integral forms can always be used to determine what happens at the boundary surface between different media.

The following statements can be made regarding the electric and magnetic fields at any surface of discontinuity:

(a) The *tangential* component of **E** is continuous at the surface. That is, it is the same *just* outside the surface as it is *just* inside the surface.

(b) The *tangential* component of **H** is continuous across a surface except at the surface of a *perfect conductor.* At the surface of a perfect conductor the *tangential* component of **H** is discontinuous by an amount equal to the surface current per unit width.

(c) The *normal* component of **B** is continuous at the surface of discontinuity.

(d) The *normal* component of **D** is continuous if there is no surface charge density. Otherwise **D** is discontinuous by an amount equal to the surface charge density.

The proof of these boundary conditions is obtained by a direct application of Maxwell's equations at the boundary between the media.

Suppose the surface of discontinuity to be the plane $x = 0$ as shown in Fig. 4-2. Consider the small rectangle of width Δx and length Δy enclosing a small portion of each of media (1) and (2).

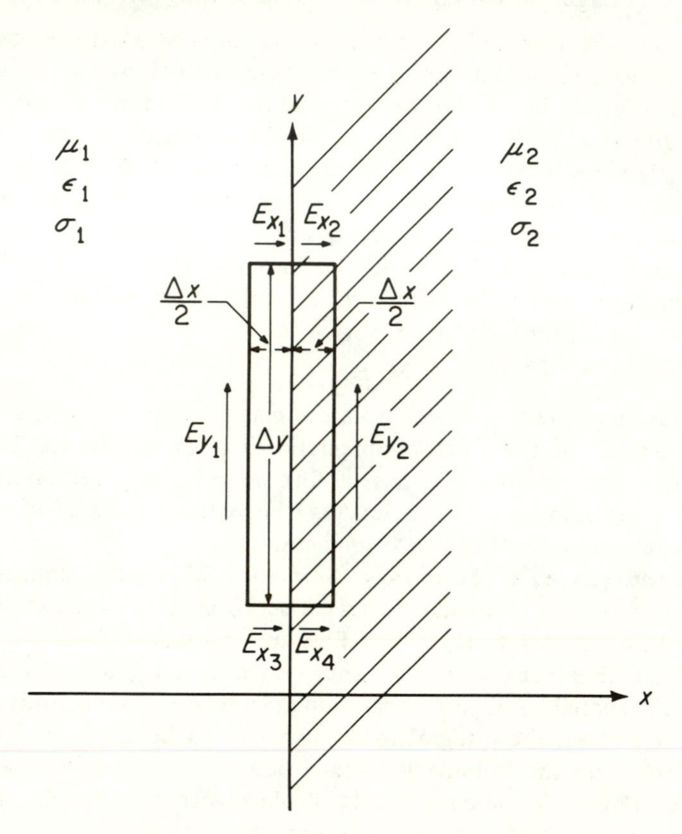

Figure 4-2. A boundary surface between two media.

The integral form of the second Maxwell equation (II) is

$$\oint \mathbf{E} \cdot d\mathbf{s} = -\int_S \dot{\mathbf{B}} \cdot d\mathbf{a}$$

For the elemental rectangle of Fig. 4-2 this becomes

$$E_{y_2}\,\Delta y - E_{x_2}\frac{\Delta x}{2} - E_{x_1}\frac{\Delta x}{2} - E_{y_1}\,\Delta y + E_{x_3}\frac{\Delta x}{2} + E_{x_4}\frac{\Delta x}{2} = -\dot{B}_z\,\Delta x\,\Delta y \tag{4-13}$$

where B_z is the average magnetic flux density through the rectangle $\Delta x\,\Delta y$. Now consider conditions as the area of the rectangle is made to approach zero by reducing the width Δx of the rectangle, always keeping the surface of discontinuity between the sides of the rectangle.

If it is assumed that B is always finite, then the right-hand side of eq. (13) will approach zero. If E is also assumed to be everywhere finite, then the $\Delta x/2$ terms of the left-hand side will reduce to zero, leaving

$$E_{y_2}\,\Delta y - E_{y_1}\,\Delta y = 0$$

for $\Delta x = 0$. Therefore

$$E_{y_1} = E_{y_2}$$

That is, *the tangential component of E is continuous.*

Similarly the integral statement of eq. (I) is

$$\oint \mathbf{H}\cdot d\mathbf{s} = \int_S (\dot{\mathbf{D}} + \mathbf{J})\cdot d\mathbf{a}$$

which becomes

$$H_{y_2}\,\Delta y - H_{x_2}\frac{\Delta x}{2} - H_{x_1}\frac{\Delta x}{2} - H_{y_1}\,\Delta y + H_{x_3}\frac{\Delta x}{2} + H_{x_4}\frac{\Delta x}{2} = (\dot{D}_z + J_z)\Delta x\,\Delta y \qquad (4\text{-}14)$$

If the rate of change of electric displacement $\dot{D}$ and current density J are both considered to be finite, then as before (14) reduces to

$$H_{y_2}\,\Delta y - H_{y_1}\,\Delta y = 0$$

or

$$H_{y_1} = H_{y_2}$$

The *tangential component of H is continuous* (for *finite* current densities; that is, for any *actual* case).

Note on a Perfect Conductor. A *perfect* conductor is one which has infinite conductivity. In such a conductor the electric field strength **E** is zero for any finite current density. Most actual conductors have a finite value for conductivity.* However, the actual conductivity may be very large and for many practical applications it is useful to assume it to be infinite. Such an assumption will lead to difficulties (because of indeterminacy) in formulating the boundary conditions unless care is taken in setting them up. As will be shown later, the depth of penetration into a conductor of an alternating electric field and of the current produced by the field decreases as the conductivity increases. Thus in a good conductor a high-frequency current will flow in a thin sheet near the surface, the depth of this sheet approaching zero as the conductivity approaches infinity. This gives rise to the useful concept of a *current sheet*. In a current sheet a finite current per unit width, J_s amperes per meter, flows in a sheet of vanishingly small depth Δx, but with the required infinitely large current density J, such that

**Superconductors* and the phenomenon of superconductivity are treated in sec. 9.07.

$$\lim_{\Delta x \to 0} J\,\Delta x = J_s \qquad \text{A/m}$$

Consider again the above example of the magnetomotive force around the small rectangle. If the current density J_z becomes infinite as Δx approaches zero, the right-hand side of eq. (14) will not become zero. Let J_s amperes per meter be the actual current per unit width flowing along the surface. Then as $\Delta x \to 0$ the eq. (14) for H becomes

$$H_{y_2}\,\Delta y - H_{y_1}\,\Delta y = J_{sz}\,\Delta y$$

Hence

$$H_{y_1} = H_{y_2} - J_{sz} \tag{4-15}$$

(Note that $\mathbf{D} = \epsilon\mathbf{E}$ remains finite and therefore $D_z\,\Delta x$ is zero for $\Delta x = 0$.)

Now, if the electric field is zero within a perfect conductor, the magnetic field must also be zero (for *alternating* fields) as the second Maxwell equation (II) shows. Then in eq. (15), H_{y_2} must be zero and so

$$H_{y_1} = -J_{sz} \tag{4-16}$$

Equation (16) states that the current per unit width along the surface of a perfect conductor is equal to the magnetic field strength H just outside the surface. The magnetic field and surface current will be parallel to the surface, but perpendicular to each other. In vector notation this is written

$$\mathbf{J}_s = \hat{\mathbf{n}} \times \mathbf{H}$$

where $\hat{\mathbf{n}}$ is the unit vector along the outward normal to the surface.

Conditions on the Normal Components of **B** *and* **D**. The remaining boundary conditions are concerned with the *normal* components of **B** and **D**. The integral form of the third field equation is

$$\oint_S \mathbf{D}\cdot d\mathbf{a} = \int_V \rho\, dV \tag{III}$$

When applied to the elementary "pill-box" volume of Fig. 4-3, eq. (III) becomes

$$D_{n_1}\,da - D_{n_2}\,da + \Psi_{\text{edge}} = \rho\,\Delta x\,da \tag{4-17}$$

In this expression da is the area of each of the flat surfaces of the pillbox, Δx is their separation, and ρ is the average charge density within the volume $\Delta x\,da$. Ψ_{edge} is the outward electric flux through the curved-edge surface of the pillbox. As $\Delta x \to 0$, that is, as the flat surfaces of the box are squeezed together, always keeping the boundary surface between them, $\Psi_{\text{edge}} \to 0$, for finite values of displacement density. Also for *finite* values of average charge density ρ, the right-hand side of (17) approaches zero, and (17) reduces to

$$D_{n_1}\,da - D_{n_2}\,da = 0$$

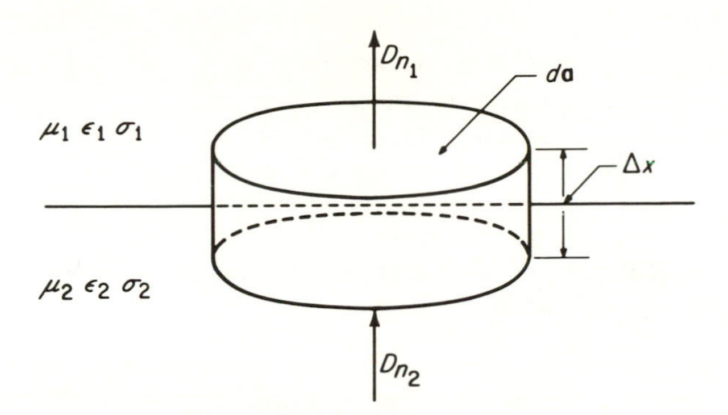

Figure 4-3. A "pill-box" volume enclosing a portion of a boundary surface.

(for $\Delta x = 0$). Then for the case of no surface charge the condition on the normal components of **D** is

$$D_{n_1} = D_{n_2} \tag{4-18}$$

That is, if there is no surface charge the normal component of **D** is continuous across the surface.

In the case of a metallic surface, the charge is considered to reside "on the surface." If this layer of surface charge has a *surface charge density* ρ_s coulombs per *square* meter, the charge density ρ of the surface layer is given by

$$\rho = \frac{\rho_s}{\Delta x} \qquad \text{coulomb/cu m}$$

where Δx is thickness of the surface layer. As Δx approaches zero, the charge density approaches infinity in such a manner that

$$\lim_{\Delta x \to 0} \rho\, \Delta x = \rho_s$$

Then in Fig. 4-3, if the surface charge is always kept between the two flat surfaces as the separation between them is decreased, the right-hand side of eq. (17) approaches $\rho_s\, da$ as Δx approaches zero. Equation (17) then reduces to

$$D_{n_1} - D_{n_2} = \rho_s \tag{4-19}$$

When there is a surface charge density ρ_s, the normal component of displacement density is *discontinuous* across the surface by the amount of the surface charge density.

For any metallic conductor the displacement density $\mathbf{D} = \epsilon\mathbf{E}$ within the conductor will be a very small quantity (it will be zero in the electrostatic case, or in the case of a perfect conductor). Then if medium (2) is a metallic conductor $D_{n_2} = 0$ and eq. (19) becomes

$$D_{n_1} = \rho_s \tag{4-20}$$

The normal component of displacement density in the dielectric is equal to the surface charge density on the conductor.

In the case of magnetic flux density **B**, since there are no isolated "magnetic charges," a similar analysis leads at once to

$$B_{n_1} = B_{n_2}$$

The normal component of magnetic flux density is always continuous across a boundary surface.

PROBLEMS

1. Show that the displacement current through the capacitor is equal to the conduction current I (Fig. 4-4).

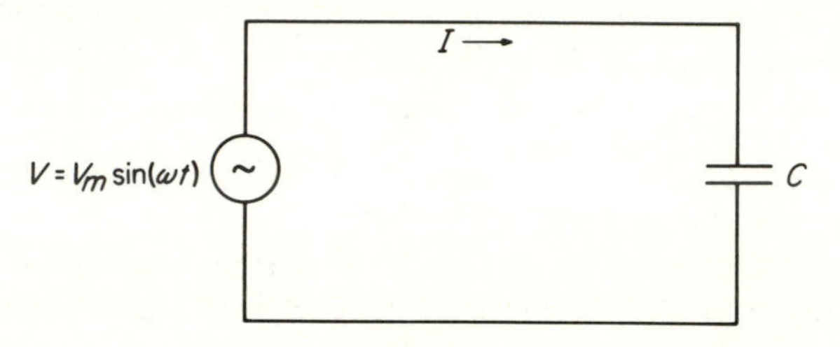

Figure 4-4.

2. *Within* a perfect conductor **E** is always zero. Using Maxwell's equations, show that **H** must also be zero for time varying fields. Can a steady (unchanging) magnetic field exist within a perfect conductor? Show that the normal component of $\dot{\mathbf{B}}$ (and therefore $\dot{\mathbf{H}}$) must be zero at the surface of a perfect conductor.

3. A "transmission line" consists of two parallel perfectly conducting planes of large extent, separated by a distance d meters, and guiding between them a uniform plane wave (Fig. 4-5). The conducting planes carry an alternating linear current density J_s A/m in the y direction, that is,

$$J_s = J_0 \cos \omega\left(t - \frac{y}{c}\right)$$

Applying Maxwell's first equation in the region between the conductors find the electric field strength, and hence the voltage between the planes, when $d = 0.1$ meter and the effective linear current density is $J_{s\,\mathrm{eff}} = 1$ A/m.

4. A square loop of wire, 20 cm by 20 cm, has a voltmeter (of infinite impedance) connected in series with one side. Determine the voltage indicated by the meter when the loop is placed in an alternating magnetic field, the maximum intensity of which is 1 ampere per meter. The plane of the loop is perpendicular to the magnetic field; the frequency is 10 MHz.

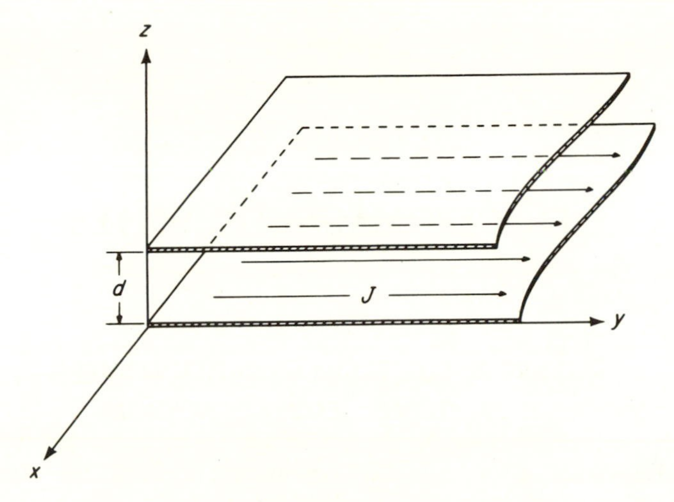

Figure 4-5. Parallel-plane "transmission" line.

5. A No. 10 copper wire carries a conduction current of 1 amp at 60 Hz. What is the displacement current in the wire? For copper assume $\epsilon = \epsilon_v$, $\mu = \mu_v$, $\sigma = 5.8 \times 10^7$.

6. The electric vector **E** of an electromagnetic wave in free space is given by the expressions

$$E_x = E_z = 0 \qquad E_y = A\cos\omega\left(t - \frac{z}{c}\right)$$

Using Maxwell's equations for free-space conditions determine expressions for the components of the magnetic vector **H**.

BIBLIOGRAPHY

Abraham, M., and R. Becker, *The Classical Theory of Electricity and Magnetism*, Blackie and Son, Ltd., London and Glasgow, 1937.

Hayt, W. H. Jr., *Engineering Electromagnetics* (2nd ed.), McGraw-Hill Book Company, New York, 1967.

Harrington, R. F., *Introduction to Electromagnetic Engineering*, McGraw-Hill Book Company, New York, 1958.

Ramo, S., J. R. Whinnery and T. Van Duzer, *Fields and Waves in Communication Electronics*, John Wiley & Sons, Inc., New York, 1965.

Schelkunoff, S. A., *Electromagnetic Fields*, Blaisdell Publishing Co., Waltham, Mass., 1963.

Skilling, H. H., *Fundamentals of Electric Waves* (2nd ed.), John Wiley & Sons, Inc., New York, 1948.

Chapter 5

ELECTROMAGNETIC WAVES

PART–I ELECTROMAGNETIC WAVES IN A HOMOGENEOUS MEDIUM

In the solution of any electromagnetic problem the fundamental relations that must be satisfied are the four field equations

$$\nabla \times \mathbf{H} = \dot{\mathbf{D}} + \mathbf{J} \tag{I}$$

$$\nabla \times \mathbf{E} = -\dot{\mathbf{B}} \tag{II}$$

$$\nabla \cdot \mathbf{D} = \rho \tag{III}$$

$$\nabla \cdot \mathbf{B} = 0 \tag{IV}$$

in which the dot superscript indicates partial differentiation with respect to time. In addition there are three relations that concern the characteristics of the medium in which the fields exist. These are the *constitutive relations*

$$\mathbf{D} = \epsilon\mathbf{E} \tag{5-1}$$

$$\mathbf{B} = \mu\mathbf{H} \tag{5-2}$$

$$\mathbf{J} = \sigma\mathbf{E} \tag{5-3}$$

where ϵ, μ, and σ are the permittivity, permeability, and conductivity of the medium, which is assumed to be homogeneous, isotropic, and source-free. A *homogeneous* medium is one for which the quantities ϵ, μ, and σ are constant throughout the medium. The medium is *isotropic* if ϵ is a scalar constant, so that **D** and **E** have everywhere the same direction. The form of Maxwell's equations, given by (I) and (II), is for *source-free* regions, that is, regions in which there are no *impressed* voltages or currents (no generators). The relations of the fields to their sources will be considered in chap. 10 and subsequent chapters.

When the relations (1), (2), and (3) are inserted in (I) and (II), Maxwell's equations become differential equations relating the electric and magnetic field strengths **E** and **H**. If they are then solved as simultaneous equations, they will determine the laws which both **E** and **H** must obey.

5.01 Solution for Free-space Conditions. Before obtaining the solution for the general case it is instructive to consider the simple, but important, particular case of electromagnetic phenomena in free space, or more generally, in a perfect dielectric containing no charges and no conduction currents. For this case the field equations become

$$\nabla \times \mathbf{H} = \dot{\mathbf{D}} \tag{5-4}$$

$$\nabla \times \mathbf{E} = -\dot{\mathbf{B}} \tag{5-5}$$

$$\nabla \cdot \mathbf{D} = 0 \tag{5-6}$$

$$\nabla \cdot \mathbf{B} = 0 \tag{5-7}$$

Differentiate (4) with respect to time. Since the curl operation is a differentiation with respect to space, the order of differentiation may be reversed, that is,

$$\frac{\partial \nabla \times \mathbf{H}}{\partial t} = \nabla \times \dot{\mathbf{H}}$$

Also since ϵ and μ are independent of time

$$\dot{\mathbf{D}} = \epsilon \dot{\mathbf{E}} \tag{5-8}$$

$$\dot{\mathbf{B}} = \mu \dot{\mathbf{H}} \tag{5-9}$$

so that there results

$$\nabla \times \dot{\mathbf{H}} = \epsilon \ddot{\mathbf{E}} \tag{5-10}$$

The symbol $\ddot{\mathbf{E}}$ means $\partial^2 \mathbf{E}/\partial t^2$.

Take the curl of both sides of (5) and using (9), obtain

$$\nabla \times \nabla \times \mathbf{E} = -\mu \nabla \times \dot{\mathbf{H}} \tag{5-11}$$

Substitute eq. (10) into (11)

$$\nabla \times \nabla \times \mathbf{E} = -\mu\epsilon \ddot{\mathbf{E}} \tag{5-12}$$

It was shown in identity (1-28) that

$$\nabla \times \nabla \times \mathbf{E} = \nabla \nabla \cdot \mathbf{E} - \nabla^2 \mathbf{E}$$

Combine this equation with (12) to obtain

$$\nabla \nabla \cdot \mathbf{E} - \nabla^2 \mathbf{E} = -\mu\epsilon \ddot{\mathbf{E}} \tag{5-13}$$

But

$$\nabla \cdot \mathbf{E} = \frac{1}{\epsilon} \nabla \cdot \mathbf{D} = 0$$

and therefore eq. (13) becomes

$$\nabla^2 \mathbf{E} = \mu\epsilon \ddot{\mathbf{E}} \tag{5-14}$$

This is the law that **E** must obey.

Differentiating (5) with respect to time and taking the curl of (4) it will be found on combining that **H** obeys the same law, viz.,

$$\nabla^2 \mathbf{H} = \mu\epsilon \ddot{\mathbf{H}} \tag{5-15}$$

Equations (14) and (15) are known as the *wave equations.* Thus the first condition on either **E** or **H** is that it must satisfy the wave equation. (Note that although **E** and **H** obey the same law, **E** is not equal to **H**.)

5.02 Uniform Plane-wave Propagation. The wave equation reduces to a very simple form in the special case where **E** and **H** are considered to be independent of two dimensions, say y and z. Then

$$\nabla^2 \mathbf{E} = \frac{\partial^2 \mathbf{E}}{\partial x^2}$$

so that (14) becomes

$$\frac{\partial^2 \mathbf{E}}{\partial x^2} = \mu\epsilon \frac{\partial^2 \mathbf{E}}{\partial t^2} \tag{5-16}$$

Vector eq. (16) is equivalent to three scalar equations, one for each of the scalar components of **E**. In general, for uniform plane wave propagation in the x direction, **E** may have components E_y and E_z, but (as will be seen later) not E_x. Without loss of generality attention can be restricted to one of the components, say E_y, knowing that results for E_z will be similar to those obtained for E_y. Then the equation to be solved has the form

$$\frac{\partial^2 E_y}{\partial x^2} = \mu\epsilon \frac{\partial^2 E_y}{\partial t^2} \tag{5-16a}$$

Equation (16a) is a second-order partial differential equation, which occurs frequently in mechanics and engineering. For example it is the differential equation for the displacement from equilibrium along a uniform string. Electrical engineers will recognize it as the differential equation for voltage or current along a lossless transmission line. Its general solution is of the form

$$E = f_1(x - v_0 t) + f_2(x + v_0 t) \tag{5-17}$$

where $v_0 = 1/\sqrt{\mu\epsilon}$ and f_1 and f_2 are *any* functions (not necessarily the same) of $(x - v_0 t)$ and $(x + v_0 t)$ respectively. The expression $f(x - v_0 t)$ means a function f of the variable $(x - v_0 t)$. Examples are: $A \cos \beta(x - v_0 t)$, $C\, e^{k(x - v_0 t)}$, $\sqrt{x - v_0 t}$, etc. All of these expressions represent wave motion.

A wave* may be defined in the following way: If a physical

*The term *wave* also has an entirely different usage, viz.: a *recurrent function of time at a point,* as in the expression *sinusoidal voltage wave.* Usually there will be no doubt as to which kind of wave is meant.

phenomenon that occurs at one place at a given time is reproduced at other places at later times, the time delay being proportional to the space separation from the first location, then the group of phenomena constitute a wave. Note that a wave is not necessarily a repetitive phenomenon in time. Those who survive a tidal wave are thankful for this.

The functions $f_1(x - v_0 t)$ and $f_2(x + v_0 t)$ describe such a wave mathematically, the variation of the wave being confined to one dimension in space. This is shown by Fig. 5-1.

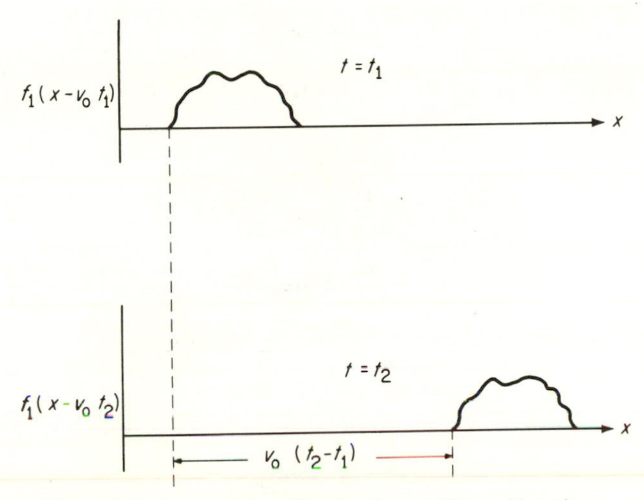

Figure 5-1. A wave traveling in the positive x direction.

If a fixed time is taken, say t_1, then the function $f_1(x - v_0 t_1)$ becomes a function of x since $v_0 t_1$ is a constant. Such a function is represented by the first curve. If another time, say t_2, is taken, another function of x is obtained, exactly the same shape as the first except that the second curve is displaced to the right by a distance $v_0(t_2 - t_1)$. This shows that the phenomenon has traveled in the positive x direction with a velocity v_0.

On the other hand, the function $f_2(x + v_0 t)$ corresponds to a wave traveling in the negative x direction. Thus the general solution of the wave equation in this case is seen to consist of two waves, one traveling to the right (away from the source), and the other traveling to the left (back toward the source). If there is no reflecting surface present to reflect the wave back to the source, the second term of (17) is zero and the solution is given by

$$E = f_1(x - v_0 t) \tag{5-18}$$

Problem 1. Verify that (17) is a solution of (16a).

Problem 2. Does the function $e^{k(x-v_0t)}$ represent a wave if k is a *real* number? Sketch it as a function of x for several instants of time.

5.03 Uniform Plane Waves. Equation (18) is a solution of the wave equation for the particular case where the electric field is independent of y and z and is a function of x and t only. Such a wave is called a *uniform plane wave*. Although this is a special case of electromagnetic wave propagation, it is a very important one practically and will be considered further.

The plane-wave equation

$$\frac{\partial^2 \mathbf{E}}{\partial x^2} = \mu\epsilon \frac{\partial^2 \mathbf{E}}{\partial t^2}$$

may be written in terms of the components of $\mathbf{E}$ as

$$\frac{\partial^2 E_x}{\partial x^2} = \mu\epsilon \frac{\partial^2 E_x}{\partial t^2} \tag{5-19a}$$

$$\frac{\partial^2 E_y}{\partial x^2} = \mu\epsilon \frac{\partial^2 E_y}{\partial t^2} \tag{5-19b}$$

$$\frac{\partial^2 E_z}{\partial x^2} = \mu\epsilon \frac{\partial^2 E_z}{\partial t^2} \tag{5-19c}$$

In a region in which there is no charge density

$$\nabla \cdot \mathbf{E} = \frac{1}{\epsilon} \nabla \cdot \mathbf{D} = 0$$

That is

$$\frac{\partial E_x}{\partial x} + \frac{\partial E_y}{\partial y} + \frac{\partial E_z}{\partial z} = 0$$

For a uniform plane wave in which $\mathbf{E}$ is independent of y and z, the last two terms of this relation are equal to zero so that it reduces to

$$\frac{\partial E_x}{\partial x} = 0$$

Therefore there is no variation of E_x in the x direction. From eq. (19a) it is seen that the second derivative with respect to time of E_x must then be zero. This requires that E_x be either zero, constant in time, or increasing uniformly with time. A field satisfying either of the last two of these conditions would not be a part of the wave motion, and so E_x can be put equal to zero. Therefore a *uniform* plane wave progressing in the x direction has no x component of $\mathbf{E}$. A similar analysis would show that there is no x component of $\mathbf{H}$. It follows, therefore, that uniform plane electromagnetic waves are transverse and have components of $\mathbf{E}$ and $\mathbf{H}$ only in directions perpendicular to the direction of propagation.

Relation between **E** *and* **H** *in a Uniform Plane Wave.* For a uniform plane wave traveling in the x direction **E** and **H** are both independent of y and z, and **E** and **H** have no x component. In this case

$$\nabla \times \mathbf{E} = -\frac{\partial E_z}{\partial x}\hat{\mathbf{y}} + \frac{\partial E_y}{\partial x}\hat{\mathbf{z}}$$

$$\nabla \times \mathbf{H} = -\frac{\partial H_z}{\partial x}\hat{\mathbf{y}} + \frac{\partial H_y}{\partial x}\hat{\mathbf{z}}$$

Then the first Maxwell Equation (I) can be written

$$-\frac{\partial H_z}{\partial x}\hat{\mathbf{y}} + \frac{\partial H_y}{\partial x}\hat{\mathbf{z}} = \epsilon\left(\frac{\partial E_y}{\partial t}\hat{\mathbf{y}} + \frac{\partial E_z}{\partial t}\hat{\mathbf{z}}\right)$$

and the second equation (II) becomes

$$-\frac{\partial E_z}{\partial x}\hat{\mathbf{y}} + \frac{\partial E_y}{\partial x}\hat{\mathbf{z}} = -\mu\left(\frac{\partial H_y}{\partial t}\hat{\mathbf{y}} + \frac{\partial H_z}{\partial t}\hat{\mathbf{z}}\right)$$

Equating the $\hat{\mathbf{y}}$ terms and then the $\hat{\mathbf{z}}$ terms yields the four relations

$$-\frac{\partial H_z}{\partial x} = \epsilon\frac{\partial E_y}{\partial t} \tag{5-20a}$$

$$\frac{\partial H_y}{\partial x} = \epsilon\frac{\partial E_z}{\partial t} \tag{5-20b}$$

$$\frac{\partial E_z}{\partial x} = \mu\frac{\partial H_y}{\partial t} \tag{5-20c}$$

$$\frac{\partial E_y}{\partial x} = -\mu\frac{\partial H_z}{\partial t} \tag{5-20d}$$

Now if $E_y = f_1(x - v_0t)$, where $v_0 = 1/\sqrt{\mu\epsilon}$, then

$$\frac{\partial E_y}{\partial t} = \frac{\partial f_1}{\partial(x - v_0t)}\frac{\partial(x - v_0t)}{\partial t} = -v_0\frac{\partial f_1}{\partial(x - v_0t)}$$

This is generally written as

$$\frac{\partial E_y}{\partial t} = f_1'(x - v_0t)\frac{\partial(x - v_0t)}{\partial t} = -v_0f_1'(x - v_0t)$$

where $f_1'(x - v_0t)$, or more simply f_1', means

$$\frac{\partial f_1(x - v_0t)}{\partial(x - v_0t)}$$

Substituting for $\partial E_y/\partial t$ in (20a) above gives

$$\frac{\partial H_z}{\partial x} = v_0\epsilon f_1'$$

Then

$$H_z = \sqrt{\frac{\epsilon}{\mu}}\int f_1'\,dx + C$$

Now

$$\frac{\partial f_1}{\partial x} = f'_1 \frac{\partial(x - v_0 t)}{\partial x} = f'_1$$

Hence

$$H_z = \sqrt{\frac{\epsilon}{\mu}} \int \frac{\partial f_1}{\partial x}\, dx + C$$

$$= \sqrt{\frac{\epsilon}{\mu}} f_1 + C$$

$$= \sqrt{\frac{\epsilon}{\mu}} E_y + C \qquad (5\text{-}21)$$

The constant of integration C that appears indicates that a field independent of x could be present. Inasmuch as this field would not be a part of the wave motion, it will be neglected and the relation between H_z and E_y becomes

$$H_z = \sqrt{\frac{\epsilon}{\mu}} E_y$$

or

$$\frac{E_y}{H_z} = \sqrt{\frac{\mu}{\epsilon}} \qquad (5\text{-}22)$$

Similarly it can be shown that

$$\frac{E_z}{H_y} = -\sqrt{\frac{\mu}{\epsilon}} \qquad (5\text{-}23)$$

Since

$$E = \sqrt{E_y^2 + E_z^2} \qquad \text{and} \qquad H = \sqrt{H_z^2 + H_y^2}$$

where E and H are the total electric and magnetic field strengths, there also results

$$\frac{E}{H} = \sqrt{\frac{\mu}{\epsilon}} \qquad (5\text{-}24)$$

Equation (24) states that in a traveling* plane electromagnetic wave there is a definite ratio between the amplitudes of E and H and that this ratio is equal to the square root of the ratio of permeability to the dielectric constant of the medium. Since the units of E are volts per meter and the units of H are amperes per meter, the ratio

*The term *traveling* wave is used to indicate that the wave is progressing in one direction and there is no *standing* wave (see section on reflection). When there is a reflected wave resulting in a standing-wave distribution, the ratio E/H can have any value between zero and infinity.

$$\frac{E}{H} = \sqrt{\frac{\mu}{\epsilon}}$$

will have the dimensions of impedance or ohms. For this reason it is customary to refer to the ratio $\sqrt{\mu/\epsilon}$ as the *characteristic impedance* or *intrinsic impedance* of the (nonconducting) medium. For free space

$$\mu = \mu_v = 4\pi \times 10^{-7} \qquad \text{henrys/m}$$

$$\epsilon = \epsilon_v \approx \frac{1}{36\pi \times 10^9} \qquad \text{f/m}$$

so that

$$\sqrt{\frac{\mu}{\epsilon}} = \sqrt{\frac{\mu_v}{\epsilon_v}} \approx 120\pi = 377 \text{ ohms}$$

For any medium, whether conducting or not, the intrinsic impedance is designated by the symbol η. When the medium is free space or a vacuum, the subscript v is used. That is, the intrinsic impedance of free space is

$$\eta_v = \sqrt{\frac{\mu_v}{\epsilon_v}} = 377 \text{ ohms}$$

The relative orientation of **E** and **H** may be determined by taking their dot product and using (22) and (23).

$$\mathbf{E}\cdot\mathbf{H} = E_y H_y + E_z H_z = \eta H_y H_z - \eta H_y H_z = 0$$

Thus in a uniform plane wave, **E** and **H** are at right angles to each other. Further information about the field vectors may be obtained from their cross-product

$$\mathbf{E} \times \mathbf{H} = \hat{\mathbf{x}}(E_y H_z - E_z H_y) = \hat{\mathbf{x}}(\eta H_z^2 + \eta H_y^2) = \hat{\mathbf{x}}\eta H^2$$

Thus the electric field vector crossed into the magnetic field vector gives the direction in which the wave travels.

Problem 3. Nonuniform plane waves also can exist under special conditions to be discussed later. Show that the function

$$F = e^{-\alpha z} \sin \frac{\omega}{v}(x - vt)$$

satisfies the wave equation $\nabla^2 F = (1/c^2)\ddot{F}$ provided that the wave velocity is given by

$$v = c\left(1 + \frac{\alpha^2 c^2}{\omega^2}\right)^{-1/2}$$

5.04 The Wave Equations for a Conducting Medium. In the foregoing sections Maxwell's equations were solved for the particular case of a perfect dielectric, such as free space, in which there were neither charges nor conduction currents. For regions in which the conductivity is not zero and conduction currents may exist, the more general solution must be obtained. It follows in a manner similar to the simpler case already considered.

Recall Maxwell's equations:

$$\nabla \times \mathbf{H} = \epsilon\dot{\mathbf{E}} + \mathbf{J} \tag{I}$$

$$\nabla \times \mathbf{E} = -\mu\dot{\mathbf{H}} \tag{II}$$

If the medium has a conductivity σ (mhos/m), the conduction current density will be given by Ohm's law:*

$$\mathbf{J} = \sigma\mathbf{E} \tag{5-25}$$

so that eq. (I) becomes

$$\nabla \times \mathbf{H} = \epsilon\dot{\mathbf{E}} + \sigma\mathbf{E} \tag{5-26}$$

Take the curl of both sides of eq. (II) and then substitute into it the time derivative of (26).

$$\begin{aligned}\nabla \times \nabla \times \mathbf{E} &= -\mu\nabla \times \dot{\mathbf{H}}\\ &= -\mu\epsilon\ddot{\mathbf{E}} - \mu\sigma\dot{\mathbf{E}}\end{aligned}$$

Recall that

$$\nabla \times \nabla \times \mathbf{E} = \nabla\nabla\cdot\mathbf{E} - \nabla^2\mathbf{E}$$

Combining these last two equations, there results

$$\nabla^2\mathbf{E} - \mu\epsilon\ddot{\mathbf{E}} - \mu\sigma\dot{\mathbf{E}} = \nabla\nabla\cdot\mathbf{E} \tag{5-27}$$

Now for any homogeneous medium in which ϵ is constant

$$\nabla\cdot\mathbf{E} = \frac{1}{\epsilon}\nabla\cdot\mathbf{D}$$

But $\nabla\cdot\mathbf{D} = \rho$, and since there is no net charge within a conductor (although there may be a charge on the surface), the charge density ρ equals zero† and therefore

$$\nabla\cdot\mathbf{D} = 0$$

*Equation (25) is the vector statement (applicable to an elemental volume) of the more familiar relation $I = V/R$.

†The statement of no *net* charge within a conductor is consistent with our notion of current flow as a drift of free negative electrons through the positive atomic lattice of the conductor. Within any macroscopic element of volume the positive and negative charges are equal in number (on the average), and the net charge is zero. It is easily shown for steady-state sinusoidal time variations that $\nabla\cdot\mathbf{D} = 0$ (and therefore $\rho = 0$) in conductors is a direct consequence of Maxwell's equations and Ohm's law (see problem 5). It can also be shown that if a charge ever were placed within a conductor (in some manner not explained) the "transient time" or "relaxation time" required for this charge to appear on the surface would be exceedingly small for any materials considered to be conductors (see problem 4).

Equation (27) then becomes

$$\nabla^2 \mathbf{E} - \mu\epsilon\ddot{\mathbf{E}} - \mu\sigma\dot{\mathbf{E}} = 0 \tag{5-28}$$

This is the wave equation for **E**. The wave equation for **H** is obtained in a similar manner.

$$\nabla \times \nabla \times \mathbf{H} = \epsilon\nabla \times \dot{\mathbf{E}} + \sigma\nabla \times \mathbf{E}$$

$$\nabla \times \mathbf{E} = -\mu\dot{\mathbf{H}}$$

$$\nabla\nabla\cdot\mathbf{H} - \nabla^2\mathbf{H} = -\mu\epsilon\ddot{\mathbf{H}} - \mu\sigma\dot{\mathbf{H}}$$

But

$$\nabla\cdot\mathbf{H} = \frac{1}{\mu}\nabla\cdot\mathbf{B} = 0$$

Therefore

$$\nabla^2\mathbf{H} - \mu\epsilon\ddot{\mathbf{H}} - \mu\sigma\dot{\mathbf{H}} = 0 \tag{5-29}$$

This is the wave equation for **H**.

Problem 4. Using $\nabla\cdot\mathbf{D} = \rho$, Ohm's law, and the equation of continuity, show that if at any instant a charge density ρ existed within a conductor, it would decrease to $1/e$ times this value in a time ϵ/σ seconds. Calculate this time for a copper conductor.

5.05 Sinusoidal Time Variations. In practice most generators produce voltages and currents, and hence electric and magnetic fields, which vary sinusoidally with time (at least approximately). Even where this is not the case any periodic variation can always be analyzed in terms of sinusoidal variations with fundamental and harmonic frequencies, so it is customary in most problems to assume sinusoidal time variations. This can be expressed by writing, for example,

$$E = E_0 \cos \omega t \tag{5-30a}$$

$$E = E_0 \sin \omega t \tag{5-30b}$$

where $f = \omega/2\pi$ is the frequency of the variation. The above expressions suggest that a sinusoidal time factor would have to be attached to every term in any equation. Fortunately this is not necessary since the time factor may be suppressed through the use of *phasor* notation. Consider the electric field strength vector as an example. The time varying field $\tilde{\mathbf{E}}(\mathbf{r}, t)$ may be expressed in terms of the corresponding phasor quantity $\mathbf{E}(\mathbf{r})$ as

$$\tilde{\mathbf{E}}(\mathbf{r}, t) = \text{Re}\,\{\mathbf{E}(\mathbf{r})e^{j\omega t}\} \tag{5-31}$$

in which the symbol ($\sim$) has been placed over the time-varying quantity to distinguish it from the phasor quantity. This distinction will be made wherever necessary from this point to the end of the book.

In order to understand phasor notation it is necessary to consider one component at a time, say the x component. The phasor E_x is defined by the relation

$$\tilde{E}_x(\mathbf{r}, t) = \operatorname{Re}\{E_x(\mathbf{r})e^{j\omega t}\} \tag{5-32}$$

In general, $E_x(\mathbf{r})$ is a complex number and thus at some fixed point $\mathbf{r}$ in space E_x may be represented as a point in the complex plane shown in Fig. 5-2. Multiplication by $e^{j\omega t}$ results in a rotation through the

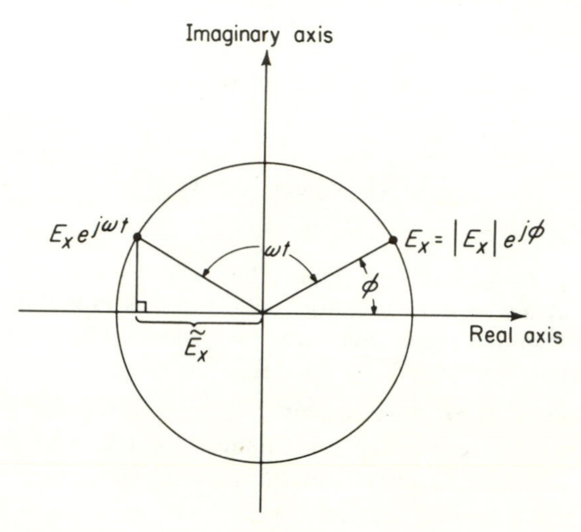

Figure 5-2. Relationship between time-varying and phasor quantities.

angle ωt measured from the angle ϕ. As time progresses, the point $E_x e^{j\omega t}$ traces out a circle with center at the origin. Taking the real part is the same as taking the projection on the real axis and this projection varies sinusoidally with time. The phase of the sinusoid is determined by the argument ϕ of the complex number E_x and thus the time-varying quantity may be expressed as

$$\begin{aligned}\tilde{E}_x &= \operatorname{Re}\{|E_x|e^{j\phi}e^{j\omega t}\} \\ &= |E_x| \cos(\omega t + \phi)\end{aligned} \tag{5-33}$$

The phasor definition used above is one of many possible definitions. For instance, one could just as easily employ the time factor $e^{-j\omega t}$ or make use of the imaginary part instead of the real part. Another possibility is to introduce a $\sqrt{2}$ factor on the right-hand side of the phasor definition, thus making $|E_x|$ the rms value of the sinusoid rather than the peak value.

Maxwell's Equations Using Phasor Notation. Having already derived the general time-varying Maxwell equations, it is desirable now to find the corresponding set of phasor equations applicable for sinusoidal time variations. Consider for instance the equation given in time-varying form by

$$\nabla \times \tilde{\mathbf{H}} = \frac{\partial \tilde{\mathbf{D}}}{\partial t} + \tilde{\mathbf{J}} \tag{5-34}$$

For the sinusoidal steady state we may substitute the phasor relations as follows:

$$\nabla \times \mathrm{Re}\{\mathbf{H}e^{j\omega t}\} = \frac{\partial}{\partial t}\mathrm{Re}\{\mathbf{D}e^{j\omega t}\} + \mathrm{Re}\{\mathbf{J}e^{j\omega t}\}$$

The operations of taking the real part and differentiation may be interchanged with the result that

$$\mathrm{Re}\{(\nabla \times \mathbf{H} - j\omega\mathbf{D} - \mathbf{J})e^{j\omega t}\} = 0$$

If the above relation is to be valid for all t, then

$$\nabla \times \mathbf{H} = j\omega\mathbf{D} + \mathbf{J} \tag{5-35}$$

which is the required differential equation in phasor form. Evidently the phasor equation may be derived from the time-varying equation by replacing each time-varying quantity with a phasor quantity and each time derivative with a $j\omega$ factor.

To summarize, for sinusoidal time variations Maxwell's equations may be expressed in phasor form as

$$\nabla \times \mathbf{H} = j\omega\mathbf{D} + \mathbf{J} \qquad \oint \mathbf{H}\cdot d\mathbf{s} = \int (j\omega\mathbf{D} + \mathbf{J})\cdot d\mathbf{a} \tag{5-36}$$

$$\nabla \times \mathbf{E} = -j\omega\mathbf{B} \qquad \oint \mathbf{E}\cdot d\mathbf{s} = -\int j\omega\mathbf{B}\cdot d\mathbf{a} \tag{5-37}$$

$$\nabla\cdot\mathbf{D} = \rho \qquad \oint \mathbf{D}\cdot d\mathbf{a} = \int \rho\, dV \tag{5-38}$$

$$\nabla\cdot\mathbf{B} = 0 \qquad \oint \mathbf{B}\cdot d\mathbf{a} = 0 \tag{5-39}$$

The above equations contain the equation of continuity.

$$\nabla\cdot\mathbf{J} = -j\omega\rho \qquad \text{or} \qquad \oint \mathbf{J}\cdot d\mathbf{a} = -\int j\omega\rho\, dV \tag{5-40}$$

The constitutive relations retain their forms, being $\mathbf{D} = \epsilon\mathbf{E}$, $\mathbf{B} = \mu\mathbf{H}$, $\mathbf{J} = \sigma\mathbf{E}$.

For sinusoidal time variations the wave equation (14) for the electric field in a lossless medium becomes

$$\nabla^2\mathbf{E} = -\omega^2\mu\epsilon\mathbf{E} \tag{5-41}$$

which is the vector *Helmholtz equation.* In a conducting medium, the wave equation (28) becomes

$$\nabla^2 \mathbf{E} + (\omega^2 \mu\epsilon - j\omega\mu\sigma)\mathbf{E} = 0 \tag{5-42}$$

Equations of the same form may be written for $\mathbf{H}$, of course.

Problem 5. Using Maxwell's equation (I) show that

$$\nabla \cdot \mathbf{D} = 0 \quad \text{in a conductor,}$$

if Ohm's law and sinusoidal time variations (i.e., as $e^{j\omega t}$) are assumed.

Wave Propagation in a Lossless Medium. For the uniform plane wave case in which there is no variation with respect to y or z, the wave equation in phasor form* may be expressed as

$$\frac{\partial^2 \mathbf{E}}{\partial x^2} = -\omega^2 \mu\epsilon \mathbf{E} \quad \text{or} \quad \frac{\partial^2 \mathbf{E}}{\partial x^2} = -\beta^2 \mathbf{E} \tag{5-43}$$

where $\beta = \omega\sqrt{\mu\epsilon}$. Considering the E_y component (as was done for time-varying fields) the solution may be written in the form

$$E_y = C_1 e^{-j\beta x} + C_2 e^{+j\beta x} \tag{5-44}$$

in which C_1 and C_2 are arbitrary complex constants. The corresponding time-varying field is

$$\begin{aligned} \tilde{E}_y(x, t) &= \mathrm{Re}\,\{E_y(x)e^{j\omega t}\} \\ &= \mathrm{Re}\,\{C_1 e^{j(\omega t - \beta x)} + C_2 e^{j(\omega t + \beta x)}\} \end{aligned} \tag{5-45}$$

By taking the real part in eq. (45) the solution is displayed directly in its sinusoidal form. For example when C_1 and C_2 are real, eq. (45) becomes

$$\tilde{E}_y(x, t) = C_1 \cos(\omega t - \beta x) + C_2 \cos(\omega t + \beta x) \tag{5-46}$$

It is evident that in a homogeneous, lossless medium the assumption of a sinusoidal time variation results in a space variation which is also sinusoidal.

Equations (45) or (46) represent the sum of two waves traveling in opposite directions. If $C_1 = C_2$, the two *traveling waves* combine to form a simple *standing wave* which does not progress. The wave velocity may be obtained by rewriting the expression for E_y as a function of $(x \pm vt)$. Clearly the velocity must be

$$v = \frac{\omega}{\beta} \tag{5-47}$$

The same result may be obtained by identifying some point in the

*In phasor form a *plane* wave is defined as one for which the *equiphase* surface is a plane. If the equiphase surface is also an equiamplitude surface, it is a *uniform* plane wave. Nonuniform plane waves are encountered in wave guides (chap. 8).

waveform and observing its velocity. For a wave traveling in the positive x direction the point is given by $\omega t - \beta x =$ a constant. Differentiation yields $v = dx/dt = \omega/\beta$ as above. This velocity of some point in the sinusoidal waveform is called the *phase velocity*. In addition, β is called the *phase-shift constant* and is a measure of the phase shift in radians per unit length.

Another important quantity is the wavelength λ, defined as that distance over which the sinusoidal waveform passes through a full cycle of 2π radians as illustrated in Fig. 5-3. Since eq. (45) indicates that

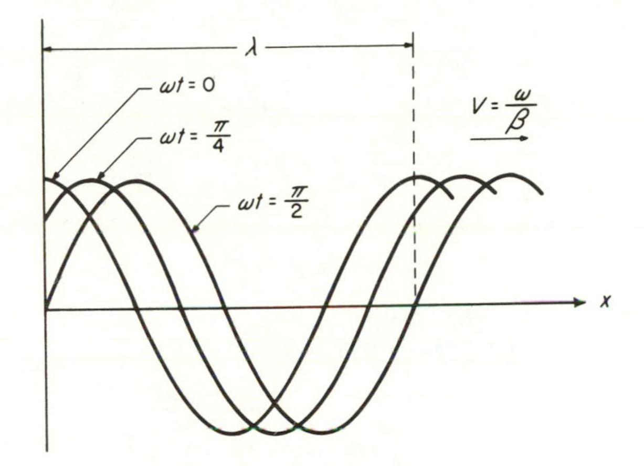

Figure 5-3. A sinusoidal traveling wave.

the fields vary in space as sinusoidal functions of βx, it is clear by the definition of wavelength that

$$\beta\lambda = 2\pi \tag{5-48}$$

Substituting for β from eq. (47), we have

$$\omega\lambda = 2\pi v$$

and since $\omega = 2\pi f$, we have finally

$$v = f\lambda \tag{5-49}$$

in which f is measured in cycles per second or *hertz*. The foregoing discussion is sufficiently general to apply for any value of β. For the value of β given in eq. (43), however, the phase velocity must be

$$v = \frac{\omega}{\beta} = \frac{1}{\sqrt{\mu\epsilon}} = v_0$$

Wave Propagation in a Conducting Medium. For sinusoidal time variations the use of phasor notation permits a straightforward solution to be obtained for the wave equation in a conducting medium. The wave equation (42) may be written in the form of a Helmholtz equation as

$$\nabla^2\mathbf{E} - \gamma^2\mathbf{E} = 0 \tag{5-50}$$

where $\gamma^2 = j\omega\mu(\sigma + j\omega\epsilon)$.

The *propagation constant* γ is a complex number having real and imaginary parts designated by α and β respectively. That is, $\gamma = \alpha + j\beta$. Actually, the square root of γ^2 has two values, but for the sake of convenience we shall agree that γ is that value of the square root which has a positive real part (given by α). It may be shown easily that if α is positive, β is also positive.

Returning once again to the uniform plane wave traveling in the x direction we note that the electric field strength must satisfy

$$\frac{\partial^2\mathbf{E}}{\partial x^2} = \gamma^2\mathbf{E} \tag{5-51}$$

which has as one possible solution

$$\mathbf{E}(x) = \mathbf{E}_0 e^{-\gamma x} \tag{5-52}$$

In time-varying form, this becomes

$$\begin{aligned}\tilde{\mathbf{E}}(x, t) &= \text{Re}\,\{\mathbf{E}_0 e^{-\gamma x + j\omega t}\} \\ &= e^{-\alpha x}\,\text{Re}\,\{\mathbf{E}_0 e^{j(\omega t - \beta x)}\}\end{aligned} \tag{5-53}$$

This is the equation of a wave traveling in the x direction and attenuated by the factor $e^{-\alpha x}$. As in the lossless cases, the phase-shift factor and the phase velocity are given by

$$\beta = \frac{2\pi}{\lambda} \qquad \text{and} \qquad v = f\lambda = \frac{\omega}{\beta}$$

In terms of the "primary" constants of the medium, that is σ, μ, and ϵ, the values of α and β are

$$\begin{aligned}\alpha &= \text{real part of } \sqrt{(j\omega\mu)(\sigma + j\omega\epsilon)} \\ &= \omega\sqrt{\frac{\mu\epsilon}{2}\left(\sqrt{1 + \frac{\sigma^2}{\omega^2\epsilon^2}} - 1\right)}\end{aligned} \tag{5-54}$$

$$\beta = \omega\sqrt{\frac{\mu\epsilon}{2}\left(\sqrt{1 + \frac{\sigma^2}{\omega^2\epsilon^2}} + 1\right)} \tag{5-55}$$

Problem 6. From the expression $\gamma = \sqrt{j\omega\mu(\sigma + j\omega\epsilon)}$ derive expressions (54) and (55).

Problem 7. For any conductive medium, begin with a uniform plane wave

$$\mathbf{E}(x) = \mathbf{E}_0 e^{-\gamma x}, \qquad \mathbf{H}(x) = \mathbf{H}_0 e^{-\gamma x}$$

and use Maxwell's equations to show that

$$\eta\mathbf{H}_0 = \hat{\mathbf{x}} \times \mathbf{E}_0, \qquad \mathbf{E}_0 = -\eta\hat{\mathbf{x}} \times \mathbf{H}_0, \qquad \hat{\mathbf{x}}\cdot\mathbf{E}_0 = 0, \qquad \hat{\mathbf{x}}\cdot\mathbf{H}_0 = 0 \tag{5-56}$$

in which $\eta^2 = j\omega\mu/(\sigma + j\omega\epsilon)$. Note that η is the root (of η^2) having a positive real part. Make use of the identities

$$\nabla \times (\psi\mathbf{A}) = \nabla\psi \times \mathbf{A} + \psi\nabla \times \mathbf{A}$$
$$\nabla\cdot(\psi\mathbf{A}) = \mathbf{A}\cdot\nabla\psi + \psi\nabla\cdot\mathbf{A}$$

5.06 Conductors and Dielectrics. In electromagnetics, materials are divided roughly into two classes: conductors and dielectrics or insulators. The dividing line between the two classes is not sharp and some media (the earth for example) are considered as conductors in one part of the radio frequency range, but as dielectrics (with loss) in another part of the range.

In Maxwell's first equation:

$$\nabla \times \mathbf{H} = \sigma\mathbf{E} + j\omega\epsilon\mathbf{E}$$

the first term on the right is conduction current density and the second term is displacement current density. The ratio $\sigma/\omega\epsilon$ is therefore just the ratio of conduction current density to displacement current density in the medium. Hence, $\sigma/\omega\epsilon = 1$ can be considered to mark the dividing line between conductors and dielectrics. For *good conductors* such as metals $\sigma/\omega\epsilon$ is very much greater than unity over the entire radio frequency spectrum. For example for copper, even at the relatively high frequency of 30,000 MHz, $\sigma/\omega\epsilon$ is about 3.5×10^8. For *good dielectrics* or insulators $\sigma/\omega\epsilon$ is very much less than unity in the radio frequency range. For example, for mica at audio or radio frequencies $\sigma/\omega\epsilon$ is of the order of 0.0002. For good conductors σ and ϵ are nearly independent of frequency, but for most materials classed as dielectrics the "constants" σ and ϵ are functions of frequency. It has been found for these materials that the ratio $\sigma/\omega\epsilon$ is often relatively constant over the frequency range of interest. For this and other reasons the properties of dielectrics are usually given in terms of the dielectric "constant" ϵ and the ratio $\sigma/\omega\epsilon$. Under these circumstances the ratio $\sigma/\omega\epsilon$ is known as the *dissipation factor D* of the dielectric. For reasonably good dielectrics, that is, those having small values of D, the dissipation factor is practically the same as the *power factor* of the dielectric. Actually, power factor is given by

$$\text{P.F.} = \sin\phi$$

where
$$\phi = \tan^{-1} D$$

Dissipation factor and power factor differ by less than 1 per cent when their values are less than 0.15.

Most materials used in radio are required either to pass conduction currents readily or to prevent the flow of conduction current as completely as possible. For this reason most materials met with in practice will fall into either the good conductor or the good insulator class. The important practical exception is the earth, which occupies an in-between position throughout most of the radio frequency spectrum. This case will be treated in detail in the chapter on propagation. For both good conductors and good dielectrics certain approximations are valid which simplify considerably the expressions for α and β.

Wave Propagation in Good Dielectrics. For this case $\sigma/\omega\epsilon \ll 1$ so that it is possible to write to a very good approximation

$$\sqrt{1 + \frac{\sigma^2}{\omega^2\epsilon^2}} \cong \left(1 + \frac{\sigma^2}{2\omega^2\epsilon^2}\right)$$

where only the first two terms of the binomial expansion have been used. Then expression (54) for α becomes

$$\alpha \cong \omega\sqrt{\frac{\mu\epsilon}{2}\left[\left(1 + \frac{\sigma^2}{2\omega^2\epsilon^2}\right) - 1\right]} = \frac{\sigma}{2}\sqrt{\frac{\mu}{\epsilon}} \tag{5-57}$$

This expression may be compared with the expression for the attenuation factor of a low-loss transmission line having zero series resistance. In that case the expression for α is

$$\alpha = \frac{G}{2}\sqrt{\frac{L}{C}} = \frac{G}{2}Z_0$$

The expression for β reduces in a similar manner

$$\beta \cong \omega\sqrt{\frac{\mu\epsilon}{2}\left[\left(1 + \frac{\sigma^2}{2\omega^2\epsilon^2}\right) + 1\right]} = \omega\sqrt{\mu\epsilon}\left(1 + \frac{\sigma^2}{8\omega^2\epsilon^2}\right) \tag{5-58}$$

$\omega\sqrt{\mu\epsilon}$ is the phase-shift factor for a perfect dielectric. The effect of a small amount of loss is to add the second term of (58) as a small correction factor. The velocity of the wave in the dielectric is given by

$$v = \frac{\omega}{\beta} = \frac{1}{\sqrt{\mu\epsilon}\left(1 + \dfrac{\sigma^2}{8\omega^2\epsilon^2}\right)}$$

$$\cong v_0\left(1 - \frac{\sigma^2}{8\omega^2\epsilon^2}\right) \tag{5-59}$$

where $v_0 = 1/\sqrt{\mu\epsilon}$ is the velocity of the wave in the dielectric when the conductivity is zero. The effect of a small amount of loss is to reduce slightly the velocity of propagation of the wave. It will be shown later that the general expression for the intrinsic or characteristic impedance of a medium which has a finite conductivity is

$$\eta = \sqrt{\frac{j\omega\mu}{\sigma + j\omega\epsilon}}$$

Using the same approximations as above, this becomes for a good dielectric

$$\eta = \sqrt{\frac{\mu}{\epsilon}\left(\frac{1}{1+\dfrac{\sigma}{j\omega\epsilon}}\right)}$$

$$\cong \sqrt{\frac{\mu}{\epsilon}}\left(1 + j\frac{\sigma}{2\omega\epsilon}\right)$$

Since $\sqrt{\mu/\epsilon}$ is the intrinsic impedance of the dielectric when $\sigma = 0$, it is seen that the chief effect of a small amount of loss is to add a small reactive component to the intrinsic impedance.

Wave Propagation in a Good Conductor. For this case $\sigma/\omega\epsilon \gg 1$ so that the expression for γ may be written

$$\gamma = \sqrt{(j\omega\mu\sigma)\left(1 + j\frac{\omega\epsilon}{\sigma}\right)}$$

$$\cong \sqrt{j\omega\mu\sigma} = \sqrt{\omega\mu\sigma}\ \underline{/45^\circ}$$

Therefore
$$\alpha = \beta = \sqrt{\frac{\omega\mu\sigma}{2}}$$

The velocity of the wave in the conductor will be

$$v = \frac{\omega}{\beta} = \sqrt{\frac{2\omega}{\mu\sigma}}$$

and the intrinsic impedance of the conductor is

$$\eta \cong \sqrt{\frac{j\omega\mu}{\sigma}} = \sqrt{\frac{\omega\mu}{\sigma}}\ \underline{/45^\circ}$$

It is seen that in good conductors where σ is very large, both α and β are also large. This means that the wave is attenuated greatly as it progresses through the conductor and the phase shift per unit length is also great. The velocity of the wave, being inversely proportional to β, is very small in a good conductor, and is of the same order of magnitude as that of a sound wave in air. The characteristic impedance is also very small and has a *reactive* component. The angle of this impedance is always 45 degrees for good conductors.

Depth of Penetration. In a medium which has conductivity the wave is attenuated as it progresses owing to the losses which occur. In a good conductor at radio frequencies the rate of attenuation is very great and the wave may penetrate only a very short distance before being reduced to a negligibly small percentage of its original strength. A term that has significance under such circumstances is the *depth of penetration.* The depth of penetration, δ, is defined as that depth in which the wave has been attenuated to $1/e$ or approximately 37 per cent of its original value. Since the amplitude decreases by the factor $e^{-\alpha x}$ it is apparent that at that distance x, which makes $\alpha x = 1$, the amplitude is only $1/e$ times its value at $x = 0$. By definition this distance is equal to δ, the depth of penetration; so

$$\alpha\delta = 1 \qquad \text{or} \qquad \delta = \frac{1}{\alpha}$$

The general expression for depth of penetration is

$$\delta = \frac{1}{\alpha} = \frac{1}{\omega\sqrt{\frac{\mu\epsilon}{2}\left(\sqrt{1 + \frac{\sigma^2}{\omega^2\epsilon^2}} - 1\right)}}$$

For a *good* conductor the depth of penetration is

$$\delta = \frac{1}{\alpha} \cong \sqrt{\frac{2}{\omega\mu\sigma}}$$

As an example of the order of magnitude of δ in metals, the depth of penetration of a one-megahertz wave into copper which has a conductivity $\sigma = 5.8 \times 10^7$ mhos per meter and a permeability approximately equal to that of free space is

$$\delta = \sqrt{\frac{2 \times 10^7}{2\pi \times 10^6 \times 4\pi \times 5.8 \times 10^7}} = 0.0667 \text{ mm}$$

At 100 MHz it is 0.00667 mm, whereas at 60 Hz it is 8.67 mm. For comparison, at one MHz the depth of penetration is 25 cm into sea water and 7.1 m into fresh water.

Problem 8. Earth is considered to be a good conductor when $\omega\epsilon/\sigma \ll 1$. Determine the highest frequency for which earth can be considered a good conductor if $\ll 1$ means less than 0.1. Assume the following constants:

$$\sigma = 5 \times 10^{-3} \text{ mho/meter} \qquad \epsilon = 10\epsilon_v$$

Problem 9. A copper wire carries a conduction current of 1 amp. Determine the displacement current in the wire at 100 MHz. (Assume that copper has about the same permittivity as free space, that is $\epsilon = \epsilon_v$. For copper $\sigma = 5.8 \times 10^7$ mhos/m.)

5.07 Polarization. The polarization of a uniform plane wave refers to the time-varying behavior of the electric field strength vector at some fixed point in space. Consider, for example, a uniform plane wave traveling in the z direction with the $\tilde{\mathbf{E}}$ and $\tilde{\mathbf{H}}$ vectors lying in the x-y plane. If $\tilde{E}_y = 0$ and only $\tilde{E}_x$ is present, the wave is said to be *polarized in the* x *direction*: A similar statement holds for polarization in the y direction. If both $\tilde{E}_x$ and $\tilde{E}_y$ are present and are *in phase*, the resultant electric field has a direction dependent on the relative magnitude of $\tilde{E}_x$ and $\tilde{E}_y$. The angle which this direction makes with the x axis is $\tan^{-1} \tilde{E}_y/\tilde{E}_x$ and this angle will be constant with time (see Fig. 5-4). In all the above cases in which the direction of the resultant vector is constant with time, the wave is said to be *linearly polarized.*

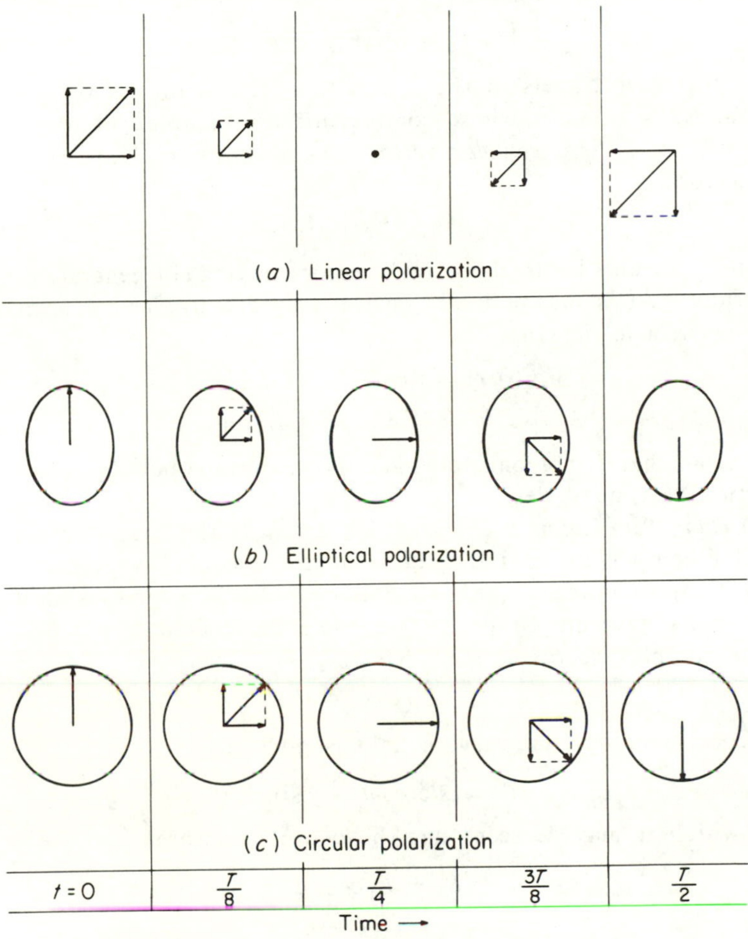

Figure 5-4. Linear, elliptical and circular polarization.

If $\tilde{E}_x$ and $\tilde{E}_y$ are not in phase, that is, if they reach their maximum values at different instants of time, then the direction of the resultant electric vector will vary with time. In this case it can be shown that the locus of the endpoint of the resultant $\tilde{\mathbf{E}}$ will be an ellipse and the wave is said to be *elliptically polarized.* In the particular case where $\tilde{E}_x$ and $\tilde{E}_y$ have equal magnitudes and a 90-degree phase difference, the locus of the resultant $\tilde{\mathbf{E}}$ is a circle and the wave is *circularly polarized.*

The polarization concept outlined above may be developed further using phasor notation. The electric field of a uniform plane wave traveling in the z direction may be expressed in phasor form as

$$\mathbf{E}(z) = \mathbf{E}_0 e^{-j\beta z}$$

and in time-varying form as

$$\tilde{\mathbf{E}}(z, t) = \text{Re}\,\{\mathbf{E}_0 e^{-j\beta z} e^{j\omega t}\} \tag{5-60}$$

Since the wave travels in the z direction, $\tilde{\mathbf{E}}$ lies in the x-y plane. In general $\mathbf{E}_0$ is a vector whose components are complex numbers, or in other words, $\mathbf{E}_0$ is a *complex vector*. As a result it may be written in the form

$$\mathbf{E}_0 = \mathbf{E}_r + j\mathbf{E}_i \tag{5-61}$$

in which $\mathbf{E}_r$ and $\mathbf{E}_i$ are both real vectors having (in general) different directions. At some point in space (say $z = 0$) the resultant time-varying electric field is

$$\begin{aligned}\tilde{\mathbf{E}}(0, t) &= \text{Re}\,\{(\mathbf{E}_r + j\mathbf{E}_i)e^{j\omega t}\}\\ &= \mathbf{E}_r \cos \omega t - \mathbf{E}_i \sin \omega t\end{aligned} \tag{5-62}$$

It is clear that $\tilde{\mathbf{E}}$ not only changes its magnitude but also changes its direction as time varies.

Circular Polarization. Consider for example the case in which the x and y components of the electric field are equal in magnitude. If the y component leads the x component by 90 degrees and if both components have amplitude E_a, then the electric field at $z = 0$ is given by the complex vector

$$\mathbf{E}_0 = (\hat{\mathbf{x}} + j\hat{\mathbf{y}})E_a$$

The corresponding time-varying field is given by

$$\tilde{\mathbf{E}}(0, t) = (\hat{\mathbf{x}} \cos \omega t - \hat{\mathbf{y}} \sin \omega t)E_a$$

from which it may be seen that the components are

$$\tilde{E}_x = E_a \cos \omega t$$

$$\tilde{E}_y = -E_a \sin \omega t$$

These components satisfy the relation

$$\tilde{E}_x^2 + \tilde{E}_y^2 = E_a^2$$

which indicates that the endpoint of $\tilde{E}(0, t)$ traces out a circle of radius E_a as time progresses. Furthermore it may be seen that the *sense* or *direction* of rotation is that of a *left-handed* screw advancing in the z direction (the direction of propagation). Thus the wave is said to be *left circularly polarized.* Similar remarks hold for right circular polarization represented by the complex vector

$$\mathbf{E}_0 = (\hat{\mathbf{x}} - j\hat{\mathbf{y}})E_a$$

It is apparent that a reversal of the sense of rotation may be obtained by a 180-degree phase shift applied either to the x component or to the y component of the electric field.

Elliptical Polarization. A somewhat more general example arises when the x and y components of the electric field differ in amplitude. Assuming again that the y component leads the x component by 90 degrees, such a field may be represented by the complex vector

$$\mathbf{E}_0 = \hat{\mathbf{x}}A + j\hat{\mathbf{y}}B$$

in which A and B are positive real constants. The corresponding time-varying field is given by

$$\tilde{\mathbf{E}}(0, t) = \hat{\mathbf{x}}A \cos \omega t - \hat{\mathbf{y}}B \sin \omega t$$

The components of the time-varying field are

$$\tilde{E}_x = A \cos \omega t$$

$$\tilde{E}_y = -B \sin \omega t$$

from which it is evident that

$$\frac{\tilde{E}_x^2}{A^2} + \frac{\tilde{E}_y^2}{B^2} = 1$$

Thus the endpoint of the $\tilde{\mathbf{E}}(0, t)$ vector traces out an ellipse and the wave is said to be *elliptically polarized.* Inspection of the equations indicates that the sense of polarization is again left-handed.

Elliptical polarization is in fact the most general form of polarization. The polarization is completely specified by the orientation and axial ratio of the polarization ellipse and by the sense in which the endpoint of the electric field vector moves around the ellipse. Various methods for representing the polarization of a wave are covered in chap. 12.

5.08 Direction Cosines. Sometimes it is necessary to write the expression for a plane wave that is traveling in some arbitrary direction with respect to a fixed set of axes. This is most conveniently done in terms of the direction cosines of the normal to the plane of the wave. By definition of a uniform plane wave the equiphase surfaces are planes. Thus in the expression

$$\mathbf{E}(x) = \mathbf{E}_0 e^{-j\beta x}$$

for a wave traveling in the x direction, the planes of constant phase are given by the equation

$$x = \text{a constant}$$

For a plane wave traveling in some arbitrary direction, say the s direction, it is necessary to replace x with an expression that, when put equal to a constant, gives the equiphase surfaces.

The equation of a plane is given by

$$\hat{\mathbf{n}}\cdot\mathbf{r} = \text{a constant}$$

where $\mathbf{r}$ is the radius vector from the origin to any point P on the plane and $\hat{\mathbf{n}}$ is the unit vector normal to the plane (sometimes called the *wave normal*). That this is so can be seen from Fig. 5-5, in which a plane perpendicular to the unit vector $\hat{\mathbf{n}}$ is seen from its side, thus appearing as the line *F-F*. The dot product $\hat{\mathbf{n}}\cdot\mathbf{r}$ is the projection of the radius vector $\mathbf{r}$ along the normal to the plane, and it is apparent that this will have the constant value OM for all points on the plane. Now the dot product of two vectors is a scalar equal to the sum of the products of the components of the vectors along the axes of the co-ordinate system. Therefore

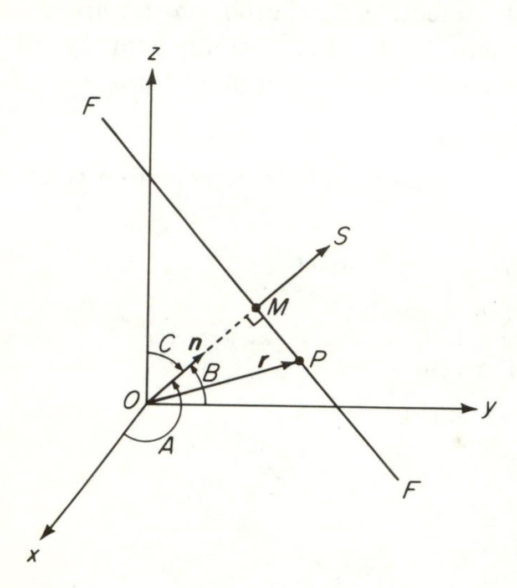

Figure 5-5.

$$\hat{\mathbf{n}}\cdot\mathbf{r} = x\cos A + y\cos B + z\cos C$$

where x, y, z, are the components of the vector $\mathbf{r}$, and $\cos A$, $\cos B$, $\cos C$ are the components of the unit vector $\hat{\mathbf{n}}$ along the x, y, and z axes. A, B and C are the angles that the unit vector $\hat{\mathbf{n}}$ makes with the positive x, y, and z axes, respectively. Their cosines are termed the *direction cosines* or *direction components* of the vector.

The equation of a plane wave traveling in the direction $\hat{\mathbf{n}}$, normal to the planes of constant phase, can now be written as

$$\begin{aligned}\mathbf{E}(\mathbf{r}) &= \mathbf{E}_0 e^{-j\beta\hat{\mathbf{n}}\cdot\mathbf{r}} \\ &= \mathbf{E}_0 e^{-j\beta(x\cos A + y\cos B + z\cos C)}\end{aligned} \tag{5-63}$$

In time-varying form and assuming $\mathbf{E}_0 = \mathbf{E}_r + j\mathbf{E}_i$, this becomes

$$\begin{aligned}\tilde{\mathbf{E}}(\mathbf{r}, t) &= \text{Re}\,\{\mathbf{E}_0 e^{-j(\beta\hat{\mathbf{n}}\cdot\mathbf{r} - \omega t)}\} \\ &= \mathbf{E}_r\cos(\beta\hat{\mathbf{n}}\cdot\mathbf{r} - \omega t) + \mathbf{E}_i\sin(\beta\hat{\mathbf{n}}\cdot\mathbf{r} - \omega t)\end{aligned} \tag{5-64}$$

Wavelength and Phase Velocity. The uniform plane wave expressions considered so far all can be broken up into functions of the form

$$e^{-jhu}$$

where h is some real constant and u represents distance measured along a straight line. Referring back to the original discussion on wavelength and phase velocity, we see that these quantities can be stated in terms of h and ω for the given direction $\hat{\mathbf{u}}$. The wavelength in the $\hat{\mathbf{u}}$ direction is given by

$$\lambda_u = \frac{2\pi}{h} \tag{5-65}$$

and the phase velocity in the $\hat{\mathbf{u}}$ direction is given by

$$v_u = \frac{\omega}{h} \tag{5-66}$$

With this information, we can determine the wavelength and phase velocity in any direction.

For the uniform plane wave already discussed, the wavelength and phase velocity in the direction of the wave normal $\hat{\mathbf{n}}$ are given by

$$\lambda = \frac{2\pi}{\beta} \qquad \text{and} \qquad v = \frac{\omega}{\beta}$$

If we look at the uniform plane wave expression in direction cosine form, it is easy to determine the wavelengths and phase velocities in the directions of the co-ordinate axes by comparison with the e^{-jhu} form discussed above. Thus we have for the x direction

$$\lambda_x = \frac{2\pi}{\beta \cos A} = \frac{\lambda}{\cos A} \qquad \text{and} \qquad v_x = \frac{\omega}{\beta \cos A} = \frac{v}{\cos A}$$

As long as the angle A is not zero, both the wavelength and the phase velocity measured along the x axis are *greater* than when measured along the wave normal. Similar statements hold for the y and z directions. These relations between the velocities and wavelengths in the various directions are shown more clearly in Fig. 5-6, which shows

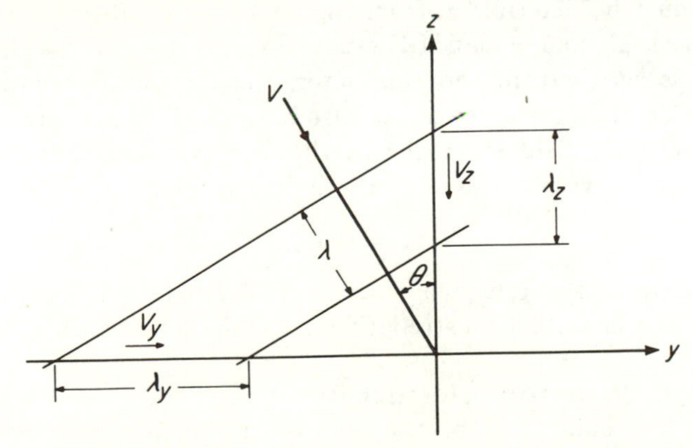

Figure 5-6. Relations between wavelengths and velocities in different directions.

successive crests of an incident wave intersecting the y and z axes. For small angles of θ it is seen that the velocity v_y, with which a crest moves along the y axis, becomes very great, approaching infinity as θ approaches zero.

PART II—REFLECTION AND REFRACTION OF PLANE WAVES

5.09 Reflection by a Perfect Conductor—Normal Incidence. When an electromagnetic wave traveling in one medium impinges upon a second medium having a different dielectric constant, permeability, or conductivity, the wave in general will be partially transmitted and partially reflected. In the case of a plane wave in air incident normally upon the surface of a perfect conductor, the wave is entirely reflected. For fields that vary with time neither **E** nor **H** can exist within a perfect conductor so that none of the energy of the incident wave can be transmitted. Since there can be no loss within a perfect conductor, none of the energy is absorbed. As a result the amplitudes of **E** and **H** in the reflected wave are the same as in the incident wave, and the only difference is in the direction of power flow. If the expression for the electric field of the *incident* wave is

$$E_i e^{-j\beta x}$$

and the surface of the perfect conductor is taken to be the $x = 0$ plane as shown in Fig. 5-7, the expression for the reflected wave will be

$$E_r e^{j\beta x}$$

where E_r must be determined from the boundary conditions. Inasmuch as the tangential component of **E** must be continuous across the boundary and **E** is zero within the conductor, the tangential component of **E** just outside the conductor must also be zero. This requires that the sum of the electric field strengths in the incident and reflected waves add to give zero resultant field strength in the plane $x = 0$. Therefore

$$E_r = -E_i$$

The amplitude of the reflected electric field strength is equal to that of the incident electric field strength, but its phase has been reversed on reflection.

The resultant electric field strength at any point a distance $-x$ from the $x = 0$ plane will be the sum of the field strengths of the

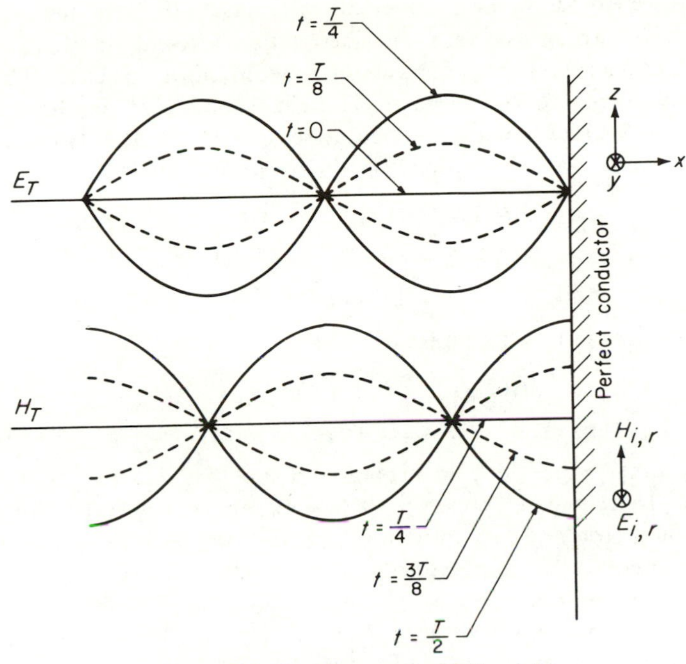

Figure 5-7. Standing waves of E and H.

incident and reflected waves at that point and will be given by

$$\begin{aligned} E_T(x) &= E_i e^{-j\beta x} + E_r e^{j\beta x} \\ &= E_i(e^{-j\beta x} - e^{j\beta x}) \\ &= -2jE_i \sin \beta x \end{aligned} \tag{5-67}$$

$$\tilde{E}_T(x, t) = \text{Re}\{-2jE_i \sin \beta x \, e^{j\omega t}\}$$

If E_i is chosen to be real,

$$\tilde{E}_T(x, t) = 2E_i \sin \beta x \sin \omega t \tag{5-67a}$$

Equation (67) shows that the incident and reflected waves combine to produce a *standing wave*, which does not progress. The magnitude of the electric field varies sinusoidally with distance from the reflecting plane. It is zero at the surface and at multiples of half wavelength from the surface. It has a maximum value of twice the electric field strength of the incident wave at distances from the surface that are odd multiples of a quarter wavelength.

Inasmuch as the boundary conditions require that the electric field strength be reversed in phase on reflection in order to produce zero resultant field at the surface, it follows that the magnetic field strength must be reflected without reversal of phase. If both magnetic and electric field strengths were reversed, there would be no reversal of direction of energy propagation, which is required in this case. Therefore, the phase of the reflected magnetic field strength H_r is the same* as that of the incident magnetic field strength H_i at the surface of reflection $x = 0$. The expression for the resultant magnetic field will be

$$\begin{aligned} H_T(x) &= H_i e^{-j\beta x} + H_r e^{+j\beta x} \\ &= H_i(e^{-j\beta x} + e^{+j\beta x}) \\ &= 2H_i \cos \beta x \end{aligned} \tag{5-68}$$

H_i is real since it is in phase with E_i.

$$\begin{aligned} \tilde{H}_T(x, t) &= \text{Re}\,\{H_T(x)e^{j\omega t}\} \\ &= 2H_i \cos \beta x \cos \omega t \end{aligned} \tag{5-68a}$$

The resultant magnetic field strength H also has a standing-wave distribution. In this case, however, it has maximum value at the surface of the conductor and at multiples of a half wavelength from the surface, whereas the zero points occur at odd multiples of a quarter wavelength from the surface. From the boundary conditions for H it follows that there must be a surface current of J_s amperes per meter, such that $J_s = H_T(\text{at } x = 0)$.

Since E_i and H_i were in time phase in the incident plane wave, a comparison of (67) and (68) shows that E_T and H_T are 90 degrees out of time phase because of the factor j in (67). This is as it should be, for it indicates no average flow of power. This is the case when the energy transmitted in the forward direction is equalled by that reflected back.

That E_T and H_T are 90 degrees apart in time phase can be seen more clearly by rewriting (67) and (68). Replacing $-j$ by its equivalent $e^{-j(\pi/2)}$ and combining this with the $e^{j\omega t}$ term to give $e^{j[\omega t-(\pi/2)]}$, eq. (67) becomes

$$\begin{aligned} \tilde{E}_T(x, t) &= \text{Re}\,\{2E_i \sin \beta x\, e^{-j\pi/2} e^{j\omega t}\} \\ &= 2E_i \sin \beta x \cos (\omega t - \pi/2) \end{aligned} \tag{5-69}$$

*An alternative way of arriving at this same result is from a consideration of current flow in the conductor. If it is assumed for the incident wave, which is traveling to the right in the positive x direction, that E_i is in the positive y direction and H_i is in the positive z direction (it will be seen later that the direction of energy propagation is always the direction of the vector $\mathbf{E} \times \mathbf{H}$), the current flow in the conductor will be in the same direction as the incident electric field, that is, in the positive y direction. This current flow produces an electric field $-E_y$ to oppose the incident field (Lenz's law) and produces a magnetic field, which is shown by application of the right-hand rule to be in the positive z direction. Therefore the magnetic field of the reflected wave has the *same* direction as in the incident wave.

Likewise rewriting (68),

$$H_T = 2H_i \cos \beta x \cos (\omega t) \tag{5-70}$$

Comparison of (69) and (70) shows that E_T and H_T differ in time phase by $\pi/2$ radians or 90 degrees.

5.10 Reflection by a Perfect Conductor—Oblique Incidence. Whenever a wave is incident obliquely on the interface between two media, it is necessary to consider separately two special cases. The first of these is the case in which the electric vector is parallel to the boundary surface or perpendicular to the plane of incidence. (The plane of incidence is the plane containing the incident ray and the normal to the surface.) This case is often termed *horizontal polarization.* In the second case the magnetic vector is parallel to the boundary surface, and the electric vector is parallel to the plane of incidence. This case is often termed *vertical polarization.* The two cases are shown in Fig. 5-8. The terms "horizontally and vertically

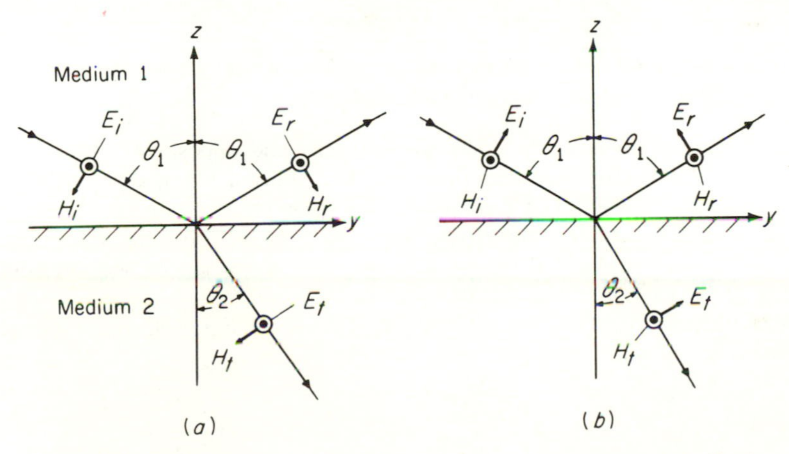

Figure 5-8. Reflection and refraction of waves having (*a*) perpendicular (horizontal) polarization and (*b*) parallel (vertical) polarization.

polarized waves" refer to the fact that waves from horizontal and vertical antennas, respectively, would produce these particular orientations of electric and magnetic vectors in waves striking the surface of the earth. However, it is seen that, whereas the electric vector of a "horizontally" polarized wave is horizontal, the electric vector of a "vertically" polarized wave is not wholly vertical but has some horizontal component. More significant designations are the terms "perpendicular" and "parallel" polarization to indicate that the electric vector is perpendicular or parallel to the plane of incidence. In waveguide work the terms *transverse electric* (TE) and *transverse magnetic*

(TM) are used to indicate that the electric or magnetic vector respectively is parallel to the boundary plane. The reason for this will be discussed later. In the present problem of a wave incident on a perfect conductor, the wave is totally reflected with the angle of incidence equal to the angle of reflection.

CASE I: E *Perpendicular to the Plane of Incidence.* Let the incident and reflected waves make angles $\theta_i = \theta_r = \theta$ with the z axes as in Fig. 5-9.

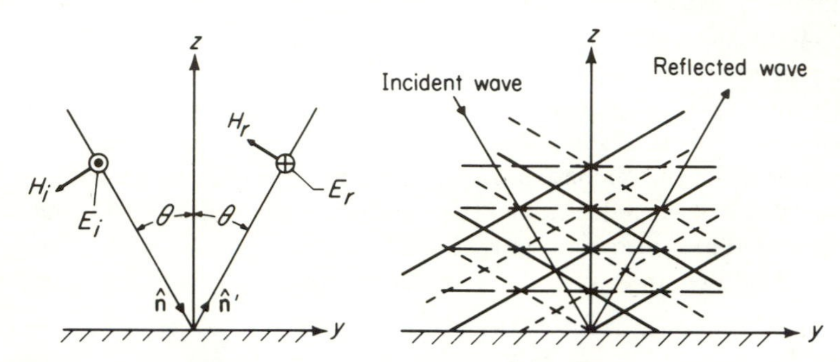

Figure 5-9. Field pattern above a reflecting plane when the wave is incident at an oblique angle (perpendicular polarization).

Because these two waves have equal wavelengths and opposite directions along the z axis, there must be a standing-wave distribution along this axis. In the y direction the incident and reflected waves both progress to the right with the same velocity and wavelength so there will be a traveling wave in the positive y direction. That these conclusions are correct can be seen by adding the expressions representing the two waves.

With the co-ordinate system chosen as shown in Fig. 5-9, the expression for the reflected wave is

$$\begin{aligned} E_{\text{reflected}} &= E_r e^{-j\beta\hat{\mathbf{n}}'\cdot\mathbf{r}} \\ &= E_r e^{-j\beta(x\cos A + y\cos B + z\cos C)} \end{aligned} \tag{5-71}$$

where E_r is the amplitude of the electric field strength of the reflected wave at the origin. For the wave normal of the reflected wave

$$\hat{\mathbf{n}}'\cdot\mathbf{r} = x\cos\frac{\pi}{2} + y\cos\left(\frac{\pi}{2} - \theta\right) + z\cos\theta = y\sin\theta + z\cos\theta$$

so that (71) becomes

$$E_{\text{reflected}} = E_r e^{-j\beta(y\sin\theta + z\cos\theta)} \tag{5-72}$$

For the incident wave

$$\hat{\mathbf{n}}\cdot\mathbf{r} = x\cos\frac{\pi}{2} + y\cos\left(\frac{\pi}{2} - \theta\right) + z\cos(\pi - \theta) = y\sin\theta - z\cos\theta$$

and

$$E_{\text{incident}} = E_i e^{-j\beta(y\sin\theta - z\cos\theta)} \tag{5-73}$$

From the boundary conditions

$$E_r = -E_i$$

Therefore the total electric field strength (sum of incident and reflected field strengths) will be

$$\begin{aligned} E &= E_i[e^{-j\beta(y \sin \theta - z \cos \theta)} - e^{-j\beta(y \sin \theta + z \cos \theta)}] \\ &= 2jE_i \sin (\beta z \cos \theta) e^{-j\beta y \sin \theta} \\ &= 2jE_i \sin \beta_z z \, e^{-j\beta_y y} \end{aligned} \tag{5-74}$$

where $\beta = \omega/v = 2\pi/\lambda$ is the phase-shift constant of the incident wave, $\beta_z = \beta \cos \theta$ is the phase-shift constant in the z direction, and $\beta_y = \beta \sin \theta$ is the phase-shift constant in the y direction. Equation (74) shows a standing-wave distribution of electric field strength along the z axis. The wavelength λ_z (twice the distance between nodal points), measured along this axis, is greater than the wavelength λ of the incident waves. The relation between the wavelengths is

$$\lambda_z = \frac{2\pi}{\beta_z} = \frac{2\pi}{\beta \cos \theta} = \frac{\lambda}{\cos \theta}$$

The planes of zero electric field strength occur at multiples of $\lambda_z/2$ from the reflecting surface. The planes of maximum electric field strength occur at odd multiples of $\lambda_z/4$ from the surface.

The whole standing-wave distribution of electric field strength is seen from eq. (74) to be traveling in the y direction with a velocity

$$v_y = \frac{\omega}{\beta_y} = \frac{\omega}{\beta \sin \theta} = \frac{v}{\sin \theta}$$

This is the velocity with which a crest of the incident wave moves along the y axis. The wavelength in this direction is

$$\lambda_y = \frac{\lambda}{\sin \theta}$$

CASE II: E *Parallel to the Plane of Incidence.* In this case E_i and E_r will have the instantaneous directions shown in Fig. 5-10 because the components

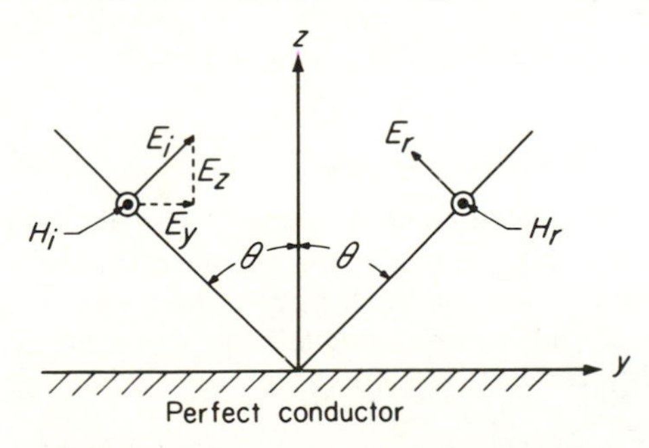

Figure 5-10. Reflection of a parallel-polarized wave.

parallel to the perfectly conducting boundary must be equal and opposite. The magnetic field strength vector **H** will be reflected without phase reversal as an examination of the direction of current flow will show. The magnitudes of **E** and **H** will be related by

$$\frac{E_i}{H_i} = \frac{E_r}{H_r} = \eta$$

For the incident wave the expression for magnetic field strength would be

$$H_{\text{incident}} = H_i e^{-j\beta(y \sin\theta - z\cos\theta)}$$

and for the reflected wave

$$H_{\text{reflected}} = H_r e^{-j\beta(y \sin\theta + z\cos\theta)}$$

and, since $H_i = H_r$, the total magnetic field strength is

$$H = 2H_i \cos\beta_z z\, e^{-j\beta_y y} \tag{5-75}$$

where, as before

$$\beta_z = \beta\cos\theta$$
$$\beta_y = \beta\sin\theta$$

The magnetic field strength has a standing-wave distribution in the z direction with the planes of maximum H located at the conducting surface and at multiples of $\lambda_z/2$ from the surface. The planes of zero magnetic field strength occur at odd multiples of $\lambda_z/4$ from the surface.

In adding together the electric field strengths of the incident and reflected waves it is necessary to consider separately the components in the y and z directions. For the incident wave

$$E_i = \eta H_i, \qquad E_z = \eta\sin\theta H_i, \qquad E_y = \eta\cos\theta H_i$$

For the reflected wave

$$H_r = H_i, \qquad E_z = \eta\sin\theta H_r, \qquad E_y = -\eta\cos\theta H_r$$

The total z component of electric field strength is

$$E_z = 2\eta\sin\theta H_i \cos\beta_z z\, e^{-j\beta_y y} \tag{5-76}$$

The total y component of electric field strength is

$$E_y = 2j\eta\cos\theta H_i \sin\beta_z z\, e^{-j\beta_y y} \tag{5-77}$$

where

$$\beta_z = \beta\cos\theta \qquad \text{and} \qquad \beta_y = \beta\sin\theta$$

Both components of the electric field strength have a standing-wave distribution above the reflecting plane. However, for the normal or z component of **E** the maxima occur at the plane and multiples of $\lambda_z/2$ from the plane whereas for the component of **E** parallel to the reflecting plane the minima occur at the plane and at multiples of $\lambda_z/2$ from the plane.

Problem 10. Derive (75), (76) and (77) using the boundary condition $E_{\tan} = 0$.

Problem 11. Sketch the planes of zero magnetic field strength, zero E_z, and zero E_y for the case of oblique reflection with **E** parallel to plane of incidence (Fig. 5-10).

5.11 Reflection by a Perfect Dielectric—Normal Incidence.

When a plane electromagnetic wave is incident normally on the surface of a perfect dielectric, part of the energy is transmitted and part of it is reflected. A *perfect* dielectric is one with zero conductivity, so that there is no loss or absorption of power in propagation through the dielectric.

As before, consider the case of a plane wave traveling in the x direction incident on a boundary that is parallel to the $x = 0$ plane. Let E_i be the electric field strength of the incident wave striking the boundary, E_r be the electric field strength of the reflected wave leaving the boundary in the first medium, and E_t be the electric field strength of the transmitted wave propagated into the second medium. Similar subscripts will be applied to the magnetic field strength H. Let ϵ_1 and μ_1 be the constants of the first medium and ϵ_2 and μ_2 be the constants of the second medium. Designating by η_1 and η_2 the ratios $\sqrt{\mu_1/\epsilon_1}$ and $\sqrt{\mu_2/\epsilon_2}$, the following relations will hold

$$E_i = \eta_1 H_i$$

$$E_r = -\eta_1 H_r$$

$$E_t = \eta_2 H_t$$

The continuity of the tangential components of **E** and **H** require that

$$H_i + H_r = H_t$$

$$E_i + E_r = E_t$$

Combining these

$$H_i + H_r = \frac{1}{\eta_1}(E_i - E_r) = H_t = \frac{1}{\eta_2}(E_i + E_r)$$

$$\eta_2(E_i - E_r) = \eta_1(E_i + E_r)$$

$$E_i(\eta_2 - \eta_1) = E_r(\eta_2 + \eta_1)$$

$$\frac{E_r}{E_i} = \frac{\eta_2 - \eta_1}{\eta_2 + \eta_1} \tag{5-78}$$

Also

$$\frac{E_t}{E_i} = \frac{E_i + E_r}{E_i} = 1 + \frac{E_r}{E_i}$$

$$= \frac{2\eta_2}{\eta_2 + \eta_1} \tag{5-79}$$

Furthermore

$$\frac{H_r}{H_i} = -\frac{E_r}{E_i} = \frac{\eta_1 - \eta_2}{\eta_1 + \eta_2} \tag{5-80}$$

$$\frac{H_t}{H_t} = \frac{\eta_1}{\eta_2}\frac{E_t}{E_t} = \frac{2\eta_1}{\eta_1 + \eta_2} \tag{5-81}$$

The permeabilities of all known insulators do not differ appreciably from that of free space, so that $\mu_1 = \mu_2 = \mu_v$. Inserting this relation the above expressions can be written in terms of the dielectric constants as follows:

$$\frac{E_r}{E_i} = \frac{\sqrt{\mu_v/\epsilon_2} - \sqrt{\mu_v/\epsilon_1}}{\sqrt{\mu_v/\epsilon_2} + \sqrt{\mu_v/\epsilon_1}}$$

$$\frac{E_r}{E_i} = \frac{\sqrt{\epsilon_1} - \sqrt{\epsilon_2}}{\sqrt{\epsilon_1} + \sqrt{\epsilon_2}} \tag{5-82}$$

Similarly

$$\frac{E_t}{E_i} = \frac{2\sqrt{\epsilon_1}}{\sqrt{\epsilon_1} + \sqrt{\epsilon_2}} \tag{5-83}$$

$$\frac{H_r}{H_i} = \frac{\sqrt{\epsilon_2} - \sqrt{\epsilon_1}}{\sqrt{\epsilon_1} + \sqrt{\epsilon_2}} \tag{5-84}$$

$$\frac{H_t}{H_i} = \frac{2\sqrt{\epsilon_2}}{\sqrt{\epsilon_1} + \sqrt{\epsilon_2}} \tag{5-85}$$

5.12 Reflection by a Perfect Insulator—Oblique Incidence. If a plane wave is incident upon a boundary surface that is not parallel to the plane containing **E** and **H**, the boundary conditions are more complex. Again part of the wave will be transmitted and part of it reflected, but in this case the transmitted wave will be refracted; that is, the direction of propagation will be altered. Consider Fig. 5-11,

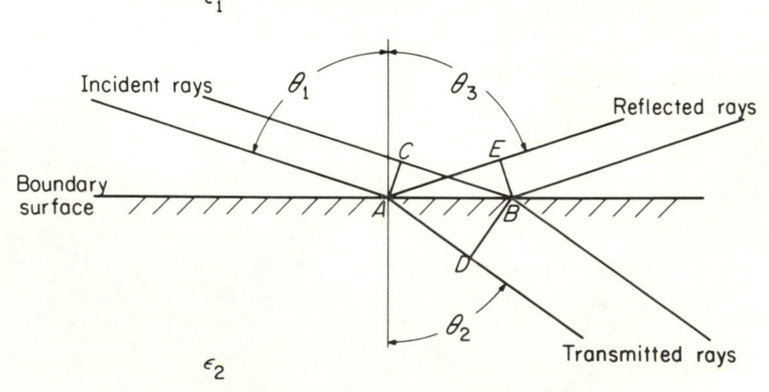

Figure 5-11. Reflection and refraction.

which shows two rays of the wave. (A ray is a line drawn normal to the equiphase surfaces.)

In the diagram incident ray (2) travels the distance CB, whereas transmitted ray (1) travels the distance AD and reflected ray (1) travels from A to E. If v_1 is the velocity of the wave in medium (1) and v_2 is the velocity in medium (2), then

$$\frac{CB}{AD} = \frac{v_1}{v_2}$$

Now $CB = AB \sin\theta_1$ and $AD = AB \sin\theta_2$, so that

$$\frac{\sin\theta_1}{\sin\theta_2} = \frac{v_1}{v_2}$$

In terms of the constants of the media, v_1 and v_2 are given by

$$v_1 = \frac{1}{\sqrt{\mu_1\epsilon_1}} = \frac{1}{\sqrt{\mu_v\epsilon_1}}$$

$$v_2 = \frac{1}{\sqrt{\mu_2\epsilon_2}} = \frac{1}{\sqrt{\mu_v\epsilon_2}}$$

Therefore

$$\frac{\sin\theta_1}{\sin\theta_2} = \sqrt{\frac{\epsilon_2}{\epsilon_1}} \tag{5-86}$$

Furthermore

$$AE = CB$$

and as a result, $\sin\theta_1 = \sin\theta_3$, or

$$\theta_1 = \theta_3 \tag{5-87}$$

The angle of incidence is equal to the angle of reflection; the angle of incidence is related to the angle of refraction by eq. (86), which in optics is known as the *law of sines*, or *Snell's law*.

In a later section it will be shown that the power transmitted per square meter in a wave is the vector product of **E** and **H**. Since **E** and **H** are at right angles to each other, in this case the power transmitted per square meter is equal to E^2/η. The power in the incident wave striking AB will be proportional to $(1/\eta_1)E_i^2 \cos\theta_1$, that reflected will be $(1/\eta_1) E_r^2 \cos\theta_1$, and that transmitted through the boundary will be $(1/\eta_2) E_t^2 \cos\theta_2$. By the conservation of energy

$$\frac{1}{\eta_1} E_i^2 \cos\theta_1 = \frac{1}{\eta_1} E_r^2 \cos\theta_1 + \frac{1}{\eta_2} E_t^2 \cos\theta_2$$

$$\frac{E_r^2}{E_i^2} = 1 - \frac{\eta_1 E_t^2 \cos\theta_2}{\eta_2 E_i^2 \cos\theta_1}$$

$$\frac{E_r^2}{E_i^2} = 1 - \frac{\sqrt{\epsilon_2}\; E_t^2 \cos\theta_2}{\sqrt{\epsilon_1}\; E_i^2 \cos\theta_1} \tag{5-88}$$

CASE I: *Perpendicular (Horizontal) Polarization.* In this case the electric vector **E** is perpendicular to the plane of incidence and parallel to the reflecting surface. Let the electric field strength E_i of the incident wave be in the positive x direction [outward in Fig. 5-8(*a*)], and let the assumed positive directions for E_r and E_t in the reflected and transmitted waves also be in the positive x direction. Then, applying the boundary condition that the tangential component of **E** is continuous across the boundary,

$$E_i + E_r = E_t$$

$$\frac{E_t}{E_i} = 1 + \frac{E_r}{E_i} \tag{5-89}$$

Insert this in eq. (88)

$$\frac{E_r^2}{E_i^2} = 1 - \sqrt{\frac{\epsilon_2}{\epsilon_1}}\left(1 + \frac{E_r}{E_i}\right)^2 \frac{\cos\theta_2}{\cos\theta_1}$$

$$1 - \left(\frac{E_r}{E_i}\right)^2 = \sqrt{\frac{\epsilon_2}{\epsilon_1}}\left(1 + \frac{E_r}{E_i}\right)^2 \frac{\cos\theta_2}{\cos\theta_1}$$

$$1 - \frac{E_r}{E_i} = \sqrt{\frac{\epsilon_2}{\epsilon_1}}\left(1 + \frac{E_r}{E_i}\right) \frac{\cos\theta_2}{\cos\theta_1}$$

$$\frac{E_r}{E_i} = \frac{\sqrt{\epsilon_1}\cos\theta_1 - \sqrt{\epsilon_2}\cos\theta_2}{\sqrt{\epsilon_1}\cos\theta_1 + \sqrt{\epsilon_2}\cos\theta_2}$$

Now from eq. (86)

$$\sqrt{\epsilon_2}\cos\theta_2 = \sqrt{\epsilon_2(1 - \sin^2\theta_2)} = \sqrt{\epsilon_2 - \epsilon_1\sin^2\theta_1}$$

therefore

$$\frac{E_r}{E_i} = \frac{\sqrt{\epsilon_1}\cos\theta_1 - \sqrt{\epsilon_2 - \epsilon_1\sin^2\theta_1}}{\sqrt{\epsilon_1}\cos\theta_1 + \sqrt{\epsilon_2 - \epsilon_1\sin^2\theta_1}} \tag{5-90}$$

$$= \frac{\cos\theta_1 - \sqrt{(\epsilon_2/\epsilon_1) - \sin^2\theta_1}}{\cos\theta_1 + \sqrt{(\epsilon_2/\epsilon_1) - \sin^2\theta_1}} \tag{5-90a}$$

Equation (90) gives the ratio of reflected to incident electric field strength for the case of a perpendicularly polarized wave.

CASE II: *Parallel (Vertical) Polarization.* In this case **E** is parallel to the plane of incidence and **H** is parallel to the reflecting surface. Again applying the boundary condition that the tangential component of **E** is continuous across the boundary in this case gives [Fig. 5-8(*b*)],

$$(E_i - E_r)\cos\theta_1 = E_t\cos\theta_2$$

$$\frac{E_t}{E_i} = \left(1 - \frac{E_r}{E_i}\right)\frac{\cos\theta_1}{\cos\theta_2}$$

Insert this in eq. (88)

$$\left(\frac{E_r}{E_i}\right)^2 = 1 - \sqrt{\frac{\epsilon_2}{\epsilon_1}}\left(1 - \frac{E_r}{E_i}\right)^2 \frac{\cos\theta_1}{\cos\theta_2}$$

$$1 - \frac{E_r^2}{E_i^2} = \sqrt{\frac{\epsilon_2}{\epsilon_1}}\left(1 - \frac{E_r}{E_i}\right)^2 \frac{\cos\theta_1}{\cos\theta_2}$$

$$1 + \frac{E_r}{E_i} = \sqrt{\frac{\epsilon_2}{\epsilon_1}}\left(1 - \frac{E_r}{E_i}\right) \frac{\cos\theta_1}{\cos\theta_2}$$

$$\frac{E_r}{E_i}\left(1 + \sqrt{\frac{\epsilon_2}{\epsilon_1}}\frac{\cos\theta_1}{\cos\theta_2}\right) = \sqrt{\frac{\epsilon_2}{\epsilon_1}}\frac{\cos\theta_1}{\cos\theta_2} - 1$$

$$\frac{E_r}{E_i} = \frac{\sqrt{\epsilon_2}\cos\theta_1 - \sqrt{\epsilon_1}\cos\theta_2}{\sqrt{\epsilon_2}\cos\theta_1 + \sqrt{\epsilon_1}\cos\theta_2}$$

$$= \frac{\sqrt{\epsilon_2}\cos\theta_1 - \sqrt{\epsilon_1(1 - \sin^2\theta_2)}}{\sqrt{\epsilon_2}\cos\theta_1 + \sqrt{\epsilon_1(1 - \sin^2\theta_2)}}$$

Recall that $\sin^2\theta_2 = \epsilon_1/\epsilon_2 \sin^2\theta_1$

$$\frac{E_r}{E_i} = \frac{(\epsilon_2/\epsilon_1)\cos\theta_1 - \sqrt{(\epsilon_2/\epsilon_1) - \sin^2\theta_1}}{(\epsilon_2/\epsilon_1)\cos\theta_1 + \sqrt{(\epsilon_2/\epsilon_1) - \sin^2\theta_1}} \tag{5-91}$$

Equation (91) gives the reflection coefficient for parallel or vertical polarization, that is, the ratio of reflected to incident electric field strength when **E** is parallel to the plane of incidence.

Problem 12. Derive E_r/E_i and E_t/E_i for both polarizations by matching both electric and magnetic fields at the boundary (instead of using conservation of energy).

Brewster Angle. Of particular interest is the possibility in eq. (91) of obtaining no reflection at a particular angle. This occurs when the numerator is zero. For this case

$$\sqrt{\frac{\epsilon_2}{\epsilon_1} - \sin^2\theta_1} = \frac{\epsilon_2}{\epsilon_1}\cos\theta_1$$

$$\frac{\epsilon_2}{\epsilon_1} - \sin^2\theta_1 = \frac{\epsilon_2^2}{\epsilon_1^2} - \frac{\epsilon_2^2}{\epsilon_1^2}\sin^2\theta_1$$

$$(\epsilon_1^2 - \epsilon_2^2)\sin^2\theta_1 = \epsilon_2(\epsilon_1 - \epsilon_2)$$

$$\sin^2\theta_1 = \frac{\epsilon_2}{\epsilon_1 + \epsilon_2}$$

$$\cos^2\theta_1 = \frac{\epsilon_1}{\epsilon_1 + \epsilon_2}$$

$$\tan\theta_1 = \sqrt{\frac{\epsilon_2}{\epsilon_1}} \tag{5-92}$$

At this angle, which is called the *Brewster angle*, there is no reflected wave when the incident wave is parallel (or vertically) polarized. If the incident wave is not entirely parallel polarized, there will be some reflection, but the reflected wave will be entirely of perpendicular (or horizontal) polarization.

Examination of eq. (90), which is for perpendicular polarization, shows that there is no corresponding Brewster angle for this polarization.

Problem 13. Prove for parallel polarization that

$$\frac{E_r}{E_i} = \frac{\tan(\theta_1 - \theta_2)}{\tan(\theta_1 + \theta_2)}$$

and for perpendicular polarization that

$$\frac{E_r}{E_i} = \frac{\sin(\theta_2 - \theta_1)}{\sin(\theta_2 + \theta_1)}$$

Problem 14. The gas laser depicted in Figure 5-12 uses "Brewster angle"

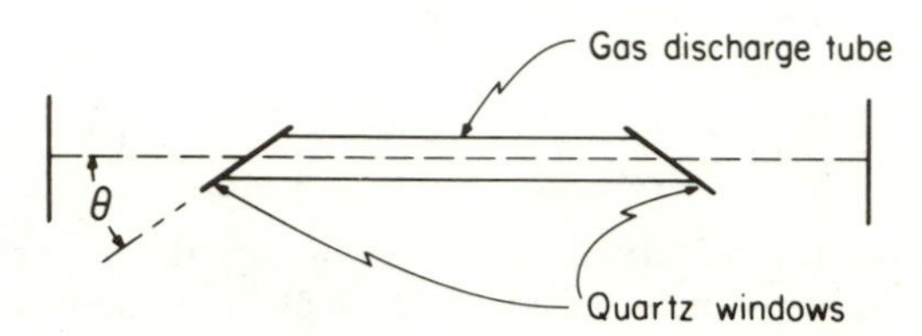

Figure 5-12. A gas laser with "Brewster angle" windows.

quartz windows on the gas discharge tube in order to minimize reflection losses. Determine the angle θ if the index of refraction for quartz at the wavelength of interest is $n = 1.45$. (For a dielectric, the index of refraction n, defined as the ratio of the velocity of light in a vacuum to the phase velocity in the dielectric medium, is equal to $\sqrt{\epsilon_r}$.) Because of these windows, the laser output is almost completely linearly polarized. What is the direction of polarization?

Total Internal Reflection. If ϵ_1 is greater than ϵ_2 both the reflection coefficients given by eqs. (90) and (91) become complex numbers when

$$\sin\theta_1 > \sqrt{\frac{\epsilon_2}{\epsilon_1}}$$

Both coefficients take on the form $(a + jb)/(a - jb)$ and thus have unit magnitude. In other words, the reflection is total provided that θ_1 is great enough and also provided that medium (1) is denser than medium (2). Total reflection does not imply that there is no field in medium (2), however. In this region the fields have the form

$$e^{-j\beta_2(y \sin\theta_2 - z \cos\theta_2)}$$

Snell's law gives the y variation as

$$e^{-j\beta_2 y\sqrt{\epsilon_1/\epsilon_2}\sin\theta_1}$$

and the z variation as

$$\begin{aligned} e^{j\beta_2 z\cos\theta_2} &= e^{j\beta_2 z[\pm\sqrt{1-\sin^2\theta_2}]} \\ &= e^{j\beta_2 z[\pm j\sqrt{(\epsilon_1/\epsilon_2)\sin^2\theta_1 - 1}]} \\ &= e^{\mp\beta_2 z\sqrt{(\epsilon_1/\epsilon_2)\sin^2\theta_1 - 1}} \end{aligned}$$

In the above expression the lower sign must be chosen so that the fields decrease exponentially as z becomes increasingly negative. That is,

$$\cos\theta_2 = -j\sqrt{\frac{\epsilon_1}{\epsilon_2}\sin^2\theta_1 - 1} = -j\sqrt{\frac{\epsilon_1}{\epsilon_2}}\sqrt{\sin^2\theta_1 - \frac{\epsilon_2}{\epsilon_1}}$$

in this problem. Thus under conditions of total internal reflection, a field does exist in the less dense medium. This field has a phase progression along the boundary but decreases exponentially away from it. It thus is an example of a *nonuniform* plane wave.

The phase velocity along the interface is given by

$$\frac{\omega}{\beta_2\sqrt{\epsilon_1/\epsilon_2}\sin\theta_1}$$

which, under conditions of total internal reflection is less than the phase velocity ω/β_2 of a uniform plane wave in medium (2). Consequently the nonuniform plane wave in medium (2) is a *slow wave*. Also, since some kind of a surface between two media is necessary to support the wave, it is called a *surface wave*.*

The reflection coefficients have already been derived. They are

$$\frac{E_r}{E_i} = \frac{\sqrt{\epsilon_1}\cos\theta_1 - \sqrt{\epsilon_2}\cos\theta_2}{\sqrt{\epsilon_1}\cos\theta_1 + \sqrt{\epsilon_2}\cos\theta_2}$$

for perpendicular polarization and

$$\frac{E_r}{E_i} = \frac{\sqrt{\epsilon_2}\cos\theta_1 - \sqrt{\epsilon_1}\cos\theta_2}{\sqrt{\epsilon_2}\cos\theta_1 + \sqrt{\epsilon_1}\cos\theta_2}$$

for parallel polarization. For total internal reflection, the value of $\cos\theta_2$ derived above may be substituted to yield

$$\frac{E_r}{E_i} = \frac{\cos\theta_1 + j\sqrt{\sin^2\theta_1 - (\epsilon_2/\epsilon_1)}}{\cos\theta_1 - j\sqrt{\sin^2\theta_1 - (\epsilon_2/\epsilon_1)}} = e^{j\phi_\perp} \qquad (5\text{-}93)$$

for perpendicular polarization and

$$\frac{E_r}{E_i} = \frac{(\epsilon_2/\epsilon_1)\cos\theta_1 + j\sqrt{\sin^2\theta_1 - (\epsilon_2/\epsilon_1)}}{(\epsilon_2/\epsilon_1)\cos\theta_1 - j\sqrt{\sin^2\theta_1 - (\epsilon_2/\epsilon_1)}} = e^{j\phi_\parallel} \qquad (5\text{-}94)$$

*This is one type of surface wave. Another type used in ground wave propagation is covered in chap. 16. In the literature the term "surface wave" has a variety of connotations depending upon the problem and the writer. A comprehensive, but by no means conclusive discussion of surface waves is given in the "Surface Waves" section of the *I. E. E. E. Transactions on Antennas and Propagation*, AP-7, Dec., 1959.

for parallel polarization. Note that $\phi_\perp$ and $\phi_\parallel$ are the phase-shift angles for the two polarizations.

Total internal reflection is widely used in binocular optics where glass prisms are employed to shorten the instrument, thus making it easier to carry. In such applications it is important to keep dust from collecting on the reflecting faces of the glass prisms since the dust particles could disturb the exponentially decreasing fields, thus causing the image to become dim or blurred.

Problem 15. The prism shown in Fig. 5-13 is one application of total

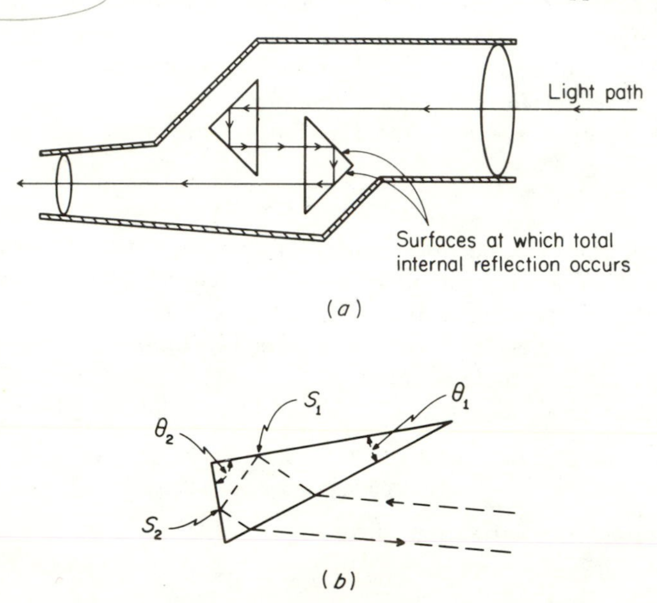

Figure 5-13. Total internal reflection as used in (*a*) binocular optics, and (*b*) laser applications.

internal reflection. The incident radiation from the right is to pass into the prism at the Brewster angle, be totally reflected at S_1 and S_2 and return parallel to the incident path to within 6 min of arc. Specify the angles for use at 6328Å (the He: Ne laser) if the prism is to be constructed from quartz. (Use the refractive index $n = \sqrt{\epsilon_r} = 1.46$.)

Problem 16. In cases of total internal reflection an incident wave polarized linearly at 45 degrees to the plane of incidence can be transformed into a circularly polarized wave after reflection. Find the angle of incidence at which this occurs for a given value of (ϵ_1/ϵ_2). Also note the minimum value of (ϵ_1/ϵ_2) required to produce the specified change in polarization.

5.13 Reflection at the Surface of a Conductive Medium. Suppose that a uniform plane wave in a medium with constants $\epsilon_1, \mu_1, \sigma_1$ is incident normally upon a second medium of infinite depth having the constants $\epsilon_2, \mu_2, \sigma_2$, as shown in Fig. 5-14. At the boundary, continuity of the tangential field components demands that

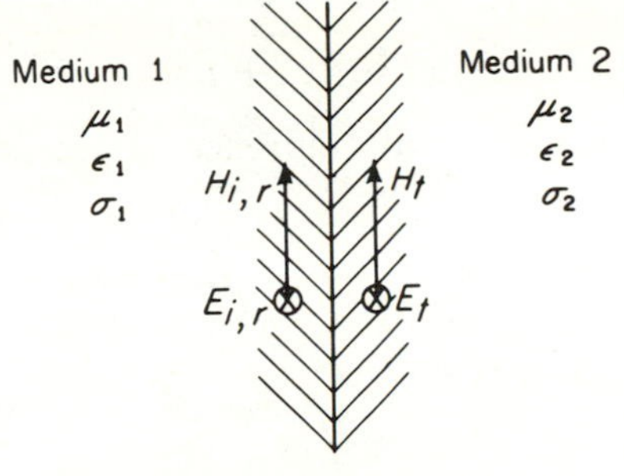

Figure 5-14. Reflection and transmission at a boundary between two conductive media.

$$E_i + E_r = E_t$$

and

$$H_i + H_r = H_t$$

The latter equation may be modified by noting that

$$\eta_1 H_i = E_i, \qquad \eta_1 H_r = -E_r, \qquad \eta_2 H_t = E_t$$

Thus if E_i is regarded as known, we have two equations in the two unknowns E_r and E_t. These equations may be expressed as

$$E_r - E_t = -E_i$$

$$\frac{E_r}{\eta_1} + \frac{E_t}{\eta_2} = \frac{E_i}{\eta_1}$$

and their solution yields the reflection and transmission coefficients

$$\frac{E_r}{E_i} = \frac{\eta_2 - \eta_1}{\eta_2 + \eta_1} \tag{5-95}$$

$$\frac{E_t}{E_i} = \frac{2\eta_2}{\eta_2 + \eta_1} \tag{5-96}$$

The reflection and transmission coefficients for the magnetic field are

$$\frac{H_r}{H_i} = \frac{\eta_1 - \eta_2}{\eta_2 + \eta_1} \tag{5-97}$$

$$\frac{H_t}{H_i} = \frac{2\eta_1}{\eta_2 + \eta_1} \tag{5-98}$$

It is interesting to evaluate expressions (95), (96), (97), and (98) for the case of an electromagnetic wave in air incident normally upon a copper sheet. A frequency of 1 MHz will be assumed. For this example

$$\mu_1 = \mu_v \qquad \mu_2 = \mu_v$$

$$\epsilon_1 = \epsilon_v \qquad \epsilon_2 = \epsilon_v$$

$$\sigma_1 = 0 \qquad \sigma_2 = 5.8 \times 10^7 \text{ mhos/m}$$

so that

$$\eta_1 = \sqrt{\frac{\mu_v}{\epsilon_v}} = 377 \text{ ohms}$$

$$\eta_2 = \sqrt{\frac{j2\pi \times 10^6 \times 4\pi \times 10^{-7}}{5.8 \times 10^7 + j2\pi \times 10^6 \times 8.854 \times 10^{-12}}} = 0.000369\underline{/45^\circ} \text{ ohms}$$

Then the ratio of reflected to incident electric field strengths, as given by eq. (95), is

$$\frac{E_r}{E_i} = \frac{3.69 \times 10^{-4} \underline{/45^\circ} - 377}{3.69 \times 10^{-4} \underline{/45^\circ} + 377}$$

$$= -0.9999986 \underline{/-0.000079^\circ}$$

$$= -\frac{H_r}{H_i}$$

It is seen that differences between these reflection coefficients for copper and the coefficients of minus and plus unity, which would be obtained for a *perfect* reflector, are indeed negligible. For most practical purposes, copper can be considered a perfect reflector of radio waves.

The relative strengths of the transmitted fields for this case are

$$\frac{E_t}{E_i} = \frac{7.38 \times 10^{-4} \underline{/45^\circ}}{3.69 \times 10^{-4} \underline{/45^\circ} + 377} = 0.00000196 \underline{/45^\circ}$$

$$\frac{H_t}{H_i} = \frac{2 \times 377}{377 + 3.69 \times 10^{-4} \underline{/45^\circ}} = 1.9999986 \underline{/-0.00004^\circ}$$

The electric field strength just inside the metal is approximately 2×10^{-6} times that of the initial wave: The magnetic field strength just inside the metal is approximately twice the magnetic field strength of the initial wave. This last result could be inferred from the fact that, since the magnetic field strength is reflected without phase reversal, the total magnetic field strength just outside the surface of the copper is approximately double that of the initial wave and therefore, because of continuity requirements, H just inside the copper is also approximately twice the magnetic field strength of the incident wave. The ratio of E to H just inside the metal is equal to η_2, the characteristic impedance of the copper. That is

$$\frac{E_t}{H_t} = \eta_2 = 0.000369 \underline{/45^\circ} \text{ ohms}$$

For many practical purposes this is sufficiently close to zero to consider the copper sheet to be a zero-impedance surface.

Problem 17. Determine the normal incidence reflection coefficients for sea water, fresh water, and "good" earth at frequencies of 60 Hz, 1 MHz, and 1 GHz. Use $\epsilon_r = 80$, $\sigma = 4$ mhos/m for sea water; $\epsilon_r = 80$, $\sigma = 5 \times 10^{-3}$ for fresh water; and $\epsilon_r = 15$, $\sigma = 10 \times 10^{-3}$ for good earth.

5.14 Surface Impedance. It has been seen that at high frequencies the current is confined almost entirely to a very thin sheet at the surface of the conductor. In many applications it is convenient to make use of a *surface impedance* defined by

$$Z_s = \frac{E_{\text{tan}}}{J_s} \tag{5-99}$$

where E_{tan} is the electric field strength parallel to, and at the surface of, the conductor and J_s is the linear current density that flows as a result of this E_{tan}. The linear current density J_s represents the total conduction current per meter width flowing in the thin sheet. If it is assumed that the conductor is a flat plate with its surface at the $y = 0$ plane (Fig. 5-15), the current distribution in the y direction will be given by

$$J = J_0 e^{-\gamma y}$$

where J_0 is the current density at the surface.

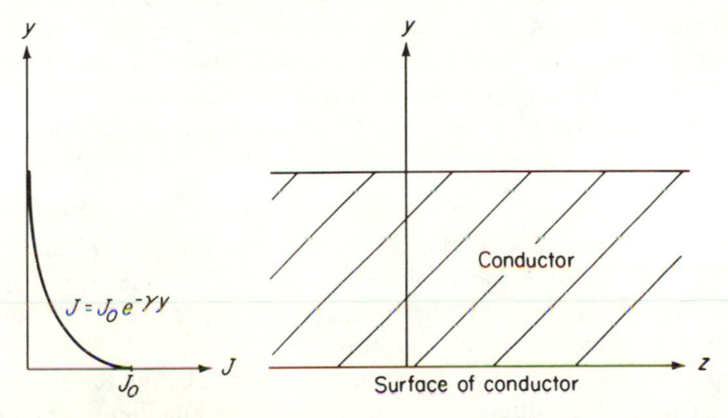

Figure 5-15. Current distribution in a thick flat-plate conductor.

It is assumed that the thickness of the conductor is very much greater than the depth of penetration, so that there is no reflection from the back surface of the conductor. The total conduction current per meter width, that is, the linear current density is

$$\begin{aligned} J_s &= \int_0^\infty J\,dy = J_0 \int_0^\infty e^{-\gamma y}\,dy \\ &= -\frac{J_0}{\gamma}\left[e^{-\gamma y}\right]_0^\infty = \frac{J_0}{\gamma} \end{aligned} \tag{5-100}$$

But J_0, the current density at the surface, is

$$J_0 = \sigma E_{\text{tan}}$$

Therefore

$$Z_s = \frac{E_{\tan}}{J_s} = \frac{\gamma}{\sigma}$$

The constant γ for propagation in a conducting medium was found to be

$$\gamma = \sqrt{j\omega\mu(\sigma + j\omega\epsilon)} \approx \sqrt{j\omega\mu\sigma}$$

This gives for a thick conductor

$$Z_s = \sqrt{\frac{j\omega\mu}{\sigma}} = \eta \text{ (for the conducting medium)} \tag{5-101}$$

It is seen that for good conductors *the surface impedance of a plane conductor that is very much thicker than the skin depth is just equal to the characteristic impedance of the conductor.* This is also the input impedance of the conductor when viewed as a transmission line conducting energy into the interior of the metal (the transmission-line analogy will be examined in sec. 5.15). When the thickness of the plane conductor is not great compared with the depth of penetration, reflection of the wave occurs at the back surface of the conductor. Under these conditions, the input impedance is approximately equal to the input impedance of a lossy line terminated in an open circuit, viz.,

$$Z_{\text{in}} = \eta \coth \gamma l \tag{5-102}$$

where l is the thickness of the conductor, and η and γ are its intrinsic impedance and propagation constant respectively. The approximation is ordinarily valid because the actual termination $\eta_v = 377$ ohms is very much greater than η of the conductor.

Surface Impedance of Good Conductors. For any material normally classed as a good conductor $\sigma \gg \omega\epsilon$, and if the conductor thickness is very much greater than the depth of penetration, the surface impedance of such a conductor is

$$Z_s \approx \sqrt{\frac{j\omega\mu}{\sigma}} = \sqrt{\frac{\omega\mu}{\sigma}}\,\underline{/45^\circ} \tag{5-103}$$

The *surface resistance* is

$$R_s \approx \sqrt{\frac{\omega\mu}{2\sigma}} \tag{5-103a}$$

and the surface reactance has the same magnitude as R_s at all frequencies

$$X_s \approx \sqrt{\frac{\omega\mu}{2\sigma}} \tag{5-103b}$$

The surface resistance defined by (103a) as the real part of the surface impedance is the high-frequency or *skin-effect resistance* per unit length of a flat conductor of unit width. (It has the dimension of ohms and its value does not depend upon the units used to measure length and width as long as they are the same.) Recalling that the expression for depth of penetration in a conductor is

$$\delta = \sqrt{\frac{2}{\omega\mu\sigma}} \tag{5-104}$$

it is seen that

$$R_s = \frac{1}{\sigma\delta} \tag{5-105}$$

The surface resistance of a flat conductor at any frequency is equal to the d-c resistance of a thickness δ of the same conductor, where δ is the depth of penetration or *skin depth.* This means that the conductor, having a thickness very much greater than δ and having the exponential current distribution throughout its depth, has the same resistance as would a thickness δ of the conductor with the current distributed uniformly throughout this thickness. From this it follows that the power loss per unit area of the plane conductor will be given by $J_{s\,\text{eff}}^2 R_s$, where R_s is its surface resistance and $J_{s\,\text{eff}}$ is the linear current density or current per meter width (effective value) flowing in the conductor. This same conclusion can be obtained from consideration of power flow, a subject that will be taken up in the next chapter.

Problem 18. Show that, when a given uniform plane wave is incident normally on a good conductor, the linear current density J_s is essentially independent of the conductivity σ.

5.15 The Transmission-line Analogy. The student familiar with ordinary transmission-line theory cannot have failed to notice the similarity between the equations of wave propagation developed in this chapter and those giving voltage and current distributions along uniform transmission lines. The similarity is especially marked in the expressions for the reflection coefficients in the two cases. This similarity is more than a coincidence. There exists a close analogy between the propagation of plane waves in a homogeneous medium and the propagation of voltage and current along a uniform transmission line. This analogy is so close that it can be used not only as an aid in obtaining an understanding of a new subject, but also to obtain the solutions to actual problems. Because of his background in transmission-line theory the engineer often finds himself able to write directly the solutions to electromagnetic-wave problems, or at any rate to set them up in terms of familiar circuit concepts. For these reasons the analogy will be considered step-by-step in some detail in order that the similarities, and the differences, may be fully understood.

For a uniform transmission line having the constants R, L, C and G per unit length, the voltage and current equations may be written in the differential form as

$$\left.\begin{aligned}\frac{\partial \tilde{V}}{\partial x} + L\frac{\partial \tilde{I}}{\partial t} + R\tilde{I} = 0 \\ \frac{\partial \tilde{I}}{\partial x} + C\frac{\partial \tilde{V}}{\partial t} + G\tilde{V} = 0\end{aligned}\right\} \tag{5-106a}$$

For a homogeneous medium Maxwell's equations are

$$\nabla \times \tilde{\mathbf{E}} = -\mu\frac{\partial \tilde{\mathbf{H}}}{\partial t}$$

$$\nabla \times \tilde{\mathbf{H}} = \epsilon\frac{\partial \tilde{\mathbf{E}}}{\partial t} + \sigma\tilde{\mathbf{E}}$$

For a uniform plane wave propagating in the x direction and having only components E_y and H_z these become

$$\left.\begin{aligned}\frac{\partial \tilde{E}_y}{\partial x} + \mu\frac{\partial \tilde{H}_z}{\partial t} = 0 \\ \frac{\partial \tilde{H}_z}{\partial x} + \epsilon\frac{\partial \tilde{E}_y}{\partial t} + \sigma\tilde{E}_y = 0\end{aligned}\right\} \tag{5-106b}$$

Inspection of these equations shows that the following quantities are analogous:

V (volt)	E (volt/m)
I (amp)	H(amp/m)
C (farad/m)	ϵ (farad/m)
L (henry/m)	μ (henry/m)
G (mho/m)	σ (mho/m)
R (ohm/m)	......

In this analogy there appears to be nothing corresponding to R. The reason for this will be seen later. If the voltages and currents, and electric and magnetic field strengths are assumed to vary sinusoidally with time, so that phasor notation is applicable, then eqs. (106) become

$$\left.\begin{aligned}\frac{\partial V}{\partial x} + (R + j\omega L)I = 0 \\ \frac{\partial I}{\partial x} + (G + j\omega C)V = 0\end{aligned}\right\} \tag{5-107a}$$

$$\left.\begin{aligned}\frac{\partial E_y}{\partial x} + (0 + j\omega\mu)H_z = 0 \\ \frac{\partial H_z}{\partial x} + (\sigma + j\omega\epsilon)E_y = 0\end{aligned}\right\} \tag{5-107b}$$

Differentiating with respect to x, these equations combine to give the following second-order differential equations:

$$\left.\begin{aligned}\frac{\partial^2 V}{\partial x^2} - (R + j\omega L)(G + j\omega C)V = 0 \\ \frac{\partial^2 I}{\partial x^2} - (R + j\omega L)(G + j\omega C)I = 0\end{aligned}\right\} \tag{5-108a}$$

$$\left.\begin{aligned}\frac{\partial^2 E_y}{\partial x^2} - (j\omega\mu)(\sigma + j\omega\epsilon)E_y = 0 \\ \frac{\partial^2 H_z}{\partial x^2} - (j\omega\mu)(\sigma + j\omega\epsilon)H_z = 0\end{aligned}\right\} \tag{5-108b}$$

A possible solution for any of these equations would be of the form

$$V, I, E_y, \text{ or } H_z = Ae^{-\gamma x} + Be^{\gamma x} \tag{5-109}$$

where
$$\gamma^2 = (R + j\omega L)(G + j\omega C)$$

for the eqs. (108a), and

$$\gamma^2 = (j\omega\mu)(\sigma + j\omega\epsilon)$$

for the eqs. (108b). It is customary to take γ to mean the particular square root (of γ^2) which has a positive real part. When the variation with time is expressed explicitly, the first term of expression (109) represents a wave traveling to the right and the second expression represents a wave traveling to the left.

An alternative solution to eqs. (108) is often used in transmission-line theory. In this solution the exponentials are combined differently and the solution appears in terms of hyperbolic functions, and can be written

$$\left.\begin{aligned} V &= A_1 \cosh \gamma x + B_1 \sinh \gamma x \\ I &= A_2 \cosh \gamma x + B_2 \sinh \gamma x \end{aligned}\right\} \qquad (5\text{-}110a)$$

Let $V = V_R$, $I = I_R$ at $x = 0$

and $V = V_S$, $I = I_S$ at $x = x_1$

Substitute these in (110a) and find for the coefficients the values

$$A_1 = V_R, \quad B_1 = -\sqrt{\frac{R + j\omega L}{G + j\omega C}}\, I_R$$

$$A_2 = I_R, \quad B_2 = -\sqrt{\frac{G + j\omega C}{R + j\omega L}}\, V_R$$

$$\left.\begin{aligned} E &= A_1 \cosh \gamma x + B_1 \sinh \gamma x \\ H &= A_2 \cosh \gamma x + B_2 \sinh \gamma x \end{aligned}\right\} \qquad (5\text{-}110b)$$

Let $E = E_R$, $H = H_R$ at $x = 0$

and $E = E_S$, $H = H_S$ at $x = x_1$

Substitute these in (110b) and evaluate the coefficients.

$$A_1 = E_R, \quad B_1 = -\sqrt{\frac{j\omega\mu}{\sigma + j\omega\epsilon}}\, H_R$$

$$A_2 = H_R, \quad B_2 = -\sqrt{\frac{\sigma + j\omega\epsilon}{j\omega\mu}}\, E_R$$

In transmission-line theory it is customary to write

$$Z = R + j\omega L \qquad Y = G + j\omega C$$

$$Z_0 = \sqrt{\frac{Z}{Y}} = \sqrt{\frac{R + j\omega L}{G + j\omega C}}$$

where Z_0 is called the *characteristic impedance* of the transmission line.

Similarly in wave theory it is customary to write

$$\eta = \sqrt{\frac{j\omega\mu}{\sigma + j\omega\epsilon}}$$

where η is called the *characteristic impedance* or *intrinsic impedance* of the medium.

In terms of these quantities, and writing $l = -x_1$, eqs. (110) become

$$\left.\begin{aligned} V_S &= V_R \cosh \gamma l + Z_0 I_R \sinh \gamma l \\ I_S &= I_R \cosh \gamma l + \frac{V_R}{Z_0} \sinh \gamma l \end{aligned}\right\} \qquad (5\text{-}111a)$$

$$\left.\begin{aligned} E_S &= E_R \cosh \gamma l + \eta H_R \sinh \gamma l \\ H_S &= H_R \cosh \gamma l + \frac{E_R}{\eta} \sinh \gamma l \end{aligned}\right\} \qquad (5\text{-}111b)$$

When the line is very long (or the homogeneous medium very thick) so that $\text{Re}\{\gamma l\}$ is a large quantity,

$$\cosh \gamma l \approx \frac{e^{\gamma l}}{2} \approx \sinh \gamma l$$

and the ratios of voltage to current and E to H are seen from eqs. (111) to be

$$\frac{V_S}{I_S} = Z_0 \qquad \frac{E_S}{H_S} = \eta$$

The characteristic impedance Z_0, and intrinsic impedance η are, respectively, the ratios of V to I on a transmission line and E to H in a uniform plane wave under conditions where there is no reflected wave from the termination, or, in other words, when the wave along the line or in the medium is a *traveling* wave. Equations (111) are the general equations for the propagation of waves along uniform transmission lines or plane waves in homogeneous media. For the special cases of a "lossless" line or a "lossless" (nonconducting) medium the following simplifications occur

$R = G = 0$ so that

$$Z_0 = \sqrt{L/C}$$

$$\gamma = \sqrt{(j\omega L)(j\omega C)} = j\omega\sqrt{LC}$$

but $\gamma = \alpha + j\beta$

Therefore $\alpha = 0$

$$\beta = \omega\sqrt{LC}$$

$\sigma = 0$ so that

$$\eta = \sqrt{\mu/\epsilon}$$

$$\gamma = \sqrt{(j\omega\mu)(j\omega\epsilon)} = j\omega\sqrt{\mu\epsilon}$$

but $\gamma = \alpha + j\beta$

Therefore $\alpha = 0$

$$\beta = \omega\sqrt{\mu\epsilon}$$

Under these circumstances, since $\cosh j\beta = \cos\beta$, and $\sinh j\beta = j\sin\beta$, the general eqs. (111) reduce to

$$\left.\begin{aligned} V_S &= V_R\cos\beta l + jZ_0 I_R \sin\beta l \\ I_S &= I_R\cos\beta l + j\frac{V_R}{Z_0}\sin\beta l \end{aligned}\right\} \qquad \text{(5-112a)}$$

$$\left.\begin{aligned} E_S &= E_R\cos\beta l + j\eta H_R \sin\beta l \\ H_S &= H_R\cos\beta l + j\frac{E_R}{\eta}\sin\beta l \end{aligned}\right\} \qquad \text{(5-112b)}$$

The quantities $1/\sqrt{LC}$ and $1/\sqrt{\mu\epsilon}$ have the dimensions of velocity and are in fact the velocities of wave propagation along the lossless line and in the lossless medium respectively. In either case, when the dielectric is air, so that $\mu = \mu_v$ and $\epsilon = \epsilon_v$,

$$\frac{1}{\sqrt{LC}} = c \approx 3 \times 10^8 \qquad \text{meter/sec}$$

$$\frac{1}{\sqrt{\mu_v\epsilon_v}} = c \approx 3 \times 10^8 \qquad \text{meter/sec}$$

It might be expected that a wave of arbitrary shape traversing a conducting medium would decrease in amplitude exponentially but would retain its wave-shape as it progresses. However, it may be shown, by decomposing the wave into its fundamental and harmonic components through Fourier analysis, that the attenuation factor α is different at different frequencies. Therefore such a waveshape becomes distorted as it progresses. Similar distortion occurs on a lossy transmission line. However in the latter case a "distortionless line" can be produced by loading the line to obtain a particular relation among the line constants R, L, G, C (see problem 27).

It has been seen that propagation of a uniform plane wave in a homogeneous medium is analogous to propagation along a uniform transmission line. If the uniform plane wave passes abruptly from one medium to another, the surface of discontinuity being a plane perpendicular to the direction of propagation, the analogy will continue to hold (see Fig. 5-16). This is so because the

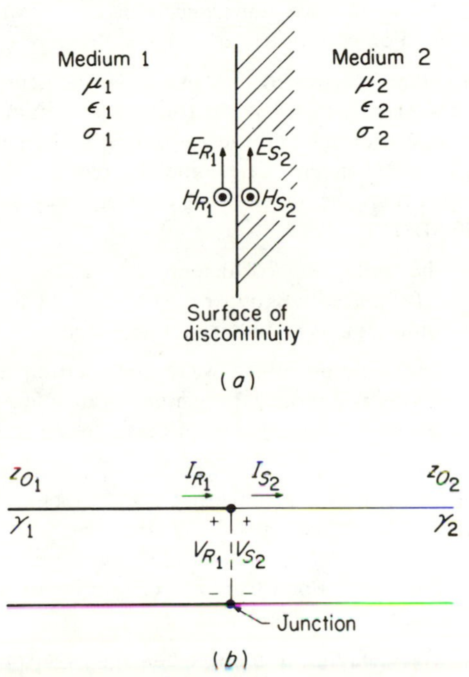

Figure 5-16. Reflection and transmission (*a*) at a boundary surface between two media and (*b*) at a junction between two transmission lines.

boundary conditions at the surface of discontinuity are the same as those existing at the junction between two transmission lines having different constants. For the latter the continuity requirements are that (1) the voltage be continuous across the junction and (2) the current be continuous across the junction. These are the same requirements that hold for the analogous quantities E and H across a boundary surface. The usefulness of the above analogy is particularly apparent in solving multi-junction problems, such as problems 20, 23, and 24.

ADDITIONAL PROBLEMS

19. The electric field strength of a uniform plane electromagnetic wave in free space is 1 volt per meter, and the frequency is 300 MHz. If a very large thick flat copper plate is placed normal to the direction of wave propagation, determine (a) the electric field strength at the surface of the plate; (b) the magnetic field strength at the surface of the plate; (c) the depth of penetration; (d) the conduction current density at the surface; (e) the conduction current density at a distance of 0.01 mm below the surface; (f) the linear current density J_s; (g) the surface impedance; (h) the power loss per square meter of surface area. For copper use $\sigma = 5.8 \times 10^7$, $\epsilon \approx \epsilon_v$, $\mu \approx \mu_v$.

20. A uniform plane electromagnetic wave is incident normally upon a sheet of dielectric material, which has the following constants: $\epsilon = 4\epsilon_v$, $\mu = \mu_v$, $\sigma = 0$. If the sheet is 2 cm thick and the amplitude of the electric field strength of the incident wave is 100 mv/m, determine the electric field strength of the wave after passing through the sheet (a) if the frequency is 3000 MHz; (b) if the frequency is 30 MHz.

21. Determine the reflection coefficients for an electromagnetic wave incident normally on (a) a sheet of copper; (b) a sheet of iron. Use $f = 1$MHz. Assume $\sigma = 1 \times 10^7$ mhos/m, $\mu = 1000\mu_v$ for the iron.

22. In the analogy between plane wave propagation in a homogeneous conducting medium and wave propagation along transmission lines, there appears to be nothing corresponding to *R*. Discuss. [Suggestion: Compare eqs. (107b).]

23. A thick brass plate is plated with a 0.0005-inch thickness of silver. What is the surface impedance at (a) 10 kHz, (b) 1 MHz, (c) 100 MHz? Compare the surface impedance of the plated brass with that of a solid silver plate and a solid brass plate. (For silver $\sigma = 6.2 \times 10^7$; for brass $\sigma = 1 \times 10^7$; for both assume that $\mu = \mu_v$, $\epsilon = \epsilon_v$.)

24. A sheet of glass, having a relative dielectric constant of 8 and negligible conductivity, is coated with a silver plate. Show that at a frequency of 100 MHz the surface impedance will be *less* for a 0.001 cm coating than it is for a 0.002 cm coating, and explain why.

25. Determine the voltmeter reading by two different methods. Assume that all the conductors are perfect and that the coaxial cable is lossless.

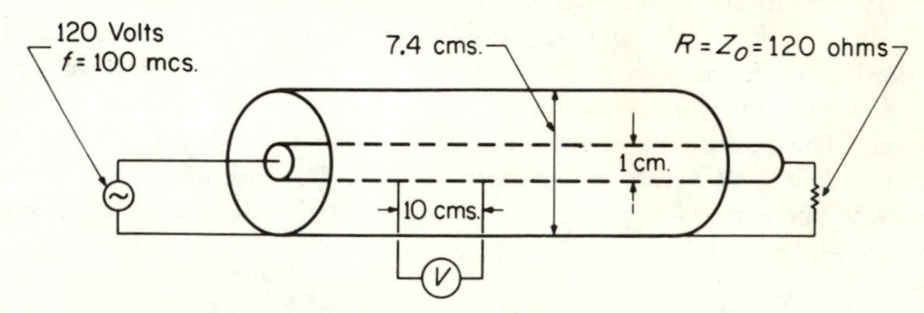

Figure 5-17. Coaxial cable for Problem 25.

26. "Free-space cloth" consists of a cloth coated with conducting material that has a surface impedance of 377 ohms per square. Show that if the thickness of the coating is much greater than the depth of penetration, the surface impedance will be complex, with a reactance equal to the resistance (assuming $\sigma/\omega\epsilon \gg 1$ for the conducting material). However, if the coating is made sufficiently thin, show that the surface impedance will be almost a pure resistance. Determine appropriate values for σ and l, where l is the thickness of the coating.

27. A transmission line is said to be *distortionless* if $L/R = C/G$. (a) Show that the distortionless condition is necessary in order to make $\tilde{V} = e^{-\alpha x} f(x - vt)$ a solution of eq. (106). It is assumed that α and v are positive, real quantities. (b) Show that for sinusoidal oscillations the distortionless condition makes the attenuation constant [that is, the real part of γ in eq. (109)] independent of frequency.

28. Owing to corrosion and tarnishing problems, metallic reflectors may prove unsuitable at optical frequencies for applications requiring exposure to the atmosphere. An alternative is the use of dielectric mismatching to build up a high reflectivity. A typical scheme illustrated in Fig. 5-18 uses alter-

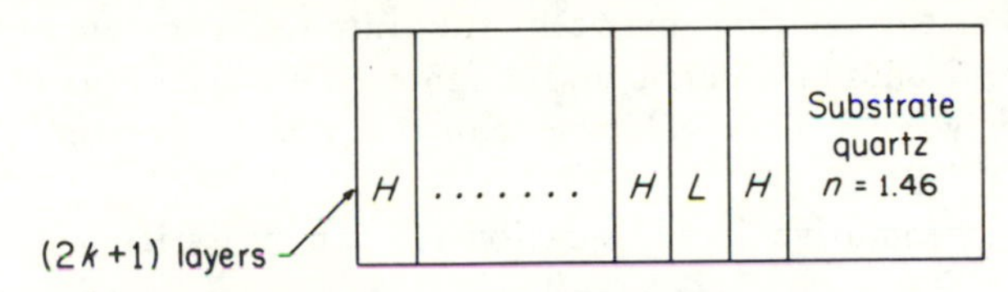

Figure 5-18. High reflectivity can be obtained through dielectric mismatching.

nating layers of high dielectric material, H, and low dielectric material, L, each layer being one-quarter wavelength thick (in the dielectric). Using refractive indices $n_H = 1.6$ and $n_L = 1.2$, find the power reflection coefficient for $k = 6$, and for $k = 25$.

29. A Fabry-Perot cavity consists of parallel plates between which an essentially uniform plane wave reflects back and forth. Such a cavity forms the feedback system for lasers. The atomic transition at 6328Å is 0.04Å wide and thus can (and does) radiate power into any cavity mode within this width. (a) If the cavity length is 100 cm, how many cavity modes are contained within this width of the atomic transition? (Because of this, a single transition laser is almost always a multiple-frequency device. Each of the resonant modes within the transition receives power from the radiating atoms and thus can act as an independent oscillator.) (b) What should the cavity length be to make the laser a single-frequency device?

BIBLIOGRAPHY

See bibliography for chap. 4.

Chapter 6

POYNTING VECTOR AND THE FLOW OF POWER

6.01 Poynting's Theorem. As electromagnetic waves propagate through space from their source to distant receiving points, there is a transfer of energy from the source to the receivers. There exists a simple and direct relation between the rate of this energy transfer and the amplitudes of electric and magnetic field strengths of the electromagnetic wave. This relation can be obtained from Maxwell's equations as follows.

The magnetomotive force equation (I) can be written

$$\mathbf{J} = \nabla \times \mathbf{H} - \epsilon\dot{\mathbf{E}} \tag{I}$$

This expresses a relation between quantities which have the dimensions of current density. If it is multiplied through by $\mathbf{E}$, there will result a relation between quantities which will have the dimensions of power per unit volume. That is

$$\mathbf{E}\cdot\mathbf{J} = \mathbf{E}\cdot\nabla \times \mathbf{H} - \epsilon\mathbf{E}\cdot\dot{\mathbf{E}} \tag{6-1}$$

Recall that for any vectors the following identity holds

$$\nabla\cdot\mathbf{E} \times \mathbf{H} = \mathbf{H}\cdot\nabla \times \mathbf{E} - \mathbf{E}\cdot\nabla \times \mathbf{H}$$

Therefore

$$\mathbf{E}\cdot\mathbf{J} = \mathbf{H}\cdot\nabla \times \mathbf{E} - \nabla\cdot\mathbf{E} \times \mathbf{H} - \epsilon\mathbf{E}\cdot\dot{\mathbf{E}} \tag{6-2}$$

Introducing the second field equation,

$$\nabla \times \mathbf{E} = -\mu\dot{\mathbf{H}} \tag{II}$$

obtain

$$\mathbf{E}\cdot\mathbf{J} = -\mu\mathbf{H}\cdot\dot{\mathbf{H}} - \epsilon\mathbf{E}\cdot\dot{\mathbf{E}} - \nabla\cdot\mathbf{E} \times \mathbf{H} \tag{6-3}$$

Now

$$\mathbf{H}\cdot\dot{\mathbf{H}} = \frac{1}{2}\frac{\partial}{\partial t}H^2 \qquad \text{and} \qquad \mathbf{E}\cdot\dot{\mathbf{E}} = \frac{1}{2}\frac{\partial}{\partial t}E^2$$

so that

$$\mathbf{E}\cdot\mathbf{J} = -\frac{\mu}{2}\frac{\partial}{\partial t}H^2 - \frac{\epsilon}{2}\frac{\partial}{\partial t}E^2 - \nabla\cdot\mathbf{E}\times\mathbf{H}$$

Integrating over a volume V,

$$\int_V \mathbf{E}\cdot\mathbf{J}\,dV = -\frac{\partial}{\partial t}\int_V \left(\frac{\mu}{2}H^2 + \frac{\epsilon}{2}E^2\right)dV - \int_V \nabla\cdot\mathbf{E}\times\mathbf{H}\,dV \quad (6\text{-}4)$$

Using the divergence theorem the last term can be changed from a volume integral to a surface integral over the surface S surrounding V, that is,

$$\int_V \nabla\cdot\mathbf{E}\times\mathbf{H}\,dV = \oint_S \mathbf{E}\times\mathbf{H}\cdot d\mathbf{a}$$

Then eq. (4) can be written

$$\int_V \mathbf{E}\cdot\mathbf{J}\,dV = -\frac{\partial}{\partial t}\int_V \left(\frac{\mu}{2}H^2 + \frac{\epsilon}{2}E^2\right)dV - \oint_S \mathbf{E}\times\mathbf{H}\cdot d\mathbf{a} \quad (6\text{-}5)$$

A physical interpretation of eq. (5) leads to some interesting conclusions. It will be considered term by term.

The term on the left-hand side represents (instantaneous) power dissipated in the volume V. This result is obtained as a generalization of Joule's law. A conductor of cross-sectional area A, carrying a current I and having a voltage drop E per unit length will have a power loss of EI watts per unit length. The power dissipated per unit volume would be

$$\frac{EI}{A} = EJ \quad \text{watts per unit volume}$$

In this case **E** and **J** are in the same direction. In general, where this may not be true, the power dissipated per unit volume would still be given by the product of **J** and the component of **E** having the same direction as **J**. That is, the power dissipated per unit volume would always be given by

$$\mathbf{E}\cdot\mathbf{J}$$

and the total power dissipated in a volume V would be

$$\int_V \mathbf{E}\cdot\mathbf{J}\,dV \quad (6\text{-}6)$$

When the **E** in this expression represents the electric field strength required to produce the current density **J** in the conducting medium, the expression (6) represents power dissipated as ohmic (I^2R) loss. However, if the **E** is an electric field strength due to a source of power, for example due to a battery, then the power represented by the integral expression (6) would be used up in driving the current against the battery voltage and hence charging the battery. If the direction of **E** were opposite to that of **J**, the "dissipated" power represented by (6) would be negative. In this case, the battery would be generating electric power.

Consider next the first term on the right-hand side of eq. (5). In the electrostatic field it was found that the quantity $\frac{1}{2}\epsilon E^2$ could be considered to represent the *energy density* or stored electric energy per unit volume of the electric field. Also for the steady magnetic field the quantity $\frac{1}{2}\mu H^2$ represented the stored energy density of the magnetic field. If it is assumed that these quantities continue to represent stored energy densities when the fields are changing with time (and there seems to be no real reason for considering otherwise), the integral represents the total stored energy in the volume V. The negative time derivative of this quantity then represents the rate at which the stored energy in the volume is decreasing.

The interpretation of the remaining term follows from the application of the law of conservation of energy. The rate of energy dissipation in the volume V must equal the rate at which the stored energy in V is decreasing, plus the rate at which energy is entering the volume V from outside. The term

$$-\oint_S \mathbf{E} \times \mathbf{H} \cdot d\mathbf{a}$$

therefore must represent the rate of flow of energy inward through the surface of the volume. Then this expression without the negative sign,

$$\oint_S \mathbf{E} \times \mathbf{H} \cdot d\mathbf{a} \tag{6-7}$$

represents rate of flow of energy *outward* through the surface enclosing the volume.

The interpretation of eq. (5) leads to the conclusion that the integral of $\mathbf{E} \times \mathbf{H}$ over any closed surface gives the rate of energy flow through that surface. It is seen that the vector

$$\mathbf{P} = \mathbf{E} \times \mathbf{H} \tag{6-8}$$

has the dimensions of watts per square meter. It is *Poynting's theorem* that the vector product $\mathbf{P} = \mathbf{E} \times \mathbf{H}$ at any point is a measure of the rate of energy flow per unit area at that point. The direction of flow is perpendicular to **E** and **H** in the direction of the vector $\mathbf{E} \times \mathbf{H}$.

EXAMPLE 1: *Power Flow for a Plane Wave.* The expression for rate of energy flow per unit area is checked very easily in the case of a uniform plane wave traveling with a velocity

$$v_0 = \frac{1}{\sqrt{\mu\epsilon}}$$

The total energy density due to electric and magnetic fields is given by

$$\tfrac{1}{2}(\epsilon E^2 + \mu H^2)$$

For a wave moving with a velocity $\mathbf{v}_0$ the rate of flow of energy per unit area would be

$$\mathbf{P} = \tfrac{1}{2}(\epsilon E^2 + \mu H^2)\mathbf{v}_0 \tag{6-9}$$

Recalling that for a plane wave the magnitudes of E and H are related by

$$\frac{E}{H} = \sqrt{\frac{\mu}{\epsilon}}$$

eq. (9) becomes

$$\begin{aligned}\mathbf{P} &= \frac{1}{2}\left(\epsilon\sqrt{\frac{\mu}{\epsilon}}\;EH + \mu\sqrt{\frac{\epsilon}{\mu}}\;EH\right)\mathbf{v}_0 \\ &= \left(\frac{EH}{v_0}\right)\mathbf{v}_0 \\ &= \mathbf{E}\times\mathbf{H}\end{aligned}$$

EXAMPLE 2: *Power Flow in a Concentric Cable.* Consider the transfer of power to a load resistance R along a concentric cable which has a d-c voltage V between conductors and a steady current I flowing in the inner and outer conductors. The conductors are assumed to have negligible resistance. The radius of the inner conductor is a and the (inside) radius of the outer conductor is b. The magnetic field strength H will be directed in circles about the axis. By Ampere's law the magnetomotive force around any of these circles will be equal to the current enclosed, that is,

$$\oint \mathbf{H}\cdot d\mathbf{s} = I$$

in the region between the conductors.

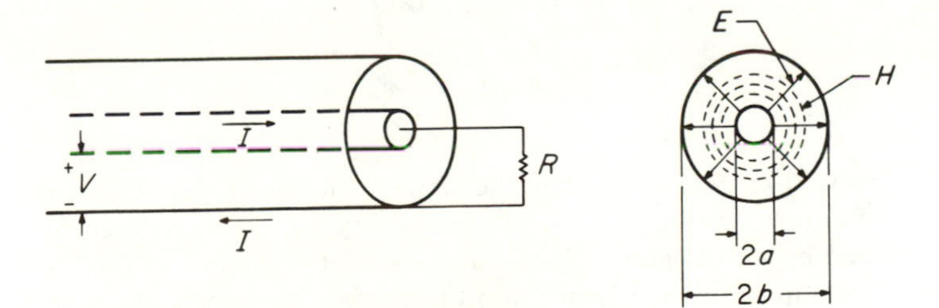

Figure 6-1.

For this case H is constant along any of the circular paths so

$$\oint \mathbf{H}\cdot d\mathbf{s} = 2\pi r H$$

where r is the radius of the circle being considered. Hence

$$H = \frac{I}{2\pi r}$$

The electric field strength **E** will be directed radially. In the example on page 52 it was shown that

$$V = \frac{q}{2\pi\epsilon} \log \frac{b}{a}$$

where q was the charge per unit length. Also it was shown that

$$E = \frac{q}{2\pi\epsilon r}$$

Therefore the magnitude of E will be given by

$$E = \frac{V}{r \log \dfrac{b}{a}}$$

The Poynting vector is

$$\mathbf{P} = \mathbf{E} \times \mathbf{H}$$

It is directed parallel to the axis of the cable. Since **E** and **H** are everywhere at right angles, the magnitude of **P** is simply

$$P = EH$$

The total power flow along the cable will be given by the integration of the Poynting vector over any cross-sectional surface. If the conductors are considered to be perfect, E will have value only in the region between them and the Poynting vector will have value only in the same region. Let the element of area be $2\pi r dr$. Then

$$\begin{aligned} W &= \int_S \mathbf{E} \times \mathbf{H} \cdot d\mathbf{a} \\ &= \int_a^b \frac{V}{r \log b/a} \left(\frac{I}{2\pi r}\right) 2\pi r\, dr \\ &= \frac{VI}{\log b/a} \int_a^b \frac{dr}{r} \\ &= VI \end{aligned}$$

This is the well-known result that the power flow along the cable is the product of the voltage and current. It is interesting to observe that this result was obtained by an integration over an area that did not include the conductors. According to this picture, for the perfect conductor case the flow of power is entirely external to the conductors. Even when the conductors have resistance, there is no contribution within the conductors to the Poynting vector in the direction parallel to the axis, for there is no value of E within a conductor at right angles to the direction of current flow. In the case of the open-wire transmission line, the fields extend throughout all space and there is a value of the Poynting vector everywhere in space, except within conducting bodies. Therefore the rather remarkable conclusion is reached that when a transmission line is used to deliver power from a generator to a load, the power transmission takes place through all the nonconducting regions of space and none of the power flows through the conductors that make up the transmission line.

EXAMPLE 3: *Conductor Having Resistance.* When a conductor having resistance carries a direct current I, there will be a value of E within the conductor. It will be parallel to the direction of the current ($\mathbf{E} = \mathbf{J}/\sigma$), so there will still be no radial component of **E**. Hence there will still be no value of the Poynting vector within the wire parallel to the axis, but there will now be a radial component of **P**. Consider a wire of length L having a voltage drop V_L along the wire. Let the wire be parallel to the z axis. Then in the wire and at its surface

$$E_z = \frac{V_L}{L}$$

The magnetic field strength **H** will be in the ϕ direction and at the surface of the wire it will have a value

$$H_\phi = \frac{I}{2\pi a}$$

where a is the radius of the wire. E_z and H_ϕ are are at right angles, so the Poynting vector will have a magnitude

$$P = E_z H_\phi$$

and will be directed radially into the wire. The total power flowing into the wire through the surface will be

$$\begin{aligned} W &= \int_0^L E_z H_\phi 2\pi a\, dz \\ &= \frac{V_L I}{L}\int_0^L dz \\ &= V_L I \end{aligned}$$

which is the usual expression for loss due to ohmic resistance. This derivation shows that the power required to supply this loss may be considered as coming from the field outside the wire, entering it through the surface of the wire.

It is interesting to observe how the power flow continues inward. Inside the wire the value of H does not vary with the radius in the same way as outside, because the current enclosed varies with r in this case. If J is the current density, the current enclosed at a radius r will be

$$I_{\text{enc}} = \pi r^2 J$$

For a wire of radius a having a total current I

$$I_{\text{enc}} = \frac{\pi r^2 I}{\pi a^2} = \frac{r^2}{a^2} I$$

Therefore inside the wire $(r < a)$

$$H = \frac{r^2 I}{2\pi r a^2}$$

The power flowing inward through an imaginary cylindrical shell of radius $r < a$ will be

$$W = \frac{V_L}{L} 2\pi r L H$$

$$= V_L I \frac{r^2}{a^2} \tag{6-10}$$

Equation (10) shows that the power dissipated within any shell is proportional to the volume enclosed by the shell through which the power is flowing. Hence the power dissipated per unit volume is uniform throughout the wire.

The configuration of the electric field about a two-wire line will appear somewhat as illustrated in Fig. 6-2 when there is a resistance drop in the conductors. The curvature near the surface of the wire is due to the voltage drop along the wire.

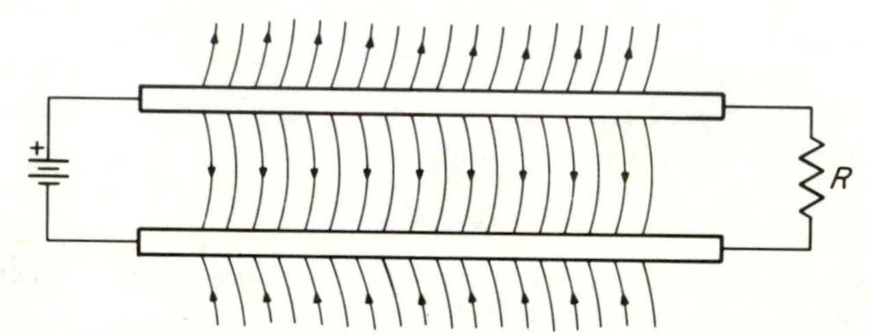

Figure 6-2. Electric-field configuration about a two-wire transmission line having resistance.

EXAMPLE 4: *Poynting Vector about a-c Lines.* When a transmission line delivers a-c power, the voltage, and therefore the electric and magnetic fields, vary with time. Also, if it is a long line, the phases of voltage and current (and E and H) will vary along the length of the line. For the simple case of a lossless line terminated in its characteristic impedance which is a pure resistance, the variation in time and along the line of both voltage and current will be given by the expression for a traveling wave, that is, they are proportional to

$$\cos \omega \left(t - \frac{z}{v}\right)$$

For any value of z and t there will be a certain distribution of the Poynting vector over a plane parallel to the x and y axes. At every point in this plane, **P** will be parallel to the z axis. The Poynting vector will be given by an expression of the form

$$\mathbf{P} = \mathbf{E} \times \mathbf{H} = A \cos^2 \omega \left(t - \frac{z}{v}\right) f(x, y)$$

The function $f(x, y)$ will not vary with z or t. For a fixed value of time, the total power passing through a plane will vary with the position of the plane, that is with z, whereas for a fixed value of z the power through the plane will vary with time. It will be noted that the power flow past a given plane is in pulses of double frequency, a fact readily appreciated when observing the flicker of a 25-cycle electric light bulb.

In a polyphase line a study of the Poynting vector shows that the power passing through a plane of fixed z will not vary as a function of time. In this case the Poynting vector distribution spirals about the line as it is propagated forward. The value of P integrated over a plane of constant z will be found to be independent of time. In such a plane where $z =$ constant, the distribution of the Poynting vector would appear to be revolving about the line.

6.02 Note on the Interpretation of $\mathbf{E} \times \mathbf{H}$. The interpretation of $\mathbf{E} \times \mathbf{H}$ as the power flow per unit area is an extremely useful concept, especially in radiation problems. For example, an integration of $\mathbf{E} \times \mathbf{H}$ over a surface enclosing a transmitting antenna gives the power radiated by the antenna. Although this interpretation of $\mathbf{E} \times \mathbf{H}$ never gives an answer which is known to be erroneous, it sometimes leads to a picture which the engineer is loath to accept. Most engineers find acceptable the concept of energy transmission through space, either with or without guiding conductors, when wave motion is present. However, for many engineers this picture becomes disturbing for transmission line propagation in the d-c case. When $\mathbf{E}$ and $\mathbf{H}$ are static fields produced by unrelated sources, the picture becomes even less credible. The classic illustration of a bar magnet on which is placed an electric charge is one which is often cited. In this example a static electric field is crossed with a steady magnetic field and a strict interpretation of Poynting's theorem seems to require a continuous circulation of energy around the magnet. This is a picture that the engineer generally is not willing to accept (although he usually does not question the theory of permanent magnetism, which requires a continuous circulation of electric currents within the magnet). Fortunately, there exists an easy way out of the dilemma posed by this last example.

First, it is observed that the surface integral in eq. (5) is over the *closed* surface surrounding the volume. If any closed surface is taken about the bar magnet, it is found that $\mathbf{E} \times \mathbf{H}$ integrated over this closed surface is always zero. In other words, the *net* power flow away from the magnet is zero as it should be. Secondly, it is noted that, even though the power flow through any closed surface is correctly given by eq. (7), it does not necessarily follow that $\mathbf{P} = \mathbf{E} \times \mathbf{H}$ represents correctly the power flow at each point. For, to the vector $\mathbf{E}$

$\times$ **H**, could be added any other vector having zero divergence (that is, any vector that is the curl of another vector) without changing the value of the integral in (7). This can be shown by applying the divergence theorem. Suppose the correct value for power flow at any point is not $\mathbf{E} \times \mathbf{H}$, but rather $\mathbf{P} = \mathbf{E} \times \mathbf{H} + \mathbf{F}$, where **F** is the curl of some other vector, say **G**. Then the net power flow through any closed surface would be

$$\oint_S (\mathbf{E} \times \mathbf{H} + \mathbf{F})\cdot d\mathbf{a} = \oint_S (\mathbf{E} \times \mathbf{H})\cdot d\mathbf{a} + \int_V \nabla\cdot\mathbf{F}\, dV$$

$$= \oint_S (\mathbf{E} \times \mathbf{H})\cdot d\mathbf{a}$$

because $\nabla\cdot\nabla \times \mathbf{G} \equiv 0$.

It is seen that even though it may be possible to write an expression that gives correctly the net flow of power through a closed surface, it is still not possible to state just where the energy is. This problem is by no means peculiar to the electromagnetic field. The total potential energy of a raised weight is a readily calculable quantity but the "distribution" of this energy is not known. Just where the potential energy of a raised weight or a charged body "resides" is a question for philosophic speculation only. It cannot be answered on the basis of any measurements that the engineer can make.

6.03 Instantaneous, Average and Complex Poynting Vector. In an a-c circuit, the instantaneous power $\tilde{W}$ is always given by the product of the instantaneous voltage $\tilde{V}$ and the instantaneous current $\tilde{I}$.

$$\tilde{W} = \tilde{V}\tilde{I}$$

The quantities $\tilde{V}$ and $\tilde{I}$ may be expressed in terms of phasors V and I as follows;

$$\tilde{V} = \text{Re}\,\{Ve^{j\omega t}\} = \text{Re}\,\{|V|e^{j\theta_v}\, e^{j\omega t}\} = |V| \cos(\omega t + \theta_v) \tag{6-11}$$

$$\tilde{I} = \text{Re}\,\{Ie^{j\omega t}\} = \text{Re}\,\{|I|e^{j\theta_i}\, e^{j\omega t}\} = |I| \cos(\omega t + \theta_i) \tag{6-12}$$

Now the instantaneous power may be expressed as

$$\tilde{W} = |V|\,|I| \cos(\omega t + \theta_v) \cos(\omega t + \theta_i)$$

$$= \frac{|V|\,|I|}{2}[\cos(\theta_v - \theta_i) + \cos(2\omega t + \theta_v + \theta_i)] \tag{6-13}$$

which consists of an average part and an oscillating part. If we define $\theta_v - \theta_i = \theta$ (see Fig. 6-3) we may write the average power as

$$W_{\text{av}} = \frac{|V|\,|I|}{2} \cos\theta \tag{6-14}$$

Another useful quantity is the "reactive power," or more accurately, the reactive volt-amperes (VAR).

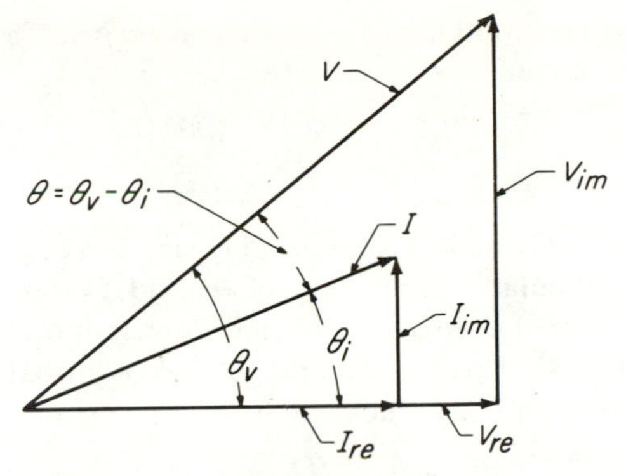

Figure 6-3.

$$W_{\text{react}} = \frac{|V|\,|I|}{2} \sin \theta \tag{6-15}$$

Since θ is the phase angle between the voltage and the current, W_{av} and W_{react} are the *in-phase* and *out-of-phase* components of the volt-ampere product. The presence of an out-of-phase or reactive component indicates that the instantaneous power changes sign (reverses direction of flow) over part of an a-c cycle.

Consider now the *complex power* W, defined as one-half the product of V and the complex conjugate of I.

$$\begin{aligned} W &= \tfrac{1}{2} VI^* = \tfrac{1}{2} |V|\, e^{j\theta_v}\, |I|\, e^{-j\theta_i} \\ &= \frac{|V|\,|I|}{2}\, e^{j\theta} \\ &= W_{\text{av}} + jW_{\text{react}} \end{aligned} \tag{6-16}$$

This shows that the average and reactive power may be recovered from the complex power by taking the real and imaginary parts.

$$W_{\text{av}} = \tfrac{1}{2}\,\text{Re}\,\{VI^*\}$$

$$W_{\text{react}} = \tfrac{1}{2}\,\text{Im}\,\{VI^*\}$$

In electromagnetic field theory there are relations similar to the above between the Poynting vector, the electric field strength and the magnetic field strength. The instantaneous power flow per square meter is

$$\tilde{\mathbf{P}} = \tilde{\mathbf{E}} \times \tilde{\mathbf{H}}$$

Following the procedure already outlined for circuit theory, we may define the complex Poynting vector $\mathbf{P}$ as

$$\mathbf{P} = \tfrac{1}{2}\, \mathbf{E} \times \mathbf{H}^* \tag{6-17}$$

from which we may obtain the average and reactive parts of the power flow per square meter.

$$\mathbf{P}_{\text{av}} = \tfrac{1}{2}\,\text{Re}\,\{\mathbf{E} \times \mathbf{H}^*\}$$

$$\mathbf{P}_{\text{react}} = \tfrac{1}{2}\,\text{Im}\,\{\mathbf{E} \times \mathbf{H}^*\}$$

The product of **E** and **H** in eq. (17) is a vector product. Only mutually perpendicular components of **E** and **H** contribute anything to power flow, and the direction of the flow is normal to the plane containing **E** and **H**. Thus in rectangular co-ordinates, the complex flow of power per unit area normal to the y-z plane is

$$P_x = \tfrac{1}{2}\,(E_y H_z^* - E_z H_y^*) \tag{6-18}$$

with corresponding expressions for the other directions. In spherical co-ordinates, the outward (radial) flow of complex power per unit area is

$$P_r = \tfrac{1}{2}\,(E_\theta H_\phi^* - E_\phi H_\theta^*) \tag{6-19}$$

The Poynting Theorem in Complex Form. Maxwell's curl equations may be expressed in phasor form as

$$\nabla \times \mathbf{H} = (\sigma + j\omega\epsilon)\,\mathbf{E} + \mathbf{J} \qquad \text{and} \qquad \nabla \times \mathbf{E} = -\,j\omega\mu\,\mathbf{H}$$

in which **J** represents nonohmic currents such as convection current or specified source current. These equations are used in the following derivation of the complex Poynting theorem.

$$\begin{aligned}\nabla\cdot(\mathbf{E} \times \mathbf{H}^*) &= \mathbf{H}^*\cdot\nabla \times \mathbf{E} - \mathbf{E}\cdot\nabla \times \mathbf{H}^* \\ &= -\,j\omega\mu\,\mathbf{H}\cdot\mathbf{H}^* - (\sigma - j\omega\epsilon)\,\mathbf{E}\cdot\mathbf{E}^* - \mathbf{E}\cdot\mathbf{J}^*\end{aligned} \tag{6-20}$$

Integrating over the volume V surrounded by surface S, we have

$$\begin{aligned}\oint_S \mathbf{E} \times \mathbf{H}^*\cdot d\mathbf{a} = &-\,j\omega \int_V (\mu\mathbf{H}\cdot\mathbf{H}^* - \epsilon\mathbf{E}\cdot\mathbf{E}^*)\,dV \\ &- \int_V \sigma\mathbf{E}\cdot\mathbf{E}^*\,dV - \int_V \mathbf{E}\cdot\mathbf{J}^*\,dV\end{aligned} \tag{6-21}$$

Note that the time-average stored energy densities (electric and magnetic) are given by

$$U_e = \tfrac{1}{4}\,\epsilon\mathbf{E}\cdot\mathbf{E}^* \qquad \text{and} \qquad U_m = \tfrac{1}{4}\,\mu\mathbf{H}\cdot\mathbf{H}^* \tag{6-22}$$

The Poynting theorem now may be separated into real and imaginary parts and written as

$$\text{Re}\oint_S \mathbf{P}\cdot d\mathbf{a} + \tfrac{1}{2}\int_V \sigma\mathbf{E}\cdot\mathbf{E}^*\,dV = -\,\tfrac{1}{2}\,\text{Re}\int_V \mathbf{E}\cdot\mathbf{J}^*\,dV \tag{6-23}$$

$$\text{Im}\oint_S \mathbf{P}\cdot d\mathbf{a} + 2\omega\int_V (U_m - U_e)\,dV = -\,\tfrac{1}{2}\,\text{Im}\int_V \mathbf{E}\cdot\mathbf{J}^*\,dV \tag{6-24}$$

Problem 1. Verify that

$$|V|\,|I|\cos\theta = V_{\mathrm{re}}I_{\mathrm{re}} + V_{\mathrm{im}}I_{\mathrm{im}} = \mathrm{Re}\;VI^*$$

and

$$|V|\,|I|\sin\theta = V_{\mathrm{im}}I_{\mathrm{re}} - V_{\mathrm{re}}I_{\mathrm{im}} = \mathrm{Im}\;VI^*$$

Problem 2. A concentric cable (assumed perfectly conducting) is one wavelength long and is terminated in its characteristic impedance, a pure resistance.

(a) Indicate the magnitude and direction of the Poynting vector along the line at successive one-eighth period intervals of time throughout a cycle.

(b) Repeat part (a) for the case where the line is terminated by a short circuit.

Problem 3. A short vertical transmitting antenna erected on the surface of a perfectly conducting earth produces effective field strength

$$E_{\mathrm{eff}} = E_{\theta\mathrm{eff}} = 100 \sin\theta \qquad \mathrm{mv/m}$$

at points a distance of 1 mile from the antenna (θ is the polar angle). Compute the Poynting vector and the total power radiated. (For the *distant* field, $H = H_\phi = E_\theta/\eta_v$.)

6.04 Power Loss in a Plane Conductor. An evaluation of the normal component of Poynting vector at the surface of a conductor will give the power flow per unit area through the surface and hence the power loss in the conductor.

Let there be a tangential component of magnetic field strength $\mathbf{H}_{\mathrm{tan}}$ at the surface of a metallic conductor (assumed for the present to be an infinitely large flat plate having a thickness very much greater than the skin depth δ). From the continuity requirements across the boundary surface the tangential component of $\mathbf{H}$ just inside the conductor will have this same value $\mathbf{H}_{\mathrm{tan}}$. Inside the conductor $\mathbf{E}_{\mathrm{tan}}$, the tangential component of $\mathbf{E}$ is related to $\mathbf{H}_{\mathrm{tan}}$ by

$$\frac{E_{\mathrm{tan}}}{H_{\mathrm{tan}}} = \eta_m$$

and

$$\eta_m = \sqrt{\frac{j\omega\mu_m}{\sigma_m + j\omega\epsilon_m}} \approx \sqrt{\frac{j\omega\mu_m}{\sigma_m}} = \sqrt{\frac{\omega\mu_m}{\sigma_m}} \qquad \underline{/45^\circ}$$

where η_m is the intrinsic impedance of the conductor. (The subscript m has been used to indicate that the quantities inside the metallic conductor are meant.) Just inside the surface of the conductor $E_{\mathrm{tan}} = \eta_m H_{\mathrm{tan}}$ and, from the continuity requirements across the boundary, the tangential component of electric field strength just outside the surface will also be E_{tan}. Then the average (or real) power flow per unit area normal to the surface will be

$$\mathbf{P}_n(\text{real}) = \tfrac{1}{2}\,\text{Re}(\mathbf{E}_{\tan} \times \mathbf{H}^*_{\tan}) \tag{6-25}$$

When $\mathbf{E}_{\tan}$ and $\mathbf{H}_{\tan}$ are at right angles, and since for any good conductor $E_{\tan}$ leads $H_{\tan}$ by 45 degrees in time phase, (25) becomes

$$\begin{aligned} P_n &= \tfrac{1}{2}\,|E_{\tan}|\,|H_{\tan}| \cos 45^\circ \\ &= \frac{1}{2\sqrt{2}}\,|\eta_m|\,|H_{\tan}|^2 \\ &= \frac{1}{2\sqrt{2}}\,\frac{|E_{\tan}|^2}{|\eta_m|} \end{aligned} \tag{6-26}$$

where the bars | | indicate the absolute magnitude of the complex quantity. For a conductor which has a thickness very much greater than the skin depth δ, the surface impedance Z_s is equal to the intrinsic impedance η_m of the conductor, so that

$$P_n = \frac{1}{2\sqrt{2}}\,|Z_s|\,|H_{\tan}|^2 = \frac{1}{2\sqrt{2}}\,\frac{|E_{\tan}|^2}{|Z_s|} \qquad \text{watt/sq m} \tag{6-27}$$

In a conductor the linear current density $\mathbf{J}_s$ is equal in magnitude to the tangential magnetic field strength at the surface, so

$$P_n = \frac{1}{2\sqrt{2}}\,|Z_s||J_s|^2 \qquad \text{watt/sq m} \tag{6-28}$$

In expressions (27) and (28), $E_{\tan}$, $H_{\tan}$, and J_s are peak values. In terms of effective values

$$\begin{aligned} P_n &= \frac{1}{\sqrt{2}}\,\frac{|E_{t(\text{eff})}|^2}{|Z_s|} = \frac{1}{\sqrt{2}}\,|Z_s|\,|H_{t(\text{eff})}|^2 \\ &= \frac{1}{\sqrt{2}}\,|Z_s|\,|J_{s(\text{eff})}|^2 \\ &= R_s J^2_{s(\text{eff})} \qquad \text{watt/sq m} \end{aligned} \tag{6-29}$$

This result agrees with that previously obtained in chap. 5.

Power Loss in a Simple Resonator. The cavity resonator is a widely used microwave component which is equivalent to the familiar "tuned circuit" used at low frequencies. The simplest type of resonator consists of a pair of parallel conducting plates between which a standing wave can exist (see Fig. 6-4).

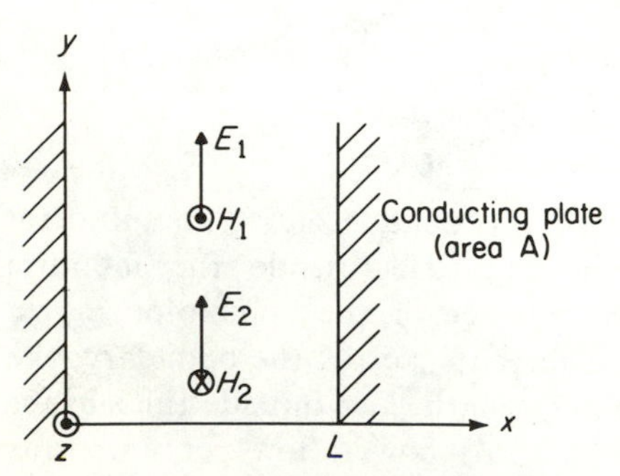

Figure 6-4. The parallel-plate resonator.

For purposes of approximate calculation, the fields between

the plates will be assumed to be made up of two uniform plane waves traveling in opposite directions and numbered 1 and 2.

$$\mathbf{E}_1 = \hat{\mathbf{y}}\, E_1\, e^{-j\beta x} \qquad \mathbf{E}_2 = \hat{\mathbf{y}}\, E_2\, e^{j\beta x}$$

$$\mathbf{H}_1 = \hat{\mathbf{z}}\, H_1\, e^{-j\beta x} \qquad \mathbf{H}_2 = -\hat{\mathbf{z}}\, H_2\, e^{j\beta x}$$

$$E_1 = \eta H_1 \qquad E_2 = \eta H_2$$

The conditions of zero tangential electric field at $x = 0$ and $x = L$ may be applied, assuming the end-plates to be perfect conductors. The boundary conditions then may be expressed as

$$E_1 + E_2 = 0$$

$$E_1\, e^{-j\beta L} + E_2\, e^{j\beta L} = 0$$

If these conditions are to be satisfied, it must be true that

$$e^{j2\beta L} = 1 \qquad \text{or} \qquad \beta L = n\pi$$

where n is a positive integer. Since $\beta = 2\pi/\lambda$ the above relation becomes

$$L = n\,\frac{\lambda}{2}$$

which states that source-free fields can exist between the plates only when the spacing is an integral number of half-wavelengths.

Any practical resonator has various forms of power loss associated with it and the effect of these losses may be estimated by calculating the Q of the resonator. Here we shall take into account only the power absorbed by the end-plates, ignoring power loss due to radiation, the presence of lossy dielectrics and loss through input and output couplings. The Q formula is given (in chap. 8) as

$$Q = \frac{\omega U_T}{W_L}$$

in which U_T is the total stored energy and W_L is the total power lost, in this case lost to the end-plates.

The stored energy is the sum of the electric and magnetic stored energies; since they are equal in magnitude, the total stored energy is just twice the electric stored energy. The total electric field strength is

$$\mathbf{E} = \mathbf{E}_1 + \mathbf{E}_2 = \hat{\mathbf{y}}\, 2j\, E_2 \sin \beta x$$

Since the electric stored energy density is given by

$$U_e = \tfrac{1}{4}\,\epsilon\mathbf{E}\cdot\mathbf{E}^*$$

the total stored energy may be calculated as follows:

$$
\begin{aligned}
U_T &= \int 2U_e\,dV \\
&= 2A\epsilon\,|E_2|^2 \int_0^{n\pi/\beta} \sin^2 \beta x\,dx \\
&= \tfrac{1}{2}\,A\epsilon\,|E_2|^2\,n\lambda
\end{aligned}
$$

The power lost is determined by applying (27) at both plates as follows:

$$
\begin{aligned}
W_L &= 2A\,R_s\,\frac{|2H_2|^2}{2} \\
&= 4A\,R_s\,\eta^{-2}\,|E_2|^2 \\
&= 2A\epsilon\,|E_2|^2 \sqrt{\frac{2\omega}{\mu\sigma}}
\end{aligned}
$$

Substitution into the formula for Q yields

$$
Q = \frac{n}{4} \cdot \frac{\lambda}{\delta}
$$

in which δ is the penetration depth. At a frequency of 10 GHz ($\lambda =$ 3 cm) and for copper end-plates, the Q for $n = 1$ is approximately 11,200.

Problem 4. A uniform plane wave having field components E_x and H_y is guided in the z direction between a pair of parallel copper planes. If the frequency is 100 MHz and the field strength of the transmitted wave is $E_x = 1$ volt/m, determine by two methods the power loss per square meter in each of the conducting planes.

BIBLIOGRAPHY

See bibliography for chap. 4.

Chapter 7

GUIDED WAVES

In the wave propagation so far discussed, only uniform plane waves, remote from any guiding surfaces, have been considered. In many actual cases, propagation is by means of *guided* waves, that is, waves that are guided along or over conducting or dielectric surfaces. Common examples of guided electromagnetic waves are the waves along ordinary parallel-wire and coaxial transmission lines, waves in wave guides, and waves that are guided along the earth's surface from a radio transmitter to the receiving point. The study of such guided waves will now be undertaken.

7.01 Waves between Parallel Planes. For purposes of study a simple illustrative example is that of an electromagnetic wave, propagating between a pair of parallel perfectly conducting planes of infinite extent in the y and z directions (Fig. 7-1). In order to determine the electromagnetic field configurations in the region between the planes, Maxwell's equations will be solved subject to the appro-

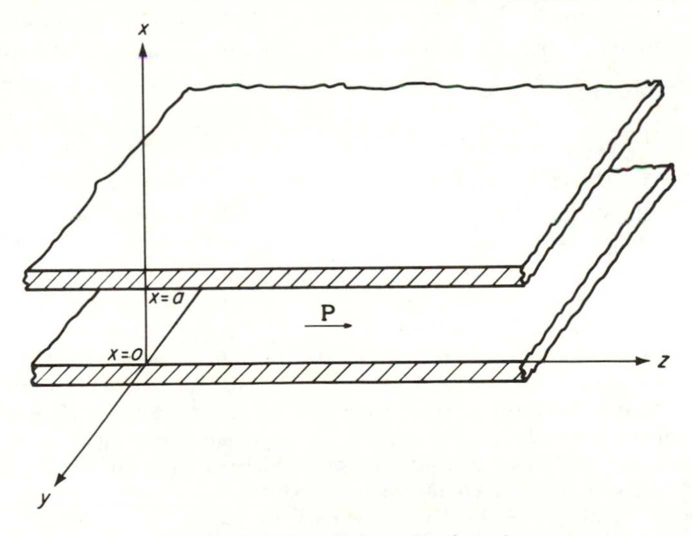

Figure 7-1. Parallel conducting planes.

priate boundary conditions. Because perfectly conducting planes have been assumed, these boundary conditions are very simple, being

$$E_{\text{tangential}} = 0, \qquad H_{\text{normal}} = 0$$

at the surfaces of the conductors.*

In general, Maxwell's curl equations and the wave equations are

$$\nabla \times \mathbf{H} = (\sigma + j\omega\epsilon)\mathbf{E} \qquad \nabla \times \mathbf{E} = -j\omega\mu\mathbf{H} \tag{7-1}$$

$$\nabla^2\mathbf{E} = \gamma^2\mathbf{E} \qquad \nabla^2\mathbf{H} = \gamma^2\mathbf{H} \tag{7-2}$$

where

$$\gamma = \sqrt{(\sigma + j\omega\epsilon)(j\omega\mu)} \tag{7-3}$$

In rectangular co-ordinates, and for the nonconducting region between the planes, these equations become

$$\left.\begin{aligned} \frac{\partial H_z}{\partial y} - \frac{\partial H_y}{\partial z} &= j\omega\epsilon E_x & \frac{\partial E_z}{\partial y} - \frac{\partial E_y}{\partial z} &= -j\omega\mu H_x \\ \frac{\partial H_x}{\partial z} - \frac{\partial H_z}{\partial x} &= j\omega\epsilon E_y & \frac{\partial E_x}{\partial z} - \frac{\partial E_z}{\partial x} &= -j\omega\mu H_y \\ \frac{\partial H_y}{\partial x} - \frac{\partial H_x}{\partial y} &= j\omega\epsilon E_z & \frac{\partial E_y}{\partial x} - \frac{\partial E_x}{\partial y} &= -j\omega\mu H_z \end{aligned}\right\} \tag{7-4}$$

$$\left.\begin{aligned} \frac{\partial^2 \mathbf{E}}{\partial x^2} + \frac{\partial^2 \mathbf{E}}{\partial y^2} + \frac{\partial^2 \mathbf{E}}{\partial z^2} &= -\omega^2\mu\epsilon\mathbf{E} \\ \frac{\partial^2 \mathbf{H}}{\partial x^2} + \frac{\partial^2 \mathbf{H}}{\partial y^2} + \frac{\partial^2 \mathbf{H}}{\partial z^2} &= -\omega^2\mu\epsilon\mathbf{H} \end{aligned}\right\} \tag{7-5}$$

It will be assumed that propagation is in the z direction, and that the variation of all field components in this direction may be expressed in the form $e^{-\bar{\gamma}z}$, where in general

$$\bar{\gamma} = \bar{\alpha} + j\bar{\beta} \tag{7-6}$$

is a complex propagation constant,† whose value is to be determined. This is a quite reasonable assumption because it is easy to show (problem 4, page 240) that for any uniform transmission line or guide the fields must obey an exponential law along the line. When the

*It is easy to show for actual conductors such as copper or brass (which have a very high, but not infinite, conductivity) that the finite conductivity has negligible effect on the field configuration. Therefore it is possible to use the fields calculated on the basis of perfectly conducting planes to determine the surface currents that must flow in these planes. The currents so calculated may then be used to compute the losses, and hence the attenuation, which occur with finitely conducting planes. This is a standard engineering approach.

†In general $\bar{\gamma}$ will not be equal to γ, defined by eq. (3), but $\bar{\gamma}$ reduces to γ in the special case of uniform plane waves.

time variation factor is combined with the z variation factor, it is seen that the combination

$$e^{j\omega t}e^{-\bar{\gamma}z} = e^{(j\omega t-\bar{\gamma}z)} = e^{-\bar{\alpha}z}e^{j(\omega t-\bar{\beta}z)} \tag{7-7}$$

represents a wave propagating in the z direction. If $\bar{\gamma}$ happens to be an imaginary number, that is if $\bar{\alpha}=0$, expression (5) represents a wave without attenuation. On the other hand, if $\bar{\gamma}$ is real so that $\bar{\beta}=0$, there is no wave motion but only an exponential decrease in amplitude.

Since the space between the planes is infinite in extent in the y direction, there are no boundary conditions to be met in this direction, and it can be assumed that the field is uniform or constant in the y direction. This means that the derivatives with respect to y in (4) can be put equal to zero. In the x direction however, there are certain boundary conditions which must be met. Therefore it is not possible to specify arbitrarily what the distribution of fields in this direction will be. This answer must come out of the solution of the differential equations when the boundary conditions are applied.

When the variation in the z direction of each of the field components is shown explicitly by writing, for example,

$$H_y = H_y^0 e^{-\bar{\gamma}z}$$

it is seen immediately that

$$\frac{\partial H_y}{\partial z} = -\bar{\gamma}H_y^0 e^{-\bar{\gamma}z} = -\bar{\gamma}H_y$$

with similar results for the z derivatives of the other components. Making use of this result and remembering that the y derivative of any component is zero, eqs. (4) and (5) become

$$\left.\begin{aligned} \bar{\gamma}H_y &= j\omega\epsilon E_x & \bar{\gamma}E_y &= -j\omega\mu H_x \\ -\bar{\gamma}H_x - \frac{\partial H_z}{\partial x} &= j\omega\epsilon E_y & -\bar{\gamma}E_x - \frac{\partial E_z}{\partial x} &= -j\omega\mu H_y \\ \frac{\partial H_y}{\partial x} &= j\omega\epsilon E_z & \frac{\partial E_y}{\partial x} &= -j\omega\mu H_z \end{aligned}\right\} \tag{7-8}$$

$$\begin{aligned} \frac{\partial^2 \mathbf{E}}{\partial x^2} + \bar{\gamma}^2\mathbf{E} &= -\omega^2\mu\epsilon\mathbf{E} \\ \frac{\partial^2 \mathbf{H}}{\partial x^2} + \bar{\gamma}^2\mathbf{H} &= -\omega^2\mu\epsilon\mathbf{H} \end{aligned} \tag{7-9}$$

In eqs. (9) it should be remembered that each of these equations is really three equations, one for each of the components of **E** or **H**. Equations (8) can be solved simultaneously to yield the following equations

$$H_x = -\frac{\bar{\gamma}}{h^2}\frac{\partial H_z}{\partial x} \qquad E_x = -\frac{\bar{\gamma}}{h^2}\frac{\partial E_z}{\partial x}$$

$$H_y = -\frac{j\omega\epsilon}{h^2}\frac{\partial E_z}{\partial x} \qquad E_y = +\frac{j\omega\mu}{h^2}\frac{\partial H_z}{\partial x} \tag{7-10}$$

where

$$h^2 = \bar{\gamma}^2 + \omega^2\mu\epsilon \tag{7-11}$$

In eqs. (10) the various components of electric and magnetic field strengths are expressed in terms of E_z and H_z. With the exception of one possibility, to be discussed later, it will be observed that there must be a z component of either **E** or **H**; otherwise all the components would be zero and there would be no fields at all in the region considered. Although in the general case both E_z and H_z could be present at the same time, it is convenient and desirable to divide the solutions into two sets. In the first of these, there is a component of **E** in the direction of propagation (E_z), but no component of **H** in this direction. Such waves are called *E waves*, or more commonly, *transverse magnetic* (TM) waves, because the magnetic field strength **H** is entirely transverse. The second set of solutions has a component of **H** in the direction of propagation, but no E_z component. Such waves are called *H waves* or *transverse electric* (TE) waves. The solutions to eqs. (8) and (9) for these two cases will now be obtained. Since the differential equations are linear, the sum of these two sets of solutions yields the most general solution.

7.02 Transverse Electric Waves ($E_z \equiv 0$). Inspection of eqs. (10) shows that when $E_z \equiv 0$, but H_z does not equal zero, the field components H_y and E_x will also equal zero, whereas, in general, there will be nonzero values for the components H_x and E_y. Since each of the field components obeys the wave equation as given by eqs. (9), the wave equation can be written for the component E_y

$$\frac{\partial^2 E_y}{\partial x^2} + \bar{\gamma}^2 E_y = -\omega^2\mu\epsilon E_y$$

This can be written as

$$\frac{\partial^2 E_y}{\partial x^2} = -h^2 E_y \tag{7-12}$$

Recalling that $E_y = E_y^0(x)\,e^{-\bar{\gamma}z}$, eq. (12) reduces to

$$\frac{d^2 E_y^0}{dx^2} = -h^2 E_y^0 \tag{7-12a}$$

where as before

$$h^2 = \bar{\gamma}^2 + \omega^2\mu\epsilon$$

Equation (12a) is the differential equation of simple harmonic motion. Its solution can be written in the form

$$E_y^0 = C_1 \sin hx + C_2 \cos hx \tag{7-13}$$

where C_1 and C_2 are arbitrary constants.

Showing the variation with time and in the z direction the expression for E_y is

$$E_y = (C_1 \sin hx + C_2 \cos hx)\, e^{-\bar{\gamma} z} \tag{7-13a}$$

The arbitrary constants C_1 and C_2 can be determined from the boundary conditions. For the parallel-plane wave guide of Fig. 7-1 the boundary conditions are quite simple. They require that the tangential component of **E** be zero at the surface of the (perfect) conductors for all values of z and time. This requires that

$$\left.\begin{aligned} E_y &= 0 \quad \text{at } x = 0 \\ E_y &= 0 \quad \text{at } x = a \end{aligned}\right\} \text{ for all values of } z \qquad \text{(boundary conditions)}$$

In order for the first of these conditions to be true, it is evident that C_2 must be zero. Then the expression for E_y is

$$E_y = C_1 \sin hx\, e^{-\bar{\gamma} z}$$

Application of the second boundary condition imposes a restriction on h. In order for E_y to be zero at $x = a$ for all values of z and t it is necessary that

$$h = \frac{m\pi}{a} \tag{7-14}$$

where $m = 1, 2, 3, \ldots$.

(The special case of $m = 0$ will be discussed later.) Therefore

$$E_y = C_1 \sin\left(\frac{m\pi}{a} x\right) e^{-\bar{\gamma} z} \tag{7-15}$$

The other components of E and H can be obtained by inserting eq. (15) in eqs. (8). When this is done, it is seen that the expression for the field strengths for transverse electric waves between parallel planes are

$$\begin{aligned} E_y &= C_1 \sin\left(\frac{m\pi}{a} x\right) e^{-\bar{\gamma} z} \\ H_z &= -\frac{m\pi}{j\omega\mu a} C_1 \cos\left(\frac{m\pi}{a} x\right) e^{-\bar{\gamma} z} \\ H_x &= -\frac{\bar{\gamma}}{j\omega\mu} C_1 \sin\left(\frac{m\pi}{a} x\right) e^{-\bar{\gamma} z} \end{aligned} \tag{7-16}$$

Each value of m specifies a particular field configuration or *mode*, and the wave associated with the integer m is designated as the TE_{m0} wave or TE_{m0} mode. The second subscript (equal to zero in this case) refers to another factor which varies with y, which is found in the general case of rectangular guides. It will be noticed that the smallest value of m that can be used in eqs. (16) is $m = 1$, because $m = 0$ makes all the fields identically zero. That is, the lowest-order mode that can exist in this case is the TE_{10} mode.

In writing expressions for the field components as in eqs. (16), the variation of all the fields in the z direction is the same for any particular value of m and is shown by the factor $e^{-\bar{\gamma}z}$. Rather than carry this factor through the entire analysis, it is sometimes convenient to make use of the zero superscript notation mentioned earlier in this chapter. Thus eqs. (16) can be written

$$
\begin{aligned}
E_y^0 &= C_1 \sin \frac{m\pi}{a} x \\
H_x^0 &= -\frac{\bar{\gamma}}{j\omega\mu} C_1 \sin \frac{m\pi}{a} x \\
H_z^0 &= \frac{-m\pi}{j\omega\mu a} C_1 \cos \frac{m\pi}{a} x
\end{aligned}
\tag{7-17}
$$

The factor $\bar{\gamma}$ is the propagation constant, which is ordinarily complex, the real part $\bar{\alpha}$ being the attenuation constant and the imaginary part $\bar{\beta}$ being the phase-shift constant. However, it will be shown in sec. 7.04 that for the present problem of waves guided by perfectly conducting walls, $\bar{\gamma}$ is either pure real or pure imaginary. In that range of frequencies where $\bar{\gamma}$ is real, $\bar{\alpha}$ has value but $\bar{\beta}$ is zero, so that there is attenuation but no phase shift and, therefore, no wave motion. In the range of frequencies where $\bar{\gamma}$ is imaginary, $\bar{\alpha}$ is zero but $\bar{\beta}$ has value, so that there is propagation by wave motion without attenuation. It is this latter range of frequencies that is of chief interest in wave guide propagation. Writing $\bar{\gamma} = j\bar{\beta}$, eqs. (16) for TE_{m0} waves in the propagation range may be written as

$$
\begin{aligned}
E_y &= C_1 \sin \left(\frac{m\pi}{a} x\right) e^{-j\bar{\beta}z} \\
H_x &= -\frac{\bar{\beta}}{\omega\mu} C_1 \sin \left(\frac{m\pi}{a} x\right) e^{-j\bar{\beta}z} \\
H_z &= \frac{jm\pi}{\omega\mu a} C_1 \cos \left(\frac{m\pi}{a} x\right) e^{-j\bar{\beta}z}
\end{aligned}
\tag{7-16a}
$$

A sketch of these field distributions at some particular instant of time is shown in Fig. 7-2 for the TE_{10} mode.

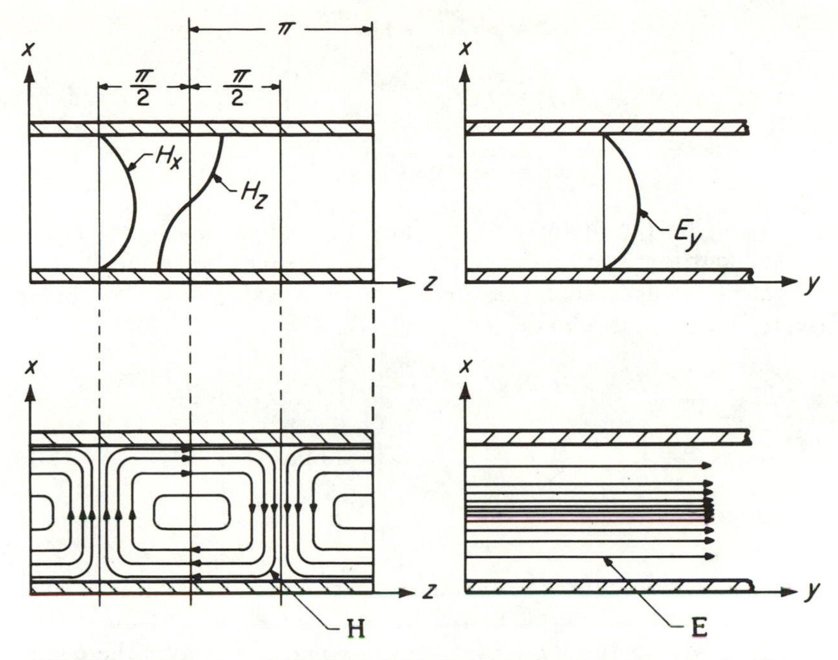

Figure 7-2. Electric and magnetic fields between parallel planes for the TE_{10} wave.

7.03 Transverse Magnetic Waves ($H_z = 0$). The case of transverse magnetic waves between parallel planes can be solved in a manner similar to that used for TE waves. In this instance H_z will be zero, and inspection of eqs. (10) shows that H_x and E_y will also be zero, while in general, E_z, E_x, and H_y will have value. Solving the wave equation for H_y, gives as before

$$H_y = (C_3 \sin hx + C_4 \cos hx)\, e^{-\bar{\gamma}z} \tag{7-18}$$

The boundary conditions cannot be applied directly to H_y to evaluate the constants C_3 and C_4, because in general the tangential component of H is not zero at the surface of a conductor. However from eqs. (8) the expressions for E_z can be obtained in terms of H_y, and then the boundary conditions applied to E_z. From eqs. (8) and (18)

$$E_z = \frac{h}{j\omega\epsilon}[C_3 \cos hx - C_4 \sin hx]\, e^{-\bar{\gamma}z}$$

Applying the boundary conditions that E_z must be zero at $x = 0$ shows that $C_3 = 0$. The second condition that E_z must be zero at $x = a$ requires that $h = m\pi/a$ where m is any integer. Then the expressions for E_z^0, H_y^0, and E_x^0 become

$$\left.\begin{aligned} E_z^0 &= -\frac{m\pi C_4}{j\omega\epsilon a}\sin\frac{m\pi}{a}x \\ H_y^0 &= C_4\cos\frac{m\pi}{a}x \\ E_x^0 &= \frac{\bar{\gamma}C_4}{j\omega\epsilon}\cos\frac{m\pi}{a}x \end{aligned}\right\} \tag{7-19}$$

Multiplying by the factor $e^{-\bar{\gamma}z}$ to show the variation in the z direction, and putting $\bar{\gamma} = j\bar{\beta}$ for the range of frequencies in which wave propagation occurs, the expressions for TM waves between parallel perfectly conducting planes are

$$\left.\begin{aligned} H_y &= C_4\cos\left(\frac{m\pi}{a}x\right)e^{-j\bar{\beta}z} \\ E_x &= \frac{\bar{\beta}}{\omega\epsilon}C_4\cos\left(\frac{m\pi}{a}x\right)e^{-j\bar{\beta}z} \\ E_z &= \frac{jm\pi}{\omega\epsilon a}C_4\sin\left(\frac{m\pi}{a}x\right)e^{-j\bar{\beta}z} \end{aligned}\right\} \tag{7-19a}$$

As in the case of transverse electric waves, there is an infinite number of modes corresponding to the various values of m from 1 to infinity. However in this case of transverse magnetic waves there is also the possibility of $m = 0$, because $m = 0$ in the above equations does

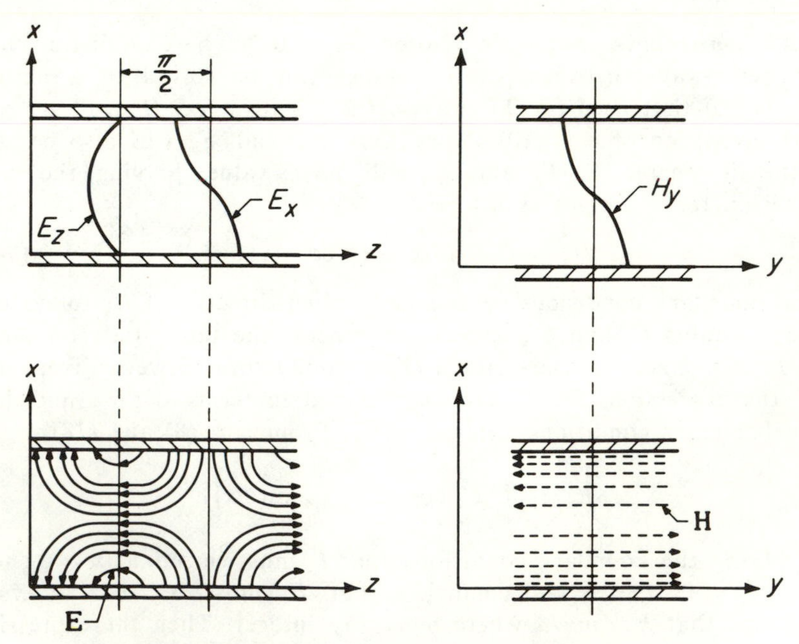

Figure 7-3. The TM_{10} wave between parallel planes.

not make all the fields vanish. This particular case of $m = 0$ will be discussed in detail in a later section. A sketch of the TM_{10} wave between parallel planes is shown in Fig. 7-3.

7.04 Characteristics of TE and TM Waves. The transverse electric and transverse magnetic waves between parallel conducting planes exhibit some interesting and rather surprising properties that seem quite different from those of uniform plane waves in free space. These properties can be studied by investigating the propagation constant $\bar{\gamma}$ for these waves.

Examination of eqs. (16) for TE waves and eqs. (19a) for TM waves shows that for each of the components of **E** or **H** there is a sinusoidal or cosinusoidal standing-wave distribution across the guide in the x direction. That is, each of these components varies in magnitude, but not in phase, in the x direction. In the y direction, by assumption, there is no variation of either magnitude or phase of any of the field components. Thus any x-y plane is an equiphase plane for each of the field components (that is, any particular component, $\tilde{E}_y$ for example, reaches its maximum value in time at the same instant for all points on the plane). Also these equiphase surfaces progress along the guide in the z direction with a velocity $\bar{v} = \omega/\bar{\beta}$, where $\bar{\beta}$, the phase-shift constant, is the imaginary part of the propagation constant $\bar{\gamma}$. Now from eq. (11), $\bar{\gamma}$ can be expressed in terms of h and frequency and the constants of the medium by

$$\bar{\gamma} = \sqrt{h^2 - \omega^2\mu\epsilon} \tag{7-20}$$

Inserting the restrictions on h imposed by eq. (14), this becomes

$$\bar{\gamma} = \sqrt{\left(\frac{m\pi}{a}\right)^2 - \omega^2\mu\epsilon} \tag{7-21}$$

Inspection of eq. (21) shows that at frequencies sufficiently high so that $\omega^2\mu\epsilon > (m\pi/a)^2$, the quantity under the radical will be negative and $\bar{\gamma}$ will be a pure imaginary equal to $j\bar{\beta}$, where

$$\bar{\beta} = \sqrt{\omega^2\mu\epsilon - \left(\frac{m\pi}{a}\right)^2} \tag{7-22}$$

Under these conditions the fields will progress in the z direction as waves, and the attenuation of these waves will be zero (for perfectly conducting planes).

As the frequency is decreased, a critical frequency $f_c = \omega_c/2\pi$ will be reached at which

$$\omega_c^2\mu\epsilon = \left(\frac{m\pi}{a}\right)^2 \tag{7-23}$$

For all frequencies less than f_c, the quantity under the radical will be positive and the propagation constant will be a real number. That is, $\bar{\alpha}$ will have value but $\bar{\beta}$ will equal zero. This means that the fields will be attenuated exponentially in the z direction and that there will be no wave motion, since the phase shift per unit length is now zero. The frequency f_c, at which wave motion ceases, is called the cut-off frequency of the guide. From eq. (23)

$$f_c = \frac{m}{2a\sqrt{\mu\epsilon}} \tag{7-24}$$

It is seen that for each value of m, there is a corresponding cut-off frequency below which wave propagation cannot occur. Above the cut-off frequency, wave propagation does occur and the attenuation of the wave is zero (for perfectly conducting planes). The phase-shift constant $\bar{\beta}$, in the range where wave propagation occurs, is given by eq. (22). It is seen that $\bar{\beta}$ varies from zero at the cut-off frequency up to the value $\omega\sqrt{\mu\epsilon}$ as the frequency approaches infinity. The distance required for the phase to shift through 2π radians is a wavelength, so that the wavelength $\bar{\lambda}$ is given in terms of $\bar{\beta}$ by

$$\bar{\lambda} = \frac{2\pi}{\bar{\beta}} \tag{7-25}$$

Also the *wave* or *phase* velocity is given by the wavelength times the frequency, so that

$$\bar{v} = \bar{\lambda}f = \frac{2\pi f}{\bar{\beta}} = \frac{\omega}{\bar{\beta}} \tag{7-26}$$

When the expression for $\bar{\beta}$ is put in eqs. (25) and (26), the wavelength and wave velocity are given by

$$\bar{\lambda} = \frac{2\pi}{\sqrt{\omega^2\mu\epsilon - (m\pi/a)^2}} \tag{7-27}$$

$$\bar{v} = \frac{\omega}{\sqrt{\omega^2\mu\epsilon - (m\pi/a)^2}} \tag{7-28}$$

It is seen that at the cut-off frequency both $\bar{\lambda}$ and $\bar{v}$ are infinitely large. As the frequency is raised above the cut-off frequency, the velocity decreases from this very large value. It approaches a lower limit,

$$\bar{v} \to v_0 = \frac{1}{\sqrt{\mu\epsilon}} \tag{7-29}$$

as the frequency becomes high enough so that $(m\pi/a)^2$ is negligible compared with $\omega^2\mu\epsilon$. When the dielectric medium between the plates is air, μ and ϵ have their free-space values μ_v and ϵ_v, and the lower limit of velocity, given by (29), is just the free-space velocity c, where as usual

$$c = \frac{1}{\sqrt{\mu_v \epsilon_v}} \approx 3 \times 10^8 \qquad \text{meter/sec}$$

Therefore the phase velocity of the wave varies from a value equal to the velocity of light in free-space up to an infinitely large value as the frequency is reduced from extremely high values down to the cut-off frequency. This *wave* velocity or *phase* velocity is different from the velocity with which the energy propagates. The distinction between these velocities will be considered in sec. 7.06.

7.05 Transverse Electromagnetic Waves. For transverse electric (TE) waves between the parallel planes, it was seen that the lowest value of m that could be used without making all the field components zero was $m = 1$. That is, the lowest-order TE wave is the TE_{10} wave. For transverse magnetic (TM) waves however, a value of m equal to zero does not necessarily require that all the fields be zero. Putting $m = 0$ in eqs. (19a) leaves

$$\left.\begin{aligned} H_y &= C_4\, e^{-j\bar{\beta} z} \\ E_x &= \frac{\bar{\beta}}{\omega\epsilon} C_4\, e^{-j\bar{\beta} z} \\ E_z &= 0 \end{aligned}\right\} \tag{7-30}$$

For this special case of transverse magnetic waves the component of **E** in the direction of propagation, that is E_z, is also zero so that the electromagnetic field is *entirely transverse*. Consistent with previous notation this wave is called the transverse electromagnetic (TEM) wave. Although it is a special case of guided-wave propagation, it is an extremely important one, because it is the familiar type of wave propagated along all ordinary two-conductor transmission lines when operating in their customary (low-frequency) manner. It is usually called the *principal wave*.

There are several interesting properties of TEM waves which follow as special cases of the more general TE or TM types of waves. For the TEM waves between the parallel planes it is seen from eqs. (30) that not only are the fields entirely transverse, but they are constant in amplitude across a cross section normal to the direction of propagation; of course, their ratio is also constant. For $m = 0$ and an air dielectric, the expressions for $\bar{\gamma}$, $\bar{\beta}$, $\bar{v}$, and $\bar{\lambda}$ reduce to

$$\left.\begin{aligned} \bar{\gamma} &\to \gamma = j\omega\sqrt{\mu_v \epsilon_v} \\ \bar{\beta} &\to \beta = \omega\sqrt{\mu_v \epsilon_v} \\ \bar{v} &\to v = \frac{1}{\sqrt{\mu_v \epsilon_v}} = c \\ \bar{\lambda} &\to \lambda = \frac{2\pi}{\omega\sqrt{\mu_v \epsilon_v}} = \frac{c}{f} \end{aligned}\right\} \tag{7-31}$$

Unlike TE and TM waves, the velocity of the TEM wave is independent of frequency and has the familiar free-space value, $c \approx 3 \times 10^8$ meter/sec. (It has this value only when the planes are perfectly conducting and the space between them is a vacuum. The effect of finite conductivity for the conducting planes is to reduce the velocity slightly. This effect will be considered in sec. 7.09.) Also from eq. (24), the cut-off frequency for the TEM wave is zero. This means that for transverse electromagnetic waves, all frequencies down to zero can propagate along the guide. The ratio of E to H between the parallel planes for a traveling wave is

$$\left|\frac{E_x}{H_y}\right| = \frac{E_x}{H_y} = \frac{\bar{\beta}}{\omega\epsilon} = \sqrt{\frac{\mu_v}{\epsilon_v}} \tag{7-32}$$

which is just the intrinsic impedance, η_v, of free space.

A sketch of the TEM wave between parallel planes is shown in Fig. 7-4.

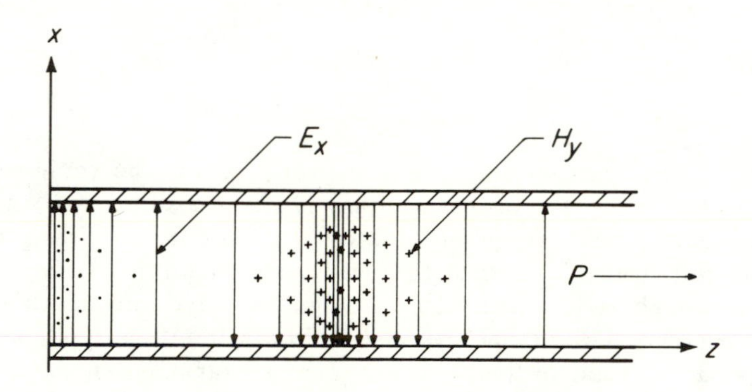

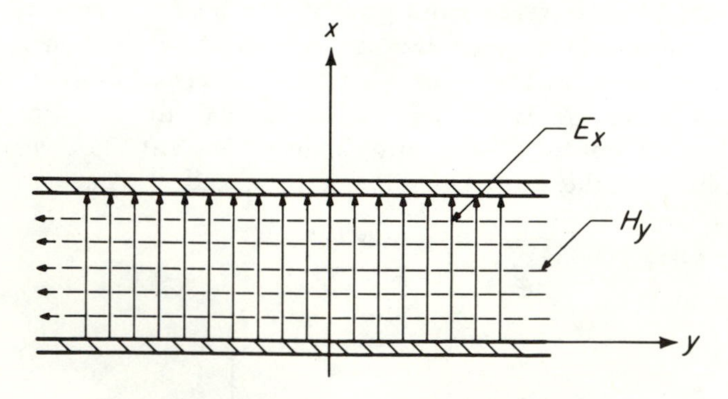

Figure 7-4. The TEM wave between parallel planes.

7.06 Velocities of Propagation. It was seen that except for the TEM wave, the velocity with which an electromagnetic wave propagates (in an air dielectric) between a pair of parallel planes is always greater than c, the free-space velocity of electromagnetic waves. In actual rectangular or cylindrical wave guides (to be considered in chap. 8), the TEM wave cannot exist and the *wave* or *phase velocity* is always greater than the free-space velocity. On the other hand, the *velocity* with which the *energy* propagates along a guide is always less than the free-space velocity. The relation between these velocities is made clear by consideration of a simple and well-known illustration. Figure 7-5 might be considered to represent water waves approaching

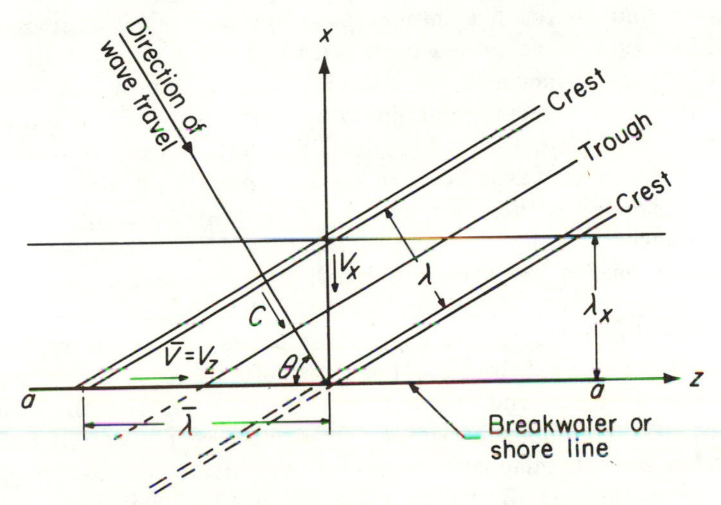

Figure 7-5. Water wave approaching a breakwater.

the shore line or a breakwater *a-a* at an angle θ. The velocity of the waves could be determined by measuring the distance λ between successive crests and recording the frequency f with which the crests passed a given observation point. The velocity c with which the waves are traveling would be given by

$$c = \lambda f$$

Alternatively, if one wished to determine the velocity c without going into the water, this could be done by measuring the angle θ and the velocity $\bar{v} = v_z$ with which the crests move *along the shore line* (in the z direction). This velocity would be given by

$$\bar{v} = \bar{\lambda} f$$

where $\bar{\lambda}$ is now the distance between crests *along the shore line.* Evidently $\bar{v}$ and $\bar{\lambda}$ are greater than c and λ respectively, and are related to them by

$$\bar{\lambda} = \frac{\lambda}{\cos \theta} \tag{7-33a}$$

$$\bar{v} = \frac{c}{\cos \theta} \tag{7-33b}$$

When the direction of wave-travel is nearly parallel to the shore, that is, when the angle θ is small, the velocity $\bar{v}$ with which the crests move along the shore line is very nearly equal to c, the free-space velocity of the waves. However, when the angle θ is near 90 degrees the velocity with which the crests advance along the shore line is very great, and approaches infinity as θ approaches 90 degrees.

Consider now wave propagation within a wave guide. It is always possible, though sometimes not too practical, to obtain the field configuration within a rectangular guide by superposing two or more plane waves in a suitable manner. For the TE_{m0} waves in rectangular guides and for these same waves between parallel planes as already considered, this separation into component waves is quite simple. It is left for the student to show (problem 2 on page 240) that two uniform plane waves having the same amplitude and frequency, but opposite phases can be added to produce the field distributions of the TE_{m0} waves. The directions of the component waves are as shown in Fig. 7-6, where the angle θ between the walls of the guide and the direction of the waves depends upon the frequency and the dimension a. For each of the component waves the electric vector **E** will be in the y direction and the magnetic vector **H** will lie in the x-z plane and will be perpendicular to the direction of travel of that wave. In order

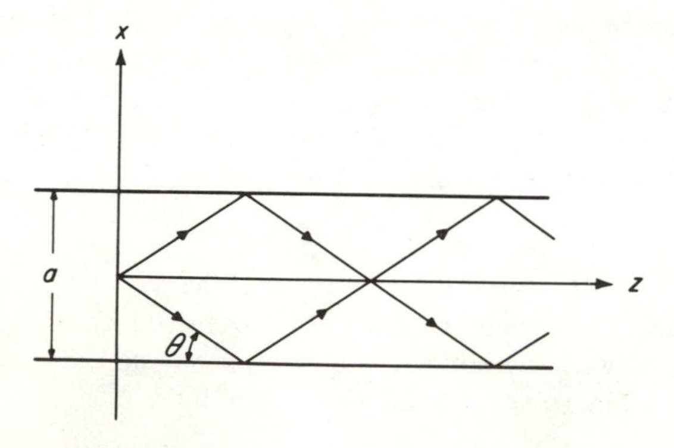

Figure 7-6. Direction of travel of the component uniform plane waves between parallel planes.

to satisfy the boundary conditions at the walls of the guide, the electric fields due to the two component waves must add to zero at those surfaces. The only way in which it is possible to have E_y equal to zero at the walls and still have values of E_y at points between the walls is to have a standing-wave distribution of E_y across the guide, with the nodal points of the standing wave occurring at the wall surfaces. This condition requires that a, the separation between the walls, must be some multiple of a half-wavelength measured in the direction perpendicular to the walls. Referring again to Fig. 7-5 the required condition is that

$$\frac{m\lambda_x}{2} = a$$

where m is an integer and where λ_x is the distance between crests measured in the x direction. Since $\lambda_x = \lambda/\sin\theta$, it is seen that the condition on θ is

$$\sin\theta = \frac{m\lambda}{2a} \tag{7-34}$$

Because the sine cannot be greater than unity, it is apparent that a, the separation between the walls must be greater than $\lambda/2$, where λ is the free-space wavelength of the wave. The wavelength for which

$$\lambda = \frac{2a}{m} \tag{7-35}$$

is the cut-off wavelength for that value of m. At the cut-off wavelength $\sin\theta$ is unity and θ is 90 degrees. That is, the waves bounce back and forth between the walls of the guide, and there is no wave motion parallel to the axis. As λ is decreased from the cut-off value, θ also decreases, so that at wavelengths much shorter than cut-off (very high frequency) the waves travel almost parallel to the axis of the guide.

The wavelength $\bar{\lambda} = \lambda_z$, *parallel* to the walls of the guide, which is the wavelength ordinarily measured in wave-guide work, is given by

$$\bar{\lambda} = \frac{\lambda}{\cos\theta} = \frac{\lambda}{\sqrt{1 - (m\lambda/2a)^2}} \tag{7-36}$$

This is the distance between equiphase points in the direction of the axis of the guide. The phase velocity in this direction is

$$\bar{v} = \frac{c}{\cos\theta} = \frac{c}{\sqrt{1 - (m\lambda/2a)^2}} \tag{7-37}$$

It is evident that because of the zig-zag path traveled by each of the component waves, the velocity,* v_g, with which the energy propagates along the axis of the guide will be less than the free-space velocity c. In terms of the angle θ, for a guide with an air dielectric, it will be

$$v_g = c \cos \theta$$

In terms of the width dimension a in wavelengths, it is

$$v_g = c\sqrt{1 - \left(\frac{m\lambda}{2a}\right)^2} \tag{7-38}$$

It will be noted that the product of the phase velocity and the velocity with which the energy propagates is equal to the square of the free-space velocity, that is,

$$\bar{v}v_g = c^2 \tag{7-39}$$

As the frequency is reduced toward the cut-off frequency, the angle θ approaches 90 degrees, so that the phase velocity $\bar{v}$ becomes very large, and the velocity with which the energy propagates becomes very small. At the cut-off frequency $\bar{v}$ is infinite, but v_g is zero, that is propagation of energy along the guide *by wave motion* ceases.†

For a (lossless) dielectric in the guide having permittivity ϵ and permeability μ, different from ϵ_v and μ_v, the velocity c must be replaced by $v_0 = 1/\sqrt{\mu\epsilon}$.

7.07 Attenuation in Parallel-plane Guides. The problem of wave propagation between parallel conducting planes has been solved for the theoretical case of *perfect* conductors, and the solutions appear as eqs. (16a), (19a), and (30) for the TE_{m0},TM_{m0} and TEM modes respectively. In actual wave guides the conductivity of the walls is usually very large, but it is rarely infinite, and there are usually some losses. These losses will modify the results obtained for the lossless case by the introduction of the multiplying factor $e^{-\alpha z}$ in eqs. (16a), (19a), and (30). The problem now is to determine this attenuation factor α that is caused by losses in the walls of the guide.‡

*This velocity, v_g, is the the group velocity. The terms *phase* velocity, *group* velocity, and *signal* velocity are discussed in more detail in Appendix I.

†This is not to say that there are no fields within the guide. In sec. 7.04 it was seen that below cut-off frequency $\bar{\gamma}$ is real, so that $\bar{\alpha}$ has value and $\bar{\beta}$ is zero. This means that the fields then penetrate into the guide with an exponential decrease in amplitude, and with no phase shift (for the infinitely long guide with perfectly conducting walls). A wave guide operated in this manner is known as an *attenuator.*

‡This attenuation factor α for the *propagating wave* is normally a small quantity. It should not be confused with the very large attenuation factor $\bar{\alpha}$ given by the real part of (21) which applies below the cut-off frequency.

In order to see how α may be evaluated for wave guides, consider the familiar problem of attenuation in ordinary two-conductor transmission lines. For any line with uniformly distributed constants, the voltage and current phasors along the line (when the line is terminated in its characteristic impedance) are

$$V = V_0 e^{-\alpha z} e^{-j\beta z} \tag{7-40}$$

$$I = I_0 e^{-\alpha z} e^{-j\beta z} \tag{7-41}$$

and the average power transmitted is

$$W_{av} = \tfrac{1}{2} \operatorname{Re} \{VI^*\}$$

$$= \tfrac{1}{2} \operatorname{Re} \{V_0 I_0^*\} e^{-2\alpha z} \tag{7-42}$$

The rate of decrease of transmitted power along the line will be

$$-\frac{\partial W_{av}}{\partial z} = +2\alpha W_{av} \tag{7-43}$$

The decrease of transmitted power per *unit length* of line is

$$-\Delta W_{av} = 2\alpha W_{av}$$

and this must be equal to the power lost or dissipated per unit length. Therefore

$$\frac{\text{Power lost per unit length}}{\text{Power transmitted}} = \frac{2\alpha W_{av}}{W_{av}} = 2\alpha$$

so that

$$\alpha = \frac{\text{Power lost per unit length}}{2 \times \text{power transmitted}} \tag{7-44}$$

Using eq. (44), the attenuation factor can be determined for more general cases of guided wave transmission where the terms "voltage" and "current" may no longer apply.

The computation of power loss in a wave guide appears at first glance to be a rather difficult problem, because the loss depends upon the field configuration within the guide, and the field configuration, in turn, depends to some extent upon the losses. The attack on this problem is one that is used quite often in engineering. It is first assumed that the losses will have negligible effect upon the field distribution within the guide. Using the field distributions calculated for the lossless case, the magnetic field strength tangential to each conducting surface is used to determine the current flow in that surface. Using this value of current and the known resistance of the walls, the losses are computed and α is determined from (44). If desired, a second and closer approximation could then be made, using a field distribution corrected to account for the calculated losses. However, for metallic conductors of high conductivity such as copper or brass, the first approximation yields quite accurate results, and a second approximation is rarely necessary.

EXAMPLE 1: *Attenuation Factor for the TEM Wave.* The expressions obtained for magnetic and electric fields between parallel perfectly conducting planes (Fig. 7-1) in the case of the TEM mode were

$$H_y = C_4\, e^{-j\beta z}$$

$$E_x = \eta C_4\, e^{-j\beta z} \tag{7-30a}$$

The linear current density in each of the conducting planes will be given by

$$\mathbf{J}_s = \hat{\mathbf{n}} \times \mathbf{H}$$

so the amplitude of the linear current density in each plane is

$$J_s = C_4$$

The loss per square meter in each conducting plane is

$$\tfrac{1}{2}\, J_s^2 R_s = \tfrac{1}{2}\, C_4^2 R_s$$

where

$$R_s = \sqrt{\frac{\omega\mu_m}{2\sigma_m}}$$

is the resistive component of the surface impedance given by the expression

$$Z_s = \sqrt{\frac{j\omega\mu_m}{\sigma_m}}$$

μ_m and σ_m refer of course to values in the metallic conductor. The total loss in the upper and lower conducting surfaces per meter length for a width b meters of the guide is

$$C_4^2 R_s b$$

The power transmitted down the guide per unit cross-sectional area is

$$\tfrac{1}{2}\,\mathrm{Re}\,(\mathbf{E} \times \mathbf{H}^*)_z \tag{7-45}$$

E_x and H_y are at right angles and in time phase and $|E_x| = \eta|H_y|$, so (45) reduces to

$$\tfrac{1}{2}\,\eta C_4^2$$

For a spacing a meters between the planes the cross-section area of a width b meters of the guide is ba square meters and the power transmitted through this area is

$$\text{Power transmitted} = \tfrac{1}{2}\,\eta C_4^2 ba$$

From (44) the attenuation factor is

$$\alpha = \frac{C_4^2 R_s b}{2 \times \frac{1}{2}\eta C_4^2 ba}$$

$$= \frac{R_s}{\eta a} = \frac{1}{\eta a}\sqrt{\frac{\omega\mu_m}{2\sigma_m}} \qquad \text{nepers/meter} \tag{7-46}$$

This expression should be compared with the corresponding expression for the attenuation factor of an ordinary transmission line (eq. 7-121), which is

$$\alpha = \frac{R}{2Z_0}$$

where R is the resistance per unit length of the line (that is *twice* the conductor resistance).

EXAMPLE 2: *Attenuation of TE Waves.* The expressions for E and H for the transverse electric modes between perfectly conducting parallel planes (Fig. 7-1) are

$$\left.\begin{aligned} E_y &= C_1 \sin\left(\frac{m\pi}{a}x\right) e^{-j\bar{\beta}z} \\ H_x &= -\frac{\bar{\beta}}{\omega\mu} C_1 \sin\left(\frac{m\pi}{a}x\right) e^{-j\bar{\beta}z} \\ H_z &= \frac{jm\pi}{\omega\mu a} C_1 \cos\left(\frac{m\pi}{a}x\right) e^{-j\bar{\beta}z} \end{aligned}\right\} \tag{7-16a}$$

The amplitude of linear current density in the conducting planes will be equal to the tangential component of H (i.e., H_z) at $x = 0$ and $x = a$

$$|J_{sy}| = |H_z| \qquad (\text{at } x = 0,\ x = a)$$

$$= \frac{m\pi C_1}{\omega\mu a}$$

It is interesting to note in passing that for these modes there is no flow of current in the direction of wave propagation. The loss in each plate is

$$\tfrac{1}{2} J_{sy}^2 R_s = \frac{m^2\pi^2 C_1^2\sqrt{\omega\mu_m/2\sigma_m}}{2\omega^2\mu^2 a^2} \tag{7-47}$$

The power transmitted in the z direction through an element of area $da = dx\,dy$ is

$$\text{Power transmitted per unit area} = \tfrac{1}{2}\,\text{Re}\,(\mathbf{E} \times \mathbf{H}^*)\cdot\hat{\mathbf{z}}$$

$$= -\tfrac{1}{2}(E_y H_x)$$

$$= \frac{\bar{\beta}C_1^2}{2\omega\mu}\sin^2\left(\frac{m\pi}{a}x\right)$$

Power transmitted in the z direction for a guide 1 meter wide with a spacing between conductors of a meters is

$$\int_{x=0}^{x=a} \frac{\bar{\beta}C_1^2}{2\omega\mu}\sin^2\left(\frac{m\pi}{a}x\right) dx = \frac{\bar{\beta}C_1^2 a}{4\omega\mu} \tag{7-48}$$

Dividing twice expression (47) by twice expression (48), the attenuation factor is

$$\alpha = \frac{2m^2\pi^2\sqrt{\omega\mu_m/2\sigma_m}}{\bar{\beta}\omega\mu a^3}$$

Recalling that $\bar{\beta} = \sqrt{\omega^2\mu\epsilon - (m\pi/a)^2}$ the expression for attenuation factor for TE waves between parallel conducting planes for frequencies above cut-off is

$$\alpha = \frac{2m^2\pi^2\sqrt{\omega\mu_m/2\sigma_m}}{\omega\mu a^3\sqrt{\omega^2\mu\epsilon - (m\pi/a)^2}} \tag{7-49}$$

The value of this expression decreases from infinity at cut-off to quite low values at higher frequencies. For frequencies very much higher than cut-off the attenuation varies inversely as the three-halves power of the frequency.

Attenuation Factor for TM Waves. The expression for the attenuation factor for TM waves between parallel conducting planes can be obtained in a similar manner. It differs from expression (49) in that the attenuation reaches a minimum at a frequency that is $\sqrt{3}$ times the cut-off frequency (f_c) and then increases with frequency. At frequencies much higher than cut-off the attenuation of the TM modes increases directly as the square root of frequency.

A sketch of variation of attenuation with frequency for different modes propagating between parallel conducting planes is shown in Fig. 7-7.

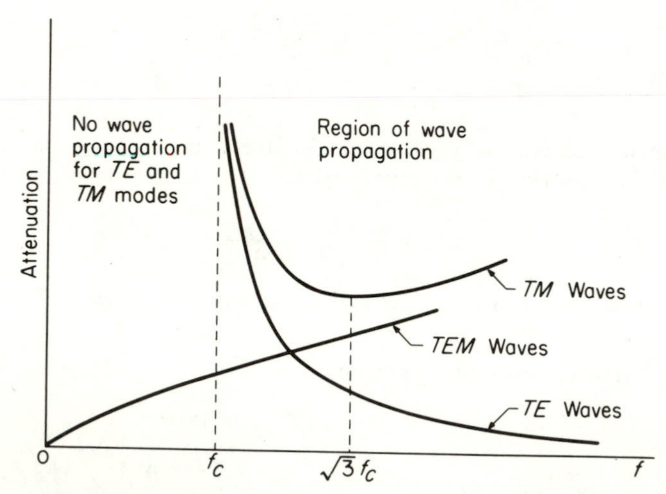

Figure 7-7. Attenuation-versus-frequency characteristics of waves guided between parallel conducting plates.

7.08 Wave Impedances. In ordinary transmission-line theory, extensive use is made of the "characteristic impedance," Z_0, of the line.

This impedance gives the ratio of voltage to current (for an infinitely long line), and its real part is a measure of the power transmitted for a given amplitude of current. In transmission-line theory power is propagated along one axis only, and only one impedance constant is involved. However, in three-dimensional wave propagation power may be transmitted along any or all of the three axes of the co-ordinate system, and consequently three impedance constants must be defined. For example, in the Cartesian co-ordinate system the complex power per unit area transmitted in the x, y, and z directions respectively is given by

$$P_x = \tfrac{1}{2}(E_y H_z^* - E_z H_y^*) \qquad P_y = \tfrac{1}{2}(E_z H_x^* - E_x H_z^*)$$
$$P_z = \tfrac{1}{2}(E_x H_y^* - E_y H_x^*)$$

The real or average Poynting vector in any of the three directions is given by the real part of the appropriate expression. It is now convenient to define the *wave impedances* at a point by the following ratios of electric to magnetic field strengths:

$$\left.\begin{aligned} Z_{xy}^+ &= \frac{E_x}{H_y} \qquad & Z_{yz}^+ &= \frac{E_y}{H_z} \qquad & Z_{zx}^+ &= \frac{E_z}{H_x} \\ Z_{yx}^+ &= -\frac{E_y}{H_x} \qquad & Z_{zy}^+ &= -\frac{E_z}{H_y} \qquad & Z_{xz}^+ &= -\frac{E_x}{H_z} \end{aligned}\right\} \tag{7-50}$$

These are the wave impedances looking along the *positive* directions of the co-ordinates, and this fact is indicated by the superscript plus sign. The impedances in the opposite directions are the negative of those given above, and the negative direction is indicated by a superscript minus sign. Thus in the directions of *decreasing* co-ordinates

$$\begin{aligned} Z_{xy}^- &= -\frac{E_x}{H_y} \qquad & Z_{yz}^- &= -\frac{E_y}{H_z} \qquad & Z_{zx}^- &= -\frac{E_z}{H_x} \\ Z_{yx}^- &= \frac{E_y}{H_x} \qquad & Z_{zy}^- &= \frac{E_z}{H_y} \qquad & Z_{xz}^- &= \frac{E_x}{H_z} \end{aligned}$$

Corresponding definitions would obtain for any orthogonal co-ordinate system. In terms of these wave impedances the x, y, and z components of the complex Poynting vector are

$$P_x = \tfrac{1}{2}(Z_{yz}^+ H_z H_z^* + Z_{zy}^+ H_y H_y^*) = -\tfrac{1}{2}(Z_{yz}^- H_z H_z^* + Z_{zy}^- H_y H_y^*)$$
$$P_y = \tfrac{1}{2}(Z_{zx}^+ H_x H_x^* + Z_{xz}^+ H_z H_z^*) = -\tfrac{1}{2}(Z_{zx}^- H_x H_x^* + Z_{xz}^- H_z H_z^*)$$
$$P_z = \tfrac{1}{2}(Z_{xy}^+ H_y H_y^* + Z_{yx}^+ H_x H_x^*) = -\tfrac{1}{2}(Z_{xy}^- H_y H_y^* + Z_{yx}^- H_x H_x^*)$$

The subscripts on the wave impedances indicate the particular components of **E** and **H** involved, and the algebraic signs of the wave impedances have been chosen so that, if the real part of any given impedance is positive, the corresponding average power flow is in the direction indicated by the impedance.

Applying these definitions to waves propagating between parallel planes the wave impedance in the direction of propagation can be found. For the TEM wave (the exceptional case where both E and H are transverse), the wave impedance is given by eq. (32), and it is seen to be equal to η_v, the same as for a uniform plane wave in free space. For TE waves, the wave impedance can be obtained from eq. (17). It is

$$Z_{yx}^{+} = -\frac{E_y}{H_x} = \frac{j\omega\mu}{\bar{\gamma}} \tag{7-51}$$

where

$$\bar{\gamma} = \sqrt{\left(\frac{m\pi}{a}\right)^2 - \omega^2\mu\epsilon}$$

The wave impedance in the z direction is constant over the cross section of the guide. For frequencies below cut-off for which $\bar{\gamma}$ is real, the impedance is a pure reactance indicating no acceptance of power by the guide and therefore no transmission down the guide. For frequencies above cut-off $\bar{\gamma}$ is a pure imaginary (under the assumption of perfectly conducting walls) and can be written

$$\bar{\gamma} = j\bar{\beta} = j\sqrt{\omega^2\mu\epsilon - \left(\frac{m\pi}{a}\right)^2}$$

so that

$$Z_{yx}^{+} = \frac{\omega\mu}{\sqrt{\omega^2\mu\epsilon - (m\pi/a)^2}} = \frac{\omega\mu}{\bar{\beta}} \tag{7-52}$$

The wave impedance is real and decreases from an infinitely large value at cut-off toward the asymptotic value of $\eta = \sqrt{\mu/\epsilon}$ as the frequency increases to values much higher than cut-off.

These results could equally well have been obtained by considering the TE wave as being made of two uniform plane waves reflected back and forth between the conducting planes and making an angle θ with the axis of propagation (Fig. 7-6). For the TE wave propagating in the positive z direction the transverse component of **E** will be E_y, whereas the transverse component of **H** will be $-H_x = -H \cos\theta$; therefore the wave impedance in the z direction is

$$Z_{yx}^{+} = -\frac{E_y}{H_x} = -\frac{E}{H\cos\theta} = \frac{\eta}{\cos\theta} \tag{7-53}$$

Making use of eq. (37), this may be written as

$$Z_{yx}^{+} = \frac{\bar{v}}{c}\eta = \frac{\omega\mu}{\bar{\beta}}$$

which is the same result as was obtained in (52).

For TM waves the transverse component of **E** will be $E_x = E\cos\theta$, whereas the transverse component of **H** will be H_y. The wave impedance for this case is

$$Z^+_{xy} = \frac{E_x}{H_y} = \frac{E\cos\theta}{H} = \eta\cos\theta \tag{7-54}$$

It varies from zero at the cut-off frequency up to the asymptotic value η for frequencies much higher than cut-off.

There is a marked resemblance between the properties of these wave impedances and the characteristic impedances of the prototype T or π sections in ordinary filter theory. For example, the wave impedance for TE waves between parallel planes may be written as

$$Z^+_{yx} = \frac{\eta}{\cos\theta} = \frac{\eta}{\sqrt{1 - \sin^2\theta}} \tag{7-55}$$

Making use of the relations

$$\sin\theta = \frac{m\lambda}{2a}, \qquad \lambda_c = \frac{2a}{m}, \qquad f_c = \frac{1}{\lambda_c\sqrt{\mu\epsilon}}$$

where λ_c and f_c are the cut-off wavelength and cut-off frequency, eq. (55) becomes

$$Z^+_{yx} = \frac{\eta}{\sqrt{1 - (f_c/f)^2}} \tag{7-56}$$

This is similar to the expression for the characteristic impedance of the prototype π section of a high-pass filter, which is

$$Z_{0\pi} = \frac{\sqrt{L/C}}{\sqrt{1 - (f_c/f)^2}} \tag{7-57}$$

Similarly the expression for the wave impedance for TM waves between parallel planes may be written as

$$Z^+_{xy} = \eta\sqrt{1 - \left(\frac{f_c}{f}\right)^2} \tag{7-58}$$

which corresponds to the expression for characteristic impedance of the prototype T section of a high-pass filter,

$$Z_{0T} = \sqrt{\frac{L}{C}}\sqrt{1 - \left(\frac{f_c}{f}\right)^2} \tag{7-59}$$

The wave impedances for waves between parallel planes are shown as functions of frequency in Fig. 7-8. In chap. 8 a general transmission-line analogy will be developed for TM and TE waves in cylindrical guides of any cross-sectional shape.

In this chapter the characteristics of waves propagating between two parallel planes have been considered in some detail. The concepts

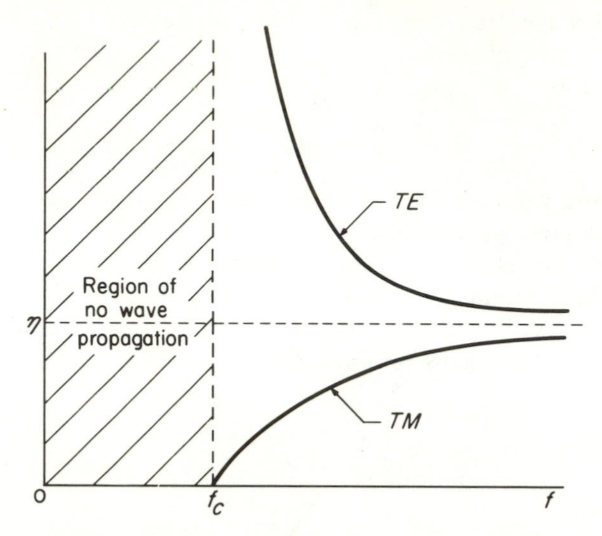

Figure 7-8. Wave impedances for waves between parallel conducting planes.

developed in the treatment of this simple illustrative system are quite general and may be extended to apply to all guided systems, including "ordinary" two-conductor transmission lines and also enclosed wave guides. Before leaving this simple system some consideration will be given to the electric field configuration and current flow within the metal walls of the guiding system.

7.09 Electric Field and Current Flow Within the Conductor. When an electromagnetic wave is guided along the surface of a conductor, currents flow in the conductor and charges appear and disappear on its surface. The current distribution within the conductor and the charge distribution on the surface can be obtained from a straightforward solution of Maxwell's equations, subject to the appropriate boundary conditions at the boundary surface between the dielectric and the conductor. However, the results are somewhat complex and require interpretation. For this reason, before obtaining the exact solution, it is advantageous to consider in a qualitative manner, and from facts already known, certain features of the problem.

In Fig. 7-9 a TEM wave is guided along the surface (in the x-z plane) of a conductor which, for the moment, will be assumed to be perfectly conducting. For the case considered the electric field strength, $\mathbf{E} = \hat{\mathbf{y}}E_y$, will be normal to the surface, and the magnetic field strength $\mathbf{H} = -\hat{\mathbf{x}}H_x$ will be parallel to the surface. There will be a surface current J_{sz}, flowing in the z direction, and related to the magnetic

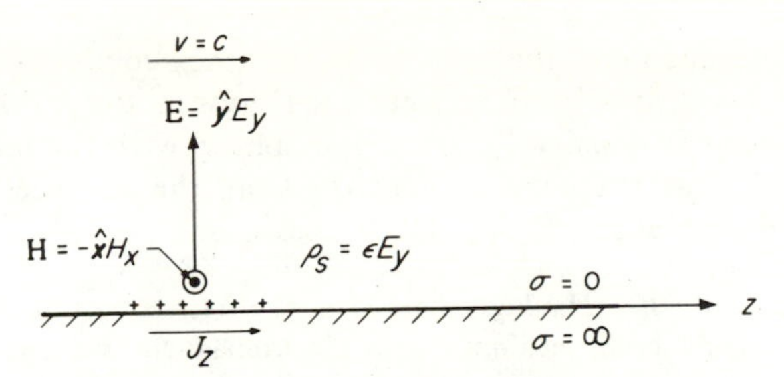

Figure 7-9. Current and surface charge on a perfect conductor guiding an electromagnetic wave.

field strength by the vector relation $\mathbf{J}_s = \mathbf{n} \times \mathbf{H}$, which in this case becomes $J_{sz} = -H_x$. Since $E_y = -\eta_v H_x$, it follows that

$$J_{sz} = \frac{E_y}{\eta_v} \tag{7-60}$$

A surface charge density appears on the surface, the value of which is given by

$$\rho_s = D_y = \epsilon_v E_y \tag{7-61}$$

From (60) and (61) it is seen that both ρ_s and J_{sz} are proportional to E_y, so that at any instant of time the position of maximum charge occurs at the same value of z as the position of maximum surface current.

When the conductivity of the conductor is reduced from infinity to a large but finite value such as obtains for ordinary metallic conductors, the situation is modified in several respects. The chief effect is the introduction of a small tangential component of $\mathbf{E}$, which is required to drive the linear current density $\mathbf{J}_s$ against the surface impedance Z_s of the conductor. Making the assumption (known to be very good) that H will not be changed appreciably by the finite rather than infinite conductivity, the tangential component of $\mathbf{E}$ can be obtained from

$$\begin{aligned} E_z &= J_{sz} Z_s = -H_x Z_s \\ &= -H_x \sqrt{\frac{j\omega\mu}{\sigma}} = -H_x \sqrt{\frac{\omega\mu}{\sigma}} \quad \underline{/45^\circ} \end{aligned}$$

The horizontal or tangential component of E is seen to lead $-H_x$ and therefore E_y by an angle of 45 degrees. The conductor is considered to be sufficiently good that the inequality $\sigma \gg \omega\epsilon$ holds for all frequencies considered. The depth of penetration, although small for good conductors, is not zero, and the linear current density J_{sz} is now distributed throughout the thickness of the conductor, with approximately two-thirds of it concentrated within the "skin depth" δ. The linear current density J_{sz} is still in phase with the magnetic field strength $-H_x$, but the current density J_z at the surface is in phase with E_z, and so leads $-H_x$ by 45 degrees.

RIGOROUS SOLUTION. Having obtained a qualitative picture of what happens within a conducting medium as an electromagnetic wave is guided along its surface, it is now in order to set up and obtain a more rigorous solution as a boundary-value problem. The problem is that of finding solutions to Maxwell's equations in regions 0 and 1 (Fig. 7-10), which will fit the boundary conditions at the surface of the conductor.

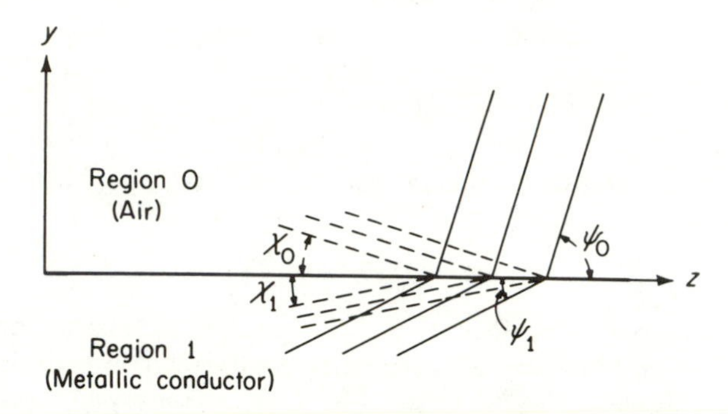

Figure 7-10. Equiphase lines (solid) and equiamplitude lines (dashed) for an electromagnetic wave guided along a conducting plane.

The following assumptions will be made:

(1) No variations in the x direction. Therefore $\partial/\partial x \equiv 0$.

(2) Variations in the z direction can be represented by $e^{-\gamma_0 z}$ in the dielectric and by $e^{-\gamma_1 z}$ in the metal. The values of γ_0 and γ_1 must come out of the solution.

(3) Variations in the y direction are as yet unknown and must be solved for.

Then Maxwell's equations become:

Above the surface (Region 0):

$$\left.\begin{aligned} \frac{\partial E_z}{\partial y} + \gamma_0 E_y &= -j\omega\mu_v H_x \\ -\gamma_0 H_x &= j\omega\epsilon_v E_y \\ -\frac{\partial H_x}{\partial y} &= j\omega\epsilon_v E_z \end{aligned}\right\} \qquad (7\text{-}62a)$$

Within the conductor (Region 1):

$$\left.\begin{aligned} \frac{\partial E_z}{\partial y} + \gamma_1 E_y &= -j\omega\mu_1 H_x \\ -\gamma_1 H_x &= (\sigma_1 + j\omega\epsilon_1)E_y \\ -\frac{\partial H_x}{\partial y} &= (\sigma_1 + j\omega\epsilon_1)E_z \end{aligned}\right\} \qquad (7\text{-}62b)$$

Combining gives

$$\frac{\partial^2 H_x}{\partial y^2} + \gamma_0^2 H_x = -\omega^2 \mu_v \epsilon_v H_x \qquad \frac{\partial^2 H_x}{\partial y^2} + \gamma_1^2 H_x = j\omega\mu_1(\sigma_1 + j\omega\epsilon_1)H_x$$

or

$$\frac{\partial^2 H_x}{\partial y^2} = h_0^2 H_x \qquad \frac{\partial^2 H_x}{\partial y^2} = h_1^2 H_x$$

where $h_0^2 = -\gamma_0^2 - \omega^2 \mu_v \epsilon_v$ $\qquad h_1^2 = -\gamma_1^2 + \gamma_m^2$

and $\gamma_m^2 = j\omega\mu_1(\sigma_1 + j\omega\epsilon_1)$

Solutions to these differential equations may be written as

$$H_x = C_1 e^{h_0 y} + C_2 e^{-h_0 y} \quad \text{(7-63a)} \qquad H_x = C_3 e^{h_1 y} + C_4 e^{-h_1 y} \quad \text{(7-63b)}$$

In taking the square root of h^2, if it is agreed that that root which has a positive real part will be used, then only the second term of (63a) need be considered. The first term represents a field which becomes infinitely large at $y = \infty$. Since this could not represent a physical field, this first term will be discarded by putting $C_1 = 0$. Similarly within the conductor, the second term represents a nonphysical field that becomes infinite at $y = -\infty$. Therefore C_4 can be put equal to zero Showing the variation in the z direction, the expressions for magnetic field strength can now be written:

Above the surface:

$$H_x = C_2 e^{-h_0 y} e^{-\gamma_0 z}$$

Below the surface:

$$H_x = C_3 e^{+h_1 y} e^{-\gamma_1 z}$$

At the surface ($y = 0$), these expressions must be equal at all instants of time and for all values of z, because H_x must be continuous across the boundary. This requires that $C_2 = C_3$ and $\gamma_0 = \gamma_1$. Then the expressions for vertical and horizontal components of electric field strength can be written:

Above the surface:

$$E_y = \frac{-\gamma_0 C_2}{j\omega\epsilon_v} e^{-h_0 y} e^{-\gamma_0 z}$$

$$E_z = \frac{h_0 C_2}{j\omega\epsilon_v} e^{-h_0 y} e^{-\gamma_0 z}$$

In the conductor:

$$E_y = -\frac{\gamma_0 C_2}{\sigma_1 + j\omega\epsilon_1} e^{h_1 y} e^{-\gamma_0 z}$$

$$E_z = -\frac{h_1 C_2}{\sigma_1 + j\omega\epsilon_1} e^{h_1 y} e^{-\gamma_0 z}$$

At $y = 0$ the expressions for E_z must be equal. Therefore

$$\frac{h_0}{j\omega\epsilon_v} = \frac{-h_1}{\sigma_1 + j\omega\epsilon_1} \approx \frac{-h_1}{\sigma_1} \qquad \text{(for metallic conductors)}$$

$$h_0 = -\frac{j\omega\epsilon_v h_1}{\sigma_1} \qquad h_0^2 = \frac{-\omega^2\epsilon_v^2}{\sigma_1^2} h_1^2$$

$$\gamma_0^2 = -\omega^2 \mu_v \epsilon_v - h_0^2 = -\omega^2 \mu_v \epsilon_v + \frac{\omega^2 \epsilon_v^2}{\sigma_1^2}(-\gamma_0^2 + \gamma_m^2)$$

From this,

$$\gamma_0 \approx \sqrt{-\omega^2 \mu_v \epsilon_v \left(1 - \frac{j\omega\mu_1\epsilon_v}{\mu_v \sigma_1}\right)}$$

$$\approx \frac{j\omega}{c}\sqrt{1 - \frac{j\omega\mu_1\epsilon_v}{\mu_v\sigma_1}}$$

For nonferrous metallic conductors $\mu_1 \approx \mu_v$, $\epsilon_1 \approx \epsilon_v$, so that

$$\gamma_0 = \gamma_1 \approx \frac{j\omega}{c}\sqrt{1 - \frac{j\omega\epsilon_v}{\sigma_1}} \approx \frac{\omega^2\epsilon_v}{2c\sigma_1} + j\frac{\omega}{c}\left(1 + \frac{\omega^2\epsilon_v^2}{8\sigma_1^2}\right)$$

$$h_0^2 \approx -\frac{\omega^2}{c^2}\left(\frac{j\omega\epsilon_v}{\sigma_1}\right) \qquad h_0 \approx -\frac{j\omega}{c}\sqrt{\frac{j\omega\epsilon_v}{\sigma_1}} = (1 - j)\frac{\omega}{c}\sqrt{\frac{\omega\epsilon_v}{2\sigma_1}}$$

$$h_1 \approx \sqrt{j\omega\mu_v\sigma_1} = (1 + j)\sqrt{\frac{\omega\mu_v\sigma_1}{2}}$$

The resultant expressions for the fields in the two regions are:

Above the conductor:

$$\left.\begin{aligned} H_x &= C_2 e^{-h_0 y} e^{-\gamma_0 z} \\ E_y &= \frac{-\gamma_0}{j\omega\epsilon_v} H_x \approx -\eta_v H_x \\ E_z &= \frac{h_0}{j\omega\epsilon_v} H_x \approx -\eta_v\sqrt{\frac{j\omega\epsilon_v}{\sigma_1}}\, H_x \\ \frac{E_y}{E_z} &= -\frac{\gamma_0}{h_0} \approx \sqrt{\frac{\sigma_1}{\omega\epsilon_v}} \quad \underline{/-45^\circ} \end{aligned}\right\} \quad (7\text{-}64a)$$

Within the conductor:

$$\left.\begin{aligned} H_x &= C_2 e^{h_1 y} e^{-\gamma_0 z} \\ E_y &= \frac{-j\omega\epsilon_v}{\sigma_1}\eta_v H_x \\ E_z &\approx -\sqrt{\frac{j\omega\mu_v}{\sigma_1}}\, H_x \\ \frac{E_y}{E_z} &\approx \sqrt{\frac{\omega\epsilon_v}{\sigma_1}} \quad \underline{/45^\circ} \end{aligned}\right\} \quad (7\text{-}64b)$$

It is seen that in the region of the air dielectric, outside the conductor, the electric field strength is almost normal to the surface. The field is elliptically polarized, the small horizontal component of **E** leading the vertical component by 45 degrees. Within the conductor the field is almost horizontal or parallel to the surface, the very small vertical component leading the horizontal component by 45 degrees.

The equiphase and equiamplitude surfaces can be obtained from the first of eqs. (64a and b). By letting

$$\gamma_0 = \alpha_0 + j\beta_0, \quad h_0 = p_0 + jq_0, \quad h_1 = p_1 + jq_1$$

these equations can be written as

$$H_x = C_2 e^{(-p_0 y - \alpha_0 z)}\, e^{j(-q_0 y - \beta_0 z)} \quad \text{(in the dielectric)}$$

and

$$H_x = C_2 e^{(p_1 y - \alpha_0 z)}\, e^{j(+q_1 y - \beta_0 z)} \quad \text{(in the conductor)}$$

Equiamplitude surfaces are obtained by setting the real exponents equal to a constant. This leads to

$$\tan\chi_0 = -\frac{y}{z} = \frac{\alpha_0}{p_0} \approx \sqrt{\frac{\omega\epsilon_v}{2\sigma_1}} \quad \text{(in the dielectric)} \qquad (7\text{-}65)$$

$$\tan\chi_1 = \frac{y}{z} = \frac{\alpha_0}{p_1} \approx \frac{\omega\epsilon_v}{\sigma_1}\sqrt{\frac{\omega\epsilon_v}{2\sigma_1}} \quad \text{(in the conductor)} \qquad (7\text{-}66)$$

Equiphase surfaces are obtained by setting the imaginary exponents equal to a constant. The slopes of the equiphase surfaces are given by

$$\tan\psi_0 = \frac{y}{z} = -\frac{\beta_0}{q_0} \approx \sqrt{\frac{2\sigma_1}{\omega\epsilon_v}} \quad \text{(in the dielectric)} \qquad (7\text{-}67)$$

$$\tan \psi_1 = \frac{y}{z} = \frac{\beta_0}{q_1} \approx \sqrt{\frac{2\omega\epsilon_v}{\sigma_1}} \quad \text{(in the conductor)} \tag{7-68}$$

The angles χ_0, χ_1, ψ_0, and ψ_1 are shown in Fig. 7-10 where the equiphase lines are shown solid and the equiamplitude lines are shown dotted. In order to show the angles, their sizes have been very much exaggerated in this diagram. It is seen from eqs. (65) and (67) that in the dielectric the equiphase and equiamplitude surfaces are mutually perpendicular. In the conductor both equiamplitude and equiphase surfaces are nearly parallel to the surface, with the equiamplitude surface making a much smaller angle than the equiphase surface. Since the equiamplitude and equiphase surfaces are not parallel, the wave solutions obtained are examples of *nonuniform plane waves.* Also, since the phase velocity above the conductor and normal to the equiphase surfaces is less than c, the wave there may be referred to as a *slow wave.* Note that the phase velocity component along the conducting surface is also less than c, being given by

$$c\left(1 + \frac{\omega^2\epsilon_v^2}{8\sigma_1^2}\right)^{-1}$$

The above analysis provides a useful description of wave propagation along and within the surface of a good conductor. The question of how such waves can be excited is beyond the scope of the elementary treatment of this chapter.

Within the conductor the conduction current density, given by

$$\mathbf{J} = \sigma\mathbf{E}$$

is seen to have both horizontal and vertical components. Using eqs. (64b) it is possible to make an instantaneous plot of current flow in the conductor as the electromagnetic wave is guided along its surface. This has been done in Fig. 7-11 for a copper conductor at 100 MHz. In order to show the current flow adequately the vertical scale has been expanded by a factor of 100,000. The current magnitudes, indicated by the lengths of the arrows, are drawn to scale, but the vertical current scale is 10^5 times the horizontal current scale. Thus, if a horizontal current density of 1 *ampere* per square meter is represented by an arrow of unit length, an arrow of the same length in the vertical direction represents only 10 *microamperes* per square meter. It is apparent from the figure that the vertical currents are very small compared with the horizontal currents. However, it is these minute vertical currents that bring to the surface the charges on which the external electric flux terminates. Since total current normal to the surface must be continuous across the boundary surface, the vertical conduction current within the conductor at the surface is equal to the displacement current normal to the surface in the dielectric (the displacement current in the conductor is negligible). These vertical currents are a maximum at those places where density on the surface is zero.

The plot of Fig. 7-11 is for a single instant of time. As time passes, the entire field configuration shown sweeps to the right with a velocity approximately equal to the velocity of light in free space.

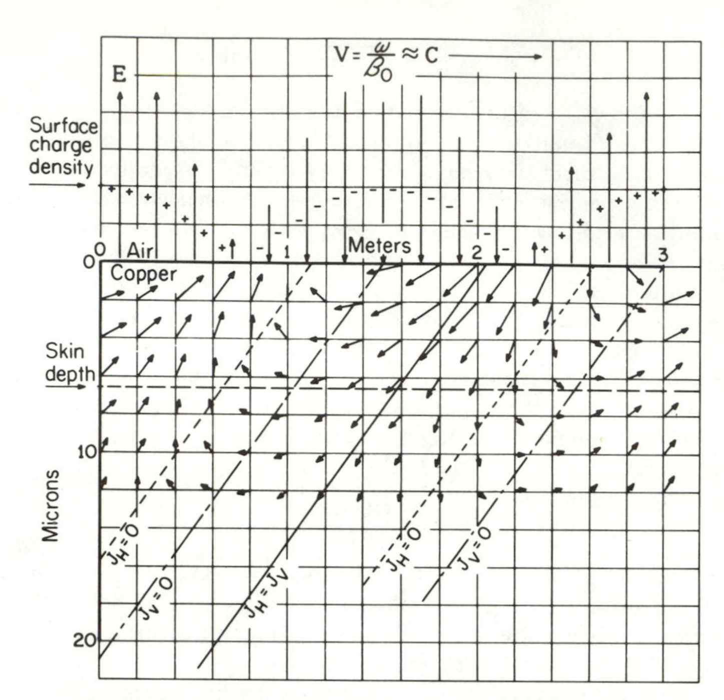

Figure 7-11. Instantaneous current distribution within a copper conductor as a 100-MHz wave is guided over its surface. The vertical scale has been expanded by a factor of 10^5 times the horizontal current scale. Length of arrow refers to magnitude at tail of arrow.

7.10 Transmission Lines. In the study of wave propagation between parallel planes, it was found that there were many possible modes or types of waves which could be propagated. Except for the special case of the transverse electromagnetic (TEM) wave, however, all of these modes require a certain minimum separation (in wavelengths) between the conductors for propagation to be possible. Only for the TEM wave could the conductor separation be small compared with a wavelength. This statement also holds for practical transmission lines, such as coaxial or parallel-wire lines, and it is for this reason that only the TEM mode need be considered at low frequencies, that is at power, audio, and radio frequencies below 200 or 300 MHz. All other modes would require impractically large cross-sectional dimensions of the guiding systems. If a system of conductors guides this low-frequency-type TEM wave, it is called a transmission line, whereas if it supports TE or TM waves, it is called a wave guide. Transmission lines are considered in this present chapter and wave guides will be studied in chap. 8. Transmission lines always consist of at least two separate conductors between which a voltage can exist, but wave guides may, and often do, involve only one conductor; for example, a hollow rectangular or circular cylinder within which the wave propagates.

Although the TEM transmission-line wave is but one special case of guided wave propagation, it is so important practically, that it is usually treated as "ordinary transmission-line theory" quite early in the training of the electrical engineer. In this treatment, circuit concepts are extended to cover this distributed-constants circuit. It is the purpose in the following sections to show how the circuit approach follows directly from Maxwell's equations.

Actual two-conductor transmission lines usually take the form of parallel-wire or coaxial lines. Before considering these practical cases, however, circuit concepts will be developed for the simpler case of a parallel-plane transmission line carrying the TEM wave.

7.11 Circuit Representation of the Parallel-plane Transmission Line. In communication engineering a transmission line carrying the principal (TEM) wave is represented as a distributed-constants network having a series impedance $Z = R + j\omega L$ per unit length and a shunt admittance $Y = G + j\omega C$ per unit length. It is instructive to draw the equivalent circuit and evaluate the constants for the parallel-plane transmission line of Fig. 7-1. For the special case of perfectly conducting planes and a perfect (lossless) dielectric, the series resistance and shunt conductance are both zero, so that the equivalent circuit representation is that of Fig. 7-12, where there is an inductance L per unit length and a capacitance C per unit length. The values of these constants in terms of the line dimensions and the constants of the medium between the planes can be obtained directly from Maxwell's equations.

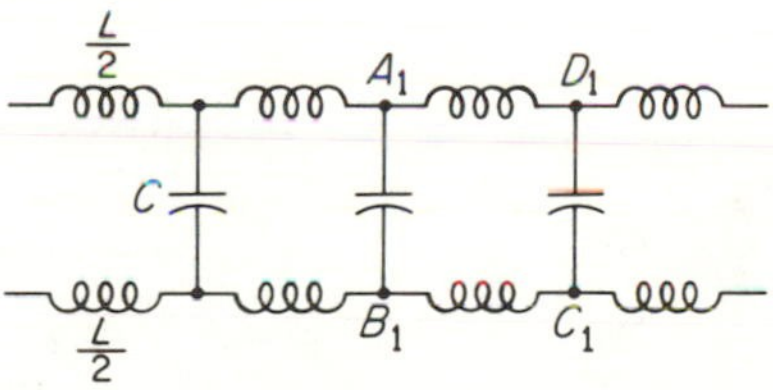

Figure 7-12. Circuit representation of a lossless line.

Consider the various sections of a parallel-plane transmission line shown in Fig. 7-13. It is assumed that the line is carrying the TEM mode in the positive z direction, so that $\mathbf{E} = \hat{\mathbf{x}}E_x$ and $\mathbf{H} = \hat{\mathbf{y}}H_y$. The linear surface current density in the lower plane is $J_{sz} = H_y$. The separation between the planes is a meters and, although they are infinite in extent in the y direction, a section b meters wide will be considered as being the transmission line. (By making this section a part of planes of infinite extent the field will not depend on y, and edge effects are eliminated.) Applying the emf equation to the closed path $ABCDA$

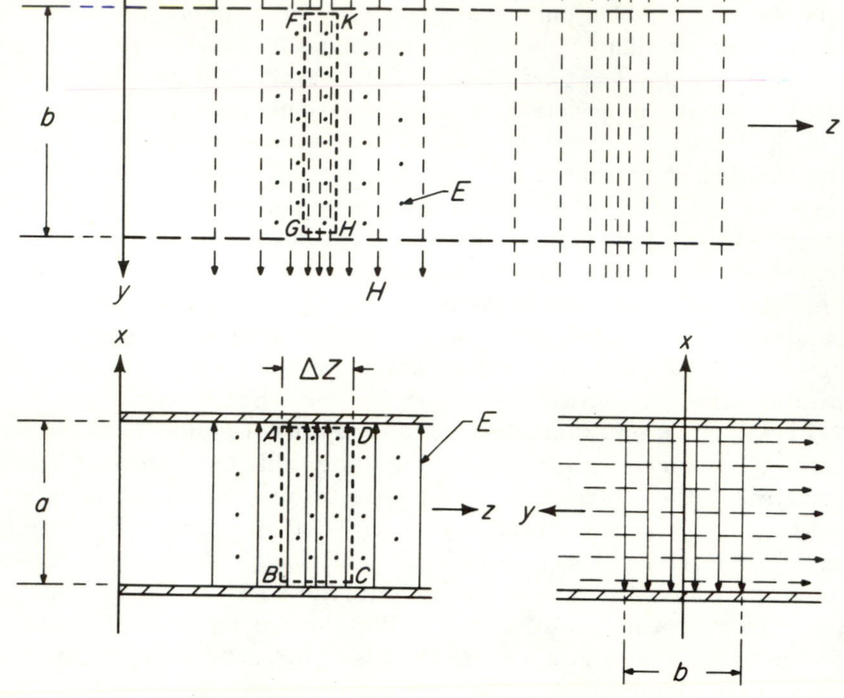

Figure 7-13. Parallel-plane transmission line.

$$\oint \mathbf{E}\cdot d\mathbf{s} = -j\omega \int_S \mathbf{B}\cdot d\mathbf{a}$$

becomes

$$V_{AB} + V_{BC} + V_{CD} + V_{DA} = -j\omega B_y a\,\Delta z$$

For perfectly conducting planes the tangential component of **E** is zero and so

$$V_{BC} = V_{DA} = 0$$

which leaves

$$V_{CD} - V_{BA} = -j\omega B_y a\,\Delta z$$

Dividing through by Δz and expressing in the differential form

$$\frac{dV}{dz} = -j\omega B_y a \tag{7-69}$$

It will be seen that

$$B_y = \mu H_y = \mu J_{sz} = \frac{\mu I}{b}$$

where I is the current flowing in the strip of width b meters. Therefore, eq. (69) becomes

$$\frac{dV}{dz} = -\frac{j\omega\mu a}{b} I \tag{7-70}$$

Comparison with the ordinary circuital form of the transmission-line equation

$$\frac{dV}{dz} = -j\omega LI$$

shows that for the parallel-plane transmission line of width b meters and spacing a meters

$$L = \mu \frac{a}{b}$$

Similarly, writing the mmf equation for the path $FGHK$ in the y-z plane gives

$$bH_{FG} - bH_{KH} = j\omega\epsilon E_x b\,\Delta z$$

which becomes

$$\frac{d(bH_y)}{dz} = -j\omega\epsilon E_x b$$

Replace bH_y by $bJ_{sz} = I$ and E_x by V/a. Then

$$\frac{dI}{dz} = -\frac{j\omega\epsilon b}{a} V \tag{7-71}$$

Comparison with the usual equation

$$\frac{dI}{dz} = -j\omega CV$$

shows that for the parallel-plane transmission line

$$C = \epsilon \frac{b}{a}$$

It is seen that for a parallel-plane transmission line the inductance per unit length is simply the permeability μ of the medium multiplied by a geometry factor a/b, which in this case is proportional to the spacing and inversely proportional to the width of the line. Also the capacitance per unit length is the dielectric constant ϵ of the medium, multiplied by a geometry factor which, in this case, is proportional to the width and inversely proportional to the spacing. The reciprocal of the square root of the product of L and C gives the velocity of wave propagation along the line. That is,

$$v = \frac{1}{\sqrt{LC}} = \frac{1}{\sqrt{\mu\epsilon}} \tag{7-72}$$

For lines of different cross section the geometry factors will of course be different. However, since the velocity of propagation is given by $v = 1/\sqrt{LC}$ for all uniform unloaded *lossless* transmission lines, and since the velocity is independent of the line geometry (whether parallel-wire, coaxial, etc.), it follows that the geometry factors for L and C must always be reciprocal (for perfect conductors). For example, for parallel-wire lines it was found in chap. 2 that the capacitance per unit length was

$$C = \frac{\pi\epsilon}{\ln \dfrac{b + \sqrt{b^2 - 4a^2}}{2a}} = \frac{\pi\epsilon}{\cosh^{-1}\left(\dfrac{b}{2a}\right)} \tag{7-73}$$

Therefore, the inductance per unit length must be

$$L = \frac{\mu \ln \dfrac{b + \sqrt{b^2 - 4a^2}}{2a}}{\pi} = \frac{\mu \cosh^{-1}\left(\dfrac{b}{2a}\right)}{\pi} \tag{7-74}$$

It is, of course, more than just a coincidence that the geometry factors for the L and C of a line are reciprocal. The significance of this relation is discussed in sec. 7.13.

A clear concept of the meaning of the permeability constant μ and dielectric constant ϵ is obtained from the parallel-plane transmission line of Fig. 7-13. If this line has unit width and unit separation, so that $a = b = 1$, then

$$L = \mu \qquad \text{and} \qquad C = \epsilon$$

Thus ϵ is the capacitance between conductors of 1 meter length of the parallel plane line, which is 1 meter wide and has a separation of 1 meter. Similarly, μ is the inductance per meter length of the same line. In terms of voltage and current, μ is a measure of the change per unit length of the transverse voltage when the current is changing at the rate of 1 amp/sec. Also the dielectric constant ϵ is a measure of the capacitive (displacement) current flow per unit length when the voltage between the planes is changing at the rate of 1 volt/sec. In terms of electric and magnetic fields, μ is a measure of the rate of change of E with distance owing to a change of H with time. Similarly, ϵ is a measure of the rate of change of H with distance owing to a change of E with time. Of course this is just the information conveyed by Maxwell's equations in their differential form.

In one respect the equivalent circuit representation of a transmission line may be misleading. In the equivalent circuit there exists a voltage drop $L(dI/dt)$ along each unit length of line. In the actual line, since E tangential to the surface of a perfect conductor is always zero, the voltage drop *along the surface of the line* is necessarily zero. Even if the conductors are imperfect so that an E parallel to the surface of the conductors is possible, the only voltage drop along the line would be that due to the current flow through the surface impedance, and this is ordinarily very small as has already been seen. The $L(dI/dt)$ drop in the equivalent circuit represents in the actual line the change per unit length of the *transverse* voltage between conductors. With a zero voltage drop along paths tangential to the (perfect) conductors, the difference of the transverse voltages AB and DC is equal to the induction voltage $-d\Phi/dt$ around the closed path $ABCDA$ (Fig. 7-13). But in the equivalent *circuit* representation of Fig. 7-12, where *fields are not considered*, the voltage around the closed path $A_1B_1C_1D_1A_1$ is zero. Therefore the induction voltage $-d\Phi/dt$ (which is responsible for the change in transverse voltage along the line) is shown as a series voltage drop, $-L\, dI/dt$, across a lumped inductive reactance.

The characteristic impedance of the lossless parallel-plane transmission line is

$$Z_0 = \sqrt{\frac{L}{C}} = \sqrt{\frac{\mu}{\epsilon}}\frac{a}{b} = \eta\frac{a}{b} \tag{7-75}$$

For the line of unit dimensions, $a = b = 1$, the characteristic impedance is just the intrinsic impedance of the dielectric medium between the plates.

7.12 Parallel-plane Transmission Lines with Loss. If the parallel-plane transmission lines have loss, the results obtained above must be modified. The loss in the line will be due to the resistance of the conductors and to any conductivity of the dielectric between them. Again applying the electromotive force equation around the path $ABCDA$ of Fig. 7-13, the voltages V_{BC} and V_{DA} will now not be zero but will each have a value

$$V_{BC} = V_{DA} = J_{sz}Z_s\,\Delta z$$

This is the voltage drop in length Δz of each conductor due to J_{sz} flowing against the surface impedance Z_s. Then around the path, the emf equation gives

$$V_{CD} - V_{BA} = -j\omega B_y a\,\Delta z - 2J_{sz}Z_s\,\Delta z \tag{7-76}$$

Writing $B_y = \mu H_y = \mu J_{sz} = \mu I/b$, and putting (76) in the differential form

$$\frac{dV}{dz} = -j\omega LI - Z'I = -(j\omega L + Z')I \tag{7-77}$$

where, as before,

$$L = \frac{\mu a}{b}$$

and

$$Z' = \frac{2Z_s}{b} \tag{7-78}$$

is the series impedance per unit length of the line (that is, twice the surface impedance of a width b of each conductor). The impedance Z' is complex and can be written as $Z' = R' + j\omega L'$ where R' will be the series resistance per unit length and $j\omega L'$ will be the surface or internal reactance per unit length. Then eq. (77) can be written

$$\frac{dV}{dz} = -[R' + j\omega(L' + L)]I \tag{7-77a}$$

If the dielectric between the conducting plates is not perfect, but has a value σ, then there will be a transverse conduction current density $\sigma\mathbf{E}$, which will modify the magnetomotive force around the rectangle *FGHK*. The mmf equation will now be

$$-(bH_{KH} - bH_{FG}) = (\sigma E_x + j\omega\epsilon E_x)b\,\Delta z$$

Then

$$\frac{d(bH_y)}{dz} = -b(\sigma + j\omega\epsilon)E_x$$

Replacing bH_y by $bJ_{sz} = I$ and E_x by V/a

$$\frac{dI}{dz} = -\left(\frac{b\sigma}{a} + j\frac{\omega\epsilon b}{a}\right)V$$

$$= -(G + j\omega C)V \tag{7-79}$$

where $C = \epsilon b/a$ is the capacitance per unit length and $G = b\sigma/a$ is the conductance per unit length of line.

Equations (77) and (79) are in the circuital form, familiar to engineers, and may be solved to yield the well-known "transmission-line equations."

7.13 E and H about Long Parallel Cylindrical Conductors of Arbitrary Cross Section. In sec. 7.11 it was found that the geometry factors for the L and C of parallel perfectly conducting cylinders were always reciprocal. As might be suspected, this interesting result is not just a coincidence, but follows as a logical consequence of the similarity that exists between all two-dimensional electric and magnetic field distributions. It is well known that lines of **E** and **H** about long parallel circular cylinders are always orthogonal, and that the magnitudes of **E** and **H** are related at all points by a constant factor that is dependent on the charge on the conductors and the current flowing

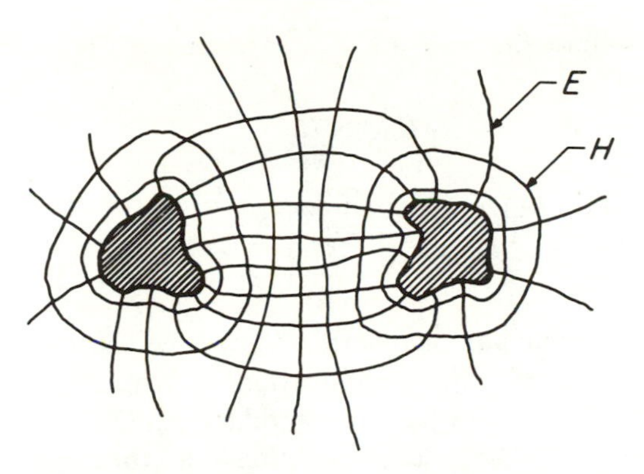

Figure 7-14. Parallel cylinders of arbitrary cross section.

through them. It is easy to show that this same correspondence between electric and magnetic fields must hold even in the more general case where the parallel cylinders have any arbitrary cross section as in Fig. 7-14.

The static electric field configuration is obtained as a solution to Laplace's equation subject to the boundary conditions of the problem. In rectangular co-ordinates, for two-dimensional fields that are independent of the z co-ordinate, Laplace's equation is

$$\frac{\partial^2 V}{\partial x^2} + \frac{\partial^2 V}{\partial y^2} = 0 \tag{7-80}$$

where V is the (electrostatic) potential, the gradient of which gives the electric field. Similarly, the magnetic field configuration can be obtained as the curl of a magnetic (vector) potential that has the direction of the current producing it. When the conductors are entirely in one direction, say the z direction, the vector potential has only one component A_z, and the components of magnetic field strength lie in the x-y plane and are given by

$$\mu H_x = \frac{\partial A_z}{\partial y} \qquad \mu H_y = -\frac{\partial A_z}{\partial x} \tag{7-81}$$

Under these conditions it can be readily shown that A_z also satisfies eq. (80). In a region in which there are no currents, Ampere's law indicates that the line integral of $\mathbf{H}$ around every closed path is zero. That is,

$$\oint \mathbf{H} \cdot d\mathbf{s} = 0$$

The differential vector statement of this law is

$$\nabla \times \mathbf{H} = 0 \tag{7-82}$$

For this case where there is no z component of $\mathbf{H}$, relation (82) becomes

$$\frac{\partial H_x}{\partial y} - \frac{\partial H_y}{\partial x} = 0$$

Inserting relations (81) gives

$$\frac{\partial^2 A_z}{\partial x^2} + \frac{\partial^2 A_z}{\partial y^2} = 0 \tag{7-83}$$

Thus for two-dimensional magnetic fields the potential A_z satisfies Laplace's equation, and the configuration of the magnetic field, obtained from (81), is always such that relation (83) is satisfied. In addition, of course, the boundary conditions of the particular problem must also be satisfied.

Now consider the problem of two parallel cylindrical conductors (assumed perfectly conducting) that carry equal and oppositely directed currents I. For the d-c case the current is uniformly distributed throughout the conductor, but for rapidly alternating fields, because of the phenomenon of skin effect, the current exists only near the surface of the conductor. Although this is a d-c field analysis, the results will be applied chiefly to the alternating field cases, so the assumption of current concentrated in a thin sheet at the surface of the conductor will be used. Except when the spacing between conductors is large compared with their diameters, the current distribution around the circumference of the conductor will not be uniform. The actual current distribution will be such that the boundary conditions at the surface of the conductor are satisfied. This is similar to the electrostatic problem where the charge distribution around the cylinders was such as to make the cylinders equipotential surfaces, and satisfy the condition that $E_{\text{tan}} = 0$.

The corresponding boundary condition for the magnetic field strength is that $H_{\text{norm}} = 0$ (for a perfect conductor). That is, the magnetic field strength at the surface is entirely *tangential*. Equations (81) indicate that if the magnetic field strength normal to the surface is zero, there can be no change of A_z in a direction tangential to the surface. Therefore the conductor must also be an "equipotential" surface for the magnetic potential A_z. Because in this case both A_z and V satisfy Laplace's equation, and in addition satisfy the same boundary conditions, it follows that the expressions for A_z and V will always be identical, except for some constant factor. Then, because

$$E_x = -\frac{\partial V}{\partial x} \qquad \mu H_x = \frac{\partial A_z}{\partial y}$$

$$E_y = -\frac{\partial V}{\partial y} \qquad \mu H_y = -\frac{\partial A_z}{\partial x} \tag{7-84}$$

it follows that $\mathbf{E}$ and $\mu\mathbf{H}$ will always be orthogonal, and that their magnitudes will be related to each other by the same factor that related V and A_z.

The electric and magnetic field configurations obtained from solutions of Laplace's equation are for the electrostatic and steady current cases, respectively. In general, it would not be expected that these same solutions would hold for alternating fields, especially at high frequencies. It turns out that for TEM fields, however, the transverse field configurations obtained for the static cases also hold for the alternating cases. This is because the general Maxwell emf and mmf equations reduce to their steady-field counterparts in the TEM case. For example, the mmf equation in the x-y plane for the region outside the conductors (no conduction current) is

$$(\nabla \times \mathbf{H})_z = j\omega\epsilon E_z \tag{7-85}$$

But for TEM fields in the x-y plane, E_z is zero so that for any path in this plane this equation reduces to eq. (82), which yielded the Laplace equation (83). Similarly the Maxwell emf equation

$$(\nabla \times \mathbf{E})_z = -\,j\omega\mu H_z$$

reduces to

$$(\nabla \times \mathbf{E})_z = 0 \tag{7-86}$$

for the TEM problem. Equation (86), the integral form of which is

$$\oint \mathbf{E}\cdot d\mathbf{s} = 0 \tag{7-87}$$

states that for any path in the x-y plane the electric field is conservative. Therefore, $\mathbf{E}$ is derivable as the gradient of a scalar potential V, and in a region in which there are no charges, the relation

$$\nabla\cdot\mathbf{E} = 0$$

leads directly to Laplace's equation.

7.14 Transmission-line Theory. For all types of TEM mode transmission lines (coaxial, parallel-wire), the basic equations have the same form as eqs. (77a) and (79) for parallel-plate lines. If R, L, G, and C are the total resistance, inductance, conductance, and capacitance per unit length, the transmission-line equations may be expressed as

$$\frac{dV}{dz} = -(R + j\omega L)I \tag{7-88}$$

$$\frac{dI}{dz} = -(G + j\omega C)V \tag{7-89}$$

Differentiating and combining gives

$$\frac{d^2V}{dz^2} = \gamma^2 V \tag{7-90}$$

$$\frac{d^2I}{dz^2} = \gamma^2 I \tag{7-91}$$

where

$$\gamma^2 = (R + j\omega L)(G + j\omega C)$$

Solutions to eqs. (90) and (91) may be written in either exponential- or hyperbolic-function form. In the exponential form, viz.,

$$V = V' e^{-\gamma z} + V'' e^{+\gamma z} \tag{7-92}$$

$$I = I' e^{-\gamma z} + I'' e^{+\gamma z} \tag{7-93}$$

the solutions are shown as the sum of two waves, one traveling in the positive z direction and the other traveling in the negative z direction. The ratio of voltage to current for the wave traveling in the positive z direction is

$$\frac{V'}{I'} = Z_0 \tag{7-94}$$

whereas for the "reflected" wave traveling in the opposite direction

$$\frac{V''}{I''} = -Z_0 \tag{7-95}$$

Z_0 is the characteristic impedance of the line and is related to the so-called primary constants R, L, C, and G by

$$Z_0 = \sqrt{\frac{R + j\omega L}{G + j\omega C}} \tag{7-96}$$

If the line is terminated in an impedance Z_R located at $z = 0$, the ratio of V to I at this point will be equal to Z_R so that

$$Z_R = \frac{V}{I} = \frac{V' + V''}{I' + I''} = \frac{Z_0(I' - I'')}{I' + I''}$$

These relations can be recombined to give the *reflection coefficients*,

$$\Gamma_R = \frac{V''}{V'} = \frac{Z_R - Z_0}{Z_R + Z_0}, \qquad \frac{I''}{I'} = \frac{Z_0 - Z_R}{Z_0 + Z_R} \tag{7-97}$$

In the hyperbolic-function form the solutions to (90) and (91) are

$$V = A_1 \cosh \gamma z + B_1 \sinh \gamma z$$

$$I = A_2 \cosh \gamma z + B_2 \sinh \gamma z \tag{7-98}$$

The constants A_1, A_2, B_1 and B_2 are evaluated by applying the boundary conditions. Let

$$V = V_R, \qquad I = I_R \qquad \text{at } z = 0$$

$$V = V_S, \qquad I = I_S \qquad \text{at } z = z_1$$

Substituting these relations in (98) and using eqs. (88) and (89),

$$V_S = V_R \cosh \gamma z_1 - Z_0 I_R \sinh \gamma z_1$$

$$I_S = I_R \cosh \gamma z_1 - \frac{V_R}{Z_0} \sinh \gamma z_1 \tag{7-99}$$

It is usual to make the location of the terminating impedance Z_R the reference point ($z = 0$), and to consider the sending end as being to the left of this reference point, that is, in the $-z$ direction as in Fig. 7-15. Then letting $l = -z_1$, eqs. (99) become

$$V_S = V_R \cosh \gamma l + Z_0 I_R \sinh \gamma l \tag{7-100}$$

$$I_S = I_R \cosh \gamma l + \frac{V_R}{Z_0} \sinh \gamma l \tag{7-101}$$

where l is measured from the receiving end of the line.

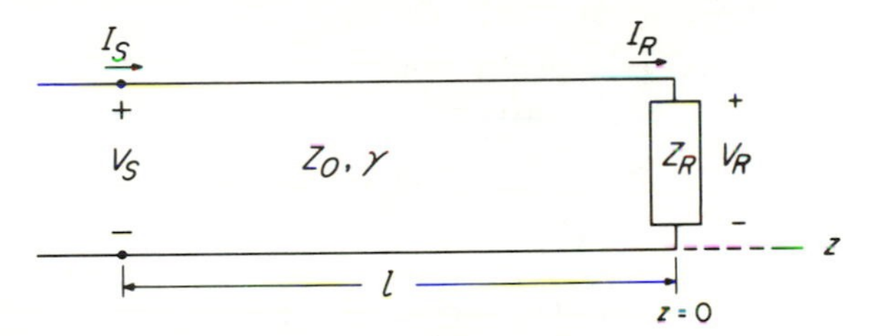

Figure 7-15.

These are the general transmission-line equations that relate the voltages and currents at the two ends of the line. The general expression for the input impedance of the line is obtained by dividing (100) by (101), that is,

$$Z_{\text{in}} = \frac{V_S}{I_S} = \frac{V_R \cosh \gamma l + Z_0 I_R \sinh \gamma l}{I_R \cosh \gamma l + (V_R/Z_0) \sinh \gamma l} \tag{7-102}$$

Certain special cases are of interest. For a line short-circuited at the receiving end, $Z_R = 0$, and therefore $V_R = 0$, and the input impedance is

$$Z_{\text{sc}} = Z_0 \tanh \gamma l \tag{7-103a}$$

On the other hand, for an open-circuited line $Z_R = \infty$, $I_R = 0$, so that the input impedance is

$$Z_{\text{oc}} = Z_0 \coth \gamma l \tag{7-103b}$$

The product of (103a) and (103b) gives

$$Z_{\text{sc}} Z_{\text{oc}} = Z_0^2 \tag{7-104}$$

7.15 Low-loss Radio Frequency and UHF Transmission Lines. The low-loss transmission line is of special interest to the engineer concerned with the transmission of energy at radio and ultrahigh frequencies. There are two reasons for this. First, most practical lines designed for use at these frequencies will be low-loss lines. Second, at ultrahigh frequencies, sections of low-loss line are used as circuit elements, and a knowledge of the operation of such "distributed-constants circuits" is of considerable importance.

A *low-loss transmission line* is one for which

$$R \ll \omega L \qquad (7\text{-}105)$$

$$G \ll \omega C$$

where R, L, C, and G are the resistance, inductance, capacitance, and conductance per unit length of the line. When the above inequalities hold, the following approximations are valid:

$$Z = R + j\omega L \approx j\omega L$$

$$Y = G + j\omega C \approx j\omega C$$

$$Z_0 = \sqrt{\frac{R + j\omega L}{G + j\omega C}} \approx \sqrt{\frac{L}{C}} \qquad (7\text{-}106)$$

$$\gamma = \sqrt{(R + j\omega L)(G + j\omega C)} \approx j\omega\sqrt{LC} \qquad (7\text{-}107)$$

Since $\gamma = \alpha + j\beta$, this last expression gives

$$\alpha \approx 0 \qquad (7\text{-}108)$$

$$\beta \approx \omega\sqrt{LC} \qquad (7\text{-}109)$$

The approximation for β is very good for low-loss lines, but occasionally the approximation of zero for α may not be good enough, even though α is very small compared with β. A closer approximation for α may be obtained by rearranging the expression for γ and using the binomial expansion. Thus

$$\gamma = j\omega\sqrt{LC}\sqrt{\left(1 + \frac{R}{j\omega L}\right)\left(1 + \frac{G}{j\omega C}\right)}$$

$$\approx j\omega\sqrt{LC}\left(1 + \frac{R}{2j\omega L}\right)\left(1 + \frac{G}{2j\omega C}\right)$$

$$\approx j\omega\sqrt{LC}\left(1 + \frac{R}{2j\omega L} + \frac{G}{2j\omega C}\right)$$

$$\approx \frac{R}{2\sqrt{L/C}} + \frac{G\sqrt{L/C}}{2} + j\omega\sqrt{LC}$$

which gives

$$\alpha \approx \frac{1}{2}\left(\frac{R}{Z_0} + GZ_0\right) \tag{7-110}$$

$$\beta \approx \omega\sqrt{LC} \tag{7-111}$$

The more correct value for α given by (110) need only be used in place of (108) when the line losses are being considered. As far as voltage and current distributions are concerned, the attenuation of most low-loss ultrahigh frequency lines is so small that the approximation $\alpha = 0$ gives satisfactory results. This may seem strange in view of the fact that R, and therefore α, *increases* with frequency, and α is not usually neglected at low (power and audio) frequencies. The explanation for this apparent paradox is that although α, the attenuation *per unit length*, increases approximately as the square root of frequency, the attenuation *per wavelength* decreases as the square root of the frequency. Transmission lines are ordinarily a few wavelengths long at most, and αl can usually be neglected (compared with βl) at the ultrahigh frequencies. Thus for many purposes, low-loss lines may be treated as though they were lossless; that is, as if $R = G = \alpha = 0$.

Using the approximate values for the secondary constants given by (106), (107), (108), and (109), the general transmission line equations become for this low-loss, high-frequency case

$$V_S = V_R \cos\beta l + jI_R Z_0 \sin\beta l \tag{7-112}$$

$$I_S = I_R \cos\beta l + j\frac{V_R}{Z_0}\sin\beta l \tag{7-113}$$

where now $Z_0 \approx \sqrt{L/C}$ is a pure resistance.

The input impedance of such a line is

$$\begin{aligned} Z_S &= \frac{V_S}{I_S} \\ &= Z_R\left(\frac{\cos\beta l + j(Z_0/Z_R)\sin\beta l}{\cos\beta l + j(Z_R/Z_0)\sin\beta l}\right) \\ &= Z_0\left(\frac{Z_R\cos\beta l + jZ_0\sin\beta l}{Z_0\cos\beta l + jZ_R\sin\beta l}\right) \end{aligned} \tag{7-114}$$

The voltage and current distributions along the line are obtained from eqs. (112) and (113) by replacing l, the length of line, by x, the distance from the terminating impedance Z_R. Since voltmeters and ammeters read magnitude without regard to phase, the magnitudes of expressions (112) and (113) have been used in Fig. 7-16 to show the standing-wave distributions for various conditions of the terminating impedance Z_R.

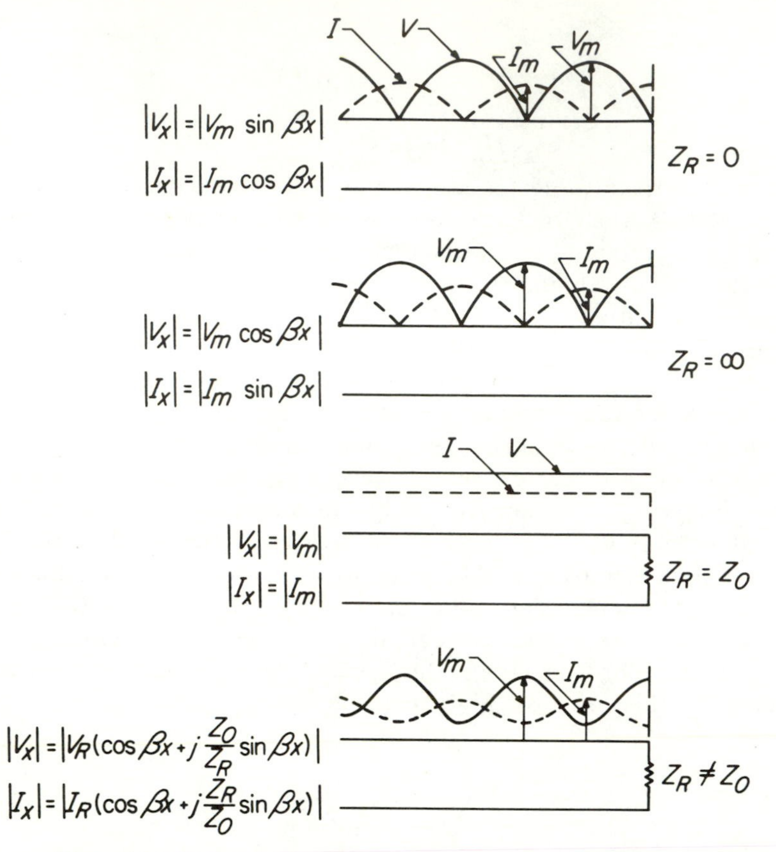

Figure 7-16. Voltage and current distribution along a lossless line.

In general the terminating impedance Z_R will be a complex impedance having both resistance and reactance, but it will be shown later that the results for the general case may be inferred from those obtained for the particular case of a pure-resistance termination. For this latter case where $Z_R = R$, eqs. (112) and (113) may be written as

$$|V_x| = V_R \sqrt{\cos^2 \beta x + (R_0/R)^2 \sin^2 \beta x} \tag{7-115}$$

$$|I_x| = I_R \sqrt{\cos^2 \beta x + (R/R_0)^2 \sin^2 \beta x} \tag{7-116}$$

For the lossless line being considered Z_0 is a pure resistance

$$Z_0 = R_0 = \sqrt{\frac{L}{C}}$$

Examination of eqs. (115) and (116) shows that the voltage and current distributions are given by the square root of the sum of a cosine-squared term and a sine-squared term. It is evident that the maximum value of voltage or current will occur at that value of x that makes the larger of these terms a maximum. In the particular case of a line terminated in R_0, that is, for which $R = R_0$, the sine and cosine terms have equal amplitudes and the square root of the sum of their squares has constant value for all values of x. That is, there are no standing waves on the line. For all other cases, however, the magnitude will vary along the length of the line. When R is less than R_0, the amplitude of the sine term of (115) will be larger than that of the cosine term and the voltage maxima will occur at those values of x that make $\sin \beta x$ a maximum, viz., at $x = \lambda/4$, $3\lambda/4$, and so on. Also the voltage minima will occur at those values of x that make the sine term a minimum, viz., $x = 0$, $\lambda/2$, and so on, also for this case of $R < R_0$, the *current* maxima will occur at $x = 0$, $\lambda/2$, and so on, and the current minima at $x = \lambda/4$, $3\lambda/4$, and so on. When the terminating resistor is larger than R_0, the conditions for both voltage and current are reversed.

One of the important measurable quantities on a transmission line is the standing-wave ratio of voltage or current. When R is less than R_0, eq. (115) shows that the voltage maximum, which occurs when $\sin \beta x = 1$, will have a value

$$V_{\text{max}} = V_R \frac{R_0}{R}$$

Also the voltage minimum, which occurs when $\sin \beta x = 0$, will have a value

$$V_{\text{min}} = V_R$$

The ratio of maximum voltage to minimum voltage is therefore

$$\frac{V_{\text{max}}}{V_{\text{min}}} = \frac{R_0}{R} \qquad (\text{for } R < R_0)$$

Similarly the standing wave of current ratio is given by

$$\frac{I_{\text{max}}}{I_{\text{min}}} = \frac{R_0}{R} \qquad (\text{for } R < R_0)$$

For $R > R_0$ these expressions are just reversed, that is

$$\frac{V_{\text{max}}}{V_{\text{min}}} = \frac{I_{\text{max}}}{I_{\text{min}}} = \frac{R}{R_0} \qquad (\text{for } R > R_0)$$

Using these expressions, the value of a terminating resistance may be determined in terms of R_0 from relative measurements of voltage or current along the line. R_0 is readily calculable from the line dimensions.

Case where Z_R is not a Pure Resistance. When the terminating impedance Z_R is *not* a pure resistance, standing-wave measurements can be still used, and in this case will yield values of both resistance and reactance of the termination. From eqs. (115) and (116) it was seen that with a resistance termination a voltage maximum or minimum always occurred right at the termination ($x = 0$). However, when the terminating impedance has reactance as well as resistance, the maximum or minimum is always displaced from the position $x = 0$, and the direction and amount of this displacement can be used to determine the sign and magnitude of the reactance of the load.

Figure 7-17 shows a transmission line terminated in an impedance

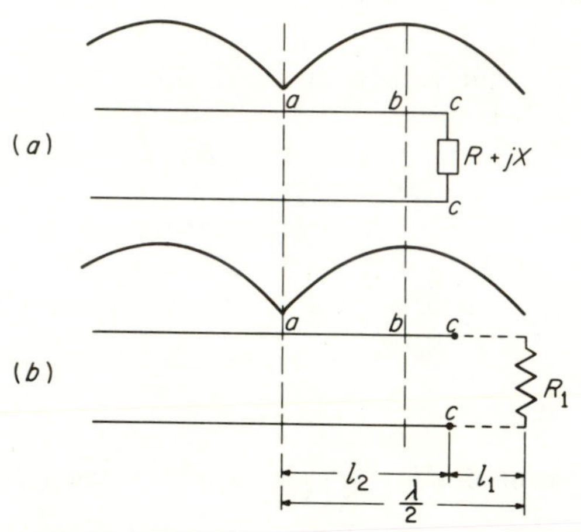

Figure 7-17. A complex terminating impedance in (*a*) is replaced by a pure resistance termination in (*b*).

that has a reactive component. The voltage distribution along the line is shown. Because the impedance is not a pure resistance, the voltage maximum (or minimum) does not occur at the termination. Now any complex impedance can be obtained by placing a pure resistance of proper value at the end of an appropriate length of (lossless) transmission line. In part (*b*) of Fig. 7-17, the complex impedance $R + jX$ has been replaced by the proper value of resistance R_1 at the end of a length l_1 of line, such that the impedance at *c-c* looking towards R_1 is equal to $R + jX$. The standing wave back from *c-c* toward the source will be unchanged and that toward R_1 will be just a continuation of it. Quite evidently the proper position for R_1 is at a distance of one-half wavelength from the minimum point *a* (or the maximum point *b* if R_1 is greater than R_0), and the proper value of R_1 is given by the standing-wave ratio on the line, that is, by

$$\frac{R_1}{R_0} = \frac{V_{\min}}{V_{\max}} \qquad \text{or} \qquad \frac{R_1}{R_0} = \frac{V_{\max}}{V_{\min}}$$

Because any resistance greater than R_0 can be obtained by a resistance less than R_0 at the end of a quarter-wave section of line (see below), it is really only necessary to consider for R_1 resistances less than or equal to R_0. It is then possible to state that any impedance whatsoever can be obtained by means of a pure resistance R_1 (not greater than R_0) at the end of a length l_1 of lossless transmission line, less than one-half wavelength long.

The value of the impedance $Z = R + jX$ is given in terms of R_1 and l_1 by eq. (114). Rationalizing and separating into real and imaginary parts, eq. (114) becomes

$$R = \frac{R_0^2 R_1}{R_0^2 \cos^2 \beta l_1 + R_1^2 \sin^2 \beta l_1} \tag{7-117}$$

$$X = \frac{R_0(R_0^2 - R_1^2) \sin \beta l_1 \cos \beta l_1}{R_0^2 \cos^2 \beta l_1 + R_1^2 \sin^2 \beta l_1} \tag{7-118}$$

Equations (117) and (118) make it possible to determine both the resistance and reactance values of a terminating impedance from standing-wave measurements on the transmission line. The sign of the reactance, that is, whether inductive (positive) or capacitive (negative) can be obtained by inspection as shown in Fig. 7-18.

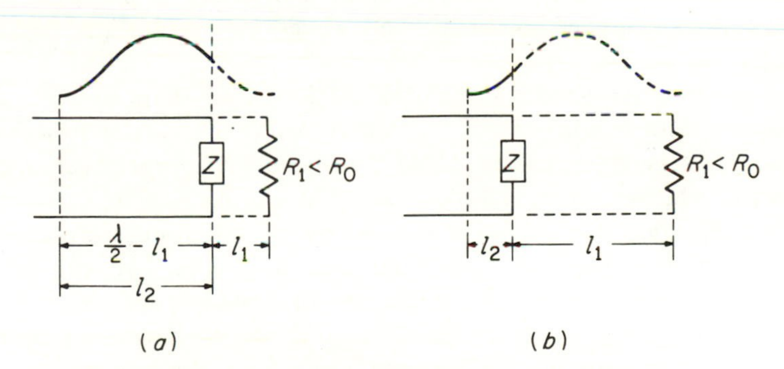

Figure 7-18. A terminating impedance that is inductive (*a*) or capacitive (*b*).

Considering the value of R_1 to be less than R_0, eq. (118) shows that when l_1 is less than one-quarter wavelength, the reactance X is positive (i.e., inductive), whereas if l_1 is between one-quarter and one-half wavelength, X will be negative (capacitive). From this results the conclusion that, if the standing wave of voltage slopes down toward the terminating impedance [Fig. 7-18(*a*)], the impedance is inductive; if the slope is up toward the impedance [Fig. 7-18(*b*)], the impedance is capacitive. Of course, if the slope is zero at the termination, the terminating impedance is a pure resistance.

In practice the measurable quantities are l_2, the distance from the termination to the minimum point *a*, and the *standing-wave ratio*

$$\rho = \frac{R_0}{R_1} = \frac{V_{\max}}{V_{\min}}$$

In terms of these measurable quantities, the resistance and reactance of the terminating impedance are given by

$$R = \frac{\rho R_0}{\rho^2 \cos^2 \beta l_2 + \sin^2 \beta l_2} \tag{7-119}$$

$$X = \frac{-R_0(\rho^2 - 1)\sin \beta l_2 \cos \beta l_2}{\rho^2 \cos^2 \beta l_2 + \sin^2 \beta l_2} \tag{7-120}$$

7.16 UHF Lines as Circuit Elements. The transfer of energy from one point to another is only one use of transmission lines. At the ultrahigh frequencies an equally important application is the use of sections of lines as circuit elements. Above 150 MHz the ordinary lumped-circuit elements become difficult to construct and, at the same time, the required physical size of sections of transmission lines has become small enough to warrant their use as circuit elements. They can be used in this manner up to about 3000 MHz where their physical size then becomes too small and wave-guide technique begins to take over.

In Fig. 7-19 are shown some line sections and their low-frequency equivalents. The magnitude of the input reactance of the first four of these sections is given by eq. (114) when the appropriate value of Z_R is inserted; that is, $Z_R = 0$ for the shorted sections and $Z_R = \infty$ for the open sections. The resistive component of the input impedance is negligible for the usual low-loss lines used at UHF. Thus it is seen that for line lengths less than a quarter of a wavelength the shorted section is equivalent to an inductance, and the open section to a capacitance. For lengths of line between a quarter and a half wavelength, the shorted section is capacitive and the open section is inductive. However, it should be noted that unlike their low-frequency equivalents, these "inductances" and "capacitances" change value with frequency.

The Quarter-wave and Half-wave Sections. For the particular case of the shorted quarter-wave line or the open half-wave line, the input reactance, given by (114), goes to infinity, and the resistive component of the input impedance must be taken into account. This corresponds

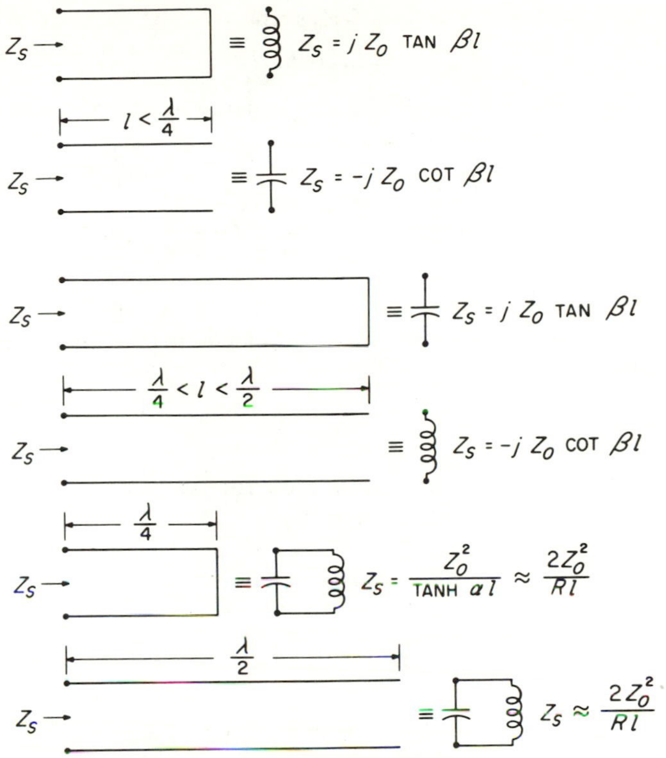

Figure 7-19. Input impedance of various transmission-line sections.

to conditions in the parallel-resonant circuit (the low-frequency analogue), which has an infinite impedance if resistance is neglected. In both cases (the quarter-wave line and the parallel-resonant circuit) the actual input impedance when the series resistance is not neglected is a pure resistance of very high value. In the case of the line its value is given approximately by

$$R_{ar} = \frac{2Z_0^2}{Rl}$$

where R_{ar} is the input resistance of the line at a resonant length and R is the series resistance per unit length of the line. l is the length of the resonant section, which will be an odd multiple of a quarter wavelength for a shorted line or an even multiple of one-quarter wavelength for an open line. This expression is obtained directly from eqs. (100) and (101) in which the actual line loss is not neglected as follows:

For a shorted line for which $V_R = 0$, eqs. (100) and (101) become

$$V_S = I_R Z_0 \sinh \gamma l$$

$$I_S = I_R \cosh \gamma l$$

Dividing the voltage equation by the current equation gives the input impedance of a short-circuited line as

$$\begin{aligned} Z_S &= Z_0 \tanh \gamma l \\ &= Z_0 \frac{\sinh \alpha l \cos \beta l + j \cosh \alpha l \sin \beta l}{\cosh \alpha l \cos \beta l + j \sinh \alpha l \sin \beta l} \end{aligned}$$

For line lengths that are an odd multiple of a quarter wavelength, $\sin \beta l = \pm 1$ and $\cos \beta l = 0$. Under these conditions the input impedance becomes

$$Z_S = Z_0 \frac{\cosh \alpha l}{\sinh \alpha l}$$

If αl is very small, as is generally true for sections of low-loss line, $\cosh \alpha l \approx 1$ and $\sin \alpha l \approx \alpha l$ so that

$$Z_S \approx \frac{Z_0}{\alpha l}$$

When $\omega L \gg R$ and $\omega C \gg G$, α is given in terms of the line constants by

$$\alpha = \frac{1}{2}\left(R \sqrt{\frac{C}{L}} + G \sqrt{\frac{L}{C}}\right) \tag{7-110}$$

For the air-dielectric lines commonly used the losses due to the conductance G are negligible, so that G can be neglected and

$$\alpha = \frac{R}{2}\sqrt{\frac{C}{L}} = \frac{R}{2Z_0} \tag{7-121}$$

Substituting this in the above expression for input impedance of a short-circuited line, whose length is an odd multiple of a quarter wavelength, gives

$$Z_S = \frac{Z_0}{\alpha l} = \frac{2Z_0^2}{Rl} \tag{7-122}$$

An identical expression is obtained for an open-ended section that is a multiple of a half-wave long.

Resonance in Line Sections. The shorted quarter-wave section has other properties of the parallel-resonant circuit. It is a *resonant* circuit and produces the resonant rise of voltage or current which exists in such circuits. The mechanism of resonance is particularly easy to visualize in this case. If it is assumed that a small voltage is induced into the line near the shorted end, there will be a voltage wave sent down the line and reflected without change of phase at the open end. This reflected wave travels back and is reflected again at the shorted end with reversal of phase. Because it required one-half cycle to travel up and back the line, this twice-reflected wave now will be in phase with the original induced voltage and so adds directly to it. Evidently those additions continue to increase the voltage (and current) in the line until the I^2R loss is equal to the power being put into the line. A voltage step-up of several hundred times is possible depending upon the Q of the line.

Input Impedance of the Tuned Line. When the quarter-wave section is tapped at some point x along its length, a further correspondence between this circuit and the simple low-frequency parallel-resonant circuit is observed. The reactance looking toward the shorted end will be inductive and of value $Z_{sc} = jZ_0 \tan \beta x$. The reactance looking toward the open end will be of equal magnitude but opposite sign, i.e., a capacitive reactance. Its value is given by $Z_{oc} = -jZ_0 \cot \beta(\lambda/4 - x) = -jZ_0 \tan \beta x$. The equal but opposite reactances are in parallel just as they are in Fig. 7-20(*b*) and the input impedance will be

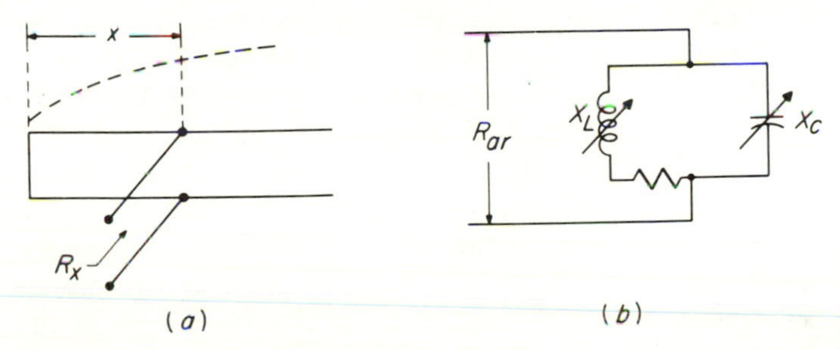

Figure 7-20. (*a*) Tapped quarter-wave line and (*b*) its equivalent circuit.

purely resistive. As the tap point is moved from the shorted end toward the open end of the line, the impedance seen at the tap point is a pure resistance that varies from zero to the quite high value already given ($R_S = 2Z_0^2/Rl$). This corresponds in the circuit of Fig. 7-20(*b*) to varying the reactances X_L and X_C from low to high values, meanwhile always keeping the circuit tuned (i.e., $X_L = X_C$).

It is of interest to know how the input resistance varies as the tap point is moved along the quarter-wave section. For the relatively high Q circuits used in such applications the voltage distribution along the line may be considered sinusoidal and it is a simple matter to determine the input resistance at any point a distance x from the shorted end. For a given magnitude of voltage and current on the quarter-wave section a certain fixed amount of power input will be required to supply the I^2R losses, regardless of where this power is fed in. This power input is equal to

$$\frac{V_S^2}{R_S} = \frac{V_S^2 Rl}{2Z_0^2}$$

where V_S and R_S are, respectively, the voltage and input resistance at the open end of the section. When the tap point of the feed line is at a distance x from the shorted end [Fig. 7-20(*a*)], the power input is given by V_x^2/R_x, where R_x is the input resistance at the point x. V_x is the voltage at this point and equals $V_S \sin \beta x$. Therefore

$$\frac{V_x^2}{R_x} = \frac{V_S^2 \sin^2 \beta x}{R_x} = \frac{V_S^2 Rl}{2Z_0^2}$$

which gives

$$R_x = \frac{2Z_0^2}{Rl} \sin^2 \beta x$$

Thus the input resistance varies as the square of the sine of the angular distance from the shorted end.

Q of Resonant Transmission-line Sections. One of the important properties of any resonant circuit is its selectivity or its ability to pass freely some frequencies, but to discriminate against others. The selectivity of a resonant circuit may be conveniently stated in terms of the ratio $\Delta f/f_0$, where f_0 is the resonant frequency and $\Delta f = f_2 - f_1$ is the frequency difference between the "half-power" frequencies. In the case of a series-resonant circuit $\Delta f/2$ represents the amount the frequency must be shifted away from the resonant frequency in order to reduce the current to 70.7 per cent of I_0, its value at the resonant frequency. (A constant voltage source is assumed.) Evidently this occurs when the reactance of the circuit becomes equal to the resistance and the phase angle of the circuit is 45 degrees. For the parallel-resonant case $\Delta f/2$ represents the frequency shift away from unity power factor resonance necessary to reduce the voltage across the parallel circuit to 70.7 per cent of its value at resonance. (A constant current source is assumed.) This occurs when the absolute magnitude of the impedance is 70.7 per cent of the impedance at resonance.

The ratio $f_0/\Delta f$ may be used to define the Q of a resonant circuit. The Q of a resonant transmission-line section can be determined as follows:

The input impedance of any shorted line section is given by

$$\begin{aligned} Z_S &= Z_0 \tanh \gamma l \\ &= Z_0 \frac{\sinh \alpha l \cos \beta l + j \cosh \alpha l \sin \beta l}{\cosh \alpha l \cos \beta l + j \sinh \alpha l \sin \beta l} \end{aligned}$$

When the frequency is a resonant frequency f_0, then $\beta l = n\pi/2$ (where n is an odd integer), $\cos \beta l = 0$ and $\sin \beta l = \pm 1$. The expression for the input impedance becomes

$$Z_S = Z_0 \frac{\cosh \alpha l}{\sinh \alpha l} = \frac{Z_0}{\tanh \alpha l} \approx \frac{Z_0}{\alpha l}$$

When the frequency is shifted off resonance by a small amount δf, that is when $f = f_0 + \delta f$, then

$$\beta l = \frac{2\pi f}{v} l = \frac{2\pi(f_0 + \delta f)}{v} l = \frac{n\pi}{2} + \frac{2\pi\, \delta f l}{v}$$

Under these conditions (with $n = 1$)

$$\cos \beta l = - \sin \frac{(2\pi\, \delta f l)}{v}$$

$$\sin \beta l = \cos \frac{(2\pi\, \delta f l)}{v}$$

and

$$Z_S = Z_0 \frac{- \sinh \alpha l \sin \left(\frac{2\pi\, \delta f l}{v}\right) + j \cosh \alpha l \cos \left(\frac{2\pi\, \delta f l}{v}\right)}{- \cosh \alpha l \sin \left(\frac{2\pi\, \delta f l}{v}\right) + j \sinh \alpha l \cos \left(\frac{2\pi\, \delta f l}{v}\right)}$$

For moderately high Q circuits the first term in the numerator is the product of two small quantities and may be neglected in comparison with other terms. Putting

$$\cosh \alpha l \approx 1, \qquad \sinh \alpha l \approx \alpha l, \qquad \cos \left(\frac{2\pi\, \delta f l}{v}\right) \approx 1,$$

$$\sin \left(\frac{2\pi\, \delta f l}{v}\right) \approx \left(\frac{2\pi\, \delta f l}{v}\right)$$

gives

$$Z_S = \frac{Z_0}{\alpha l + j \left(\frac{2\pi\, \delta f l}{v}\right)}$$

When the imaginary term in the denominator is equal to the real term, the impedance Z_S will be 70.7 per cent of its value for a resonant length, and the frequency shift required to make this true will be $\Delta f/2$. Therefore

$$\frac{2\pi\, \Delta f l}{2v} = \alpha l$$

$$\Delta f = \frac{\alpha v}{\pi} = \frac{2\alpha f_0}{\beta}$$

The Q of the resonant section is

$$Q = \frac{f_0}{\Delta f} = \frac{\beta}{2\alpha} \tag{7-123}$$

Alternative forms of this expression are

$$Q = \frac{\pi f_0}{\alpha v} = \frac{2\pi f_0 Z_0}{Rv} = \frac{\omega L}{R} \tag{7-124}$$

The Q is independent of the number of quarter wavelengths in the resonant section as long as αl is a small quantity. It is interesting to observe that the Q of a resonant section of transmission line is equal to the ratio of inductive reactance per unit length to resistance per unit length.

A similar analysis could be carried through for an open-ended resonant section (for which the length would be some multiple of a half wave-length). The expression for Q in this case would be identical with the above.

The Quarter-wave Line as a Transformer. When a section of transmission line is used as a reactance, or as a resonant circuit, it is a two-terminal network. The input terminals of the section are connected across the generator or load and the other terminals are left open or shorted as the case may be. However, a section of line is often used as a four-terminal network, in which case it is inserted in series between generator and load. Because the input impedance is in general different from the load impedance connected across the output terminals, the line section is an impedance-transforming network. This is true for all lengths of line, but the quarter-wave section has certain particular properties that make it very useful in this respect.

For any impedance termination Z_R, the input impedance of a section of lossless line is given by eq. (114) as

$$Z_S = Z_R\left(\frac{\cos\beta l + jZ_0/Z_R \sin\beta l}{\cos\beta l + jZ_R/Z_0 \sin\beta l}\right)$$

For the particular case of a quarter-wave section, $\beta l = \pi/2$, and this reduces to

$$Z_S = \frac{Z_0^2}{Z_R}$$

For the case under consideration, where Z_0 is a pure resistance R_0 this is

$$Z_S = \frac{R_0^2}{Z_R} \tag{7-125}$$

Thus the quarter-wave section is an impedance transformer, or more correctly an impedance inverter. Whatever the terminating impedance may be, the inverse impedance will appear at the input. If the output impedance consisted of a resistance R_2 in series with an inductive reactance X_{L_2}, the input impedance would be given by a resistance R_1 in *parallel* with a capacitive reactance X_{C_1}, where

$$R_1 = \frac{R_0^2}{R_2} \qquad \text{and} \qquad X_{C_1} = \frac{R_0^2}{X_{L_2}}$$

A pure resistance termination R is transformed into a pure resistance of value R_0^2/R.

This property of matching any two impedances Z_1, Z_2 such that $Z_1 Z_2 = Z_0^2$ finds many practical applications. It can be used to join together, without impedance mismatch, lines having different characteristic impedances; it is only necessary to make the characteristic impedance of the quarter-wave matching section the geometric mean of the Z_0's to be matched. By means of the quarter-wave section a pure resistance load can be matched to a generator having a generator impedance that is resistive so long as the geometric mean between the resistances gives a value for the required characteristic impedance that is practicable to obtain.

Voltage Step-up of the Quarter-wave Transformer. As long as the quarter-wave transforming section is considered as being lossless, the ratio between input and output voltages will just be the square root of the ratio of the input and output impedances being matched. From the voltage equation (112), for the quarter-wave section

$$\frac{V_S}{V_R} = \frac{jI_R Z_0}{V_R} = \frac{jZ_0}{Z_R} = j\sqrt{\frac{Z_S}{Z_R}}$$

or calling V_R/V_S the voltage step-up

$$\left|\frac{V_R}{V_S}\right| = \sqrt{\frac{Z_R}{Z_S}}$$

For the infinite impedance termination, that is an open circuit, this simple relation indicates an infinite voltage step-up, and it becomes necessary to resort to the exact eqs. (100) and (101) for the correct answer in this case. For the quarter-wave section the voltage equation of (100) becomes

$$V_S = jV_R \sinh \alpha l + jI_R Z_0 \cosh \alpha l$$

In open circuit I_R is zero and the voltage step-up is

$$\left|\frac{V_R}{V_S}\right| = \frac{1}{\sinh \alpha l} \approx \frac{1}{\alpha l} = \frac{2Z_0}{Rl}$$

For the quarter-wave section this may be written

$$\left|\frac{V_R}{V_S}\right| = \frac{8Z_0}{R\lambda} = \frac{8Z_0 f}{Rv}$$

while for a three-quarter-wave section the voltage step-up would be

$$\left|\frac{V_R}{V_S}\right| = \frac{8Z_0 f}{3Rv}$$

7.17 Transmission-line Charts. Many transmission-line problems may be solved very easily by the use of graphical procedures. Some of these procedures emphasize the voltage reflection coefficient at the load, Γ_R, as defined in (97). In order to develop these procedures it is necessary to use Γ_R and rewrite the voltage and current expressions (92) and (93) in the following form (setting $z = -l$):

$$V = V'(e^{+\gamma l} + \Gamma_R e^{-\gamma l})$$
$$= V' e^{+\gamma l}(1 + \Gamma_R e^{-2\gamma l}) \tag{7-126}$$
$$I = \frac{V'}{Z_0}(e^{+\gamma l} - \Gamma_R e^{-\gamma l})$$
$$= \frac{V'}{Z_0} e^{+\gamma l}(1 - \Gamma_R e^{-2\gamma l}) \tag{7-127}$$

The input impedance Z_{in} at a distance l from the load is thus given by

$$Z_{\text{in}} = \frac{V}{I} = Z_0 \frac{1 + \Gamma_R e^{-2\gamma l}}{1 - \Gamma_R e^{-2\gamma l}} \tag{7-128}$$

which is identical to (102). The impedance expression (128) may be simplified by defining the normalized impedance z:

$$z = \frac{Z_{\text{in}}}{Z_0}$$

A further simplification may be realized if one defines Γ to be the coefficient at a point on the line distant l from the load. It is easy to show that

$$\Gamma = \Gamma_R e^{-2\gamma l}$$

and thus that (128) may be expressed as

$$z = \frac{1 + \Gamma}{1 - \Gamma} \tag{7-129}$$

The quantities z and Γ are complex numbers with real and imaginary parts given by

$$z = r + jx$$
$$\Gamma = u + jv$$

Frequently it is convenient to write Γ in polar co-ordinates,

$$\Gamma = K e^{j\theta}$$

and at the load,

$$\Gamma_R = K_R e^{j\theta_R}$$

Since

$$\gamma = \alpha + j\beta$$

it is clear that

$$\theta = \theta_R - 2\beta l$$
$$K = K_R\, e^{-2\alpha l}$$

Equation (129) is of central importance because it states the relationship between the normalized impedance and the reflection coefficient at any distance from the load. The quantities z and Γ may be plotted in their respective complex planes having axes r, x and u, v. Equation (129) gives the relationship or *mapping* between points in the two complex planes. Since (129) is a *bilinear transformation*,* it has the property that circles map into circles (remember that a straight line is a degenerate circle) and furthermore it is a *conformal* transformation which means that the angle between two line segments is preserved in mapping between the z plane and the Γ plane.

The mapping between the z plane and the Γ plane is illustrated in Fig. 7-21. The reader should verify for himself that the points *ABCDE*

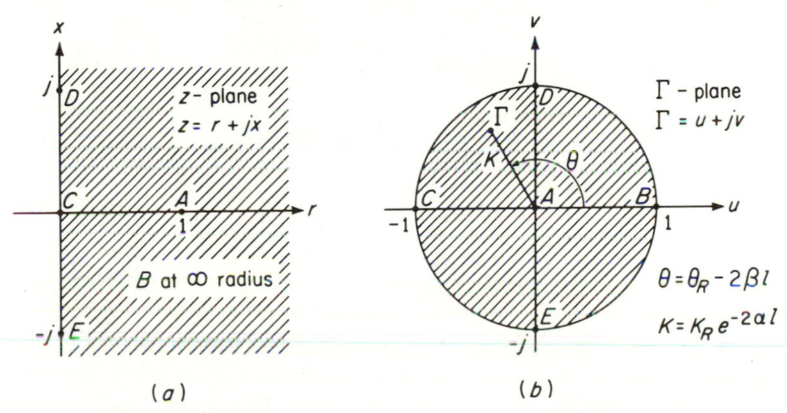

Figure 7-21. The mapping between the normalized impedance plane (*a*) and the reflection coefficient plane (*b*).

map as shown and that the region $r > 0$ in the impedance plane maps into the interior of the unit circle in the reflection coefficient plane. The mapping of the constant-resistance and constant-reactance lines into the Γ plane may be studied by first equating the real and imaginary parts of (129) to give

$$r = \frac{1 - u^2 - v^2}{(1-u)^2 + v^2}, \qquad x = \frac{2v}{(1-u)^2 + v^2} \tag{7-130}$$

which may be written as follows:

$$\left(u - \frac{r}{r+1}\right)^2 + v^2 = \left(\frac{1}{r+1}\right)^2 \tag{7-131}$$

*See, for instance, R. V. Churchill, *Introduction to Complex Variables and Applications*, McGraw-Hill Book Company, New York, 1948.

$$(u - 1)^2 + \left(v - \frac{1}{x}\right)^2 = \frac{1}{x^2} \tag{7-132}$$

Equation (131) represents a circle with center $[r/(r+1), 0]$ and radius $1/(r+1)$ while (132) represents a circle with center $(1, 1/x)$ and radius $1/|x|$. These circles of constant resistance and constant reactance are evident in Fig. 7-22 which shows them in the interior of the unit circle

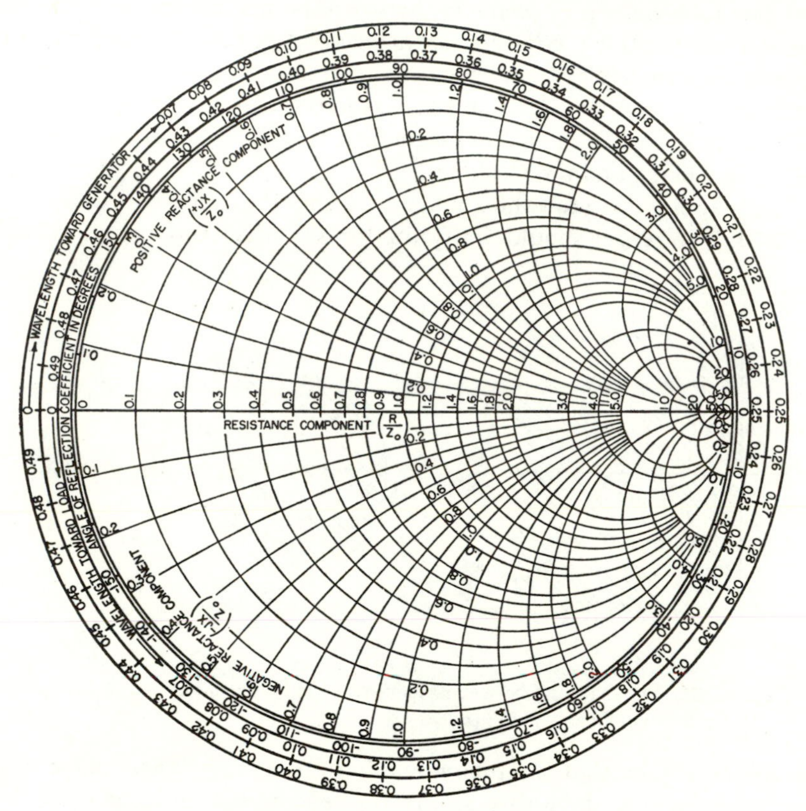

Figure 7-22. Smith chart for transmission-line calculations.

in the reflection coefficient plane. The chart of Fig. 7-22 is useful in transmission-line calculations and is known as the *Smith chart*.* There are several types of transmission-line charts but the Smith chart is the most widely used.

The admittance at any point on the line may be dealt with just as easily as the impedance. The input admittance, characteristic admittance,

*P. H. Smith, "An Improved Transmission-Line Calculator," *Electronics*, **17**, 130 (Jan.,1944); also **12**, 29 (Jan., 1939).

and normalized admittance are $Y = 1/Z$, $Y_0 = 1/Z_0$ and $y = 1/z$. Using these definitions, (129) may be expressed as

$$y = \frac{1 - \Gamma}{1 + \Gamma} \tag{7-133}$$

Note that replacing Γ by $-\Gamma$ in (133) results in the same expression as in (129). This means that the normalized admittance may be found by rotating the appropriate point on the Smith chart by 180 degrees and using the same numerical chart scales as are used for normalized impedance. For the sake of completeness, it should be mentioned that (129) and (133) have inverse forms given respectively by

$$\Gamma = \frac{z - 1}{z + 1} \tag{7-134}$$

$$\Gamma = \frac{1 - y}{1 + y} \tag{7-135}$$

Equations (134) and (135) are frequently useful when mapping from the Γ plane to the z plane or the y plane.

The variation of voltage and current magnitudes along a transmission line also may be studied graphically. This is especially easy to do for a lossless line whose voltage and current magnitudes from (126) and (127) are

$$|V| = |V'|\,|1 + \Gamma| \tag{7-136}$$

$$|I| = \left|\frac{V'}{Z_0}\right|\,|1 - \Gamma| \tag{7-137}$$

Fig. 7-23 shows how the Smith chart may be used to obtain the expressions $|1 + \Gamma|$ and $|1 - \Gamma|$. This diagram is sometimes referred to as the "crank diagram" because of the crank-like clockwise rotation which occurs as the observation point moves along the line away from the load. Inspection of the diagram reveals that the voltage standing-wave ratio is given by

$$\rho = \frac{1 + |\Gamma|}{1 - |\Gamma|} = \frac{1 + K}{1 - K} \tag{7-138}$$

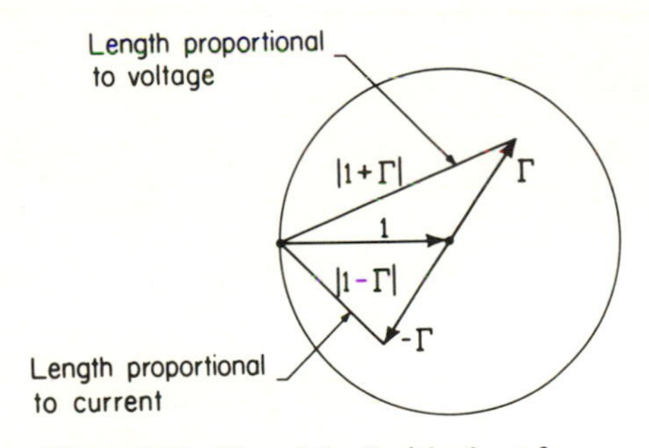

Figure 7-23. Use of the Smith chart for calculation of voltage and current on a lossless line.

and the discussion of sec. 7.15 indicates that the numerical value of ρ may be read from the Smith chart resistance scale to the right of the origin.

7.18 Impedance Matching by Means of Stub Lines. When a line is terminated in an impedance other than its characteristic impedance Z_0, reflection will occur and there will be standing waves of voltage and current along the line which may be very large if there is considerable "mismatch." In general, these standing waves are undesirable because they prevent maximum transfer of power and because they increase the line losses. It is possible to obtain an impedance match between the line and its load by use of a properly located "stub line."

Consider the problem of matching a known load admittance on the end of a lossless line using a single, shorted shunt stub as illustrated in Fig. 7-24. Varying the stub length L varies only the shunt susceptance so the line length l must be so chosen that the input admittance may be brought to the center of the chart by the addition of a shunt susceptance. Thus transformation over the distance l changes the admittance to y_1 which lies on the constant-conductance circle passing through the origin. The stub input admittance y_s (susceptance) is determined by the

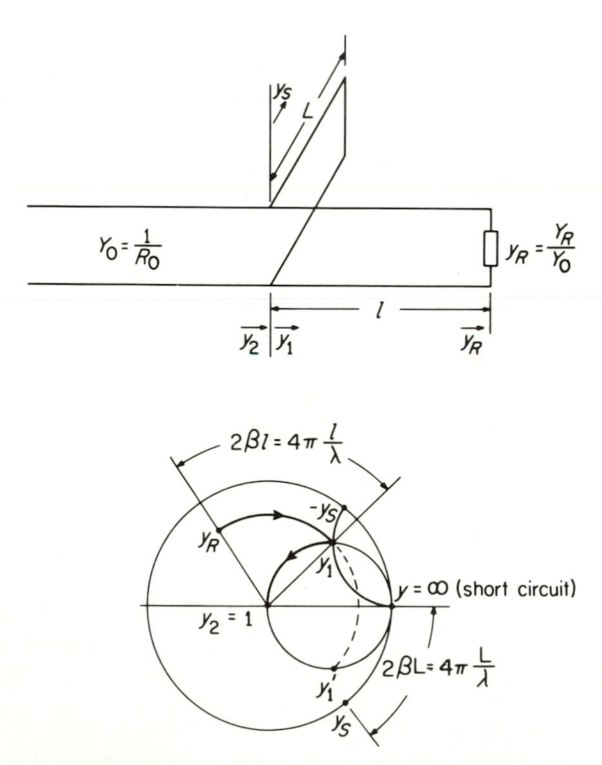

Figure 7-24. Matching a lossless line by means of a single, shorted shunt stub.

length L chosen as shown to give $y_2 = 1$. Note that the line may also be matched using the point y_1' instead of y_1. The standing-wave ratio between load and stub may be read off the resistance (conductance) scale at the point where it is crossed by the dashed arc $y_1 - y_1'$.

The single stub device described above has the advantage that it will match any load but it has the disadvantage that the line length l needs to be adjustable, a requirement which is often very awkward particularly for coaxial lines. One solution to this problem is to use the double stub tuner shown in Fig. 7-25. The stub of length L_1 adds susceptance to bring the point y_1 over to y_2 on the circle lying in the lower half of the Smith chart; this circle is the unity-conductance circle rotated by $\frac{3}{8}\lambda$. Transformation by $\frac{3}{8}\lambda$ brings the admittance to the point y_3 which can then

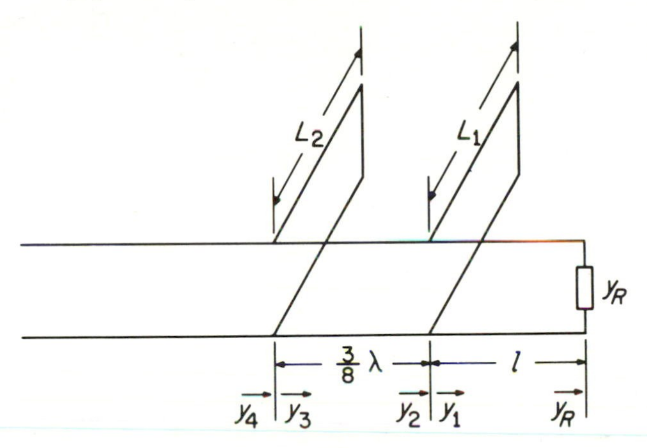

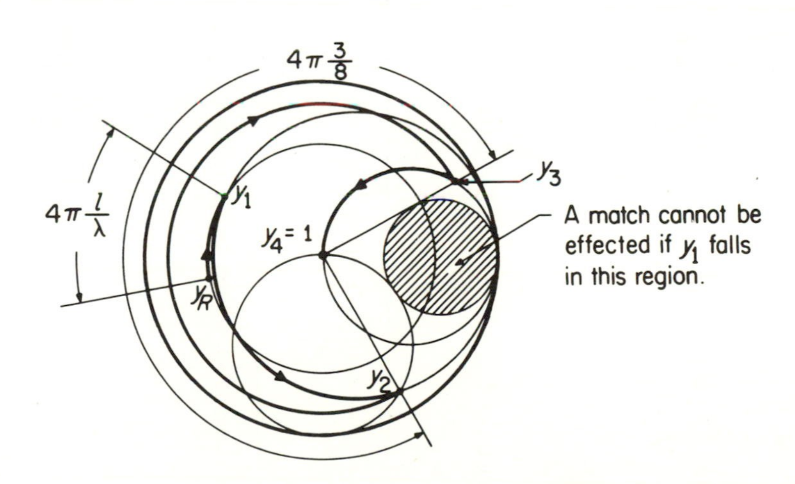

Figure 7-25. The double stub tuner on a lossless line.

Table 7-1

NUMERICAL DATA ON UHF TRANSMISSION LINES

	Parallel Wire Lines Conductor radius $= a$ Conductor spacing (between centers) $= b$	*Coaxial Lines* Outer radius of inner conductor $= a$ Inner radius of outer conductor $= b$
Inductance L (henry/m length of line)	$\frac{\mu_v}{\pi}\cosh^{-1}\frac{b}{2a}$ or $\frac{\mu_v}{\pi}\ln\frac{b}{a}$ for $b \gg a$	$\frac{\mu_v}{2\pi}\ln\frac{b}{a}$
Capacitance C (F/m length of line)	$\frac{\pi\epsilon}{\cosh^{-1} b/2a}$ or $\frac{\pi\epsilon}{\ln b/a}$ for $b \gg a$	$\frac{2\pi\epsilon}{\ln b/a}$
Resistance R (ohms/unit length of line)	$\frac{R_s}{\pi a} = \frac{1}{\pi a}\sqrt{\frac{\omega\mu_v}{2\sigma}}$ ohms/m For copper lines $R = \frac{8.31 \times 10^{-8} f^{1/2}}{a}$ ohms/m $\approx \frac{\sqrt{f_{\text{MHz}}}}{a_{\text{in}}}$ ohms/1000 ft	$\frac{R_s}{2\pi}\left(\frac{1}{a}+\frac{1}{b}\right) = \frac{1}{2\pi}\sqrt{\frac{\omega\mu_v}{2\sigma}}\left(\frac{1}{a}+\frac{1}{b}\right)$ ohms/m For copper lines $R = 4.16 \times 10^{-8} f^{1/2}\left(\frac{1}{a}+\frac{1}{b}\right)$ ohms/m
Conductance G (mhos/m length of line)	ωC (dissipation factor) $\approx \omega C$ (power factor of dielectric)	

Table 7-1—*Continued*

	Parallel Wire Lines Conductor radius $= a$ Conductor spacing (between centers) $= b$	*Coaxial Lines* Outer radius of inner conductor $= a$ Inner radius of outer conductor $= b$
Characteristic impedance $Z_0 = \sqrt{\dfrac{R + j\omega L}{G + j\omega C}} \approx \sqrt{\dfrac{L}{C}}$ (air dielectric)	$120 \cosh^{-1} \dfrac{b}{2a}$ or $276 \log_{10} \dfrac{b}{a}$ for $b \gg a$	$\dfrac{\eta}{2\pi} \ln \dfrac{b}{a} = \dfrac{138}{\sqrt{\epsilon_r}} \log_{10} \dfrac{b}{a}$
Attenuation constant α (neper/m)	$\alpha = \dfrac{R}{2Z_0} + \dfrac{GZ_0}{2}$	
Phase-shift constant β (radians/m)	$\beta = \dfrac{2\pi}{\lambda} = \dfrac{\omega}{v_0}$	
Phase velocity v_0 (meter/sec)	$v_0 \approx \dfrac{1}{\sqrt{LC}} \approx \dfrac{3 \times 10^8}{\sqrt{\mu_r \epsilon_r}}$ for low-loss lines	

For air $\mu_r \approx 1$; $\epsilon_r \approx 1$

For copper $\mu_r \approx 1$; $\epsilon_r \approx 1$; $\sigma = 5.8 \times 10^7$

$\mu_v = 4\pi \times 10^{-7}$ $\epsilon_v \approx \dfrac{1}{36\pi \times 10^9}$

be changed to 1 by the addition of the susceptance due to the stub of length L_2. If a particular load cannot be matched (i.e., if y_1 falls within the shaded region in Fig. 7-25), a change in the length l will make a match possible. The stub lengths and the standing-wave ratios on the line may be found in much the same way as for the case of the single stub tuner. For the double stub tuner it should be noted that there are two sets of stub lengths which will match the load to the line. The reader should work out the alternative procedure in the same way as was done in Fig. 7-25.

PROBLEMS

1. A TEM wave is guided between two perfectly conducting parallel planes (Fig. 7-26). The frequency is 300 MHz. Determine the voltage reading of the (infinite impedance) voltmeter (a) by using Maxwell's electromotive force law (Faraday's induction law); (b) in terms of voltages induced in conductors which are parallel to the electric field.

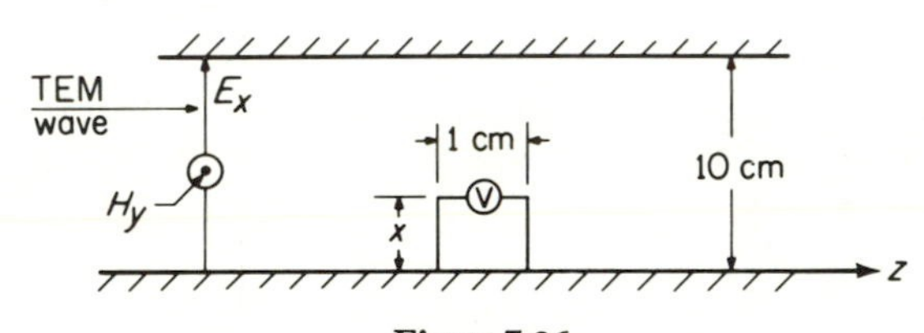

Figure 7-26.

2. Show that the field configuration of the TE_{m0} wave between parallel planes can be obtained by superposing two plane waves that are reflected back and forth between the walls of the guide as indicated in Fig. 7-6.

3. (a) Derive an expression for the attenuation factor for the TM_{10} wave between parallel conducting planes.

(b) Verify that the attenuation is a minimum at a frequency which is $\sqrt{3}$ times the cut-off frequency.

4. For any uniform transmission line, for which R, L, C, and G per unit length are independent of position along the line (and, of course, independent of the magnitude of voltage and current), show that variation along the line of V and I can always be represented by an exponential law.

5. Use Maxwell's equations to show that it is impossible for the TEM wave to exist within any single-conductor wave guide (such as an ordinary rectangular or circular guide).

HINT: For $\oint \mathbf{H} \cdot d\mathbf{s}$ to have value in the transverse plane, there must be a longitudinal flow of current (conduction or displacement).

6. A plane wave propagating in a dielectric medium of permittivity ϵ_1 and permeability $\mu_1 = \mu_v$ is incident at an angle θ_1 upon a second dielectric of permittivity ϵ_2 and permeability $\mu_2 = \mu_v$. The wave is polarized parallel to the plane of incidence. Then, if the electric and magnetic field strengths of the incident wave are E_1 and H_1, the component of E_1 parallel to the boundary surface will be $E_1 \cos \theta_1$ and the component of H_1 parallel to the surface will be H_1, so that the "wave impedance" of medium (1) in a direction normal to the surface would be $E_1 \cos \theta_1/H_1 = \eta_1 \cos \theta_1$. Similarly the "wave impedance" for the refracted ray in medium (2) in the direction normal to the surface would be $E_2 \cos \theta_2/H_2 = \eta_2 \cos \theta_2$. It would be expected when these impedances normal to the boundary surface are equal that there would be no reflection at the surface. Show that the condition that these impedances be equal is the same condition that led to the Brewster angle in eq. (5-91).

7. (a) In chap. 8 (eq. 8-56), the expression for phase velocity in a rectangular guide of any cross section is shown to be $\bar{v} = v_0/\sqrt{1 - \omega_c^2/\omega^2}$ where ω_c is a constant which depends upon the dimensions of the guide. Show that the group velocity defined by $v_g = d\omega/d\bar{\beta}$ is given by $v_g = v_0\sqrt{1 - \omega_c^2/\omega^2}$.

(b) Using the definition $v_g = d\omega/d\bar{\beta}$, show that eq. (7-38) follows from (7-37).

8. The analysis in sec. 7.09 results in expressions for the fields both inside and outside the conductor. Find the fields outside a good conductor by equating the wave impedance with the surface impedance of the conductor (see sec. 5.13). Compare with the results of sec. 7.09.

9. Show how to develop parallel-plate wave-guide theory without assuming the $e^{-\bar{\gamma}z}$ variation in the direction of propagation (use the separation of variables technique).

10. Derive an expression for the inductance and capacitance per unit length of a coaxial transmission line.

11. Repeat for a parallel-wire line (assume perfect conductors).

12. Compute the line "constants" per unit length, R, L, C, G, α, and β, and the characteristic impedance, Z_0, for each of the following lines at the frequencies indicated.

(a) No. 12 wires (diameter = 0.0808 in.) spaced 3 in. apart at 10 MHz; at 100 MHz.

(b) $\frac{3}{8}$-in. diameter rods spaced 1 in. apart at 100 MHz; at 1000 MHz.

(c) A coaxial line having a $\frac{1}{8}$-in. diameter inner conductor and $\frac{3}{8}$-in. outer conductor, at 1000 MHz.

13. (a) A dipole antenna is fed by a transmission line consisting of No. 12 wires at 3-in. spacing. The measured ratio $V_{max}/V_{min} = 4$, and the location of a voltage minimum is 2.8 meters from the antenna feed point. $f = 112$ MHz. Determine the antenna impedance.

(b) If a current indicator is used instead of a voltage indicator, where will the maximum and minimum readings be obtained, and what will be their ratio?

14. A shorted length of a parallel-rod transmission line is connected between grid and plate of a tube to make a UHF oscillator. What should be the length of the line to tune to 300 MHz, if the effective capacitance between grid and plate is 3 pF? The rods are $\frac{3}{8}$ in. in diameter and are spaced 1 in. apart.

15. A lossless transmission line has a characteristic impedance of 300 ohms and is one-quarter wavelength long. What will be the voltage at the open-circuited receiving end, when the sending end is connected to a generator which has 50-ohm internal impedance and a generated voltage of 10 volts?

16. For low-loss transmission-line sections which are much shorter than one-quarter wavelength show that the input reactance can be represented by

$$X_{\text{in}} \approx \omega L_{\text{in}} = \omega L l$$

when the line is shorted, and

$$X_{\text{in}} \approx \frac{1}{\omega C_{\text{in}}} = \frac{1}{\omega C l}$$

when the line is open.

$$L = \frac{Z_0}{v} \qquad \text{and} \qquad C = \frac{1}{vZ_0}$$

are the inductance and capacitance per unit length of the line.

17. Derive the expression for Q [eq. (123)] for an open-end half-wave line.

18. Show that a coaxial line having an outer conductor of radius b will have minimum attenuation when the radius a of the inner conductor satisfies the ratio $b/a = 3.6$.

19. In the Γ plane any straight line passing through $u = 2$, $v = 0$ may be written in the form $u = kv + 2$ where k is any real number. In the z plane this line maps into a circle; find its equation, its center, and its radius and illustrate your result with a rough sketch. In the sketch, show typical circles and indicate the parts of the circles which map into the interior of the Smith-chart unit circle.

20.

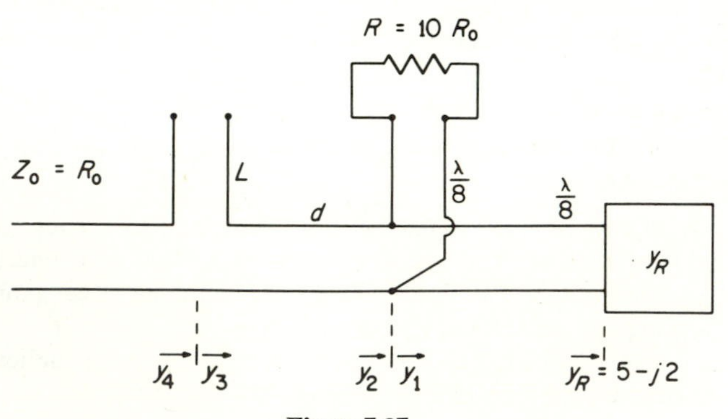

Figure 7-27.

In the stub-matching problem shown in Fig. 7-27, use the Smith chart to find the values of d and L which will result in a match ($y_4 = 1$), making d as short as possible. On the chart mark the points y_1, y_2, y_3, writing them in the form $y = g + jb$. What is the VSWR on the line section of length d?

BIBLIOGRAPHY

Everitt, W. L., *Communication Engineering*, McGraw-Hill Book Company, New York, 1937.

Johnson, W. C., *Transmission Lines and Networks*, McGraw-Hill Book Company, New York, 1950.

Moore, R. K., *Traveling-Wave Engineering*, McGraw-Hill Book Company, New York, 1960.

Chapter 8

WAVE GUIDES

8.01 Rectangular Guides. Practical wave guides usually take the form of rectangular or circular cylinders. Other cross-sectional shapes are possible, but in general these other shapes offer no electrical advantages over the simpler forms in use and are more expensive to manufacture.

In order to determine the electromagnetic field configuration within the guide, Maxwell's equations are solved subject to the appropriate boundary conditions at the walls of the guide. Again assuming perfect conductivity for the walls of the guide, the boundary conditions are simply that E_{tan} and H_{norm} will be zero at the surface of the conductors. For rectangular guides Maxwell's equations and the wave equations are expressed in rectangular co-ordinates and the solution follows almost exactly as for waves between parallel planes. Assuming that variations in the z direction may be expressed as $e^{-\bar{\gamma}z}$, where $\bar{\gamma} = \bar{\alpha} + j\bar{\beta}$, Maxwell's equations become (for the loss-free region within the guide)

$$\begin{aligned} \frac{\partial H_z}{\partial y} + \bar{\gamma} H_y &= j\omega\epsilon E_x & \frac{\partial E_z}{\partial y} + \bar{\gamma} E_y &= -j\omega\mu H_x \\ \frac{\partial H_z}{\partial x} + \bar{\gamma} H_x &= -j\omega\epsilon E_y & \frac{\partial E_z}{\partial x} + \bar{\gamma} E_x &= j\omega\mu H_y \\ \frac{\partial H_y}{\partial x} - \frac{\partial H_x}{\partial y} &= j\omega\epsilon E_z & \frac{\partial E_y}{\partial x} - \frac{\partial E_x}{\partial y} &= -j\omega\mu H_z \end{aligned} \qquad (8\text{-}1)$$

and the wave equations for E_z and H_z are

$$\left.\begin{aligned} \frac{\partial^2 E_z}{\partial x^2} + \frac{\partial^2 E_z}{\partial y^2} + \bar{\gamma}^2 E_z &= -\omega^2 \mu\epsilon E_z \\ \frac{\partial^2 H_z}{\partial x^2} + \frac{\partial^2 H_z}{\partial y^2} + \bar{\gamma}^2 H_z &= -\omega^2 \mu\epsilon H_z \end{aligned}\right\} \qquad (8\text{-}2)$$

Equations (1) can be combined into the form

$$H_x = -\frac{\bar{\gamma}}{h^2}\frac{\partial H_z}{\partial x} + j\frac{\omega\epsilon}{h^2}\frac{\partial E_z}{\partial y}$$

$$
\left.\begin{aligned}
H_y &= -\frac{\bar{\gamma}}{h^2}\frac{\partial H_z}{\partial y} - j\frac{\omega\epsilon}{h^2}\frac{\partial E_z}{\partial x} \\
E_x &= -\frac{\bar{\gamma}}{h^2}\frac{\partial E_z}{\partial x} - j\frac{\omega\mu}{h^2}\frac{\partial H_z}{\partial y} \\
E_y &= -\frac{\bar{\gamma}}{h^2}\frac{\partial E_z}{\partial y} + j\frac{\omega\mu}{h^2}\frac{\partial H_z}{\partial x}
\end{aligned}\right\} \qquad (8\text{-}3)
$$

where

$$h^2 = \bar{\gamma}^2 + \omega^2\mu\epsilon$$

These equations give the relationships among the fields within the guide. It will be noticed that, if E_z and H_z are both zero, *all* the fields within the guide will vanish. Therefore, for wave-guide transmission (no inner conductor) there must exist either an E_z or an H_z component. As in the case of waves between parallel planes, it is convenient to divide the possible field configurations within the guide into two sets, transverse magnetic (TM) waves for which $H_z \equiv 0$, and transverse electric (TE) waves for which $E_z \equiv 0$. For the rectangular guide shown in Fig. 8-1 the boundary conditions are:

$$E_x = E_z = 0 \qquad \text{at } y = 0 \text{ and } y = b$$

$$E_y = E_z = 0 \qquad \text{at } x = 0 \text{ and } x = a$$

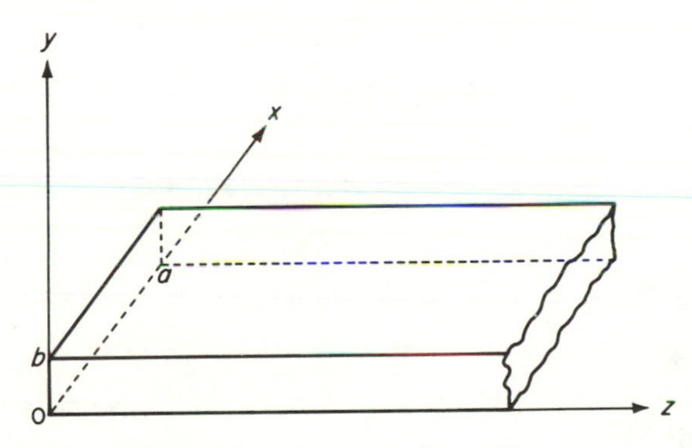

Figure 8-1. A rectangular guide.

8.02 Transverse Magnetic Waves in Rectangular Guides. The wave equations (2) are partial differential equations that can be solved by the usual technique of assuming a product solution. This procedure leads to two ordinary differential equations, the solutions of which are known. Noting that

$$E_z(x, y, z) = E_z^0(x, y)\, e^{-\bar{\gamma}z}$$

let

$$E_z^0 = XY \qquad (8\text{-}4)$$

where X is a function of x alone, and Y is a function of y alone. Inserting (4) in (2) gives

$$Y\frac{d^2X}{dx^2} + X\frac{d^2Y}{dy^2} + \bar{\gamma}^2 XY = -\omega^2 \mu\epsilon XY$$

Putting $h^2 = \bar{\gamma}^2 + \omega^2\mu\epsilon$ as before, this becomes

$$Y\frac{d^2X}{dx^2} + X\frac{d^2Y}{dy^2} + h^2 XY = 0$$

Divide by XY,

$$\frac{1}{X}\frac{d^2X}{dx^2} + h^2 = -\frac{1}{Y}\frac{d^2Y}{dy^2} \tag{8-5}$$

Equation (5) equates a function of x alone to a function of y alone. The only way in which such a relation can hold for all values of x and y is to have each of these functions equal to some constant, say A^2. Then

$$\frac{1}{X}\frac{d^2X}{dx^2} + h^2 = A^2 \tag{8-6}$$

$$\frac{1}{Y}\frac{d^2Y}{dy^2} = -A^2 \tag{8-7}$$

A solution of eq. (6) is

$$X = C_1 \cos Bx + C_2 \sin Bx$$

where

$$B^2 = h^2 - A^2$$

The solution of eq. (7) is

$$Y = C_3 \cos Ay + C_4 \sin Ay$$

This gives

$$\begin{aligned} E_z^0 = XY &= C_1C_3 \cos Bx \cos Ay + C_1C_4 \cos Bx \sin Ay \\ &+ C_2C_3 \sin Bx \cos Ay + C_2C_4 \sin Bx \sin Ay \end{aligned} \tag{8-8}$$

The constants $C_1, C_2, C_3, C_4, A,$ and B must now be selected to fit the boundary conditions, viz.,

$$E_z^0 = 0 \qquad \text{when } x = 0, x = a, y = 0, y = b$$

If $x = 0$, the general expression (8) becomes

$$E_z^0 = C_1C_3 \cos Ay + C_1C_4 \sin Ay$$

For E_z^0 to vanish (for all values of y) it is evident that C_1 must be zero. Then the general expression for E_z^0 will be

$$E_z^0 = C_2C_3 \sin Bx \cos Ay + C_2C_4 \sin Bx \sin Ay \tag{8-9}$$

When $y = 0$, eq. (9) reduces to

$$E_z^0 = C_2C_3 \sin Bx$$

For this to be zero for all values of x it is possible to have either C_2 or C_3 equal to zero (assuming $B \neq 0$). Putting $C_2 = 0$ in (9) would make E_z^0 identically zero, so instead C_3 will be put equal to zero. Then the general expression (9) for E_z^0 reduces to

$$E_z^0 = C_2 C_4 \sin Bx \sin Ay \tag{8-10}$$

In addition to the amplitude constant $C = C_2 C_4$, there are still two unknown constants, A and B. However, there are two more boundary conditions to be applied.

If $x = a$

$$E_z^0 = C \sin Ba \sin Ay$$

In order for this to vanish for all values of y (and assuming $A \neq 0$, because $A = 0$ would make E_z^0 identically zero) the constant B must have the value

$$B = \frac{m\pi}{a} \qquad \text{where } m = 1, 2, 3, \ldots$$

Again if $y = b$,

$$E_z^0 = C \sin \frac{m\pi}{a} x \sin Ab$$

and for this to vanish for all values of x, A must have the value

$$A = \frac{n\pi}{b} \qquad \text{where } n = 1, 2, 3, \ldots$$

Therefore the final expression for E_z^0 is

$$E_z^0 = C \sin \frac{m\pi}{a} x \sin \frac{n\pi}{b} y \tag{8-11}$$

Making use of eqs. (3) and putting $\bar{\gamma} = j\bar{\beta}$ (as in sec. 7.02) for frequencies above the cut-off frequency, the following expressions are obtained:

$$\left.\begin{aligned} E_x^0 &= \frac{-j\bar{\beta}C}{h^2} B \cos Bx \sin Ay \\ E_y^0 &= \frac{-j\bar{\beta}C}{h^2} A \sin Bx \cos Ay \\ H_x^0 &= \frac{j\omega\epsilon C}{h^2} A \sin Bx \cos Ay \\ H_y^0 &= \frac{-j\omega\epsilon C}{h^2} B \cos Bx \sin Ay \end{aligned}\right\} \tag{8-12}$$

where $$B = \frac{m\pi}{a} \qquad \text{and} \qquad A = \frac{n\pi}{b} \tag{8-13}$$

These expressions show how each of the components of electric and magnetic field strengths varies with x and y. The variation with time and along the axis of the guide, that is in the z direction, is shown by putting back into each of these expressions the factor $e^{j\omega t-\bar{\gamma}z}$ and then taking the real part.

In the derivation of the fields it was found necessary to restrict the constants A and B to the values given by expressions (13). In these expressions a and b are the width and height of the guide, and m and n are integers. Now, by definition,

$$A^2 + B^2 = h^2$$

and

$$h^2 = \bar{\gamma}^2 + \omega^2 \mu\epsilon$$

Therefore,

$$\begin{aligned}\bar{\gamma} &= \sqrt{h^2 - \omega^2\mu\epsilon} \\ &= \sqrt{A^2 + B^2 - \omega^2\mu\epsilon} \\ &= \sqrt{\left(\frac{m\pi}{a}\right)^2 + \left(\frac{n\pi}{b}\right)^2 - \omega^2\mu\epsilon}\end{aligned} \tag{8-14}$$

Equation (14) defines the propagation constant for a rectangular guide for TM waves. For low frequencies, where $\omega^2\mu\epsilon$ is small, $\bar{\gamma}$ will be a real number. The propagation constant met with in ordinary transmission-line theory is a complex number, that is $\bar{\gamma} = \bar{\alpha} + j\bar{\beta}$, where $\bar{\alpha}$ is the attenuation constant (attenuation per unit length) and $\bar{\beta}$ is the phase-shift constant (phase shift per unit length). If $\bar{\gamma}$ is real, $\bar{\beta}$ must be zero, and there can be no phase shift along the tube. This means there can be no wave motion along the tube for low frequencies. However, as the frequency is increased, a value for ω will be reached that will make the expression under the radical in (14) equal to zero. If this value of ω is called ω_c, then for all values of ω greater than ω_c, the propagation constant $\bar{\gamma}$ will be imaginary and will have the form $\bar{\gamma} = j\bar{\beta}$. For the case under consideration (perfectly conducting walls) the attenuation constant $\bar{\alpha}$ is zero for all frequencies such that $\omega > \omega_c$. For these frequencies

$$\bar{\beta} = \sqrt{\omega^2\mu\epsilon - \left(\frac{m\pi}{a}\right)^2 - \left(\frac{n\pi}{b}\right)^2} \tag{8-15}$$

The value of ω_c is given by

$$\omega_c = \frac{1}{\sqrt{\mu\epsilon}}\sqrt{\left(\frac{m\pi}{a}\right)^2 + \left(\frac{n\pi}{b}\right)^2} \tag{8-16}$$

The cut-off frequency, that is the frequency below which wave propagation will not occur, is

$$f_c = \frac{1}{2\pi\sqrt{\mu\epsilon}}\sqrt{\left(\frac{m\pi}{a}\right)^2 + \left(\frac{n\pi}{b}\right)^2} \tag{8-17}$$

and the corresponding cut-off wavelength is

$$\lambda_c = \frac{2}{\sqrt{\left(\frac{m}{a}\right)^2 + \left(\frac{n}{b}\right)^2}} \tag{8-18}$$

from which it is clear that

$$f_c\lambda_c = v_0$$

The velocity of wave propagation will be given by

$$\bar{v} = \frac{\omega}{\bar{\beta}} = \frac{\omega}{\sqrt{\omega^2\mu\epsilon - \left(\frac{m\pi}{a}\right)^2 - \left(\frac{n\pi}{b}\right)^2}} \tag{8-19}$$

This last expression indicates that the velocity of propagation of the wave in the guide is greater than the phase velocity in free space. As the frequency is increased above cut-off, the phase velocity decreases from an infinitely large value and approaches v_0 ($v_0 = c$ in free space) as the frequency increases without limit.

Since the wavelength in the guide is given by $\bar{\lambda} = \bar{v}/f$, it will be longer than the corresponding free-space wavelength. From the expression for $\bar{v}$

$$\bar{\lambda} = \frac{2\pi}{\sqrt{\omega^2\mu\epsilon - \left(\frac{m\pi}{a}\right)^2 - \left(\frac{n\pi}{b}\right)^2}} \tag{8-20}$$

In the above expressions the only restriction on m and n is that they be integers. However from eqs. (12) and (13) it is seen that if either m or n is zero the fields will all be identically zero. Therefore the lowest possible value for either m or n (for TM waves) is unity. From eq. (17) it is evident that the lowest cut-off frequency will occur for $m = n = 1$. Substituting these values in eqs. (13) gives the fields for the lowest frequency TM wave which can be propagated through the guide. This particular wave is called the TM_{11} wave for obvious reasons. Higher-order waves (larger values of m and n) require higher frequencies in order to be propagated along a guide of given dimensions.

8.03 Transverse Electric Waves in Rectangular Guides. The equations for transverse electric waves ($E_z = 0$) can be derived in a manner similar to that for transverse magnetic waves. This is left as an exercise for the student. H_z^0 will be found to have the same general form as eq. (8). This is differentiated with respect to x and y to find E_x^0, E_y^0, H_x^0, and H_y^0. The boundary conditions are then applied to E_x^0 and E_y^0 to give the resulting expressions:

$$\left.\begin{aligned}
H_z^0 &= C \cos Bx \cos Ay \\
H_x^0 &= \frac{j\bar{\beta}}{h^2} CB \sin Bx \cos Ay \\
H_y^0 &= \frac{j\bar{\beta}}{h^2} CA \cos Bx \sin Ay \\
E_x^0 &= \frac{j\omega\mu}{h^2}\, CA \cos Bx \sin Ay \\
E_y^0 &= -\frac{j\omega\mu}{h^2} CB \sin Bx \cos \mathrm{Ay} \\
B &= \frac{m\pi}{a} \qquad A = \frac{n\pi}{b}
\end{aligned}\right\} \tag{8-21}$$

In the above expression $\bar{\gamma}$ has been put equal to $j\bar{\beta}$, which is valid for frequencies above cut-off.

For TE waves the equations for $\bar{\beta}, f_c, \lambda_c, \bar{v}$, and $\bar{\lambda}$ are found to be identical to those for TM waves. However, in eqs. (21) for TE waves it will be found possible to make either m or n (but not both) equal to zero without causing all the fields to vanish. That is, a lower order is possible than in the TM wave case. The lowest-order TE wave in rectangular guides is therefore the TE_{10} wave. This wave which has the lowest cut-off frequency is called the *dominant* wave.

It is seen that the subscripts m and n represent the number of half-period variations of the field along the x and y co-ordinates respectively. By convention,* the x co-ordinate is assumed to coincide with the larger transverse dimension, so the TE_{10} wave has a lower cut-off frequency than the TE_{01}.

For practical reasons in most experimental work with rectangular guides the dominant TE_{10} wave is used. For this wave, substituting $m = 1$ and $n = 0$, the fields are

$$\left.\begin{aligned}
H_z^0 &= C \cos \frac{\pi x}{a} \\
H_x^0 &= \frac{j\bar{\beta}aC}{\pi} \sin \frac{\pi x}{a} \\
E_y^0 &= \frac{-j\omega\mu aC}{\pi} \sin \frac{\pi x}{a} \\
E_x^0 &= H_y = 0
\end{aligned}\right\} \tag{8-22}$$

$$\bar{\beta} = \sqrt{\omega^2 \mu\epsilon - \left(\frac{\pi}{a}\right)^2}$$

*"Definition of Terms Relating to Wave Guides," *IRE Standards on Radio Wave Propagation*, 1945.

$$f_c = \frac{c}{2a} \qquad \lambda_c = 2a \qquad h = \frac{\pi}{a}$$

For the TE_{10} wave the cut-off frequency is that frequency for which the corresponding (free-space) half wavelength is equal to the width of the

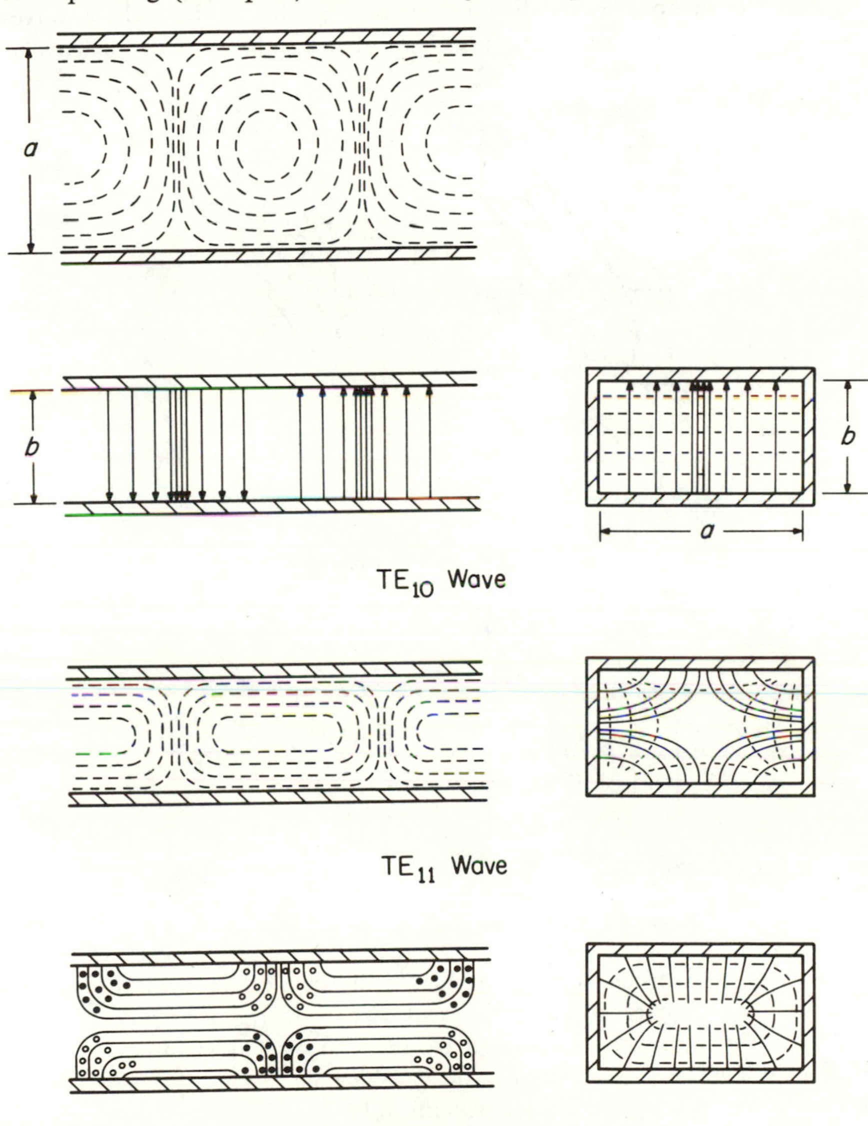

Figure 8-2. Electric (solid) and magnetic (dashed) field configurations for the lower-order modes in a rectangular guide.

guide. For the TE_{10} wave the cut-off frequency is independent of the dimension b.

In Fig. 8-2 are sketched the field configurations for the lower-order TE and TM waves in rectangular guides.

Possible methods for feeding rectangular guides so that these waves may be initiated are shown in Fig. 8-3. In order to launch a particular

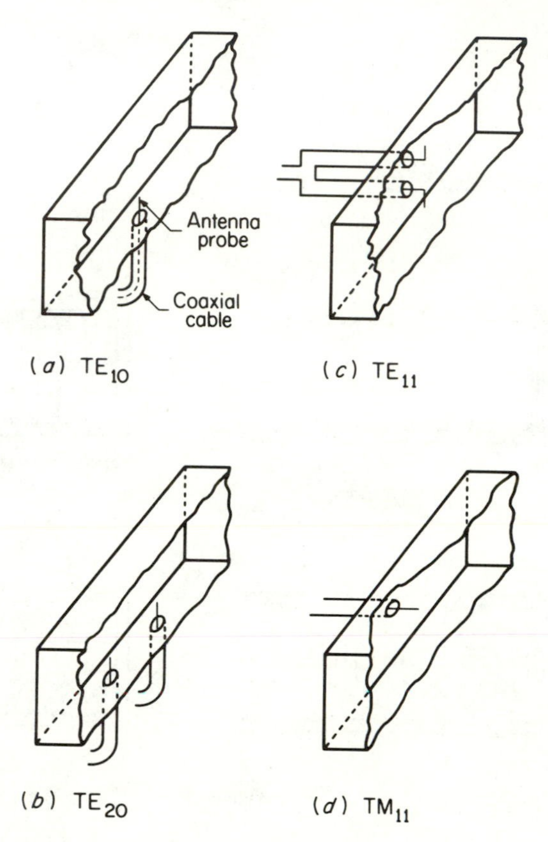

Figure 8-3. Excitation methods for various modes.

mode, a type of probe is chosen which will produce lines of E and H that are roughly parallel to the lines of E and H for that mode. Thus in Fig. 8-3(a) the probe is parallel to the y axis and so produces lines of E in the y direction and lines of H which lie in the x-z plane. This is the correct field configuration for the TE_{10} mode. In (b), the parallel probes fed with opposite phase tend to set up the TE_{20} mode. In (d) the probe parallel to the z axis produces magnetic field lines in the x-y plane, which is correct for the TM modes. The field configuration due to probes and antennas is the subject of chap. 10.

It is possible for several modes to exist simultaneously in a guide if the frequency is above cut-off for those particular modes. However the guide dimensions are often chosen so that only the dominant mode can exist.

Problem 1. A rectangular guide has cross-section dimensions

$$a = 7 \text{ cm} \qquad b = 4 \text{ cm}$$

Determine all the modes which will propagate at a frequency of (a) 3000 MHz, (b) 5000 MHz.

Problem 2. Starting with expressions (8-16) and (8-20) derive the relation

$$\lambda = \frac{\bar{\lambda}\lambda_c}{\sqrt{\bar{\lambda}^2 + \lambda_c^2}}$$

where λ is the free-space wavelength ($\lambda = c/f$), $\bar{\lambda}$ is the wavelength measured in the guide, and λ_c is the cut-off wavelength.

Problem 3. Show by means of arrows the directions of the instantaneous Poynting vector for the TE_{10} wave in a rectangular guide.

Problem 4. (a) Indicate the (instantaneous) directions of current flow in all the walls of a rectangular guide carrying a TE_{10} wave.

(b) Where in the guide could slots be cut without affecting operation?

Problem 5. Starting with eqs. (8-2) and (8-3) derive expressions (8-21) for TE waves.

8.04 Impossibility of TEM Wave in Wave Guides. The waves that will propagate inside hollow rectangular cylinders have been divided into two sets: the transverse magnetic waves of eqs. (11) and (12) which have no z component of H, and the transverse electric waves of eqs. (21) that have no z component of E. It will be found that corresponding sets of TM and TE waves can also propagate within circular wave guides, or indeed, in cylindrical guides of any cross-sectional shape. It is easily shown, however, that the familiar TEM wave, for which there is no axial component of either E or H, cannot possibly propagate within a single-conductor wave guide.

Suppose a TEM wave is assumed to exist within a hollow guide of any shape. Then lines of H must lie entirely in the transverse plane. Also in a nonmagnetic material,

$$\nabla \cdot \mathbf{H} = 0$$

which requires that the lines of H be *closed* loops. Therefore, if a TEM wave exists in the guide, the lines of H will be closed loops in plane perpendicular to the axis. Now by Maxwell's first equation the magnetomotive force around each of these closed loops must equal the axial current (conduction or displacement) through the loop. In the case of a "guide" with an inner conductor, e.g., a coaxial transmission line, this axial current through the H loops is the conduction current in the inner conductor. However, for a hollow wave guide having no inner conductor, this axial current must be a displacement current. But an axial displacement current requires an axial component of E, something not present in a TEM wave. Therefore the TEM wave cannot exist in a single-conductor wave guide.

8.05 Bessel Functions. In solving for the electromagnetic fields within guides of circular cross section, a differential equation known as Bessel's equation is encountered. The solution of the equation leads to *Bessel Functions*. These functions will be considered briefly in this section in preparation for the following section on circular wave guides. These same functions can be expected to appear in any two-dimensional problem in which there is circular symmetry. Examples of such problems are the vibrations of a circular membrane, the propagation of waves within a circular cylinder, and the electromagnetic field distribution about an infinitely long wire.

The differential equation involved in these problems has the form

$$\frac{d^2P}{d\rho^2} + \frac{1}{\rho}\frac{dP}{d\rho} + \left(1 - \frac{n^2}{\rho^2}\right)P = 0 \tag{8-23}$$

where n is any integer.* One solution to this equation can be obtained by assuming a power-series solution

$$P = a_0 + a_1\rho + a_2\rho^2 + \dots \tag{8-24}$$

Substitution of this assumed solution back into (23) and equating the coefficients of like powers leads to a series solution for the differential equation. For example in the special case where $n = 0$, eq. (23) is

$$\frac{d^2P}{d\rho^2} + \frac{1}{\rho}\frac{dP}{d\rho} + P = 0 \tag{8-25}$$

When the power series (24) is inserted in (25) and the sums of the coefficients of each power of ρ are equated to zero, the following series is obtained

$$\begin{aligned} P = P_1 &= C_1\left[1 - \left(\frac{\rho}{2}\right)^2 + \frac{(\frac{1}{2}\rho)^4}{(2!)^2} - \frac{(\frac{1}{2}\rho)^6}{(3!)^2} + \dots\right] \\ &= C_1\left[1 - \frac{\rho^2}{2^2} + \frac{\rho^4}{2^2\cdot 4^2} - \frac{\rho^6}{2^2\cdot 4^2\cdot 6^2} + \dots\right] \\ &= C_1\sum_{r=0}^{\infty}(-1)^r\frac{(\frac{1}{2}\rho)^{2r}}{(r!)^2} \end{aligned} \tag{8-26}$$

This series is convergent for all values of ρ, either real or complex. It is called Bessel's function of the *first kind* of *order zero* and is denoted by the symbol

$$J_0(\rho)$$

The zero order refers to the fact that it is the solution of (23) for the

*If n is not restricted to integral values, the symbol ν is used. See Appendix II.

case of $n = 0$. The corresponding solutions for $n = 1, 2, 3$, etc., are designated $J_1(\rho)$, $J_2(\rho)$, $J_3(\rho)$, where the subscript n denotes the order of the Bessel function. Since eq. (23) is a second-order differential equation, there must be two linearly independent solutions for each value of n. The second solution may be obtained in a manner somewhat similar to that used for the first, but starting with a slightly different series that is suitably manipulated to yield a solution.* This second solution is known as Bessel's function of the *second kind*, or Neumann's function, and is designated by the symbol†

$$N_n(\rho)$$

where again n indicates the order of the function. For the zero order of this solution of the second kind, the following series is obtained

$$N_0(\rho) = \frac{2}{\pi}\left\{\ln\left(\frac{\rho}{2}\right) + \gamma\right\}J_0(\rho) - \frac{2}{\pi}\sum_{r=1}^{\infty}(-1)^r\frac{(\frac{1}{2}\rho)^{2r}}{(r!)^2}\left(1 + \frac{1}{2} + \frac{1}{3} + \ldots + \frac{1}{r}\right) \tag{8-27}$$

The complete solution of (25) is then

$$P = AJ_0(\rho) + BN_0(\rho) \tag{8-28}$$

A plot of $J_0(\rho)$ and $N_0(\rho)$ is shown in Fig. 8-4. Because all the Neumann functions become infinite at $\rho = 0$, these second solutions cannot be

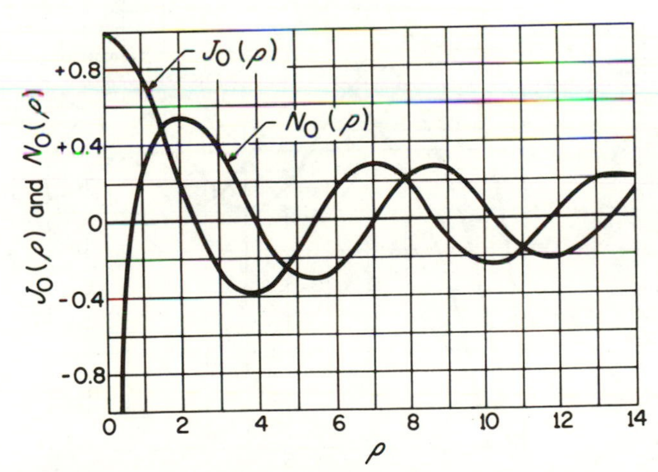

Figure 8-4. Zero-order Bessel functions of the first and second kinds.

*N.W. McLachlan, *Bessel Functions for Engineers*, Oxford University Press, New York, 1934.

†The symbol $Y(\rho)$ is used in some texts and tables. It should be noted that there are other forms for this second solution which differ by a constant from the one given.

used for any physical problem in which the origin is included, as for example the hollow wave-guide problem.

It is apparent that [except near the origin for $N_0(\rho)$] these curves bear a marked similarity to damped cosine and sine curves. Indeed, for large values of (ρ) these functions do approach the sinusoidal forms.

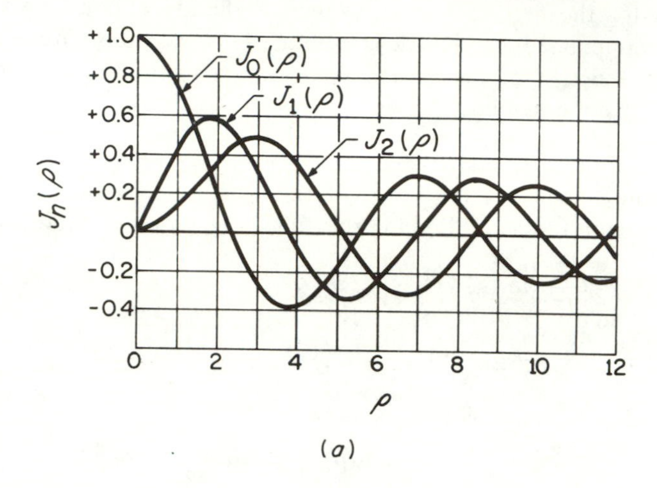

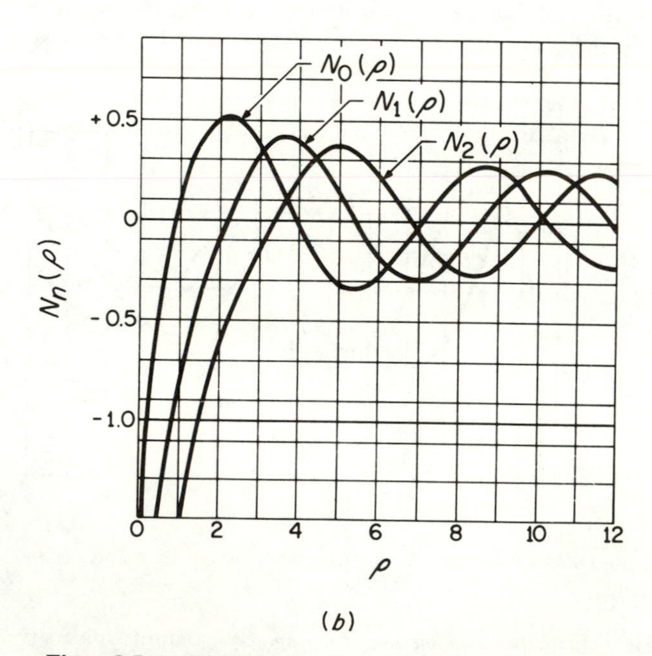

Figure 8-5. (*a*) Higher-order Bessel functions of the first kind. (*b*) Higher-order Bessel functions of the second kind (Neumann functions).

As ρ becomes very large

$$J_0(\rho) \to \sqrt{\frac{2}{\pi\rho}} \cos\left(\rho - \frac{\pi}{4}\right) \tag{8-29}$$

$$N_0(\rho) \to \sqrt{\frac{2}{\pi\rho}} \sin\left(\rho - \frac{\pi}{4}\right) \tag{8-30}$$

Figs. 8-5(*a*) and (*b*) show Bessel functions of the first and second kinds for the higher orders. A further discussion of these functions is given in Appendix II.

8.06 Solution of the Field Equations: Cylindrical Co-ordinates. The method of solution of the electromagnetic equations for guides of circular cross section is similar to that followed for rectangular guides. However, in order to simplify the application of the boundary conditions (electric field tangential to the surface equals zero), it is expedient to express the field equations and the wave equations in the cylindrical co-ordinate system.

In cylindrical co-ordinates in a nonconducting region (and again assuming the variation with z to be given by $e^{-\bar{\gamma}z}$), Maxwell's equations are

$$\left.\begin{aligned}
\frac{\partial H_z}{\rho\partial\phi} + \bar{\gamma}H_\phi &= j\omega\epsilon E_\rho \\
\frac{\partial E_z}{\rho\partial\phi} + \bar{\gamma}E_\phi &= -j\omega\mu H_\rho \\
-\bar{\gamma}H_\rho - \frac{\partial H_z}{\partial\rho} &= j\omega\epsilon E_\phi \\
-\bar{\gamma}E_\rho - \frac{\partial E_z}{\partial\rho} &= -j\omega\mu H_\phi \\
\frac{1}{\rho}\left(\frac{\partial(\rho H_\phi)}{\partial\rho} - \frac{\partial H_\rho}{\partial\phi}\right) &= j\omega\epsilon E_z \\
\frac{1}{\rho}\left(\frac{\partial(\rho E_\phi)}{\partial\rho} - \frac{\partial E_\rho}{\partial\phi}\right) &= -j\omega\mu H_z
\end{aligned}\right\} \tag{8-31}$$

These equations can be combined to give

$$\left.\begin{aligned}
h^2 H_\rho &= j\frac{\omega\epsilon}{\rho}\frac{\partial E_z}{\partial\phi} - \bar{\gamma}\frac{\partial H_z}{\partial\rho} \\
h^2 H_\phi &= -j\omega\epsilon\frac{\partial E_z}{\partial\rho} - \frac{\bar{\gamma}}{\rho}\frac{\partial H_z}{\partial\phi} \\
h^2 E_\rho &= -\bar{\gamma}\frac{\partial E_z}{\partial\rho} - j\frac{\omega\mu}{\rho}\frac{\partial H_z}{\partial\phi} \\
h^2 E_\phi &= -\frac{\bar{\gamma}}{\rho}\frac{\partial E_z}{\partial\phi} + j\omega\mu\frac{\partial H_z}{\partial\rho} \\
h^2 &= \bar{\gamma}^2 + \omega^2\mu\epsilon
\end{aligned}\right\} \tag{8-32}$$

The wave equation in cylindrical co-ordinates for E_z is

$$\frac{\partial^2 E_z}{\partial \rho^2} + \frac{1}{\rho^2}\frac{\partial^2 E_z}{\partial \phi^2} + \frac{\partial^2 E_z}{\partial z^2} + \frac{1}{\rho}\frac{\partial E_z}{\partial \rho} = -\omega^2 \mu\epsilon E_z \qquad (8\text{-}33)$$

Proceeding in a manner similar to that followed in the rectangular case, let

$$E_z = P(\rho)Q(\phi)e^{-\bar{\gamma}z} = E_z^0 e^{-\bar{\gamma}z} \qquad (8\text{-}34)$$

where $P(\rho)$ is a function of ρ alone and $Q(\phi)$ is a function of ϕ alone. Substituting the expression for E_z in the wave equation gives

$$Q\frac{d^2 P}{d\rho^2} + \frac{Q}{\rho}\frac{dP}{d\rho} + \frac{P}{\rho^2}\frac{d^2 Q}{d\phi^2} + PQ\bar{\gamma}^2 + \omega^2\mu\epsilon PQ = 0$$

Divide by PQ,

$$\frac{1}{P}\frac{d^2 P}{d\rho^2} + \frac{1}{\rho P}\frac{dP}{d\rho} + \frac{1}{Q\rho^2}\frac{d^2 Q}{d\phi^2} + h^2 = 0 \qquad (8\text{-}35)$$

As before, eq. (35) can be broken up into two ordinary differential equations

$$\frac{d^2 Q}{d\phi^2} = -n^2 Q \qquad (8\text{-}36)$$

$$\frac{d^2 P}{d\rho^2} + \frac{1}{\rho}\frac{dP}{d\rho} + \left(h^2 - \frac{n^2}{\rho^2}\right) P = 0 \qquad (8\text{-}37)$$

where n is a constant. The solution of eq. (36) is

$$Q = (A_n \cos n\phi + B_n \sin n\phi) \qquad (8\text{-}38)$$

Dividing through by h^2, eq. (37) is transformed into

$$\frac{d^2 P}{d(\rho h)^2} + \frac{1}{(\rho h)}\frac{dP}{d(\rho h)} + \left[1 - \frac{n^2}{(\rho h)^2}\right] P = 0 \qquad (8\text{-}39)$$

This is a standard form of Bessel's equation in terms of (ρh). Using only the solution that is finite at $(\rho h) = 0$, gives

$$P(\rho h) = J_n(\rho h) \qquad (8\text{-}40)$$

where $J_n(\rho h)$ is Bessel's function of the first kind of order n. Substituting the solutions (38) and (40) in (34),

$$E_z = J_n(\rho h)(A_n \cos n\phi + B_n \sin n\phi)e^{-\bar{\gamma}z} \qquad (8\text{-}41)$$

The solution for H_z will have exactly the same form as for E_z and can therefore be written

$$H_z = J_n(\rho h)(C_n \cos n\phi + D_n \sin n\phi)e^{-\bar{\gamma}z} \qquad (8\text{-}42)$$

For TM waves the remaining field components can be obtained by inserting (41) into (32). For TE waves (42) must be inserted into the set corresponding to (32).

8.07 TM and TE Waves in Circular Guides. As in the case of rectangular guides, it is convenient to divide the possible solutions for circular guides into transverse magnetic and transverse electric waves. For the TM waves H_z is identically zero and the wave equation for E_z is used. The boundary conditions require that E_z must vanish at the surface of the guide. Therefore, from (41)

$$J_n(ha) = 0 \tag{8-43}$$

where a is the radius of the guide. There is an infinite number of possible TM waves corresponding to the infinite number of roots of (43). As before $h^2 = \bar{\gamma}^2 + \omega^2\mu\epsilon$, and, as in the case of rectangular guides, h^2 must be less than $\omega^2\mu\epsilon$ for transmission to occur. This means that h must be small or else extremely high frequencies will be required. This in turn means that only the first few roots of (43) will be of practical interest. The first few roots are

$$\left.\begin{aligned} (ha)_{01} &= 2.405 \qquad & (ha)_{11} &= 3.85 \\ (ha)_{02} &= 5.52 \qquad & (ha)_{12} &= 7.02 \end{aligned}\right\} \tag{8-44}$$

The first subscript refers to the value of n and the second refers to the roots in their order of magnitude. The various TM waves will be referred to as TM_{01}, TM_{12}, etc.

Since $\bar{\gamma} = \sqrt{h^2 - \omega^2\mu\epsilon}$, this gives for $\bar{\beta}$

$$\bar{\beta}_{nm} = \sqrt{\omega^2\mu\epsilon - h^2_{nm}}$$

The cut-off or critical frequency below which transmission of a wave will not occur is

$$f_c = \frac{h_{nm}}{2\pi\sqrt{\mu\epsilon}}$$

where

$$h_{nm} = \frac{(ha)_{nm}}{a}$$

The phase velocity is

$$\bar{v} = \frac{\omega}{\bar{\beta}} = \frac{\omega}{\sqrt{\omega^2\mu\epsilon - h^2_{nm}}}$$

From eqs. (32) the various components of TM waves can be computed in terms of E_z. The expressions for TM waves in circular guides are

$$E_z^0 = A_n J_n(h\rho)\cos n\phi$$

$$\left.\begin{aligned} H_\rho^0 &= -\frac{jA_n\omega\epsilon n}{h^2\rho}J_n(\rho h)\sin n\phi \\ H_\phi^0 &= -\frac{jA_n\omega\epsilon}{h}J_n'(\rho h)\cos n\phi \\ E_\rho^0 &= \frac{\bar\beta}{\omega\epsilon}H_\phi^0 \\ E_\phi^0 &= -\frac{\bar\beta}{\omega\epsilon}H_\rho^0 \end{aligned}\right\} \tag{8-45}$$

The variations of each of these field components with time and in the z direction are shown by multiplying each of the expressions of (45) by the factor $e^{j(\omega t-\bar\beta z)}$ and taking the real part. In the original expression (41) for E_z, the arbitrary constant B_n has been put equal to zero. The relative amplitudes of A_n and B_n determine the orientation of the field in the guide, and for a circular guide and any particular value of n, the $\phi = 0$ axis can always be oriented to make either A_n or B_n equal to zero.

For *transverse electric* waves E_z is identically zero and H_z is given by eq. (42). By substituting (42) into eqs. (32), the remaining field components can be found. The expressions for TE waves in circular guides are

$$\left.\begin{aligned} H_z^0 &= C_nJ_n(h\rho)\cos n\phi \\ H_\rho^0 &= \frac{-j\bar\beta C_n}{h}J_n'(h\rho)\cos n\phi \\ H_\phi^0 &= \frac{jn\bar\beta C_n}{h^2\rho}J_n(h\rho)\sin n\phi \\ E_\rho^0 &= \frac{\omega\mu}{\bar\beta}H_\phi^0 \\ E_\phi^0 &= -\frac{\omega\mu}{\bar\beta}H_\rho^0 \end{aligned}\right\} \tag{8-46}$$

The boundary conditions to be met for TE waves are that $E_\phi = 0$ at $\rho = a$. From (32) E_ϕ is proportional to $\partial H_z/\partial\rho$, and therefore to $J_n'(h\rho)$, where the prime denotes the derivative with respect to $(h\rho)$. Therefore, for TE waves the boundary conditions require that

$$J_n'(ha) = 0 \tag{8-47}$$

and it is the roots of (47) which must be determined. The first few of these roots are

$$\left.\begin{aligned} (ha)_{01}' &= 3.83 \qquad (ha)_{11}' = 1.84 \\ (ha)_{02}' &= 7.02 \qquad (ha)_{12}' = 5.33 \end{aligned}\right\} \tag{8-48}$$

The corresponding TE waves are referred to as TE_{01}, TE_{11}, and so on.

The equations for $f_c, \bar{\beta}, \bar{\lambda}$, and $\bar{v}$ are identical to those for the TM waves. It is understood, of course, that the roots of eq. (47) are to be used in connection with TE waves only.

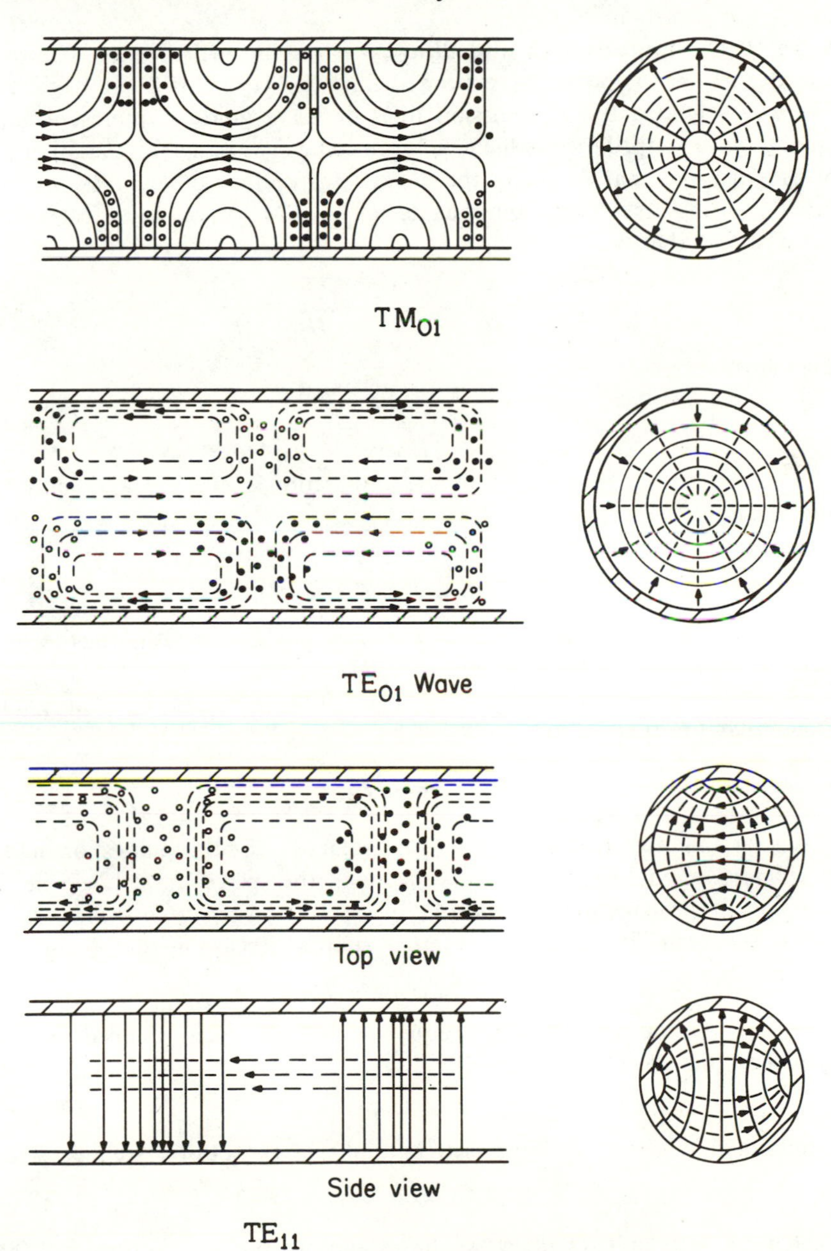

Figure 8-6. TE and TM waves in circular guides.

Inspection of eqs. (44) and (48) shows that the wave having the lowest cut-off frequency is the TE_{11} wave. The wave having the next lowest cut-off frequency is the TM_{01}. Some representative TM and TE waves are shown in Fig. 8-6.

8.08 Wave Impedances and Characteristic Impedances. The wave impedances at a point have been defined by eqs. (7-50) on page 197. For waves guided by transmission lines or wave guides, interest centers on the wave impedance which is seen when looking in the direction of propagation, that is, along the z axis. Inspection of expressions (12) for the transverse field components of a TM wave in a rectangular guide shows that

$$\frac{E_x}{H_y} = -\frac{E_y}{H_x} = \frac{\sqrt{E_x^2 + E_y^2}}{\sqrt{H_x^2 + H_y^2}} = \frac{\bar{\beta}}{\omega\epsilon}$$

Therefore

$$Z_{xy} = Z_{yx} = \frac{\bar{\beta}}{\omega\epsilon} = Z_z \tag{8-49}$$

The wave impedances looking in the z direction are equal and may be put equal to Z_z, where

$$Z_z = \frac{E_{\text{trans}}}{H_{\text{trans}}} = \frac{\sqrt{E_x^2 + E_y^2}}{\sqrt{H_x^2 + H_y^2}} \tag{8-50}$$

is the ratio of the total transverse electric field strength to the total transverse magnetic field strength.

A similar inspection of eqs. (45) for TM waves in circular guides shows that for them also

$$Z_z = Z_{\rho\phi} = -Z_{\phi\rho} = \frac{\bar{\beta}}{\omega\epsilon} \tag{8-51}$$

It is seen that for TM waves in rectangular or circular guides, or indeed in cylindrical guides of any cross section, the wave impedance in the direction of propagation is *constant over the cross section of the guide*, and is the same for guides of different shapes. Recalling that

$$\bar{\beta} = \sqrt{\omega^2 \mu\epsilon - h^2}$$

and that the cut-off angular frequency ω_c has been defined as that frequency that makes

$$\omega_c^2 \mu\epsilon = h^2$$

it follows that $\bar{\beta}$ can be expressed in terms of the cut-off frequency by

$$\bar{\beta} = \omega\sqrt{\mu\epsilon}\sqrt{1 - (\omega_c^2/\omega^2)} \tag{8-52}$$

Then from (49) or (51) the wave impedance in the z direction for TM waves is

$$Z_z(\text{TM}) = \sqrt{\frac{\mu}{\epsilon}}\sqrt{1 - (\omega_c^2/\omega^2)}$$

$$= \eta\sqrt{1 - (\omega_c^2/\omega^2)} \tag{8-53}$$

Thus for any cylindrical guide the wave impedance for TM waves is dependent only on the intrinsic impedance of the dielectric and the ratio of the frequency to the cut-off frequency.

For TE waves the same conclusion can be reached. However for TE waves it is found that

$$Z_z(\text{TE}) = \frac{\omega\mu}{\bar{\beta}} = \frac{\eta}{\sqrt{1 - (\omega_c^2/\omega^2)}} \tag{8-54}$$

For TEM waves between parallel planes or on ordinary parallel-wire or coaxial transmission lines the cut-off frequency is zero, and the wave impedance reduces to

$$Z_z(\text{TEM}) = \eta \tag{8-55}$$

The dependence of $\bar{\beta}$ on the ratio of frequency to cut-off frequency as shown by (52) affects the phase velocity and the wavelength in a corresponding manner. Thus the phase or wave velocity in a cylindrical guide of any cross section is given by

$$\bar{v} = \frac{\omega}{\bar{\beta}} = \frac{1}{\sqrt{\mu\epsilon}\sqrt{1 - (\omega_c^2/\omega^2)}} = \frac{v_0}{\sqrt{1 - (\omega_c^2/\omega^2)}} \tag{8-56}$$

where $v_0 = 1/\sqrt{\mu\epsilon}$, and μ and ϵ are the constants of the dielectric. The wavelength in the guide, measured in the direction of propagation, is

$$\bar{\lambda} = \frac{\bar{v}}{f} = \frac{2\pi}{\bar{\beta}} = \frac{1}{f\sqrt{\mu\epsilon}\sqrt{1 - (\omega_c^2/\omega^2)}}$$

$$= \frac{\lambda_0}{\sqrt{1 - (\omega_c^2/\omega^2)}} \tag{8-57}$$

where λ_0 is the wavelength of a TEM wave of frequency f in a dielectric having the constants μ and ϵ. Since $\omega_c^2/\omega^2 = \lambda_0^2/\lambda_c^2$ it follows that

$$\bar{\lambda} = \frac{\lambda_0\lambda_c}{\sqrt{\lambda_c^2 - \lambda_0^2}}$$

or

$$\lambda_0 = \frac{\bar{\lambda}\lambda_c}{\sqrt{\lambda_c^2 + \bar{\lambda}^2}} \tag{8-58}$$

A quantity of great usefulness in connection with ordinary two-conductor transmission lines is the (integrated) *characteristic impedance*, Z_0, of the line. For such lines, Z_0 can be defined in terms of the voltage-current ratio or in terms of the power transmitted for a given voltage or a given current. That is, for an infinitely long line

$$Z_0 = \frac{V}{I}; \qquad Z_0 = \frac{2W}{II^*}; \qquad Z_0 = \frac{VV^*}{2W} \tag{8-59}$$

where V and I are peak phasors. For ordinary transmission lines these definitions are equivalent, but for wave guides they lead to three values that depend upon the guide dimensions in the same way, but which differ by a constant.

For example, consider the three definitions given by (59) for the case of the TE_{10} mode in a rectangular guide (Fig. 8-1). The voltage will be taken as the maximum voltage from the lower face of the guide to the upper face. This occurs at $x = a/2$ and has a value

$$V_m = \int_0^b E_y(\max)\, dy = bE_y(\max) = \frac{-j\omega\mu baC}{\pi} \tag{8-60}$$

The longitudinal linear current density in the lower face is

$$J_z = -H_x = -\frac{j\bar{\beta}aC}{\pi} \sin \frac{\pi x}{a} \tag{8-61}$$

The total longitudinal current in the lower face is

$$I = \int_0^a J_z\, dx = \frac{-j2a^2\bar{\beta}C}{\pi^2}$$

Then the "integrated" characteristic impedance by the first definition is

$$Z_0(V, I) = \frac{\pi b}{2a}\frac{\omega\mu}{\bar{\beta}} = \frac{\pi b}{2a} Z_z = \frac{\pi b \eta}{2a\sqrt{1 - (f_c^2/f^2)}} \tag{8-62}$$

In terms of the second definition, the characteristic impedance for the TE_{10} wave in a rectangular guide is found to be

$$Z_0(W, I) = \frac{\pi^2 b}{8a} Z_z = \frac{\pi}{4} Z_0(V, I) \tag{8-63}$$

In terms of the third definition the integrated characteristic impedance is

$$Z_0(W, V) = \frac{2b}{a} Z_z = \frac{4}{\pi} Z_0(V, I) \tag{8-64}$$

In the next section the utility of the concept of characteristic impedance for cylindrical wave guides will be demonstrated.

8.09 Transmission-line Analogy for Wave Guides. There exists a useful analogy between the electric and magnetic field strengths of TM and TE waves and the voltages and currents on suitably loaded transmission lines. This analogy enables the engineer to draw "equivalent circuits," which are often helpful to him in dealing with unfamiliar electromagnetic problems.

For TM waves ($H_z = 0$) in rectangular co-ordinates the field equations are

$$\left.\begin{aligned} \frac{\partial H_y}{\partial z} &= -j\omega\epsilon E_x & \frac{\partial E_z}{\partial y} - \frac{\partial E_y}{\partial z} &= -j\omega\mu H_x \\ \frac{\partial H_x}{\partial z} &= j\omega\epsilon E_y & \frac{\partial E_x}{\partial z} - \frac{\partial E_z}{\partial x} &= -j\omega\mu H_y \\ \frac{\partial H_y}{\partial x} - \frac{\partial H_x}{\partial y} &= j\omega\epsilon E_z & \frac{\partial E_y}{\partial x} - \frac{\partial E_x}{\partial y} &= -j\omega\mu H_z \\ & \frac{\partial H_x}{\partial x} + \frac{\partial H_y}{\partial y} = 0 \end{aligned}\right\} \tag{8-65}$$

Now since $H_z = 0$,

$$(\nabla \times \mathbf{E})_z = 0$$

That is, in the x-y plane the electric field has no curl (the voltage around a closed path is zero) and so in this plane E may be written as the gradient of some scalar potential V. Then

$$E_x = -\frac{\partial V}{\partial x} \qquad E_y = -\frac{\partial V}{\partial y} \tag{8-66}$$

From the first equations of (65) and (66) and using (3)

$$\frac{\partial}{\partial z}\left(\frac{j\omega\epsilon}{h^2}\frac{\partial E_z}{\partial x}\right) = -j\omega\epsilon\frac{\partial V}{\partial x}$$

whence

$$\frac{\partial}{\partial z}\left(\frac{j\omega\epsilon}{h^2}E_z\right) = -j\omega\epsilon V \tag{8-67}$$

From the fifth of (65) and the first of (66) and using (3)

$$\frac{\partial E_x}{\partial z} - \frac{\partial E_z}{\partial x} = -\frac{\omega^2\mu\epsilon}{h^2}\frac{\partial E_z}{\partial x}$$

whence

$$\begin{aligned} \frac{\partial V}{\partial z} &= \left(\frac{\omega^2\mu\epsilon}{h^2} - 1\right)E_z \\ &= -\left(j\omega\mu + \frac{h^2}{j\omega\epsilon}\right)\left(\frac{j\omega\epsilon}{h^2}E_z\right) \end{aligned} \tag{8-68}$$

The quantity $j\omega\epsilon E_z$ is the longitudinal displacement current density and $1/h^2$ has the dimensions of area, so $j\omega\epsilon E_z/h^2$ represents a current in the z direction and will be designated by I_z. Then (67) and (68) become

$$\frac{\partial I_z}{\partial z} = -j\omega\epsilon V \qquad \frac{\partial V}{\partial z} = -\left[j\omega\mu + \frac{h^2}{j\omega\epsilon}\right]I_z \tag{8-69}$$

These are the differential equations for a lossless transmission line having a series impedance per unit length $Z = j\omega\mu + (h^2/j\omega\epsilon)$ and a shunt admittance per unit length $Y = j\omega\epsilon$. The "equivalent circuit" for such a transmission line is that shown in Fig. 8-7.

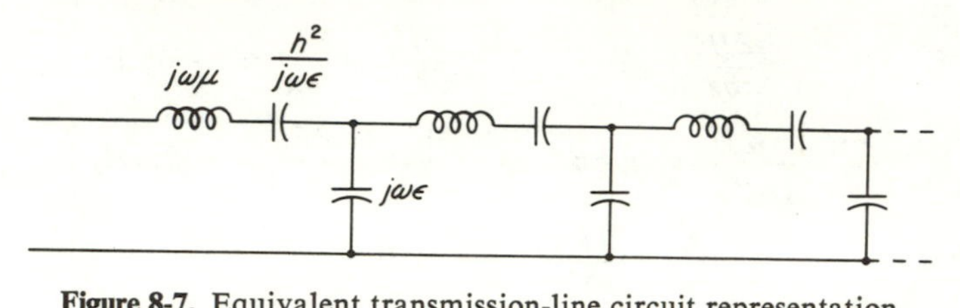

Figure 8-7. Equivalent transmission-line circuit representation for TM waves.

For TE waves the two equations of interest from the set corresponding to (65) are

$$\frac{\partial E_y}{\partial z} = j\omega\mu H_x \qquad \frac{\partial H_z}{\partial y} - \frac{\partial H_y}{\partial z} = j\omega\epsilon E_x \tag{8-70}$$

Since $E_z = 0, (\nabla \times \mathbf{H})_z = 0$; then in the x-y plane it is possible to define a scalar (magnetic) potential U such that

$$H_x = -\frac{\partial U}{\partial x} \qquad H_y = -\frac{\partial U}{\partial y} \tag{8-71}$$

From (70) and (71) and using eqs. (3)

$$\frac{\partial}{\partial z}\left(\frac{j\omega\mu}{h^2}\frac{\partial H_z}{\partial x}\right) = -j\omega\mu\frac{\partial U}{\partial x} \qquad \frac{\partial H_z}{\partial y} - \frac{\partial H_y}{\partial z} = \frac{\omega^2\mu\epsilon}{h^2}\frac{\partial H_z}{\partial y}$$

whence

$$\frac{\partial}{\partial z}\left(\frac{j\omega\mu}{h^2}H_z\right) = -j\omega\mu U \qquad \frac{\partial U}{\partial z} = -\left(\frac{h^2}{j\omega\mu} + j\omega\epsilon\right)\left(\frac{j\omega\mu}{h^2}H_z\right) \tag{8-72}$$

The quantity $j\omega\mu H_z/h^2$ has the dimensions of voltage and U has the dimensions of current, so (72) may be written

$$\frac{\partial V_1}{\partial z} = -ZI_1 \qquad \frac{\partial I_1}{\partial z} = -YV_1 \tag{8-73}$$

where now

$$V_1 = \frac{j\omega\mu H_z}{h^2} \qquad I_1 = U$$

$$Z = j\omega\mu \qquad Y = j\omega\epsilon + \frac{h^2}{j\omega\mu}$$

The "equivalent circuit" for TE waves is shown in Fig. 8-8.

The "loaded" transmission-line circuits of Figs. 8-7 and 8-8 have high-pass filter characteristics. The cut-off frequency for the line of

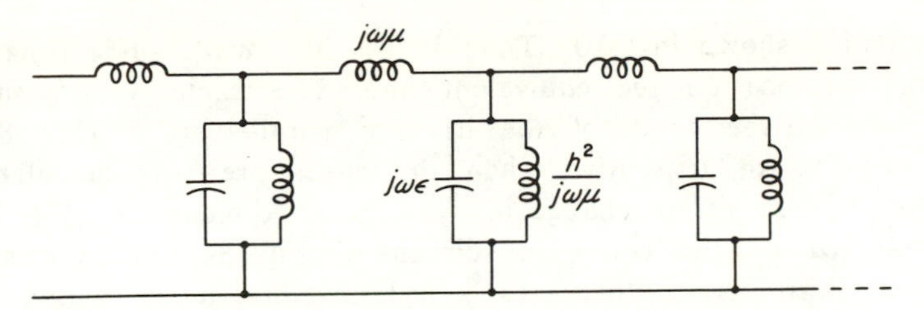

Figure 8-8. Equivalent transmission-line circuit representation for TE waves.

Fig. 8-7 occurs when the series reactance equals zero, whereas for the line of Fig. 8-8, the cut-off frequency is that which makes the shunt susceptance equal to zero. Both of these equalities require that

$$h^2 = \omega_c^2 \mu\epsilon$$

as was already obtained from wave theory. The characteristic impedance of the line of Fig. 8-7 is

$$Z_0(\mathrm{TM}) = \sqrt{\frac{Z}{Y}} = \sqrt{\frac{j\omega\mu + (h^2/j\omega\epsilon)}{j\omega\epsilon}} = \sqrt{\frac{\mu}{\epsilon}}\sqrt{1 - \frac{\omega_c^2}{\omega^2}} = Z_z(\mathrm{TM}) \tag{8-74}$$

The characteristic impedance of the line of Fig. 8-8 is

$$Z_0(\mathrm{TE}) = \sqrt{\frac{j\omega\mu}{j\omega\epsilon + (h^2/j\omega\mu)}} = \sqrt{\frac{\mu}{\epsilon}}\sqrt{\frac{1}{1 - (\omega_c^2/\omega^2)}} = Z_z(\mathrm{TE}) \tag{8-75}$$

The characteristic impedances of the equivalent transmission lines are equal to the corresponding wave impedances as would be expected.

The concept of a wave guide as an equivalent transmission line with a certain characteristic impedance and propagation constant is a powerful tool in the solution of many wave-guide problems because it enables the engineer to obtain the solution by means of well-known circuit and transmission-line theory. For example, the wave-guide problems, illustrated in Fig. 8-9, can be solved in terms of the "equiv-

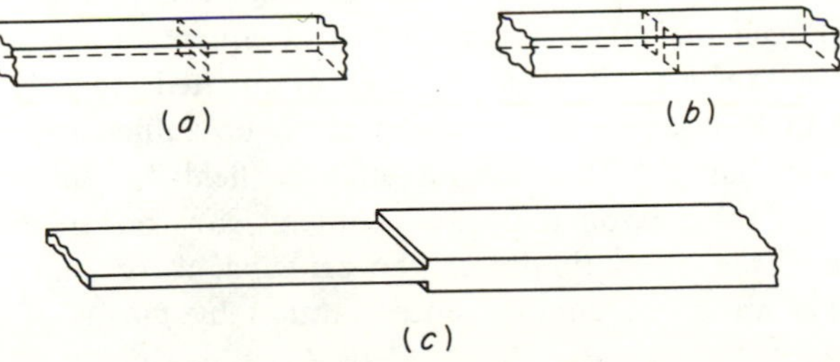

Figure 8-9. Typical discontinuities in wave guides: (*a*) Iris with edges perpendicular to **E**; (*b*) Iris with edges parallel to **E**; (*c*) Change of wave-guide dimensions.

alent circuits" shown in 8-10. Thus an iris in a wave guide behaves as a shunt reactance on the equivalent line. The reactance is positive or inductive when the edges of the iris are parallel to **E** [Fig. 8-9(*b*)]; it is negative or capacitive when the edges are perpendicular to **E** [Fig. 8-9(*a*)]. An abrupt change in wave-guide dimensions [Fig. 8-9(*c*)] is represented by the equivalent circuit of Fig. 8-10(*c*), in which two equivalent transmission lines are joined together, with an appropriate reactance shunted across the junction.

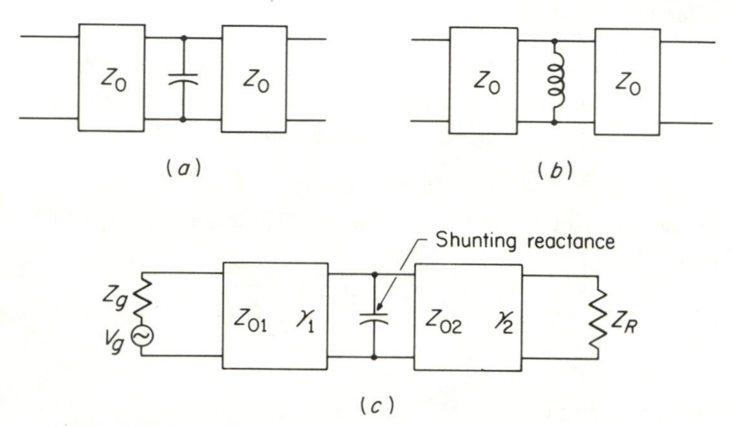

Figure 8-10. "Equivalent circuits" for wave-guide discontinuities illustrated in Fig. 8-9.

In these examples the calculation of the actual value of shunting reactance to be used in any particular case is, of course, a field problem. However, it is a field problem which can be solved in a fairly straightforward manner by matching solutions at the boundary.* The procedure is to represent the field at the junction or discontinuity by the sum of principal and higher-order waves, the relative amplitudes of which are obtained by matching the tangential components of **E** and **H** at the boundary. The higher-order waves are set up by the discontinuity and are required in order to meet the boundary conditions. However, in general, they have cut-off frequencies higher than the frequency of transmission and so are attenuated rapidly. The load impedance and the generator are assumed to be sufficiently far removed from the iris or junction to be out of the field of these higher-order waves. It is for this reason that the problem can be treated in terms of the effect of the discontinuity on the *principal wave only*, which fact, in turn, makes valid the circuit representation by means of an ordinary

*J. R. Whinnery and H. W. Jamieson, "Equivalent Circuits for Discontinuities in Transmission Lines," *Proc. IRE*, **32**, 98–114, February, 1944; S. A. Schelkunoff. *Electromagnetic Waves*, D. Van Nostrand Co., Inc., Princeton, N. J., 1943, p. 492.

transmission line. (Otherwise, a transmission line having a different set of "constants" for each mode would be required.) It is found that the higher-order modes make no contribution to the voltage at the junction, and therefore the total voltage is just the voltage of the principal wave. However, in the region of the junction the higher-order waves do make contributions to the current. Although the *total* current must be continuous across the junction, the principal-wave current is discontinuous by the amount of the higher-order mode current. This discontinuity of principal-wave current is accounted for by the effect of the equivalent shunting reactance.

The values for equivalent shunting reactances have been calculated in terms of the iris or junction dimensions for many cases and may be found in handbooks.* Using this known reactance, together with the known characteristic impedance for the guide [as given, for example, by eq. (75) for the TE_{10} wave], the wave-guide problems of Fig. 8-9 are readily solved in terms of the well-known circuit problems of Fig. 8-10.

8.10 Attenuation Factor and Q of Wave Guides. In solving Maxwell's equations for the region within rectangular or circular wave guides, the assumptions were made that the dielectric was lossless and that the walls of the guide were perfectly conducting. Under these conditions an expression was obtained for the propagation constant $\bar{\gamma}$ which was

$$\bar{\gamma} = \sqrt{h^2 - \omega^2 \mu\epsilon}$$

The quantity h^2 is a real number, the value of which depends upon the guide dimensions and the order of the mode being considered. For example, for rectangular guides h^2 is given by

$$h^2 = \left(\frac{m\pi}{a}\right)^2 + \left(\frac{n\pi}{b}\right)^2$$

For frequencies below cut-off $\omega^2\mu\epsilon$ is less than h^2, and $\bar{\gamma}$ is a real number, which is then put equal to $\bar{\alpha}$. That is, below cut-off,

$$\bar{\gamma} = \bar{\alpha} = \sqrt{h^2 - \omega^2 \mu\epsilon}$$

In general, for frequencies well below cut-off $\bar{\alpha}$ is a large number, and the fields decrease exponentially at a rapid rate. At the cut-off frequency $\bar{\gamma}$ becomes equal to zero; for all frequencies above cut-off $\bar{\gamma}$ is a pure imaginary and the attenuation constant $\bar{\alpha}$ is zero. This result is correct for the assumed conditions. However, whereas the dielectric within the guide may be very nearly lossless (air, for example), the walls of an actual guide do have some loss. Therefore a finite, though perhaps

*N. Marcuvitz, *Waveguide Handbook*, Radiation Laboratory Series, Vol. 10, McGraw-Hill Book Company, New York, 1948.

small, value of attenuation would be expected in the range of frequencies above cut-off.

The actual attenuation factor for waves propagating within cylindrical guides may be calculated to a very good approximation by the method already outlined for parallel-plane guides. In this approach it is assumed that the finite conductivity of the walls will have only a small effect on the configurations within the guide: In particular the magnetic field tangential to the wall is expected to depend only slightly on the conductivity of the walls. This is very nearly true as long as the conductivity is high, as it is for metals. Then the tangential magnetic field strength computed for perfectly conducting walls is used to determine the linear current density in the walls. This linear current density, squared and multiplied by the actual surface resistance of the walls, gives the actual power loss per unit area in the walls. The attenuation factor *in the range of propagation* is then given by

$$\alpha = \frac{\text{power lost per unit length}}{2 \times \text{power transmitted}} \tag{8-76}$$

The power transmitted is obtained by integrating the axial component of the Poynting vector over the cross section of the guide. Because the transverse components of E and H have been found to be in phase and normal to each other, the axial Poynting vector is given simply as

$$P_z = \tfrac{1}{2}\,|E_{\text{trans}}||H_{\text{trans}}|$$

Using (50), this may be written

$$P_z = \tfrac{1}{2}\,Z_z|H_{\text{trans}}|^2$$

or

$$P_z = \frac{1}{2Z_z}\,|E_{\text{trans}}|^2$$

The total power transmitted is

$$W = \tfrac{1}{2}Z_z\int_{\text{area}} |H_{\text{trans}}|^2\,da \tag{8-77}$$

where the integration is over the cross-section area of the guide. The power lost per unit length of guide is

$$\begin{aligned} W_{\text{lost}} &= \tfrac{1}{2}R_s\int_{\text{surf}} |J|^2\,da \\ &= \tfrac{1}{2}R_s\int_{\text{surf}} |H_{\tan}|^2\,da \end{aligned} \tag{8-78}$$

where the integration is taken over the wall surface of a unit length of the guide.

Formulas for attenuation factors for rectangular and circular guides computed by this or equivalent methods can be found in many textbooks and handbooks. For the dominant TE_{10} mode in rectangular guides the result is

$$\left.\begin{aligned} \alpha &= \frac{R_s}{b\eta_0 K}\left[1 + \frac{2bf_c^2}{af^2}\right] \quad \text{nepers/m} \\ &= \frac{8.7R_s}{b\eta_0 K}\left[1 + \frac{2bf_c^2}{af^2}\right] \quad \text{db/m} \end{aligned}\right\} \tag{8-79}$$

where

$$K = \sqrt{1 - \frac{f_c^2}{f^2}} \qquad R_s = \sqrt{\frac{\omega\mu_m}{2\sigma_m}}$$

Attenuation vs. frequency curves are sketched for typical rectangular brass guides in Fig. 8-11, and for circular guides in Fig. 8-12. The attenuation is very high near the cut-off frequency, but decreases to a quite low value at frequencies somewhat above cut-off. For TM

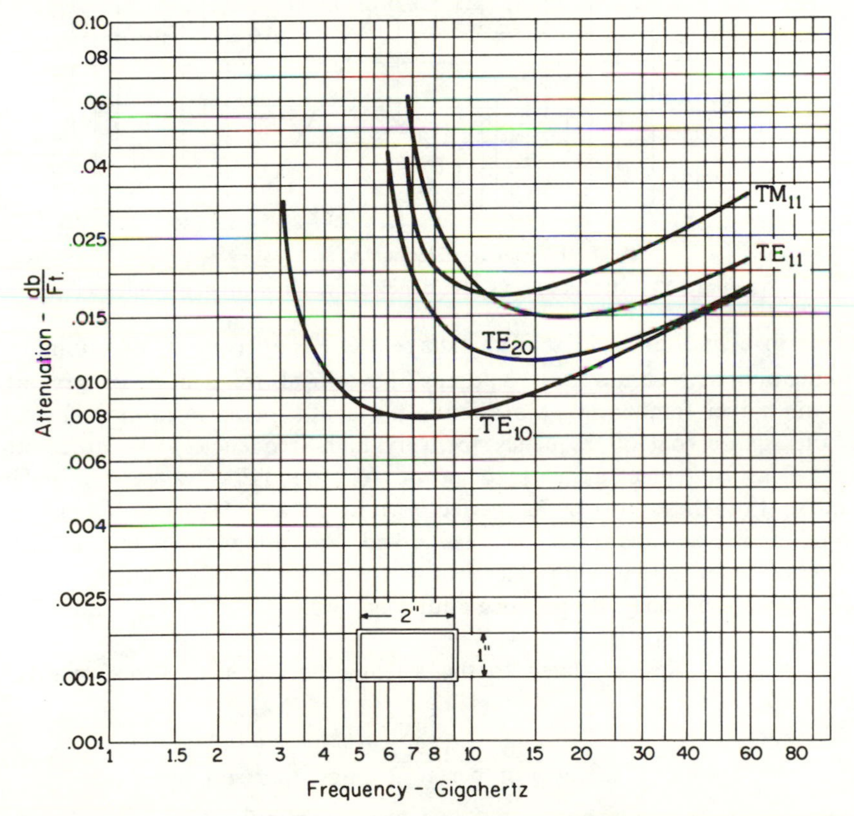

Figure 8-11. Attenuation vs. frequency curves for various modes in a typical rectangular brass guide.

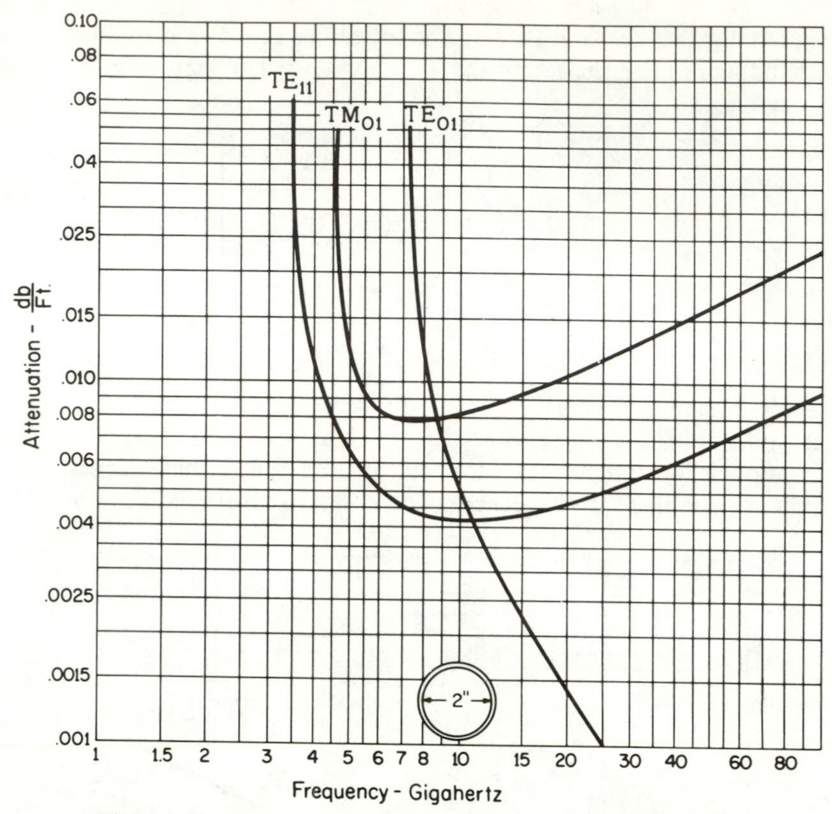

Figure 8-12. Attenuation vs. frequency curves in a circular guide.

waves in cylindrical guides of any shape, there is a frequency of minimum attenuation that occurs at $\sqrt{3}$ times the critical frequency. In general, for still higher frequencies, the attenuation again increases (approximately as the square root of frequency for very high frequencies). An exception to this last statement appears to occur for the TE_{0m} waves in circular guides. For these waves in perfectly circular guides the wall currents decrease as the frequency increases, and the attenuation theoretically decreases indefinitely with increasing frequency. Unfortunately, slight deformations of the guide produce additional wall currents that nullify this desirable characteristic.

A quantity closely related to the attenuation factor α is the "quality" factor Q.

$$Q = \omega \frac{\text{energy stored per unit length}}{\text{energy lost per unit length per second}} \tag{8-80}$$

Expression (80) may be considered as a general definition for Q, applicable to wave guides and to ordinary transmission lines. It should be compared with the circuit definition of Q, which may be stated as

$$Q = \omega \frac{\text{energy stored in circuit}}{\text{energy lost per second}} \tag{8-81}$$

For the TEM wave on the lossless or distortionless transmission line, the velocity v represented both the phase velocity and the group velocity. For wave guides these two velocities are different, the group velocity, v_g, being related to the phase velocity $\bar{v}$ by

$$v_g = \frac{v_0^2}{\bar{v}} \qquad \text{where } v_0 = \frac{1}{\sqrt{\mu\epsilon}} \tag{8-82}$$

For a wave guide,

energy transmitted per second

$$= v_g \times (\text{energy stored per unit length})$$

or

$$\text{energy stored per unit length} = \frac{1}{v_g} \times \text{power transmitted} \tag{8-83}$$

Using (80), (83), and (76), the Q of a wave guide is given by

$$Q = \frac{\omega}{v_g}\left(\frac{\text{power transmitted}}{\text{power lost per unit length}}\right)$$

$$= \frac{\omega}{2\alpha v_g} \tag{8-84}$$

This also may be written in the following equivalent forms

$$Q = \frac{\omega\bar{v}}{2\alpha v_0^2} = \frac{\omega}{2\alpha v_0\sqrt{1 - \omega_c^2/\omega^2}} \tag{8-84a}$$

Because of the low attenuation factors obtainable with wave guides compared to transmission lines, it is possible to construct wave-guide sections having extremely high Q's. This is of importance when such sections are used as resonators, or as the elements of wave-guide filters.

8.11 Dielectric Slab Wave Guide. The discussion of total internal reflection in chap. 5 suggests that a dielectric slab or rod may be able to support guided wave modes similar to those in completely enclosed metallic wave guides. Under conditions of total internal reflection and with a reflection coefficient of suitable phase shift, it seems possible that waves could propagate unattenuated by "bouncing" between the two surfaces of a dielectric slab. This is true in fact as will be shown in the following analysis of wave propagation along a flat dielectric slab in a vacuum (see Fig. 8-13).

The problem is similar to that of propagation over a conducting plane and therefore the method of solution used will be the same as

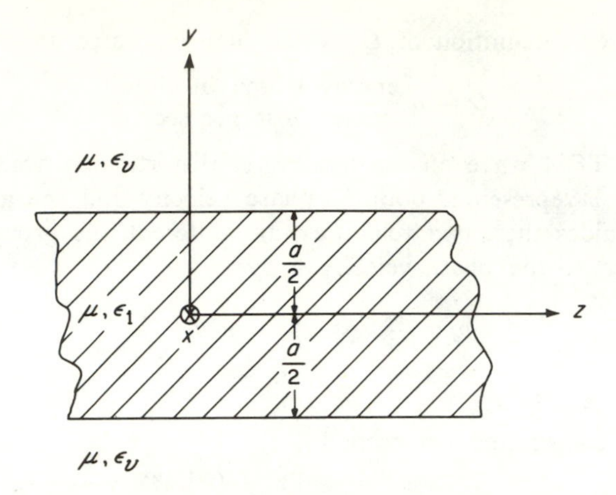

Figure 8-13. Dielectric wave guide.

in sec. 7.09. For TM modes the analysis proceeds as follows:

Outside the slab ($y > a/2$):

$$H_x = H_x^0 e^{-\gamma_0 z}$$

$$\frac{\partial^2 H_x^0}{\partial y^2} = h_0^2 H_x^0$$

where $h_0^2 = -\gamma_0^2 - \omega^2 \mu \epsilon_v$

$$H_x^0 = Ce^{-h_0 y}$$

$$E_y^0 = \frac{j\gamma_0}{\omega\epsilon_v} H_x^0$$

$$E_z^0 = \frac{j}{\omega\epsilon_v}\frac{\partial H_x^0}{\partial y} = -j\frac{h_0}{\omega\epsilon_v} Ce^{-h_0 y}$$

Inside the slab:

$$H_x = H_x^0 e^{-\gamma_1 z}$$

Note: $\gamma_1 = \gamma_0$ in order for fields to match across surface

$$\frac{\partial^2 H_x^0}{\partial y^2} = -k_1^2 H_x^0$$

where $k_1^2 = \gamma_0^2 + \omega^2 \mu \epsilon_1$ (8-85)

$H_x^0 = \sin k_1 y$ or $H_x^0 = \cos k_1 y$

The surface impedance looking down on the surface of the slab is given by

$$Z_s = -\frac{E_z}{H_x} = j\frac{h_0}{\omega\epsilon_v} \tag{8-86}$$

If h_0 is positive real (which is necessary for the existence of a guided wave), Z_s is a positive reactance. Thus in general, a surface wave can exist over a *reactive* surface and in particular, a TM surface wave can exist over an *inductive* surface. For the dielectric slab, further information can be obtained only by matching tangential field components at the dielectric surface.

Case I: $H_x^0 = \sin k_1 y$

Continuity of tangential magnetic field gives

$$C = e^{h_0 a/2} \sin\left(\frac{k_1 a}{2}\right)$$

and continuity of tangential electric field gives

$$-\frac{\epsilon_v k_1}{\epsilon_1} \cot\left(\frac{k_1 a}{2}\right) = h_0 \tag{8-87}$$

Equations (85) for h_0 and k_1 may be combined to give

$$h_0^2 + k_1^2 = \omega^2 \mu(\epsilon_1 - \epsilon_v) \tag{8-88}$$

Finally h_0 may be eliminated from (87) and (88) to give a transcendental equation which may be solved graphically to yield values of k_1, the transverse phase-shift constant within the dielectric. The longitudinal propagation constant γ_0 and the transverse attenuation constant h_0 then may be obtained from (85).

CASE II: $H_x^0 = \cos k_1 y$

Continuity of tangential magnetic field gives

$$C = e^{h_0 a/2} \cos\left(\frac{k_1 a}{2}\right)$$

and continuity of tangential electric field gives

$$\frac{\epsilon_v k_1}{\epsilon_1} \tan\left(\frac{k_1 a}{2}\right) = h_0 \tag{8-89}$$

which may be combined with (88) to give a transcendental equation for k_1 as before. It is important to note that in this case the z component of electric field is zero at $y = 0$ so that the solution applies as well to a dielectric slab of thickness $a/2$ over a perfectly conducting plane.

The dielectric slab is an example of a continuous reactive surface which can support a guided wave; a similar guided wave effect is present in the dielectric rod "light pipes" used to guide light around corners. Surface waves are also supported by corrugated periodic surfaces such as the surface consisting of thin, vertical conducting strips attached to a horizontal conducting plate. If the strips are close together they may be regarded as shorted parallel-plate transmission lines. Thus it is easy to see that such a corrugated surface could approximate a continuous surface with a reactive surface impedance and thus could support a surface wave.

Problem 6. Derive the TE mode field equations for wave propagation along a dielectric slab. Discuss the applicability of the above results to the problem of wave propagation over a dielectric-coated, perfectly conducting plane. Show that the TE surface impedance is capacitive and that the guided waves are slow waves.

ADDITIONAL PROBLEMS

7. Verify the results obtained in eqs. (63) and (64).

8. Show that for a coil the definition for Q given by (81) reduces to $Q = \omega L/R$.

9. Show that at frequencies much higher than the cut-off frequency, the Q of a rectangular guide carrying the dominant TE_{10} wave approaches the value

$$Q \to b\alpha_m$$

where $\alpha_m = \sqrt{\omega\mu_m\sigma_m/2}$ is the attenuation factor for a wave propagating in the metal of the guide walls. (Note: Assume $\mu_m \approx \mu_v$.)

BIBLIOGRAPHY

Southworth, G. C., "Hyper-frequency Wave Guides—General Considerations and Experimental Results," *BSTJ*, **15**, 284 (1936).

Barrow, W. L., "Transmission of Electromagnetic Waves in Hollow Tubes of Metal," *Proc. IRE*, **24**, 1298 (1936).

Schelkunoff. S. A., "Transmission Theory of Plane Electromagnetic Waves," *Proc. IRE*, **25**, 1457 (1937).

Chu, L. J. and W. L. Barrow, "Electromagnetic Waves in Hollow Metal Tubes of Rectangular Cross Section," *Proc. IRE*, **26**, 1520 (1938).

Clavier, A. G., "Attenuation and Q Factors in Wave Guides," *Electrical Communication*, **23**, No. 4 (1946).

Schelkunoff, S. A., *Electromagnetic Waves*, D. Van Nostrand Co., Inc., Princeton, N. J.. 1943.

Bronwell, A. B. and R. E. Beam, *Theory and Application of Microwaves*, McGraw-Hill Book Company, New York, 1947.

Collin, R. E., *Field Theory of Guided Waves*, McGraw-Hill Book Company, New York, 1960.

Harrington, R. F., *Time-harmonic Electromagnetic Fields*, McGraw-Hill Book Company, New York, 1961.

Chapter 9

INTERACTION OF FIELDS AND MATTER

Up to this point the electromagnetic properties of matter have been given in terms of ϵ, μ and σ; these are respectively the permittivity, permeability and conductivity of the material and they may be found tabulated in reference books for an enormous number of different materials. The quantities ϵ, μ and σ express the *macroscopic* properties of a substance and in many cases they may be used without giving any thought to the *microscopic* (atomic and molecular) effects which are responsible for the macroscopic behavior. However, the increasing importance of properties of materials in electrical engineering has made it highly desirable that the engineer have some understanding of how these properties arise and why they differ in different materials. This understanding requires a consideration of the interaction of fields and matter on a microscopic scale.

It has been known for many years that electric charge plays a very important part in the constitution of matter. The Bohr model of the atom postulates negatively charged electrons whirling about a nucleus consisting of positively charged protons and uncharged neutrons; in this model the Coulomb attractive force between opposite charges is balanced by the outward centrifugal force to maintain the electrons in stable orbits. If external energy is available (ultraviolet light, for instance) the atoms of a gas may become *ionized*, that is, electrons may be freed from an atom, leaving it as an *ion* with a net positive charge. On the other hand, electrons may attach themselves to an atom, thus forming a negative ion. In a metal, some of the electrons are free to move from one atom to the next, giving rise to the *conduction* of electricity; heating of the metal may cause some of these electrons to be given off into space, as in the case of radio tubes with thermionic cathodes.

Since matter is made up largely of charged "particles," external electric and magnetic fields must exert some kind of influence on matter. This influence will be present whether the particles are free to move about (such as electrons in a conductor, or electrons and ions in a dilute gas) or are tightly bound together (as in the atoms or molecules of a dielectric). Each charged particle is subject not only to external fields but also to the fields of the other particles including the long-range Coulomb fields of charged particles and the short-range fields of neutral particles. This brief discussion suggests that a study of fields and matter should begin by considering the manner in which charged particles respond to the application of electric and magnetic fields. To do this, it is necessary to examine the *equation of motion* which states precisely how the motion of a particle changes under the influence of an applied force.

9.01 Charged-particle Equation of Motion. The velocity $\mathbf{v}$ of a particle of mass m under the influence of an external force $\mathbf{F}$ is governed by the equation of motion,

$$\mathbf{F} = \frac{d}{dt}(m\mathbf{v}) \tag{9-1}$$

which states that the rate of change of momentum is equal to the applied force. The velocity $\mathbf{v}$ is defined as the rate of change of the position vector $\mathbf{r}$ which denotes the location of the particle. That is,

$$\mathbf{v} = \frac{d\mathbf{r}}{dt} = \dot{\mathbf{r}} \tag{9-2}$$

Under nonrelativistic conditions, m is a constant and (1) reduces to the familiar force equation

$$\mathbf{F} = m\frac{d\mathbf{v}}{dt} = m\dot{\mathbf{v}} \tag{9-3}$$

The external force $\mathbf{F}$ could be electric, magnetic, gravitational or mechanical and it is through this force that the external world acts upon the particle. The acceleration $\dot{\mathbf{v}}$ represents the response of the particle to the applied force. Before seeking the response to specific physical forces, it is necessary to examine the possible forms of $\dot{\mathbf{v}}$. A good deal of information about the acceleration may be uncovered simply by expressing it first in rectangular co-ordinates and then in cylindrical co-ordinates.

Rectangular Co-ordinates. In terms of their rectangular components, force, velocity and position may be expressed as

$$\begin{aligned} \mathbf{F} &= \hat{\mathbf{x}}F_x + \hat{\mathbf{y}}F_y + \hat{\mathbf{z}}F_z && \text{(a)} \\ \mathbf{v} &= \hat{\mathbf{x}}v_x + \hat{\mathbf{y}}v_y + \hat{\mathbf{z}}v_z && \text{(b)} \\ \mathbf{r} &= \hat{\mathbf{x}}x + \hat{\mathbf{y}}y + \hat{\mathbf{z}}z && \text{(c)} \end{aligned} \tag{9-4}$$

The unit vectors $\hat{\mathbf{x}}$, $\hat{\mathbf{y}}$ and $\hat{\mathbf{z}}$ are constant in both magnitude and direction so that substitution of the position vector into (2) gives

$$\mathbf{v} = \hat{\mathbf{x}}\dot{x} + \hat{\mathbf{y}}\dot{y} + \hat{\mathbf{z}}\dot{z} \tag{9-5}$$

and substitution of (5) into (3) gives

$$\mathbf{F} = m[\hat{\mathbf{x}}\dot{v}_x + \hat{\mathbf{y}}\dot{v}_y + \hat{\mathbf{z}}\dot{v}_z] = m[\hat{\mathbf{x}}\ddot{x} + \hat{\mathbf{y}}\ddot{y} + \hat{\mathbf{z}}\ddot{z}] \tag{9-6}$$

for the force equation. For a given force two successive integrations of (6) give the orbit of the particle.

Cylindrical Co-ordinates. In terms of their cylindrical components, the force, velocity and position may be expressed as

$$\begin{aligned} \mathbf{F} &= \hat{\boldsymbol{\rho}}F_\rho + \hat{\boldsymbol{\phi}}F_\phi + \hat{\mathbf{z}}F_z \qquad &\text{(a)} \\ \mathbf{v} &= \hat{\boldsymbol{\rho}}v_\rho + \hat{\boldsymbol{\phi}}v_\phi + \hat{\mathbf{z}}v_z \qquad &\text{(b)} \\ \mathbf{r} &= \hat{\boldsymbol{\rho}}\rho + \hat{\mathbf{z}}z \qquad &\text{(c)} \end{aligned} \tag{9-7}$$

The co-ordinate directions are as shown in Fig. 9-1. Expression of the

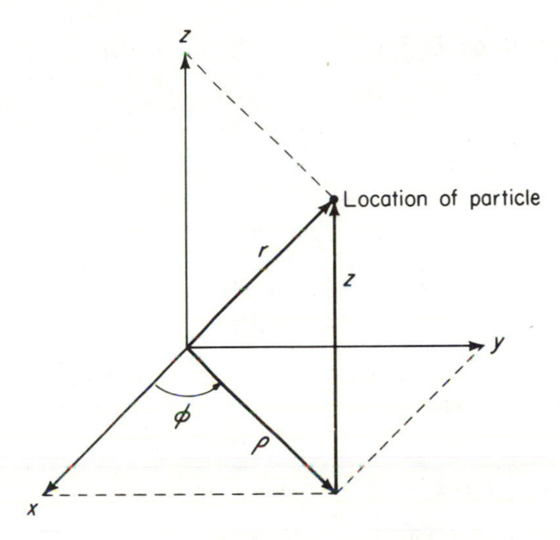

Figure 9-1.

velocity in terms of the co-ordinates is complicated by the fact that the direction of $\hat{\boldsymbol{\rho}}$ is not constant. Thus, substitution of the position vector (7c) into (2) gives

$$\begin{aligned} \mathbf{v} &= \frac{d}{dt}(\hat{\boldsymbol{\rho}}\rho + \hat{\mathbf{z}}z) \\ &= \hat{\boldsymbol{\rho}}\frac{d\rho}{dt} + \rho\frac{d\hat{\boldsymbol{\rho}}}{dt} + \hat{\mathbf{z}}\frac{dz}{dt} \end{aligned} \tag{9-8}$$

The time derivative of $\hat{\boldsymbol{\rho}}$ may be expressed as

$$\frac{d\hat{\boldsymbol{\rho}}}{dt} = \frac{\partial\hat{\boldsymbol{\rho}}}{\partial\rho}\frac{d\rho}{dt} + \frac{\partial\hat{\boldsymbol{\rho}}}{\partial\phi}\frac{d\phi}{dt} + \frac{\partial\hat{\boldsymbol{\rho}}}{\partial z}\frac{dz}{dt}$$

The first and last terms on the right-hand side are zero because the direction of $\hat{\boldsymbol{\rho}}$ remains fixed as either ρ or z is varied. The change in

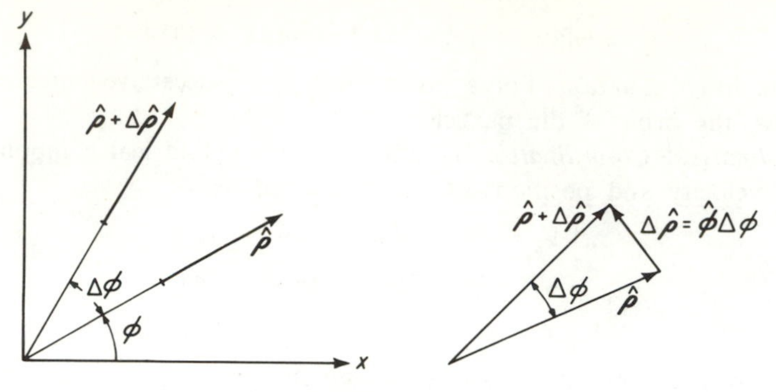

Figure 9-2.

$\hat{\boldsymbol{\rho}}$ with ϕ is depicted in Fig. 9-2 which shows that

$$\frac{\partial\hat{\boldsymbol{\rho}}}{\partial\phi}=\lim_{\Delta\phi\to 0}\frac{\Delta\hat{\boldsymbol{\rho}}}{\Delta\phi}=\hat{\boldsymbol{\phi}} \tag{9-9}$$

and consequently that

$$\frac{d\hat{\boldsymbol{\rho}}}{dt}=\hat{\boldsymbol{\phi}}\frac{d\phi}{dt} \tag{9-10}$$

Substitution of (10) into (8) gives finally

$$\mathbf{v}=\hat{\boldsymbol{\rho}}\dot{\rho}+\hat{\boldsymbol{\phi}}\rho\dot{\phi}+\hat{\mathbf{z}}\dot{z} \tag{9-11}$$

The quantity $\dot{\phi}$ is the angular velocity and is usually written as ω; thus from (11) one may write

$$v_\phi=\rho\dot{\phi}=\rho\omega \tag{9-12}$$

The time derivative of velocity may be treated similarly as follows:

$$\begin{aligned}\dot{\mathbf{v}}&=\frac{d}{dt}(\hat{\boldsymbol{\rho}}v_\rho+\hat{\boldsymbol{\phi}}v_\phi+\hat{\mathbf{z}}v_z)\\&=\hat{\boldsymbol{\rho}}\frac{dv_\rho}{dt}+v_\rho\frac{d\hat{\boldsymbol{\rho}}}{dt}+\hat{\boldsymbol{\phi}}\frac{dv_\phi}{dt}+v_\phi\frac{d\hat{\boldsymbol{\phi}}}{dt}+\hat{\mathbf{z}}\frac{dv_z}{dt}\end{aligned} \tag{9-13}$$

The time derivative of $\hat{\boldsymbol{\phi}}$ may be expressed as

$$\frac{d\hat{\boldsymbol{\phi}}}{dt}=\frac{\partial\hat{\boldsymbol{\phi}}}{\partial\rho}\frac{d\rho}{dt}+\frac{\partial\hat{\boldsymbol{\phi}}}{\partial\phi}\frac{d\phi}{dt}+\frac{\partial\hat{\boldsymbol{\phi}}}{\partial z}\frac{dz}{dt}$$

and the first and last terms on the right-hand side are zero as before. The change in $\hat{\boldsymbol{\phi}}$ may be depicted as in Fig. 9-3 which shows that

$$\frac{\partial\hat{\boldsymbol{\phi}}}{\partial\phi}=\lim_{\Delta\phi\to 0}\frac{\Delta\hat{\boldsymbol{\phi}}}{\Delta\phi}=-\hat{\boldsymbol{\rho}} \tag{9-14}$$

and consequently that

$$\frac{d\hat{\boldsymbol{\phi}}}{dt}=-\hat{\boldsymbol{\rho}}\frac{d\phi}{dt}=-\hat{\boldsymbol{\rho}}\frac{v_\phi}{\rho} \tag{9-15}$$

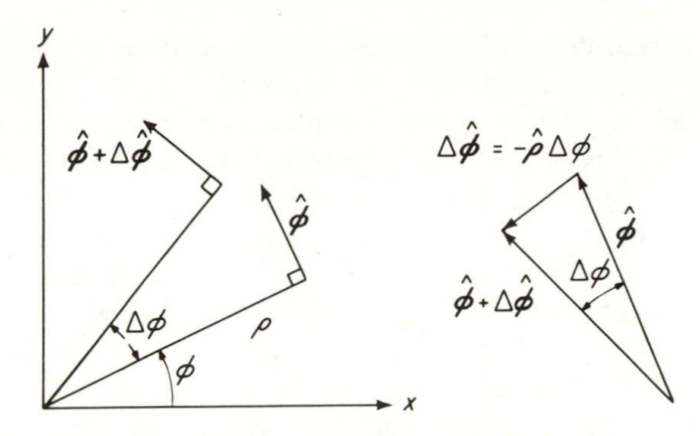

Figure 9-3.

Substitution of (15) and (10) into (13) gives

$$\dot{\mathbf{v}} = \hat{\boldsymbol{\rho}}\left(\dot{v}_\rho - \frac{v_\phi^2}{\rho}\right) + \hat{\boldsymbol{\phi}}\left(\dot{v}_\phi + \frac{v_\rho v_\phi}{\rho}\right) + \hat{\mathbf{z}}\dot{v}_z \tag{9-16}$$

Equation (16) may be expressed in terms of the space co-ordinates ρ, ϕ, z using the relations

$$\begin{aligned} \dot{v}_\rho &= \ddot{\rho} \\ \dot{v}_\phi &= \frac{d}{dt}\left(\rho\frac{d\phi}{dt}\right) = \rho\ddot{\phi} + \dot{\rho}\dot{\phi} \\ \dot{v}_z &= \ddot{z} \end{aligned} \tag{9-17}$$

Thus (16) becomes

$$\dot{\mathbf{v}} = \hat{\boldsymbol{\rho}}(\ddot{\rho} - \rho\dot{\phi}^2) + \hat{\boldsymbol{\phi}}(\rho\ddot{\phi} + 2\dot{\rho}\dot{\phi}) + \hat{\mathbf{z}}\ddot{z} \tag{9-18}$$

$$= \hat{\boldsymbol{\rho}}(\ddot{\rho} - \rho\omega^2) + \hat{\boldsymbol{\phi}}\frac{1}{\rho}\frac{d}{dt}(\rho^2\omega) + \hat{\mathbf{z}}\ddot{z} \tag{9-19}$$

Expressions (16), (18) and (19) all are used frequently for the acceleration in the nonrelativistic equation of motion.

The acceleration expressed in cylindrical co-ordinates evidently is much more complicated than when expressed in rectangular co-ordinates. Nevertheless, the cylindrical co-ordinate expression displays clearly certain effects of rotational motion which are hidden by use of rectangular co-ordinates. These effects become evident when the equation of motion (3) is written as follows:

$$\mathbf{F} + \hat{\boldsymbol{\rho}}m\rho\omega^2 - \hat{\boldsymbol{\phi}}mv_\rho\omega = m(\hat{\boldsymbol{\rho}}\dot{v}_\rho + \hat{\boldsymbol{\phi}}\dot{v}_\phi + \hat{\mathbf{z}}\dot{v}_z) \tag{9-20}$$

Equation (20) is the equation of motion written with two of the acceleration terms brought over to the force side of the equation. The right-hand side of (20) is the mass times the acceleration that would be measured in a frame of reference rotating with the particle at the angular velocity ω. Thus, the left-hand side represents the force on the particle as measured in the rotating frame of reference. Two of the force terms are not external but are due solely to the rotation of the particle; as a result they are sometimes referred to as "fictitious" forces despite the fact that their existence is by no means fictitious. The quantity $\hat{\boldsymbol{\rho}} m\rho\omega^2$ is the familiar *centrifugal force* directed outward from the center of rotation. The quantity $-\hat{\boldsymbol{\phi}} m v_\rho \omega$ is the less familiar *Coriolis force* which tends to give an angular acceleration to a particle with an inward velocity component (negative v_ρ) and an angular deceleration to a particle with an outward velocity component (positive v_ρ). The Coriolis force is used regularly by figure skaters who begin spins with arms and one leg outstretched and draw them in close to the body to accelerate the spin; the resulting rapid spin is stopped simply by once again extending the arms.

The equation of motion states that the rate of change of momentum is proportional to the applied force; this statement may be applied not only to the equation as a whole, but also to its vector components. In the case of the $\hat{\boldsymbol{\phi}}$ component in cylindrical co-ordinates, a slightly different statement applies when the force equation is multiplied by the radius ρ. Using (19) for the acceleration, this procedure gives

$$\rho F_\phi = \frac{d}{dt}(m\rho^2\omega)$$

which states that the rate of change of *angular momentum* is equal to the applied *torque*.

9.02 Force and Energy. The principal forces which act on a charged particle are electric and magnetic. The electric force is proportional to the electric field strength and the magnetic force is proportional to the magnetic flux density as shown by the force relation

$$\mathbf{F} = e(\mathbf{E} + \mathbf{v} \times \mathbf{B}) \qquad (9\text{-}21)$$

in which e is the charge on the particle (e is negative for an electron). It is important to note that the magnetic force acts only on a particle in motion and it acts in a direction perpendicular to the direction of motion. Insertion of (21) into (3) gives the nonrelativistic equation of motion

$$e(\mathbf{E} + \mathbf{v} \times \mathbf{B}) = m\dot{\mathbf{v}} \qquad (9\text{-}22)$$

In sec. 9.01, $\dot{\mathbf{v}}$ was expressed in rectangular and cylindrical co-ordinates. The force also may be expressed quite simply in the two co-ordinate systems. In rectangular co-ordinates, the determinant form of the cross-product

$$\mathbf{v} \times \mathbf{B} = \begin{vmatrix} \hat{\mathbf{x}} & \hat{\mathbf{y}} & \hat{\mathbf{z}} \\ v_x & v_y & v_z \\ B_x & B_y & B_z \end{vmatrix}$$

is helpful in writing the force as

$$\begin{aligned}\mathbf{F} = e[\hat{\mathbf{x}}(E_x + v_y B_z - v_z B_y) + \hat{\mathbf{y}}(E_y + v_z B_x - v_x B_z) \\ + \hat{\mathbf{z}}(E_z + v_x B_y - v_y B_x)]\end{aligned} \tag{9-23}$$

In cylindrical co-ordinates, the cross-product may be expressed as

$$\mathbf{v} \times \mathbf{B} = \begin{vmatrix} \hat{\rho} & \hat{\phi} & \hat{\mathbf{z}} \\ v_\rho & v_\phi & v_z \\ B_\rho & B_\phi & B_z \end{vmatrix}$$

from which the force may be written in the form

$$\begin{aligned}\mathbf{F} = e[\hat{\rho}(E_\rho + v_\phi B_z - v_z B_\phi) + \hat{\phi}(E_\phi + v_z B_\rho - v_\rho B_z) \\ + \hat{\mathbf{z}}(E_z + v_\rho B_\phi - v_\phi B_\rho)]\end{aligned} \tag{9-24}$$

Expressions (23) and (24) when substituted into eqs. (6) and (20) give the complete equation of motion in both rectangular and cylindrical co-ordinates; in these forms the equation of motion may be used in a wide variety of situations involving charged particles. The simplest of these situations is the one in which an external field (**E**, **B**) is applied and the number of charged particles is so small that their presence does not alter the field; in such a case **F** is a given function and two successive integrations of the equation of motion (with respect to time) give the velocity and position of the particle as functions of time. Then, elimination of time from the position relations gives the orbit of the charged particle. However, in many situations, the charged-particle density is sufficient to alter the applied field, thus making it necessary to solve the equation of motion simultaneously with Maxwell's equations.

Before proceeding to specific examples of particle motion, it must be noted that work is done (energy is expended) when a charged particle is moved from one point to another in the presence of an external field. Suppose a field **F** acts on a charged particle as the particle moves from point (1) to point (2). The work done by the field on the particle is expressed as

$$W_{12} = \int_1^2 \mathbf{F} \cdot d\mathbf{s} \tag{9-25}$$

Substitution of (3) into (25) gives

$$\begin{aligned}W_{12} &= m \int_1^2 \frac{d\mathbf{v}}{dt} \cdot d\mathbf{s} = m \int_1^2 d\mathbf{v} \cdot \frac{d\mathbf{s}}{dt} = m \int_1^2 \mathbf{v} \cdot d\mathbf{v} = \tfrac{1}{2} m \int_1^2 d(\mathbf{v} \cdot \mathbf{v}) \\ &= \tfrac{1}{2} m \int_1^2 d(v^2) = \tfrac{1}{2} m(v_2^2 - v_1^2)\end{aligned} \tag{9-26}$$

Equation (26) states that the work done is equal to the increase in *kinetic energy.*

In order to see how the work done is shared between the electric and magnetic fields, substitute (21) into (25).

$$W_{12} = e\int_1^2 (\mathbf{E} + \mathbf{v} \times \mathbf{B})\cdot d\mathbf{s} \tag{9-27}$$

It is important to note that

$$\mathbf{v} \times \mathbf{B}\cdot d\mathbf{s} = d\mathbf{s}\cdot\mathbf{v} \times \mathbf{B} = d\mathbf{s} \times \mathbf{v}\cdot\mathbf{B} = 0$$

because $d\mathbf{s}$ and $\mathbf{v}$ are in the same direction. Thus

$$W_{12} = e\int_1^2 \mathbf{E}\cdot d\mathbf{s} \tag{9-28}$$

which states that the work is done entirely by the electric field.

In many cases of interest the electric field is either steady or varying so slowly that electromagnetic effects are negligible. Under such conditions, the electric field strength may be written as the gradient of a scalar potential, that is

$$\mathbf{E} = -\nabla V \tag{9-29}$$

Substitution of (29) into (28) gives

$$W_{12} = -e(V_2 - V_1) \tag{9-30}$$

which states that the work done is proportional to the potential difference between points (2) and (1). A particularly useful relation may be obtained by combining (30) with (26):

$$\tfrac{1}{2}m(v_2^2 - v_1^2) = -e(V_2 - V_1) \tag{9-31}$$

Equation (31) indicates that the increase in kinetic energy from point (1) to point (2) is proportional to the potential difference between the two points.

Problem 1. Two parallel metal plates lie in the planes $z = 0$ and $z = d$. The plates are at the potential $V = 0$ and $V = -V_0$ respectively. An electron enters the region between the plates at the origin and at $t = 0$. The initial velocity is

$$\mathbf{v} = \hat{\mathbf{y}}v_a + \hat{\mathbf{z}}v_b$$

Find the velocity and position as functions of time. Determine the orbit and show that it is in the form of a parabola. Find the point of deepest penetration into the parallel-plate region both from the orbit and from energy considerations.

9.03 Circular Motion in a Magnetic Field. Consider the motion of a charged particle in a steady magnetic field given by

$$\mathbf{B} = \hat{\mathbf{z}}B \tag{9-32}$$

Suppose that the velocity is constant in time and is entirely in the $\hat{\boldsymbol{\phi}}$ direction. Under these conditions, the equation of motion has only a $\hat{\boldsymbol{\rho}}$

component obtained by substituting (24) into (20):

$$ev_\phi B + m\rho\omega^2 = 0 \tag{9-33}$$

Equation (33) states the radial force condition under which uniform circular motion may exist. This condition may be simplified using (12) with the result

$$eB + m\omega = 0 \tag{9-34}$$

stating that a particle with a negative charge must rotate in the $\hat{\phi}$ direction. This rotation is at a fixed angular frequency ω_c, called the *cyclotron frequency* because of its importance in the design of cyclotrons for nuclear research. The cyclotron frequency is given by

$$\omega_c = -\frac{eB}{m} \tag{9-35}$$

Equation (12) states that the radius of the circular orbit is proportional to the velocity (since the frequency is a constant).

The acceleration of a moving electron in a steady magnetic field is utilized in some cathode ray tubes to produce deflection of an electron beam. After emission from the cathode and after acceleration to a fixed velocity, the electrons enter a region in which there is a nearly uniform, transverse magnetic field. The electrons then follow a circular orbit until they leave the magnetic field region, whereupon they move in a straight line which is at an angle to their original path. Deflection of an electron beam also may be accomplished if the beam is passed through a nearly uniform, transverse electric field such as that between a pair of parallel, flat-plate electrodes having a potential difference between them. Between the electrodes the electron path is approximately parabolic but beyond the electric field region the electron path reverts to a straight line at an angle to the original path. Electron beam deflection by electric and magnetic fields is also fundamental to the operation of the electron microscope. In this instrument regions of nonuniform electric or magnetic fields act as *lenses* which have essentially the same effects on electron beams as glass lenses have on light waves. In an electron microscope the electron beam passes through the specimen and is then focused by field lenses to form a real image on a photographic plate. For further discussion of electric and magnetic lenses the reader is referred to the literature.*

Problem 2. A positively charged particle has a velocity $\mathbf{v} = -\hat{\mathbf{x}}v_0$ at the origin in the presence of the magnetic field given by (32). Find the rectangular components of velocity and position as functions of time by integrating the equation of motion in rectangular co-ordinates. Also find the orbit in rectangu-

*See W. W. Harman, *Fundamentals of Electronic Motion*, McGraw-Hill Book Company, New York, 1953.

lar co-ordinates and check your result with the orbit you would expect from the foregoing discussion of circular motion.

9.04 Crossed-field Motion of a Charged Particle. An interesting and useful form of charged-particle motion occurs under the influence of crossed electric and magnetic fields. Suppose that the crossed fields are steady and uniform, given by

$$\mathbf{B} = \hat{\mathbf{z}}B \qquad \text{and} \qquad \mathbf{E} = \hat{\mathbf{y}}E$$

Suppose also that a positively charged particle of e coulombs and mass m is initially at the origin and traveling in the $\hat{\mathbf{x}}$ direction with a velocity v_0 as shown in Fig. 9-4. The problem is to find the velocity and position of the particle at all subsequent times.

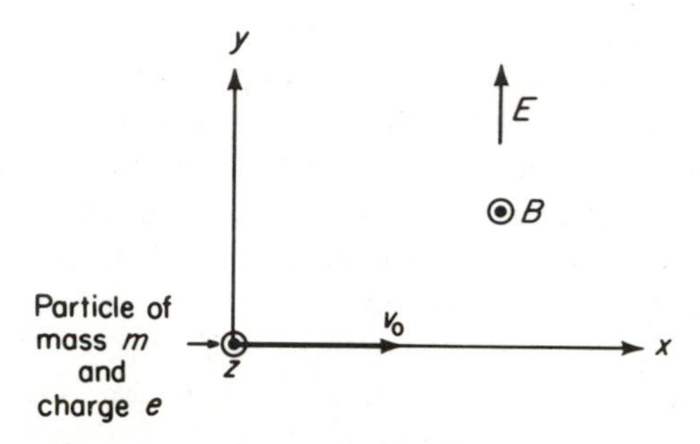

Figure 9-4. Initial conditions for crossed-field motion.

In rectangular co-ordinates, the equation of motion may be expressed as

$$m\dot{\mathbf{v}} = e[\hat{\mathbf{x}}v_y B + \hat{\mathbf{y}}(E - v_x B)] \tag{9-36}$$

The cyclotron frequency definition $\omega_c = -e(B/m)$ permits expression of (36) in component form as

$$\dot{v}_x = -\omega_c v_y \tag{9-37}$$

$$\dot{v}_y = \frac{e}{m}E + \omega_c v_x \tag{9-38}$$

$$\dot{v}_z = 0 \tag{9-39}$$

The z component may be dealt with immediately. Since there is no initial velocity in the $\hat{\mathbf{z}}$ direction, $v_z = 0$. Also, since the particle starts at the origin, $z = 0$ at all points along its path.

Equations (37) and (38) are coupled and thus must be combined to obtain a differential equation in one unknown. Taking the time derivative of (38) and eliminating v_x gives

$$\ddot{v}_y = -\omega_c^2 v_y \tag{9-40}$$

which has the general solution

$$v_y = C_1 \sin \omega_c t + C_2 \cos \omega_c t \tag{9-41}$$

The arbitrary constant C_2 must be zero since $v_y = 0$ at $t = 0$. The x component of velocity may be found by substituting (41) into (38). This gives

$$\omega_c v_x = -\frac{e}{m} E + \omega_c C_1 \cos \omega_c t$$

Since $v_x = v_0$ at $t = 0$, the arbitrary constant C_1 must be given by

$$C_1 = \frac{e}{m\omega_c} E + v_0 = -\frac{E}{B} + v_0$$

Thus, the velocity at any time t may be expressed in component form as

$$v_x = \frac{E}{B} + \left(v_0 - \frac{E}{B}\right) \cos \omega_c t \tag{9-42}$$

$$v_y = \left(v_0 - \frac{E}{B}\right) \sin \omega_c t \tag{9-43}$$

The position of the particle may be found by integrating the velocity components with respect to time. Integrating (42) gives

$$x = \frac{E}{B} t + \frac{1}{\omega_c}\left(v_0 - \frac{E}{B}\right) \sin \omega_c t \tag{9-44}$$

with the arbitrary constant of integration set equal to zero since $x = 0$ at $t = 0$. Integrating (43) gives

$$y = -\frac{1}{\omega_c}\left(v_0 - \frac{E}{B}\right) \cos \omega_c t + C_3$$

The arbitrary constant C_3 is determined by the condition $y = 0$ at $t = 0$ which gives

$$y = \frac{1}{\omega_c}\left(v_0 - \frac{E}{B}\right)(1 - \cos \omega_c t) \tag{9-45}$$

With $v_0 = 0$, (44) and (45) are the parametric equations of a cycloid with t as the parameter; a cycloid is the path traced by a point on the rim of a rolling wheel. The cycloidal trajectory of the particle is shown in Fig. 9-5 along with the trajectories (known as trochoids) for several initial velocities different from $v_0 = 0$. It is especially interesting to note the focusing effect which causes the charged particles to pass through a succession of fixed points on the x axis for any value of initial velocity v_0. These focal points are given by $\cos \omega_c t = 1$ or $\omega_c t = n2\pi$ where n is a positive or negative integer or zero; the focal point positions are thus given by

$$x = n \frac{2\pi E}{\omega_c B}$$

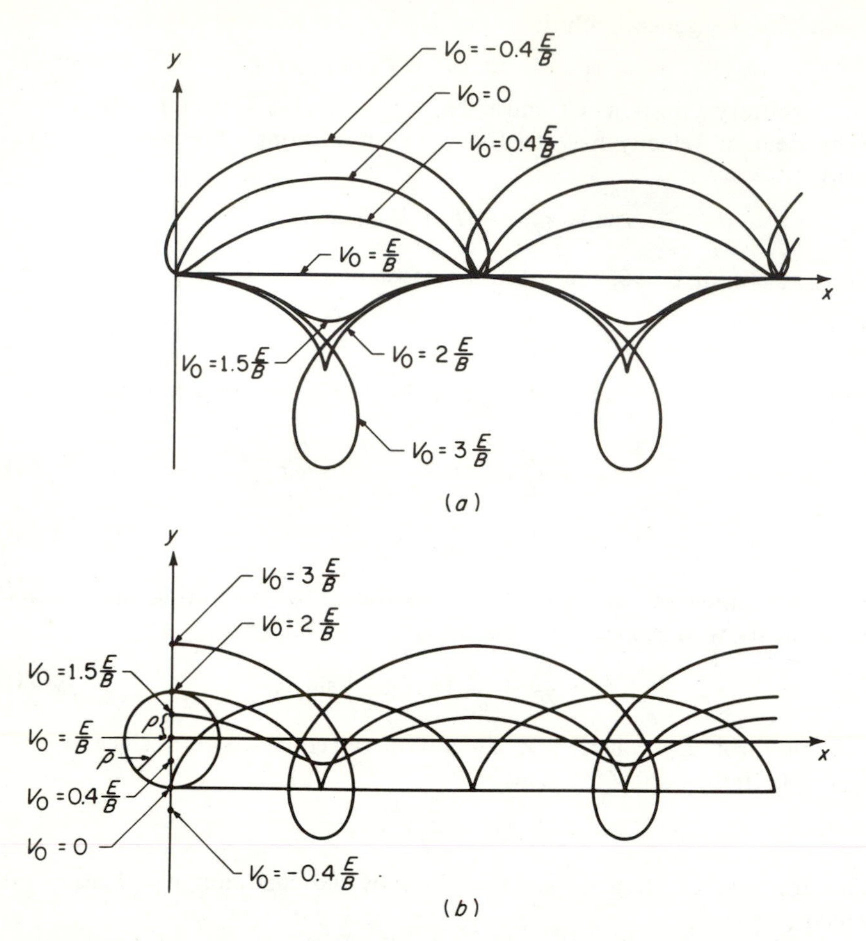

Figure 9-5. Crossed-field particle trajectories (*a*) as calculated and (*b*) as determined by tracing the paths of points on the spoke of a rolling wheel.

Further points of interest are that the time of flight between focal points is independent of v_0 and that the condition $v_0 = E/B$ gives a straight-line trajectory. The properties of crossed-field trajectories are utilized in the design of traveling-wave magnetrons and microwave photomultipliers.*

There is a simple alternative approach to the motion of charged particles in crossed electric and magnetic fields that provides additional insight and an easy means for obtaining quick answers on what at first

*O.L. Gaddy and D.F. Holshouser. "A Microwave Frequency Dynamic Crossed-Field Photomultiplier," *Proc. IEEE*, **51**, No. 1, pp. 153–162, January, 1963.

appears to be a rather complicated set of trajectories. From eqs. (42) and (43) it is noted that if the electron is injected with an initial velocity $v_0 = \bar{v}_0 = E/B$, the particle continues at velocity v_0 on a straight-line path along $y = 0$. This is because the electric force Ee is just cancelled by the y-directed magnetic force ev_xB, and all accelerating forces are reduced to zero. For an injection velocity v_0 different from $\bar{v}_0$, the magnetic force can be considered as the sum of $e\bar{\mathbf{v}}_0 \times \mathbf{B}$ and $e(\mathbf{v} - \bar{\mathbf{v}}_0) \times \mathbf{B}$. The first of these components cancels the effect of the electric field, and the second force component produces the typical circular motion of a particle in a magnetic field (sec. 9.03), with an angular frequency $\omega_c = -e(B/m)$ and a radius $\rho = m(v_0 - \bar{v}_0)/eB$. Hence the particle motion is the sum of a straight-line motion along the x axis at a velocity $\bar{v}_0$ and the circular motion just described. Such motion is exactly that described by a point on the spoke of a wheel of radius $\bar{\rho} = m\bar{v}_0/eB = mE/eB^2$ rolling along a straight line as depicted in Fig. 9-5(*b*). Different injection velocities v_0 provide the trajectories shown.

Problem 3. Assume that the crossed-field motion discussed above takes place between parallel plates lying in the $y = 0$ and $y = d$ planes. The plates are at potentials $V = 0$ and $V = -V_0$, respectively. Use (31) to find the value of B such that the charged particle will just graze the plate at $y = d$. Check your result using the trajectory equations (44) and (45).

Problem 4. A magnetron consists of two coaxial cylindrical electrodes with radii $\rho = a$ and $\rho = b\,(a < b)$ in a region permeated by a magnetic field $\mathbf{B} = \hat{\mathbf{z}}B$. Suppose that the inner electrode (cathode) is at zero potential and the outer electrode (anode) is at potential V. Find the value of B such that electrons leaving the cathode with zero initial velocity will just graze the surface of the anode.

9.05 Space-charge-limited Diode. In the problems discussed so far, attention has been confined to the motion of a single particle. The conclusions reached also apply to a number of charged particles provided the number is not so great as to modify the field. However, if the charged-particle density is great enough, the field acting on any one particle will be modified by the presence of all the others. Under these latter conditions, the equation of motion must be solved simultaneously with Maxwell's equations subject to the appropriate boundary conditions.

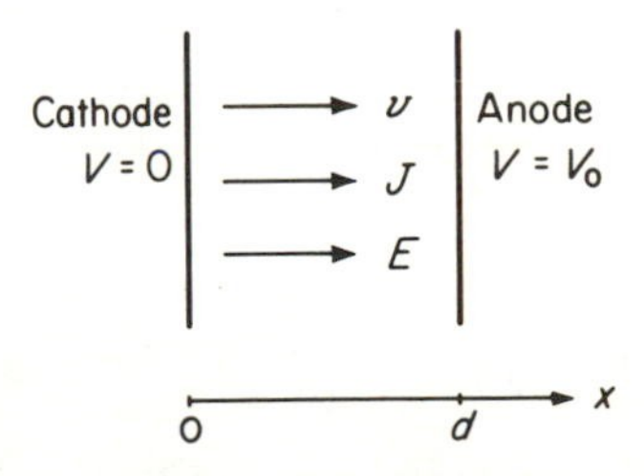

Figure 9-6. Parallel-plane diode.

This method of analysis may be used with the parallel-plane d-c diode depicted in Fig. 9-6. It will be assumed that electrons are given off by the cathode and that they have zero initial velocity. Thus, for any value of x, one may write (31) as

$$\tfrac{1}{2}mv^2 = -eV \tag{9-46}$$

Equation (46) makes use of the equation of motion plus the approximation that $\mathbf{E} = -\nabla V$ which is valid for steady fields. The notion of a steady field might well be questioned for the case of the diode since the field at any point must surely fluctuate as electrons shoot by. Because of the large number of electrons present, however, the total effect of the electrons moving from cathode to anode is essentially that of a continuous, charged fluid flowing in a steady stream. The current density in this stream is

$$\mathbf{J} = \rho\mathbf{v} = Ne\mathbf{v} \tag{9-47}$$

in which ρ is the charge density and N is the number density of the charged particles.

Maxwell's equations now may be brought into the picture. Under steady-field conditions, they contain Poisson's equation

$$\nabla^2 V = -\frac{\rho}{\epsilon} \tag{9-48}$$

and the equation of continuity which may be written in differential and integral form as

$$\nabla\cdot\mathbf{J} = 0 \qquad \text{and} \qquad \oint_S \mathbf{J}\cdot d\mathbf{a} = 0 \tag{9-49}$$

If the surface S is a rectangular box located between the cathode and the anode, (49) indicates that the current entering the left face of the box must be equal to the current leaving the right face, a condition requiring that the current density be constant everywhere between the two electrodes; thus we may write $J(x) = J_0$.

Poisson's equation now may be rewritten using (46) and (47), noting that the field quantities vary only in the x direction.

$$\begin{aligned} \frac{d^2V}{dx^2} &= -\frac{J_0}{\epsilon v} \\ &= -\frac{J_0}{\epsilon}\left(-\frac{m}{2eV}\right)^{1/2} \\ &= KV^{-(1/2)} \end{aligned} \tag{9-50}$$

in which
$$K = -\frac{J_0}{\epsilon}\left(-\frac{m}{2e}\right)^{1/2}$$

In order to solve (50) note that

$$\frac{d}{dx}\left(\frac{dV}{dx}\right)^2 = 2\frac{dV}{dx}\frac{d^2V}{dx^2}$$

and that

$$\frac{d}{dx}(V^{1/2}) = \tfrac{1}{2}V^{-(1/2)}\frac{dV}{dx}$$

Thus (50), when multiplied by dV/dx, becomes

$$\frac{d}{dx}\left(\frac{dV}{dx}\right)^2 = 4K\frac{d}{dx}V^{1/2}$$

and integration gives

$$\left(\frac{dV}{dx}\right)^2 = 4KV^{1/2} + C_1 \tag{9-51}$$

Evaluation of the arbitrary constant C_1 requires a consideration of the effect of space charge on the potential gradient dV/dx. In Fig. 9-7, curve (1) is a straight line representing the potential with no

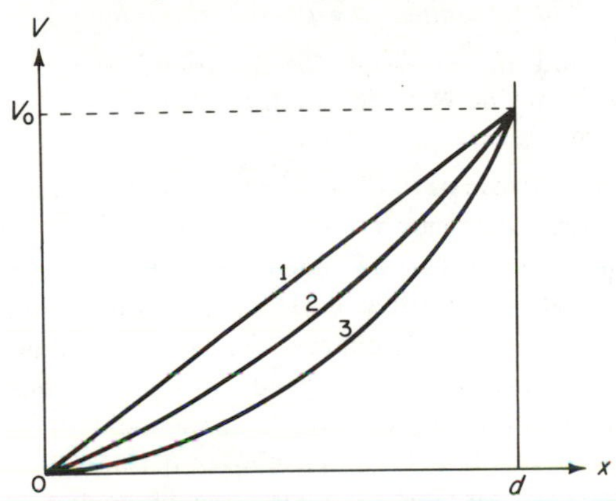

Figure 9-7. The effect of space charge on potential.

charges present. Some negative charge between the plates will tend to repel electrons near the cathode (decrease the potential gradient) and accelerate electrons near the anode (increase the potential gradient), thus causing the potential to vary as in curve (2). As the number of electrons increases, a limiting case [curve (3)] is reached in which there is zero potential gradient at the cathode. Further change in the potential curve would produce a potential barrier near the cathode which could be overcome only if the electrons had a finite initial velocity. This discussion leads to the *space-charge-limited* boundary condition:

$$\frac{dV}{dx} = 0 \qquad \text{at} \qquad x = 0 \tag{9-52}$$

Thus, the arbitrary constant in (51) is zero and we have

$$\frac{dV}{dx} = 2K^{1/2}V^{1/4} \tag{9-53}$$

Equation (53) may be integrated to give

$$V^{3/4} = \tfrac{3}{2} K^{1/2} x \tag{9-54}$$

after setting the arbitrary constant to zero since $V = 0$ at $x = 0$. From (54) it is clear that

$$\frac{V}{V_0} = \left(\frac{x}{d}\right)^{4/3}$$

giving curve (3) of Fig. 9-7. In addition, if (54) is evaluated at $x = d$ and the substitution for K is made, then it may be noted that

$$|J_0| \propto V_0^{3/2} \tag{9-55}$$

Equation (55) gives the current-voltage curve for the diode and it is known as the *Child-Langmuir law* or the *three-halves-power law.*

Problem 5. Find an expression for the transit time (time taken for an electron to pass from one electrode to the other) for a space-charge-limited, parallel-plane diode.

9.06 Plasma Oscillations. The word *plasma* as used here denotes an assembly of charged particles in which the time-average charge density is zero. That is, the number of negative particles per unit volume is equal to the number of positive particles per unit volume if all particles have the same magnitude of charge. Thus, a plasma forms whenever the atoms in a gas are ionized to produce equal numbers of ions and electrons, as for example in the earth's ionosphere or in certain regions of a gas discharge tube. If a plasma is disturbed, powerful restoring forces can be set up, tending to make the particles go into oscillatory motions about their equilibrium positions.

By far the most elementary plasma motion is the oscillation of an infinite plasma slab. Suppose that the positive and negative particles are displaced from each other a small distance x as shown in Fig. 9-8. If the thickness d is small, the time-varying field E within the slab may be calculated using concepts from electrostatics. The excess charges on either side of the slab constitute charge sheets and if the charged-particle density (either positive or negative) is N, then the surface charge density on each sheet is

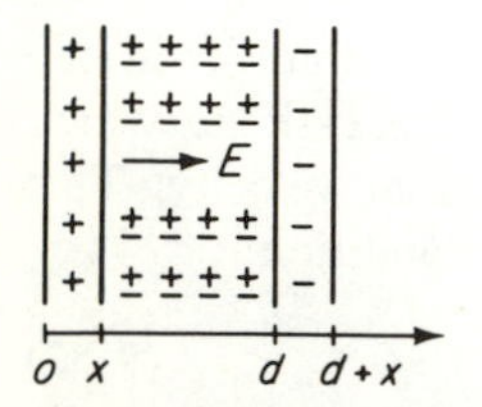

Figure 9-8. Displacement of positive and negative particles in a plasma slab.

$$\rho_s = Nex \qquad \text{coulomb/m}^2$$

in which e is positive for the left-hand sheet and negative for the right-hand sheet. Application of Gauss's law gives the field inside the slab as

$$E = \frac{N|e|x}{\epsilon_v} \tag{9-56}$$

which states that the restoring force is proportional to the displacement, the condition for *simple harmonic motion.*

In a plasma the positive particles are usually ions and because their mass is much greater than that of the negative electrons, they may be considered to be stationary. Thus for the electrons, the equation of motion is of the form

$$m\ddot{x} = -\frac{Ne^2}{\epsilon_v}x \tag{9-57}$$

Equation (57) is valid provided the charged particles do not collide with each other or with any neutral particles which may be present. This equation describes oscillations occurring at a fixed frequency ω_p called the *plasma frequency* and given by

$$\omega_p^2 = \frac{Ne^2}{m\epsilon_v} \tag{9-58}$$

Equation (58) permits (57) to be written in the form

$$\ddot{x} = -\omega_p^2 x \tag{9-59}$$

which has the solution

$$x = C_1 \sin \omega_p t + C_2 \cos \omega_p t \tag{9-60}$$

This solution gives the displacement x for *free* oscillations of the plasma slab, oscillations which can exist only at the plasma frequency. Oscillations can occur at other frequencies but only if the slab is *driven* by an external oscillating field.

9.07 Wave Propagation in a Plasma. Electromagnetic waves propagate through a plasma in much the same manner as through any dielectric material. To begin an analysis of this wave propagation, let us assume that the plasma is made up of electrons and ions and that the ions are relatively immobile (due to their mass) in comparison with the electrons. Maxwell's equations in time-varying form must be combined with the equation of motion for electrons in order to analyze the oscillations which are possible. The electron motions enter Maxwell's equations in the electron current density and electron charge density. These are given by

$$\tilde{\mathbf{J}} = \tilde{n}e\tilde{\mathbf{v}} \tag{9-61}$$

and

$$\tilde{\rho} = \tilde{n}e \tag{9-62}$$

in which $\tilde{n}$ and $\tilde{\mathbf{v}}$ are time-varying electron density and electron velocity. If it is assumed that the electron oscillations are small in amplitude and sinusoidal in form, $\tilde{n}$ and $\tilde{\mathbf{v}}$ may be written as

$$\tilde{n} = N + \text{Re}\{ne^{j\omega t}\} \tag{9-63}$$

$$\tilde{\mathbf{v}} = \text{Re}\,\{\mathbf{v}e^{j\omega t}\} \tag{9-64}$$

in which N is the steady (or *ambient*) electron density; the velocity is assumed to have no steady (or *drift*) component. Since the oscillations are small in amplitude, (61) may be approximated by

$$\tilde{\mathbf{J}} = Ne\,\text{Re}\,\{\mathbf{v}e^{j\omega t}\} \tag{9-65}$$

in which the product of two small quantities is assumed to be negligible. Similar relations also may be written for the ions but since they are assumed to be immobile, they contribute no current-density term to Maxwell's equations. However, while the ions do contribute a steady charge density term to Maxwell's equations, this term serves only to cancel out the steady part of the electron charge density in Gauss's law, thus assuring us that no steady electric fields will arise within the plasma.

The equation of motion and Maxwell's equations now may be written for sinusoidal oscillations by the use of (65), (62) and (63). These may be expressed in phasor form by a straightforward manipulation. With the phasor velocity $\mathbf{v}$ and the phasor electron-density variation n, the equation of motion and Maxwell's equations with zero external magnetic field are

$$e\mathbf{E} = j\omega m\mathbf{v} \tag{9-66}$$

$$\nabla \times \mathbf{H} = j\omega\epsilon_v\mathbf{E} + Ne\mathbf{v} \tag{9-67}$$

$$\nabla \times \mathbf{E} = -j\omega\mu_v\mathbf{H} \tag{9-68}$$

$$\nabla\cdot\mathbf{E} = \frac{ne}{\epsilon_v} \tag{9-69}$$

$$\nabla\cdot\mathbf{H} = 0 \tag{9-70}$$

Equations (67) and (69) contain the equation of continuity,

$$\nabla\cdot(N\mathbf{v}) = -j\omega n \tag{9-71}$$

The above equations may be solved directly for any given problem. However, a major simplification may be achieved by first substituting the equation of motion into Maxwell's equations. Eliminating v from (66) and (67) gives

$$\nabla \times \mathbf{H} = j\omega\epsilon_v\left(1 - \frac{Ne^2}{\omega^2 m\epsilon_v}\right)\mathbf{E} \tag{9-72}$$

which suggests that the plasma can be represented by a dielectric permittivity given by

$$\epsilon = \epsilon_v\left(1 - \frac{\omega_p^2}{\omega^2}\right) \tag{9-73}$$

in which ω_p^2 is as given by (58). In ionospheric work the relative permittivity usually is written as

$$\frac{\epsilon}{\epsilon_v} = K_0 = 1 - X \qquad \text{where} \qquad X = \frac{\omega_p^2}{\omega^2} \tag{9-74}$$

The permittivity concept also enters into Gauss's law (69) as may be shown by eliminating $\mathbf{v}$ from (66) and (71) and then using this relation to eliminate n from (69). This procedure permits (69) to be expressed as

$$\nabla \cdot (\epsilon \mathbf{E}) = 0 \tag{9-75}$$

which is completely consistent with the idea that a plasma can be represented by the new permittivity given by (73).

The great advantage of the plasma-permittivity concept is that we can use all the existing solutions for wave propagation in a dielectric in order to study wave propagation in a plasma. For instance, the phasor Helmholtz equation

$$\frac{\partial^2 E_y}{\partial x^2} = -\omega^2 \mu\epsilon E_y \qquad \left(\text{with } \frac{\partial}{\partial y} = \frac{\partial}{\partial z} = 0\right) \tag{9-76}$$

has the solution

$$E_y = C_1 e^{-j\omega(\mu\epsilon)^{1/2}x} + C_2 e^{+j\omega(\mu\epsilon)^{1/2}x} \tag{9-77}$$

For electrons in a vacuum, (77) may be expressed as

$$E_y = C_1 e^{-j\beta_v(1-X)^{1/2}x} + C_2 e^{+j\beta_v(1-X)^{1/2}x} \tag{9-78}$$

in which $\beta_v = \omega(\mu_v\epsilon_v)^{1/2}$. Equation (78) demonstrates that wave propagation in a plasma is similar to wave propagation in an ordinary dielectric provided X is less than unity, that is provided $\omega > \omega_p$. If X is greater than unity, that is if $\omega < \omega_p$, then (78) carries more meaning if written as

$$E_y = C_3 e^{-\beta_v(X-1)^{1/2}x} + C_4 e^{+\beta_v(X-1)^{1/2}x} \tag{9-79}$$

Thus when $\omega < \omega_p$, electromagnetic waves are *attenuated* by a plasma. By analogy with a wave guide, ω_p may be regarded as the *cut-off frequency* of a plasma.

One must appreciate that attenuation in a cut-off plasma as described above is not accompanied by the absorption of power and its subsequent conversion into heat; in this respect the plasma is like a cut-off wave guide. Thus a wave passing from free space into a cut-off plasma is not absorbed or transmitted but rather *totally reflected*. It is this type of reflection which causes radio waves to "bounce" off the ionosphere, thus making long-range radio communication possible.

Of course, some power loss is always present in a plasma due to the fact that the electrons frequently collide with gas molecules, ions, or even other electrons. Such collisions cause some of the power carried by an electromagnetic wave to be transformed into heat. At frequencies above the plasma frequency these collisional losses cause a propagating wave to be attenuated, and below the plasma frequency the losses cause total reflection to become partial reflection.

Problem 6. Collisional effects in a plasma may be taken into account approximately by including in the equation of motion the frictional force term $-m\nu\tilde{\mathbf{v}}$ in which ν is the collision frequency of the electrons with other particles. Show that the concept of a plasma permittivity is valid with collisions included and also show that the relative permittivity is given by

$$K_0 = 1 - \frac{X}{U} \tag{9-80}$$

in which $U = 1 - jZ$ and $Z = \nu/\omega$

Application to Superconductors. In a metal there exists a very high density N of free electrons. At room temperature these electrons collide very frequently with the metal atoms, causing power loss and the familiar *resistance* property of metals. At temperatures near absolute zero, however, some metals exhibit a complete loss of resistance. This phenomenon is termed *superconductivity.* Superconductivity was first observed in Holland in 1911 by Kamerlingh Onnes. In 1933 Meissner and Ochsenfeld discovered that superconducting metals could not be penetrated by a magnetic field; furthermore, they observed that when a normal metal permeated by a magnetic field is cooled to the superconducting state, the metal does not "freeze in" the magnetic flux but acts to expel it. Theoretical explanations of superconductivity began with the macroscopic theory of London and London* in 1935 and culminated in 1957 with the highly successful microscopic theory of Bardeen, Cooper and Schrieffer.†

Recent theories of superconductivity are very involved and quite beyond the scope of the present discussion. However, a very simple picture of the superconducting state results when one assumes that the electrons in a superconductor are able to move about freely without collisions; this theoretical model is essentially the collisionless plasma model already analyzed. Because of the exceedingly high electron density in a metal, the plasma is cut off and waves can penetrate only a very short distance into the metal (a distance of the order of 200Å). Thus, viewing the superconductor as a cut-off plasma provides a crude explanation for its ability to prevent magnetic fields from penetrating.

*F. London and H. London, *Proc. Roy. Soc.* (London), **A 149**, 71 (1935); *Physica,* **2**, 341 (1935).

†J. Bardeen, L.N. Cooper and J.R. Schrieffer, *Phys. Rev.*, **108**, 1175 (1957).

From (79) a wave entering a superconductor would be attenuated with a "skin depth" given by

$$\delta = [\beta_v(X - 1)^{1/2}]^{-1} \tag{9-81}$$

$$\approx [\beta_v X^{1/2}]^{-1} \qquad \text{for large } N$$

$$\approx \left[\frac{m}{Ne^2 \mu_v}\right]^{1/2} \tag{9-82}$$

which states that for a large electron density the skin depth is independent of frequency. This suggests that a superconductor would tend to exclude a magnetic field even if the field were applied very slowly.

9.08 Polarization of Dielectric Materials. Dielectrics differ from plasmas and conducting media in that dielectrics contain no free charges, but rather charges which are tightly bound together to form atoms and molecules. The application of a steady external electric field causes a small but significant separation of the bound charges so that each infinitesimal element of volume behaves as if it were an electrostatic dipole. The induced dipole field tends to oppose the applied field, as indicated in Fig. 9-9. The "volume element" indicated in the figure could represent an atom, a molecule, or any small region in the material. A dielectric material which exhibits this separation of bound charge is said to be *polarized.*

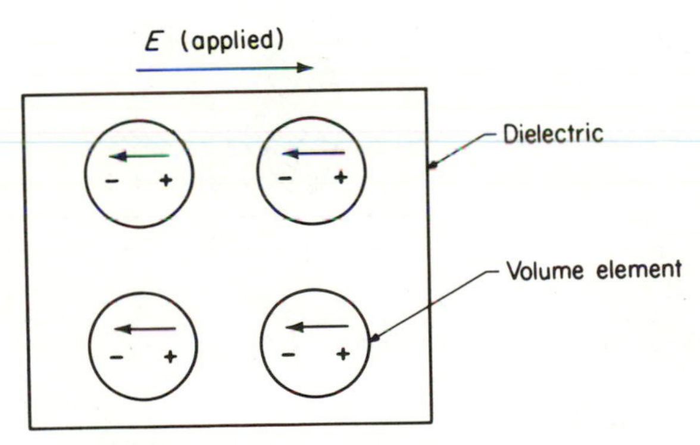

Figure 9-9. Induced electric fields (small arrows) opposing the applied field.

On a microscopic scale, the type of polarization is determined by the material. In most materials polarization occurs only in the presence of an applied field, although a few materials exhibit permanent polarization; *ferroelectric* crystals exhibit a spontaneous, permanent polarization while certain other materials called *electrets* become permanently polarized if allowed to solidify in a strong electric field. On an atomic scale, charge separation can occur due to the displacement of the negative electron cloud relative to the positive nucleus; this is called *electronic* polarization. On a molecular scale, *ionic* and *orientational* polarization are important. Ionic polarization results from the separation of positive and negative ions in molecules held together by ionic bonds. Orientational polarization arises in materials whose molecules are permanently polarized but randomly oriented; the application of an external electric field causes the molecules to align themselves with the applied field, thus producing a net polarization in the material. On a still larger scale, one encounters *space-charge* polarization which arises when free-conduction electrons are present but are prevented from moving great distances by barriers such as grain boundaries; when a field is applied, the electrons "pile-up" against these barriers, producing the separation of charge required to polarize the material. The four types of polarization mentioned above are indistinguishable in a steady field but in a time-varying field each type exhibits a different characteristic response time.

Electronic polarization may be understood to a first order of approximation by regarding an atom as a point nucleus surrounded by a spherical electron cloud of uniform charge density as shown in Fig. 9-10. With an applied steady field the nucleus and the electron cloud

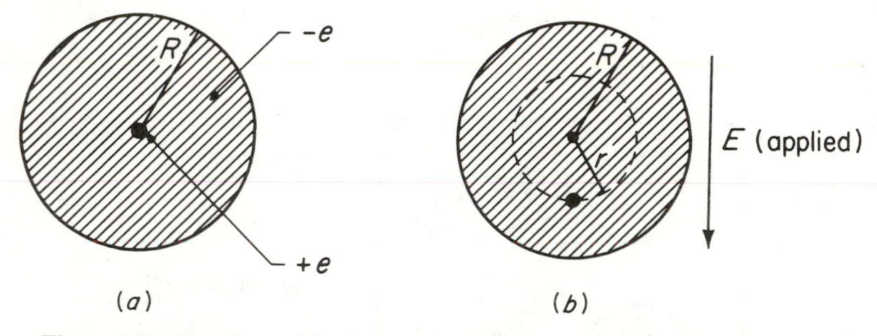

Figure 9-10. Simple model of an atom (*a*) with no applied field and (*b*) with an applied electric field.

are displaced relative to each other until their mutual attractive force is just equal to the force due to the applied field. This force F may be calculated by applying Gauss's law to the sphere of radius r:

$$F = eE = \frac{e}{\epsilon_v}\frac{1}{4\pi r^2}\frac{r^3}{R^3}e = \frac{e^2}{\epsilon_v}\frac{r}{4\pi R^3} \tag{9-83}$$

The magnitude of the dipole moment is given by

$$p = er = 4\pi\epsilon_v R^3\, E \tag{9-84}$$

which also may be written in vector form by giving the scalar moment the direction of a line drawn from the negative charge to the positive charge; thus

$$\mathbf{p} = 4\pi\epsilon_v R^3 \mathbf{E} = \alpha_e \mathbf{E} \text{ where } \alpha_e = 4\pi\epsilon_v R^3 \tag{9-85}$$

The quantity α_e in (85) is called the *electronic polarizability* of the atom; it is the constant of proportionality between the dipole moment and the electric field strength. It should be noted that this linear relationship between **p** and **E** holds only for small applied fields.

Any molecule develops a dipole moment proportional to the applied electric field for small field magnitudes. This proportionality may be expressed as

$$\mathbf{p} = \alpha \mathbf{E} \tag{9-86}$$

in which the total polarizability α is the sum of the polarizabilities arising from each of the different types of polarization (electronic, ionic, etc.). If (86) is multiplied by N_0, the number of molecules per cubic meter, the result is the *dipole moment per unit volume* or the *polarization* and it is represented by the letter $\mathscr{P}$:

$$\mathscr{P} = N_0 \mathbf{p} = N_0 \alpha \mathbf{E} \tag{9-87}$$

Equation (87) is often written as

$$\mathscr{P} = \epsilon_v \chi \mathbf{E} \qquad \text{where} \qquad \epsilon_v \chi = N_0 \alpha \tag{9-88}$$

The quantity χ (chi) is known as the *susceptibility* of the material.

9.09 Equivalent Volume and Surface Charges. According to the picture of polarization developed so far, the application of an external electric field to a dielectric causes the formation of dipoles within the material. The dipoles in turn produce an electric field of their own. Thus, in order to understand dielectrics, it is necessary to calculate the field of a volume distribution of infinitesimal dipoles.

As shown in chap. 2, a single z-directed dipole of moment $\mathbf{p} = \hat{\mathbf{z}}p$ located at the origin has a potential given by

$$\begin{aligned} 4\pi\epsilon_v V(\mathbf{r}) &= \frac{p \cos\theta}{r^2} \\ &= \frac{\mathbf{p} \cdot \hat{\mathbf{r}}}{r^2} \\ &= -\mathbf{p} \cdot \nabla\left(\frac{1}{r}\right) \end{aligned} \tag{9-89}$$

Equation (89) is valid for any orientation of the moment **p** and it may be adapted easily for an arbitrary position $\mathbf{r}'$ of the dipole. Thus if R is the distance from the source point (dipole) to the observation point, the potential may be expressed as

$$4\pi\epsilon_v V(\mathbf{r}) = -\mathbf{p}\cdot\nabla\left(\frac{1}{R}\right)$$

$$= \mathbf{p}\cdot\nabla'\left(\frac{1}{R}\right) \tag{9-90}$$

Consider now a large number of infinitesimal dipoles distributed throughout a given volume as shown in Fig. 9-11. Since $\mathscr{P}$ is the dipole

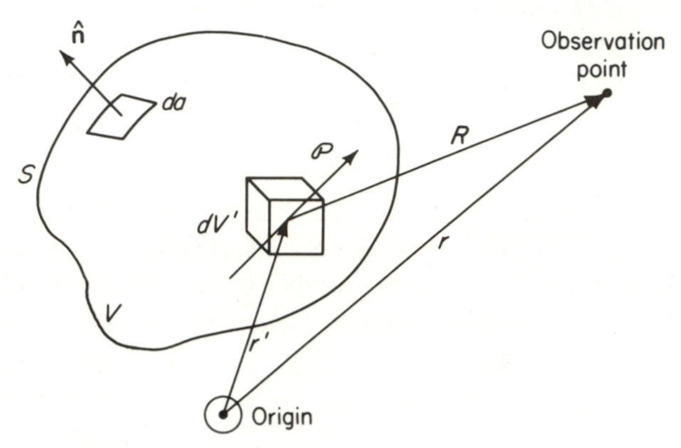

Figure 9-11. A volume distribution of polarization.

moment per unit volume, the moment of the volume dV' must be $\mathscr{P}\,dV'$. Thus the potential at the observation point must be given by

$$4\pi\epsilon_v V(\mathbf{r}) = \int_V \mathscr{P}\cdot\nabla'\left(\frac{1}{R}\right)\,dV' \tag{9-91}$$

Equation (91) may be rewritten using the identity (1-31) for the divergence of the product of a scalar and a vector. Thus

$$4\pi\epsilon_v V(\mathbf{r}) = \int_V \nabla'\cdot\left(\frac{\mathscr{P}}{R}\right)dV' - \int_V \frac{\nabla'\cdot\mathscr{P}}{R}\,dV' \tag{9-92}$$

The divergence theorem allows (92) to be expressed as

$$4\pi\epsilon_v V(\mathbf{r}) = \oint_S \frac{\mathscr{P}\cdot\hat{\mathbf{n}}}{R}\,da' - \int_V \frac{\nabla'\cdot\mathscr{P}}{R}\,dV' \tag{9-93}$$

in which the relation $d\mathbf{a} = \hat{\mathbf{n}}\,da$ has been used. Equation (93) is particularly important because, by comparison with the formula for the potential of a given charge density, it shows that a volume distribution of dipoles may be represented as a surface charge density plus a volume charge density. These are given by

$$\rho_s = \mathscr{P}\cdot\hat{\mathbf{n}} \tag{9-94}$$

and

$$\rho = -\nabla\cdot\mathscr{P} \tag{9-95}$$

Such a result is not surprising if one returns to the simple picture of

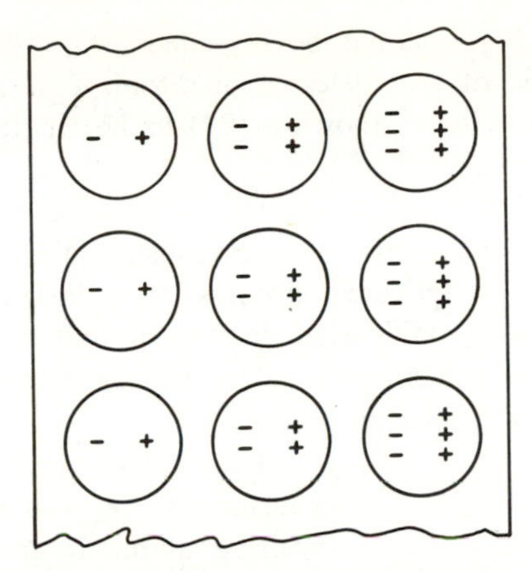

Figure 9-12. Nonuniform polarization in a dielectric slab.

molecular polarization as shown in Fig. 9-12, which depicts the nonuniform polarization arising, for example, in a nonuniform dielectric. It is easy to see that incomplete cancellation of the dipole "end charges" would produce a volume charge density and that the uncancelled charges at the surface of the material would produce a surface charge density.

9.10 The Permittivity Concept. In the interior of a dielectric material, the surface charge density is zero and the volume charge density is given by (95). Since no free charge is present the bound charge may be represented by ρ and thus Gauss's law becomes

$$\nabla\cdot(\epsilon_v \mathbf{E}) = \rho = -\nabla\cdot\mathscr{P}$$

which may be written in the form

$$\nabla\cdot(\epsilon_v \mathbf{E} + \mathscr{P}) = 0 \tag{9-96}$$

This suggests the use of a displacement density (or flux density) vector $\mathbf{D}$ defined as

$$\mathbf{D} = \epsilon_v \mathbf{E} + \mathscr{P} \tag{9-97}$$

which permits (96) to be written in the familiar form

$$\nabla\cdot\mathbf{D} = 0 \tag{9-98}$$

Of course, if free charge were present in the dielectric, a suitable charge density would have to replace the zero on the right-hand side of (98).

The theory of dielectrics may be simplified even further by recalling the proportionality between polarization and electric field strength. Substitution of (88) into (97) gives

$$\mathbf{D} = \epsilon_v(1 + \chi)\mathbf{E} \tag{9-99}$$

which suggests that $1 + \chi$ could be regarded as a *relative permittivity* ϵ_r which is characteristic of the dielectric material. This in turn leads to the use of a total permittivity ϵ, allowing (99) to be written in the form

$$\mathbf{D} = \epsilon\mathbf{E} \tag{9-100}$$

in which $\epsilon = \epsilon_v\epsilon_r = \epsilon_v(1 + \chi)$. This establishes the exceedingly useful result that the rather complicated process of polarization may be represented by a simple modification in the permittivity. The use of a modified permittivity is an alternative to the representation consisting of a collection of charge dipoles in space.

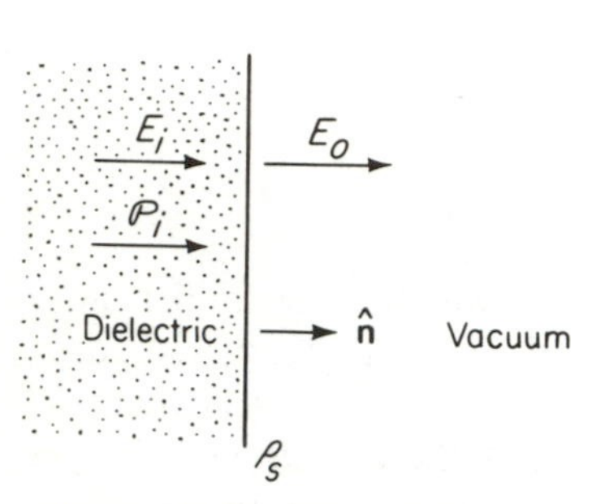

Figure 9-13. The surface of a polarized dielectric.

The above result applies to the interior of a dielectric but reveals nothing about boundary conditions at its surface. Consider the dielectric surface represented by the surface charge density ρ_s as depicted in Fig. 9-13. With a surface charge present, the boundary condition is known to be

$$\epsilon_v E_0 - \epsilon_v E_i = \rho_s \tag{9-101}$$

Substitution of (94) into (101) leads to

$$\epsilon_v E_0 - (\epsilon_v E_i + \mathscr{P}_i) = 0$$

or

$$D_0 - D_i = 0 \tag{9-102}$$

which states that the normal component of the displacement density is continuous at the surface of a dielectric on which there is no free surface-charge density. The equivalent surface charge does not affect the tangential component of $\mathbf{E}$ so that the tangential electric field strength must be continuous at the boundary.

The great advantage of the permittivity concept is that it permits easy solution of problems that would be difficult to formulate using the dipole-distribution concept. Naturally, one must be careful to keep the two concepts entirely separate, representing the dielectric *either* by its permittivity *or* by its dipole charge distribution (its equivalent surface and volume charges).

Problem 7. A conducting sphere of radius a is held at the potential V_0 (zero potential is taken to be at infinite radius). The sphere is surrounded by a spherical shell of uniform dielectric (ϵ_r) with inner radius a and outer radius b. First find the potential and the electric field strength everywhere outside the spherical conductor. Then find a surface charge and volume charge representation for the dielectric shell. Check your equivalent-charge distributions by using them to recalculate the potential and electric field strength.

9.11 Magnetic Polarization. The Bohr model of the atom pictures electrons rotating around a nucleus, a charge motion which may be regarded as a current loop. A small current loop has associated with it a *magnetic moment* given by the product of the current and the area of the loop. Thus, it is reasonable to associate with an atom the property of magnetic moment designated by the letter m. In reality, this moment arises not only from the rotation of the electrons, but also from their "spin." Furthermore, some atoms possess permanent moments while others have moments which are induced by external magnetic fields.

When a magnetic field is applied to an atom possessing a permanent magnetic moment, there is a resultant torque which tends to rotate the atom. If the atom is imagined to be represented by a current loop, it is easy to see that the atom will orient itself so that its own magnetic field will add to the applied magnetic field. Materials whose atoms respond so as to increase the magnetic field are said to be *paramagnetic*. In general this effect is very small, but in *ferromagnetic* materials it is extremely large. Some atoms do not have permanent dipole moments, but their induced dipole moments form in such a way that the induced magnetic field opposes the applied field. Materials whose atoms respond so as to reduce the magnetic field are said to be *diamagnetic*.

Whatever the mechanism of dipole formation, one may assign a vector dipole moment $\mathbf{m}$ to an atom and multiply it by the atomic density N_0 to obtain the *magnetic moment per unit volume* or the *magnetization* represented by the letter $\mathcal{M}$:

$$\mathcal{M} = N_0 \, \mathbf{m} \tag{9-103}$$

For very small applied fields the magnetic moment is proportional to the applied magnetic field strength $\mathbf{H}$ in all materials except ferromagnetic materials. This proportionality may be expressed as

$$\mathcal{M} = \chi_m \, \mathbf{H} \tag{9-104}$$

which defines the *magnetic susceptibility* χ_m. In paramagnetic materials χ_m is positive and in diamagnetic materials it is negative.

9.12 Equivalent Volume and Surface Currents. According to the picture of magnetic polarization outlined above, the application of an external magnetic field to a material causes the formation of magnetic moments within the material. These magnetic moments may be represented by infinitesimal current loops or in other words, *magnetic dipoles*. Clearly, the understanding of magnetization requires first of all the calculation of the field of a volume distribution of magnetic dipoles.

As shown in sec. 3.13, a small loop with scalar moment m and vector moment $\mathbf{m} = \hat{\mathbf{z}}m$ centered at the origin and lying in the x-y plane produces the vector potential $\mathbf{A}$ given by

$$\frac{4\pi}{\mu}\mathbf{A}(\mathbf{r}) = \hat{\boldsymbol{\phi}}\frac{m\sin\theta}{r^2}$$
$$= \frac{\hat{\mathbf{z}} \times \hat{\mathbf{r}}m}{r^2}$$
$$= \frac{\mathbf{m} \times \hat{\mathbf{r}}}{r^2}$$
$$= -\mathbf{m} \times \nabla\left(\frac{1}{r}\right) \tag{9-105}$$

Equation (105) is valid for any orientation of $\mathbf{m}$. Furthermore, it may be generalized to apply to a loop at any point $\mathbf{r}'$ by replacing r with R. For the general case,

$$\frac{4\pi}{\mu}\mathbf{A}(\mathbf{r}) = -\mathbf{m} \times \nabla\left(\frac{1}{R}\right)$$
$$= \mathbf{m} \times \nabla'\left(\frac{1}{R}\right) \tag{9-106}$$

If a large number of dipoles is contained within a volume V (the situation is similar to that of Fig. 9-11), the magnetic moment of a volume dV' must be $\mathcal{M}(\mathbf{r}')\,dV'$. Thus the potential of the entire assembly of magnetic dipoles is given by

$$\frac{4\pi}{\mu}\mathbf{A}(\mathbf{r}) = \int_V \mathcal{M} \times \nabla'\left(\frac{1}{R}\right)dV' \tag{9-107}$$

Equation (107) may be rewritten using identity number (1-33) for the curl of the product of a scalar and a vector. The result is

$$\frac{4\pi}{\mu}\mathbf{A}(\mathbf{r}) = \int_V \frac{\nabla' \times \mathcal{M}}{R}dV' - \int_V \nabla' \times \left(\frac{\mathcal{M}}{R}\right)dV' \tag{9-108}$$

The second volume integral in (108) may be converted into a surface integral using identity (1–54) giving

$$\frac{4\pi}{\mu}\mathbf{A}(\mathbf{r}) = \int_V \frac{\nabla' \times \mathcal{M}}{R}dV' + \oint_S \frac{\mathcal{M} \times \hat{\mathbf{n}}}{R}da' \tag{9-109}$$

When (109) is compared with the expression for the vector potential due to a current distribution, it becomes apparent that the volume distribution of magnetic dipoles is equivalent to a surface current density and a volume current density given by

$$\mathbf{J}_s = \mathcal{M} \times \hat{\mathbf{n}} \tag{9-110}$$

and

$$\mathbf{J} = \nabla \times \mathcal{M} \tag{9-111}$$

This result is to be expected since the magnetic properties may be derived by regarding the material as being made up of many small current loops.

In the interior of the material, incomplete cancellation of the currents in adjacent loops results in a net volume current density. Similarly, at the surface of the material, the currents are not cancelled, giving a net equivalent surface current density. These remarks are illustrated by Fig. 9-14, which could represent, for example, nonuniform magnetic material having as a result nonuniform magnetization.

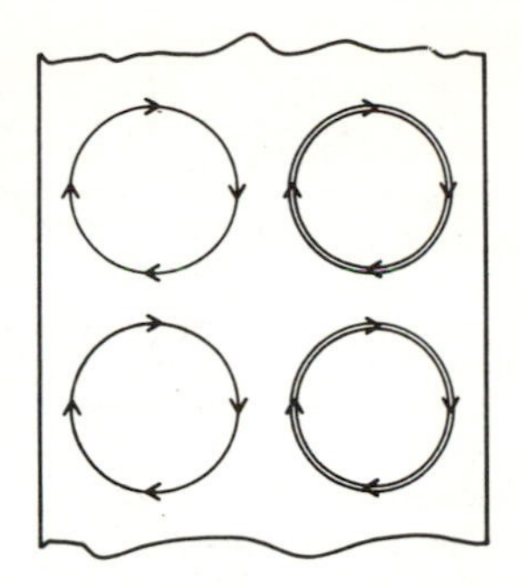

Figure 9-14. Representation of atomic magnetic moments by current loops.

9.13 The Permeability Concept. Inside a magnetically polarized material, the magnetic field is related to the polarization current density through the differential form of Ampere's circuital law which may be expressed as

$$\nabla \times \frac{\mathbf{B}}{\mu_v} = \mathbf{J} = \nabla \times \mathscr{M} \tag{9-112}$$

when there is no current due to the motion of free charges. This may be written in the form

$$\nabla \times \left(\frac{\mathbf{B}}{\mu_v} - \mathscr{M}\right) = 0 \tag{9-113}$$

which suggests the use of a magnetic field strength $\mathbf{H}$ defined as

$$\mathbf{H} = \frac{\mathbf{B}}{\mu_v} - \mathscr{M} \tag{9-114}$$

This allows (113) to be written in the familiar form

$$\nabla \times \mathbf{H} = 0 \tag{9-115}$$

If a nonpolarization current density were present, it would replace the zero on the right-hand side of (115).

It should be recalled at this point that the magnetization is proportional to the magnetic field strength. Substitution of (104) into (114) gives

$$\mathbf{B} = \mu_v(\mathbf{H} + \mathscr{M}) \tag{9-116}$$

$$= \mu_v(1 + \chi_m)\mathbf{H}$$

$$= \mu\mathbf{H} \qquad \text{where} \qquad \mu = \mu_v\mu_r = \mu_v(1 + \chi_m) \tag{9-117}$$

Equation (117) defines a property of the material called the *relative permeability* μ_r and also makes use of the total permeability μ.

The equivalent magnetic polarization surface current density establishes the boundary condition on the tangential fields. For the boundary shown in Fig. 9-15, the boundary condition is known to be

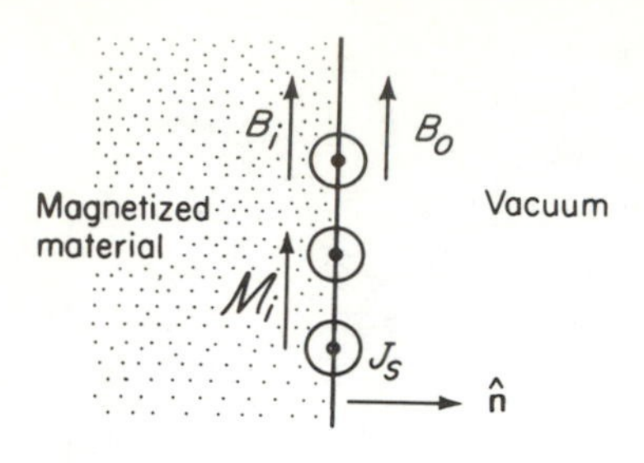

Figure 9-15. The surface of a magnetized material.

$$\frac{B_0}{\mu_v} - \frac{B_t}{\mu_v} = J_s \qquad \text{(9-118)}$$

where J_s represents the magnetic polarization surface current density, and it is assumed that there are no surface currents due to the motion of free charges. Substitution of (110) into (118) leads to

$$\frac{B_0}{\mu_v} - \left(\frac{B_t}{\mu_v} - \mathcal{M}_t\right) = 0$$

or

$$H_0 - H_t = 0$$

which states the well-known boundary condition that the tangential component of the magnetic field strength is continuous across the boundary in the absence of free-charge surface currents.

Problem 8. A long, straight wire of radius a carries a direct current I. The wire is surrounded by a uniform sheath of material with relative permeability μ_r. The sheath carries no current and has inner radius b and outer radius c, such that $a < b < c$. Find **B** and **H** everywhere and use this result to obtain an equivalent magnetic polarization current representation for the sheath. Check your result by recalculation of the field from the polarization currents.

9.14 Frequency Response of Dielectric Materials. Since the polarization of a material requires the displacement of particles, each having a finite mass, it is clear that the strength of the polarization must decrease with increasing frequency. That is, the particle's inertia tends to prevent it from "following" rapid oscillations in the applied field. The existence of several different types of polarization makes a complete analysis somewhat involved. For a qualitative understanding of polarization, however, it is sufficient to consider only electronic polarization and to note that the other types exhibit a similar frequency response.

In the analysis of electronic polarization, it is convenient to regard the heavy positive nucleus as being fixed in space and surrounded by an electron cloud of charge e and mass m which oscillates back and forth in response to an applied electric field E, as illustrated in Fig. 9-16. The electron cloud moves in response to three forces: the force due to the applied field, the restoring force due to the coulomb attraction of the positive nucleus, and frictional forces which result in dissipation of energy and consequent damping of oscillations. The applied force is simply $e\tilde{E}_x$. The restoring force may be obtained from (83) and expressed as

$$-k\tilde{x} \qquad \text{where} \qquad k = \frac{e^2}{4\pi\epsilon_v R^3}$$

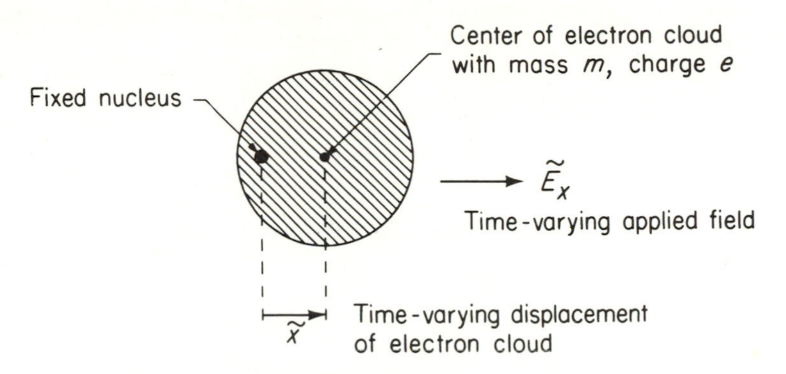

Figure 9-16. Electronic polarization in a time-varying field.

The frictional force is generally proportional to the particle mass and to the velocity and so may be expressed as

$$-mg\dot{\tilde{x}}$$

in which g is a positive constant. The complete equation of motion now may be expressed as

$$m\ddot{\tilde{x}} = -mg\dot{\tilde{x}} - k\tilde{x} + e\tilde{E}_x \tag{9-119}$$

Equation (119) may be simplified greatly by considering only sinusoidal oscillations and making use of phasor notation. A phasor displacement x and a phasor electric field E_x may be defined through the relations

$$\tilde{x} = \text{Re}\{xe^{j\omega t}\} \tag{9-120}$$

and

$$\tilde{E}_x = \text{Re}\{E_x e^{j\omega t}\} \tag{9-121}$$

This allows (119) to be transformed into

$$-\omega^2 mx = -j\omega mgx - kx + eE_x \tag{9-122}$$

which may be solved readily to give the displacement,

$$x = \frac{(e/m)E_x}{\omega_0^2 - \omega^2 + j\omega g} \tag{9-123}$$

in which $\omega_0^2 = k/m$. Once the displacement is known, the dipole moment, polarization, polarizability, susceptibility and relative permittivity all follow directly. They are, respectively,

$$\begin{aligned}
p_x &= ex = \alpha E_x \\
\mathscr{P}_x &= N_0 p_x = N_0 \alpha E_x = N_0 ex = \epsilon_v \chi E_x \\
\alpha &= e\left(\frac{x}{E_x}\right) \\
\chi &= \frac{N_0 e}{\epsilon_v}\left(\frac{x}{E_x}\right) \\
\epsilon_r &= 1 + \chi
\end{aligned} \tag{9-124}$$

all of which are, in general, complex numbers (phasors). The susceptibility may be separated into real and imaginary parts as follows:

$$\chi = \frac{N_0 e^2}{m\epsilon_v}\left[\frac{1}{\omega_0^2 - \omega^2 + j\omega g}\right]$$

$$= \frac{N_0 e^2}{m\epsilon_v}\left[\frac{\omega_0^2 - \omega^2}{(\omega_0^2 - \omega^2)^2 + \omega^2 g^2} - j\frac{\omega g}{(\omega_0^2 - \omega^2)^2 + \omega^2 g^2}\right] \tag{9-125}$$

This leads to a separation of the relative permittivity into real and imaginary parts in the form

$$\epsilon_r = \epsilon_r' - j\epsilon_r'' \tag{9-126}$$

in which
$$\epsilon_r' = 1 + \frac{N_0 e^2}{m\epsilon_v}\left[\frac{\omega_0^2 - \omega^2}{(\omega_0^2 - \omega^2)^2 + \omega^2 g^2}\right]$$

and
$$\epsilon_r'' = \frac{N_0 e^2}{m\epsilon_v}\left[\frac{\omega g}{(\omega_0^2 - \omega^2)^2 + \omega^2 g^2}\right]$$

Formulas of this type are often referred to as *dielectric dispersion formulas.* It is clear that the imaginary part of the permittivity is associated with power loss or dissipation within the dielectric. Furthermore, at any given frequency ϵ_r'' produces the same type of macroscopic effect as the conductivity σ and thus for practical purposes the two are indistinguishable. In macroscopic theory, it is customary to use ϵ_r'' for dielectrics and σ for metals. Frequently dielectric losses are described using the loss tangent $\tan\delta$, defined as

$$\tan\delta = \frac{\epsilon_r''}{\epsilon_r'} \tag{9-127}$$

Examination of (126) shows that the properties of a dielectric change most rapidly at frequencies near the resonance frequency ω_0. This is illustrated in Fig. 9-17. For electronic polarization the resonance frequency is very high—usually in the ultraviolet range.

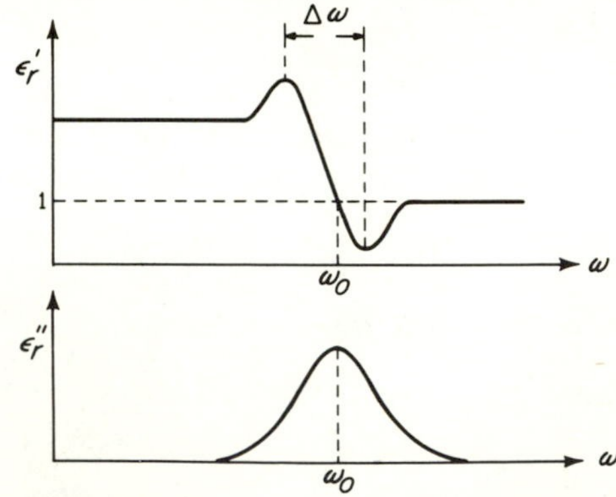

Figure 9-17. The effect of electronic polarization on the permittivity of a dielectric.

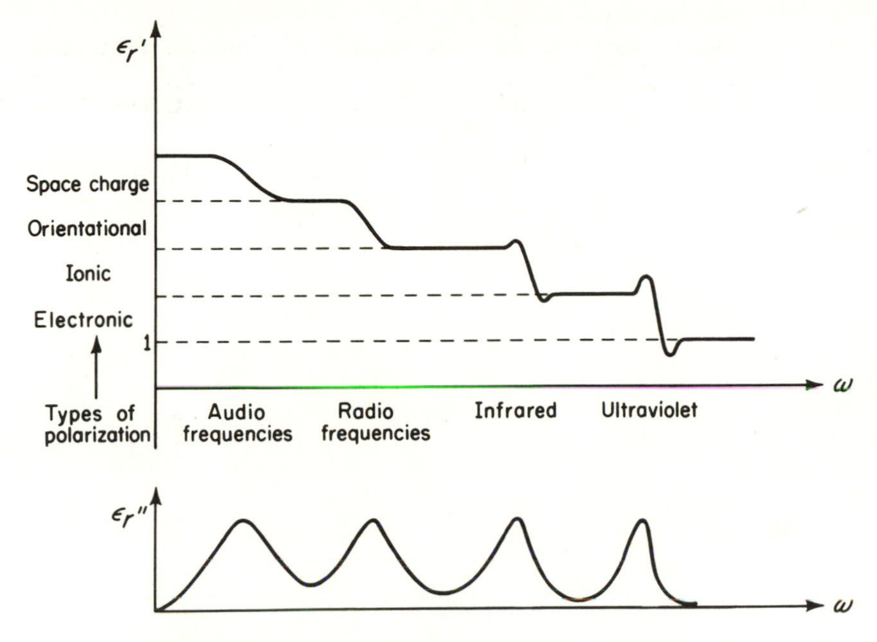

Figure 9-18. The variation of permittivity with frequency.

The other types of polarization exhibit effects qualitatively similar to electronic polarization, but with the difference that they have much lower resonance frequencies. This behavior is sketched in Fig. 9-18. It is evident that the resonance frequencies are associated with the effective masses of the microscopic bodies contributing to the polarization effect; the greater the effective mass, the lower the resonance frequency.

Problem 9. Consider the electronic polarization phenomenon for the case in which the two "peaks" in ϵ_r' (Fig. 9-17) are close to ω_0. Find the frequency separation $\Delta\omega$ between the peaks.

Problem 10. In problem 6, the relative permittivity of a plasma (80) was derived using the concept of particle *currents* in the medium. Derive the relative permittivity expression using the concept of *polarization* of the plasma. Assume that a displaced electron leaves a positive "hole" at its original position and that the only forces acting on the electron are those due to collisions and to the applied field.

BIBLIOGRAPHY

Harman, W.W., *Fundamentals of Electronic Motion*, McGraw-Hill Book Company, New York, 1953.

Nussbaum, A., *Electromagnetic and Quantum Properties of Materials*, Prentice-Hall, Inc., Englewood Cliffs, N.J., 1966.

Nussbaum, A., *Electronic and Magnetic Behavior of Materials*, Prentice-Hall, Inc., Englewood Cliffs, N.J., 1966.

Spangenberg, K.R., *Fundamentals of Electron Devices*, McGraw-Hill Book Company, New York, 1957.

Spangenberg, K.R., *Vacuum Tubes*, McGraw-Hill Book Company, New York, 1948.

Chapter 10

RADIATION

10.01 Potential Functions and the Electromagnetic Field. In the previous chapters, relations which exist among the electromagnetic field vectors have been studied, but so far no consideration has been given to the means by which the fields are generated. We shall now consider how these fields are related to their *sources*, that is, the charges and currents that produce them. Although it is possible in theory to obtain expressions for the electric and magnetic field strengths **E** and **H** directly in terms of the charge and current densities ρ and **J**, such a derivation is, in general, quite difficult. It will be recalled that in the study of the electrostatic field and the steady magnetic field it was found possible, and often simpler, first to set up *potentials* in terms of the charges or currents, and then to obtain the electric or magnetic fields from these potentials. Similarly, in the electromagnetic field it turns out to be much simpler first to set up potentials that are related to the charges and currents, and then to obtain **E** and **H** from these potentials.

The problem of determining suitable potentials for the electromagnetic field may be approached in several different ways, all of which require a certain amount of educated guesswork. A simple way, using the heuristic approach, is to generalize the potentials already developed for the steady electric and magnetic fields, and hope that these generalized potentials will satisfy Maxwell's equations and the requirements of the problem. A second method is to start with Maxwell's equations and from these derive differential equations that the potentials must satisfy. Once again the potentials are guessed and then shown to satisfy the differential equations. The third method is to attempt a direct solution of the derived differential equations for the potentials. Because of the importance of understanding the use of potentials in electromagnetic theory, all three approaches will be developed; the first two will be considered in this section, and in sec. 10.10 the phasor form of the differential equations will be solved.

1. *Heuristic Approach.* In electrostatics a scalar potential V was set

up, related to the electric charge distribution through

$$V(\mathbf{r}) = \frac{1}{4\pi\epsilon}\int \frac{\rho(\mathbf{r}')}{R}\,dV' \tag{2-26}$$

From this potential the electric field was obtained through

$$\mathbf{E} = -\nabla V \tag{2-24}$$

For the steady magnetic field in a homogeneous medium, a vector potential $\mathbf{A}$ was found and related to the source currents through

$$\mathbf{A}(\mathbf{r}) = \frac{\mu}{4\pi}\int \frac{\mathbf{J}(\mathbf{r}')}{R}\,dV' \tag{3-28}$$

From this magnetic potential the magnetic field was obtained through

$$\mu\mathbf{H} = \nabla \times \mathbf{A} \tag{3-29}$$

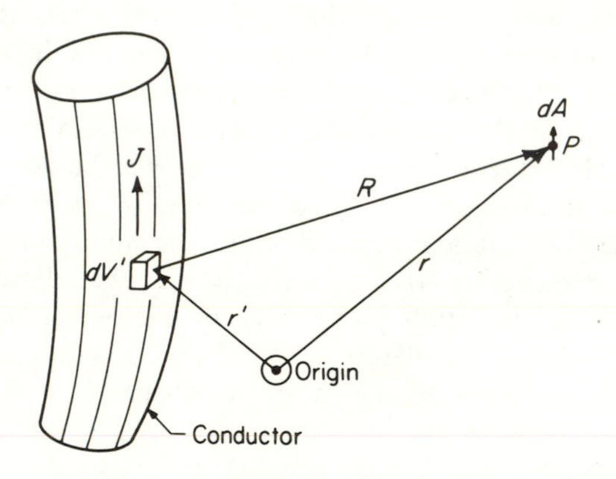

Figure 10-1.

The sources of the *electromagnetic* field are current and charge distributions that vary with time, so it is reasonable to try these same potentials, generalized for time variations. That is, we might expect to be able to write

$$\mathbf{A}(\mathbf{r}, t) = \frac{\mu}{4\pi}\int \frac{\mathbf{J}(\mathbf{r}', t)}{R}\,dV' \tag{10-1}$$

$$V(\mathbf{r}, t) = \frac{1}{4\pi\epsilon}\int \frac{\rho(\mathbf{r}', t)}{R}\,dV' \tag{10-2}$$

in which $R = |\mathbf{r} - \mathbf{r}'|$. As written above the strengths of these potentials vary instantaneously with the strengths of the sources. However, the principal result of the time-varying theory (wave theory) discussed so far is the appearance of a finite propagation time for electromagnetic waves. It is logical that the potentials from which the fields are to be derived should also display finite propagation time. This can be readily accomplished by modifying the above relations to give

$$\mathbf{A}(\mathbf{r}, t) = \frac{\mu}{4\pi}\int \frac{\mathbf{J}(\mathbf{r}', t - R/v)}{R}\, dV' \tag{10-3}$$

$$V(\mathbf{r}, t) = \frac{1}{4\pi\epsilon}\int \frac{\rho(\mathbf{r}', t - R/v)}{R}\, dV' \tag{10-4}$$

In these expressions a time delay of R/v seconds has been introduced, so that now the potentials have been delayed or retarded by this amount. For this reason they are often called *retarded potentials.*

Having guessed the potentials, one might expect to derive the electric and magnetic fields through relations $\mu\mathbf{H} = \nabla \times \mathbf{A}$ (3-29) and $\mathbf{E} = -\nabla V$ (2-24). It would be discovered that while the first relation is correct, the second is incomplete. The reason for this is that the electric and magnetic fields obtained are not independent but must be related through Maxwell's equations, just as the source distributions (currents and charges) cannot be specified independently but are related through the equation of continuity. These relationships are automatically satisfied in the next approach which starts with the requirement that the potentials and derived fields satisfy Maxwell's equations.

2. *Maxwell's Equations Approach.* Maxwell's equations in time-varying, differential form (for constant, scalar ϵ and μ) are

$$\nabla \times \mathbf{H} = \epsilon\dot{\mathbf{E}} + \mathbf{J} \tag{10-5}$$

$$\nabla \times \mathbf{E} = -\mu\dot{\mathbf{H}} \tag{10-6}$$

$$\nabla\cdot\mathbf{E} = \frac{\rho}{\epsilon} \tag{10-7}$$

$$\nabla\cdot\mathbf{H} = 0 \tag{10-8}$$

$\mathbf{J}$ and ρ are the source current density and the source charge density, respectively, and are related by the equation of continuity

$$\nabla\cdot\mathbf{J} = -\dot{\rho} \tag{10-9}$$

Equation (8) is satisfied if $\mathbf{H}$ is represented as the curl of some vector. This leads to the following definition for the *vector potential* $\mathbf{A}$:

$$\mu\mathbf{H} = \nabla \times \mathbf{A} \tag{10-10}$$

Substitution of (10) into (6) gives

$$\nabla \times (\mathbf{E} + \dot{\mathbf{A}}) = 0 \tag{10-11}$$

Equation (11) is satisfied if $\mathbf{E} + \dot{\mathbf{A}}$ is represented as the gradient of a scalar. Setting $\mathbf{E} + \dot{\mathbf{A}}$ equal to $-\nabla V$ defines the *scalar potential* V in such a way that for slowly varying fields it is identical to the scalar potential used in chap. 2. Thus the electric field strength may be expressed as

$$\mathbf{E} = -\nabla V - \dot{\mathbf{A}} \tag{10-12}$$

Equations (10) and (12) show how the field quantities $\mathbf{E}$ and $\mathbf{H}$ may be expressed in terms of a vector potential $\mathbf{A}$ and a scalar potential V. For static fields both (10) and (12) are identical to the field expressions already derived in chaps. 2 and 3.

Expressions (10) and (12) for $\mathbf{H}$ and $\mathbf{E}$ satisfy two of Maxwell's equations, (6) and (8). The remaining two of Maxwell's equations may be used to derive differential equations for the potential functions. Substitution of (10) and (12) into (5) gives

$$\frac{1}{\mu}\nabla \times \nabla \times \mathbf{A} = -\epsilon\nabla\dot{V} - \epsilon\ddot{\mathbf{A}} + \mathbf{J}$$

The identity $\nabla \times \nabla \times \mathbf{A} = \nabla\nabla\cdot\mathbf{A} - \nabla^2\mathbf{A}$ permits the above to be written as

$$\nabla^2\mathbf{A} - \mu\epsilon\ddot{\mathbf{A}} = -\mu\mathbf{J} + \mu\epsilon\nabla\dot{V} + \nabla\nabla\cdot\mathbf{A} \tag{10-13}$$

This is one of the required differential equations; the second one may be obtained by substituting (12) into (7).

$$\nabla^2 V + \nabla\cdot\dot{\mathbf{A}} = -\frac{\rho}{\epsilon} \tag{10-14}$$

Equations (13) and (14) are complicated by the fact that they are coupled, that is, the unknowns $\mathbf{A}$ and V appear in both equations. A further complication is the fact that (13) and (14) do not yield a unique solution for the potentials. The truth of the latter statement may be inferred from the *Helmholtz theorem: Any vector field due to a finite source is specified uniquely if both the curl and the divergence of the field are specified.* The curl of $\mathbf{A}$ is given in (10) but its divergence remains unspecified. Moreover, from the defining relation (11), it is evident that there is an arbitrariness about the scalar potential as well. Examination of eqs. (13) and (14) shows that if the divergence of $\mathbf{A}$ is set equal to $-\mu\epsilon\dot{V}$, these equations become uncoupled and reduce to standard wave equations with source terms included, viz.,

$$\nabla^2\mathbf{A} - \mu\epsilon\ddot{\mathbf{A}} = -\mu\mathbf{J} \tag{10-15}$$

$$\nabla^2 V - \mu\epsilon\ddot{V} = -\frac{\rho}{\epsilon} \tag{10-16}$$

The particular choice used above, that is

$$\nabla\cdot\mathbf{A} = -\mu\epsilon\dot{V} \tag{10-17}$$

is known as the *Lorentz gauge condition.* Other gauge conditions are possible,* but if it is required that the solutions for **A** and V represent retarded electromagnetic potentials it can be shown† that (17) must be satisfied, and will be satisfied provided that the source charge and current distributions are related by the equation of continuity.

Equations similar to (15) and (16) have already been derived in the chapters on static electric and magnetic fields. The corresponding static equations are

$$\nabla^2 \mathbf{A} = -\mu \mathbf{J} \tag{10-18}$$

$$\nabla^2 V = -\frac{\rho}{\epsilon} \tag{10-19}$$

and they have the solutions already given in (10-1) and (10-2). These solutions may be used as a guide in an attempt to guess the solutions to (15) and (16), and of course such guesses have to be verified by substitution. On the other hand, a direct solution of (15) and (16) could be attempted. The latter approach is generally somewhat involved,‡ but it will be discussed later in this chapter for the special case of sinusoidal oscillations.

The potential functions may be guessed following the lines of reasoning already discussed [see discussion of eqs. (3) and (4)]. The functions are

$$\mathbf{A}(\mathbf{r}, t) = \frac{\mu}{4\pi} \int \frac{\mathbf{J}(\mathbf{r}', t - R/v)}{R}\, dV' \tag{10-20}$$

$$V(\mathbf{r}, t) = \frac{1}{4\pi\epsilon} \int \frac{\rho(\mathbf{r}', t - R/v)}{R}\, dV' \tag{10-21}$$

Equations (20) and (21) are the solutions of (15) and (16) as may be verified by direct substitution.§ Since a time delay is involved, **A** and V as defined above are the *retarded potentials*. In general it is not necessary to evaluate (21) since **E** may be obtained using one of Maxwell's equations (5); the procedure is to obtain **H** from **A** using (10) and then to obtain **E** from **H** by taking its curl and integrating with respect to time.

*For example the Coulomb gauge condition sets the divergence of **A** equal to zero; this results in reducing (14) to Poisson's equation but has the disadvantage of giving potentials which do not exhibit retardation.

†J. Grosskopf, "On the Application of the Two Methods of Solution of Maxwell's Equations in the Calculation of the Electromagnetic Fields of Radiating Conductors," *Hochfrequenz. Technik und Electro-akustic*, **49**, 205–211 (1937); also J. A. Stratton, *Electromagnetic Theory*, McGraw-Hill Book Company, New York, 1941, p. 429.

‡Solution of the time-dependent wave equation is discussed by J. D. Jackson, *Classical Electrodynamics*, John Wiley & Sons, Inc., New York, 1962, pp. 183–186.

§This process, which is usually quite straightforward, is rather involved in this case, and reference should be made to one of the following texts: H. A. Lorentz, *Theory of Electrons*, pp. 17–19; M. Mason and W. Weaver, *The Electromagnetic Field*, University of Chicago Press, Chicago, 1929 (p. 282).

It should be noted that $\mathbf{J} = 0$ in (5) when the field is evaluated in free space, that is external to the region of source currents.

Problem 1a. Show that for a current along the z axis the expression $\mu\mathbf{H} = \nabla \times \mathbf{A}$ reduces to

$$\mu H_\phi = -\sin\theta \frac{\partial A_z}{\partial r}$$

when only the distant field is considered.

(NOTE: This is an important result, for it covers the majority of practical antennas.)

Problem 1b. Verify that for a current *element* the expression in 1a gives correctly the *total* magnetic field.

10.02 Potential Functions for Sinusoidal Oscillations. In the sinusoidal steady state, calculations with the potential functions are greatly simplified. Using phasor notation, (15) and (16) may be expressed as

$$\nabla^2\mathbf{A} + \omega^2\mu\epsilon\mathbf{A} = -\mu\mathbf{J} \tag{10-22}$$

$$\nabla^2 V + \omega^2\mu\epsilon V = -\frac{\rho}{\epsilon} \tag{10-23}$$

both equations having the form of the Helmholtz wave equation with a source term included. As before, a solution may be guessed by generalizing the solutions for the static cases to take into account the phase delay due to the distance R between a source point and the observation point. It is known that a uniform plane wave traveling in the $\mathbf{R}$ direction would have a phase variation given by the factor $e^{-j\beta R}$. This suggests that the potential integrals should be modified by the inclusion of an exponential phase factor as follows:

$$\mathbf{A}(\mathbf{r}) = \frac{\mu}{4\pi}\int \mathbf{J}(\mathbf{r}')\frac{e^{-j\beta R}}{R}\,dV' \tag{10-24}$$

$$V(\mathbf{r}) = \frac{1}{4\pi\epsilon}\int \rho(\mathbf{r}')\frac{e^{-j\beta R}}{R}\,dV' \tag{10-25}$$

This solution may be verified by substitution into (22) and (23)* or (as will be shown in sec. 10.10) it may be obtained by solving (22) and (23) directly.

An important point to be made here is that the gauge condition relationship between $\mathbf{A}$ and V makes it unnecessary to evaluate V using the

*The verification is carried out in R. Plonsey and R. E. Collin, *Principles and Applications of Electromagnetic Fields*, McGraw-Hill Book Company, New York, 1961, pp. 324–325.

integral (25). Using phasor notation the expressions for the magnetic and electric field strengths are

$$\mathbf{H} = \frac{1}{\mu} \nabla \times \mathbf{A} \tag{10-26}$$

$$\mathbf{E} = -\nabla V - j\omega\mathbf{A} \tag{10-27}$$

where the vector and scalar potentials are related by the Lorentz gauge condition,

$$\nabla \cdot \mathbf{A} = -j\omega\mu\epsilon V \tag{10-28}$$

The scalar potential V may be eliminated from (27) using (28) to give

$$\mathbf{E} = \frac{\nabla\nabla \cdot \mathbf{A}}{j\omega\epsilon\mu} - j\omega\mathbf{A} \tag{10-29}$$

Equations (26) and (29) demonstrate that **E** and **H** may be expressed entirely in terms of **A**.

An alternative form of (29) may be derived from one of the Maxwell curl equations,

$$\nabla \times \mathbf{H} = j\omega\epsilon\mathbf{E} + \mathbf{J} \tag{10-30}$$

Substitution of (26) into (30) gives

$$\mathbf{E} = \frac{1}{j\omega\epsilon\mu}(\nabla \times \nabla \times \mathbf{A} - \mu\mathbf{J}) \tag{10-31}$$

in which **E** is expressed in terms of **A** only. Of course $\mathbf{J} = 0$ if **E** is evaluated outside the current distribution.

Problem 2. Show that (31) is identical to (29).

Problem 3. Consider a source ($\mathbf{J}$, ρ) in a medium whose properties are represented by μ, ϵ and σ. Beginning with Maxwell's equations in phasor form, derive the differential equations

$$\nabla^2\mathbf{A} - \gamma^2\mathbf{A} = -\mu\mathbf{J} \tag{10-32}$$

$$\nabla^2 V - \gamma^2 V = -\frac{\rho}{\epsilon} \tag{10-33}$$

in which $\gamma^2 = j\omega\mu(\sigma + j\omega\epsilon)$. It is necessary in the derivation to redefine the Lorentz gauge condition.

10.03 The Alternating Current Element (or Oscillating Electric Dipole). An excellent example of the use of the retarded vector potential occurs in the calculation of the electromagnetic field of an alternating current element (or oscillating electric dipole) in free space. A *current element $I\,dl$* refers to a filamentary current I flowing along an elemental length dl. This is approximated when a current I flows in a very short length of thin wire, if the length dl considered is so short that the current is essentially constant along the length. Although an isolated current element may appear to be a very unreal concept, any physical circuit or antenna carrying current may be considered to consist of a large number of such elements joined end to end. Therefore, if the electromagnetic field of this "building block" is known, the electromagnetic field of any actual antenna having a specified current distribution may be calculated.

Figure 10-2 shows an alternating-current element $I\,dl \cos \omega t$ located at the origin of a spherical co-ordinate system. The problem is to calculate the electromagnetic field at an arbitrary point P.

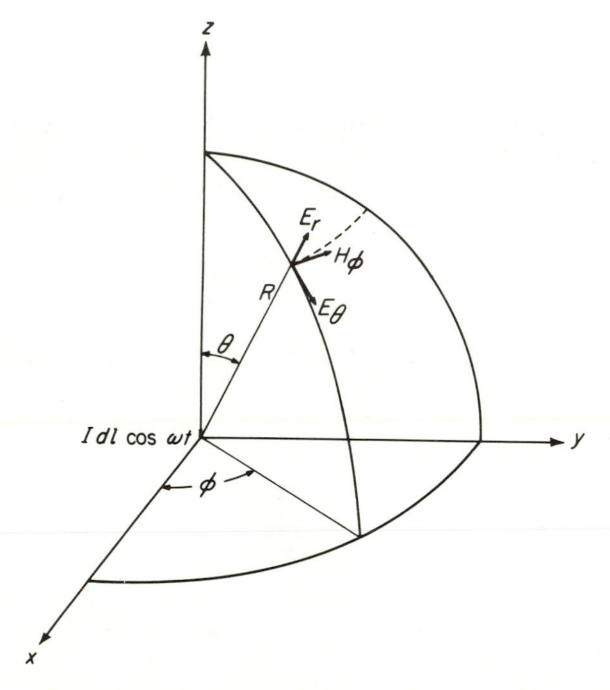

Figure 10-2. A current element at the center of a spherical co-ordinate system.

The first step is to obtain the vector potential **A** at P. The general expression for **A** is given by

$$\mathbf{A}(\mathbf{r}) = \frac{\mu}{4\pi}\int \frac{\mathbf{J}\left(t - \frac{R}{v}\right)}{R}\,dV' \tag{10-34}$$

The integration over the volume in (34) consists of an integration over the cross-sectional area of the wire and an integration along its length. The current density **J**, integrated over the cross-sectional area of the wire, is just the current I, and because this is assumed to be constant along the length, integration over the length gives $I\,dl$. Therefore in this simple example the expression for vector potential becomes*

$$A_z = \frac{\mu}{4\pi}\frac{I\,dl\cos\omega\left(t - \dfrac{r}{v}\right)}{r} \tag{10-35}$$

The vector potential has the same direction as the current element, in this case the z direction, and is retarded in time by r/v seconds. The magnetic field strength **H** is obtained through the relation

$$\mu\mathbf{H} = \nabla \times \mathbf{A}$$

Reference to the expressions in chap. 1, showing the curl in spherical co-ordinates, gives the components of **H** in terms of A_r, A_θ, and A_ϕ. From Fig. 10-2 it is seen that for this case

$$A_r = A_z\cos\theta; \qquad A_\theta = -A_z\sin\theta; \qquad A_\phi = 0 \tag{10-38}$$

Then from expressions (1-45a, b, and c) on page 16, and noting that be-cause of symmetry, $\partial/\partial\phi = 0$,

$$\mu H_r = (\nabla \times \mathbf{A})_r = 0$$

$$\mu H_\theta = (\nabla \times \mathbf{A})_\theta = 0$$

$$\begin{aligned} H_\phi &= \frac{1}{\mu r}\left[\frac{\partial}{\partial r}(rA_\theta) - \frac{\partial A_r}{\partial\theta}\right] \\ &= \frac{I\,dl}{4\pi r}\left\{\frac{\partial}{\partial r}\left[-\sin\theta\cos\omega\left(t - \frac{r}{v}\right)\right] - \frac{\partial}{\partial\theta}\left[\frac{\cos\theta}{r}\cos\omega\left(t - \frac{r}{v}\right)\right]\right\} \\ &= \frac{I\,dl\sin\theta}{4\pi}\left[\frac{-\omega\sin\omega\left(t - \dfrac{r}{v}\right)}{rv} + \frac{\cos\omega\left(t - \dfrac{r}{v}\right)}{r^2}\right] \end{aligned} \tag{10-39}$$

The electric field strength **E** can be obtained from **H** through Max-well's first equation, which at the point P (in free space) is

$$\nabla \times \mathbf{H} = \epsilon\dot{\mathbf{E}}$$

$$\mathbf{E} = \frac{1}{\epsilon}\int \nabla \times \mathbf{H}\,dt \tag{10-40}$$

Taking the curl of (39) and then integrating with respect to time (the order is immaterial) gives for the components of **E**,

$$E_\theta = \frac{I\,dl\sin\theta}{4\pi\epsilon}\left(\frac{-\omega\sin\omega t'}{rv^2} + \frac{\cos\omega t'}{r^2 v} + \frac{\sin\omega t'}{\omega r^3}\right) \tag{10-41}$$

*Equation (35) also may be derived by writing the current density **J** in terms of the Dirac deltas introduced in chap. 1 and used in chaps. 2 and 3. Suitable delta representations are

$$\mathbf{J} = \hat{\mathbf{z}}\,I\,dl\,\cos\omega t\;\delta(\mathbf{r}) \tag{10-36}$$

and

$$\mathbf{J} = \hat{\mathbf{z}}\,I\,dl\,\cos\omega t\;\delta(x)\,\delta(y)\,\delta(z) \tag{10-37}$$

$$E_r = \frac{2I\,dl \cos\theta}{4\pi\epsilon}\left(\frac{\cos\omega t'}{r^2 v} + \frac{\sin\omega t'}{\omega r^3}\right) \tag{10-42}$$

and rewriting (39)

$$H_\phi = \frac{I\,dl \sin\theta}{4\pi}\left(\frac{-\omega\sin\omega t'}{rv} + \frac{\cos\omega t'}{r^2}\right) \tag{10-43}$$

where $t' = (t - r/v)$.

It is somewhat surprising to find that something so apparently simple as a current element should give rise to an electromagnetic field as complicated as that given by (41), (42), and (43). However, a study of these expressions soon shows the significance and necessity for each of the terms.

Consider first the expression for H_ϕ. It is seen to consist of two terms, one of which varies inversely as r and the other inversely as r^2. The second of these, called the *induction* field, will predominate at points close to the current element where r is small, whereas at great distances, where r is large, the second term becomes negligible compared with the first. This first inverse-distance term is called the *radiation* or *distant* field. The two fields will have *equal amplitudes* at that value of r, which makes

$$\frac{1}{r^2} = \frac{\omega}{rv}$$

that is, at

$$r = \frac{v}{\omega} = \frac{\lambda}{2\pi} \approx \frac{\lambda}{6}$$

Except for the fact that t' has replaced t, the induction term

$$\frac{I\,dl \sin\theta \cos\omega t'}{4\pi r^2}$$

is just the magnetic field strength that would be given by a direct application of the Biot-Savart law (or Ampere's law for the current element) if this were extended to cover the case of an alternating current $I \cos\omega t$. The fact that the true field is a function of $t' = t - (r/v)$ instead of t, accounts for the finite time of propagation. However, at points close to the element where the induction term predominates, r/v is a very small quantity and $t' \approx t$.

The inverse-distance (radiation) term is an extra term, not present for steady currents. It results from the fact of the finite time of propagation, which is of no account in the steady-field case. It will be shown later that this radiation term contributes to a flow of energy away from the source (the current element), whereas the induction term contributes to energy that is stored in the field during one quarter of a cycle and returned to the circuit during the next.

Examination of the expressions for E_θ and E_r shows that E_θ has an induction ($1/r^2$) term and a radiation ($1/r$) term, and E_r has a $1/r^2$ term. In addition, both components of **E** have a term that varies as $1/r^3$. From their similarity with the components of the field of an electrostatic dipole (see chap. 2, example 1), these $1/r^3$ terms are called the *electrostatic* field terms (or sometimes just *electric* field terms).

Relation between a Current Element and an Electric Dipole. It is not just a coincidence that the expressions for the electric field of the alternating current element should contain terms that correspond to the field of an oscillating electric dipole. Although only *current* was specified in setting up the hypothetical current element, the equation of continuity (or conservation of charge) requires that there be an accumulation of charge at the ends of the element, which is given by

$$\frac{dq}{dt} = I \cos \omega t$$

That is, the charge at one end is increasing and at the other end decreasing, by the amount of the current flow (coulombs per second). In order to obtain a physical approximation of an isolated current element, one could terminate the current element in two small spheres or disks on which the charges could accumulate. If the wire is very thin compared with the radius of the spheres (so that its distributed capacitance is negligible compared with the capacitance between the spheres), the current in the wire will be uniform. In addition, the radii of the spheres should be small compared with dl, their distance apart, and in turn dl should be very much shorter than a wavelength. The arrangement then is essentially that of the original Hertzian oscillating electric dipole (Fig. 10-3).

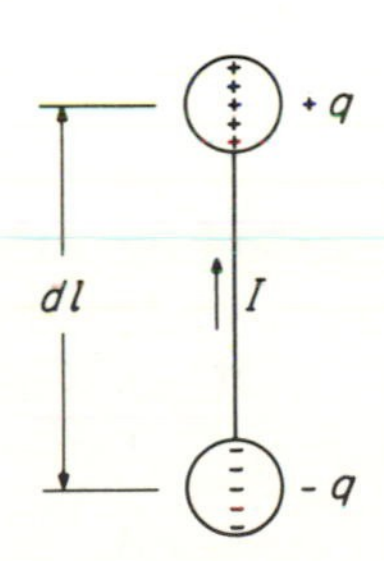

Figure 10-3. Hertzian dipole.

$$\frac{dq}{dt} = I \cos \omega t$$

$$q = \frac{I \sin \omega t}{\omega}$$

From comparison with the electrostatic dipole, the electric field strength that would be expected to result from the separated charges at the ends of the current element would be

$$E_\theta = \frac{q\, dl \sin \theta}{4\pi \epsilon r^3} = \frac{I\, dl \sin \theta \sin \omega t'}{4\pi \epsilon \omega r^3} \tag{10-44}$$

$$E_r = \frac{2q\, dl \cos \theta}{4\pi \epsilon r^3} = \frac{2I\, dl \cos \theta \sin \omega t'}{4\pi \epsilon \omega r^3} \tag{10-45}$$

These are exactly the $1/r^3$ terms that automatically appeared in the solution for the electromagnetic field of the current element.

When a current element forms part of a complete circuit there is no accumulation of charge at its ends if the current is uniform throughout the circuit, for the current from one element flows into the next. In this case, as would be expected, the $1/r^3$ terms due to accumulated charge vanish, leaving only induction and radiation fields. In terms of a chain of Hertzian dipoles (Fig. 10-4), the positive charge at the end of one dipole is just cancelled by an equal amount of negative charge at the opposite end of the adjacent dipole. However, if the current along the circuit or antenna is not uniform along its length, but is distributed as, for example, in Fig. 10-5, this could be represented as a chain of current

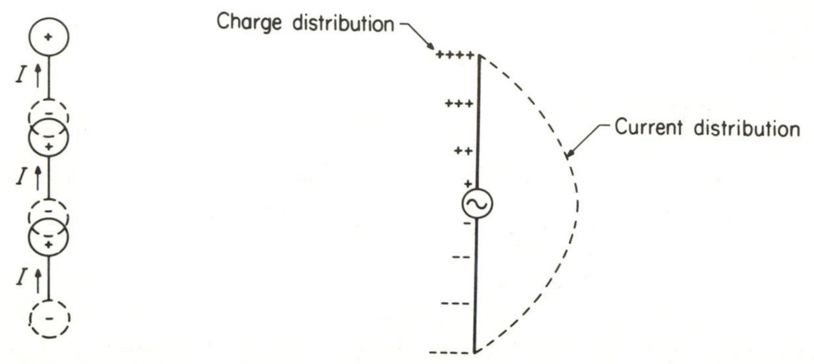

Figure 10-4. Chain of Hertzian dipoles.

Figure 10-5. Current and charge distribution on a linear antenna.

elements, or Hertzian dipoles, having slightly different amplitudes. In this case the adjacent charges do not completely cancel, and there is an accumulation of charge on the surface of the wire, as indicated in Fig. 10-5. These surface charges are responsible for a relatively strong component of electric field strength normal to the surface of the wire.

Problem 4. Obtain expressions (41) and (42) for E_θ and E_r due to a current element through the alternative relation

$$\mathbf{E} = -\dot{\mathbf{A}} - \nabla V$$

(NOTE: Obtain V from $\mathbf{A}$ through $\nabla \cdot \mathbf{A} = -\mu\epsilon\dot{V}$; alternatively write V directly from the charges that can be obtained from I through the equation of continuity.)

Problem 5. Starting with the expression $I\,dl$ for a current element, show that the phasor expressions for vector potential and field strengths will be

$$A_z = \frac{\mu I\,dl}{4\pi r} e^{-j\beta r}$$

$$H_\phi = \frac{I\,dl \sin\theta\, e^{-j\beta r}}{4\pi r}\left(j\beta + \frac{1}{r}\right)$$

$$E_\theta = \frac{\eta I\, dl \sin\theta\, e^{-j\beta r}}{4\pi r}\left(j\beta + \frac{1}{r} + \frac{1}{j\beta r^2}\right)$$

$$E_r = \frac{\eta I\, dl \cos\theta\, e^{-j\beta r}}{4\pi r}\left(\frac{2}{r} + \frac{2}{j\beta r^2}\right)$$

where $\beta = \omega/v = 2\pi/\lambda$, $\eta = \sqrt{\mu/\epsilon} = \mu v$.

10.04 Power Radiated by a Current Element. The power flow per unit area at the point P will be given by the Poynting vector at that point. The instantaneous Poynting vector is given by $\tilde{\mathbf{E}} \times \tilde{\mathbf{H}}$ and it will have both θ and r components. Replacing v by $c \approx 3 \times 10^8$ for free-space propagation, the θ component of the instantaneous Poynting vector will be

$$\begin{aligned} P_\theta &= -E_r H_\phi \\ &= \frac{I^2 dl^2 \sin 2\theta}{16\pi^2 \epsilon}\left(\frac{\sin^2 \omega t'}{r^4 c} - \frac{\cos^2 \omega t'}{r^4 c} - \frac{\sin \omega t' \cos \omega t'}{\omega r^5} + \frac{\omega \sin \omega t' \cos \omega t'}{r^3 c^2}\right) \\ &= \frac{I^2 dl^2 \sin 2\theta}{16\pi^2 \epsilon}\left(\frac{-\cos 2\omega t'}{r^4 c} - \frac{\sin 2\omega t'}{2\omega r^5} + \frac{\omega \sin 2\omega t'}{2r^3 c^2}\right) \end{aligned} \quad (10\text{-}46)$$

The average value of $\sin 2\omega t'$ or $\cos 2\omega t'$ over a complete cycle is zero. Therefore, for any value of r, the average of P_θ over a complete cycle is zero. P_θ represents only a surging back and forth of power in the θ direction without any net or average flow. The radial Poynting vector is given by

$$\begin{aligned} P_r &= E_\theta H_\phi \\ &= \frac{I^2 dl^2 \sin^2\theta}{16\pi^2 \epsilon}\left(\frac{\sin \omega t' \cos \omega t'}{\omega r^5} + \frac{\cos^2 \omega t'}{r^4 c} - \frac{\omega \sin \omega t' \cos \omega t'}{r^3 c^2}\right. \\ &\qquad \left. - \frac{\sin^2 \omega t'}{r^4 c} - \frac{\omega \sin \omega t' \cos \omega t'}{r^3 c^2} + \frac{\omega^2 \sin^2 \omega t'}{r^2 c^3}\right) \\ &= \frac{I^2 dl^2 \sin^2\theta}{16\pi^2 \epsilon}\left(\frac{\sin 2\omega t'}{2\omega r^5} + \frac{\cos 2\omega t'}{r^4 c} - \frac{\omega \sin 2\omega t'}{r^3 c^2} + \frac{\omega^2(1 - \cos 2\omega t')}{2r^2 c^3}\right) \end{aligned} \quad (10\text{-}47)$$

The *average* value of radial Poynting vector over a cycle will be due to part of the final term only and is

$$\begin{aligned} P_{r(\text{av})} &= \frac{\omega^2 I^2 dl^2 \sin^2\theta}{32\pi^2 r^2 c^3 \epsilon} \\ &= \frac{\eta}{2}\left(\frac{\omega I\, dl \sin\theta}{4\pi r c}\right)^2 \quad \text{watts/sq m} \end{aligned} \quad (10\text{-}48)$$

None of the terms in the expressions for the Poynting vector represents an *average* power flow except that of eq. (48). The only terms of **E** and **H** that contribute to this average power flow are the radiation or inverse-distance terms. At a large distance from the source these radiation terms are the only ones that have appreciable value, but even close to the current element, where the induction and electric fields predominate, only $1/r$ terms contribute to an average outward flow of power.

That it is only the inverse-distance terms that can contribute to an outward flow of power from the source can be proven by simple reasoning. If one considers two concentric shells of radii r_1 and r_2 enclosing the source, then the average outward rate of energy flow through shell r_2 must be the same as through shell r_1, if there is to be no continuous accumulation (or decrease) of energy stored in the region between them. This requires that the power density decrease as $1/r^2$ since the area of the shells increases as r^2. The power density is proportional to E times H (or to E^2 or H^2), so E and H, which contribute to an average radial power flow, must be proportional to $1/r$. Components of E or H, which are inversely proportional to r^2 or r^3, can contribute to an instantaneous flow of energy into the region between the shells, but this energy must be returned to the source because it cannot be stored permanently in a finite volume of the medium.

From eqs. (41), (42), and (43) the amplitudes of the *radiation* fields of an electric current element $I\,dl$ are

$$\begin{aligned} E_\theta &= \frac{\omega I\,dl \sin\theta}{4\pi\epsilon v^2 r} \\ &= \frac{\eta I\,dl\sin\theta}{2\lambda r} \\ &= \frac{60\pi I\,dl\sin\theta}{r\lambda} \end{aligned} \tag{10-49}$$

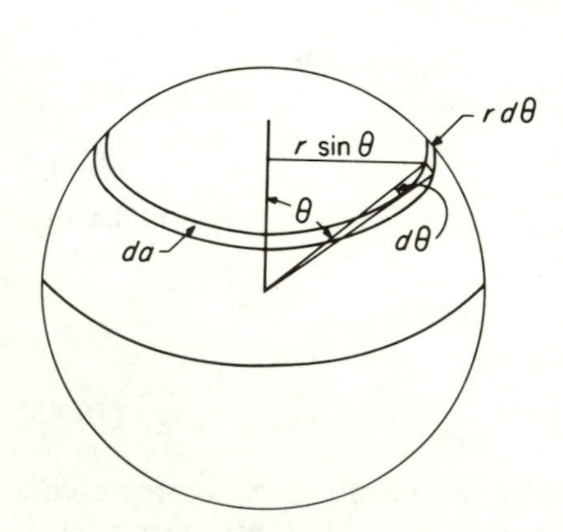

Figure 10-6. Element of area on a spherical surface.

$$\begin{aligned} H_\phi &= \frac{\omega I\,dl\sin\theta}{4\pi v r} \\ &= \frac{I\,dl\sin\theta}{2\lambda r} \end{aligned} \tag{10-50}$$

The radiation terms of E_θ and H_ϕ are in time phase and are related by

$$\frac{E_\theta}{H_\phi} = \eta \tag{10-51}$$

The total power radiated by the current element can be computed by integrating the radial Poynting vector over a spherical surface centered at the element. P_r is independent of the azimuthal angle ϕ, so the element of area on the spherical shell will be taken as the strip da where

$$da = 2\pi r^2 \sin\theta \, d\theta$$

Then the total power radiated is

$$\begin{aligned} \text{Power} &= \oint_{\text{surface}} P_{r(\text{av})} \, da = \int_0^\pi \frac{\eta}{2} \left(\frac{\omega I \, dl \sin\theta}{4\pi r c}\right)^2 2\pi r^2 \sin\theta \, d\theta \\ &= \frac{\eta\omega^2 I^2 dl^2}{16\pi c^2} \int_0^\pi \sin^3\theta \, d\theta \\ &= \frac{\eta\omega^2 I^2 dl^2}{16\pi c^2} \left[\frac{-\cos\theta}{3}(\sin^2\theta + 2)\right]_0^\pi \\ &= \frac{\eta\omega^2 I^2 dl^2}{12\pi c^2} \qquad \text{watts} \end{aligned} \tag{10-52}$$

In this expression I is maximum or peak current. In terms of effective current the power radiated is

$$\begin{aligned} \text{Power} &= \frac{\eta\omega^2 I_{\text{eff}}^2 \, dl^2}{6\pi c^2} \\ &= 80\pi^2 \left(\frac{dl}{\lambda}\right)^2 I_{\text{eff}}^2 \end{aligned}$$

The coefficient of I_{eff}^2 has the dimensions of resistance and is called the *radiation resistance* of the current element. Then, for a current element,

$$R_{\text{rad}} = 80\pi^2 \left(\frac{dl}{\lambda}\right)^2 \tag{10-53}$$

10.05 Application to Short Antennas. The hypothetical current element is a useful tool for theoretical work, but it is not a practical antenna. The *practical* "elementary dipole" is a center-fed antenna having a length that is very short in wavelengths. The current amplitude on such an antenna decreases uniformly from a maximum at the center to zero at the ends [Fig. 10-7(*a*)]. For the same current I (at the terminals) the (short) practical dipole of length l will radiate only one-quarter as much power as the current element of the same length, which has the current I throughout its entire length. (The field strengths at every point are reduced to one-half, and the power density will be reduced to one-quarter.) Therefore, the radiation resistance of a practical short dipole is one-quarter that of the current element of the same length. That is

$$\begin{aligned} R_{\text{rad}}\,(\text{short dipole}) &= 20\pi^2 \left(\frac{l}{\lambda}\right)^2 \\ &\approx 200 \left(\frac{l}{\lambda}\right)^2 \qquad \text{ohms} \end{aligned} \tag{10-54a}$$

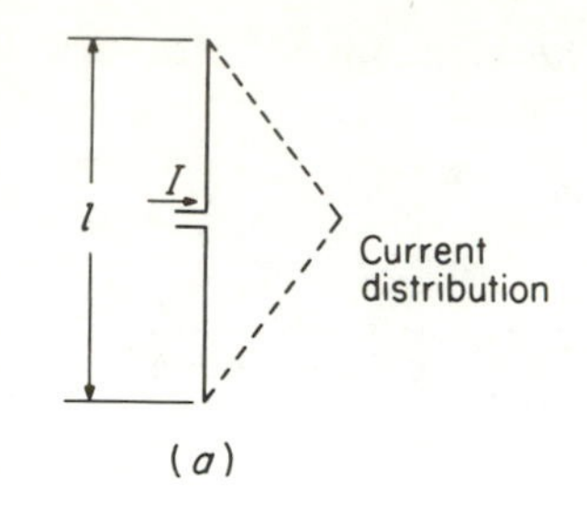

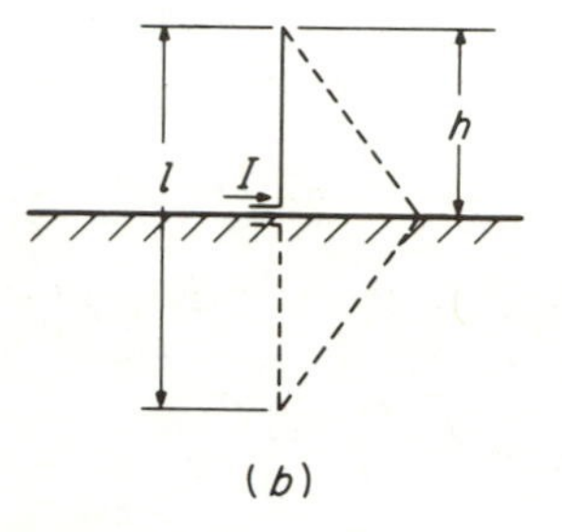

Figure 10-7. Current distribution on short antennas: (*a*) short dipole; (*b*) short monopole.

The monopole of height h [Fig. 10-7(*b*)], or short vertical antenna mounted on a reflecting plane, produces the same field strengths above the plane as does the dipole of length $l = 2h$ when both are fed with the same current. However, the short vertical antenna radiates only through the hemispherical surface above the plane, so its radiated power is only one-half that of the corresponding dipole. Therefore the radiation resistance of the monopole of height $h = l/2$ is

$$R_{\text{rad}}\,(\text{monopole}) = 10\pi^2 \left(\frac{l}{\lambda}\right)^2 = 40\pi^2 \left(\frac{h}{\lambda}\right)^2 \approx 400 \left(\frac{h}{\lambda}\right)^2 \text{ ohms} \qquad (10\text{-}54\text{b})$$

These formulas hold strictly for very short antennas* only, but they are good approximations for dipoles of lengths up to quarter wavelength, and monopoles of heights up to one-eighth wavelength.

10.06 Assumed Current Distribution. In order to calculate the electromagnetic fields of longer antennas, it is necessary to know the current distribution along the antennas. This information should be obtainable by solving Maxwell's equations subject to the appropriate boundary conditions along the antenna. However, for the cylindrical antenna, this is a comparatively difficult problem, and it is only in quite recent years that satisfactory solutions have been obtained. One of these will be considered in chap. 14. In the absence of a known antenna current, it is possible to *assume* a certain distribution and from that to calculate approximate field distributions. The accuracy of the fields so calculated will, of course, depend upon how good an assumption was made for current distribution. By thinking of the center-fed antenna as an open-circuited transmission line that has been opened out, a sinusoidal current distribution with current nodes at the ends is suggested. The fact that it is known from Abraham's work on

*Lower-case symbols, l and h, have been used in eq. (54) for these *short* antennas. For general-length antennas, discussed in the next section, the symbols L and H are used.

thin ellipsoids that the current will be truly sinusoidal for the infinitely thin case, makes this assumption more justifiable. It turns out to be a very good assumption for thin antennas, sufficiently good in fact, that even with more accurate (and much more complicated) formulas available, the sinusoidal distribution is still used for much of the work in the antenna field. When greater accuracy is desired, and in those particular cases where the sinusoidal assumption breaks down entirely, it is necessary to use a distribution that is closer to the true one.

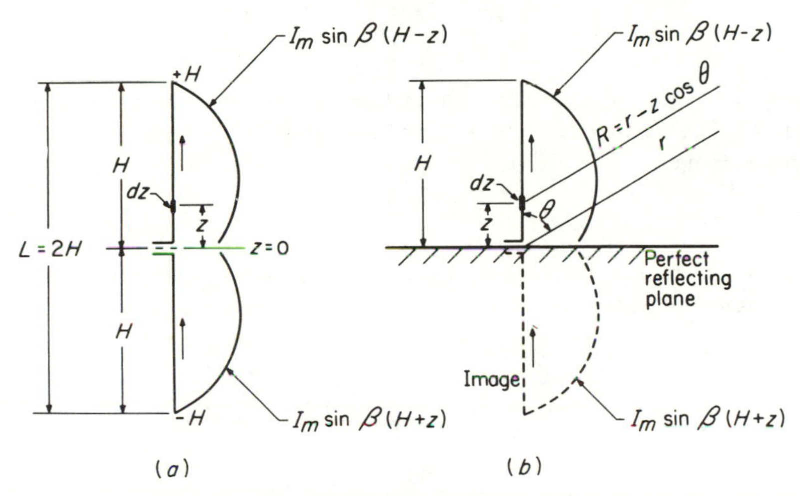

Figure 10-8. (*a*) Center-fed dipole with assumed sinusoidal current distribution; (*b*) corresponding monopole.

Figure 10-8(*a*) shows a center-fed dipole with a sinusoidal current distribution, and Fig. 10-8(*b*) shows the corresponding monopole. "A *dipole antenna** is a straight radiator, usually fed in the center, and producing a maximum of radiation in the plane normal to the axis. The length specified is the overall length." The vertical antenna (of height $H = L/2$) fed against an infinitely large perfectly conducting plane has the same radiation characteristics above the plane as does the dipole antenna of length L in free space. This is because the fields due to a current element $I\,dz$ when reflected from the plane, appear to originate at an image element located beneath the plane. Moreover, the impedance of the vertical antenna fed against the reflecting plane is just one-half that of the corresponding dipole of length $L = 2H$. Thus the dipole of Fig. 10-8(*a*) and the corresponding base-fed vertical antenna of Fig. 10-8(*b*) can be solved conveniently as one problem. The base-fed vertical antenna of Fig. 10-8(*b*) will be referred to simply as

**IRE Standards on Antennas*, 1948.

a *monopole* of height H, the infinitely large, perfectly reflecting plane being understood, unless otherwise stated. The corresponding center-fed dipole will be referred to as a dipole of *length* $L = 2H$ or as a dipole of *half-length* H.

The image principle used here is discussed further in chap. 11.

10.07 Radiation from a Quarter-wave Monopole or Half-wave Dipole. It will be assumed that the current is sinusoidally distributed as shown in Fig. 10-8. Then

$$I = I_m \sin \beta(H - z) \qquad z > 0$$

$$I = I_m \sin \beta(H + z) \qquad z < 0$$

where I_m is the value of current at the current *loop* or current *maximum*. The expression for the vector potential at a point P due to the current element $I\,dz$ will be

$$dA_z = \frac{\mu I\, e^{-j\beta R}\, dz}{4\pi R}$$

where R is the distance from the current element to the point P. The total vector potential at P due to all the current elements will be

$$A_z = \frac{\mu}{4\pi}\int_{-H}^{0} \frac{I_m \sin \beta(H + z)\, e^{-j\beta R}\, dz}{R} + \frac{\mu}{4\pi}\int_{0}^{H} \frac{I_m \sin \beta(H - z)\, e^{-j\beta R}\, dz}{R} \tag{10-55}$$

Because only the distant or radiation fields are required in this problem, it is possible to make some simplifying approximations. For the inverse-distance factor (the R in the denominator) it is valid to write

$$R \approx r$$

However, for the R in the phase factor in the numerator, it is the *difference* between R and r that is important. For very large values of R the lines to the point P are essentially parallel and for the R in the phase factor one can write approximately

$$R = r - z\cos\theta$$

Then the expression for A_z becomes

$$A_z = \frac{\mu I_m\, e^{-j\beta r}}{4\pi r}\left[\int_{-H}^{0} \sin \beta(H + z)\, e^{j\beta z \cos\theta}\, dz + \int_{0}^{H} \sin \beta(H - z)\, e^{j\beta z \cos\theta}\, dz\right] \tag{10-56}$$

For the particular case of $H = \lambda/4$,

$$\sin \beta(H + z) = \sin \beta(H - z) = \cos \beta z$$

and the integral becomes

$$
\begin{aligned}
A_z &= \frac{\mu I_m e^{-j\beta r}}{4\pi r}\int_0^H \cos\beta z(e^{j\beta z\cos\theta} + e^{-j\beta z\cos\theta})\,dz \\
&= \frac{\mu I_m e^{-j\beta r}}{4\pi r}\int_0^H [\cos\{\beta z(1+\cos\theta)\} + \cos\{\beta z(1-\cos\theta)\}]\,dz \\
&= \frac{\mu I_m e^{-j\beta r}}{4\pi r}\left\{\frac{\sin[\beta z(1+\cos\theta)]}{\beta(1+\cos\theta)} + \frac{\sin[\beta z(1-\cos\theta)]}{\beta(1-\cos\theta)}\right\}_0^{\lambda/4} \\
&= \frac{\mu I_m e^{-j\beta r}}{4\pi\beta r}\left[\frac{(1-\cos\theta)\cos\left(\frac{\pi}{2}\cos\theta\right) + (1+\cos\theta)\cos\left(\frac{\pi}{2}\cos\theta\right)}{\sin^2\theta}\right] \\
&= \frac{\mu I_m e^{-j\beta r}}{2\pi\beta r}\left[\frac{\cos\left(\frac{\pi}{2}\cos\theta\right)}{\sin^2\theta}\right] \qquad (10\text{-}57)
\end{aligned}
$$

We recall from problem 1a that when the current is entirely in the z direction,

$$\mu H_\phi = \frac{-\partial A_z}{\partial r}\sin\theta$$

The expression for magnetic field strength at a distant point will be

$$H_\phi = \frac{jI_m e^{-j\beta r}}{2\pi r}\left[\frac{\cos\left(\frac{\pi}{2}\cos\theta\right)}{\sin\theta}\right] \qquad (10\text{-}58)$$

where only the inverse-distance term has been retained. The electric field strength for the radiation field will be

$$
\begin{aligned}
E_\theta &= \eta H_\phi \\
&= \frac{j60 I_m e^{-j\beta r}}{r}\left[\frac{\cos\left(\frac{\pi}{2}\cos\theta\right)}{\sin\theta}\right] \qquad (10\text{-}59)
\end{aligned}
$$

The magnitude of the electric field strength for the radiation field of a half-wave dipole or quarter-wave monopole is

$$E_\theta = \frac{60 I_m}{r}\left[\frac{\cos\left(\frac{\pi}{2}\cos\theta\right)}{\sin\theta}\right] \quad \text{V/m} \qquad (10\text{-}60)$$

E_θ and H_ϕ are in time phase so the maximum value in time of the Poynting vector is just the product of the peak values of E_θ and H_ϕ, and the average value in time of the Poynting vector will be one-half the peak value. Then

$$P_{\text{av}} = \frac{\eta I_m^2}{8\pi^2 r^2}\left[\frac{\cos^2\left(\frac{\pi}{2}\cos\theta\right)}{\sin^2\theta}\right]$$

The total power radiated through a hemispherical surface of radius r (Fig. 10-6) will equal

$$\oint P_{\text{av}}\, da = \frac{\eta I_m^2}{4\pi} \int_0^{+\pi/2} \frac{\cos^2\left(\dfrac{\pi}{2}\cos\theta\right)}{\sin\theta}\, d\theta \tag{10-61}$$

It is necessary to evaluate this integral. Most of the difficulty in radiation problems is usually in connection with the evaluation of an integral. The following substitutions are typical.

$$\int_0^{\pi/2} \frac{\cos^2\left(\dfrac{\pi}{2}\cos\theta\right)}{\sin\theta}\, d\theta = \frac{1}{2}\int_0^{\pi/2} \frac{1+\cos(\pi\cos\theta)}{\sin\theta}\, d\theta$$

Let

$$u = \cos\theta$$

$$du = -\sin\theta\, d\theta$$

$$\frac{d\theta}{\sin\theta} = \frac{-du}{\sin^2\theta} = -\frac{du}{1-u^2}$$

$$\int_0^{\pi/2} \frac{\cos^2\left(\dfrac{\pi}{2}\cos\theta\right)}{\sin\theta}\, d\theta = -\frac{1}{2}\int_1^0 \frac{(1+\cos\pi u)}{1-u^2}\, du$$

$$= \frac{1}{4}\int_0^1 (1+\cos\pi u)\left(\frac{1}{1+u} + \frac{1}{1-u}\right) du$$

$$= \frac{1}{4}\int_{-1}^{+1} \frac{1+\cos\pi u}{1+u}\, du$$

Let

$$v = \pi(1+u)$$

$$dv = \pi\, du$$

$$\frac{dv}{v} = \frac{du}{1+u}$$

$$\pi u = v - \pi$$

$$\cos\pi u = \cos v\cos\pi + \sin v\sin\pi = -\cos v$$

Therefore

$$\int_0^{\pi/2} \frac{\cos^2\left(\dfrac{\pi}{2}\cos\theta\right)}{\sin\theta}\, d\theta = \frac{1}{4}\int_0^{2\pi} \frac{1-\cos v}{v}\, dv$$

$$= \frac{1}{4}\int_0^{2\pi}\left(\frac{v^1}{2!} - \frac{v^3}{4!} + \frac{v^5}{6!} - \frac{v^7}{8!} + \ldots\right) dv$$

$$= \frac{1}{4}\left(\frac{v^2}{2\cdot 2!} - \frac{v^4}{4\cdot 4!} + \frac{v^6}{6\cdot 6!} - \frac{v^8}{8\cdot 8!} + \ldots\right)_0^{2\pi} \tag{10-62}$$

The series of eq. (10-62) can be evaluated by substitution. It does not converge rapidly and so a number of terms must be used. This evaluation is shown below.

$$v = 2\pi = 6.2832 \qquad \log_{10} v = 0.79818$$

$$2\cdot 2! = 4 \qquad \log_{10} v^2 = 1.59636$$

$$\log_{10} 2 \cdot 2! = 0.60206$$

$$\log_{10} \frac{v^2}{2 \cdot 2!} = .99430 \qquad \frac{v^2}{2 \cdot 2!} = 9.870$$

The other terms are found in a similar manner. Using eight terms, the sum of the positive terms is 26.878 and the sum of the negative terms is 24.441. Therefore

$$\int_0^{\pi/2} \frac{\cos^2\left(\frac{\pi}{2}\cos\theta\right)}{\sin\theta}\, d\theta = 0.6093 \tag{10-63}$$

It is also possible to integrate such a function as

$$\int_0^{\pi/2} \frac{\cos^2\left(\frac{\pi}{2}\cos\theta\right)}{\sin\theta}\, d\theta$$

graphically or by Simpson's or the trapezoidal rule. For example by the trapezoidal rule if θ is taken in increments of 5°, then the following table is constructed:

θ in degrees	0	5	10	15	20	25	30	35	40	45
$\frac{\cos^2\left(\frac{\pi}{2}\cos\theta\right)}{\sin\theta}$	0	0	.003	.011	.028	.050	.086	.138	.201	.280

θ in degrees	50	55	60	65	70	75	80	85	90
$\frac{\cos^2\left(\frac{\pi}{2}\cos\theta\right)}{\sin\theta}$	.369	.468	.578	.688	.788	.875	.942	.980	1.00

Now

$$\int_0^{\pi/2} \frac{\cos^2\left(\frac{\pi}{2}\cos\theta\right)}{\sin\theta}\, d\theta = \frac{\pi}{2} \times \frac{1}{18}\left[\frac{1.000 + 0}{2} + \sum_{\theta = 5^\circ}^{\theta = 85^\circ} \frac{\cos^2\left(\frac{\pi}{2}\cos\theta\right)}{\sin\theta}\right]$$

$$= \frac{\pi}{36} \times 6.987$$

$$= 0.609$$

This method of numerical integration shows that in a given antenna problem, if the current distribution is known, the radiation resistance may always be found by straightforward methods, although the integration may be tedious. The power radiated through the hemispherical surface is obtained by inserting the value of the integral in (61). Then

$$\text{Radiated power} = \frac{0.609\eta I_m^2}{4\pi}$$

In this expression I_m is peak current. In terms of effective current the radiated power would be

$$\text{Radiated power} = \frac{0.609\eta I^2_{m(\text{eff})}}{2\pi}$$

$$= 36.5 I^2_{m(\text{eff})} \tag{10-64}$$

Therefore the radiation resistance of a quarter-wave monopole antenna is 36.5 ohms.

For the half-wave dipole antenna in free space, power would be radiated through a complete spherical surface. Therefore, for the same current the power radiated would be twice as much, and the radiation resistance for the half-wave dipole is

$$R_{\text{rad}} = 73 \text{ ohms}$$

Problem 6. Derive the expression for the radiation term of the electric field of a half-wave dipole [eq. (10-60)] without the use of the vector potential, that is, by adding directly the (distant) fields owing to the current elements.

Problem 7. Derive the general expression corresponding to eq. (59) for the (distant) electric field of a dipole antenna of any half-length *H*. It is

$$E_\theta = \frac{j60 I_m\, e^{-j\beta r}}{r}\left[\frac{\cos(\beta H \cos\theta) - \cos\beta H}{\sin\theta}\right]$$

Problem 8. Write the expression for the power radiated through a spherical surface by the dipole of half-length *H*.

10.08 Sine Integral and Cosine Integral. The series of eq. (62) that resulted from the integral

$$\int_0^x \frac{1 - \cos v}{v}\, dv$$

has been evaluated and can be found in tables. It is designated as $S_1(x)$. This series also occurs in the integral

$$\text{Ci}(x) = -\int_x^\infty \frac{\cos v}{v}\, dv \tag{10-65}$$

This latter integral is called the *cosine integral* of x, and is abbreviated as shown. A companion integral defined by

$$\text{Si}(x) = \int_0^x \frac{\sin v}{v}\, dv \tag{10-66}$$

is known as the *sine integral of* x.

The cosine integral of x is related to $S_1(x)$ by

$$\text{Ci}(x) = \ln x + C - S_1(x)$$

where $C = 0.5772157$ is Euler's constant, and

$$S_1(x) = \int_0^x \frac{1 - \cos v}{v}\, dv = \left(\frac{x^2}{2\cdot 2!} - \frac{x^4}{4\cdot 4!} + \frac{x^6}{6\cdot 6!} - + \dots\right) \quad (10\text{-}67)$$

These integrals occur frequently in radiation problems. They have been studied extensively, and tables giving their values may be found in several books.*

Problem 9. Integrate the expression of problem 7 and show that the general expression for the radiation resistance of a dipole of half-length H is

$$R_{\text{rad}} = 30\,\{S_1(b) - [S_1(2b) - S_1(b)]\cos b + [S_1(2b) - \text{Si}\,(b)]\sin b + [1 + \cos b]\,S_1(b) - \sin b\,\text{Si}\,(b)\}$$

where

$$b = 2\beta H$$

10.09 Electromagnetic Field Close to an Antenna. In section 10.07 and problem 7 expressions for the radiation or distant fields of an antenna were derived. For some purposes, for example to determine the mutual impedance between antennas, it is necessary to know the electric and magnetic fields in *the neighborhood* of the antenna. In this region the field is often called the *near* field, in contrast to the distant or radiation field. Because the near field will include induction and electric as well as radiation fields, it can be expected to be more complex than the distant field. The answer of most interest will be the component of electric field strength parallel to the antenna, that is E_z. For this reason it is convenient to use a *cylindrical* co-ordinate system, or actually a combination of cylindrical, spherical, and rectangular co-ordinate systems. This is shown in Fig. 10-9, where the antenna is assumed to extend along the z axis. In this figure the following relations will hold:

$$R = \sqrt{(z - h)^2 + y^2}$$
$$R_1 = \sqrt{(z - H)^2 + y^2}$$
$$R_2 = \sqrt{(z + H)^2 + y^2}$$
$$r = \sqrt{z^2 + y^2}$$

The co-ordinates of the point P are (ρ, ϕ, z) in cylindrical co-ordinates (or r, θ, ϕ in spherical co-ordinates). However, because of symmetry, there are no variations in the ϕ direction and so, without loss of generality, the point P may be located in the y-z plane ($\phi = 90$ degrees plane). In this case $\rho = y$, and the distances from various points along the antenna will be as indicated in the figure.

*E. Jahnke and F. Emde, *Tables of Functions*, B. G. Teubner, Leipzig 1933; F. E. Terman, *Radio Engineers Handbook*, McGraw-Hill Book Company, New York, 1943.

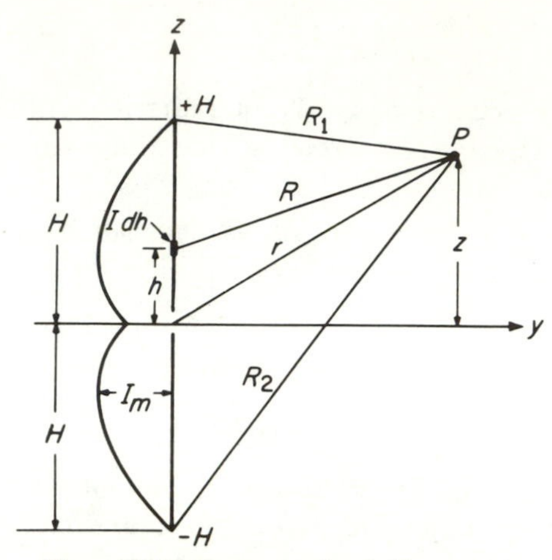

Figure 10-9. Geometry for fields near the antenna.

Again assuming a sinusoidal distribution of current, the antenna current will be

$$I = I_m \sin \beta(H - h) \qquad h > 0$$

$$I = I_m \sin \beta(H + h) \qquad h < 0$$

The expression for vector potential at the point P will be

$$A_z = \frac{\mu I_m}{4\pi}\left[\int_0^H \frac{\sin \beta(H-h)e^{-j\beta R}}{R}\,dh + \int_{-H}^0 \frac{\sin \beta(H+h)\,e^{-j\beta R}}{R}\,dh\right]$$

Replacing $\sin \beta(H - h)$ with

$$\frac{e^{j\beta(H-h)} - e^{-j\beta(H-h)}}{2j}$$

and using a corresponding expression for $\sin \beta(H + h)$, there results

$$A_z = \frac{\mu I_m}{8\pi j}\left[e^{j\beta H}\int_0^H \frac{e^{-j\beta(R+h)}}{R}\,dh - e^{-j\beta H}\int_0^H \frac{e^{-j\beta(R-h)}}{R}\,dh\right.$$

$$\left. + e^{j\beta H}\int_{-H}^0 \frac{e^{-j\beta(R-h)}}{R}\,dh - e^{-j\beta H}\int_{-H}^0 \frac{e^{-j\beta(R+h)}}{R}\,dh\right] \qquad (10\text{-}68)$$

In cylindrical co-ordinates the magnetic field strength at the point P will be given by

$$\mu H_\phi = (\nabla \times \mathbf{A})_\phi = -\frac{\partial A_z}{\partial \rho}$$

With the point P in the y-z plane, this can be written as

$$\mu H_\phi = -\mu H_x = -\frac{\partial A_z}{\partial y}$$

$$H_\phi = -\frac{I_m}{8\pi j}\left[e^{j\beta H}\int_0^H \frac{\partial}{\partial y}\left(\frac{e^{-j\beta(R+h)}}{R}\right) dh - e^{-j\beta H}\int_0^H \frac{\partial}{\partial y}\left(\frac{e^{-j\beta(R-h)}}{R}\right) dh\right.$$
$$\left. + e^{j\beta H}\int_{-H}^0 \frac{\partial}{\partial y}\left(\frac{e^{-j\beta(R-h)}}{R}\right) dh - e^{-j\beta H}\int_{-H}^0 \frac{\partial}{\partial y}\left(\frac{e^{-j\beta(R+h)}}{R}\right) dh\right] \quad (10\text{-}69)$$

Consider the first term only,

$$e^{j\beta H}\int_0^H \frac{\partial}{\partial y}\left(\frac{e^{-j\beta(R+h)}}{R}\right) dh = e^{j\beta H}\int_0^H \left[\frac{-j\beta y\, e^{-j\beta(R+h)}}{R^2} - \frac{y\, e^{-j\beta(R+h)}}{R^3}\right] dh \quad (10\text{-}70)$$

The integrand turns out to be a perfect differential. Integrating gives

$$e^{j\beta H}\left[\frac{y\, e^{-j\beta(R+h)}}{R(R+h-z)}\right]_{h=0}^{h=H} = y\, e^{j\beta H}\left[\frac{e^{-j\beta(R_1+H)}}{R_1(R_1+H-z)} - \frac{e^{-j\beta r}}{r(r-z)}\right]$$
$$= y\, e^{j\beta H}\left[\frac{(R_1 - H + z)\, e^{-j\beta(R_1+H)}}{R_1[R_1^2 - (H-z)^2]} - \frac{(r+z)\, e^{-j\beta r}}{r(r^2 - z^2)}\right]$$

But $$R_1^2 - (H-z)^2 = r^2 - z^2 = y^2$$

so that the first term becomes

$$\frac{e^{j\beta H}}{y}\left[\left(1 - \frac{H-z}{R_1}\right) e^{-j\beta(R_1+H)} - \left(1 + \frac{z}{r}\right) e^{-j\beta r}\right]$$

Similarly the second, third, and fourth terms of eq. (10-69) are

$$\frac{e^{-j\beta H}}{y}\left[\left(1 + \frac{H-z}{R_1}\right) e^{-j\beta(R_1-H)} - \left(1 - \frac{z}{r}\right) e^{-j\beta r}\right]$$
$$\frac{e^{j\beta H}}{y}\left[\left(1 - \frac{H+z}{R_2}\right) e^{-j\beta(R_2+H)} - \left(1 - \frac{z}{r}\right) e^{-j\beta r}\right]$$
$$\frac{e^{-j\beta H}}{y}\left[\left(1 + \frac{H+z}{R_2}\right) e^{-j\beta(R_2-H)} - \left(1 + \frac{z}{r}\right) e^{-j\beta r}\right]$$

Adding these four terms, the magnetic field strength can be obtained. It is

$$H_\phi = -\frac{I_m}{4\pi j}\left(\frac{e^{-j\beta R_1}}{y} + \frac{e^{-j\beta R_2}}{y} - \frac{2\cos\beta H\, e^{-j\beta r}}{y}\right) \quad (10\text{-}71)$$

The electric field can be obtained from the magnetic field by recalling that in free space

$$\mathbf{E} = \frac{1}{j\omega\epsilon}\nabla \times \mathbf{H}$$

In the $x = 0$ plane

$$E_z = \frac{1}{j\omega\epsilon}(\nabla \times \hat{\boldsymbol{\phi}} H_\phi)_z = \frac{1}{j\omega\epsilon y}\frac{\partial}{\partial y}(yH_\phi) \quad (10\text{-}72)$$

$$E_y = \frac{1}{j\omega\epsilon}(\nabla \times \hat{\boldsymbol{\phi}} H_\phi)_y = -\frac{1}{j\omega\epsilon}\frac{\partial}{\partial z}(H_\phi) \quad (10\text{-}73)$$

Substituting the expression for H_ϕ in these equations gives

$$E_z = \frac{-j\beta I_m}{4\pi\omega\epsilon y}\left(\frac{y\,e^{-j\beta R_1}}{R_1} + \frac{y\,e^{-j\beta R_2}}{R_2} - 2\cos\beta H\frac{y\,e^{-j\beta r}}{r}\right)$$

which reduces to

$$E_z = -j30I_m\left(\frac{e^{-j\beta R_1}}{R_1} + \frac{e^{-j\beta R_2}}{R_2} - 2\cos\beta H\frac{e^{-j\beta r}}{r}\right) \qquad (10\text{-}74)$$

Similarly,

$$E_y = j30I_m\left(\frac{z-H}{y}\cdot\frac{e^{-j\beta R_1}}{R_1} + \frac{z+H}{y}\cdot\frac{e^{-j\beta R_2}}{R_2} - \frac{2z\cos\beta H}{y}\frac{e^{-j\beta r}}{r}\right) \qquad (10\text{-}75)$$

and rewriting the expression for magnetic field strength

$$H_\phi = \frac{j30I_m}{\eta y}\left(e^{-j\beta R_1} + e^{-j\beta R_2} - 2\cos\beta H\,e^{-j\beta r}\right) \qquad (10\text{-}76)$$

Equations (74), (75), and (76) give the electric and magnetic field strengths both near to and far from an antenna carrying a sinusoidal current distribution. It is quite remarkable that something so complex as the electromagnetic field close to an antenna should be expressible by such simple relations. The secret of this result lies in the integral of eq. (70), the integrand of which turned out to be a perfect differential. This happy circumstance occurred only because the current distribution was stipulated to be *sinusoidal* (although it also occurs for an unattenuated traveling-wave distribution, two of which can be combined to give the sinusoidal distribution). The result becomes even more remarkable when the expression for the important parallel component E_z is considered and interpreted. Examining eq. (74), it is seen that the first term represents a spherical wave originating at the top of the antenna. The numerator is the phase factor and the denominator is the inverse-distance factor. Similarly, the second term represents a spherical wave of equal amplitude originating at the other end of the antenna, or, in the case of a monopole antenna on a reflecting plane, at the image point of the top of the antenna. Finally, the third term represents a wave originating at the center of the antenna (at the base in the case of a monopole antenna). The amplitude of this latter wave depends upon the antenna half-length H. It is zero for $H = \lambda/4$, that is, for a half-wave dipole or quarter-wave monopole. The sources of the spherical waves represented by the terms of eq. (74) are isotropic, that is they radiate uniformly in all directions, as is the case, for example, for a point source of sound. If a point source of sound is situated above a perfect (acoustic) reflecting plane, an equal image source is automatically obtained. Thus it is seen that the pressure field of a single, point source of sound located one-quarter wavelength above a reflecting plane will give a true representation of both magnitude and phase of the parallel component of electric field strength about a quarter-wave monopole or half-wave dipole carrying a sinusoidally distributed current. Antennas of other lengths may be represented by two point sources. This fact has been used* to give experimental data from which the (electrical) mutual impedance between antennas can be computed.

Equations (74) and (75) can be used to calculate the parallel and normal components of electric field strength in the immediate neighborhood of an antenna. Figures 10-10 and 10-11 show the calculated

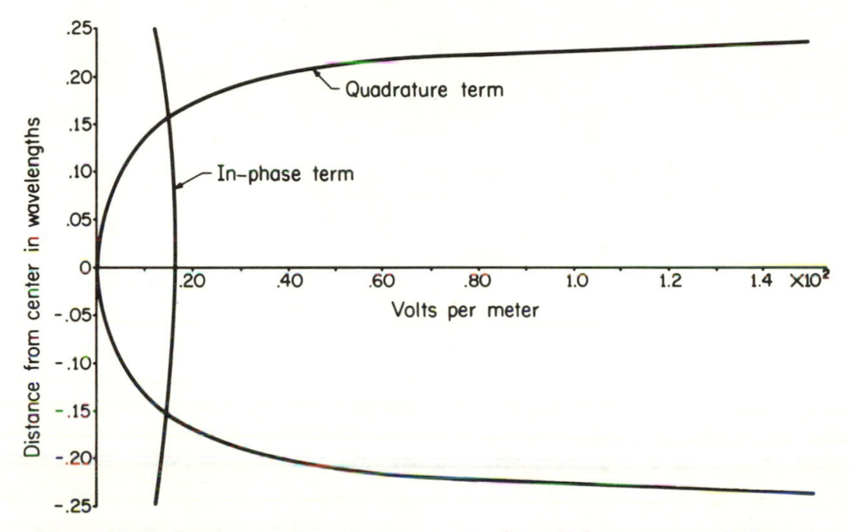

Figure 10-10. In-phase and quadrature terms of parallel component of electric field strength along a half-wave dipole. (Calculated for a number 4 wire at 20 MHz carrying an assumed sinusoidal current distribution.)

values† of the components of **E** at the surface of a half-wave dipole for the assumed sinusoidal current distribution. Figure 10-10 shows the relative magnitudes of the in-phase and quadrature terms of E_z, the parallel component of **E**. Figure 10-11 compares the relative magnitudes of the quadrature terms of parallel and perpendicular components of **E**. It is seen that except very near the ends of the antenna the normal component is very large compared with the parallel component.

*E. C. Jordan and W. L. Everitt, "Acoustic Models of Radio Antennas," *Proc. IRE*, **29**, 4, 186 (1941).

†P. S. Carter, "Circuit Relations in Radiating Systems," *Proc. IRE*, **20**, 6, 1004 (1932).

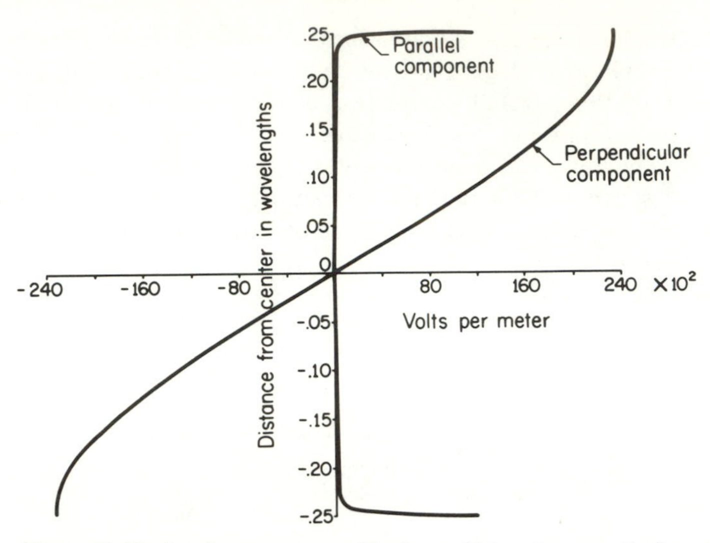

Figure 10-11. Quadrature terms of both parallel and perpendicular components of electric field strength along a half-wave dipole. (Calculated for a number 4 wire at 20 MHz carrying an assumed sinusoidal current distribution.)

Problem 10. Verify that

$$\int\left[\frac{-j\beta y\, e^{-j\beta(R+h)}}{R^2} - \frac{y\, e^{-j\beta(R+h)}}{R^3}\right] dh = \frac{y\, e^{-j\beta(R+h)}}{R(R+h-z)} \tag{10-77}$$

where R, h, z, and y are as indicated in Fig. 10-9.

10.10 Solution of the Potential Equations. Earlier in this chapter differential equations for the vector and scalar potentials were derived and their solutions were given with neither verification nor derivation. In order to complete the discussion of radiation fundamentals a solution will be carried out for the phasor Helmholtz equations

$$\nabla^2 \mathbf{A} - \gamma^2 \mathbf{A} = -\mu \mathbf{J} \tag{10-78}$$

$$\nabla^2 V - \gamma^2 V = -\rho/\epsilon \tag{10-79}$$

already derived in problem 3. The propagation constant γ is taken to have a positive real part and is given by

$$\gamma = \alpha + j\beta = \sqrt{j\omega\mu(\sigma + j\omega\epsilon)} \tag{10-80}$$

Equations (78) and (79) are four scalar equations (for V and the three rectangular components of $\mathbf{A}$) but only one need be solved since all have the same form. Of all the equations having this form, the simplest is the one having a unit point source represented by a three-dimensional Dirac delta. The function G which satisfies this equation is called the *Green's function* and the equation itself is

$$\nabla^2 G - \gamma^2 G = -\delta \tag{10-81}$$

It will be shown that the function G is an important part of the solutions to (78) and (79) and thus it is necessary first of all to find a solution to (81).

Let it be assumed for simplicity that the Dirac delta in (81) is located at the origin, making the solution G spherically symmetric about the origin. Thus at all points other than the origin, G must satisfy

$$\frac{1}{r^2}\frac{d}{dr}\left(r^2\frac{dG}{dr}\right) - \gamma^2 G = 0$$

which may be written in the more convenient form

$$\frac{1}{r}\frac{d^2}{dr^2}(rG) - \gamma^2 G = 0 \tag{10-82}$$

Equation (82) may be multiplied by r and integrated directly to give

$$G = C\frac{e^{\pm\gamma r}}{r} \tag{10-83}$$

in which C is an arbitrary constant. The solution having the negative sign in the exponent will be used since it represents a wave whose amplitude decreases with increasing distance from the source. The value of C is determined by the strength of the source and thus C may be found by returning to (81), which contains the Dirac delta source function. The procedure is to substitute (83) into (81) and then to integrate over a small spherical volume V surrounding the origin. The radius of the spherical volume is taken to be so small that everywhere inside it, $e^{-\gamma r} \approx 1$. Thus the integral is

$$C\int_V \nabla^2\left(\frac{1}{r}\right) dV - \gamma^2 C\int_V \frac{1}{r}\, dV = -\int_V \delta(\mathbf{r})\, dV \tag{10-84}$$

The first term may be evaluated by noting that the integrand appears in Poisson's equation for a point charge at the origin

$$\nabla^2\left(\frac{q}{4\pi\epsilon r}\right) = -\frac{q}{\epsilon}\delta(r)$$

Using Poisson's equation, the first term in (84) is

$$C\int_V \nabla^2\left(\frac{1}{r}\right) dV = C\int_V [-4\pi\delta(r)]\, dV = -4\pi C$$

The second term in (84) approaches zero as the radius of the sphere approaches zero and the third term is minus one, giving finally $C = (4\pi)^{-1}$. Thus the Green's function is

$$G = \frac{e^{-\gamma r}}{4\pi r} \tag{10-85}$$

and it represents an outgoing spherical wave with a unit source at the origin.

The Green's function is used in obtaining a solution for equations (78) and (79). Consider one component of (78), say the x component:

$$\nabla'^2 A_x(\mathbf{r}') - \gamma^2 A_x(\mathbf{r}') = -\mu J_x(\mathbf{r}') \tag{10-86}$$

(Primes have been attached to the co-ordinates in order to simplify the final result). The corresponding Green's function equation is

$$\nabla'^2 G(\mathbf{r}', \mathbf{r}) - \gamma^2 G(\mathbf{r}', \mathbf{r}) = -\,\delta(\mathbf{r}' - \mathbf{r}) \tag{10-87}$$

which by comparison with (85) has the solution

$$G(\mathbf{r}', \mathbf{r}) = \frac{e^{-\gamma R}}{4\pi R} \tag{10-88}$$

in which

$$R = \sqrt{(x' - x)^2 + (y' - y)^2 + (z' - z)^2} = |\mathbf{r}' - \mathbf{r}|.$$

If (86) is multiplied by G and (87) is multiplied by A_x, the difference between the two equations is

$$\delta(\mathbf{r}' - \mathbf{r})\, A_x(\mathbf{r}') - \mu J_x(\mathbf{r}')\, G(\mathbf{r}', \mathbf{r}) = G(\mathbf{r}', \mathbf{r})\, \nabla'^2 A_x(\mathbf{r}') - A_x(\mathbf{r}')\, \nabla'^2 G(\mathbf{r}', \mathbf{r}) \tag{10-89}$$

This equation now may be integrated over an arbitrary volume V surrounded by a surface S. Application of Green's theorem (1-51) gives

$$\int_V \Big[\delta(\mathbf{r}' - \mathbf{r})\, A_x(\mathbf{r}') - \mu J_x(\mathbf{r}')\, G(\mathbf{r}', \mathbf{r})\Big]\, dV' = \int_S \Big[G(\mathbf{r}', \mathbf{r})\, \nabla' A_x(\mathbf{r}') - A_x(\mathbf{r}')\, \nabla' G(\mathbf{r}', \mathbf{r})\Big] \cdot d\mathbf{a}' \tag{10-90}$$

If the surface S is a sphere with radius r, then (90) becomes

$$\int_V \Big[\delta(\mathbf{r}' - \mathbf{r})\, A_x(\mathbf{r}') - \mu J_x(\mathbf{r}')\, G(\mathbf{r}', \mathbf{r})\Big]\, dV' = \int_S \Big[G(\mathbf{r}', \mathbf{r}) \frac{\partial A_x(\mathbf{r}')}{\partial r'} - A_x(\mathbf{r}') \frac{\partial G(\mathbf{r}', \mathbf{r})}{\partial r'}\Big]\, da' \tag{10-91}$$

If the radius r is made to approach infinity (so that V includes all of space) the quantities G and A_x on S may be written as far-field approximations. Thus the Green's function has the general form

$$G(\mathbf{r}', \mathbf{r}) = M(\theta', \phi', \mathbf{r}) \frac{e^{-\gamma r'}}{r'} \tag{10-92}$$

It is reasonable to assume that A_x also has the form of an outgoing spherical wave:

$$A_x(\mathbf{r}') = N(\theta', \phi') \frac{e^{-\gamma r'}}{r'} \tag{10-93}$$

It is easy to show that (92) and (93) cause the integrand on the right-hand side of (91) to go to zero. The left-hand side of (91) thus becomes

$$A_x(\mathbf{r}) = \mu \int J_x(\mathbf{r}')\, G(\mathbf{r}', \mathbf{r})\, dV' \tag{10-94}$$

The y and z components of $\mathbf{A}$ may be treated similarly and added vectorially to give a single expression for the vector potential. A similar expression for the scalar potential also may be derived. Thus the expressions for the potentials in terms of their sources are

$$\mathbf{A}(\mathbf{r}) = \mu \int \mathbf{J}(\mathbf{r}')\, G(\mathbf{r}', \mathbf{r})\, dV' \tag{10-95}$$

$$V(\mathbf{r}) = \frac{1}{\epsilon}\int \rho(\mathbf{r}')\, G(\mathbf{r}', \mathbf{r})\, dV' \tag{10-96}$$

for which G is given by (88). If the medium is assumed to be lossless then $\gamma = j\beta$ and (95) and (96) reduce to (24) and (25).

10.11 Far-field Approximation. Electromagnetic fields far from a source are especially important since they are the radiation fields. A far-field approximation was applied to the case of a linear antenna in sec. 10.07 and this approximation need be changed only slightly in order to be applicable to any current distribution. Consider an arbitrary current distribution $\mathbf{J}$ located in the vicinity of an origin of co-ordinates as shown in Fig. 10-12. If the distance r is large compared to

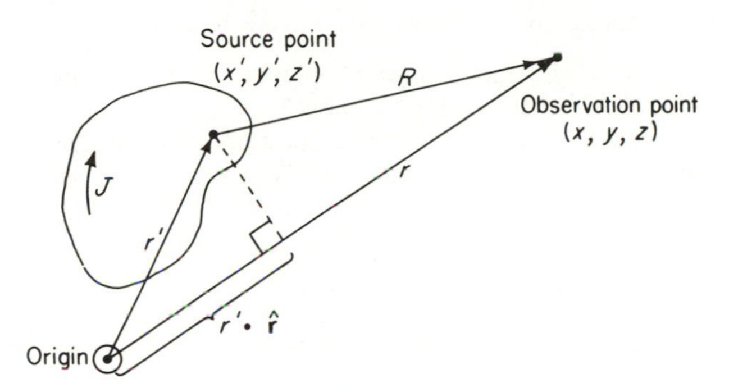

Figure 10-12.

a wavelength and also large compared to the largest dimension of the source, then the observation point is said to be in the far field. Under these conditions $\mathbf{R}$ is approximately parallel to $\mathbf{r}$ and the distance R is approximately r reduced by the projection of $\mathbf{r}'$ on $\mathbf{r}$. That is, to a first approximation

$$R \approx r - \mathbf{r}' \cdot \hat{\mathbf{r}} \tag{10-97}$$

This approximation is appropriate for the exponential in the Green's function (88) but for the denominator it is sufficient to replace R with r. These approximations permit (95) to be expressed as

$$\mathbf{A}(\mathbf{r}) \approx \frac{\mu e^{-\gamma r}}{4\pi r}\, \mathbf{f}(\theta, \phi) \tag{10-98}$$

in which

$$\mathbf{f}(\theta, \phi) = \int \mathbf{J}(\mathbf{r}')\, e^{\gamma \mathbf{r}' \cdot \hat{\mathbf{r}}}\, dV' \tag{10-99}$$

The magnetic and electric field strengths now may be expressed in very simple forms using (98).

$$\mathbf{H} = \frac{1}{\mu} \nabla \times \mathbf{A} \approx \frac{\gamma\, e^{-\gamma r}}{4\pi r} (\hat{\boldsymbol{\theta}} f_\phi - \hat{\boldsymbol{\phi}} f_\theta) \tag{10-100}$$

$$\mathbf{E} = \frac{\nabla \times \mathbf{H}}{\sigma + j\omega\epsilon} \approx -\frac{\gamma\eta e^{-\gamma r}}{4\pi r}(\hat{\boldsymbol{\theta}} f_\theta + \hat{\boldsymbol{\phi}} f_\phi) \tag{10-101}$$

where $\gamma = \sqrt{j\omega\mu(\sigma + j\omega\epsilon)}$ and $\eta = \sqrt{j\omega\mu/(\sigma + j\omega\epsilon)}$

In the derivation of (100) and (101) all terms proportional to r^{-2} and r^{-3} have been left out so that only the radiation terms remain; the details of the derivation are left as an exercise for the reader.

ADDITIONAL PROBLEMS

11. Obtain expressions for the far field of a half-wave dipole using (100) and (101), and check your answers against the results of sec. 10.07.

12. (a) At what distance from a 60-cycle circuit is the radiation field approximately equal to the induction field?

(b) Approximately what are the relative amplitudes of the radiation, induction, and electric fields at a distance of 1 wavelength from a Hertzian dipole?

13. (a) Show that the unattenuated radiation field at the surface of the earth of a quarter-wave monopole is given by the formula

$$E = \frac{6.14}{r}\sqrt{W} \qquad \text{mv/m effective}$$

where r is in miles and W is the power radiated in watts. The *unattenuated* field is the value of field strength that would exist if the earth were perfectly conducting.

(b) Derive the corresponding expression for a short monopole ($H \ll \lambda$).

14. A half-wave dipole is located parallel to and one-quarter wavelength from a plane metallic reflecting sheet. Sketch the lines of current flow in the sheet. (Suggestion: Use the image principle and the relation $\mathbf{J} = \hat{\mathbf{n}} \times \mathbf{H}$.)

15. Short vertical monopole antennas that are suitably "top-loaded" with a capacitive load, have an essentially uniform current along their whole length. Set up the vector potential and derive an expression for the average value (in time) of the Poynting vector at large distances from such an antenna.

16. (a) Set up an expression for the vector potential due to a traveling-wave current distribution

$$I(z) = I_m\, e^{-j\beta z}$$

along a terminated wire antenna of length L.

(b) Show that the distant field of such an antenna is given by

$$E_\theta = \frac{30 I_m \sin\theta}{r(1 - \cos\theta)}\,[2 - 2\cos\beta(1 - \cos\theta)\,L]^{1/2}$$

17. (a) A toroidal coil has a large number of closely wound turns on a core of high permeability so that at d-c virtually all of the magnetic flux is confined to the core (Fig. 10-13). When an alternating current flows in the

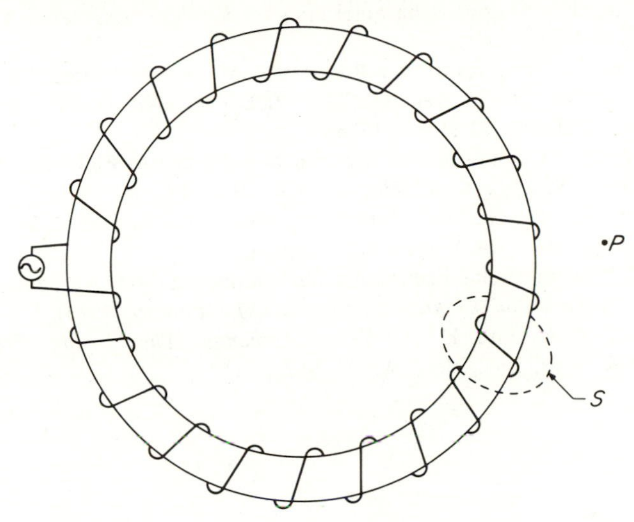

Figure 10-13. A toroidal coil on a high-permeability core.

winding, is there at the point P (1) a value of vector potential $\mathbf{A}$? (2) a value of magnetic field $\mathbf{H}$? (3) a value of electric field $\mathbf{E}$?

Assuming a current of 1 amp at 60 Hz through a 1000-turn winding, and $\mu = 1000\mu_v$:

(b) Approximately what is the voltage around the path s?

(c) What is the order of magnitude of each of the vectors of parts (1), (2) and (3) above?

18. It is possible to define a single function, called the Hertzian vector, from which both electric and magnetic field strengths may be derived. This vector is

$$\mathbf{Z} = \frac{1}{\mu\epsilon}\int \mathbf{A}\,dt$$

Using this function, show that:

(a) $\mathbf{E} = -\mu\epsilon\ddot{\mathbf{Z}} + \nabla(\nabla\cdot\mathbf{Z})$

(b) $\mathbf{H} = \epsilon\nabla \times \dot{\mathbf{Z}}$

(c) $\nabla^2\mathbf{Z} = \mu\epsilon\ddot{\mathbf{Z}}$ (in free space, where $\nabla \times \mathbf{H} = \dot{\mathbf{D}}$)

BIBLIOGRAPHY

Pistolkors, A. A., "The Radiation Resistance of Beam Antennas," *Proc. IRE*, **17**, 562 (March, 1929).

Carter, P. S., "Circuit Relations in Radiating Systems," *Proc. IRE*, **20**, 1004 (June, 1932).

Brown, G. H., "Directional Antennas," *Proc. IRE*, **25**, 78 (January 1937).

Burgess, R. E., "Aerial Characteristics," *Wireless Engineer*, **21**, 247, 154 (April, 1944).

Moullin, E. B., "The Radiation Resistance of Aerials Whose Length is Comparable with the Wavelength," *J. IEE*, **78**, 540 (1936); also *Wireless Section, IEE*, **11**, 93 (June, 1936).

Bechmann, R., "Calculation of Electric and Magnetic Field Strengths of any Oscillating Straight Conductors," *Proc. IRE*, **19**, 461 (March, 1931).

Bechmann, R., "On the Calculation of the Radiation Resistance of Beam Antennas," *Proc. IRE*, **19**, 1471 (August, 1931).

Labus, J., "Rechnerische Ermittlung der Impedanz von Antennen," *Hochfrequenz Technik und Elektroakustik*, **41**, 17 (January, 1933).

Schelkunoff, S. A., and H. T. Friis, *Antennas, Theory and Practice*, John Wiley & Sons, Inc., New York, 1952.

Chapter 11

ANTENNA FUNDAMENTALS

11.01 Introduction. The first ten chapters of this book have been devoted to the basic theory of electromagnetic fields. By the very nature of his profession the engineer cannot be content with theory, however interesting, but must eventually apply his knowledge to the solution of practical problems in the real world. In the antenna field this transition from theory to practice is greatly facilitated by a knowledge of the basic properties of antennas, as well as the role of the antenna element in the overall system. It is truly astonishing how many significant practical problems can be solved using only the simplest concepts. It is the purpose of this chapter to develop some of these elementary concepts, and to show how the antenna takes its place as a system element.

The important properties of an antenna as an element in a system are its directional characteristics, including directive gain and polarization, and its impedance. Although these characteristics will be determined for the antenna as a transmitting element, the results also hold for reception, as can be readily demonstrated by means of the reciprocity theorem (secs. 11.02 and 13.06).

11.02 Network Theorems. In ordinary circuit theory certain network theorems have proven valuable in the solution of a great many problems. The validity of these theorems is based upon the linearity and/or the bilateralism of the networks. In electromagnetic field theory the solution of any antenna problem can be obtained (at least in theory) by application of Maxwell's equations and the appropriate boundary conditions. The field equations themselves are linear, and as long as the "constants" μ, ϵ, and σ of the media involved are truly constant, that is, do not vary with the magnitude of the signal (linearity) nor with direction (bilateralism), the same theorems can be applied. The usefulness of such theorems in antenna work is evidenced by the fact that with their aid nearly all the properties of a receiving antenna can be deduced from the known transmitting properties of the same antenna. A few of the theorems that are most commonly used in antenna problems are the following:

Superposition Theorem. In a network of generators and linear impedances, the current flowing at any point is the sum of the currents that would flow if each generator were considered separately, all other generators being replaced at the time by impedances equal to their internal impedances.

This fundamental principle follows directly from the linearity of Ohm's law and the field equations. When a network of impedances is linear, a given increase in voltage produces an increase in current that is independent of the magnitude of current already flowing. Therefore the effect of each generator can be considered separately and independently of whether or not other generators are actually generating.

Thevenin's Theorem. If an impedance Z_R be connected between any two terminals of a linear network containing one or more generators, the current which flows through Z_R will be the same as it would be if Z_R were connected to a simple generator whose generated voltage is the open-circuit voltage that appeared at the terminals in question and whose impedance is the impedance of the network looking back from the terminals, with all generators replaced by impedances equal to the internal impedances of these generators.

This theorem follows from the principle of superposition. A proof of it can be found in any of the references on circuit theory.

Maximum-power Transfer Theorem. An impedance connected to two terminals of a network will absorb maximum power from the network when the impedance is equal to the conjugate of the impedance seen looking back into the network from the two terminals.

Corollary: The maximum power that can be absorbed from a network equals $V_{oc}^2/4R$, where V_{oc} is the open-circuit voltage at the output terminals and R is the resistive component of the impedance looking back from the output terminals.

*Compensation Theorem.** Any impedance in a network may be replaced by a generator of zero internal impedance, whose generated voltage at every instant is equal to the instantaneous potential difference that existed across the impedance because of the current flowing through it.

Reciprocity Theorem. In any system composed of linear bilateral impedances, if a voltage V is applied between any two terminals and the current I is measured in any branch, the ratio of V to I, called the transfer impedance, will be the same as the ratio of V to I obtained when the positions of generator and ammeter are interchanged.

*For an application to fields see G. D. Monteath, "Application of the Compensation Theorem to Certain Radiation and Propagation Problems," *Proc. IEE,* **98,** Part IV, No. 1, Oct., 1951, pp. 23–30.

In the above statement of this theorem the generator and ammeter are assumed to have zero impedance. It may be readily shown that the theorem also holds if the generator and ammeter have impedances which are equal. A particularly useful form of the theorem results when the generator and ammeter impedances are made very large ($Z_g = Z_a \rightarrow \infty$). The generator then becomes a constant-current source and the "ammeter" becomes an infinite-impedance voltmeter. The reciprocity theorem then states that if a current is applied at one pair of terminals and the open-circuit voltage is measured at a second pair of terminals, the ratio of voltage to current remains the same when the positions of current source and voltmeter are interchanged.

The reciprocity theorem is one of the most powerful theorems in both circuit and field theory. It was originally stated by Rayleigh in a form somewhat similar to the above. A generalized statement of the theorem, suitable for application to fields as well as circuits, will be considered in detail in sec. 13.06.

The circuit concept of reciprocity can be applied directly to deduce some of the "terminal" properties of antennas. Suppose that an antenna system has two "terminal pairs" or "ports" as shown in Fig. 11-1(*a*); for example, the ports could be the terminals of the dipoles in Fig. 11-1(*b*).

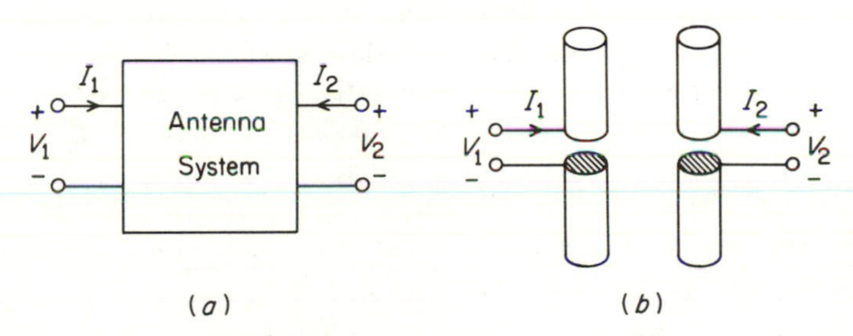

Figure 11-1. Antenna system as circuits.

Circuit theory for linear systems tells us that the voltages and currents are related by the equations

$$V_1 = Z_{11}I_1 + Z_{12}I_2 \qquad I_1 = Y_{11}V_1 + Y_{12}V_2$$

$$\text{or by} \tag{11-1}$$

$$V_2 = Z_{21}I_1 + Z_{22}I_2 \qquad I_2 = Y_{21}V_1 + Y_{22}V_2$$

The reciprocity theorem indicates that

(a) $Z_{21} = Z_{12}$ or (b) $Y_{21} = Y_{12}$ or (c) $Z'_{21} = Z'_{12}$

These are equivalent but different statements of the reciprocity theorem for circuits. Z_{12}, Z_{21} are the *mutual impedances* of the circuit; $Z_{12} = V_1/I_2$ is a measure of the open-circuit voltage produced across terminals (1) by current I_2 at terminals (2). Y_{12}, Y_{21} are the *transfer admittances;* Y_{12} is a measure of the short-circuit current at terminals (1) produced by voltage V_2 at terminals (2). Z'_{12}, Z'_{21} are the *transfer impedances;* $Z'_{12} = V_1/I_2$ is the ratio of the voltage applied at terminals (1) to the current through terminals (2) when short-circuited. Thus $Z'_{12} = 1/Y_{21}$ and $Z'_{21} = 1/Y_{12}$. The word statements of the reciprocity theorem given above correspond to (b or c) and (a) respectively.

The reciprocity theorem may be demonstrated as follows with reference to the pair of dipoles in Fig. 11-1(*b*). Consider two cases indicated by primes (′) and double primes (″), the first case having the terminals of antenna (2) shorted and a voltage V'_1 applied to antenna (1) and the second case having the terminals of antenna (1) shorted and a voltage V''_2 applied to antenna (2). In the first case $I'_2 = Y_{21}V'_1$ and in the second case $I''_1 = Y_{12}V''_2$. If $Y_{21} = Y_{12}$, we may deduce that

$$\frac{V'_1}{I'_2} = \frac{V''_2}{I''_1} \tag{11-2}$$

which is one form of the reciprocity statement for networks applied directly to antennas. In sec. (13.06) the reciprocity theorem for electromagnetic fields will be derived, and shown to reduce to the circuit statement at the antenna terminals.

The reciprocity theorem has proven a powerful and useful tool in circuit and field theory, and many corollary statements have been derived from it. Some of these corollaries, especially those concerning reciprocity of powers, have been derived under special conditions. If applied in circumstances where these special conditions are violated, the corollary statements may break down. The reciprocity theorem itself is perfectly general, and always gives the correct answer as long as only linear bilateral circuits or media are involved.

Application of Network Theorems to Antennas. From the above theorems can be deduced several very useful antenna theorems, relating the properties of transmitting and receiving antennas. So far, most of the analysis has been concerned with transmitting antennas for which the assumption of sinusoidal current distribution is known to yield results of acceptable accuracy. On the other hand, for antennas excited as receiving antennas, the current distribution varies with the direction of arrival of the received field and is not even approximately sinusoidal, except for the resonant lengths (half-wave dipole, etc.). This being the case, it is by no means obvious that the directional and impedance properties of an antenna should be identical for the transmitting and receiving conditions. Because, in the general case, the current distribution is not sinusoidal for the receiving antenna, direct computation of its properties is usually a relatively complicated problem. The following theorems make it possible to infer the properties of a receiving antenna from its properties as a transmitting antenna, and vice versa.

Equality of Directional Patterns. The directional pattern of a receiving antenna is identical with its directional pattern as a transmitting antenna.

Proof: This theorem results directly from an application of the reciprocity theorem. The directional pattern of a transmitting antenna is the polar characteristic that indicates the strength of the radiated field at a fixed distance in different directions in space. The directional pattern of a receiving antenna is the polar characteristic that indicates the response of the antenna to unit field strength from different directions. The pattern as a transmitting antenna could be measured as indicated in Fig. 11-2 by means of a short exploring dipole moved about on the surface of

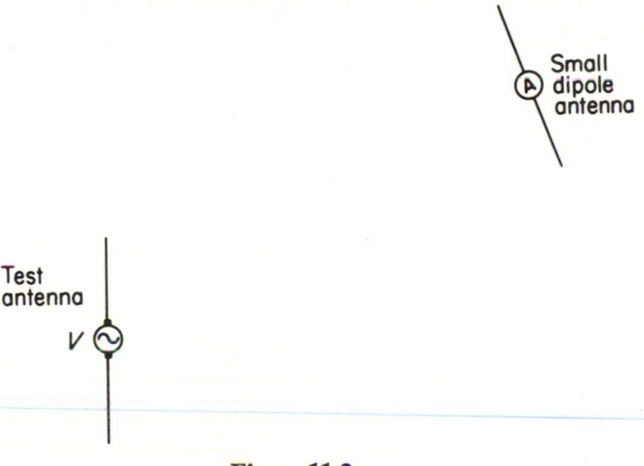

Figure 11-2.

a large sphere centered at the antenna under test. (For the case of linear polarization, the exploring dipole is always oriented so as to be perpendicular to the radius vector and parallel to the electric vector.) A voltage V is applied to the test antenna, and the current I flowing in the short dipole antenna will be a measure of the electric field at the position of the dipole antenna. If then the voltage V is applied to the dipole and the test antenna current is measured, the receiving pattern of the test antenna can be obtained. But by the reciprocity theorem, for every location of the probe antenna, the ratio of V to I is the same as before. Therefore the radiation pattern as a receiving antenna will be identical with the pattern as a transmitting antenna.

When the test antenna radiates an elliptically polarized wave, that is, when the radiated electric field strength has two components, E_θ and E_ϕ, that are not in time phase, the radiation patterns for the θ polarization and ϕ polarization are shown separately. The pattern for the particular polarization specified is obtained by keeping the exploring dipole parallel to that polarization. It follows as before that for each of the polarizations the radiation patterns for transmitting and receiving will be the same.

Equivalence of Transmitting and Receiving Antenna Impedances. The impedance of an isolated antenna when used for receiving is the same as when used for transmitting.

Proof: This theorem is particularly easy to prove for the case of two antennas which are widely separated. If antenna (2) is far from antenna (1), the self-impedance Z_{S1} of antenna (1) is given by

$$Z_{S1} = \frac{V_1}{I_1} = Z_{11}$$

Because of the wide separation of the two antennas, the mutual impedance Z_{12} may be ignored when antenna (1) is used for transmitting. However, when antenna (1) is used for receiving, Z_{12} cannot be ignored since it provides the only coupling between the two antennas. For this case one may consider a load impedance Z_L attached to antenna (1) and also one may represent the voltage $Z_{12}I_2$ as a generator as shown in the equivalent circuit of Fig. 11-3. If the two antennas are far apart, varying

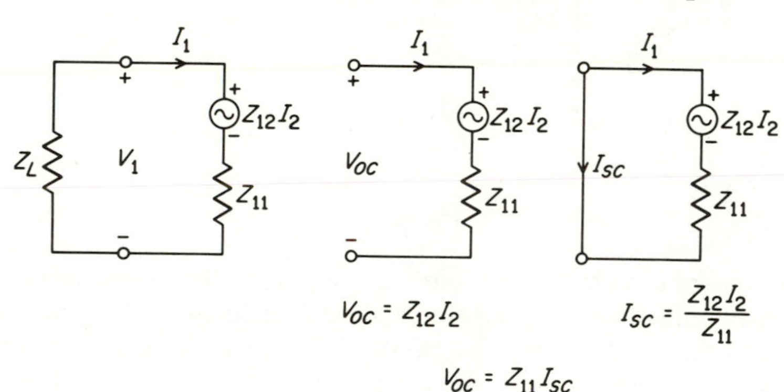

Figure 11-3. Equivalent circuit of a receiving antenna under loaded, open-circuit and short-circuit conditions.

Z_L cannot cause I_2 to vary and thus the generator in Fig. 11-3 behaves as an ideal zero-impedance, constant-voltage generator. Under these conditions, antenna (1) exhibits the terminal behavior of a generator with internal impedance Z_{11} and thus the receiving impedance is equal to the transmitting impedance.

The discussion above for one distant antenna applies equally well to any number of antennas provided they are far from the antenna whose impedance is being considered. If the other antennas are sufficiently close that the mutual impedance is not negligible (compared with the self-impedance), then the receiving impedance will be the self-impedance of the antenna plus the impedance coupled-in owing to the presence of the other antennas. However, even under these conditions the equality of the receiving and transmitting impedances will hold if in the transmitting case the other radiators are left in position and connected to impedances equal to the impedances of the generators that excited them.

Equality of Effective Lengths. The *effective length*, l_{eff}, of an antenna is a term used to indicate the effectiveness of the antenna as a radiator or collector of electromagnetic energy. The significance of the term as applied to transmitting antennas is illustrated in Fig. 11-4. The effective

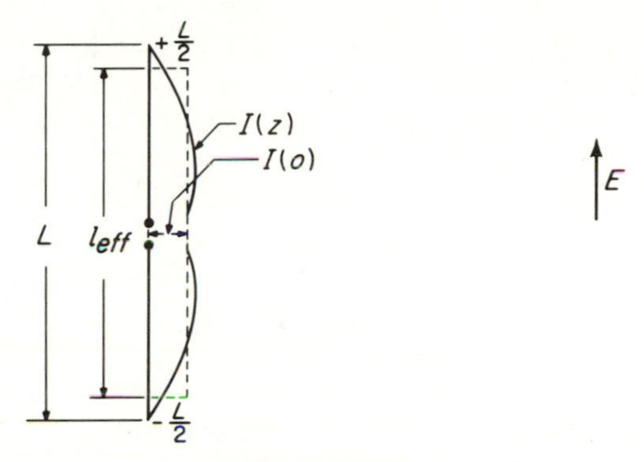

Figure 11-4.

length of a transmitting antenna is that length of an equivalent linear antenna that has a current $I(0)$ at all points along its length and that radiates the same field strength as the actual antenna in the direction perpendicular to its length. $I(0)$ is the current at the terminals of the actual antenna. That is, for transmitting

$$I(0)l_{\text{eff}}(\text{trans}) = \int_{-L/2}^{+L/2} I(z)\, dz$$

or

$$l_{\text{eff}}(\text{trans}) = \frac{1}{I(0)}\int_{-L/2}^{+L/2} I(z)\, dz \tag{11-3}$$

The effective length of a receiving antenna is defined in terms of the open-circuit voltage developed at the terminals of the antenna for a given received field strength **E**. That is, for receiving

$$l_{\text{eff}}(\text{rec}) = -\frac{V_{\text{oc}}}{E} \tag{11-4}$$

where V_{oc} is the open-circuit voltage at the antenna terminals produced by a uniform exciting field E volts per meter parallel to the antenna. [The minus sign is used in (4) to conform with conventional polarity markings (upper terminal positive).]

The equality of transmitting and receiving effective lengths may be deduced by application of the reciprocity theorem to the two cases as shown in Fig. 11-5.

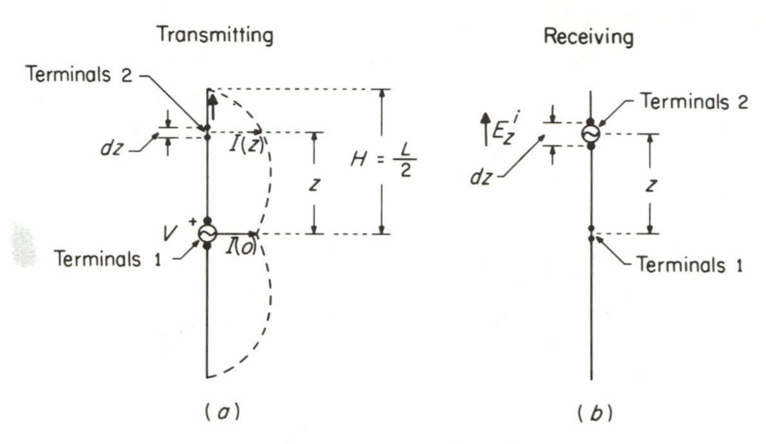

Figure 11-5.

First consider the transmitting antenna case [Fig. 11-5(*a*)]. A voltage V applied at the terminals produces a current $I(0) = V/Z_a$ at the terminals and a current $I(z)$ at any point z along the antenna. Z_a is the antenna impedance measured at the terminals. This is the "prime" situation.

Next consider the same antenna for the receiving case [Fig. 11-5(*b*)] in which an electromagnetic field E_z^i is incident upon the antenna. As a result of this field a voltage $E_z^i\,dz$ is impressed at or induced in the element dz. Because this impressed voltage is independent of the current that flows in the antenna, it can be represented by a zero-impedance generator of voltage $E_z^i\,dz$ in series at the point z. With the antenna base terminals short-circuited, the voltage $E_z^i\,dz$ at z will produce a current dI_{sc} at the terminals. This is the "double-prime" situation.

Putting $V = V_1'$, $I(z) = I_2'$, $E_z^i\,dz = V_2''$, and $dI_{sc} = I_1''$,

the reciprocity theorem (eq. 11-2) gives

$$\frac{V}{I(z)} = \frac{E_z^i\,dz}{dI_{sc}} \qquad \text{or} \qquad dI_{sc} = \frac{E_z^i\,dz}{V}\,I(z)$$

By superposition the total short-circuit current produced at the antenna terminals by all of the differential voltages impressed along the entire length of the antenna will be

$$I_{sc} = \frac{1}{V}\int E_z^i I(z)\,dz \tag{11-5}$$

Knowing the short-circuit current, the open-circuit voltage at the terminals can be determined from Thevenin's theorem:

$$V_{oc} = -I_{sc}Z_a = -\frac{Z_a}{V}\int E_z^i I(z)\,dz = -\frac{1}{I(0)}\int E_z^i I(z)\,dz \tag{11-6}$$

where again the minus sign is chosen to conform with conventional polarity markings (upper terminal positive). For an incident field $E_z^i = E_z$ constant along the length of the antenna

$$V_{oc} = -\frac{E_z}{I(0)}\int I(z)\,dz$$

so that

$$\frac{V_{oc}}{E_z} = -\frac{1}{I(0)}\int I(z)\,dz$$

Therefore from (3) and (4) the effective length of an antenna for receiving is equal to its effective length as a transmitting antenna.

In this section, by applying well-known circuit theorems, it has been demonstrated that the directive properties, impedance, and effective length of an antenna are the same whether the antenna is used for transmitting or receiving. One characteristic that is different in the two cases is the current distribution along the antenna. However, even here the reciprocity theorem can be used to determine the current distribution for reception when the current distribution for transmission is known or assumed.

11.03 Directional Properties of Dipole Antennas. Radio antennas have a two-fold function. The first of these functions is to "radiate" the radio frequency energy that is generated in the transmitter and guided to the antenna by the transmission line. In this capacity the antenna acts as an impedance-transforming device to match the impedance of the transmission line to that of free space. The other function of the antenna is to direct the energy into desired directions, and what is often more important, to suppress the radiation in other directions where it is not wanted. This second function of the antenna will be considered first under the general heading of directional characteristics.

A completely nondirectional or omnidirectional radiator radiates uniformly in all directions and is known as an isotropic radiator or a unipole. A point source of sound is an example of an isotropic radiator in acoustics. There is no such thing as an isotropic radiator of electromagnetic energy, since all radio antennas have some directivity. However, the notion of a completely nondirectional source is useful, especially for gain comparison purposes.

The radiation pattern of an antenna is a graphical representation of the radiation of the antenna as a function of direction. When the radiation is expressed as field strength, E volts per meter, the radiation pattern is a field-strength pattern. If the radiation in a given direction is expressed in terms of power per unit solid angle, the resulting pattern is a power pattern. A power pattern is proportional to the square of the field strength pattern. Unless otherwise specified, the radiation patterns referred to in this book will be field-strength patterns.

The co-ordinate system generally used in the specification of antenna radiation patterns is the spherical co-ordinate system (r, θ, ϕ), shown in Fig. 1-9 on page 16. The antenna is located at or near the origin of this system, and the field strength is specified at points on the spherical surface of radius r (or on a hemispherical surface in the case of ground-based antennas). The shape of the radiation pattern is independent of r, as long as r is chosen sufficiently large (r must be very much greater than the wavelength and very much greater than the largest dimension of the antenna system). When this is true, the magnitude of the field strength in any direction varies inversely with r, and so needs to be stated for only one value of r. For example, in broadcast antenna work it is customary to state the field strength at a radius of 1 mile. Often only the *relative* radiation pattern is used. This gives the relative field strengths in various directions, usually referred to unity in the direction of maximum radiation.

For the radiation field, the direction of **E** is always tangential to the spherical surface. For a vertical dipole **E** is in the θ direction, whereas for a horizontal loop **E** is in the ϕ direction. In general, the radiation field strength may have both E_θ and E_ϕ components, which may or may not be in time phase. The radiation characteristics are then shown by separate patterns for the theta and phi polarizations. The terms, *theta polarization* and *phi polarization* are synonymous with and replace the older terms *vertical* polarization and *horizontal* polarization, respectively. The older terms were confusing in that a theta or vertically polarized signal is not always vertical (however it is always in the vertical plane through the radius vector), although a phi or horizontally polarized signal is always horizontal.

A complete radiation pattern gives the radiation for all angles of ϕ and θ and really requires three-dimensional presentation. This is overcome by showing cross sections of the pattern in planes of interest. Cross sections in which the radiation patterns are most frequently given are the horizontal ($\theta = 90°$) and vertical ($\phi =$ constant) planes. These are called the *horizontal pattern* and *vertical patterns*, respectively. For linearly polarized antennas having patterns of simple shape the terms *E plane pattern* and *H plane pattern* are also in common use.

The E plane and H plane are respectively the planes passing through the antenna in the direction of beam maximum and parallel to the far-field **E** and **H** vectors.

For an elementary dipole antenna $I\,dl$ the magnitude of the radiation term for the field strength is

$$E = E_\theta = \frac{60\pi I\,dl}{r\lambda}\sin\theta \qquad \text{V/m} \quad (11\text{-}7)$$

where θ is the angle between the axis of the dipole and the radius vector to the point where the field strength is measured. When the dipole is vertical, the horizontal radiation pattern is a circle [Fig. 11-6(*a*)] because in this plane ($\theta = 90°$) the radiation is uniform. In any vertical plane through the axis the field strength varies as $\sin\theta$ and the vertical patterns are all the same, having the figure-eight shape shown in Fig. 11-6(*b*). When the dipole is horizontal, the horizontal pattern has the figure-eight shape, but the vertical pattern depends upon the angle which the vertical plane makes with the horizontal axis. The two vertical planes of chief interest are those perpendicular and parallel to the axis of the dipole. The vertical radiation pattern is a circle for the former and a figure eight for the latter. These two vertical patterns and the horizontal pattern are known as the *principal-plane* patterns.

As the length of a dipole is increased beyond the point where it may be considered short in terms of a wavelength, the radiation pattern in the planes through the axis changes as indicated in Fig. 11-6. Figure 11-6(*c*) shows the vertical radiation pattern of a center-fed half-wave vertical dipole, and Fig. 11-6(*d*) shows the same pattern when the dipole is one wavelength long. The expression for the magnitude of the radiation field strength due to a half-wave dipole is

$$E = \frac{60I}{r}\left[\frac{\cos\left(\frac{\pi}{2}\cos\theta\right)}{\sin\theta}\right] \qquad \text{V/m} \quad (11\text{-}8)$$

The more general expression for a center-fed dipole antenna of any length $L = 2H$ is

$$E = \frac{60I}{r}\left[\frac{\cos\beta H - \cos(\beta H\cos\theta)}{\sin\theta}\right] \qquad \text{V/m} \quad (11\text{-}9)$$

Expressions (7) and (8) were derived in chap. 10 and expression (9) was obtained in a problem in the same chapter. The radiation patterns, as given by (9), are shown in Fig. 11-6(*e*) and (*f*) for antenna lengths of $1\frac{1}{2}$ and 2 wavelengths.

The vertical radiation patterns of Fig. 11-6 also apply to the corresponding grounded vertical antennas when mounted on a perfectly

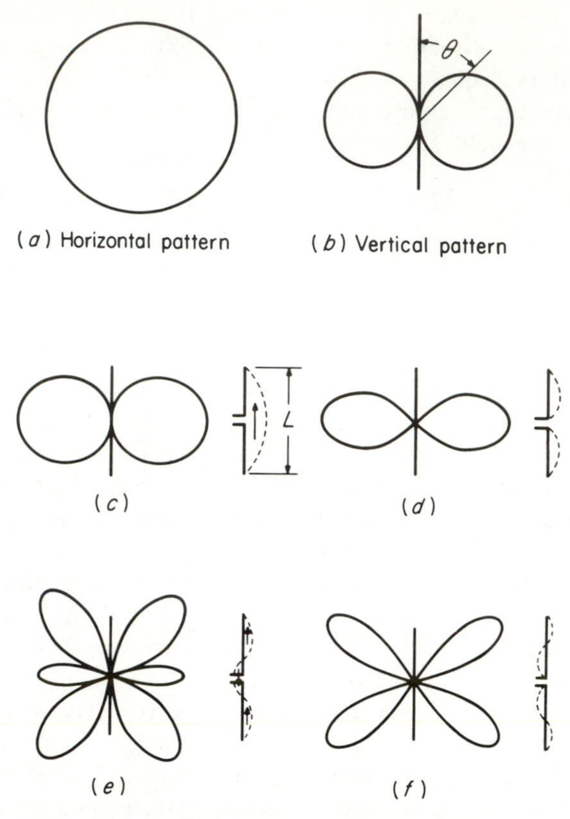

Figure 11-6. Radiation patterns of center-fed vertical dipoles: (*a*) horizontal pattern for all lengths, (*b*) vertical pattern for short dipole. Vertical patterns for dipole lengths: (*c*) one-half wavelength; (*d*) one wavelength; (*e*) one-and-a-half wavelengths; (*f*) two wavelengths. The assumed current in each case is shown dashed.

conducting ground plane. The length of the grounded vertical antenna is just one-half the length of the corresponding dipole (the image forms the other half), and of course only the top half of the pattern applies.

11.04 Traveling-wave Antennas and Effect of the Point of Feed on Standing-wave Antennas. The patterns of Fig. 11-6 are for unterminated antennas that are assumed to have a standing-wave distribution of current. Sometimes antennas are terminated to make them *aperiodic* or *nonresonant* and, in this case, they have a traveling-wave distribution. The directional pattern of a traveling-wave antenna having a length of 6 wavelengths is shown in Fig. 11-7. Assuming negligible attenuation of the wave along the antenna and a wave velocity equal

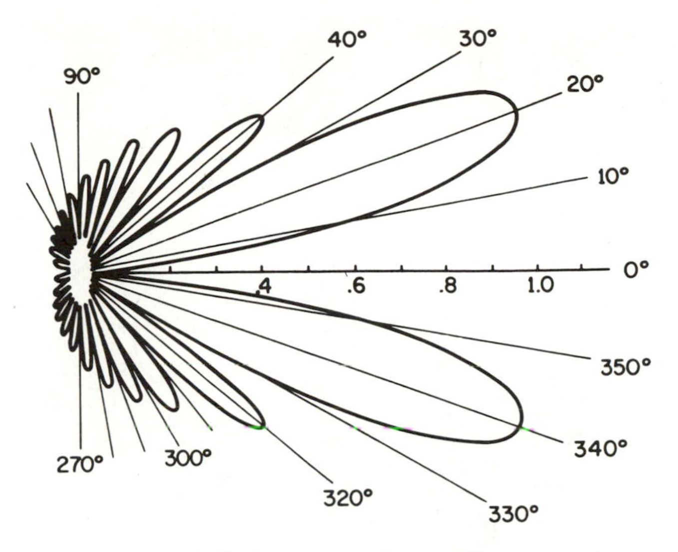

Figure 11-7. Radiation pattern of a traveling-wave antenna six wavelengths long.

to c, the expression for the pattern of a traveling-wave antenna (see problem 16, chap. 10) is

$$E = \frac{30I_m \sin\theta}{r(1 - \cos\theta)}\{2 - 2\cos[\beta L(1 - \cos\theta)]\}^{1/2} \tag{11-10}$$

It is seen that with a traveling current wave the pattern is no longer symmetrical about the $\theta = 90$ degrees plane, but instead the radiation tends to "lean" in the direction of the current wave. The angle θ between the axis of the antenna and the direction of maximum radiation becomes smaller as the antenna becomes longer.

In the case of an unterminated antenna the actual current distribution in general is a combination of standing wave and traveling wave. However, except for very long antennas, the standing wave is predominant and the traveling-wave component of current is usually neglected in pattern calculations. For the standing-wave current distribution, the pattern is always symmetrical about the $\theta = 90$ degrees plane. For *center-fed* antennas the pattern of the traveling-wave component of current is also symmetrical about $\theta = 90$ degrees, so that the effect of this latter component of current tends to be obscured. This is especially true when the angles of maximum radiation for the two current distributions nearly coincide, as is often the case. However, when an antenna is unsymmetrically fed, the pattern due to the traveling-wave current distribution is no longer symmetrical about $\theta = 90$ degrees, and its effect on the resultant pattern becomes more pronounced. This is

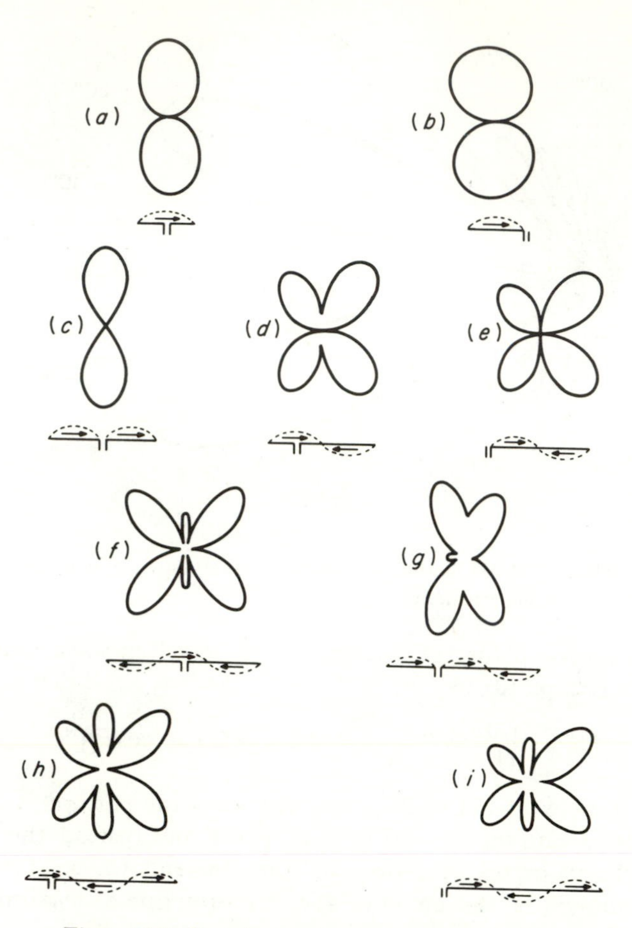

Figure 11-8. Experimental patterns of wire antennas having different locations for feed points: (*a*) half-wave center-fed, (*b*) half-wave end-fed, (*c*) full-wave center-fed, (*d*) full-wave fed one-quarter wavelength from the end, (*e*) full-wave end-fed; (*f*), (*g*), (*h*) and (*i*) one-and-one-half wave, fed as indicated.

evident in Fig. 11-8, which shows some experimental patterns of wire antennas having different locations for the feed points. The asymmetry due to the traveling-wave current shows up when the feed point is moved away from center.

Changing the location of the feed point has another, even more important, effect when the antenna is longer than a half-wavelength. This is the effect on the standing-wave current distribution, which may be quite different for different locations of the feed point, as is also illustrated in Fig. 11-8 for full-wave and one-and-one-half wavelength antennas. The effects of these different current distributions are clearly evident in the measured patterns.

11.05 Two-element Array. When greater directivity is required than can be obtained by a single antenna, antenna arrays are used. An *antenna array* is a system of similar antennas, similarly oriented. Antenna arrays make use of wave-interference phenomena that occur between the radiations from the different elements of the array. Consider the two-element array of Fig. 11-9 in which the antennas (0) and (1)

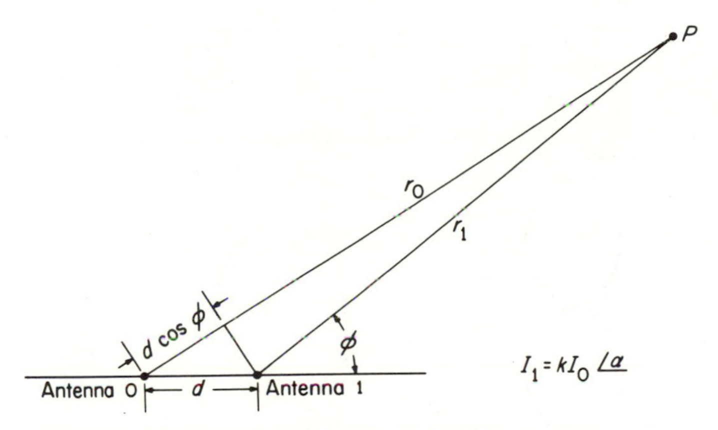

Figure 11-9. A two-element array of nondirectional radiators.

are nondirectional radiators in the plane under consideration. (For example, they could be vertical radiators when the horizontal pattern is being considered.) When the point P is sufficiently remote from the antenna system, the radius vectors to the point can be considered parallel, and it is possible to write

$$r_1 = r_0 - d\cos\phi$$

in the phase factor of the fields, and

$$\frac{1}{r_1} = \frac{1}{r_0}$$

as far as the magnitudes of the fields are concerned. The phase difference between the radiations from the two antennas will be

$$\psi = \beta d\cos\phi + \alpha$$

where $\beta d = (2\pi/\lambda)d$ is the path difference in radians and α is the phase angle by which the current I_1 leads I_0. The phasor sum of the fields will be

$$E = E_0(1 + ke^{j\psi})$$

where E_0 is the field strength due to antenna (0) alone, and where k

is the ratio of the magnitudes of I_1 and I_0. The magnitude of the total field strength is given by

$$\begin{aligned} E_T &= |E_0(1 + ke^{j\psi})| \\ &= |E_0(1 + k\cos\psi + jk\sin\psi)| \\ &= E_0\sqrt{(1 + k\cos\psi)^2 + k^2\sin^2\psi} \end{aligned}$$

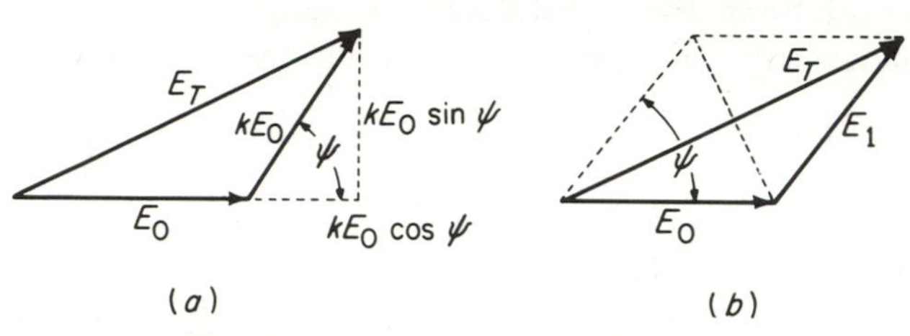

Figure 11-10. Phasor addition of fields.

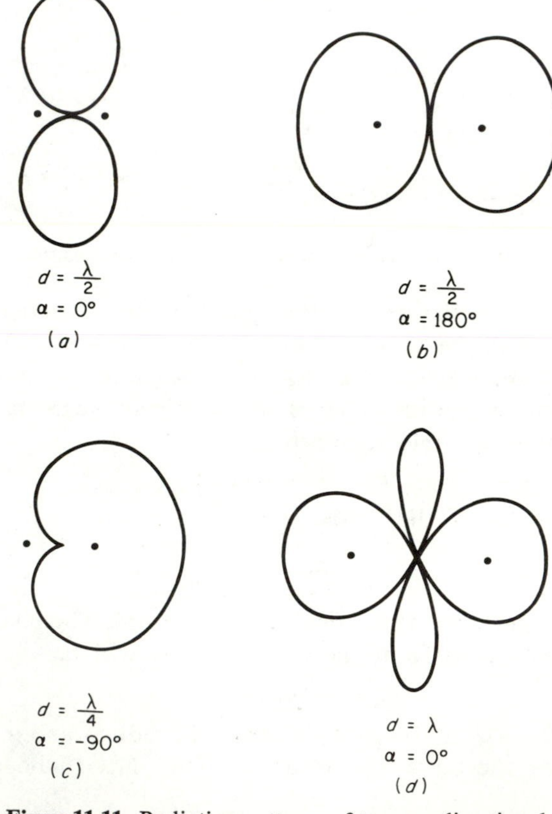

Figure 11-11. Radiation patterns of two nondirectional radiators fed with equal currents at the phasings shown.

In the particular but important case where the antenna currents have equal magnitudes, this becomes [see Fig. 11-10(*b*)]

$$E_T = 2E_0 \cos \frac{\psi}{2}$$

$$= 2E_0 \cos \left(\frac{\pi d \cos \phi}{\lambda} + \frac{\alpha}{2}\right) \tag{11-11}$$

The radiation patterns resulting from the expression for commonly used spacings and phasings are sketched in Fig. 11-11. These patterns are for the case where each of the antennas, when radiating alone, is a nondirectional or *point-source* radiator; that is, it has a circle for its radiation pattern in the plane under consideration. This is quite evidently true for vertical antennas when the horizontal pattern is being considered.

11.06 Horizontal Patterns in Broadcast Arrays. Two-element arrays are limited in the type and variety of radiation patterns that they can produce, and for broadcast arrays three or more antennas are often used. With a two-antenna array the pattern must always be symmetrical about the plane through the antennas, and the position of only two nulls can be specified. A three-element array, in which antenna configurations and spacing as well as current magnitudes and phases are all variables under the control of the designer, permits a larger number of different antenna pattern types. For a three-element array as in Fig. 11-12 the resultant horizontal pattern is given by

$$E_T = |E_0(1 + k_1 e^{j\psi_1} + k_2 e^{j\psi_2})| \tag{11-12}$$

where

$$\psi_1 = (\beta d_1) \cos \phi_1 + \alpha_1$$
$$\psi_2 = (\beta d_2) \cos \phi_2 + \alpha_2$$

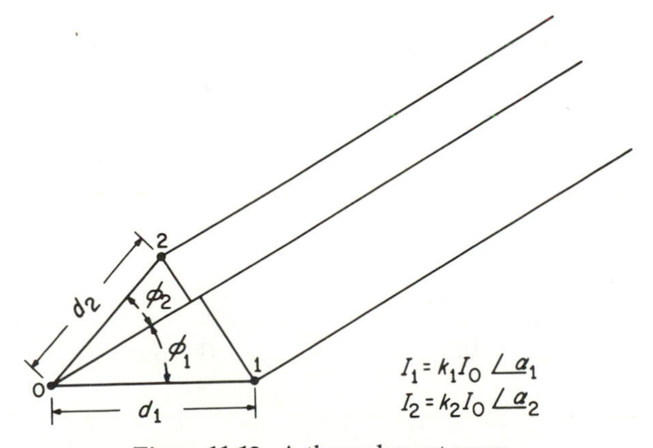

Figure 11-12. A three-element array.

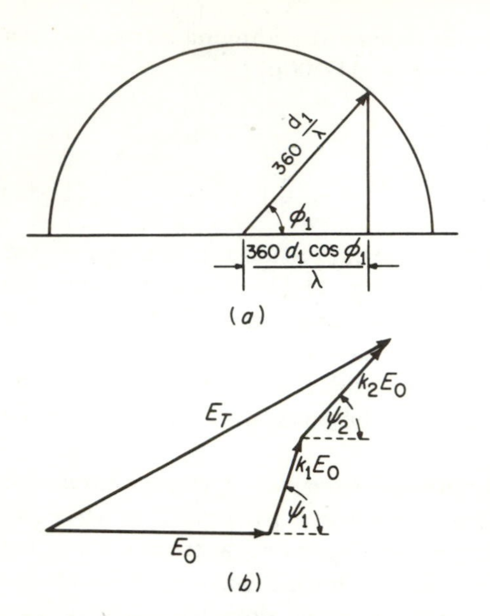

Figure 11-13. Graphical method of obtaining antenna patterns.

The evaluation of expression (12) is straightforward but rather time-consuming when a large number of points must be plotted. A graphical method which is sometimes used for evaluating an expression such as (12) is shown in Fig. 11-13. The antenna spacing is expressed in degrees and a semicircle is drawn with this spacing as radius. For each angle ϕ_1 the value of $\beta d \cos \phi$ is read off directly in degrees, and ψ_1 is obtained by adding the angle α. ψ_2 is obtained in a like manner and the vector addition of Fig. 11-13(*b*) gives the resultant E_T.

The design of an array to produce a desired pattern was once a tedious cut-and-try process, but the use of modern high-speed electronic computers with direct display of the calculated patterns has greatly facilitated this task.

11.07 Linear Arrays. For point-to-point communication at the higher frequencies the desired radiation pattern is a single narrow lobe or beam. To obtain such a characteristic (at least approximately) a multielement linear array is usually used. An array is *linear* when the elements of the array are spaced equally along a straight line (Fig. 11-15). In a *uniform linear array* the elements are fed with currents of equal magnitude and having a uniform progressive phase shift along the line. The pattern of such an array can be obtained as before by adding vectorially the field strengths due to each of the elements. For a uniform array of nondirectional elements the field strength would be

$$E_T = E_0|1 + e^{j\psi} + e^{j2\psi} + e^{j3\psi} + \cdots + e^{j(n-1)\psi}| \tag{11-13}$$

where

$$\psi = \beta d \cos \phi + \alpha$$

and α is the progressive phase shift between elements. (α is the angle by which the current in any element *leads* the current in the preceding element.)

For the purpose of computing the pattern of the linear array, eq. (13) may be viewed as a geometric progression and written in the form

$$\frac{E_T}{E_0} = \left|\frac{1 - e^{jn\psi}}{1 - e^{j\psi}}\right|$$

$$= \left|\frac{\sin \frac{n\psi}{2}}{\sin \frac{\psi}{2}}\right| \tag{11-14}$$

The maximum value of this expression is n and occurs when $\psi = 0$. This is the *principal maximum* of the array.* Since $\psi = \beta d \cos \phi + \alpha$ the principal maximum occurs when

$$\cos \phi = -\frac{\alpha}{\beta d}$$

For a *broadside* array the maximum radiation occurs perpendicular to the line of the array at $\phi = 90$ degrees, so $\alpha = 0$ degrees. For an *end-fire* array the maximum radiation is along the line of the array at $\phi = 0$, so $\alpha = -\beta d$ for this case.

The expression (14) is zero when

$$\frac{n\psi}{2} = \pm k\pi \qquad k = 1, 2, 3, \ldots$$

These are the *nulls* of the pattern. Secondary maxima occur approximately midway between the nulls, when the numerator of expression (14) is a maximum, that is when

$$\frac{n\psi}{2} = \pm(2m + 1)\frac{\pi}{2} \qquad m = 1, 2, 3, \ldots$$

The first secondary maximum occurs when

$$\frac{\psi}{2} = \frac{+3\pi}{2n}$$

(note that $\psi/2 = \pi/2n$ does not give a maximum). The amplitude of the first secondary lobe is

$$\left|\frac{1}{\sin(\psi/2)}\right| = \left|\frac{1}{\sin(3\pi/2n)}\right|$$

$$\approx \frac{2n}{3\pi} \qquad \text{for large } n$$

The amplitude of the principal maximum was n so the amplitude ratio of first secondary maximum to principal maximum is $2/3\pi = 0.212$. This means that the first secondary maximum is about 13.5 db below the principal maximum, and this ratio is *independent* of the number of elements in the uniform array, as long as the number is large. Note

*If the spacing d is equal to or greater than λ, there may be more than one principal maximum.

that for all *odd* values of n, the smallest lobe (at $\psi = \pi$) has an amplitude of unity.

The width of the principal lobe, measured between the first nulls, is twice the angle between the principal maximum and first null. This latter angle is given by

$$\frac{n\psi_1}{2} = \pi \qquad \text{or} \qquad \psi_1 = \frac{2\pi}{n}$$

For a broadside array $\cos\phi = \psi/\beta d$, and the principal maximum occurs at $\phi = \pi/2$. The first null occurs at an angle $[(\pi/2) + \Delta\phi]$ where

$$\cos\left(\frac{\pi}{2} + \Delta\phi\right) = \frac{\psi_1}{\beta d} = \frac{\lambda}{dn}$$

If $\Delta\phi$ is small, it is given approximately by

$$\Delta\phi = \frac{\lambda}{nd}$$

and the width of the principal lobe is

$$2\Delta\phi = \frac{2\lambda}{nd} \tag{11-15a}$$

For a *uniform broadside* array the width of the principal lobe (in radians) is approximately twice the reciprocal of the array length in wavelengths.

For the end-fire array $\psi = \beta d(\cos\phi - 1)$. The principal maximum is at $\phi = 0$, and the first null is at $\phi_1 = \Delta\phi$ where

$$\psi_1 = \beta d(\cos\phi_1 - 1) = -\frac{2\pi}{n}$$

or

$$(\cos\Delta\phi) - 1 = -\frac{\lambda}{nd}$$

For $\Delta\phi$ small, there results approximately

$$\frac{(\Delta\phi)^2}{2} = \frac{\lambda}{nd}$$

$$2\Delta\phi = 2\sqrt{\frac{2\lambda}{nd}} \tag{11-15b}$$

The width of the principal lobe of a *uniform end-fire* array, as given approximately by expression (15b) is greater than that for a uniform broadside array of the same length.

Fig. 11-14(a) shows plots of the expression $|(\sin n\psi/2)/(\sin\psi/2)|$ for several values of n, and Fig. 11-14(b) displays a sample pattern calculation for $n = 3$, $\alpha = \pi/2$, $\beta d = \pi/2$.

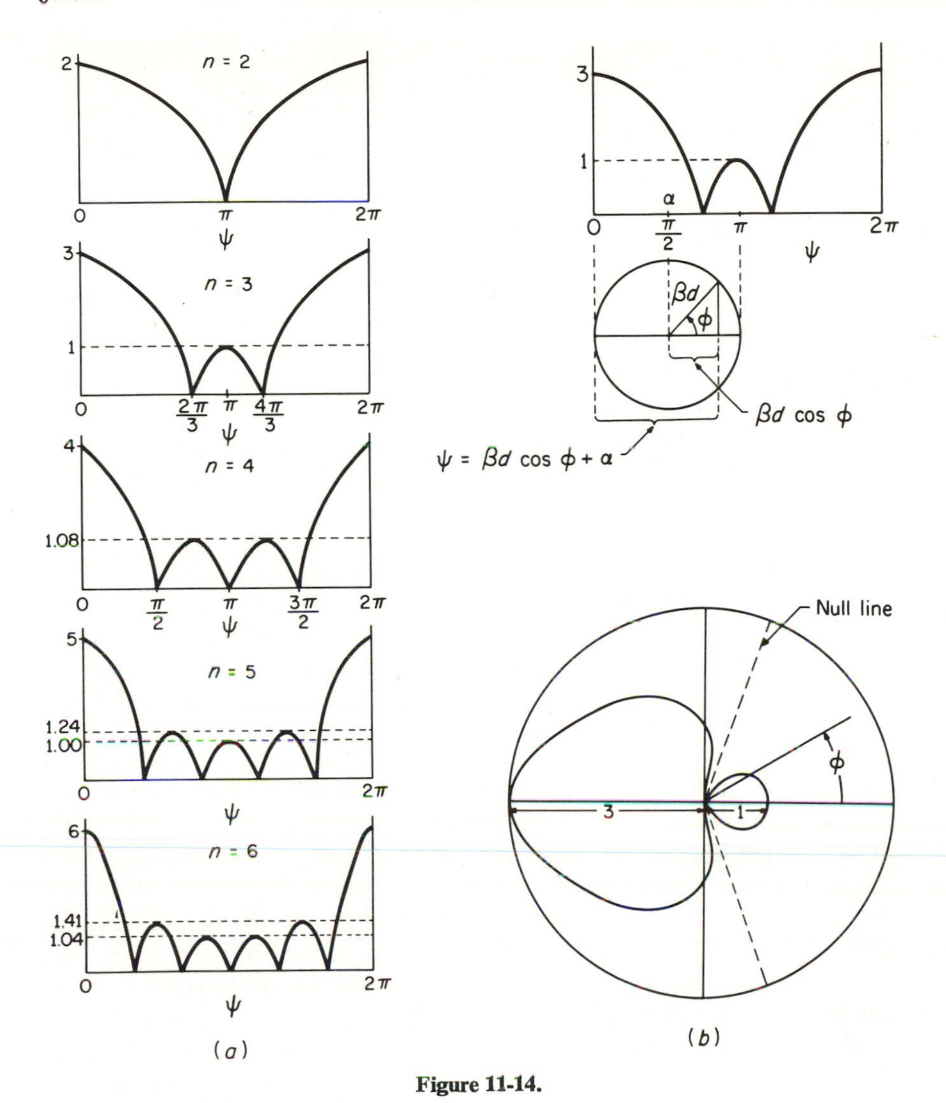

Figure 11-14.

11.08 Multiplication of Patterns. The methods of the preceding section provide straightforward means for determining the radiation patterns of uniform linear arrays. However, for such arrays there is also available another method for obtaining these same patterns. This second method, when it can be used, has the great advantage that it makes it possible to sketch rapidly, almost by inspection, the patterns of complicated arrays. Because of this fact, the method is a useful tool in the design of arrays.

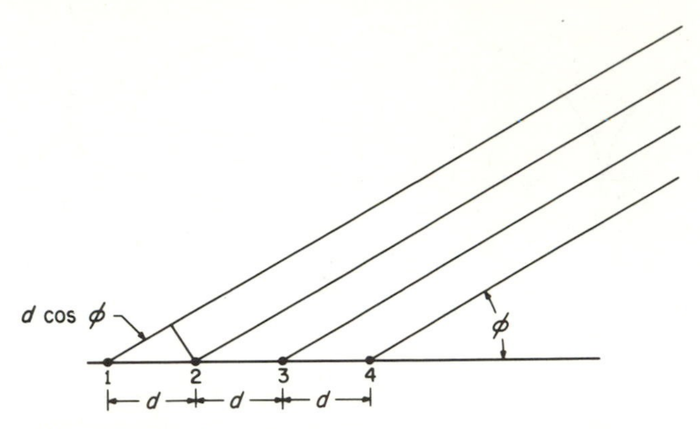

Figure 11-15. A four-element linear array of nondirectional radiators.

Consider a four-element array of antennas in Fig. 11-15, in which the spacing between units is $\lambda/2$ and the currents are in phase ($\alpha = 0$). The pattern can be obtained directly by adding the four electric fields due to the four antennas. However the same radiation pattern can be obtained from the following considerations. The pattern of antennas (1) and (2) operating as a unit, that is, two antennas spaced $\lambda/2$ and fed in phase, has been shown in Fig. 11-11(*a*) on page 360. Also antennas (3) and (4) may be considered as another similar unit with the same pattern of Fig. 11-11(*a*). As far as the resultant radiation pattern is concerned antennas (1) and (2) could be replaced by a single antenna located at a point midway between them and having as its directional characteristic the "figure eight" of Fig. 11-11(*a*). Antennas (3) and (4) could similarly be replaced by a single antenna having the figure-eight pattern. The problem is then reduced to that of determining the radiation pattern of two similar antennas that are spaced a wavelength apart and each of which has a figure-eight directional pattern. Now the pattern of two nondirectional radiators spaced λ and fed in phase is already known and is that of Fig. 11-11(*d*). For the case of Fig. 11-11(*d*) each of the antennas alone radiates equally in all directions in the plane being considered. When these antennas are replaced by radiators that radiate different amounts in different directions, the pattern of Fig. 11-11(*d*) must be modified accordingly. The resultant pattern for the original four-element array is obtained as the *product* of the pattern of Fig. 11-11(*d*) by the pattern of the unit, Fig. 11-11(*a*). This multiplication of patterns is illustrated in Fig. 11-16.

The application of this principle to more complicated arrays follows quite readily. For example the pattern of a broadside array of eight elements spaced one-half wavelength and fed in phase would be obtained

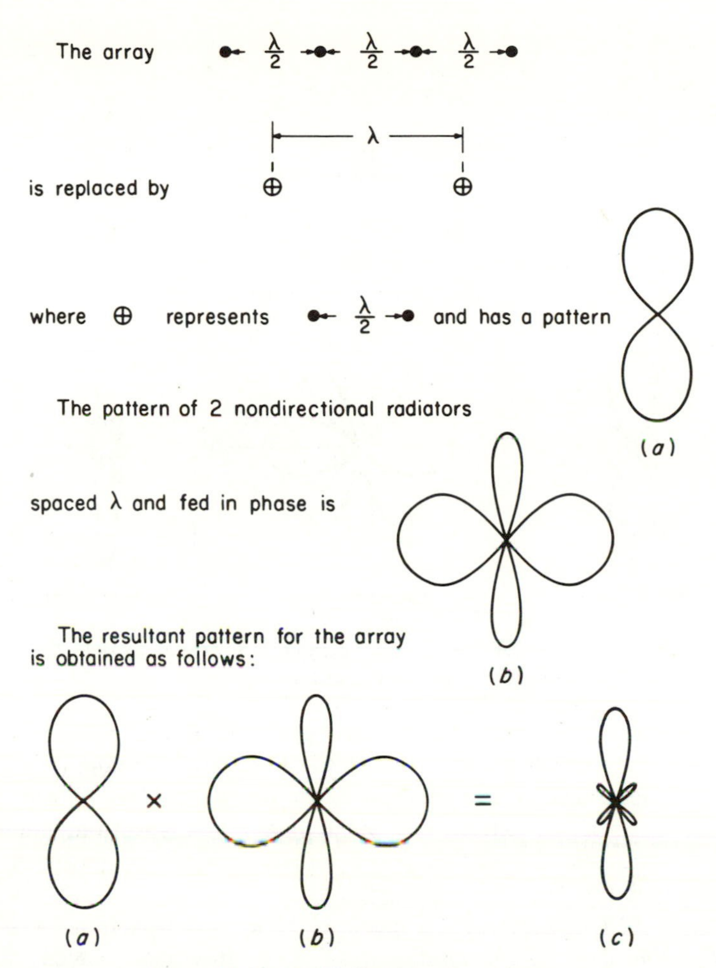

Figure 11-16. Multiplication of patterns.

by considering four elements as a unit and finding the pattern of two such units spaced a distance of two wavelengths. This is shown in Fig. 11-17. The resultant pattern is the product of the unit pattern for four elements (already obtained in Fig. 11-16) by the pattern for two nondirectional radiators spaced two wavelengths apart [calculated from eq. (13)].

This procedure provides a means for rapidly determining what the resultant pattern of a complicated array will look like without making lengthy computations, since the approximate pattern can be arrived at by inspection. The width of the principal lobe (between nulls) is the same as the width of the corresponding lobe of the group pattern. The number of secondary lobes can be determined from the number of nulls

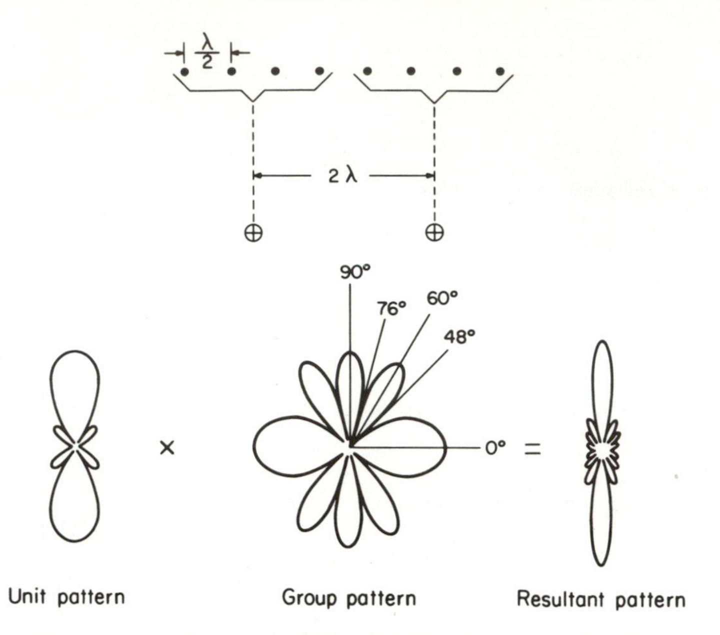

Figure 11-17. Pattern for an eight-element uniform array obtained by principle of multiplication of patterns.

in the resultant pattern, which is just the sum of the nulls in the unit and group patterns (assuming none of the nulls are coincident). Although the chief usefulness of the method is in being able to obtain an approximate idea of the pattern of a large array by inspection, the method itself is exact, and a point-by-point multiplication of patterns yields the exact pattern for the resultant.

Patterns in Other Planes. Figure 11-11(a) is the pattern, *in the plane normal to the axes of the antennas*, of two antennas spaced one-half wavelength apart and fed *in phase*. In this plane* the antennas are nondirectional or uniform radiators. If the pattern in the plane containing the antennas is desired (in which plane the antennas are *directional*), it is necessary to multiply the pattern of Fig. 11-11(a) (that is, the array factor) by the directional pattern of the antenna in the plane being considered. For half-wave dipole antennas this latter pattern will be the "figure-eight" pattern of Fig. 11-6(c). This is shown as the unit pattern in Fig. 11-18.

The resultant pattern is then obtained as a multiplication of the group pattern by the unit pattern [Fig. 11-18(a)]. If the antennas are fed 180 degrees out of phase (end-fire array) the directions of maxima

*In this plane the radiation pattern is the space factor or array factor of the array (see sec. 12.01)

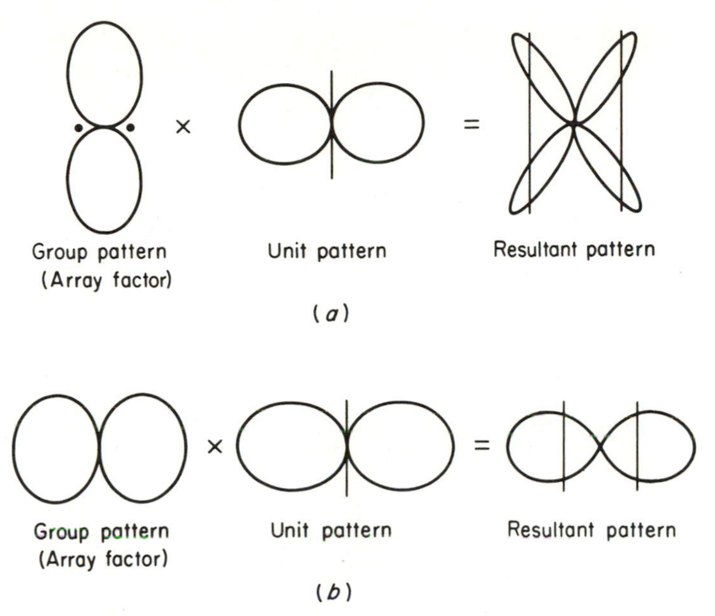

Figure 11-18. Radiation pattern (in the plane containing the axes of the two antennas) of a two-element array of half-wave dipoles: (*a*) fed in phase; (*b*) fed 180 degrees out of phase.

of group and unit patterns coincide and the desirable directional characteristic of Fig. 11-18(*b*) results.

11.09 Effect of the Earth on Vertical Patterns. The radiation patterns shown so far have been obtained on the assumption that the antenna or antenna array was situated in free space far removed from any other conducting bodies or reflecting surfaces. In practice, antennas are nearly always erected either right at, or within a few wavelengths of, the surface of the earth, or some other reflecting surface. Under these conditions currents flow in the reflecting surface, and the radiation pattern is modified accordingly. The magnitudes and phases of these induced currents will of course be dependent to some extent upon the frequency and the σ and ϵ of the reflector surface. However, for practical purposes it often is adequate to compute the resultant fields on the assumption that the surfaces are perfectly conducting. This is true, for example, for the earth at low and medium frequencies, and for metallic reflectors at any radio frequency.

In Fig. 11-19 are shown horizontal and vertical antennas located above the earth (assumed perfectly conducting). The boundary conditions to be satisfied at the surface of the perfectly conducting plane are that the tangential component of **E** and the normal component of **H** must vanish. That is, at the surface **E** is normal and **H** tangential.

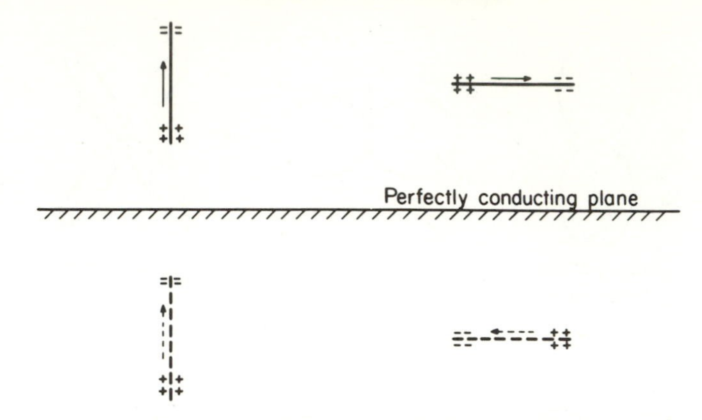

Figure 11-19. Image charges and currents replace the charges and currents induced in the conducting plane.

Charges will distribute themselves and currents will flow on the conducting surface in such a manner that these boundary conditions are satisfied. The total electric and magnetic fields will be due not only to the charges and currents on the antenna, but also to these "induced" charges and currents. As far as the electric and magnetic fields in the region above the conducting plane are concerned, the same results can be obtained with the conducting plane removed and replaced with suitably located "image" charges and currents, as shown in Fig. 11-19. The image charges will be "mirror images" of the actual charges, but will have opposite sign. The currents in actual and image antennas will have the same directions for vertical antennas, but opposite directions for horizontal antennas.

For perfectly conducting planes these same results are also given in terms of simple ray theory* as pictured in Fig. 11-20. The resultant field is considered as made up of direct and reflected waves, the image antenna being the virtual source of the reflected wave. The vertical component of electric field for the incident wave is reflected without phase reversal, whereas the horizontal component has a 180-degree phase reversal. It is seen that the phase delay due to path-length differences (that is, the effect of retardation) is automatically taken care of.

The use of the image principle makes it a simple matter to take into account the effect of the presence of the earth on the radiation patterns. The earth is replaced by an image antenna, located at a distance $2h$ below the actual antenna, where h is the height above the

*Ray theory is easily justified for the receiving case, where the incident field is a uniform plane wave. Application of the reciprocity theorem then demonstrates its validity for the transmitting case.

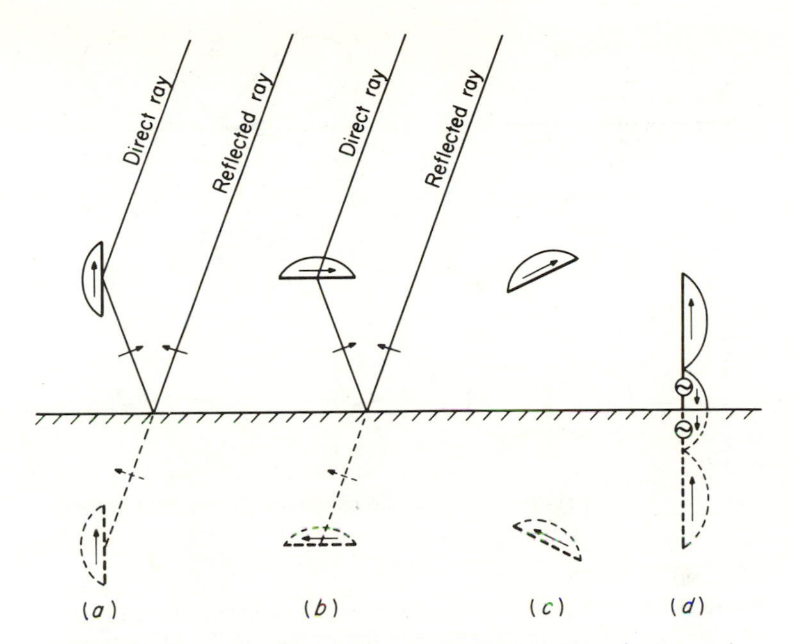

Figure 11-20. Image antennas act as virtual sources for the reflected waves.

ground of the actual antenna. The field of this image antenna is added to that of the actual antenna to yield the resultant field. The shape of the horizontal pattern will remain unchanged (its absolute value changes), but the vertical pattern is affected greatly.

For simple arrays above a reflecting surface the principle of multiplication of patterns can be used to obtain the resultant vertical patterns. The vertical pattern of the antenna (or array) is multiplied by the vertical pattern of two nondirectional or point-source radiators having equal amplitudes and spaced one above the other a distance $2h$ apart. For vertical antennas the nondirectional radiators would be considered to have the same phase, whereas for horizontal antennas the nondirectional radiators would have opposite phases. Examples of this method are shown in Fig. 11-21. Of course, only the upper half of the resultant pattern actually exists. When the antenna is sloping as in Fig. 11-20(*c*), the pattern cannot be obtained by this multiplication process but the image principle can still be used to obtain the resultant field. This same statement also applies for vertical antennas mounted at the surface of the earth when the antenna is not a multiple of one-half wavelength long [Fig. 11-20(*d*)].

When the finite conductivity of the earth must be considered, the idea of images is still valid, but the simple ray theory used here is no

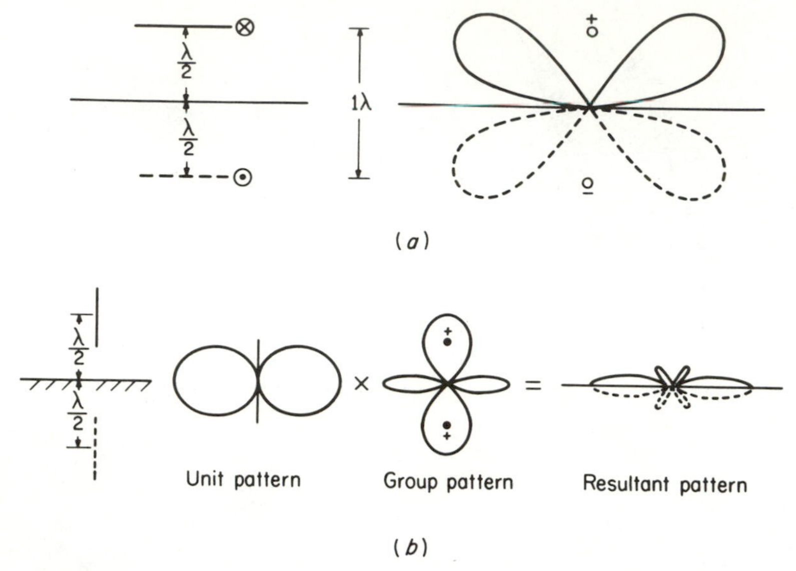

Figure 11-21. (*a*) Vertical pattern of a horizontal antenna above the earth, obtained by considering the pattern of the antenna and its negative image. (*b*) Vertical pattern of a vertical antenna above the earth, obtained by using the principle of images and the principle of multiplication of patterns.

longer adequate. It is then necessary to return to field theory for accurate answers. The effect of an imperfect earth on the radiated fields is considered in chap. 16 on ground-wave propagation. The effect of the presence of the earth on the *impedance* of elevated antennas is discussed in sec. 11.15.

11.10 Binomial Array. An example of the usefulness of the principle of multiplication of patterns is given in the derivation of the so-called binomial array. With a uniform linear array it is found that, as the array length is increased in order to increase the directivity, secondary or minor lobes always appear in the pattern. For some applications a single narrow lobe without minor lobes is desired. A study of the uniform array, using the principle of multiplication of patterns, shows that secondary lobes appear in the resultant pattern whenever the elements that produce the unit pattern or the elements that produce the group or space-factor pattern have a spacing greater than one-half wavelength. Thus in the uniform four-element array of Fig. 11-16, the secondary lobes appear in the resultant because the group pattern has four lobes. The group pattern has four lobes because the effective sources producing the group pattern are spaced a full wavelength. Reduction of the spacing of the elements of the group to one-half wavelength results in a two-lobed figure-eight pattern for the group pattern, and a resultant pattern that has only primary lobes. The antenna arrangement that will result in half-wavelength separation of the elements of the group is shown in Fig. 11-22 along with the resultant patterns. In this case antennas (2) and (3) coincide so they would be replaced with a single antenna carrying double the current in the other elements. In other words, the result is a three-element array that has the current ratios 1 : 2 : 1. The resultant pattern is shown in Fig. 11-22. Since this pattern is the product of two figure-eight patterns, it can be called a "figure-eight squared" pattern.

Using this three-element array as a unit with a second similar unit spaced one-half wavelength from it results in the four-element array shown in Fig. 11-23. The current ratios of this array are

$$1 : 3 : 3 : 1$$

and the pattern is the "figure-eight squared" pattern of the unit times a figure-eight group pattern that results in a "figure-eight cubed" pattern. This process may be continued to obtain a pattern having

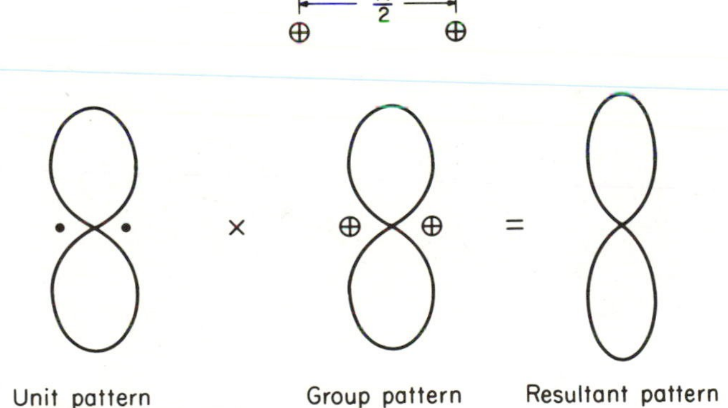

Figure 11-22. An array that produces a pattern without secondary lobes.

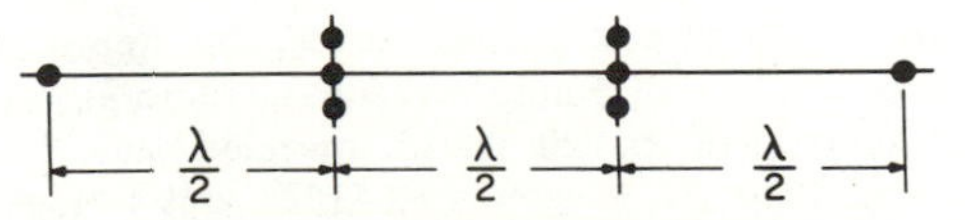

Figure 11-23. A four-element array with a "figure-eight cubed" pattern.

any desired degree of directivity and no secondary lobes.* The numbers that give the current ratios will be recognized as the binomial coefficients. For an array n half-wavelengths long the relative current in the rth element from one end is given by:

$$\frac{n!}{r!(n-r)!} \tag{11-16}$$

where

$$r = 0, 1, 2, 3, \ldots$$

11.11 Antenna Gain. The ability of an antenna or antenna array to concentrate the radiated power in a given direction, or conversely to absorb effectively incident power from that direction, is specified variously in terms of its (antenna) gain, power gain, directive gain or directivity. The precise significance of each of these terms is most readily stated by first defining a quantity known as radiation intensity.

The power radiated per unit area in any direction is given by the Poynting vector **P**. For the distant or radiation field for which **E** and **H** are orthogonal in a plane normal to the radius vector, and for which $E = \eta_v H$, the power flow per unit area is given by†

$$P = \frac{E^2}{\eta_v} \qquad \text{watts/sq m} \tag{11-17}$$

Referring to Fig. 11-24, noting that there are r^2 square meters of

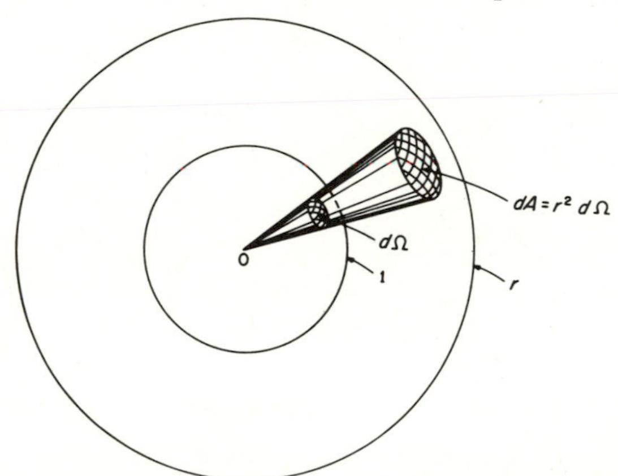

Figure 11-24.

*The binomial array was derived in 1938 using the approach described herein (E. C. Jordan, "Acoustic Models of Radio Antennas," Ohio State Experiment Station Bulletin No. 108). A subsequent patent search revealed that it had been invented by John Stone Stone in 1929. (U. S. Patents 1,643,323 and 1,715,433)

†In this and the following section all field strengths and currents are taken to be effective (or *rms*) values. Also, linear polarization has been assumed.

surface area per unit solid angle (or sterradian), and defining the *radiation intensity* $\Phi(\theta, \phi)$ in a given direction as the power per unit solid angle in that direction, we see that

$$\Phi(\theta, \phi) = r^2 P = \frac{r^2 E^2}{\eta_v} \qquad \text{watts/unit solid angle} \tag{11-18}$$

It should be noted that radiation intensity is independent of r. The total power radiated is

$$W_r = \int \Phi \, d\Omega \qquad \text{watts}$$

and since there are 4π sterradians in the total solid angle, the average power radiated per unit solid angle is

$$\frac{W_r}{4\pi} = \Phi_{av} \qquad \text{watts/sterradian}$$

Φ_{av} represents the radiation intensity that would be produced by an *isotropic* radiator (one that radiates uniformly in all directions) radiating the same total power W_r.

The *directive gain* g_d, *in a given direction*, is defined as the ratio of the radiation intensity in that direction to the average radiated power. That is,

$$g_d(\theta, \phi) = \frac{\Phi(\theta, \phi)}{\Phi_{av}} = \frac{\Phi(\theta, \phi)}{W_r/4\pi} = \frac{4\pi\Phi(\theta, \phi)}{W_r} \tag{11-19a}$$

$$= \frac{4\pi\Phi(\theta, \phi)}{\int \Phi \, d\Omega} \tag{11-19b}$$

When expressed in decibels the directive gain is denoted by G_d where

$$G_d = 10 \log_{10} g_d \tag{11-20}$$

The *directivity*, D, of an antenna is its maximum directive gain. Whereas the directive gain is a function of angles (θ, ϕ) which should be specified, the directivity is a constant, having been specified for a particular direction. However, in common usage g_d (without specification of angle) is used interchangeably with D to designate directive gain in the direction of maximum radiation.

If total input power W_t is used in expression (11-19a) instead of radiated power W_r, the result is power gain rather than directive gain. The *power gain*, g_p, is defined by

$$g_p = \frac{4\pi\Phi}{W_t} \tag{11-21}$$

where $W_t = W_r + W_l$, W_l being the ohmic losses in the antenna. It is evident that

$$\frac{g_p}{g_d} = \frac{W_r}{W_r + W_l} \tag{11-22}$$

is a measure of the efficiency of the antenna. For many well-constructed antennas the efficiency is nearly 100 per cent, so that power gain and directive gain are nearly equal, a fact which has led to the rather loose use of the term *antenna gain* (designated as g without subscript) for either g_d or g_p. For electrically small antennas and for super-directive antennas, power gain may be very much less than directive gain, and in these circumstances careful distinction between the two should be made.

Although the above definitions* have been framed by considering a transmitting antenna, they are applied to the antenna regardless of its particular function. That is, the gain of an antenna when used for receiving is the same as its gain when used for transmitting. Of course, the gain thus defined can be realized on a receiving antenna only when it is properly matched and in the presence of an appropriately polarized field.

The directivity of an antenna is easily computed when its effective length and radiation resistance are known. For example, for a current element $I\,dl$, the distant field in the direction of maximum radiation is

$$E = \frac{60\pi}{r} I \left(\frac{dl}{\lambda}\right) \tag{11-23}$$

From (10-53) the current required to radiate 1 watt is

$$I = \frac{\lambda}{\sqrt{80}\,\pi\, dl} \qquad \text{amp}$$

with a corresponding field strength in the direction of maximum radiation of

$$E = \frac{60}{r\sqrt{80}} \qquad \text{V/m} \tag{11-24}$$

Using (11-18) and (11-24) the radiation intensity is

$$\Phi = \frac{60^2}{80 \times 120\pi} = \frac{3}{8\pi}$$

so that the directivity or maximum directive gain of the current element is

$$g_d(\max) = 4\pi\Phi = 1.5 \tag{11-25a}$$

*From the *IRE Antenna Standards* (1948). See also *IEEE* No. 149, "Test Procedure for Antennas," Jan., 1965; also printed in *IEEE Transactions on Antennas and Propagation*, Vol. AP-13, No. 3, pp. 437–466, May, 1965.

or

$$G_d(\text{max}) = 10 \log_{10} 1.5 = 1.76 \text{ db} \tag{11-25b}$$

For a half-wave dipole the computed directivity is 1.64 or 2.15 db [problem 2(b)]. Thus the maximum directive gain of a half-wave dipole is only 0.39 db greater than for a current element (or for a very short dipole). However, the *power* gain for the short dipole, short monopole, or small loop may be considerably less because of low radiation resistance, and consequent higher ohmic losses (problems 4 and 6).

11.12 Effective Area. A term that has special significance for receiving antennas is effective area (sometimes called effective aperture). The *effective area* or *effective aperture* of an antenna is defined in terms of the directive gain of the antenna through the relation

$$A = \frac{\lambda^2}{4\pi} g_d \tag{11-26}$$

Using this relation it can be shown (problem 3) that the effective area is the ratio of power available at the antenna terminals to the power per unit area of the *appropriately polarized* incident wave. That is

$$W_R = PA \qquad \text{watts} \tag{11-27}$$

where W_R is the received power and P is the power flow per square meter for the incident wave. When directive gain g_d is used in (11-26) it is assumed that all of the available power is delivered to the load. This is the case for a 100 per cent efficient, correctly matched receiving antenna with the proper polarization characteristics. For a lossy antenna, g_p should be used in (11-26) and the effective area so calculated determines through (11-27) the useful power delivered to the load. For electrically small antennas this useful power may be much less than that calculated by using directive gain g_d, in (11-26). (See problems 4 and 6.)

That relation (11-27) holds for the hypothetical current element receiving antenna (assumed lossless) can be demonstrated quite simply. For an effective field strength E parallel to the antenna the power per square meter in the linearly polarized received wave is

$$P = \frac{E^2}{\eta_v} = \frac{E^2}{120\pi} \qquad \text{watts/sq m}$$

The power absorbed in a properly matched load connected to the antenna would be

$$W_R = \frac{V_{oc}^2}{4R_{rad}} = \frac{E^2 l_{eff}^2}{4R_{rad}} \tag{11-28}$$

For the current element $l_{\text{eff}} = dl$ and therefore

$$W_R = \frac{E^2\,dl^2}{4R_{\text{rad}}} \tag{11-29}$$

Inserting the value for R_{rad} from (10-53),

$$W_R = \frac{E^2\lambda^2}{320\pi^2}$$

Using (11-27) the maximum effective area is

$$A = \frac{W_R}{P} = 1.5\,\frac{\lambda^2}{4\pi}$$

which agrees with definition (11-26).

Problem 1. Compute the effective area of a half-wave dipole.

Relations between g_d (or A), l_{eff}, and R_{rad}. Only two of the three quantities, directive gain (or effective area), effective length, and radiation resistance are required to specify the radiation characteristics of an antenna that emits or receives linearly polarized waves. When two of these quantities are known the third can be derived.

Problem 2. (a) Considering an antenna for transmitting, and using (10-49), (11-18) and (11-19a) show that

$$g_d = \frac{120\pi^2}{R_{\text{rad}}}\left(\frac{l_{\text{eff}}}{\lambda}\right)^2 \tag{11-30}$$

(b) Compute g_d for a half-wave dipole. Equation (11-30) relates g_d, l_{eff} and R_{rad}.

Problem 3. Considering the antenna for receiving and using (11-30) and (11-28) show that power received in the presence of an appropriately polarized received wave for any antenna is

$$W_R = \left(\frac{g_d\lambda^2}{4\pi}\right)P = PA$$

Application to Aperture Antennas. The term effective area (or effective aperture) has particular significance when applied to electromagnetic horns and reflector antennas which have well-defined physical apertures. For these antennas the ratio of the effective aperture to the actual aperture is a direct measure of the antenna's effectiveness in radiating or receiving the power to or from the desired direction. Normal values of this ratio for reflector antennas range between 45 per cent and 75 per cent, depending upon antenna type and design, with 65 per cent being considered rather good for the extensively used parabolic reflector antenna. The theory of aperture antennas is treated in chap. 13.

Measured Gain. The directive gain of an antenna can be obtained by calculation from its known or measured radiation pattern. Although pattern measurements normally give only *relative* values of transmitted or received field strengths, these values can be used to calculate g_d through expression (11-19). Detailed accounts of such calculations are given in various books and antenna handbooks.* Power gain normally requires measurement. A common method is to compare the gain with that of a half-wave dipole or other standard antenna. Methods of power gain measurement are treated in the *IEEE Standards on Methods of Antenna Measurement*.†

Problem 4. A 10 meter high monopole is to be used as a portable transmitting antenna at 1.5 MHz. Its measured base reactance is $-j350$ ohms. Assuming the ohmic losses in the ground system and tuning coil (which has a $Q = 100$) are approximately equal, estimate antenna efficiency and power gain g_p in direction of maximum radiation. Determine its effective area.

Problem 5. Calculate approximately the radiation resistance of a 1 meter square loop antenna at a frequency of 3 MHz ($\lambda = 100$ m).

The radiation resistance can be calculated accurately from eq. (10-101), but this is a rather involved computation. Alternatively the radiation resistance can be determined to a good approximation from eq. (11-30) and the following considerations.

An electrically small current loop is the equivalent of a magnetic dipole (chap. 13) and has the same radiation pattern as an electric dipole. Therefore it has the same maximum directive gain g_d value of 1.5. The effective length l_{eff} of the loop may be determined from (10-49) and the knowledge that the field strength E in the direction of maximum radiation will be the phasor sum of two current elements phased 180 degrees and with a spacing L/λ wavelengths. Then using (11-30), R_{rad} can be determined.

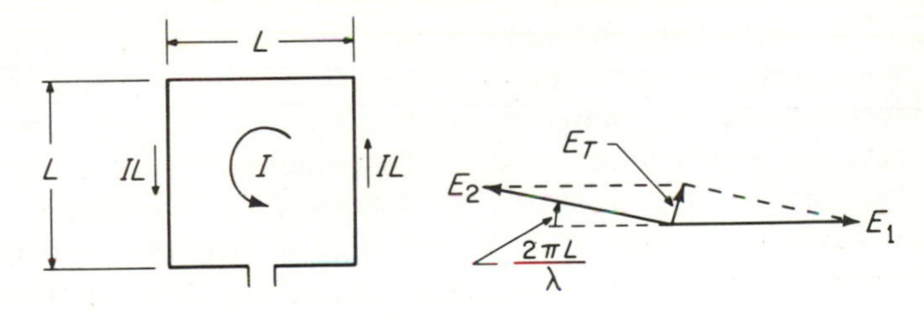

Figure 11-25.

Problem 6. (a) Determine the effective area of the loop of problem 5 at a frequency of 3 MHz.

(b) If the received field strength is 100 millivolts per meter how much power is "received" by the loop?

*See, for example, H. Jasik, *Antenna Handbook*, McGraw-Hill Book Company, New York, 1961.

†"IEEE Test Procedure for Antennas," *IEEE Trans. Antennas and Propagation*, Vol. AP-13, No. 3, pp. 437–466, May, 1965.

(c) Assuming an ohmic resistance of 1 ohm for the loop and its tuning capacitor, how much of the "received" power is actually absorbed in a matched load?

(d) If effective area had been defined in terms of power gain, so that $A_p = g_p\lambda^2/4\pi$, what would be the size of the "actual" effective area A_p?

11.13 Antenna Terminal Impedance. To the communication engineer interested in the overall design of a radio communication system, the antenna is but one link in the complicated chain that leads from the microphone to the loudspeaker. It is natural for him to consider the antenna simply as another circuit element that must be properly matched to the rest of the network for efficient power transfer. From this point of view the input or terminal impedance of the antenna is of primary concern. The input impedance of an antenna is a complicated function of frequency, which cannot be described in any simple analytical form. Nevertheless, at a single frequency, the antenna terminal impedance may be accurately represented by a resistance in series with a reactance. Over a small band of frequencies such representation can still be used, but it is now only approximate. If, as is often the case, the band of frequencies is centered about the "resonant frequency" of the antenna, a better approximation is obtained by representing the antenna as a series R, L, C circuit. If the range of operation extends over a wider band of frequencies, this representation is no longer adequate. It can be improved by adding elements to the "equivalent" network, but the number of elements required for reasonably good representation becomes very large as the frequency range is extended. A better approximation applicable to linear antennas involves replacing the equivalent lumped-constant network with a distributed-constant network, such as an open-circuited transmission line, the input impedance of which will represent reasonably well the input impedance of the antenna over a wide range of frequencies.

Lumped-constant Representation of Antenna Input Impedance. For an antenna whose half-length H is shorter than a quarter-wavelength, the input impedance can be represented over a narrow band of frequencies by a resistance R in series with a capacitive reactance X. The resistance R is the radiation resistance (and loss resistance, if any) of the antenna, referred to the terminal or base current. It is given in Fig. 14-7 (sec. 14.06), or it can be obtained from Fig. 14-5 by dividing the values of radiation resistance referred to loop current by $\sin^2 \beta H$. Approximate values of the capacitive reactance X are also given in Fig. 14-7. For antennas which have a half-length greater than about a quarter-wavelength (actually nearer 0.23λ, the exact point depending on the thickness of the antenna), the input impedance becomes inductive and would be represented by a resistance in series with an inductive reactance. For both transmitting and receiving, an antenna is often operated at its resonant frequency, that is, at the center frequency of the narrow band of operation where the antenna input impedance is a pure resistance. Below this center frequency the antenna reactance is capacitive, and above this frequency the reactance is inductive. The input impedance can then be represented approximately by a series R, L, C circuit. Quantities of interest are the required values of the equivalent R_a, L_a, and C_a, and the Q of the antenna.

Figure 11-26 illustrates this representation as a simple R, L, C circuit and shows the variation of impedance and admittance in the vicinity of resonance ($X = 0$) for such a circuit.

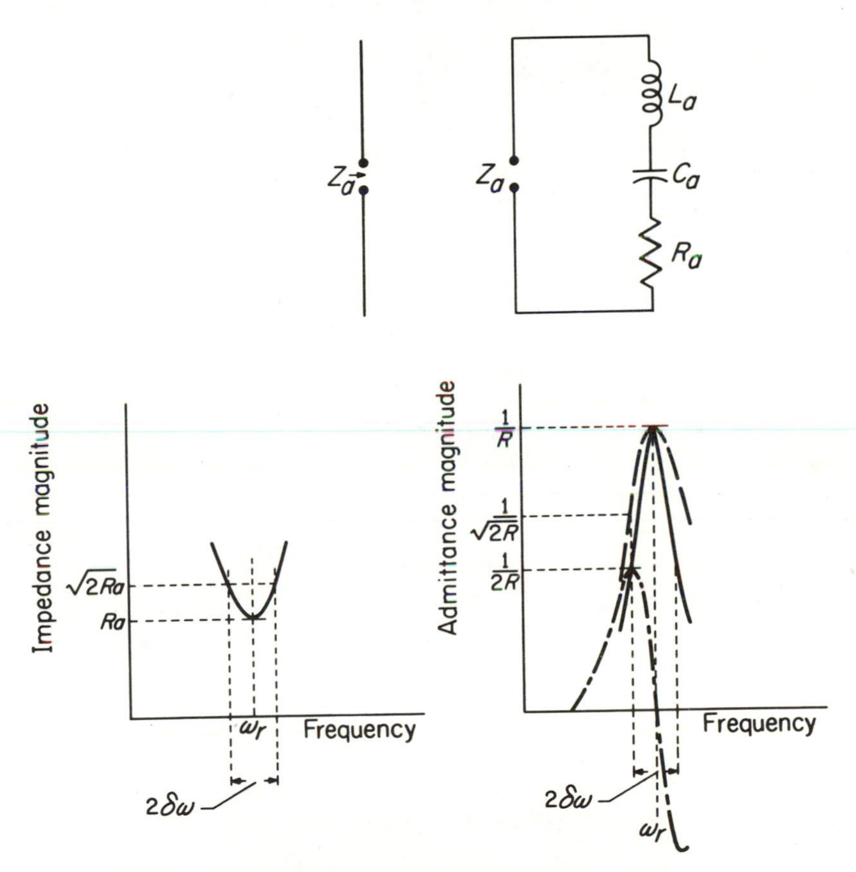

Figure 11-26. Approximate representation of antenna input impedance by simple R, L, C circuit with corresponding impedance and admittance curves.

The general expression for the impedance is

$$Z_a = R_a + j\left(\omega L_a - \frac{1}{\omega C_a}\right) \tag{11-31}$$

At the resonant frequency $f = f_r$,

$$\omega_r L_a = \frac{1}{\omega_r C_a} \qquad Z_a|_{\omega=\omega_r} = Z_r = R_a \tag{11-32}$$

For a small angular frequency increment, $\delta\omega$, from the resonance frequency, the impedance increment is

$$\delta Z_a = j\left(\delta\omega L_a + \frac{\delta\omega}{\omega^2 C_a}\right) \tag{11-33}$$

Therefore the per unit increase in impedance is

$$\begin{aligned}\frac{\delta Z_a}{Z_r} = \frac{\delta Z_a}{R_a} &= j\left(\frac{\delta\omega L_a}{R_a} + \frac{\delta\omega}{R_a\omega^2 C_a}\right)\\ &= j\left(\frac{Q}{\omega_r}\,\delta\omega + \frac{Q}{\omega_r}\,\delta\omega\right)\\ &= j2Q\,\frac{\delta\omega}{\omega_r}\end{aligned} \tag{11-34}$$

where

$$Q = \frac{\omega_r L_a}{R_a} = \frac{1}{\omega_r C_a R_a} \tag{11-35}$$

From eq. (34)

$$\left|\frac{\delta Z_a}{R_a}\right| = 2Q\,\frac{\delta\omega}{\omega_r}$$

or

$$\frac{\delta\omega}{\omega_r} = \frac{1}{2Q}\left|\frac{\delta Z_a}{R_a}\right| \tag{11-36}$$

When $\delta Z_a = R_a$, the current has dropped to $1/\sqrt{2}$ times its value at resonance and the power has dropped to one-half. The angular frequency difference between half-power points is

$$\Delta\omega = 2\delta\omega = \frac{\omega_r}{Q} \tag{11-37}$$

The frequency difference between half-power points is the band width of the circuit, and the *relative band width* is

$$\frac{\Delta\omega}{\omega_r} = \frac{1}{Q} \tag{11-38}$$

To the extent that the antenna impedance may be represented by the simple circuit of Fig. 11-26, this can be considered to be the band width of the antenna (unloaded). A more general definition for antenna band width is given in chap. 14.

When precise impedance or admittance measurements are made on an actual antenna in the neighborhood of the resonant frequency, it is found that the curves have a somewhat different shape from those of the simple R, L, C circuit shown in Fig. 11-26. This difference is due to the fact that the equivalent R_a, L_a, and C_a of the antenna are really functions of frequency, and not constants as in Fig. 11-26. A much better representation of the antenna impedance may be obtained with the series R, L, C circuit by assuming L_a and C_a to be constant as before, but taking into account the variation of R_a with frequency. For this purpose, the variation of R_a with frequency can be assumed to be linear over the frequency range of interest, and so the resistance may be written as

$$R_a = R_r\left(1 + \rho\frac{\delta\omega}{\omega_r}\right) \tag{11-39}$$

where R_r is the resistance at resonance and ρ is a positive constant. The expression for impedance for this case, illustrated in Fig. 11-27, is (for $\omega = \omega_r + \delta\omega$)

$$\begin{aligned} Z_a &= R_a + j\left(\omega L_a - \frac{1}{\omega C_a}\right) \\ &= R_r\left(1 + \rho\frac{\delta\omega}{\omega_r}\right) + j\left(\omega_r L_a + \delta\omega L_a - \frac{1}{\omega_r C_a} + \frac{\delta\omega}{\omega_r^2 C_a}\right) \end{aligned} \tag{11-40}$$

Then, remembering that $\omega_r L_a = 1/\omega_r C_a$,

$$\frac{Z_a}{R_r} = 1 + \frac{\delta\omega}{\omega_r}\left(\rho + j2\frac{\omega_r L_a}{R_r}\right) \tag{11-41}$$

$$= 1 + \frac{\delta\omega}{\omega_r}(\rho + j2Q) \tag{11-42}$$

The impedance and admittance curves for this case are shown in Fig. 11-27. It is seen that the impedance minimum or admittance maximum no longer occurs at resonance, but rather at some frequency below resonance. The frequency at which the impedance is a minimum can be found by minimizing the absolute value of expression (41) (or its square) with respect to $\delta\omega$.

Putting

$$\frac{d}{d(\delta\omega)}\left|\frac{Z_a}{R_r}\right|^2 = 0$$

gives

$$\frac{\delta\omega}{\omega_r} = \frac{-\rho}{\rho^2 + 4Q^2} \qquad \text{(for the minimum)}$$

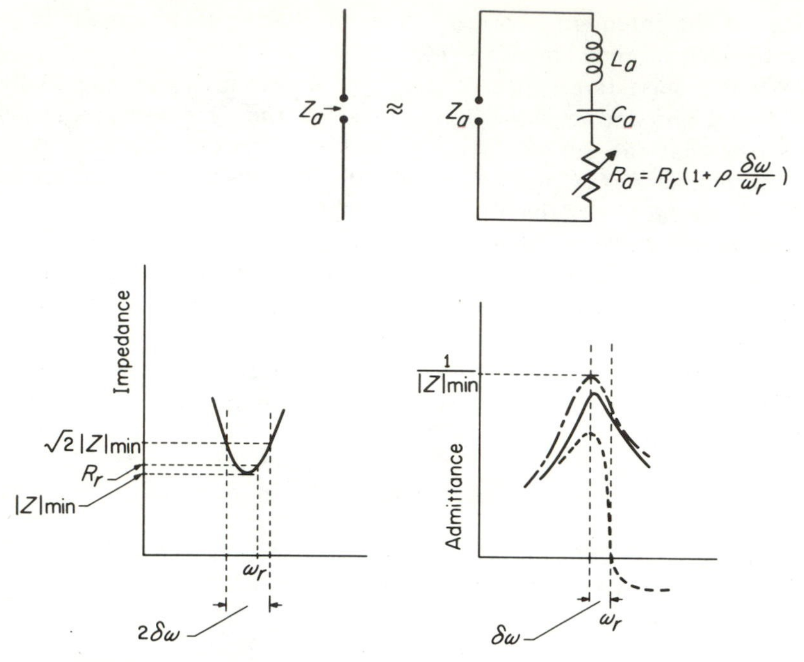

Figure 11-27. Impedance and admittance curves about resonance when the variation of radiation resistance with frequency is considered.

The impedance at the minimum will be given by

$$Z_{\min} = R_r\left[1 - \frac{\rho}{\rho^2 + 4Q^2}(\rho + j2Q)\right]$$

$$= R_r\left[1 - \frac{\rho^2}{\rho^2 + 4Q^2} - j\frac{2\rho Q}{\rho^2 + 4Q^2}\right] \qquad (11\text{-}43)$$

In order to make use of the equivalent circuit for antenna input impedance in computations, it is necessary to know, or to be able to obtain, values of the equivalent L_a, C_a, and R_a in terms of antenna dimensions. When curves such as those of Fig. 14-7 are available in the range of interest, values of these quantities may be determined from the curves. Otherwise, L_a, C_a, and Q may be calculated in terms of a quantity called the average characteristic impedance of the antenna.

Characteristic Impedance of Antennas. A quantity that has considerable usefulness in connection with antennas is the *average characteristic impedance* of the antenna. The significance of this term as applied to cylindrical antennas can be understood by first considering a biconical transmission line or a biconical antenna. The characteristic impedance of a transmission line is defined as the voltage-current ratio existing on the line when the line is infinitely long. For a uniform transmission line, this ratio is constant along the line. It is easily shown that the transmission line, formed by two infinitely long coaxial conical conductors having a common apex (Fig. 11-28), is a uniform line, and that the ratio of voltage to current along the line will remain constant, that is, independent of r. (The voltage V is applied across an infinitesimal gap at the apex and the current I flows out of one cone and into the other.) In secs. 14.13 and 14.14 the general solution for a finite length of such a transmission line (or antenna) will be obtained. It will be found that such a structure can support the TEM wave, as well as higher-order TM waves. For the outgoing TEM wave (which alone is excited on the infinitely long line) the expressions for the fields are

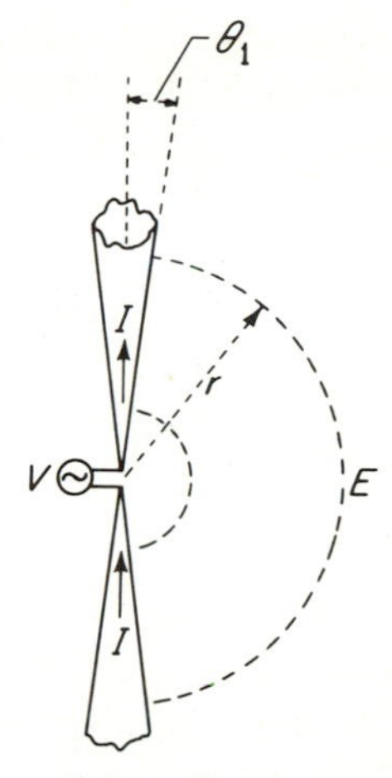

Figure 11-28. Input to a conical transmission line (or a biconical antenna).

$$\left.\begin{aligned} H_\phi &= \frac{A}{r\sin\theta}e^{-j\beta r} \\ E_\theta &= \frac{A\eta_v}{r\sin\theta}e^{-j\beta r} \\ E_r = E_\phi &= H_\theta = H_r = 0 \end{aligned}\right\} \tag{11-44}$$

Maxwell's equations in spherical co-ordinates for the case of no variation in the ϕ direction are

$$\left.\begin{aligned} \frac{1}{r}\frac{\partial(rE_\theta)}{\partial r} - \frac{1}{r}\frac{\partial E_r}{\partial\theta} &= -j\omega\mu H_\phi \\ \frac{1}{r\sin\theta}\frac{\partial}{\partial\theta}(\sin\theta H_\phi) &= j\omega\epsilon E_r \\ -\frac{1}{r}\frac{\partial(rH_\phi)}{\partial r} &= j\omega\epsilon E_\theta \end{aligned}\right\} \tag{11-45}$$

Direct substitution of (44) into (45) shows that Maxwell's equations are satisfied. In addition, because $E_r = 0$, the boundary conditions are automatically satisfied.

It will be noted that the electric field distribution is just that corresponding to the static case (problem 8, chap. 2) and that the magnitude of the voltage between the cones at any distance r from the apex is constant. That is

$$\begin{aligned} V &= \int_{\theta_1}^{\pi-\theta_1} E_\theta r\,d\theta = \eta_v A e^{-j\beta r}\int_{\theta_1}^{\pi-\theta_1}\frac{1}{\sin\theta}\,d\theta \\ &= 2\eta_v A \ln\cot\frac{\theta_1}{2}e^{-j\beta r} \end{aligned} \tag{11-46}$$

Also the amplitude of the current flowing in the cones is constant along the line and is given by

$$I = 2\pi r \sin\theta_1 H_\phi$$
$$= 2\pi A e^{-j\beta r} \tag{11-47}$$

Therefore, the characteristic impedance for a biconical transmission line or biconical antenna is constant and is

$$Z_0 = \frac{V}{I} = \frac{\eta_v}{\pi} \ln\cot\frac{\theta_1}{2}$$
$$= 120 \ln\cot\frac{\theta_1}{2} \tag{11-48}$$

Because the electric and magnetic field configurations are the same as for the stationary-fields case, this same result could have been obtained directly by using the *static* capacitance per unit length as calculated in chap. 2. Thus

$$Z_0 = \sqrt{\frac{L}{C}} = \frac{1}{cC}$$

where

$$C = \frac{\pi\epsilon}{\ln\cot\frac{\theta_1}{2}}$$

and

$$\frac{1}{\sqrt{LC}} = c = \frac{1}{\sqrt{\mu_v\epsilon_v}} = 3\times 10^8 \qquad \text{meter/sec}$$

Then

$$Z_0 = \frac{\eta_v}{\pi}\ln\cot\frac{\theta_1}{2} = 120\ln\cot\frac{\theta_1}{2}$$

For *thin* antennas, that is when θ_1 is small, the characteristic impedance is given approximately by

$$Z_0 = 120\ln\left(\frac{2}{\theta_1}\right) = 120\ln\left(\frac{2r}{a}\right) \tag{11-49}$$

where a is the cone radius at a distance r from the apex.

When a cylindrical antenna is treated in a like manner, it is evident that the corresponding "bi-cylindrical" transmission line will be non-uniform, with a capacitance per unit length and characteristic impedance that vary along the line. However, for *thin* antennas the elements dr can be considered as elements of a *biconical* line which has a cone angle $\theta_1 = a/r$, where a is the radius of the cylinder and r is the distance from the origin to the element dr (Fig. 11-29). Then the characteristic impedance at a distance r will be

$$Z_0(r) = 120 \ln \left(\frac{2}{\theta_1}\right) = 120 \ln \left(\frac{2r}{a}\right) \qquad (11\text{-}50)$$

It is seen that the characteristic impedance of a cylindrical antenna varies along the antenna, being larger near the ends. For a center-fed cylindrical antenna of half-length H, an "average" characteristic impedance can be defined by

$$Z_0(\text{av}) = \frac{1}{H}\int_0^H Z_0(r)\,dr \qquad (11\text{-}51)$$

$$= \frac{1}{H}\int_0^H 120 \ln \left(\frac{2r}{a}\right) dr$$

$$= 120\left[\ln \left(\frac{2H}{a}\right) - 1\right] \qquad (11\text{-}52)$$

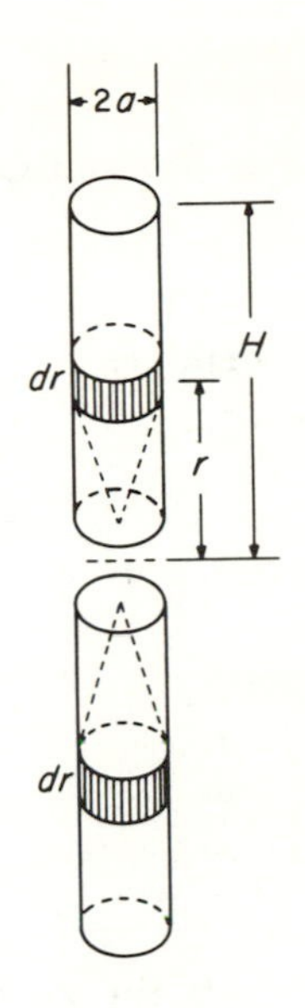

Figure 11-29.

Equivalent L_a, C_a, and Q in terms of Z_0 (av). Use of the average characteristic impedance of an antenna makes it possible to obtain values for the equivalent R, L, C circuit of the antenna in terms of the antenna dimensions. This is done by first finding a Q for the antenna by comparison with ordinary transmission-line theory, and then determining L_a and C_a in terms of this Q and the known R_a.

From transmission-line theory for low-loss lines, the Q of a transmission line is

$$Q = \frac{\omega L}{R} = \frac{\omega Z_0}{Rv} = \frac{2\pi Z_0}{\lambda R} \qquad (11\text{-}53)$$

where R, L, and C are the resistance, inductance, and capacitance per unit length of line and

$$Z_0 \approx \sqrt{\frac{L}{C}} \qquad v \approx \frac{1}{\sqrt{LC}}$$

for low-loss lines. For a resonant length ($l = \lambda/4$) of open-circuited line, it is easily shown that the input impedance is a pure resistance of value

$$R_{\text{in}} = \frac{Rl}{2} = \frac{R\lambda}{8} \qquad (11\text{-}54)$$

and this must be equal to R_a in the equivalent lumped-circuit representation of antenna input resistance (Fig. 11-26). That is,

$$\frac{R\lambda}{8} = R_a = R_{\text{rad}}$$

For the equivalent lumped-constant circuit

$$Q = \frac{\omega_r L_a}{R_a}$$

Therefore

$$\omega_r L_a = R_a Q = \frac{R\lambda}{8} \cdot \frac{2\pi Z_0}{\lambda R} = \frac{\pi Z_0}{4}$$

$$\omega_r C_a = \frac{4}{\pi Z_0}$$

and

$$\left.\begin{aligned} L_a &= \frac{Z_0}{8f_r} \qquad & C_a &= \frac{2}{\pi^2 f_r Z_0} \\ Q &= \frac{\pi Z_0}{4R_a} \qquad & R_a &= R_{\text{rad}} \end{aligned}\right\} \tag{11-55}$$

where the Z_0 in the case of a cylindrical antenna will be the average characteristic impedance Z_0(av) defined in the preceding section. The Q of the antenna as given by eq. (55) will be the *unloaded* Q. When the antenna circuit is loaded by a properly matched generator impedance or load impedance, the total Q of the circuit is one-half the Q of the antenna above. That is

$$Q_{\text{loaded}} = \tfrac{1}{2} Q_{\text{unloaded}}$$

It is this Q_{loaded} which is of chief concern in band-width considerations.

Problem 7. (a) Using the curves of Fig. 14-7, determine an approximate value for the Q of a half-wave dipole antenna constructed with No. 12 copper wire, at a frequency of 100 MHz; (b) compute Q for the same antenna using eq. (11-55).

Problem 8. From the point of view of band width, discuss the suitability of each of the following antennas for (a) an FM receiving antenna, (b) a television receiving antenna: a half-wave dipole constructed of (1) No. 12 wire, (2) 1-cm diameter rods, (3) 1-in. diameter pipes, (4) a biconical cage arrangement with a total cone angle of 10 degrees.

NOTE: The FM band covers from 88–108 MHz. The low-frequency television channels are 6 MHz wide and, at present, are 54–60, 60–66, 66–72, 76–82, 82–88. In both cases it is desired, if practical, that a single antenna should cover the entire band with a decrease of received power of less than 3 db.

11.14 The Antenna as an Opened-out Transmission Line. Using an assumed sinusoidal current distribution, the power radiated from an antenna can be calculated, and approximate values for input resistance and reactance can be obtained for very thin antennas. However, for thicker antennas, such as tower antennas for broadcast use, especially when fed near a current node, the sinusoidal-current assumption fails to give sufficiently good answers. Faced with the necessity of having to feed such antennas, and in the absence of a rigorous solution to the antenna problem, it was natural that engineers should look for a better assumption than the sinusoidal for the current distribution. Because the input impedance of an antenna goes through variations somewhat similar to the input impedance of an open-circuited transmission line, it was also natural to attempt to treat the antenna as an opened-out transmission line. This attack had the advantage of using transmission-line theory with which the engineer was already familiar, and it provided him with a simple expression for current distribution and input impedance which, although only approximate, could always be "adjusted" in the light of measured values. Although in more recent years the cylindrical antenna problem has been solved to a fairly good approximation and accurate values for input impedance are now available in the form of curves or tables, it is often still very convenient to have available simple analytic expressions, which will indicate the correct order of magnitude of input impedance and current distribution. For this reason, and because it is instructive to compare the antenna with the transmission line, one of the methods developed for treating the antenna as a transmission line will be outlined.

The most extensive study of the transmission-line representation of an antenna has been made in a series of papers* by Siegel and Labus, and their results have been used in the broadcast antenna field for many years. The method treats the antenna as an opened-out transmission line that is open-circuited at the end. An antenna differs from a transmission line in two important respects. An antenna radiates power, whereas transmission-line theory assumes negligible radiation of power. Ordinary transmission-line theory deals with *uniform* lines for which L, C, and Z_0 are constant along the line (except very close to the end). For the nonuniform line representing the antenna, L, C, and Z_0 all vary along the line, and indeed it becomes necessary to define what is meant by these quantities under such conditions. Siegel and Labus assume that the radiated power can be accounted for by introducing an equal amount of ohmic loss distributed along the transmission line. Knowing this loss, an attenuation factor can be calculated,

*J. Labus, "Mathematical Calculation of the Impedance of Antennas," *Hochf. und Elek.*, **41,** 17 (1933); E. Siegel and J. Labus, "Apparent Resistance of Antennas," *Hochf. und Elek.*, **43,** 166 (1934); E. Siegel and J. Labus, "Transmitting Antennas," *Hochf. und Elek.*, **49,** 87 (1937).

and this factor can be used to give a better approximation for the current distribution. In addition the variable characteristic impedance of the antenna is replaced by an average value. Because the value used for this average characteristic impedance determines very largely what the input impedance will be, considerable effort was expended in obtaining a truly significant expression for Z_0. The essentials of the Siegel and Labus method (with slight modifications) are outlined below.

Input Impedance by Transmission-line Analogy. Figure 11-30 illustrates the representation of a center-fed dipole antenna as an opened-out transmission line. The "equivalent" transmission line has a length equal to the half-length H of the dipole antenna. The diameter of the antenna or transmission-line conductors is $2a$. The problem of a ground-based antenna of height H, erected on a perfect reflecting plane [Fig. 11-30(*d*)], is the same as that of the dipole antenna of half-length H [Fig. 11-30(*c*)], except that values of characteristic and input impedances will be just one-half those obtained for the corresponding dipole antenna.

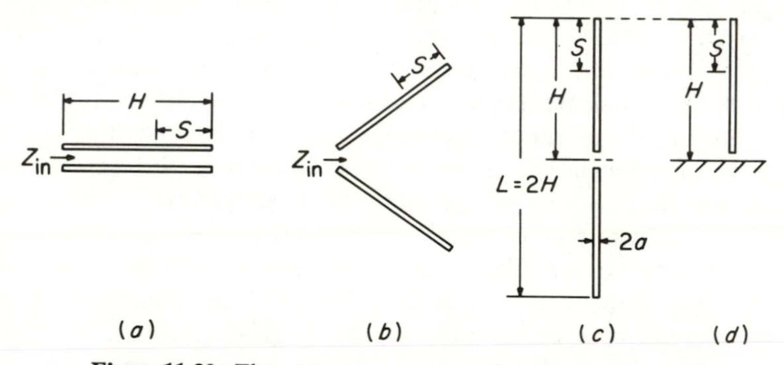

Figure 11-30. The antenna as an opened-out transmission line.

The first step is to obtain an expression for Z_c (av), the average characteristic impedance of the antenna. In sec. 11.13 a simple expression for average characteristic impedance was developed from elementary considerations. Siegel and Labus have developed a somewhat different expression in quite another way. The scalar potential for an antenna carrying a sinusoidal current distribution was set up and compared with the usual expression for the scalar potential along a uniform parallel-wire line. Comparison of these expressions yielded an expression for the characteristic impedance $Z_0(s)$ at each point s along the antenna. From this, an average characteristic impedance for the total length was defined by

$$Z_c\,(\text{av}) = \frac{1}{H}\int_0^H Z_0(s)\,ds \tag{11-56}$$

The final expressions obtained for Z_c (av) were

(a) For the center-fed dipole antenna of half-length H [Fig. 11-30(*c*)]

$$Z_c\,(\text{av}) = 120\left(\ln\frac{H}{a} - 1 - \frac{1}{2}\ln\frac{2H}{\lambda}\right) \qquad \text{ohm} \tag{11-57}$$

(b) For the antenna of height H, fed against a perfect reflecting plane [Fig. 11-30(*d*)],

$$Z_c\,(\text{av}) = 60\left(\ln\frac{H}{a} - 1 - \frac{1}{2}\ln\frac{2H}{\lambda}\right) \qquad \text{ohm} \quad (11\text{-}58)$$

The notation $Z_c\,(\text{av})$ has been used to distinguish this average characteristic impedance from the $Z_0\,(\text{av})$ given by eq. (52). For the special case of $H = \lambda/4$ (half-wave dipole or quarter-wave ground-based antenna), the expressions for $Z_c\,(\text{av})$ become

(c) Half-wave dipole,

$$Z_c\,(\text{av}) = 120\left(\ln\frac{H}{a} - 0.65\right) \qquad \text{ohm} \quad (11\text{-}59)$$

(d) Quarter-wave ground-based antenna,

$$Z_c\,(\text{av}) = 60\left(\ln\frac{H}{a} - 0.65\right) \qquad \text{ohm} \quad (11\text{-}60)$$

Having a value for $Z_c\,(\text{av})$, the expressions for the voltage and current distributions and input impedance may be written down from transmission-line theory. For the open-circuited line the expressions are

$$V_S = V_R \cosh \gamma s \qquad (11\text{-}61)$$

$$I_S = \frac{V_R}{Z_c\,(\text{av})} \sinh \gamma s \qquad (11\text{-}62)$$

$$Z_S = \frac{V_S}{I_S} = Z_c\,(\text{av}) \coth \gamma s \qquad (11\text{-}63)$$

$$Z_{\text{in}} = Z_c\,(\text{av}) \coth \gamma H \qquad (11\text{-}64)$$

In these expressions V_S and I_S are the voltage and current, respectively, at a distance s from the open end of the line and V_R is the voltage at the end of the line. $\gamma = \alpha + j\beta$ is the complex propagation constant for the line. Its imaginary part $\beta = 2\pi/\lambda$ is the phase-shift constant, and its real part α is the attenuation constant, which is still to be determined.

The expression for current distribution [eq. (62)] may be written in terms of the current at the loop or maximum, I_m, as follows: Expanding $\sinh \gamma s$, and taking the absolute magnitude, it is seen that, if the attenuation is not too great, the maximum amplitude of current (the loop current) will occur approximately at $s = \lambda/4$, and will be given by

$$I_m = I_{s=\lambda/4} = j\frac{V_R}{Z_c\,(\text{av})} \cosh \alpha\frac{\lambda}{4} \sin \beta\frac{\lambda}{4}$$

$$\approx j\frac{V_R}{Z_c\,(\text{av})}$$

Then eq. (62) may be written as

$$\begin{aligned} I_S &= (-j) I_m \sinh \gamma s \\ &= I_m(\cosh \alpha s \sin \beta s - j \sinh \alpha s \cos \beta s) \end{aligned} \qquad (11\text{-}65)$$

The next step is to determine the attenuation factor α for the "equivalent" transmission line. On the basis of a sinusoidal current distribution, the power radiated by the antenna can be computed by the Poynting vector method or the induced-emf method. This gives a value for R_{rad}, the radiation resistance, referred to the loop current. By definition,

$$\text{Power radiated} = |I_m|^2 R_{rad} \tag{11-66}$$

R_{rad} for ground-based antennas on a perfect reflecting plane is plotted as a function of antenna height in Fig. 14-5. Alternatively, for H greater than 0.2λ, R_{rad} may be obtained with good accuracy from the approximate formula

$$R_{rad} = 15\left[-\frac{\pi}{2}\sin\frac{4\pi H}{\lambda} + \left(\ln\frac{2H}{\lambda} + 1.722\right)\cos\frac{4\pi H}{\lambda} + 4.83 + 2\ln\frac{2H}{\lambda}\right] \tag{11-67}$$

For the corresponding center-fed dipole ($L = 2H$) in free space, the values of R_{rad} given by Fig. 14.5 or eq. (67) must be doubled.

In the transmission-line representation the radiated power is replaced by an equal amount of power dissipated as ohmic loss along the line. This power loss may be assumed to be due to a series resistance r ohms per unit length, shunt conductance g mhos per unit length, or both. It turns out that if the power loss is considered to be due to both series resistance and shunt conductance of such values that the I^2r loss per unit length at a current loop is equal to the V^2g loss per unit length at a voltage loop, simpler expressions result. There is the added advantage that input impedances calculated from these simpler expressions seem to be in better agreement with values determined by other means. This is especially true for short antennas where the series-resistance assumption (used by Siegel and Labus) leads to values of input resistance that are consistently too high, whereas the assumption used here leads to correct values.

Assuming both series resistance r per unit length and shunt conductance g per unit length, the total power loss along the line is

$$\begin{aligned} W &= \int_0^H (|I|^2 r + |V|^2 g)\, ds \\ &= \int_0^H (I_m^2 r|\sinh \gamma s|^2 + V_m^2 g|\cosh \gamma s|^2)\, ds \end{aligned} \tag{11-68}$$

The values of r and g are so chosen that

$$I_m^2 r = V_m^2 g \tag{11-69}$$

Then the expression (68) for power dissipated becomes

$$\begin{aligned} W = I_m^2\, r \int_0^H (&|\sinh \alpha s \cos \beta s + j \cosh \alpha s \sin \beta s|^2 \\ &+ |\cosh \alpha s \cos \beta s + j \sinh \alpha s \sin \beta s|^2)\, ds \end{aligned}$$

which reduces to

$$\begin{aligned} W &= I_m^2\, r \int_0^H \cosh 2\alpha s\, ds \\ &= I_m^2\, rH \frac{\sinh 2\alpha H}{2\alpha H} \end{aligned} \tag{11-70}$$

For small values of $2\alpha H$, this becomes approximately

$$W \approx I_m^2 \, rH$$

That is, the power loss per unit length is approximately constant and is equal to

$$I_m^2 \, r = V_m^2 \, g \tag{11-71}$$

The total power dissipated must equal the power that is actually radiated, so

$$I_m^2 \, rH = I_m^2 \, R_{\text{rad}}$$

Therefore

$$r = \frac{R_{\text{rad}}}{H}$$

Also, using (71),

$$g = \frac{I_m^2 \, r}{V_m^2} = \frac{r}{Z_c^2\,(\text{av})} = \frac{R_{\text{rad}}}{HZ_c^2\,(\text{av})}$$

For any low-loss transmission line the attenuation factor is given by

$$\alpha = \frac{1}{2}\left(\frac{r}{Z_0} + gZ_0\right)$$

so for this "equivalent" line

$$\alpha = \frac{1}{2}\left(\frac{r}{Z_c\,(\text{av})} + \frac{R_{\text{rad}}}{HZ_c\,(\text{av})}\right) = \frac{R_{\text{rad}}}{HZ_c\,(\text{av})} \tag{11-72}$$

The total attenuation for the length H is

$$\alpha H = \frac{R_{\text{rad}}}{Z_c\,(\text{av})} \tag{11-73}$$

The input impedance can now be obtained from eq. (64). For purposes of computation eq. (64) can be expanded into more suitable forms:

$$Z_{\text{in}} = Z_c\,(\text{av}) \left[\frac{\cosh\,(\alpha H + j\beta H)}{\sinh\,(\alpha H + j\beta H)}\right]$$

$$= \frac{Z_c\,(\text{av})}{2}\left(\frac{\sinh 2\alpha H - j \sin 2\beta H}{\cosh^2 \alpha H - \cos^2 \beta H}\right) \tag{11-74}$$

$$= Z_c\,(\text{av})\left(\frac{\sinh 2\alpha H - j \sin 2\beta H}{\cosh 2\alpha H - \cos 2\beta H}\right) \tag{11-75}$$

The input resistance and input reactance are, respectively,

$$R_{\text{in}} = \frac{Z_c\,(\text{av})}{2}\left(\frac{\sinh 2\alpha H}{\cosh^2 \alpha H - \cos^2 \beta H}\right) \tag{11-76}$$

$$X_{\text{in}} = \frac{Z_c\,(\text{av})}{2}\left(\frac{-\sin 2\beta H}{\cosh^2 \alpha H - \cos^2 \beta H}\right) \tag{11-77}$$

The current distribution is given by eq. (65). When only the magnitude of the current is of interest, this may be obtained from

$$I_S = I_m|\sinh \gamma s|$$

$$= I_m\sqrt{\sinh^2 \alpha s + \sin^2 \beta s} \tag{11-78}$$

Correction for the End-effect. Equation (77) for the input reactance for an antenna indicates that the reactance goes through zero for lengths of line that are integral multiples of a quarter-wavelength. It is known from experiment that this is not the case, and that, in fact, the reactance zeros occur for physical lengths of antennas that are somewhat less than multiples of $\lambda/4$. This effect, which also occurs on open-ended transmission lines, is known as *end-effect.* It is due to a decrease in L and an increased C near the end of the line. This results in a decrease in Z_0 and an increase in current near the end of the line over that given by the sinusoidal distribution. With transmission lines the magnitude of the effect depends upon the line spacing in wavelengths. The region in which the change in the line "constants" occurs is known as the *terminal zone.* In the case of the transmission line, the terminal zone extends back a distance approximately equal to the line spacing. In the case of antennas, the end-effect produces an apparent lengthening of the antenna, the amount of which depends in a rather complicated manner on the characteristic impedance, the length, and the configuration of the antenna. The effect is somewhat greater for antennas of low characteristic impedance (large cross section) than it is for thin-wire antennas. Siegel has investigated* the end-effect on both transmission lines and antennas and has computed the following table, which shows numerical values for the amount by which the apparent electrical length of the antenna exceeds its physical length measured in wavelengths. It is seen that the rule of thumb often used

Table 11-1

INCREASE IN APPARENT LENGTH OF ANTENNAS DUE TO END-EFFECT

	Per cent Increase	
H_0/λ	*Tower antenna* $Z_c\,(av) = 220$ *ohms*	*Wire antenna* $Z_c\,(av) = 500$ *ohms*
$\frac{1}{4}$	5.4	4.5
$\frac{3}{8}$	5.3	4.3
$\frac{1}{2}$	5.2	2.2
0.59	5.1	1.9

in the field, by which the physical length is made 5 per cent less than the desired electrical length, is a rather good approximation for tower antennas. In computing input impedance by this method it is the apparent electrical length that should be used for H.

EXAMPLE 1: Determine an approximate value for the input impedance at anti-resonance of a full-wave ($L = 2H \approx \lambda$) cylindrical dipole antenna having a diameter of 2 cm. The frequency is 150 MHz.

*Ernest M. Siegel, *Wavelength of Oscillations Along Transmission Lines and Antennas,* University of Texas Publication No. 4031, Aug. 15, 1940.

Physical half-length

$$H_0 \approx \frac{2 \times 0.95}{2} = 0.95 \text{ meter}$$

$$\frac{H_0}{a} = 95$$

$$Z_c\,(\text{av}) = 120(\ln 95 - 1 - \tfrac{1}{2}\ln 0.95) = 430 \text{ ohms}$$

$$\alpha H = \frac{R_{\text{rad}}}{Z_c\,(\text{av})} = \frac{210}{430} = 0.489$$

$$\beta H = \pi$$

$$Z_{\text{in}} = R_{\text{in}} = 430\left(\frac{\sinh 0.978}{\cosh 0.978 - 1}\right) = 948 \text{ ohms}$$

EXAMPLE 2: A uniform-cross-section tower antenna is 400 ft high and 7 ft square. Calculate the base impedance at a frequency of 1300 kHz.

(a) Characteristic impedance

Equivalent* radius $a = 0.5902 \times 7 \qquad = 4.14$ ft

Physical height $H_0 \qquad = 400$ ft

Then $H_0/a = 96.6 \qquad 2H_0/\lambda = 1.06$

$$Z_c\,(\text{av}) = 60\left(\ln\frac{H_0}{a} - 1 - \frac{1}{2}\ln\frac{2H_0}{\lambda}\right) \qquad = 210 \text{ ohms}$$

(b) Attenuation factor

$$\text{Electrical height } \beta H = \frac{2\pi \times 1.05 \times 400}{231 \times 3.281} = 3.48 \text{ radians}$$

$$\alpha H = \frac{R_{\text{rad}}}{Z_0} = \frac{81}{210} \qquad = 0.386 \text{ nepers}$$

(c) Theoretical base impedance

$$R = \frac{210}{2}\left(\frac{0.851}{1.155 - 0.883}\right) \qquad = 329 \text{ ohms}$$

$$X = -\frac{210}{2}\left(\frac{0.643}{0.272}\right) \qquad = -248 \text{ ohms}$$

$$Z_{\text{base}} = 329 - j248 \text{ ohms}$$

*E. Hallén has shown that the correct value for the equivalent radius of a noncircular cylinder is obtained by finding the radius of the infinitely long circular cylinder that has the same capacitance per unit length as does an infinite length of the noncircular cylinder. For the square cross section of side length d he obtains for the equivalent radius, $a = 0.5902d$. "Theoretical Investigations into the Transmitting and Receiving Qualities of Antennae," *Nova Acta Upsal* 4, **11, 7** (1938). Also, "Admittance Diagrams for Antennas and the Relation between Antenna Theories," Cruft Laboratory Report No. 46, June, 1938. Also, equivalent radii for elliptical and polyhedral cross sections have been determined by Y. T. Lo, "A Note on the Cylindrical Antenna of Noncircular Cross Section," *J. Appl. Phys.*, **24**, pp. 1338–1339, October, 1953.

EXAMPLE 3: The antenna of example 2 is to be used as antenna (1) in a two-element directional array, with a quarter-wave tower as antenna (2). The loop current of antenna (2) is equal to, but leads, the loop current of antenna (1) by 90 degrees, that is,

$$I_2\,(\text{loop}) = I_1\,(\text{loop})\,\underline{/90^\circ}$$

Assume Z_{11} is nearly equal to the self-impedance of antenna (1) and use

$$Z_{12} = 20\,\underline{/-25^\circ} \quad \text{(referred to current loops)}$$

$$Z_{22} = 36 + j20$$

Determine the driving-point impedances:

(a) Refer the mutual impedance to the base of antenna (1). The mutual impedance Z_{21} referred to the *base* current $I_1(0)$ of antenna (1) is

$$Z_{21}\,(\text{base}) = \frac{V_{21}}{I_1(0)} = \frac{I_1\,(\text{loop})}{I_1(0)}\,\frac{V_{21}}{I_1\,(\text{loop})} = \frac{I_1\,(\text{loop})}{I_1(0)}\,Z_{21}\,(\text{loop})$$

where V_{21} is the open-circuit voltage at the terminals of (2) due to the current flow in (1).

From example 2,

$$\frac{I_1\,(\text{loop})}{I_1\,(\text{base})} = \frac{\sinh \gamma\lambda/4}{\sinh \gamma H} = j\,\frac{\cosh \alpha\lambda/4}{\sinh 0.386 \cos 200^\circ + j \cosh 0.386 \sin 200^\circ}$$

$$\approx \frac{1\,\underline{/90^\circ}}{0.523\,\underline{/224.8^\circ}} = 1.91\,\underline{/-134.8^\circ}$$

then

$$Z_{21}\,(\text{base}) = 20\,\underline{/-25^\circ} \times 1.91\,\underline{/-134.8^\circ} = 38.2\,\underline{/-159.8^\circ}$$

(b) Driving-point impedances

$$Z_1' = Z_{11} + \frac{I_2}{I_1} Z_{12} \quad \text{(all referred to base)}$$

$$\frac{I_2\,(\text{base})}{I_1\,(\text{base})} = \frac{I_2\,(\text{base})}{I_1\,(\text{loop})} \times \frac{I_1\,(\text{loop})}{I_1\,(\text{base})}$$

$$= 1\,\underline{/90^\circ} \times 1.91\,\underline{/-134.8^\circ} = 1.91\,\underline{/-44.8^\circ}$$

$$Z_1' = 329 - j248 + 1.91\,\underline{/-44.8^\circ} \times 38.2\,\underline{/-159.8^\circ}$$

$$= 263 - j218$$

For antenna (2), base current equals loop current.

$$Z_2' = Z_{22} + \frac{I_1}{I_2} Z_{12} \quad \text{(loop or base)}$$

$$= 36 + j20 + \frac{38.2\,\underline{/-159.8^\circ}}{1.91\,\underline{/-44.8^\circ}}$$

$$= 36 + j20 + 20\,\underline{/-115.0^\circ}$$

$$= 27.5 + j1.9$$

11.15 Practical Antennas and Methods of Excitation. The directional and impedance characteristics of antennas have been considered briefly in the preceding sections, and will be treated in more depth in later chapters. In this section several practical considerations relating to the feeding and matching of antennas and arrays will be discussed.

Effect of the Earth on the Impedance of Elevated Antennas. A practical "half-wave dipole" refers to a *resonant-length* dipole for which the reactance is zero. Because of the end-effect (sec. 11.14) and as can be seen from the impedance curves of Figs. 14-7, 14-12, and 14-19, a resonant-length antenna is always shorter than a half-wavelength (by an amount depending on the antenna thickness) and the radiation resistance at resonance may be about 65 to 72 ohms, which is somewhat less than the theoretical value of 73 ohms for the infinitely thin dipole.

In addition to the above effect, an actual dipole is always located above the ground or other supporting and reflecting surface, so that the theoretical free-space conditions do not apply. The effect of the presence of a perfectly conducting ground on the input impedance of the antenna can most readily be accounted for by replacing the ground by an appropriately located image antenna carrying an equal current in proper phase, and then computing the driving-point impedance under those conditions by the methods of chap. 14. Figure 11-31 shows how the input impedance of a (theoretical) half-wave dipole varies with distance above a ground plane that is assumed perfectly conducting. In general the input impedance now has a reactive component as well as resistance, and the magnitude of the resistance oscillates about the free-space value of 73 ohms. For a practical dipole a similar effect could be expected, with the input resistance oscillating about the actual free-space value. The effect of a *finitely* conducting ground could be

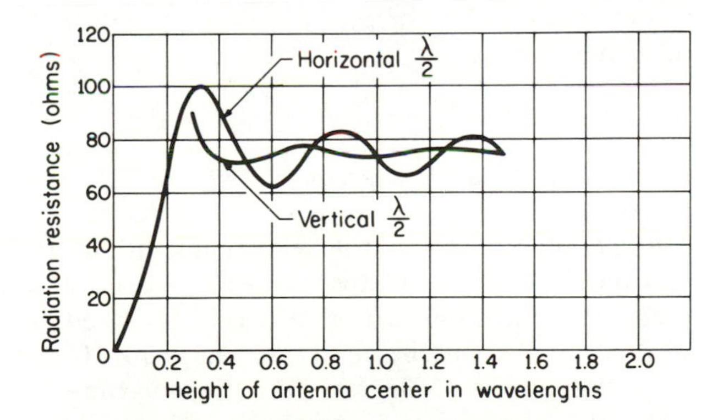

Figure 11-31. Variation of the radiation resistance of a theoretical half-wave dipole with height above a perfectly conducting earth.

determined in a similar manner, the only difference being that the image antenna would carry a current, the magnitude and phase of which would depend upon the actual reflection coefficients of the earth (see chap. 16). The final results would be similar to those of Fig. 11-31, except that the amplitude of oscillation about the free-space impedance would be slightly less, with a slight shift in the actual heights above ground at which the maximum and minimum impedances occurred. Because of irregularities of the ground itself, and unknown reflections from buildings, trees, and other surrounding objects, some deviation from the theoretical values must always be expected.

The effect of the presence of the ground on the vertical radiation pattern can also be obtained by use of the image principle. In Figs. 16-7 through 16-11 are shown the vertical patterns of short vertical and horizontal dipoles above earths of various conductivities. For greater heights above the earth the vertical pattern becomes multi-lobed. The approximate location of the maxima and minima of the pattern can be obtained by considering the perfect ground case and using the principle of multiplication of patterns as in sec. 11.08. For horizontal antennas in the plane perpendicular to the axis of the antenna the factor by which the free-space pattern must be multiplied to account for the effect of the ground is

$$2 \sin \left(\frac{2\pi h}{\lambda} \sin \psi\right) \tag{11-79}$$

where h is the height of the center of the antenna above ground and ψ is the angle of elevation above the horizontal. The first maximum in this pattern occurs at an angle ψ_{ml} which is given by

$$\sin \psi_{ml} = \frac{\lambda}{4h} \qquad \left(h > \frac{\lambda}{4}\right)$$

For a vertical antenna the corresponding ground-effect factor of a perfectly conducting earth is

$$2 \cos \left[\frac{2\pi h}{\lambda} \sin \psi\right] \tag{11-80}$$

For a finitely conducting earth, expression (79) should give a good approximation to the actual multiplying factor for all angles of ψ because, for horizontal polarization, the reflection factor of an imperfect earth is always nearly equal to minus one (Fig. 16-3). However, for vertical antennas, expression (80) will give reasonably good results only for large angles of ψ. As will be shown in chap. 16 (Fig. 16-4), for angles of ψ less than about 15 degrees (the "pseudo-Brewster angle"), the phase of the reflection factor is nearer to 180 degrees than it is to zero, and the use of (80) for low angles of ψ would lead to erroneous results.

In the vertical plane *parallel* to the axis of a horizontal antenna, the radiation is *vertically* polarized, and it is the reflection factor for vertical polarization that must be used in determining the effect of the ground. This means that for large angles of ψ, the reflection factor will be approximately plus one, but for small angles (below about 15 degrees) it will more nearly approximate minus one. This result has an effect of some practical importance in connection with low-angle radiation or reception off the end of a horizontal antenna. Because the vertical components of the direct and reflected waves are oppositely directed as they leave the horizontal antenna, they will have the same direction (or phase) after reflection of the reflected wave by the imperfect ground. Therefore, at low angles, direct and reflected waves will tend to add instead of cancel, resulting in a relatively strong vertically polarized signal off the end of a horizontal antenna.

Methods of Excitation. Several common methods of feeding half-wave antennas are illustrated in Fig. 11-32. Figure 11-32(*a*) shows the balanced-line type of center feed. Because of the mismatch between the high characteristic impedance of open-wire lines and the low input resistance of a resonant dipole, this manner of excitation results in standing waves on the feed line as indicated in the figure. However

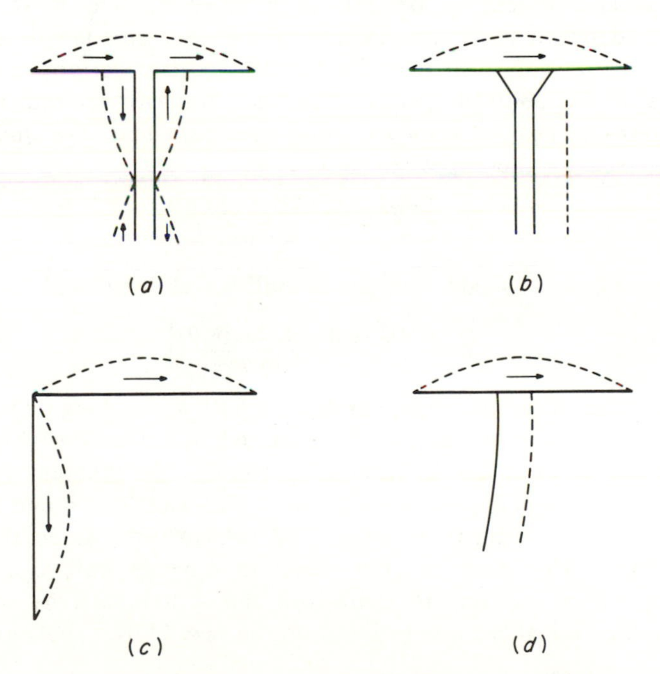

Figure 11-32. Common methods of exciting high-frequency antennas.

with solid-dielectric, low-impedance lines this mismatch can be almost completely eliminated, but there is now some loss in the dielectric. The "delta-match" or "shunt-feed" arrangement of Fig. 11-32(*b*) can result in a good impedance match and low standing waves on the feed-line if the various dimensions are properly chosen.

Probably the simplest of all possible methods of excitation is the single-wire line "end-fed" arrangement of Fig. 11-32(*c*). In this case the vertical "transmission line" also radiates energy, a result that may or may not be desired. By tapping on the vertical wire at a lower impedance point along the horizontal antenna as in (*d*), a better impedance match and lower standing-wave ratio on the feed line can be obtained.* This results in smaller radiation from the vertical wire, which now carries a traveling-wave current distribution. Optimum dimensions for the types of feed shown in (*b*) and (*d*) are dependent on the height of the antenna above ground and upon the conductivity of the ground. They may be determined by trial in each case.

Long-wire Antennas. The single-wire feeds of Figs. 11-32(*c*) and (*d*) are also suited to the excitation of horizontal long-wire antennas. When such antennas are unterminated (that is, open at the far end), the current distribution is chiefly that of a standing wave, and the antenna should preferably be cut to a resonant length, so that the input impedance is resistive. However, with the resonant-line feed of Fig. 11-32(*c*) it is usually possible to tune out a certain amount of reactance at the point of coupling between transmitter and feed line. The patterns of *end-fed* resonant long-wire antennas are multi-lobed patterns given by the expressions

$$E = \frac{60I}{r}\left[\frac{\cos\,(\pi L/\lambda \cos\theta)}{\sin\theta}\right] \tag{11-81a}$$

for wires that are an odd number of half-waves long, and

$$E = \frac{60I}{r}\left[\frac{\sin\,(\pi L/\lambda \cos\theta)}{\sin\theta}\right] \tag{11-81b}$$

for wires that are an even number of half-waves long. Qualitative patterns for these antennas can be obtained by inspection through use of the principle of multiplication of patterns. The theoretical current distribution and calculated pattern for a two-wavelength end-fed long-wire antenna are shown in Figs. 11-33(*a*) and (*b*). Since the actual current distribution consists of a traveling wave as well as a standing wave (because of loss due to radiation), the actual patterns will differ from the theoretical as was pointed out in sec. 11.04. The chief effect

*W. L. Everitt and J. F. Byrne, "Single-wire Transmission Lines for Short-wave Radio Antennas," *Proc. IRE,* **17,** 1840 (1929).

of this difference is to tilt the lobes toward the unfed end. This difference between actual patterns and theoretical patterns (based on standing waves only) is much less in the case of center-fed antennas.

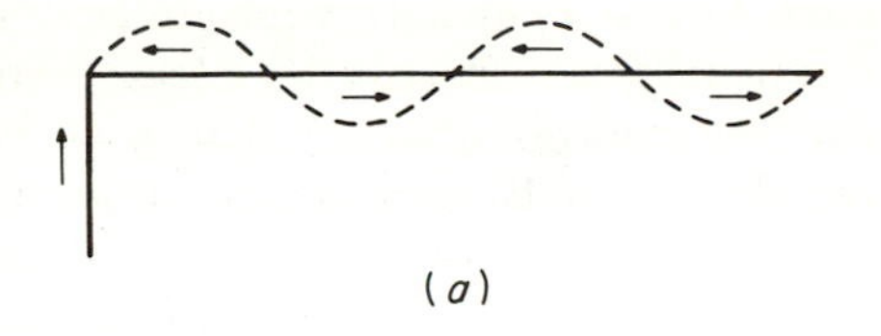

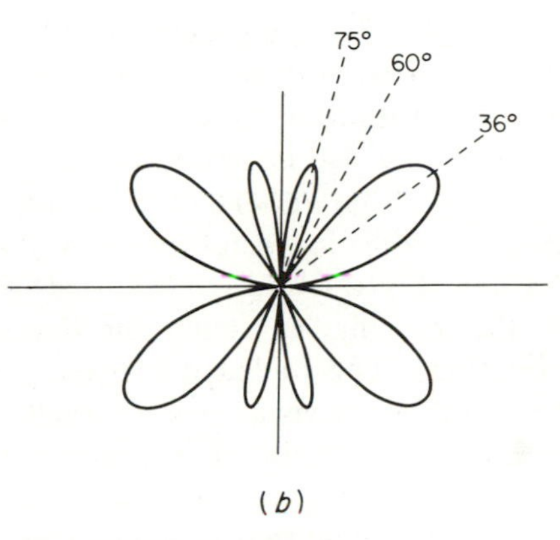

Figure 11-33. (*a*) Theoretical current distribution and (*b*) radiation pattern of a two-wavelength end-fed antenna.

If a long wire is terminated in a resistance equal to its characteristic impedance the current distribution along it is essentially that of a traveling wave, and its pattern will have the general shape of that shown in Fig. 11-7. The longer the wire, the smaller will be the angle between the wire and the first or main lobe. The important difference between the patterns of terminated and unterminated wire antennas is the absence of large rear lobes in the former.

Stub-matched and Folded Dipoles. A simple dipole has an input impedance that is too low for direct connection to an ordinary open-wire transmission line, and some sort of impedance-matching arrangement is required if the desirable condition of no standing waves on the line is to be attained. An easy way of obtaining this match is by means of the *stub-matched* dipole shown in Fig. 11-34. By making the length $L = 2H$ somewhat less than a half-wavelength, the input imped-

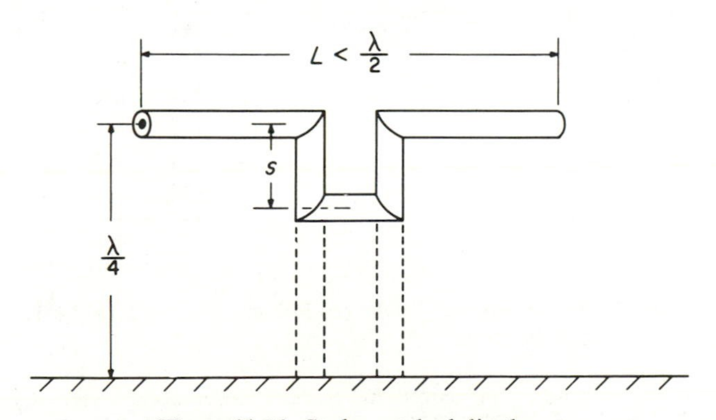

Figure 11-34. Stub-matched dipole.

ance will be a capacitive reactance in series with the radiation resistance. For a length s of stub line less than $\lambda/4$, the input impedance of the shorted transmission line will be an inductive reactance, the magnitude of which can be adjusted to tune out the capacitance of the antenna. The resulting impedance at the terminals *a-b* will be a pure resistance, the resistance of the parallel-resonant circuit. By proper choice of L and s this resistance can be adjusted to almost any value desired. The arrangement is good mechanically, because by extending the lines of the stub back beyond the shorting bar, the antenna can be mounted on, and a quarter-wavelength in front of, a reflecting ground screen, without the use of insulators.

An alternative way of obtaining a high-impedance input is by means of the *folded dipole* described by Carter* and shown in Fig. 11-35. This method has the added advantage that it also increases the band width of the antenna, an important consideration in FM and television applications. The folded half-wave dipole consists essentially of two half-wave radiators very close to each other and connected together at top and bottom. As far as the antenna currents or radiating currents are concerned, the two elements are in parallel, and if their diameters are the same, the currents in the elements will be equal and in the same direction. If 1 amp flows in each element (at the center) the total effective current is 2 amp, and the power radiated will be $(2I_1)^2 R_{\text{rad}} \approx 4 \times 73 I_1^2$ or 4 times that radiated by a single element carrying 1 ampere. However, the current that is required to be delivered by the generator at the terminals *a-b* is only 1 ampere, so that the input resistance is

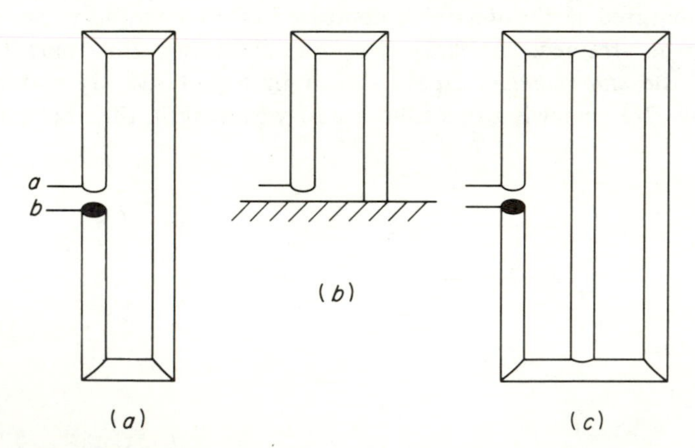

Figure 11-35. (*a*) Folded dipole. (*b*) Folded monopole. (*c*) Multi-element folded dipole.

*P. S. Carter, "Simple Television Antennas," *RCA Rev.*, **4,** 168, October 1939; W. Van B. Roberts, "Input Impedance of a Folded Dipole," *RCA Rev.*, **8,** 289 (1947).

seen to be 4 times that of a simple dipole. If there are three elements of equal diameter connected together as in Fig. 11-34(*c*), the input resistance will approximately be 9 times that of a simple dipole. If the elements are of unequal diameters, the currents will divide unequally between the elements. If it is assumed that the currents divide inversely as the characteristic impedances Z_{01} and Z_{02}, so that

$$\frac{I_2}{I_1} = \frac{Z_{01}}{Z_{02}}$$

then if element 1 is the driven element, the input resistance will be

$$R_{\text{in}} = \frac{(I_1 + I_2)^2 R_{\text{rad}}}{I_1^2} = R_{\text{rad}}\left(1 + \frac{I_2}{I_1}\right)^2 = R_{\text{rad}}\left(1 + \frac{Z_{01}}{Z_{02}}\right)^2 \qquad (11\text{-}82)$$

As before, R_{rad} is the radiation resistance of a simple half-wave dipole (theoretically $R_{\text{rad}} = 73$ ohms, actually $R_{\text{rad}} < 73$ ohms).

In order to understand the increased band width that results with the folded-dipole arrangement, consider the simple half-wave (resonant-length) dipole of Fig. 11-36(*a*) which is connected in parallel at its terminals with a shorted quarter-wavelength line. At the resonant frequency the dipole resistance is in parallel with the input impedance of the transmission line, which is a resistance of very high value. Below resonance, the antenna impedance becomes capacitive, but the transmission-line impedance becomes inductive, and the parallel combination tends to remain nearer unity power factor than does the antenna alone. Conversely, above resonance the antenna impedance becomes inductive and the line impedance becomes capacitive so that compensation is again obtained. Although compensation is far from

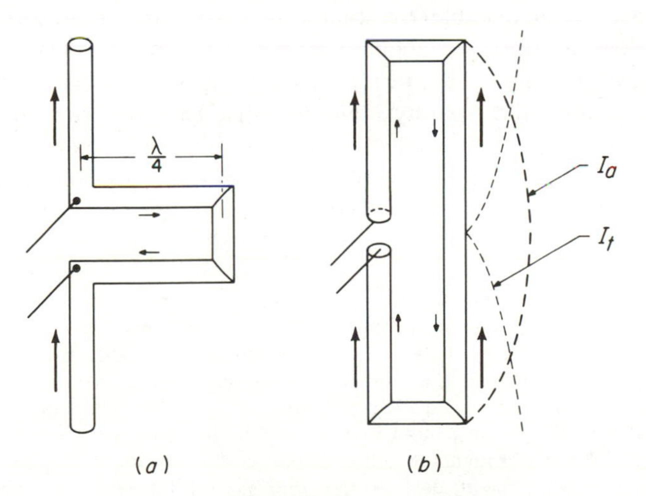

Figure 11-36.

perfect, because the susceptances are not equal and opposite, if the frequency is shifted far enough in either direction from the resonant frequency of the dipole, a point of perfect susceptance compensation (where the input impedance is a pure resistance) is again obtained. Below resonance this occurs for the same conditions that led to the stub-matched dipole of Fig. 11-34. Above resonance the point of perfect susceptance compensation occurs when the capacitive susceptance of the stub is just sufficient to tune out the inductive susceptance of the antenna. Of course, the input resistance at these stub-matched or anti-resonant points will be considerably higher than at the resonant frequency, but for some purposes the resulting standing-wave ratio is good enough over the range that the effective band width may be said to extend from one anti-resonant point to the other. This represents almost a two-to-one frequency range.

The above considerations apply directly to the folded dipole of Fig. 11-36, which has the stub line (actually two stub lines in series) as a built-in feature. The elements of the folded dipole then carry both the antenna currents, which are in the same direction in the two elements, and the transmission-line currents, which are in opposite directions in the two elements of the dipole [Fig. 11-36(*b*)]. At the resonant frequency the antenna currents are relatively large [$I_a = (V/R_{\text{in}}) = (V/4R_{\text{rad}})$ in each element, at the center], whereas the transmission-line currents are zero at the center, but have the value $I_t = V/2Z_{0t}$ at the ends, where Z_{0t} is the characteristic impedance of each of the two shorted quarter-wave transmission-line sections.

For FM broadcast reception a common type antenna is a folded dipole made of flexible solid dielectric "twin-lead." For a transmission line made of such cable the phase velocity, and hence the length of a wavelength, is only about 80 per cent of the free-space value. Therefore an electrical quarter-wavelength section is only 0.8 times $\lambda/4$ physically, and the physical line must be made shorter than would be the case with a free-space dielectric. On the other hand, the thin dielectric covering on the cable has almost negligible effect* on the apparent phase velocity and wavelength of the antenna currents, so that for resonance the physical length of the antenna should still be

*The reason for the difference in the two cases is as follows: As a transmission line the return displacement currents flow from one wire to the adjacent wire, mostly through the solid dielectric that is in parallel with the surrounding air dielectric. As an antenna the "return" displacement currents flow from both wires in parallel, through the solid dielectric and air dielectric in series to the opposite arm of the dipole (or to ground in the case of the monopole). In this latter case the length of path through the solid dielectric is short compared with the remainder of the path through the air dielectric so that the solid dielectric has negligible effect.

approximately $L \approx 0.95\lambda/2$ (that is $H \approx 0.95\lambda/4$). The method for satisfying these two conditions simultaneously is indicated in Fig. 11-37. The two elements are cut to 0.95 times $\lambda/2$, but the shorting connections are spaced only 0.8 times $\lambda/2$ apart.

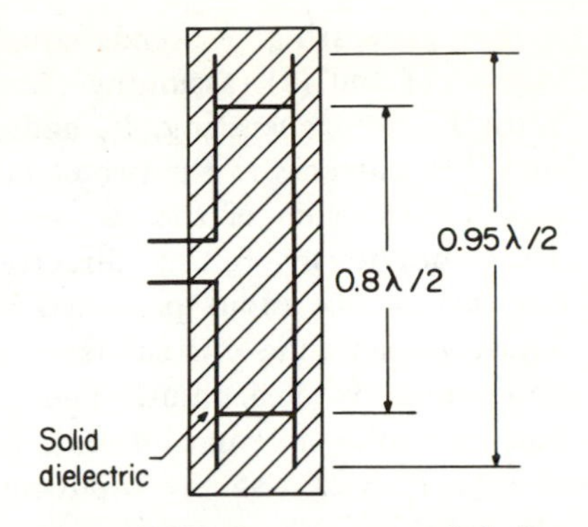

Figure 11-37. Common type of "built-in" antenna for FM reception.

An added insight into the operation of the folded dipole (and also certain of the "baluns" described in the next part of this section) is provided by the superposition principle. Consider the operation of the folded monopole of Fig. 11-35(*b*), which operation is identical with that of the corresponding folded dipole of Fig. 11-35(*a*). The single zero-impedance generator may be replaced by three equivalent generators having equal voltages, zero internal impedances, and connections and polarity as shown in Fig. 11-38(*b*). If $V_1 = V_2 = V_3 = V/2$, then a

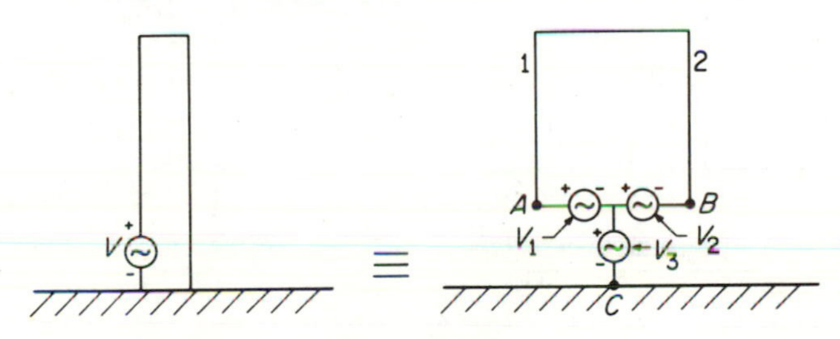

Figure 11-38.

quick check shows that as far as the currents in the two elements are concerned, the operation of 11-38(*b*) is identical with that of 11-38(*a*). In (*b*) generator V_3 causes equal antenna currents to flow in the same direction in elements (1) and (2). Generators V_1 and V_2 in series cause equal and opposite transmission-line currents to flow in elements (1) and (2). Because points *B* and *C* have the same voltage at all times they could be joined together without affecting the operation (when all generators are generating) and the circuit of Fig. 11-38(*a*) would result. Now consider the superposition principle applied to the equivalent circuit (*b*). By this principle the total currents flowing in any branch with all generators generating is just the sum of the individual currents produced by each of the generators alone, the other generators being replaced by their internal impedances. With V_1 and

V_2 not generating, V_3 sends equal antenna currents into the two elements (1) and (2) (assuming these elements to have equal diameters). With V_3 not generating, V_1 and V_2 send equal and opposite transmission-line currents in the two elements. The total currents that actually flow are the sum of the two sets of currents. The impedance relations given previously follow directly. At resonance the transmission-line currents at the input produced by V_1 and V_2 are approximately zero. The total antenna current is V_3/R_{rad}, where R_{rad} for a $\lambda/4$ monopole is approximately 36.5 ohms. The antenna current in element (1) is one-half the total antenna current, or $I_{a1} = V_3/2R_{\text{rad}}$. The actual applied voltage $V = 2V_3$, so the input impedance at resonance is $R_{\text{in}} = V/I_{a1} = 4R_{\text{rad}}$.

Baluns. An ordinary dipole is a *balanced* load in the sense that for equal currents in the two arms, the arms should have the same impedance to ground. Such a load should be fed by a transmission line such as a two-wire line, which itself is "balanced to ground."

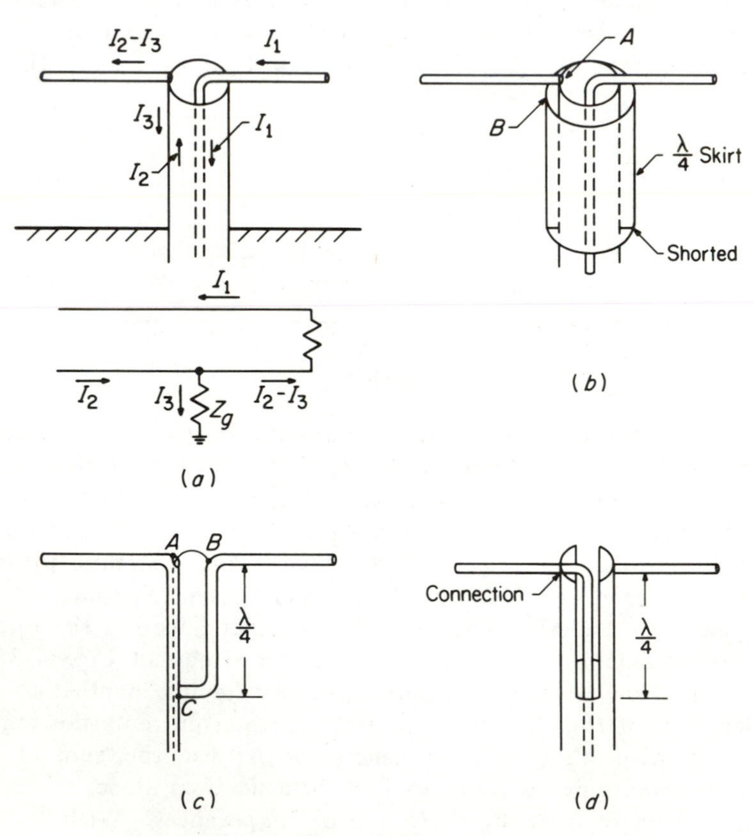

Figure 11-39. Baluns.

However, at very high and ultra-high frequencies unbalanced coaxial lines are nearly always used, so that we encounter the problem of transforming from an unbalanced to a balanced system or vice versa. The device that accomplishes this balance-to-unbalance transformation is called a *balun.* There are many different types of baluns and four of the most common are shown in Figs. 11-39 and 11-40. In Fig. 11-39(a), a balanced dipole antenna is shown connected directly to the end of an unbalanced (coaxial) line. The currents I_1 and I_2 must be equal and opposite. At the junction A, current I_2 divides into I_3, which flows down the outside of the outer conductor of the line and $I_2 - I_3$ which flows on the second arm of the dipole. The current I_3 depends upon the effective "impedance to ground" Z_g provided by this path along the outside of the conductor. This impedance can be made very high, thus making I_3 very small, by the addition of a quarter-wave skirt around the outer conductor as in Fig. 11-39(b). With the skirt shorted to the conductor at the bottom, the impedance between the points A and B (and therefore between A and the effective ground part wherever it may be) is extremely large, being limited only by the Q of the shorted quarter-wave section. In the arrangement of 11-39(c) the dipole feed is balanced by making the impedance of the shunt paths from A and B to the common point C equal. When the stub length is approximately a quarter of a wavelength, these equal shorting impedances are very large, and, in addition, the arrangement exhibits the desirable broadband characteristics discussed in connection with Fig. 11-36(a), to which it is exactly equivalent. Figure 11-39(d) is a more practical version of the arrangement shown in (c). Similar broadband characteristics are obtained with the balun of Fig. 11-40(a) for which the equivalent circuit of Fig. 11-40(b) may be drawn. The impedances Z_g, which shunt each side of the balanced load, are given by

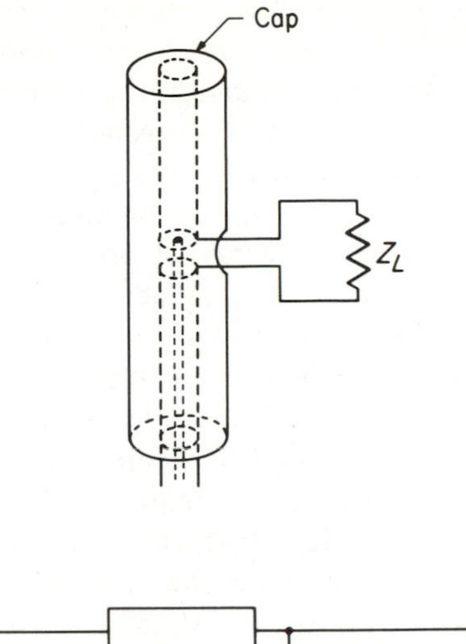

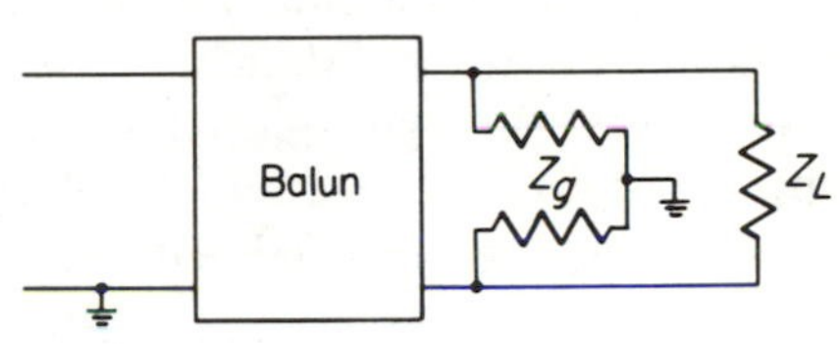

Figure 11-40. Balun and "equivalent circuit."

$$Z_g = jZ_0 \tan \beta s$$

where Z_0 is the characteristic impedance of each of the parallel stubs.

VHF Antenna Arrays. For point-to-point communication and applications such as radar the gain or directivity that can be achieved by the use of arrays is usually desirable and sometimes necessary. At these frequencies, line arrays and rectangular arrays become practical. A common array is the "mattress" antenna (Fig. 11-41) which consists of a rectangular array of coplanar elements, mounted a quarter wavelength in front of a reflecting screen. The patterns of such arrays are easily obtainable by the methods of secs. 7 and 8, but the methods of feeding the arrays to obtain the desired currents in the elements have not yet been considered. Figures 11-42 and 11-43 illustrate two methods for feeding a number of elements with equal currents, or with any specified currents, as required, for example in the binomial array of sec. 10. In Fig. 11-42 the points $A, B, C,$ and D are spaced half a wavelength apart on the transmission line, so that the voltages at these points are always equal in amplitude. By feeding the antennas from these points through quarter-wave sections, the current amplitudes depend only upon these equal voltages and the characteristic impedances of the sections, and will be independent of the antenna driving-

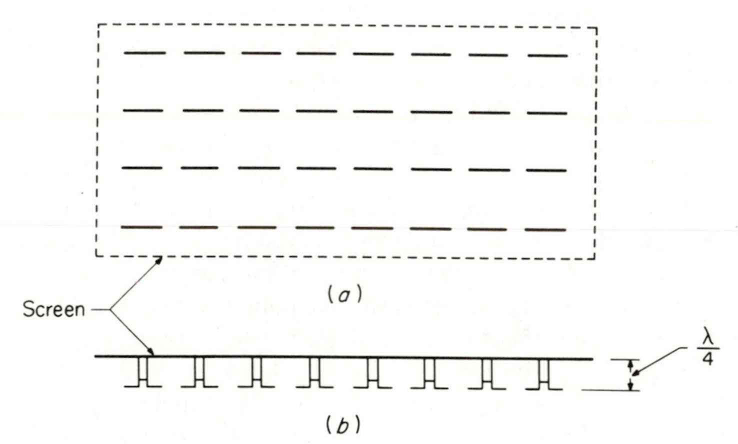

Figure 11-41. Typical rectangular array of co-planar elements.

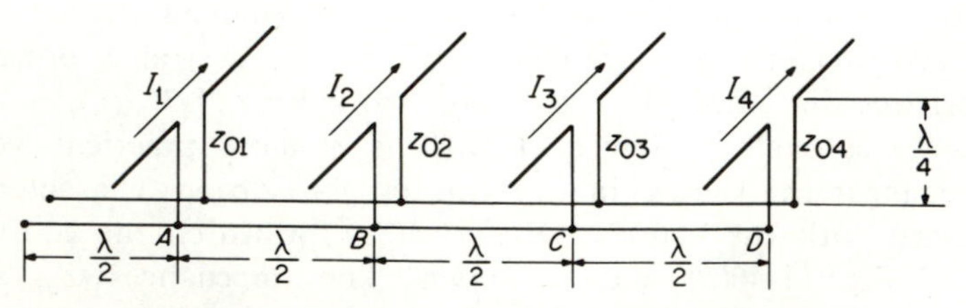

Figure 11-42. A method of feeding antennas in an array with specified currents: $I_1 = V/Z_{01}$, $I_2 = V/Z_{02}$, $I_3 = V/Z_{03}$, $I_4 = V/Z_{04}$.

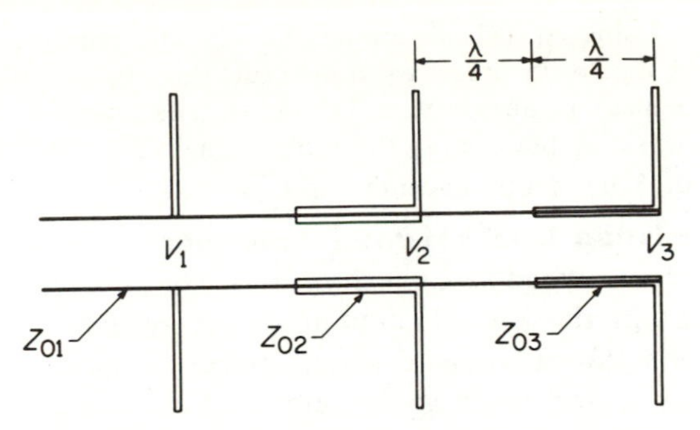

Figure 11-43. A second method of feeding the elements of an array with specified currents.

point impedances. This technique provides a simple way of circumventing the bothersome effects of unequal and (difficult to calculate) mutual impedances between antennas in an array. For similar quarter-wave sections the antenna currents will be equal, although they can be made to have almost any ratio by suitable choice of characteristic impedances for the quarter-wave sections. In the arrangement of Fig. 11-43, it is easy to show that the amplitudes of the driving voltages at the antenna terminals will be related by

$$\frac{V_2}{V_1} = \frac{Z_{02}}{Z_{01}} \qquad \frac{V_3}{V_2} = \frac{Z_{03}}{Z_{01}} \tag{11-83}$$

In practice the quarter-wave sections are made by slipping copper tubing of the correct diameter over the main feed line Z_{01}, and soldering the tubing to the inner line at both ends.

VHF and SHF Antennas. The antennas and arrays used in the VHF band, for the most part can be and are used in the UHF range also. However, at the upper end of the ultrahigh-frequency band, and especially at superhigh frequencies (3–30 GHz), the size of the elements becomes impractically small. It is then convenient to use "current-sheet" radiators such as paraboloids, horns, and slot antennas. The radiation from such sheet radiators can be computed from the fields of the individual current elements, exactly as is done for ordinary linear radiators, providing that the current distribution on the conducting sheets is known or can be estimated. However, in most cases the current distribution is neither known nor readily estimated, so that other methods of determining the radiation must be sought. A quite powerful method consists of determining the radiation from the antenna in terms of the fields that exist across the "aperture" of the antenna. Radiation from aperture antennas is the subject of chap. 13.

Problem 9. Verify the relation given in eq. (83).

Problem 10. Using eq. (82), determine the wire size required for the excited arm of a folded dipole, in order to transform the input resistance of 21.5 ohms, for an element in an array, to 150 ohms. The second arm of the dipole should be of the same diameter as the other elements ($1\frac{1}{2}$ in.), and a spacing between arms of 3 in., center-to-center, is suggested.

11.16 Transmission Loss between Antennas. Up to this point the antenna has been studied as an entity. It is now necessary to examine it as an element in the overall communication system. It is interesting to compare* the transmission of signal power in guided-wave systems such as coaxial cables, wave guides, etc., and in radiated-wave systems, such as microwave links. In the former the transmission loss per unit length is independent of distance, and usually increases with frequency. In the latter the transmission loss per unit length *decreases* with increasing distance, and the frequency variation depends upon the application. The notion of transmission loss is common in guided-wave systems but it also has direct application in radio transmission† where it is simply a measure of the ratio of power transmitted or radiated to the power received, expressed in decibels.

For *free-space* transmission, the transmission loss is easily determined. For a power W_T radiated from an isotropic antenna the power density at a distance d is $W_T/4\pi d^2$. If the transmitting antenna has a gain g_1 in the desired direction the power density is increased to $g_1 W_T/4\pi d^2$. The effective area of the receiving antenna is $A = \lambda^2 g_2/4\pi$ so the power received will be

$$W_R = \frac{g_1 W_T}{4\pi d^2}\frac{\lambda^2 g_2}{4\pi} = \frac{\lambda^2 g_1 g_2}{(4\pi d)^2} W_T$$

Hence the ratio of received to transmitted power is

$$\frac{W_R}{W_T} = \frac{\lambda^2 g_1 g_2}{(4\pi d)^2} \tag{11-84}$$

The *basic transmission loss*, L_b, is defined as the reciprocal of this ratio, expressed in decibels, for transmission between isotropic antennas ($g_1 = g_2 = 1$). That is

$$L_b = 10 \log_{10} \frac{(4\pi d)^2}{\lambda^2} \tag{11-85}$$

The actual transmission loss L will be less than L_b by the amount of the antenna gains G_1 and G_2, expressed in decibels.

$$L = L_b - G_1 - G_2 \tag{11-86}$$

*A more comprehensive discussion appears as chap. 16, "The Channel in Electronic Systems," by E. C. Jordan, in *Foundations of Future Electronics*, D. B. Langmuir and W. D. Hershberger, eds., McGraw-Hill Book Company, New York, 1961.

†K. A. Norton, "Transmission Loss in Radio Propagation," *Proc. IRE*, **41**, 146–152 (1953); also Nat. Bur. Stand. Tech. Note 12, June, 1959.

Problem 11. Determine the basic transmission loss between a ground-based antenna and an antenna on an aircraft (assuming free-space conditions) at distances of 1, 10, 100 and 200 miles

(a) at a frequency $f = 300$ MHz

(b) at a frequency $f = 3000$ MHz

(c) How high should the aircraft be flying to be within line-of-sight at distances of 100 and 200 miles? (see sec. 16.06)

Problem 12. If the ground-based antenna of problem 11 is a 25-ft-diameter paraboloid with an effective area equal to 60 per cent of actual area, and the aircraft antenna is essentially isotropic, what is the transmission loss at both frequencies?

Problem 13. If the ground-based antenna of problem 12 is also required to be azimuthally omnidirectional so that its directive gain is approximately 2 at both frequencies, what is the transmission loss at 300 MHz and 3000 MHz?

Transmission Loss as a Function of Frequency. Problems 11, 12, and 13 have emphasized that the variation with frequency of transmission loss depends upon the circumstances of the problem. Three particular cases are of special interest:

(1) For vehicular communication, air-to-ground links and navigational systems it is normally required that both antennas have omnidirectional coverage. (An "omnidirectional" antenna is one that has essentially uniform coverage in the azimuthal plane but which may have some directivity in the vertical plane. A monopole or vertical dipole is a typical omnidirectional antenna.) Under these circumstances the directional gain is fixed and independent of frequency. From (11-84) it is evident that under these circumstances of transmission between *fixed-gain* antennas, the received power is proportional to wavelength squared, or *inversely proportional* to the square of the operating frequency.

(2) For transmission between a fixed-gain antenna and a fixed-area antenna the transmission loss is *independent* of frequency. An example occurs in transmission between a satellite antenna and a ground-based antenna. It is often desired to make the satellite antenna behave as closely as possible to an isotropic radiator to allow for spinning. The size of the ground based antenna is set by cost considerations, so its effective area, $A_2 = (\lambda^2/4\pi)g_2$, being proportional to actual area, is essentially fixed. From (11-84) the received power under these circumstances is given by

$$\frac{W_R}{W_T} = \frac{g_1 A_2}{4\pi d^2}$$

and is *independent* of frequency.

(3) For ordinary microwave links both antennas are made directional with a size that is limited by cost considerations. In this case, for

which

$$\frac{W_R}{W_T} = \frac{A_1 A_2}{\lambda^2 d^2}$$

the received power is *directly proportional* to the square of frequency.

It is important to bear in mind that the conclusions reached above apply only for the ideal case of *free-space* transmission. In practice the propagation path almost always requires that allowances be made for the effects of the ground, the ionosphere, or the troposphere on the transmission loss. Moreover the important quantity is not received power, but rather received signal-to-noise ratio, so it is the variation with frequency of this latter factor that is most significant. Signal-to-noise ratio is considered in sec. 11.17 and the effects of the ground, troposphere and ionosphere on propagation characteristics are covered in chaps. 16 and 17.

11.17 Antenna Temperature and Signal-to-noise Ratio. The basic problem in the design of a communication channel is to receive a signal strong enough to yield an acceptable signal-to-noise ratio. If the noisiness of the system is low, the received signal can be very small indeed, and still produce a quite adequate transmission channel. The design of an antenna receiving system for satisfactory signal-to-noise output requires a knowledge of effects of both antenna noise and receiver noise.

Although noise power can be expressed in watts, or watts per hertz of band width, a more convenient measure is given in terms of temperature through the Nyquist relation. According to this relation* the noise power available from a resistor R at absolute temperature $T°$K is

$$P_a = kTB \tag{11-87}$$

where B is band width in hertz and k is Boltzmann's constant ($k = 1.37 \times 10^{-23}$ joule/°K). Because this noise power is independent of the value of R and directly proportional to T, the absolute temperature makes a convenient specification for noise power per hertz of band width. From (11-87) it is easy to calculate the thermal noise voltage across the resistor. Knowing that the (maximum) power available from any source having resistance R is $P_a = V^2/4R$, it follows that the thermal noise voltage across R at temperature $T°$K is given by

$$V = 2\sqrt{kTBR} \tag{11-88}$$

When an antenna or other source is connected to a receiver of gain G, both signal and source noise are amplified, and in addition noise

*Nyquist, H., "Thermal Agitation of Electric Charge in Conductors," *Phys. Rev.*, **32,** 110 (1928).

is added by the receiver, so that the input signal-to-noise ratio is degraded (decreased) at the output. There are several factors which can be used to describe* this result, but attention here will be confined to the approach that is particularly suitable in antenna work where very low-noise receivers are frequently utilized.

The design objective for antenna receiving systems is to produce a specified output signal-to-noise ratio with a minimum received or input signal. Considering an input signal S_A generated by a source (in this case the receiving antenna) at temperature T_A, the output signal power S from an amplifier of power gain G will be $S_A G$. The output noise power will be the sum of the amplified antenna noise and the receiver noise P_N. That is, the output noise power N is given by

$$N = kT_A BG + P_N = kT_A BG + kT_e BG$$

where T_e is the effective noise temperature† of the receiver network, referred to the input terminals. Then the output signal-to-noise ratio is given by

$$\frac{S}{N} = \frac{S_A}{(T_A + T_e)kB} \tag{11-89}$$

Hence the input signal required to produce a specified output S/N ratio is proportional to the effective antenna noise temperature T_A plus the effective noise temperature T_e of the network. From the point of view of signal-to-noise ratio, the system design calls for a maximization of $S_A/(T_A + T_e)$.

In view of expression (89) it is worthwhile to examine the range of antenna noise temperatures to be expected in practice. Fig. 11-44 shows the limits of antenna noise temperatures in the frequency range from 100 kHz to 100 MHz. It is seen that in the HF band and below, that is below 30 MHz, antenna noise temperatures can range from a low of 100° K at 2 MHz to a high of (10^{15})° K at 100 kHz. It seems hardly necessary to point out that these figures bear no relation to the

*See bibliographical references to Heffner, Susskind, Hayden and Hansen at end of chapter.

†Effective noise temperature T_e and *noise figure* (or noise factor) F_N of a network are related as follows:

F_N is the ratio of actual noise power out (when connected at input to a source at standard temperature $T_0 = 290$°K) to the noise power out that would exist for the same input, if the network were noiseless. Hence

$$F_N = \frac{kT_0 BG + P_N}{kT_0 BG} = \frac{kT_0 BG + kT_e BG}{kT_0 BG} = 1 + \frac{T_e}{T_0}$$

Note that $F_N = 1$, or 0 db corresponds to $T_e = 0$°K,
$F_N = 2$, or 3 db corresponds to $T_e = 290$°K,
and so on.

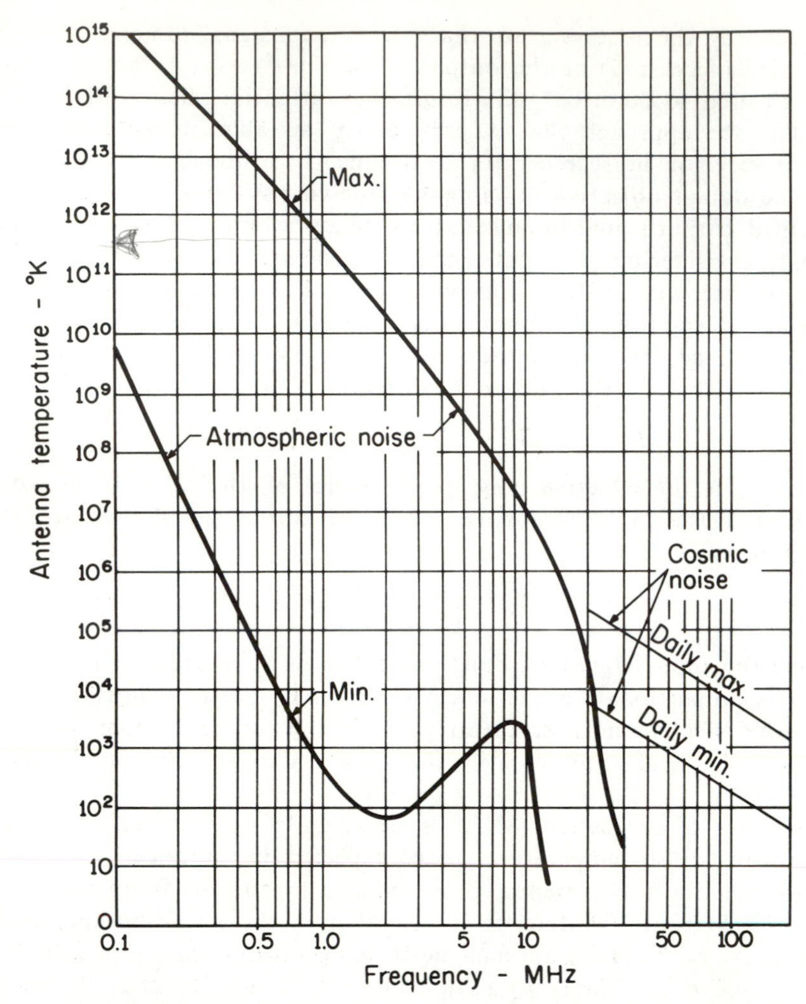

Figure 11-44. Noise temperatures at medium and high frequencies (after E. C. Hayden).

ambient temperature of the antenna but are simply a measure of the noise power (in this case atmospheric noise) available from the antenna (assumed over this frequency range to be a lossless monopole). The extraordinarily high temperatures shown are indicative of the high temperatures that would be required of a resistor to produce an equivalent amount of thermal noise according to the relation $P_a = kTB$. It is evident that over most of this frequency range the antenna noise temperature T_A is so large that it would be pointless to expend any great effort in designing a low-noise receiver to keep T_e small. A quite different situation exists, however, at higher frequencies, as can be seen

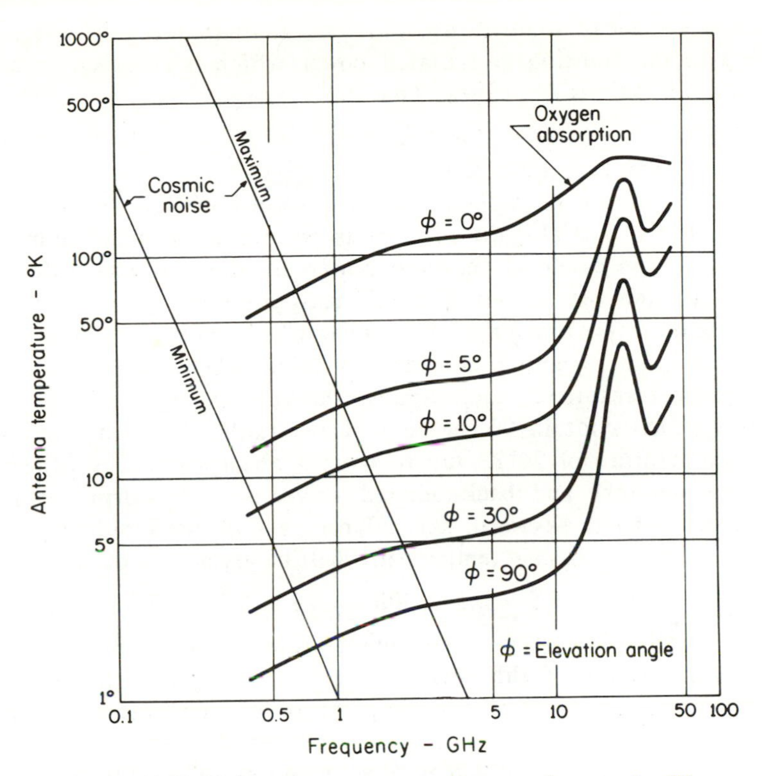

Figure 11-45. Noise temperatures at microwave frequencies (Pierce and Kompfner—after D. C. Hogg).

from Fig. 11-45. Between about 50 and 500 MHz atmospheric noise becomes negligibly small and one is concerned with cosmic or galactic noise, the daily limits of which are shown in both Figs. 11-44 and 11-45. Above about 1000 MHz cosmic noise becomes negligible and there remains only a contribution due to oxygen absorption which has a resonance peak at 60 GHz. Over this frequency range highly directive reflector-type antennas are used. In Fig. 11-45 several curves are shown corresponding to various elevation angles (above the horizon) at which the highly directive antenna is pointing. When the antenna is pointing straight up, the absorption in the thin blanket of the earth's atmosphere is small, but as the antenna points more toward the horizon, the relative thickness of the absorbing layer increases. Using the principle of detailed balancing it can be shown that the temperature "seen" by the antenna looking at a region having a temperature T is αT, where α is the fraction of the total radiation from the antenna (when transmitting) which is absorbed in the region. Also, if an antenna illuminates several regions or bodies, the antenna temperature T_A on receiving is the mean temperature of the illuminated region weighted according to the fraction of radiated power which is absorbed in each of the several regions or bodies. That is,

$$T_A = \sum \alpha_k T_k$$

where T_k = temperature of kth body

α_k = fraction of radiated power absorbed by kth body

It follows that if a high-gain antenna is pointed at a signal source in a "cold" sky it is essential that the power in the antenna side and back lobes (which see a "warm" earth at approximately $T_0 = 290°\text{K}$) be kept small if the low-temperature characteristics of the antenna are to be realized. For example, consider an antenna beamed upward at a satellite-borne transmitter, and suppose that the main beam and first side lobes, which contain 90 per cent of the radiated power, "see" an average temperature of 20°K due to oxygen absorption. The other 10 per cent of side-lobe and back-radiated power sees a warm earth at 300°K which reflects skyward, say, 50 per cent of the incident power. Then the effective antenna temperature will be given approximately by

$$T_A = 0.9 \times 20° + 0.5 \times 0.1 \times 300° + 0.5 \times 0.1 \times 20° = 34°\text{K}$$

As was indicated above, the temperature of an ideal (lossless) antenna is a measure of the noise power received by the antenna, and bears no relation to the ambient temperature of the antenna. When the antenna has ohmic loss, the situation is different. The ohmic resistance of the antenna is effectively in series with the antenna's radiation resistance and can be expected to contribute thermal noise. Since the mean-square noise voltage across resistors in series is the sum of the individual mean-square voltages, that is

$$V^2 = V_1^2 + V_2^2$$

it follows using expression (11-88) that $TR = T_1R_1 + T_2R_2$ from which the effective noise temperature of the combination is obtained as

$$T = T_1\frac{R_1}{R} + T_2\frac{R_2}{R}$$

where $R = R_1 + R_2$. Hence the temperature of each resistor adds to the total temperature in proportion to the relative magnitude of the resistance. If R_1 is the radiation resistance of an antenna which "sees" a temperature T_A, and if R_2 is the ohmic resistance of the antenna at an ambient temperature T_0, the total effective antenna temperature T_a will be

$$T_a = T_A\frac{R_1}{R} + T_0\frac{R_2}{R} = T_A\eta + T_0(1 - \eta) \tag{11-90}$$

where $\eta = R_1/(R_1 + R_2)$ is the antenna efficiency.

11.18 Space Communications. The design of space communication systems provides the engineer with challenging problems that can be made to yield quantitative answers. Unlike most other applications, the conditions of space communication can usually be approximated by "free-space propagation" assumptions, uncomplicated by the influences of the ground and ionosphere, which are often difficult to evaluate quantitatively. An example that will serve to illustrate the type of problem encountered is that of communication between two points on the earth via a 24-hour, "synchronous" satellite.

The advent of earth satellites has extended the range of line-of-sight propagation paths and made possible transoceanic transmission of microwaves with their potentiality for large band widths. The transmission path can be extended in any of several ways. The satellite, in the form of a metallic-coated plastic balloon, can be used as a passive reflector, in which case no equipment is required on the satellite. It has been estimated* that 24 such passive reflectors in random 3000-mile polar orbits can result in transatlantic transmission that would be interrupted less than 1 per cent of the time. As a second possibility the satellite can be used as an active microwave repeater, relaying the signal to the receiving location, either instantly or after storage and waiting until the satellite is near the receiving station. In this latter case the channel capacity is necessarily limited. With the satellite in a near orbit, say less than 5000 miles, the transmission loss is quite reasonable, but the ground-station antennas should be capable of scanning almost from horizon to horizon. If the satellite is placed in a 24-hour equatorial orbit (at a radius of approximately 25,000 miles) it will appear to remain fixed above some point on the equator. Three such synchronous satellites equispaced around the equator would provide nearly world-wide coverage. With the satellite "fixed" in position with respect to the earth, and stabilized in orientation, large but relatively economical antennas can be used on the ground, and an antenna with modest directivity can be used on the satellite.

EXAMPLE 4: *Synchronous Satellite Relay.* It is required to estimate the performance capabilities of a 24-hour "fixed" satellite microwave relay. From the standpoint of antenna size, and keeping in view the noise curves of Figs. 11-44 and 11-45, choose frequencies in the 1 to 10 GHz range, say $f_1 = 3$ GHz for the ground-to-satellite link, and $f_2 = 6$ GHz for the satellite-to-ground link. Then the basic transmission loss for this 25,000-mile path will be

At $f_1 = 3$ GHz or $\lambda = 0.1$ m,

*Pierce, J. R., and Kompfner, "Transoceanic Communication by Means of Satellites," *Proc. IRE*, **47**, No. 3, pp. 372–380 (1959).

$$L_b = 10 \log \left(\frac{4\pi \times 25{,}000 \times 1609}{0.1}\right)^2 = 194.5 \text{ db}$$

At $f_2 = 6$ GHz, $L_b = 200.5$ db

Using 30-meter (100-ft)-diameter antennas on the ground and a 0.3-meter-diameter on the satellite, and assuming effective areas equal to 0.65 times areas gives

$$A = 0.65\pi(15)^2 = 460 \text{ m}^2 \qquad \text{(for the ground antennas)}$$

$$A = 0.046 \text{ m}^2 \qquad \text{(for the satellite antenna)}$$

The corresponding maximum directive gains are

(a) For the ground antennas

$$G_1 = 57.5 \text{ db at } 3 \text{ GHz} \qquad G_2 = 63.5 \text{ db at } 6 \text{ GHz}$$

(b) For the satellite antenna

$$G_1 = 17.5 \text{ db at } 3 \text{ GHz} \qquad G_2 = 23.5 \text{ db at } 6 \text{ GHz}$$

With a ground transmitter power of 5 kw or 37 dbw, the power received at the satellite receiver will be

$$37 + 57.5 + 17.5 - 194.5 = -82.5 \text{ dbw}$$

This received power (0.6×10^{-8} watts) is more than adequate (see below). For the solar-battery-powered satellite transmitter the power requirements should be kept as small as possible. Using a transmitter power of 1 watt (0 dbw), the received power will be

$$0 + 23.5 + 63.5 - 200.5 \approx -113 \text{ dbw}$$

The noise power at the receiver input is

$$N = k(T_A + T_e)B \text{ watts}$$

where

$$k = 1.38 \times 10^{-23} \text{ watt/deg./hertz}$$

$$= -229 \text{ dbw for } 1\,^\circ\text{K and 1 hertz band width}$$

$$B = \text{band width, hertz}$$

From Fig. 11-45, T_A will be about 20°K if the antenna never points closer than 10 degrees to the horizon. For a parametric amplifier at this frequency T_e will be about 30°K, whereas for a maser amplifier it will be negligible. Assuming a maser amplifier and using $T_A = 20$°K the noise power per megahertz is

$$N = -229 + 13 + 60 = -156 \text{ dbw/MHz band width}$$

For a 10-MHz rf band width the noise power at the receiver input would be -146 dbw with a resulting 33-db signal-to-noise ratio. For a 100-MHz rf band width the signal-to-noise ratio would be 23 db. For an FM feedback system* an output S/N ratio of 40 db can be obtained with a signal input of 32 kTb (or 15.1 db above kTb), where b is the base modulation band width.

*Pierce and Kompfner, *loc. cit.*

For a modulation band width $b = 10$ MHz, a 310-MHz rf band width is required but the available received power need be only

$$15.1 - 146 = -131 \text{ dbw}$$

Hence the transmitter power could be reduced by nearly 18 db, that is to about 15 mw, or alternatively, smaller antennas could be used.

Problem 14. Determine approximately the minimum satellite transmitter power required for an active satellite repeater in a near orbit for which the maximum distance from satellite to ground station has been computed to be 5000 miles. Use a frequency $f = 5$ GHz, and assume the same type of FM feedback modulation as in example 4. Assume an isotropic antenna on the satellite and a 20-meter diameter ground antenna. Make any other necessary (but justifiable) assumptions.

Problem 15. It is desired to establish a transoceanic communications link using passive reflectors in 3000-mile polar orbits, for which the maximum satellite to ground station distance is 5240 miles. Using $\pi D^2/4$ for the scattering cross section of a conducting sphere of diameter D, determine the transmitter power required to receive a signal having a base band width of 5 MHz if 30-meter diameter spherical reflectors and 30-meter diameter ground antennas are used. Use $f = 6$ GHz and assume a noise temperature of 20°K.

(NOTE: *Scattering cross section,** discussed further in chap. 13, may be regarded as the area from which power would have to be extracted from the incident plane wave in order to give the same radiation intensity in a specified direction as does the obstacle, if the extracted power were re-radiated isotropically. A perfectly reflecting sphere which is large compared with a given wavelength scatters an incident plane wave approximately isotropically.)

Problem 16. By measuring the Faraday rotation of the plane of polarization of a radio wave reflected from the moon it is possible to determine the electron content of the ionosphere (chap. 17). Determine the expected signal-to-noise ratio of the received signal if the following conditions apply:

$f = 400$ MHz
Power radiated = 50 kw
Band width $B = 1$ kHz
Antenna diameters = 20 meters
Diameter of moon $D = 2160$ miles
Distance to moon $d = 225{,}000$ miles
Cosmic noise temp. $T_A = 100$°K
Receiver noise figure $F_N = 5$ db ($T_e = 630$°K)

Because the moon is neither smooth nor perfectly conducting its scattering cross section cannot be calculated. However, experimental evidence suggests an effective cross section

**IEEE Test Procedure for Antennas*, No. 149, Jan., 1965.

$$\sigma = \frac{0.5\pi D^2}{4}$$

at $f = 400$ MHz. Allow 10 db for fading from various causes (wave interference between reflections from different parts of the rough moon, moon libration, turbulence of the ionosphere, etc.)

ADDITIONAL PROBLEMS

17. Derive an expression for the radiation pattern of an antenna of length L which has a traveling-wave current distribution represented by $I = I_0 e^{-(\alpha + j\beta)l}$. The phase-shift factor β is equal to $2\pi/\lambda$ where λ is assumed to be equal to the free-space wavelength.

18. Using the principle of multiplication of patterns, sketch the following radiation patterns:

(a) The horizontal pattern of four vertical antennas spaced one-half wavelength apart and fed with equal currents, but with 180-degree phasing between adjacent elements.

(b) Same as part (2), but for eight elements.

(c) The horizontal pattern of four vertical radiators spaced one-quarter wavelength and having a progressive phase shift of 90 degrees between elements.

(d) The free-space vertical patterns (obtained for the array remote from the earth) of each of the arrays of parts (a), (b), and (c):

(1) In the plane of the array

(2) In the plane perpendicular to the plane of the array.

19. An elevated antenna is one wavelength long and is fed a quarter wavelength from one end. Assuming a sinusoidal current distribution [*not* the distribution of Fig. 11-6(*d*)] calculate its free-space radiation pattern and its radiation resistance.

20. A resonant-length dipole ($L = 2H$, slightly less than $\lambda/2$) has a free-space input impedance of $73 + j0$ ohms. What is its input impedance when placed parallel to, and a quarter-wavelength from, a large perfectly conducting screen?

21. A parasitic (unfed and short-circuited) dipole has a length of 108 cm and a radius of 0.5 cm. Determine the magnitude and phase of the current in it when placed parallel to and 0.1 wavelength from a half-wave dipole carrying 1 amp. Frequency = 150 MHz. From curves, find $Z_{12} = 68 + j10$ approximately. What is the input impedance of the driven dipole (antenna 1) if it is assumed that $Z_{11} \approx 73$ ohms?

BIBLIOGRAPHY

Hansen, R. C., "Low Noise Antennas," *Microwave J.*, **2**, No. 6, 1959, pp. 19-24.

Hayden, E. C. *et al.*, "Excitation of Currents on Aircraft," Final Report

W 33-038-ac 20778, Antenna Laboratory, Department of Electrical Engineering, University of Illinois, February, 1953.

Heffner, H., "Low Noise Devices," chap. 15, in *Foundations of Future Electronics*, D. B. Langmuir and W. D. Hershberger, eds., McGraw-Hill Book Company, New York, 1961, p. 407.

Susskind, Charles, *Encyclopedia of Electronics*, Reinhold Publishing Corp., New York, 1962, pp. 550–554.

Chapter 12

ANTENNA ARRAYS

12.01 The Mathematics of Linear Arrays. The binomial array discussed in chap. 11 is but one example of a large class of linear arrays, having special current distributions by means of which the radiation patterns can be made to have almost any prescribed shape. Schelkunoff has shown* that linear arrays can be represented as polynomials and that this representation becomes a very useful tool in the analysis and synthesis of antenna arrays.

For a general linear array of equally spaced elements (Fig. 12-1), the relative amplitude of the radiated field strength is given by

$$|E| = |a_0 e^{j\alpha_0} + a_1 e^{j\psi + j\alpha_1} + a_2 e^{j2\psi + j\alpha_2} + \dots \\ + a_{n-2} e^{j(n-2)\psi + j\alpha_{n-2}} + e^{j(n-1)\psi}| \tag{12-1}$$

where $$\psi = \beta d \cos\phi + \alpha, \qquad \beta = \frac{2\pi}{\lambda}$$

In this expression d is the spacing between elements. The coefficients a_0, a_1, a_2, etc., are proportional to the current amplitudes in the respective elements. α is the progressive phase shift (lead) from left to right; α_1, α_2,

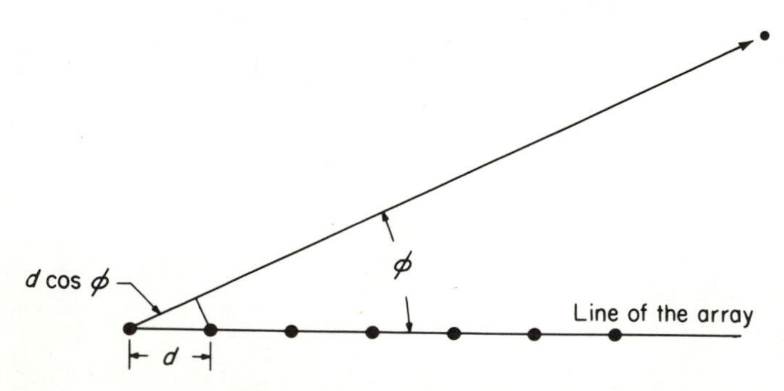

Figure 12-1. A linear array.

*S. A. Schelkunoff, "A Mathematical Theory of Linear Arrays," *BSTJ* **22**, 1, 80–107 (1943).

etc., are the deviations from this progressive phase shift. Expression (1) may be written

$$|E| = |A_0 + A_1 z + A_2 z^2 + \ldots + A_{n-2} z^{n-2} + z^{n-1}| \tag{12-2}$$

where $z = e^{j\psi}, \qquad A_m = a_m e^{j\alpha_m}$

The coefficients A_1, A_2, etc., are now complex and indicate the amplitude of current in each element and the phase deviation of that current from the progressive phase shift of the array. If any of the coefficients are zero, the corresponding element of the array will be missing, and the actual separation between adjacent elements can be greater than the "apparent separation" d. The apparent separation is the greatest common measure of the actual separations.

The following fundamental theorems are due to Schelkunoff, and lay the foundations for the method:

THEOREM I: Every linear array with commensurable separations between the elements can be represented by a polynomial, and every polynomial can be interpreted as a linear array.

Since the product of two polynomials is a polynomial, a corollary to Theorem I is:

THEOREM II: There exists a linear array with a space factor equal to the product of the space factors of two linear arrays.

THEOREM III: The space factor of a linear array of n apparent elements is the product of $(n - 1)$ virtual couplets with their null points at the zeros of E. [eq. (2)]

The *space factor* of an array is defined as the radiation pattern of a similar array of *nondirective* or isotropic elements. The degree of the polynomial which represents an array is always one less than the apparent number of elements. The actual number of elements is at most equal to the apparent number. The total length of the array is the product of the apparent separation and the degree of the polynomial.

Consider a simple two-element array in which the currents in the elements are equal in magnitude. The relative radiation field strength is represented by

$$|E| = |1 + z|$$

where

$$z = e^{j(\beta d \cos \phi + \alpha)} \tag{12-3}$$

Making use of Theorem II, a second array can be constructed which will have a radiation pattern that is the square of that given by (3), that is,

$$|E| = |1 + z|^2 = |1 + 2z + z^2|$$

It is seen that the array that will produce this pattern is a three-element array having the current ratios

$$1 : 2 : 1$$

The current in the center element will lead the left-hand element by α, and the current in the right-hand element will lead that in the left-hand element by 2α.

If the polynomial of (3) is raised to the mth power, there results the general binomial array already discussed. When the element spacing d is not greater than $\lambda/2$, such an array produces a pattern with no secondary lobes. However, the principal lobe is considerably broader than that produced by a uniform array having the same number of elements. An array having a narrower principal lobe than that given by the binomial distribution and smaller secondary lobes than that given by the uniform distribution can be obtained by raising the polynomial of the uniform array of n elements (where $n > 2$) to any desired power.

For an n-element uniform array

$$|E| = |1 + z + z^2 + \ldots + z^{n-1}| \tag{12-4}$$

It has already been shown that when n, the number of elements, is large, the ratio of the principal maximum to the first secondary maximum is approximately independent of n and is 13.5 db for the uniform array. If an array is formed to produce a pattern that is the *square* of that given by (4), the ratio of the principal to first secondary maximum will be 27 db. This second array is given by

$$\begin{aligned} |E| &= |1 + z + z^2 + \ldots + z^{n-1}|^2 \\ &= |1 + 2z + 3z^2 + \ldots + nz^{n-1} + (n-1)z^n + \ldots \\ &\quad + 2z^{2n-3} + z^{2n-2}| \end{aligned} \tag{12-5}$$

The current ratios for this array have the triangular distribution

$$1, 2, 3 \ldots (n-1), n, (n-1), \ldots 3, 2, 1$$

Raising the uniform array to a still higher power would, of course, increase still further the ratio of principal to secondary lobes. The respective patterns for the uniform, binomial, and triangular distribution are shown in Fig. 12-2.

The significance of Theorem III, the decomposition theorem, can be understood by studying the variable z, where

$$z = e^{j\psi} \qquad \psi = \beta d \cos \phi + \alpha$$

Since ψ is real, $j\psi$ is a pure imaginary, and the absolute value of z is always unity. Plotted in the complex plane, z is always on the circumference of the unit circle (Fig. 12-3).

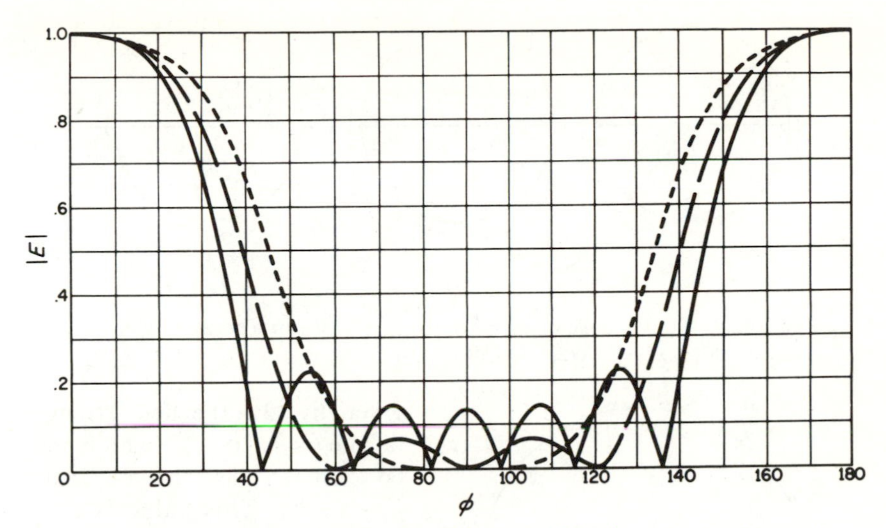

Figure 12-2. Radiation patterns for uniform (solid), triangular (dashed) and binomial (short dashed) amplitude distributions. (Courtesy *Bell System Technical Journal*)

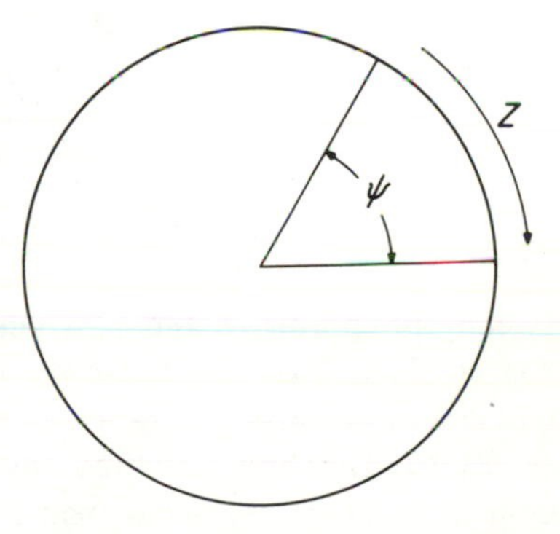

Figure 12-3. As ϕ increases from 0 degrees to 180 degrees, z moves in a clockwise direction on the unit circle.

As ϕ increases from zero to 180 degrees ψ decreases from $\beta d + \alpha$ to $-\beta d + \alpha$ and z moves in a clockwise direction. Because of symmetry the range of ϕ to be considered is from zero to 180 degrees. Thus the range* of ψ described by z is $\psi = 2\beta d$ radians. For example, for a separation between elements of $\lambda/4$, ψ varies through π radians as ϕ goes from zero to 180 degrees, and z describes a semicircle. (z retraces its path to the starting point as ϕ goes from 180 degrees to 360 degrees, and the pattern is symmetrical about the 0–180-degree line.) For $d = \lambda/2$ the range of ψ is 2π radians and z describes a complete circle as ϕ varies from zero to 180 degrees. If d is greater than $\lambda/2$, the range of ψ is greater than 2π, and z will overlap itself. The geometrical representation of Fig. 12-4 makes it a simple matter to observe the radiation characteristics as z moves around the circle within its range of operation. For example, for the simple two-element uniform array

*This is the "visible" range of ψ. Operation in the "invisible" region, outside of the visible range, is discussed later.

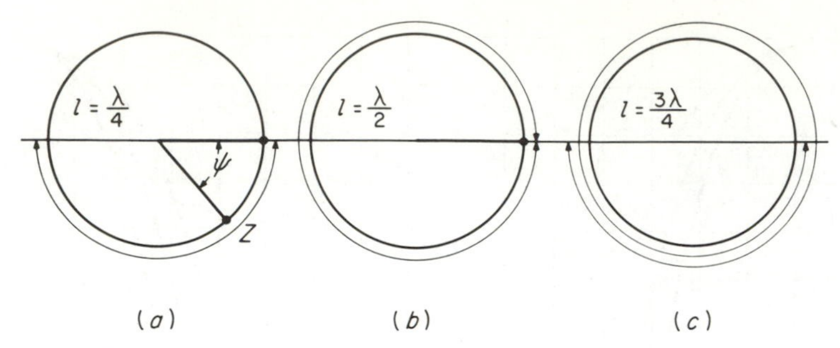

Figure 12-4. Active range of z (shown for $\alpha = -\beta l$) for a separation between elements of (*a*) $\lambda/4$, (*b*) $\lambda/2$, (*c*) $3\lambda/4$.

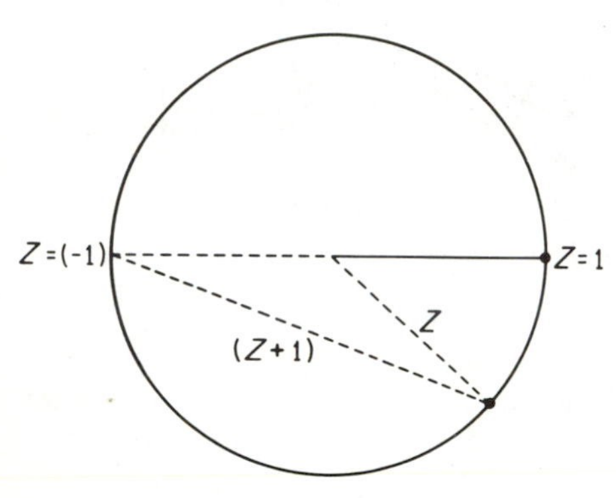

Figure 12-5.

given by (3), the field strength is the *sum* $|z + 1|$, which may be written as the difference $|z - (-1)|$. This value is given geometrically by the distance between z and the point -1 (Fig. 12-5). For the more general case of unequal amplitudes, where the source intensities are proportional to 1 and $-t$, the radiated field strength pattern is given by $|z - t|$ which geometrically is the distance between the points z and t. Since z is always on the unit circle, the pattern will have a zero only when t is also on the unit circle, and when t is within the range of z.

By the fundamental theorem of algebra, a polynomial of the $(n - 1)$th degree has $(n - 1)$ zeros (some of which may be multiple zeros) and can be factored into $(n - 1)$ binomials. Thus

$$|E| = |(z - t_1)(z - t_2) \ldots (z - t_{n-1})| \tag{12-6}$$

from which Theorem III follows directly.

It is evident that the relative radiation field in any direction is given by the products of the distances from z (corresponding to the chosen direction) to the null points of the array.

EXAMPLE 1: *Uniform Array.* Consider the case of the uniform array that is represented by

$$|E| = |1 + z + z^2 + \ldots + z^{n-1}|$$
$$= \left|\frac{1 - z^n}{1 - z}\right| = \left|\frac{z^n - 1}{z - 1}\right| \tag{12-7}$$

The null points of such an array, given by the roots of (7), are in this case the nth roots of unity (excluding $z = 1$, which is the principal maximum). In the complex plane the roots of unity all lie on the unit circle, and divide the circle into n equal parts (Fig. 12-6). The roots are

$$z = e^{-j(2\pi/n)}, e^{-j2(2\pi/n)}, \dots e^{-jm(2\pi/n)}$$

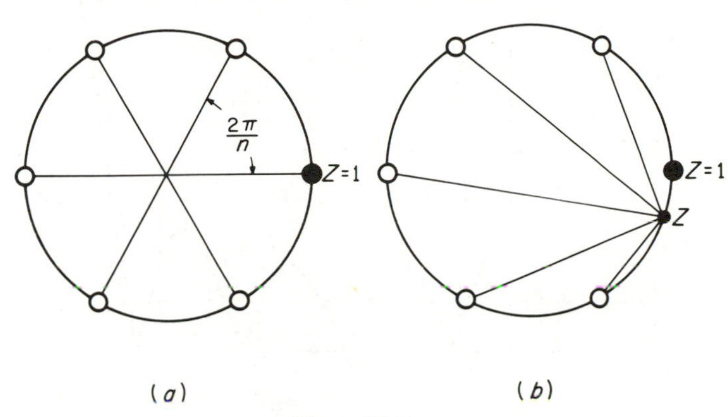

Figure 12-6.

It is seen that the null points of the array are given by $\psi_m = -m2\pi/n$ where $m = 1, 2, 3, \dots (n-1)$. Since $\psi = \beta d \cos\phi + \alpha$, the null points of the radiation pattern are given in terms of the angle ϕ by

$$\cos\phi_m = -\frac{\alpha}{\beta d} - \frac{2\pi m}{n\beta d}$$

When $z = 1$, $|E|$ has a principal maximum. Other maxima occur approximately midway between the nulls. As z moves around the circle the radiation pattern is given by the product of the lines connecting the null points to z. A plot of $|E|$ as a function of ψ is shown in Fig. 12-7. Using

$$\phi = \cos^{-1}\left(\frac{\psi - \alpha}{\beta d}\right)$$

$|E|$ can be drawn as a function of ϕ.

EXAMPLE 2: *Four-element Broadside.* A simple array, which has already been considered is the four-element broadside having half-wave spacing between elements and equal currents fed in phase. For this case

$$\beta d = \pi \qquad \alpha = 0 \qquad \psi = \pi \cos\phi$$

The range of z is $\psi = 2\beta d = 2\pi$.

The relative field strength pattern is given by

$$\begin{aligned}|E| &= |1 + z + z^2 + z^3| \\ &= \left|\frac{z^4 - 1}{z - 1}\right| \\ &= |(z - e^{-j(\pi/2)})(z - e^{-j\pi})(z - e^{-j(3\pi/2)})| \qquad (12\text{-}8)\end{aligned}$$

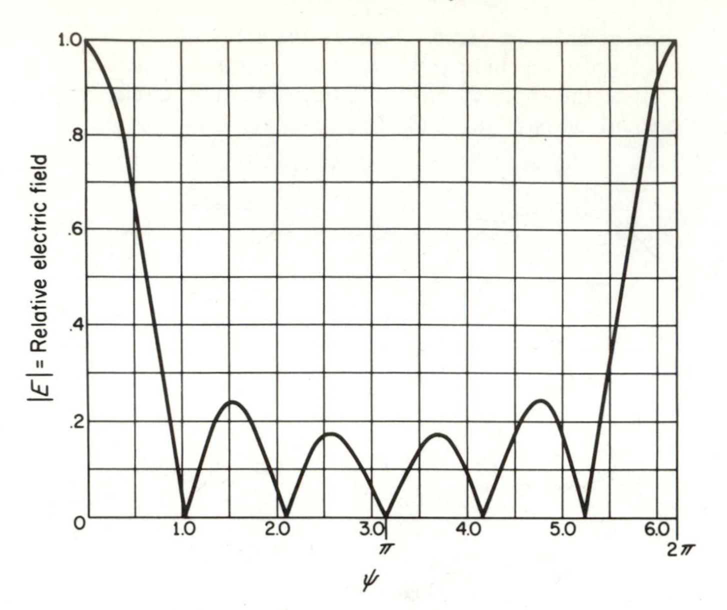

Figure 12-7. Relative field strength $|E|$ as a function of ψ.

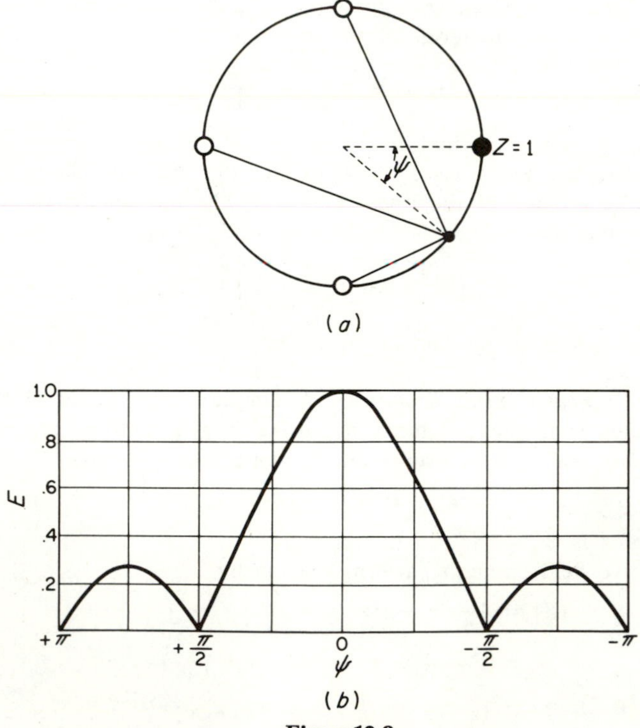

Figure 12-8.

The nulls are spaced equally on the unit circle as shown in Fig. 12-8(*a*). As ϕ increases from 0 to 180 degrees, ψ decreases from π through zero to $-\pi$ and the curve of Fig. 12-8(*b*) results. This is plotted in polar co-ordinates as a function of ϕ in Fig. 12-9.

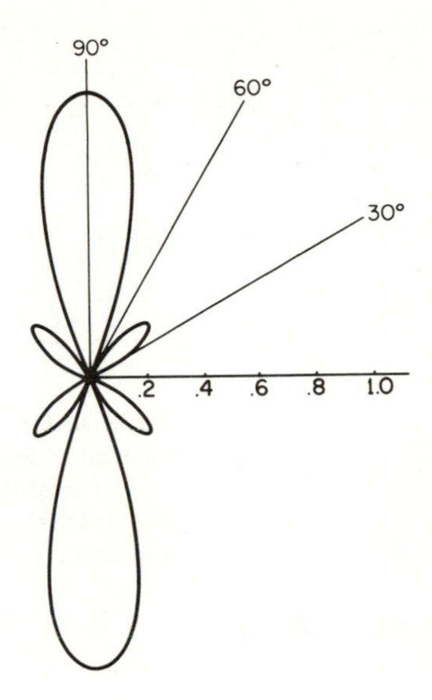

Figure 12-9. Relative field strength as a function of ϕ.

EXAMPLE 3: *Four-element End-fire.* Consider a uniform four-element end-fire array having an element spacing of one-quarter wavelength and a progressive phase shift of $-\pi/2$ radians. For this array

$$\beta d = \frac{\pi}{2} \qquad \alpha = -\frac{\pi}{2}$$

and $$\psi = \beta d \cos \phi + \alpha = \frac{\pi}{2}(\cos \phi - 1)$$

The range of ψ is π radians.

As before, the expression for $|E|$ is given by eq. (8) and the three nulls occur at $\psi = -\pi/2, -\pi, -3\pi/2$. However, in this case the range of ψ is only from $\psi = 0$ to $\psi = -\pi$ (Fig. 12-10), so the null at $-3\pi/2$ obviously has very little effect on the pattern. An improved pattern (that is, one with a narrower principal lobe and smaller secondary lobes) can be obtained with the same number of elements by spacing the nulls equally in the range of ψ. This gives rise to the array that has the

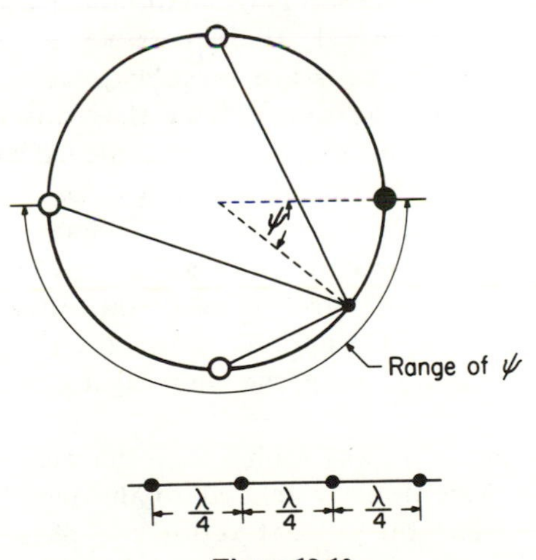

Figure 12-10.

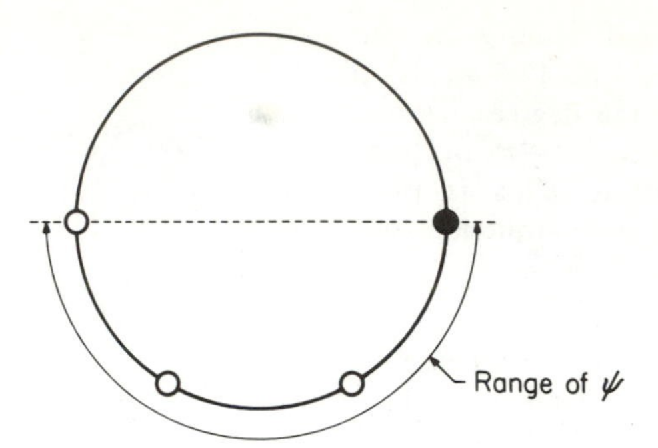

Figure 12-11. Circle diagram for a four-element array having nulls equispaced in the range of ψ. For an element spacing of one-quarter wavelength the range of ψ is π radians.

circle diagram of Fig. 12-11 and the pattern given by

$$\begin{aligned} |E| &= |(z - e^{-j(\pi/3)})(z - e^{-j(2\pi/3)})(z - e^{-j\pi})| \\ &= |z^3 + (1 - e^{-j(\pi/3)} - e^{-j(2\pi/3)})z^2 + (-1 - e^{-j(\pi/3)} - e^{-j(2\pi/3)})z - 1| \\ &= |1 + 2e^{-j(\pi/3)}z + 2e^{-j(2\pi/3)}z^2 + e^{-j\pi}z^3| \end{aligned} \tag{12-9}$$

The current amplitudes of the array are

$$1:2:2:1$$

and the progressive phase shift between elements will be $-\pi/2 - \pi/3 = -5\pi/6$ radians. The resultant pattern as a function of ϕ is shown as curve B in Fig. 12-12.

If the overall length of the array is maintained constant, but the number of elements is increased, it is possible to improve the directivity still further if the nulls are properly spaced in the range of operation. (However, it will be shown later that this improvement in directivity over that of a uniform array is achieved at the expense of other factors of antenna performance.) Curve C of Fig. 12-12 shows the pattern that results when the number of elements is increased to seven with the spacing reduced to one-eighth wavelength, so that the overall length is still $\frac{3}{4}\lambda$. To obtain this result the nulls were equispaced in the range $\psi = 2\beta d = \pi/2$. Curve D shows the pattern obtained for 13 elements at $\lambda/16$ spacing, with the nulls again equispaced in the range of ψ.

For the uniform array it was found that the maximum directivity and gain obtainable were directly related to the length of the array. In contrast to this, when the current ratios and phasings are properly chosen, it appears possible to obtain arbitrarily sharp directivity with an array of fixed length by using a sufficiently large number of

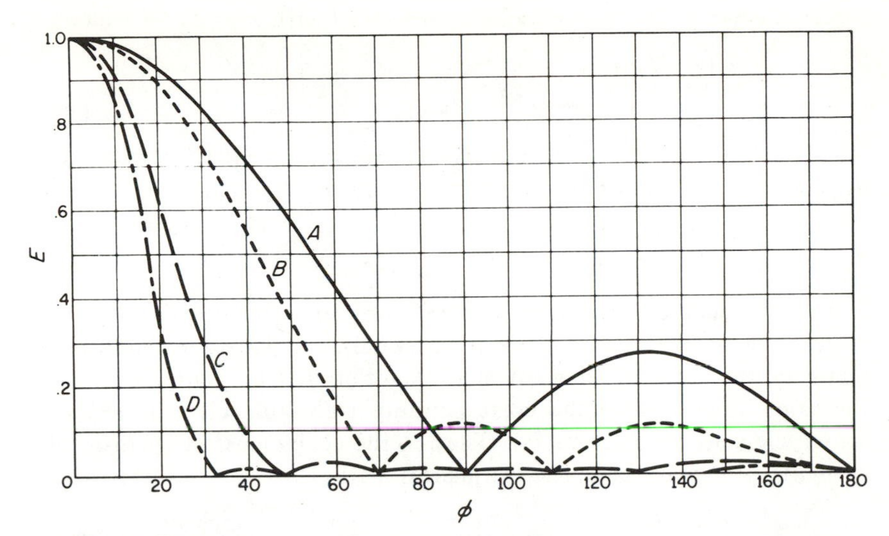

Figure 12-12. Radiation patterns for several end-fire arrays, all having an array length of $3\lambda/4$. Shown at *A* is a four-element uniform array ($d = \lambda/4$); *B*, four elements with nulls equispaced in the range of ψ ($d = \lambda/4$); *C*, seven elements with nulls equispaced in the range of ψ ($d = \lambda/8$); *D*, thirteen elements with nulls equispaced in the range of ψ ($d = \lambda/16$). (Courtesy *Bell System Technical Journal*)

elements. However, it will be found also that with the phase relations and close spacings between the elements required to obtain this result, the radiation resistance is reduced to extremely low values. That is, extremely large currents are required to produce radiation fields of appreciable strength. Associated with these large currents are large amounts of stored or reactive energy. With actual antennas that have a finite ohmic resistance, the antenna efficiency enters the picture to limit the directivity and gain that can be obtained from an array of given length.

Problem 1. (a) Draw the circle diagram and sketch the pattern of a three-element, uniform, end-fire array, using $d = \lambda/2$. (b) Using the same number of elements and same spacing, redesign the array to have nulls at $\phi = 90$ degrees and $\phi = 60$ degrees.

12.02 Antenna Synthesis. It is a simple and straightforward job to compute the radiation pattern of an array having specified configuration and antenna currents. A somewhat more difficult problem is the design of an array to produce a prescribed radiation pattern. Making use of Fourier analysis, the methods of the preceding sections may be extended to accomplish this result.

It is convenient to consider an array having an odd number of elements with a certain symmetry of current distribution about the center element. The polynomial for an array with $n = 2m + 1$ elements is

$$|E| = |A_0 + A_1 z + A_2 z^2 + \ldots + A_m z^m + A_{m+1} z^{m+1} + \ldots + A_{2m} z^{2m}| \tag{12-10}$$

Now the absolute value of z is always unity, so equation (10) can be divided by z^m without changing the value of $|E|$. That is

$$|E| = |A_0 z^{-m} + A_1 z^{-m+1} + \ldots + A_{m-1} z^{-1} + A_m + A_{m+1} z + \ldots + A_{2m} z^m| \tag{12-11}$$

It is now specified that the currents in corresponding elements on either side of the center element be equal in magnitude, but that the phase of the left-side element shall lag that of the center element by the same amount that the corresponding right-side element leads the center element (or vice versa). That is, the coefficients of corresponding elements are made complex conjugates with

$$A_m = a_0 \qquad A_{m-k} = a_k - jb_k \qquad A_{m+k} = a_k + jb_k$$

Then the sum of terms of two corresponding elements may be written

$$\begin{aligned} A_{m-k} z^{-k} + A_{m+k} z^k &= a_k(z^{+k} + z^{-k}) + jb_k(z^k - z^{-k}) \\ &= 2a_k \cos k\psi - 2b_k \sin k\psi \end{aligned}$$

since

$$z^k = e^{jk\psi}$$

The expression for $|E|$ is now*

$$\begin{aligned} |E| &= 2[\tfrac{1}{2}a_0 + a_1 \cos \psi + \ldots + a_m \cos m\psi \\ &\quad - (+b_1 \sin \psi + \ldots + b_m \sin m\psi)] \\ &= 2\left\{\frac{a_0}{2} + \sum_{k=1}^{k=m} [a_k \cos k\psi + (-b_k) \sin k\psi]\right\} \end{aligned} \tag{12-12}$$

These are the first $2m + 1$ terms of a Fourier series in which the coefficients of the cosine terms are the a_k's, and the coefficients of the sine terms are ($-b_k$'s). Now any radiation pattern specified as a function $f(\psi)$ may be expanded as a Fourier series with an infinite number of terms. Such a pattern may be approximated to any desired accuracy by means of the finite series (12). When this is done the required current distribution of the array can be written down directly. From the theory of Fourier series, this approximation is in the least-mean-square sense; i.e., the mean-square difference between the desired and the approximate pattern for ψ from 0 to 2π is minimized.

*Since $\cos m\psi$ and $\sin m\psi$ can be expressed as polynomials of $\cos \psi$ and $\sin \psi$, it may be stated, as a corollary to Schelkunoff's theorem, that for a symmetrically excited cophasal array, E can be expressed in terms of a trigonometric polynomial. Because ψ is real, the visible range is on the real axis instead of a unit circle as in the previous case. This fact will be used later in sec. 12.04 for relating the pattern function to the Tchebyscheff polynomial.

EXAMPLE 4: *Synthesized Bidirectional Array.* Design an array that will produce approximately a pattern of Fig. 12-13. This pattern is defined by

$$f(\phi) = 1 \qquad 0 < \phi < \frac{\pi}{3}$$

$$f(\phi) = 0 \qquad \frac{\pi}{3} < \phi < \frac{2\pi}{3}$$

$$f(\phi) = 1 \qquad \frac{2\pi}{3} < \phi < \pi$$

It will, of course, by symmetrical about the line of the array, $\phi = 0$. If the spacing is chosen to be $\lambda/2$, then $\psi = \pi \cos \phi + \alpha$. The corresponding ψ function is

$$F(\psi) = 1 \qquad \pi + \alpha > \psi > \frac{\pi}{2} + \alpha$$

$$F(\psi) = 0 \qquad \frac{\pi}{2} + \alpha > \psi > -\frac{\pi}{2} + \alpha$$

$$F(\psi) = 1 \qquad -\frac{\pi}{2} + \alpha > \psi > -\pi + \alpha$$

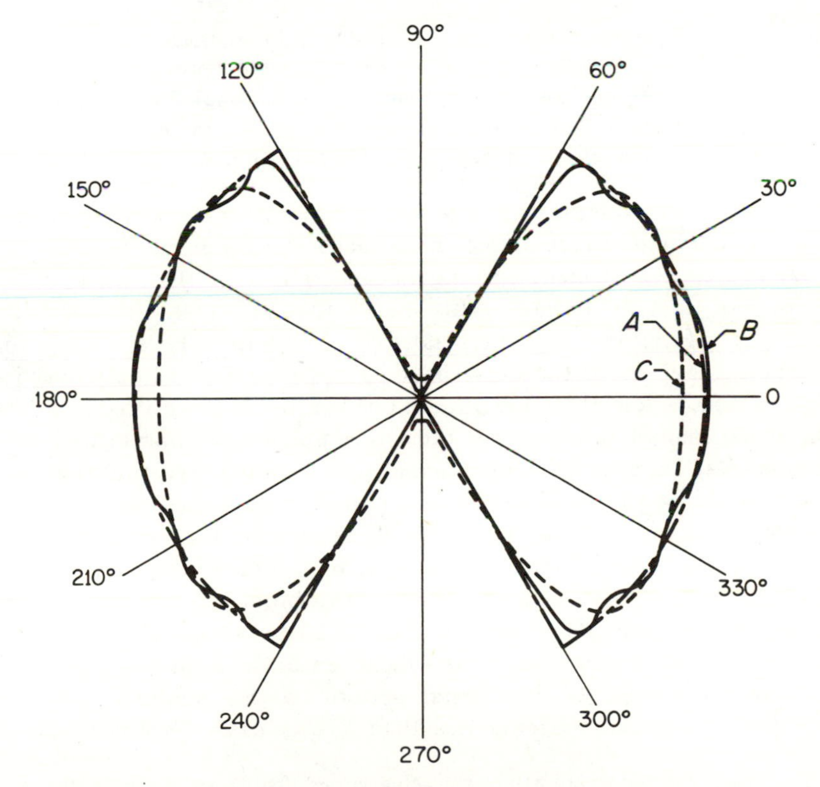

Figure 12-13. A prescribed pattern, *A*, and approximations to it, obtained with an eleven-element array, *B*, and a five-element array, *C*.

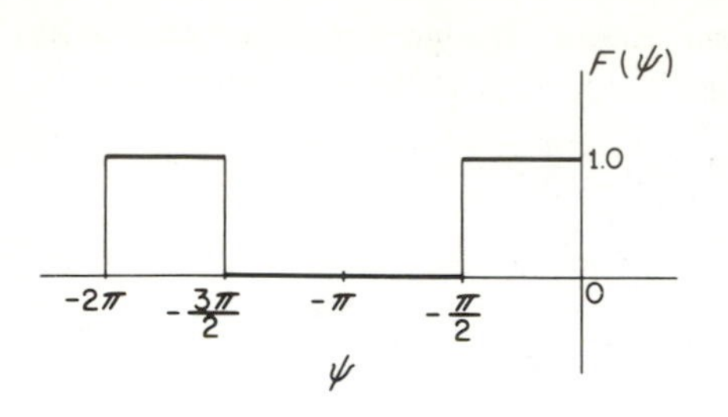

Figure 12-14.

Choosing $\alpha = -\pi$ for an end-fire array results in the function shown in Fig. 12-14. The Fourier series expansion for this function is

$$F(\psi) = \left(\frac{1}{2} + \frac{2}{\pi}\sum_{k=1}^{k=\infty}\frac{1}{k}\sin\frac{k\pi}{2}\cos k\psi\right)$$

Comparison with (12) determines the coefficients

$$a_0 = \frac{1}{2}$$

$$a_k = \frac{1}{k\pi}\sin\frac{k\pi}{2}$$

$$b_k = 0 \quad k \neq 0$$

The pattern obtained using the value of $m = 4$, is given from eq. (11) as

$$|E| = \frac{1}{\pi}\left|-\frac{1}{3}z^{-3} + z^{-1} + \frac{\pi}{2} + z - \frac{1}{3}z^3\right| \tag{12-13}$$

This is a five-element array having the current ratios indicated and an overall length of three wavelengths (the apparent spacing between elements is one-half wavelength, but four of the elements are missing). The pattern produced by this array is shown in Fig. 12-13. Also shown in this figure is the pattern obtained with an 11-element array formed using $m = 9$ in the series.

In the above example the apparent element spacing was arbitrarily chosen as one-half wavelength, which made the range of ψ equal to 2π radians. If the element spacing is less than $\lambda/2$, the range of ψ will be less than 2π radians. This means that although the radiation pattern as a function of ϕ, that is $f(\phi)$, is completely specified for the whole range of ϕ, the corresponding $f(\psi)$ is specified only over its range, which is less than the interval of 2π radians required for the Fourier expansion. It is possible then to complete the interval with any function that satisfies Dirichlet's conditions. Naturally, the function chosen would be one which would simplify the series as much as possible or make it converge rapidly. It is evident that when the apparent spacing is less than $\lambda/2$ there is an unlimited number of solutions that will satisfy the conditions of the problem. If the apparent spacing of the elements is greater than $\lambda/2$, the range of ψ is more than 2π radians. Except for some special cases* it is then not possible to obtain the prescribed directional pattern by this method.

When an apparent spacing less than $\lambda/2$ is used, $f(\psi)$ is specified

*Examples of cases where a larger spacing is permissible are given in the following article: Irving Wolff, "Determination of the Radiating System Which Will Produce a Specified Directional Characteristic," *Proc. IRE,* **25**, 5, 630 (1937).

over only a portion of the required 2π radians, and the function used to fill in the remainder of the interval can be chosen at will by the designer. A judicious choice of a "fill-in" function will produce a desirable pattern with a minimum number of elements, and conversely a poor choice of function may result in a poor pattern or in high Q factor for the array. An example will illustrate this point.

EXAMPLE 5: *Synthesized Unidirectional Array.* Let it be required to design an end-fire array that will have an approximately semicircular pattern given by

$$f(\phi) = 1 \qquad 0 < \phi < \frac{\pi}{2}$$

$$f(\phi) = 0 \qquad \frac{\pi}{2} < \phi < \pi$$

The apparent spacing is to be $\lambda/4$.

Then, for this problem,

$$\psi = \frac{\pi}{2} \cos \phi + \alpha$$

and the range of ψ is π radians. By choosing different values of α, the range of ψ which is used can be shifted anywhere in the interval of 2π radians, which is required for the Fourier expansion. This is shown in Fig. 12-15 for

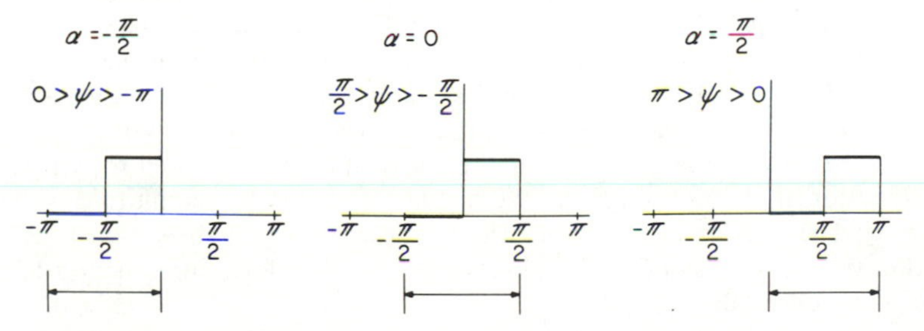

Figure 12-15. Range of ψ for $0 < \phi < \pi$ for three different values of α. (Range used is indicated by double arrows.)

three values of α. In this example there is a finite discontinuity within the range of ψ, so the coefficients of the series will decrease at a rate that is of the order of $1/n$. (They will be less in absolute magnitude than c/n, where c is some positive constant.*) If the functions were continuous in the range, the series would converge at a rate that would be at least of the order of $1/n^2$. Because no choice of a fill-in function can remove this discontinuity within the range of ψ, it is anticipated that, in this case, the fill-in function may not have much effect on the number of terms required. However, it is interesting to examine some actual cases.

*Doherty and Keller, *Mathematics of Modern Engineering,* John Wiley & Sons, Inc., New York, 1936, **1**, p. 89.

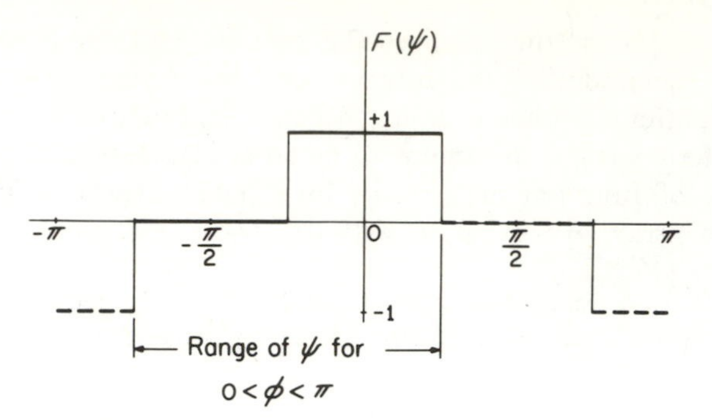

Figure 12-16. A possible choice of fill-in function (shown dashed) for $\alpha = -\pi/4$.

CASE 1: A possible choice for α and for the fill-in function is illustrated in Fig. 12-16. This choice would appear to be good, because it results in the following conditions:

(1) $f(\pi + \psi) = -f(\psi)$. Therefore, only odd harmonics will be present.

(2) $f(\psi)$ is an even function, so the coefficients of the sine terms will be zero.

(3) $f(\psi)$ has an average value of zero, so the d-c term (or center element) is eliminated.

The antenna array resulting from the choice used in Fig. 12-16 is shown in Fig. 12-17(*a*) for $m = 7$ (eight elements), and the corresponding pattern is shown in Fig. 12-17(*b*). This pattern has two serious defects. It approaches a relative value of 0.5 at $\phi = 0$ where it should be unity, and it also approaches 0.5 at $\phi = \pi$ where it should be zero. Using more terms of the series will not remedy these defects, which are inherent in the particular function used in Fig. 12-16. This function is discontinuous at the values of ψ corresponding to $\phi = 0$ and $\phi = \pi$, and the series converges to the average of the values taken by the function on the two sides of the discontinuity. Therefore, the function of Fig. 12-16 is an unsuitable choice.

CASE 2: The discontinuities at values of ψ corresponding to $\phi = 0$ and $\phi = \pi$ can be eliminated by a different choice of α, and a suitable fill-in function. A possible function is that shown in Fig. 12-18(*a*). The corresponding antenna array and resulting pattern are also shown, and it is seen that this function is a suitable one.

Whereas many other types of fill-in functions are possible, it is found for this example, where the apparent spacing is fixed at $\lambda/4$, that none of them results in appreciable improvement over the design of Fig. 12-18. However, if the apparent spacing d is permitted to have other values, this puts one more variable under the control of the designer. For a given number of antenna elements it is in general possible, with this additional control, to improve the pattern obtained. In the present example for a given number of elements, an apparent spacing of $3\lambda/8$ instead of $\lambda/4$ results in a closer approach to the ideal pattern in the critical regions, $\phi = \pm\pi/2$.

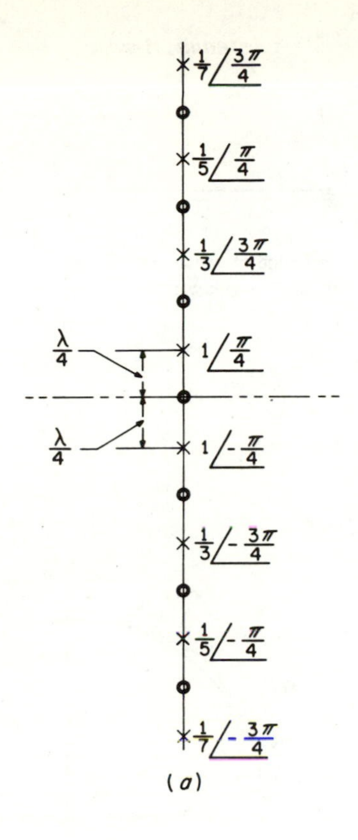

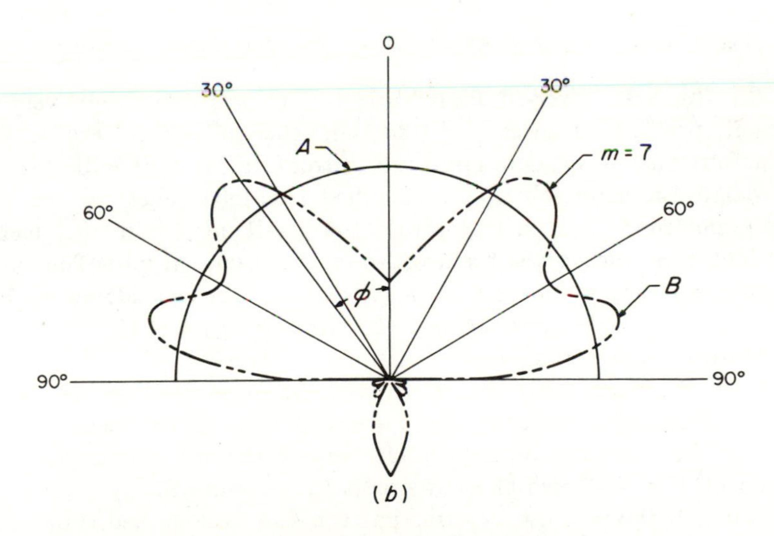

Figure 12-17. (*a*) Array and (*b*) pattern corresponding to Fig. 12-16. [Circles in (*a*) indicate elements which drop out because of zero current.]

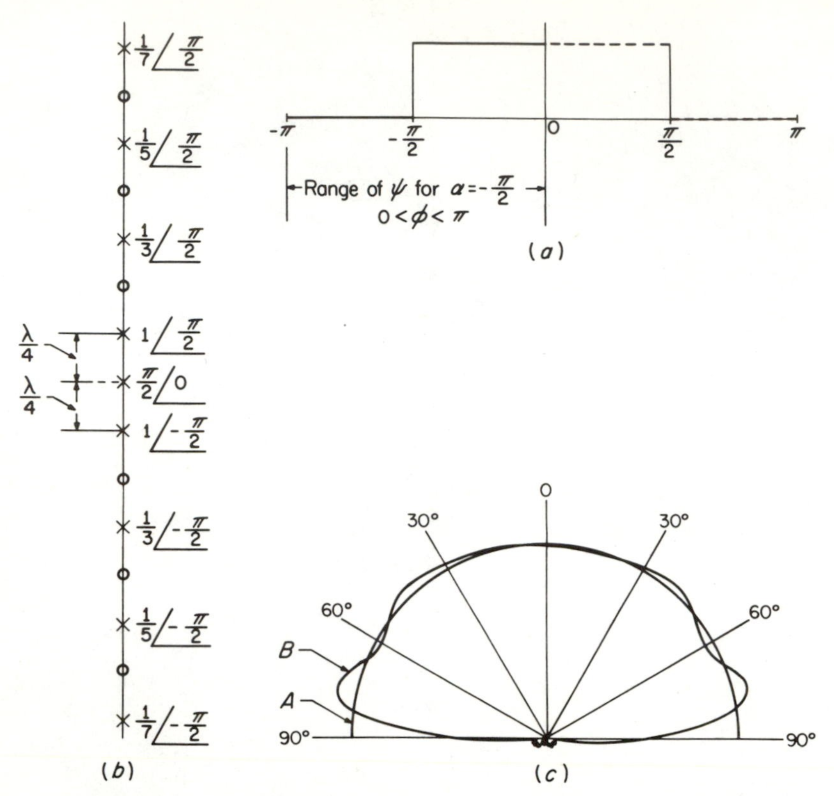

Figure 12-18. A better design (*a*) results in the array (*b*) and the pattern (*c*).

12.03 The Tchebyscheff Distribution. A particular, but very important, problem in antenna synthesis is the following: For a given linear antenna array, determine the current ratios that will result in the narrowest main lobe, for a specified side-lobe level; or, in other words, determine the current ratios that will result in the smallest side-lobe level for a given beam-width of the principal lobe. The current distribution that produces such a pattern will be considered as being the optimum *in the above sense.* The directive gain of this "optimum" distribution is less than that of a uniform array.

From the material of sec. 1, it will be recalled that a desirable pattern (but not necessarily the optimum) can be obtained by equispacing the nulls on the appropriate arc of the unit circle. An examination of Fig. 12-7, which shows a pattern obtained by equispacing the nulls, indicates how a better pattern can be obtained. For a given width of principal lobe, the first secondary lobe can be decreased by moving the second null closer to the first. Of course, this increases the second side lobe, but that is permissible as long as it does not exceed the first. It is evident that the *optimum* pattern is obtained when all the side lobes have the same level. The problem is simply that of finding the spacing of nulls which makes this true. The answer is given in terms of the Tchebyscheff polynomials.

The *Tchebyscheff* polynomials* occur quite frequently in design and synthesis problems. They are defined† by

$$T_m(x) = \cos(m \cos^{-1} x) \qquad -1 < x < +1$$

$$T_m(x) = \cosh(m \cosh^{-1} x) \qquad |x| > 1$$

The general shape of $T_m(x)$ is shown in Fig. 12-19 for both m even and m odd.

By inspection, $\qquad T_0(x) = 1, \qquad T_1(x) = x$

The higher-order polynomials can be derived as follows:

$$T_2(x) = \cos(2 \cos^{-1} x) = \cos 2\delta$$

where $\qquad \delta = \cos^{-1} x \qquad$ or $\qquad x = \cos \delta$

Now since $\qquad \cos 2\delta = 2 \cos^2 \delta - 1$

$$T_2(x) = 2x^2 - 1$$

Similarly, it can be shown that

$$T_{m+1}(x) = 2T_m(x)T_1(x) - T_{m-1}(x)$$

so that

$$T_3(x) = 4x^3 - 3x$$

$$T_4(x) = 8x^4 - 8x^2 + 1$$

and so on.

The important characteristic of the Tchebyscheff polynomials, as far as antenna pattern synthesis is concerned, is evident from Fig. 12-19. As x is allowed to vary from some point c up to a value x_0 and then back to its starting point, the function $T_m(x)$ traces out a pattern consisting of several small side lobes and one major lobe. The secondary lobes will all be of equal amplitude (unity) and will be down from the main lobe by the ratio $1/b$. This ratio can be chosen at will by suitable choice of x_0. Such a pattern will be called the optimum or Tchebyscheff pattern.‡ Since a technique for obtaining the pattern is available once the positions of the nulls on the unit circle are known (sec. 1), all that is required from the Tchebyscheff polynomials is information on the proper distribution of the nulls. This information can

*An alternative spelling of this Russian name is Chebyshev.

†Courant-Hilbert, *Methoden der Mathematischen Physik*, Julius Springer, Berlin, 1931, **1**, p. 75.

‡C. L. Dolph, "A Current Distribution for Broadside Array which Optimizes the Relationship between Beam Width and Side-lobe Level," *Proc. IRE*, **34**, No. 6, p. 335 (1946); also H. J. Riblet, *Proc. IRE*, **35**, No. 5, p. 489 (1947).

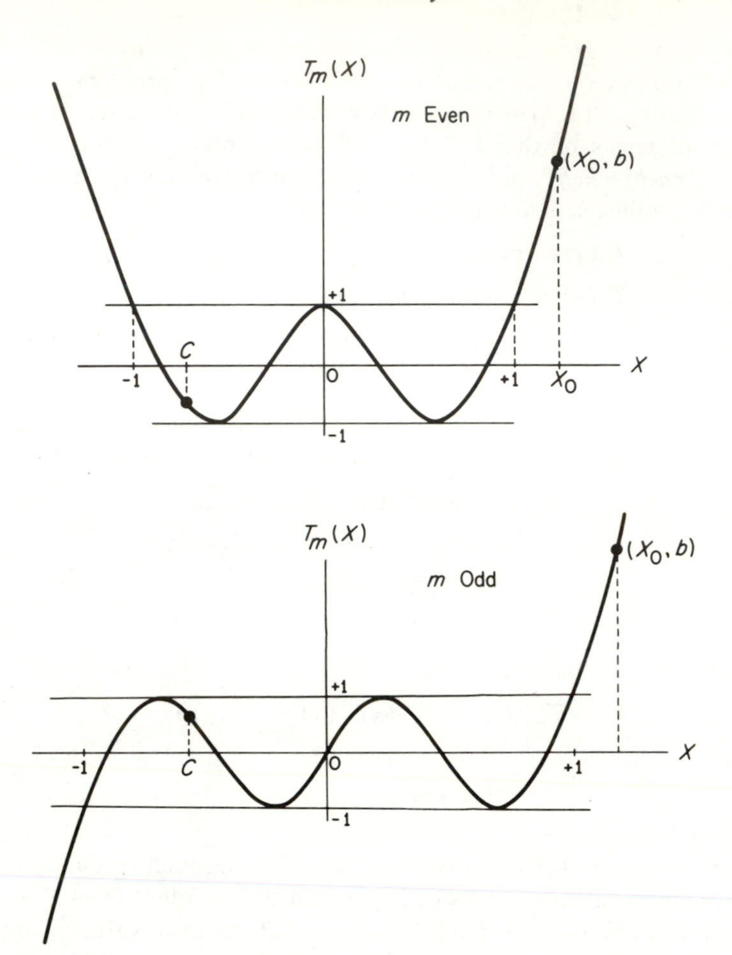

Figure 12-19. Tchebyscheff polynomials, $T_m(x)$, for m even and m odd.

be obtained by causing x to trace out the desired portion of the Tchebyscheff polynomial (of correct degree) as the variable ψ moves over its range on the unit circle. This is accomplished as follows:

Consider the Tchebyscheff polynomial of mth degree

$$T_m(x) = \cos(m \cos^{-1} x) = \cos(m\delta)$$

where

$$\cos \delta = x$$

The nulls of the pattern are given by the roots

$$\cos(m\delta) = 0$$

that is, by

$$\delta_k^0 = \frac{(2k-1)\pi}{2m} \qquad k = 1, 2, \ldots m$$

Next consider the function ψ. For a broadside array for which $\alpha = 0$,

$$\psi = \beta d \cos \phi$$

As ϕ varies from 0 to $\pi/2$ to π, ψ goes from βd to 0 to $-\beta d$ and the range of ψ is $2\beta d$.

Now let $x = x_0 \cos \psi/2$. Then, as ϕ varies from 0 through $\pi/2$ to π, ψ varies from βd through zero to $-\beta d$, and x will vary from $x_0 \cos \pi d/\lambda$ to x_0 back to $x_0 \cos(-\pi d/\lambda) = x_0 \cos \pi d/\lambda$. For example, if $d = \lambda/2$, ψ will range from π through zero to $-\pi$, and x will range from 0 to $+x_0$ and back to zero. Again, if $d = \lambda$, ψ will range twice around the circle from 2π through 0 to -2π, (two major lobes) and x will range from $-x_0$ to x_0 and back to $-x_0$. This is the correspondence desired.

The nulls in the Tchebyscheff pattern occur at values of x given by

$$x_k^0 = \cos \delta_k^0$$

so the corresponding position for the nulls on the unit circle will be given by

$$x_k^0 = x_0 \cos \frac{\psi_k^0}{2}$$

or

$$\psi_k^0 = 2 \cos^{-1} \left[\frac{x_k^0}{x_0} \right]$$

$$= 2 \cos^{-1} \left[\frac{\cos \delta_k^0}{x_0} \right] \tag{12-14}$$

where

$$\delta_k^0 = \frac{(2k-1)\pi}{2m} \qquad k = 1, 2, \ldots m$$

Equation (14) gives the required spacing of the nulls on the unit circle for a pattern whose side lobes are all equal. The degree m of the polynomial used will be equal to the number of nulls on the unit circle, and this will be one less than n, the apparent number of elements. The value of x_0 is determined by the desired ratio b of principal to side-lobe amplitudes. The value of x_0 is given in terms of b by

$$T_m(x_0) = b$$

It can be calculated by noting that if $b = \cosh \rho$, then

$$x_0 = \cosh(\rho/m)$$

The graphical-analytical method for obtaining the pattern of the array from the location of the nulls yields a detailed and accurate plot of relative field strength versus the defined angle ψ. For many purposes a rough sketch of the pattern may be adequate, and this can be obtained directly from the known properties of the Tchebyscheff polynomial. Thus, knowing the location of the nulls in the pattern and the amplitude of all the side lobes relative to the principal lobe, the pattern as a function of ψ may be sketched in with good accuracy. The pattern as a function of the azimuthal angle ϕ is then determined using the transform $\phi = \cos^{-1} \psi/\beta d$. The binomial expansion method of calculating the required current distribution from the location of the nulls proves satisfactory for small arrays, but tends to become unwieldy for larger arrays. For large multi-element arrays, alternative procedures requiring less labor have been developed.*

EXAMPLE 6: Design a four-element broadside array having a spacing $d = \lambda/2$ between elements. The pattern is to be optimum with a side lobe level which is 19.1 db down ($b = 9.0$).

For $d = \lambda/2$, the range of operation is $2\beta d = 2\pi$. Since there will be 3 nulls, use $T_3(x) = 4x^3 - 3x$. Then

$$T_3(x_0) = 4x_0^3 - 3x_0 = 9$$

Solving for x_0,

$$\rho = \cosh^{-1} b = 2.887$$

$$x_0 = \cosh\left(\frac{2.887}{3}\right)$$

$$= 1.5$$

The nulls are given by $\cos(m\delta) = \cos(3\delta) = 0$. Therefore

$$\delta_k^0 = \frac{(2k-1)\pi}{2m} = \frac{(2k-1)\pi}{6} \qquad k = 1, 2, 3, \ldots$$

Then $\quad \delta_1^0 = \frac{\pi}{6} \qquad \delta_2^0 = \frac{3\pi}{6} \qquad \delta_3^0 = \frac{5\pi}{6}$

k	δ_k^0	$x_k^0 = \cos\delta_k^0$	x_k^0/x_0	$\psi_k^0 = 2\cos^{-1}\frac{x_k^0}{x_0}$	ψ_k^0 (radians)
1	$\pi/6$	0.866	0.577	109.5°	1.910
2	$3\pi/6$	0	0	180°	π
3	$5\pi/6$	−0.866	−0.577	250.5°	4.37

The polynomial representing the array is

$$|E| = |(z - e^{j1.91})(z - e^{j\pi})(z - e^{j4.37})|$$

$$= |z^3 + 1.667z^2 + 1.667z + 1|$$

*T. J. van der Maas, "A Simplified Calculation for Dolph-Tchebyscheff Arrays," *J. Appl. Phys.*, **25**, pp. 121–124, Jan., 1954. Also, R. J. Stegen, "Excitation Coefficients and Beamwidths of Tchebyscheff Arrays," *Proc. IRE*, **41**, pp. 1671–1674, Nov., 1952.

The required relative currents in the elements are

$$1 : 1.667 : 1.667 : 1$$

In concluding this section it is pertinent to note that the Tchebyscheff "optimum" design is strictly applicable only for a line array of isotropic sources. Actual antennas are not isotropic and the effect of element directivity should not be forgotten. Moreover, instead of the typical fan-beam pattern of an equiphase line array, the more usual optimization problem calls for a maximum in only *one* direction in space, with the peak field strength in all other directions minimized.

12.04 Superdirective Arrays. In any engineering problem a major goal is the determination of an optimum solution, a process which usually requires the maximization (or minimization) of some parameter, while holding other parameters fixed. In antenna engineering a key problem is often the maximization of antenna gain or directivity. An early example was the determination by Hansen and Woodyard of the optimum phase of an end-fire line source with continuous constant-magnitude excitation to produce maximum directive gain.* Another example was the attempt by La Paz and Miller to determine the optimum current distribution on a vertical monopole for maximum gain in the horizontal plane.† Although this attempt did not yield an optimum distribution, it did yield distributions that model studies‡ confirmed as being superior to uniform or the normal sinusoidal distributions. It also led Bouwkamp and De Bruijn,§ using a function-theoretic argument, to prove that if practical considerations like radiation resistance are ignored, there is *no* upper limit on the maximum directivity that can be attained from a (continuous) current distribution. Antennas or arrays designed to yield directive gains appreciably greater than those obtainable from uniform distributions have become known as *supergain* or *superdirective* arrays. Although the term "supergain" is in common use, the correct term is "superdirective" because it is directive gain that is maximized. Indeed, as will be shown, any design that yields appreciable superdirectivity will have very low efficiency, and hence low power gain.

The Bouwkamp and De Bruijn paper triggered a series of discussions

*W. H. Hansen and J. R. Woodyard, "A new principle in directional antenna design," *Proc. IRE*, **26**, pp. 333–345, March, 1938.

†L. La Paz and G. Miller, "Optimum current distribution on vertical antennas," *Proc. IRE*, **31**, p. 214, 1943.

‡E. C. Jordan and W. L. Everitt, "Acoustic models of radio antennas," *Proc. IRE*, **29**, No. 4, p. 186, 1941.

§C. J. Bouwkamp and N. G. De Bruijn, "The problem of optimum antenna current distribution," Philips Research Reports, **1**, R11, pp. 135–158, 1945/6.

which continues even to the present. Although it is relatively easy to show in specific instances that superdirective designs are highly impractical, so strong is the urge to get the most for the least, that attempts to achieve even a small amount of superdirectivity prove almost irresistible, as evidenced by the extensive literature on the subject.* It is the purpose of this section to display by means of specific examples the nature of the difficulties associated with superdirectivity and to indicate a general solution to the maximum directivity problem, both with and without the constraints that arise from practical considerations.

Superdirective Tchebyscheff Arrays. In the case of end-fire arrays, the technique of equispacing the nulls in the range of ψ yields desirable patterns even when the spacing between the elements becomes small in wavelengths. Indeed, it was seen in sec. 12.01 that if the number of elements is increased as the spacing is decreased, so that the overall length of the array remains fixed, the technique of equispacing the nulls leads to designs having arbitrarily sharp directivity—that is, to superdirective arrays. When this same technique of equispacing the nulls is applied to broadside arrays having small spacings, it is found that as the spacings are made smaller the patterns become progressively poorer. However, if the nulls are distributed in the range of ψ according to a Tchebyscheff distribution, desirable patterns having small side lobes and arbitrarily sharp principal lobes result. The design procedure for superdirective arrays is indicated below.

Referring to Fig. 12-19, the range of the Tchebyscheff polynomial, which is used, lies between the points c and x_0. The position of the starting point c depends upon the element spacing and is given by

$$c = x_0 \cos \frac{\pi d}{\lambda}$$

For a spacing d equal to $\lambda/2$, the point c is at origin. For $d > \lambda/2$, c is negative as shown in Fig. 12-19, whereas for spacings less than $\lambda/2$, c is positive, approaching x_0 as the spacing approaches zero. Since the radiation pattern is determined entirely by that portion of the Tchebyscheff curve lying between c and x_0, it is evident that for small spacings (c near x_0) full use is not being made of the pattern control available. This failing can be remedied by compressing the desired range of Tchebyscheff curve into the region that will be used. This result can be accomplished by a simple change of scale on the abscissa.

*A. Bloch, R. G. Medhurst, S. D. Pool, "A New Approach to the Design of Superdirective Aerial Arrays," *Proc. IEE* (London), Part III, **100**, pp. 303–314, Sept., 1953. (Gives 28 references.) See also "Superdirectivity," by the same authors in *Proc. IRE*, **48**, p. 1164, 1960, which gives 31 additional references.

The method will be illustrated by an example to be solved as a problem by the student.

Problem 2. Design a five-element broadside array having a total array length of $\lambda/4$ (spacing between the elements $d = \lambda/16$). The side-lobe level is to be 25.8 db down, or 1/19.5 times the main-lobe level.

For the above problem there should be four nulls on the unit circle within the range of ψ, and four nulls in the range of the Tchebyscheff pattern that is used. Thus it would be possible to select $T_4(x)$ and use the range from 0 to x_0 and back (again four nulls). Because it is simpler to work with the lower-degree polynomials, it is suggested that $T_2(x)$ be used. Start by letting $x = x_0 \cos \psi$, where for this case,

$$\psi = \frac{\pi}{8} \cos \phi$$

Then as ϕ ranges from 0 through $\pi/2$ to π, ψ will range from $(\pi/8)$ to $(-\pi/8)$, and since $\cos(\pi/8) = 0.92388$, x will vary from $0.92388x_0$ to x_0 and back to $0.92388x_0$. By a simple translation and change of scale of the abscissa, the portion of curve between -1 and x_0 can be compressed within the range $0.92388x_0$ to x_0. When the problem is solved* in this manner, it will be found that the nulls on the unit circle will be placed at

$$\psi^\circ = \pm 17^\circ\, 17\tfrac{5}{9}' \qquad \text{and} \qquad \pm 21^\circ\, 41\tfrac{5}{11}'$$

Using these values in an expression such as eq. (9) gives the pattern of Fig. 12-20(*a*) and the current ratios which will be found to be

$$1 : -3.7680 : 5.5488 : -3.7680 : 1$$

In the above example it will be noted that to obtain the pattern by the semigraphical method of multiplying together the distances from z to the null points of the array, as indicated in eq. (6), three-figure accuracy is adequate. However, if it is desired to obtain the pattern from the simple phasor addition of the fields due to the individual elements. as in eq. (1), it is necessary to compute the current ratios with great accuracy if a reasonably accurate pattern is to be obtained.

The reason for the extreme accuracy required in this case becomes evident when the fields are added to determine the resultant field in the direction of the maximum, broadside to the array. In this direction, there is no phase difference due to difference in path lengths and so the field strength is proportional to the simple arithmetic sum of all the currents. Adding these currents, with due regard for sign, we have

$$1.0000 - 3.7680 + 5.5488 - 3.7680 + 1.0000 = 0.0128$$

*Details of the method of solution are given by Nicholas Yaru, "A Note on Supergain Antenna Arrays," *Proc. IRE*, **39**, No. 9, pp. 1081–1085, Sept., 1951.

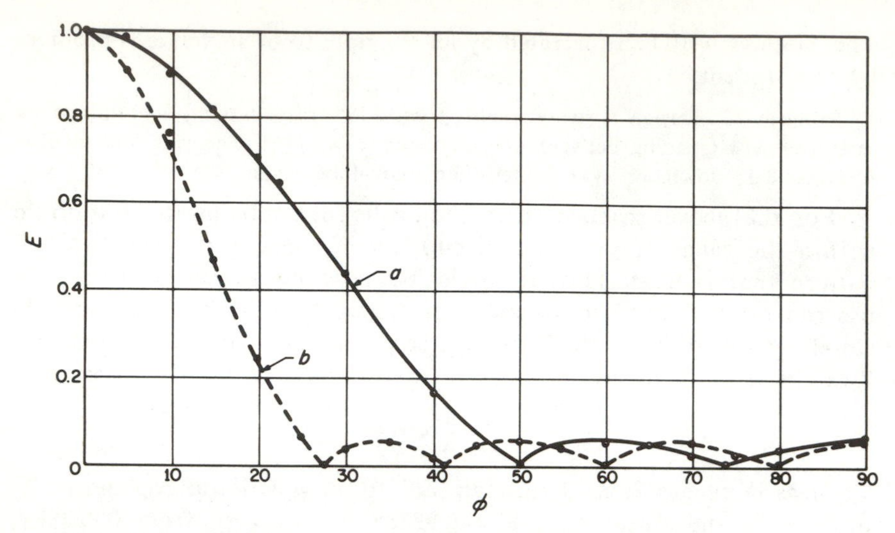

Figure 12-20. Broadside "super gain" patterns for arrays that have an overall length of one-quarter wavelength: (*a*) five-element array; (*b*) nine-element array.

It is seen that the "effective current" radiating in the direction of the maximum is only about one-fifth of 1 per cent of the current in the center element. This low value of radiation results from the fact that the array acquires its superdirective properties by virtue of the addition of the radiation from elements carrying large, almost equal and opposite currents. Furthermore, a slight error of the order of 1 per cent in the setting of any one of the currents would change the resultant by several hundred per cent, and so completely destroy the supergain pattern.

These effects are demonstrated even more clearly in the next example,* which is the case of a nine-element array having the same overall length (one-quarter wavelength) as the array of the previous example. The pattern of this array is shown in Fig. 12-20(*b*). For this array, using a null spacing corresponding to the distribution of the nulls in $T_4(x)$, the calculated current ratios are

$I_1 = I_9$	260,840.2268
$I_2 = I_8$	−2,062,922.9994
$I_3 = I_7$	7,161,483.1266
$I_4 = I_6$	−14,253,059.7032
I_5	17,787,318.7374
Total	00,000,000.0390

*See the article by Yaru, which also discusses efficiency and required accuracy.

It is seen that, if a current of the order of 17 million amperes is fed to the center element, with corresponding currents in the other elements, the total effective current radiating broadside to the array (the direction of the maximum) is equivalent to a current of 39 milliamperes in a single antenna!

From the above examples it seems reasonable to conclude that although superdirective arrays are possible in theory, they are quite impractical. However, it has been argued,* with some justification, that specific examples such as the above do not constitute a general proof of impracticability. More recently a completely general solution of the problem of maximum directivity of an array of a *given number of elements* has been obtained by Lo, et al.† Their results show maximum directivity and Q-factor as a function of element spacing. The Q-factor is equal to the ratio of the complex input power to the power radiated by the array (and hence is inversely proportional to efficiency). As would be expected, for the optimum excitation the Q-factor increases sharply as the spacing d is decreased into the superdirective region. For example, for a 10-element array excited for maximum directivity the Q-factor increases from 3.4 to a value of the order of 10^4 as d decreases from 0.45λ to 0.3λ; for smaller spacings the Q-factor increases to astronomically large values. The referenced paper gives other examples, as well as a thorough discussion of the sensitivity factor which provides a measure of the accuracy requirements on element excitations. From their results it is possible to obtain a quantitative measure of the deleterious effects of any stated amount of superdirectivity. It is also concluded that from a practical point of view the uniform current distribution yields a directivity near to the optimum. However, this conclusion may be invalid when the antenna is used for reception where *signal-to-noise ratio* is to be optimized.

12.05 Radiation from a Current Sheet. The antenna arrays considered so far in this chapter have been represented by discrete current filaments lying in a plane. However, there are many other problems in electromagnetic theory for which it is advantageous to use instead, a representation by continuous current distributions lying in a plane. For instance, a continuous current distribution can be used to represent the induced currents in a metal sheet or grating when it is excited by an incident wave; if the induced current distribution can be estimated, then the scattered field can be calculated approximately. In addition, a closely spaced array of discrete elements can be represented approximately by a continuous current distribution.

*A. Bloch, R. G. Medhurst, S. D. Pool, *loc. cit.* (1953).

†Y. T. Lo, S. W. Lee, and Q. H. Lee, "Optimization of Directivity and Signal-to-Noise Ratio of an Arbitrary Antenna Array," *Proc. IEEE*, **54**, 1033–1045, Aug., 1966.

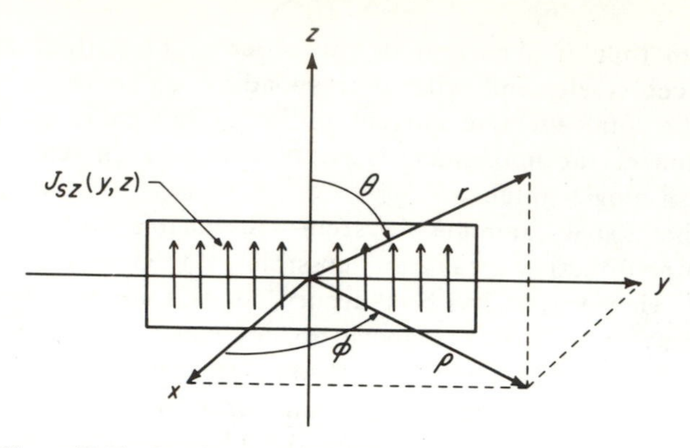

Figure 12-21. Continuous distribution of electric current lying in the y-z plane.

Consider a sheet of electric current lying in the y-z plane and flowing in the z direction as shown in Fig. 12-21. The electric field strength far from this source may be deduced using eqs. (10-99) and (10-101). The result is

$$E_\theta = \frac{-j\beta\eta}{4\pi}\frac{e^{-j\beta r}}{r} f_\theta(\theta, \phi) \tag{12-15}$$

in which

$$f_\theta = -\sin\theta\, f_z$$

and

$$f_z = \int_{\text{all space}} J_z(\mathbf{r}')e^{j\beta\hat{\mathbf{r}}\cdot\mathbf{r}'}\, dV'$$

With the observation point $\mathbf{r}$ in spherical co-ordinates and the source point $\mathbf{r}'$ in rectangular co-ordinates, the dot product in the volume integral is given by

$$\hat{\mathbf{r}}\cdot\mathbf{r}' = x'\sin\theta\cos\phi + y'\sin\theta\sin\phi + z'\cos\theta$$

For currents lying in the y-z plane, $x' = 0$ and thus (15) may be expressed as

$$E_\theta = \frac{j\beta\eta}{4\pi}\frac{e^{-j\beta r}}{r}\sin\theta \iint_{-\infty}^{\infty} J_{sz}(y', z')e^{j\beta(y'\sin\theta\sin\phi + z'\cos\theta)}\, dy'\, dz' \tag{12-16}$$

The integral in (16) has the same mathematical form as a two-dimensional Fourier transform and thus it is clear that the *far field of a planar current distribution may be expressed as the Fourier transform of the surface current density.*

In many cases the surface current density is separable, that is, it may be written in the form

$$J_{sz}(y', z') = J_a(y')J_b(z')$$

so that (16) becomes

$$E_\theta = \frac{j\beta\eta}{4\pi}\frac{e^{-j\beta r}}{r}\sin\theta\int_{-\infty}^{\infty} J_a(y')e^{j\beta y' \sin\theta\sin\phi}\,dy'\int_{-\infty}^{\infty} J_b(z')e^{j\beta z' \cos\theta}\,dz' \quad (12\text{-}17)$$

The absolute value of the integral over z' multiplied by $\sin\theta$ constitutes the *element factor* in the radiation field, an element being any one of the current filaments making up the total current distribution. The integral over y' is the *space factor* (or *array factor*) due to the arrangement of current filaments and thus (17) provides an example of the principle of multiplication of patterns discussed in chap. 11.

The radiation field in the x-y plane may be expressed as

$$E(\nu) = \frac{E_\theta}{E_0} = \int_{-\infty}^{\infty} J_a(y)e^{j2\pi\nu y}\,dy \quad (12\text{-}18)$$

in which

$$E_0 = \frac{j\beta\eta}{4\pi}\frac{e^{-j\beta r}}{r}\int_{-\infty}^{\infty} J_b(z)\,dz$$

and

$$\nu = \frac{\beta}{2\pi}\sin\phi = \frac{1}{\lambda}\sin\phi$$

Note that the primes have been dropped because the source and observation points may be distinguished easily without them. Equation (18) is in the form of a one-dimensional Fourier transform and thus it is very convenient for the computation of radiation fields. The exhaustive literature on the Fourier transform provides us with a large number of computational techniques and algebraic manipulations which may be applied directly to antenna problems.

As an example, consider a current strip lying in the y-z plane and having a uniform z-directed current distribution of length L in the y direction. The radiation pattern in the x-y plane is to be calculated and for this reason the current distribution in the z direction need not be specified. Such a current distribution would be a first approximation to the current on a flat metal strip excited by a normally-incident uniform plane wave or to the currents in a large number of closely spaced dipoles fed in phase. As illustrated in Fig. 12-22, the y variation in surface current density may be represented by

$$J_a(y) = \frac{1}{L}R^L(y)$$

in which $R^L(y)$ represents the unit-amplitude gating function of width L centered at the origin. The far field in the x-y plane may be expressed as

$$E(\nu) = \frac{1}{L}\int_{-L/2}^{L/2} e^{j2\pi\nu y}\,dy = \frac{\sin\pi\nu L}{\pi\nu L} \quad (12\text{-}19)$$

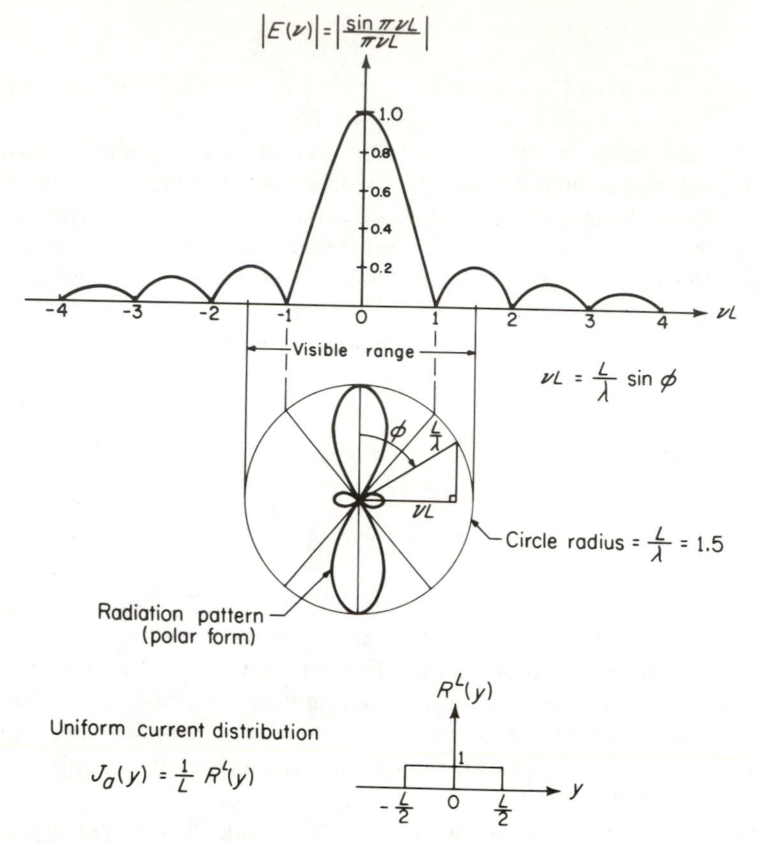

Figure 12-22. Radiation pattern of a uniform current distribution 1.5 wavelengths long.

and the radiation pattern is given by $|E(\nu)|$. A method of sketching the polar radiation pattern is illustrated in Fig. 12-22 for the particular case $L = 1.5\lambda$. It should be noted that in (19) the electric field strength is expressed as a function of ν which has the range $-\infty \leq \nu \leq \infty$ if it is regarded as a Fourier transform variable. However, the radiation pattern is obtained only from that part of the infinite range defined by $-\lambda^{-1} \leq \nu \leq \lambda^{-1}$ which is known as the *visible range* of ν (the same kind of behavior was discussed in sec. 12.02 with respect to the variable ψ).

Because the field expression (18) is given as a Fourier transform, it might (at first glance) seem possible to deduce $J_a(y)$ from $E(\nu)$ by calculating the inverse Fourier transform

$$J_a(y) = \int_{-\infty}^{\infty} E(\nu) e^{-j2\pi\nu y}\, d\nu$$

Such a procedure* would be valid only if $E(v)$ were specified or known over the whole infinite range of v. In synthesis problems (sec. 12.02) $E(v)$ is specified in the visible range; outside the visible range $E(v)$ is chosen in such a way as to permit practical realization of a suitable antenna array.

Problem 3. Side-lobe level may be reduced at the cost of increased major lobe beamwidth by "tapering" or "smoothing" the current distribution. Discuss this statement with reference to uniform, cosine, and cosine-squared current distributions.

Effect of a Uniform Phase Progression. Suppose a current distribution $J_a(y) = A(y)$ has a far field given by

$$E(v) = \int_{-\infty}^{\infty} A(y)e^{j2\pi vy}\,dy \tag{12-20}$$

If a progressive phase shift $e^{-j2\pi ky}$ is imposed on this current distribution, the new current distribution is $J_a'(y) = A(y)e^{-j2\pi ky}$ and the new far field is given by

$$\begin{aligned} E'(v) &= \int_{-\infty}^{\infty} A(y)e^{j2\pi(v-k)y}\,dy \\ &= E(v-k) \end{aligned} \tag{12-21}$$

Thus the progressive phase shift causes a shift in the radiation pattern function which in turn causes the pattern lobes to swing in the direction of the phase progression. This effect is illustrated in Fig. 12-23 for the case $kL = 1/2$ and the uniform current distribution of Fig. 12-22.

EXAMPLE 7: *Scattering by a Grating.* An elementary grating consists of a flat grid of very fine wires as shown in Fig. 12-24(a). A uniform plane wave incident on the grating causes currents to flow in the wires and these induced currents are the sources for the scattered field. Far from the grating the

*Because in all practical cases $J_a(y)$ must be zero except for a *finite* interval in y, it follows from the Wiener-Paley theorem that $E(v)$ as defined by (18) is an analytic function. From the theory of functions of a complex variable an analytic function is uniquely determined by its value over an interval. These facts seem to suggest that the knowledge (including phase) of $E(v)$ over the visible region would determine the source $J_a(y)$ uniquely. However, in practice this is not possible because $E(v)$ is not known with unlimited accuracy and, as it turns out, an immeasurably small deviation of $E(v)$ in the visible region can lead to an entirely different source function $J_a(y)$. The above misconception has appeared repeatedly in the literature in articles covering a wide variety of problems including superdirective antennas, information retrieval, optical image reconstruction and radio astronomy. For a further discussion of the matter see Y. T. Lo, "On the Theoretical Limitation of a Radio Telescope in Determining the Sky Temperature Distribution," *J. App. Phys.*, **32**, pp. 2052–2054, Oct., 1961.

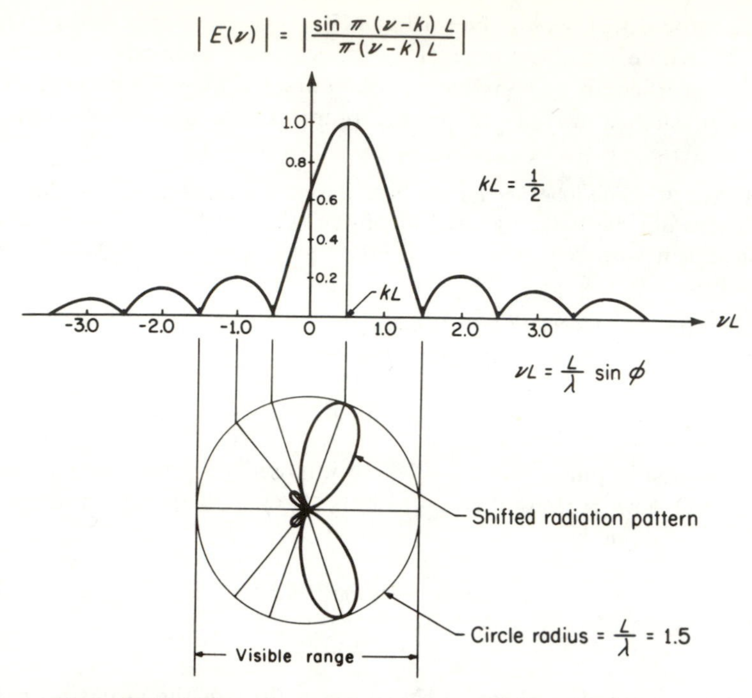

Figure 12-23. Effect of a progressive phase shift on the pattern of a uniform current distribution.

scattered (or re-radiated) field may be found from the Fourier transform of the induced current distribution.

Consider, for example, a grating "illuminated" by a normally-incident uniform plane wave whose electric vector is in the z direction. If the grating wire has a diameter much smaller than a wavelength, the surface current density in the y-z plane may be represented approximately by a row of Dirac deltas extending from $(-L/2)$ to $(+L/2)$. Such a current density function could be represented as the product of a "gating function" of length L and a "comb function," $C^d(y) = \sum_{n=-\infty}^{\infty} \delta(y - nd)$, consisting of an infinite sequence of Dirac deltas spaced a distance d as shown in Fig. 12-24(b). A current distribution of this type has a y variation of the form

$$J_a(y) = \frac{1}{L} R^L(y) d C^d(y) \tag{12-22}$$

The radiation field in the x-y plane may be found by taking the Fourier transform of J_a as indicated in (18). The transform of the gating function is known from (19) and the transform of the comb function is another comb function with a reciprocal spacing,

$$\int_{-\infty}^{\infty} d C^d(y) e^{j2\pi\nu y}\, dy = \sum_{n=-\infty}^{\infty} \delta\left(\nu - \frac{n}{d}\right) = C^{1/d}(\nu) \tag{12-23}$$

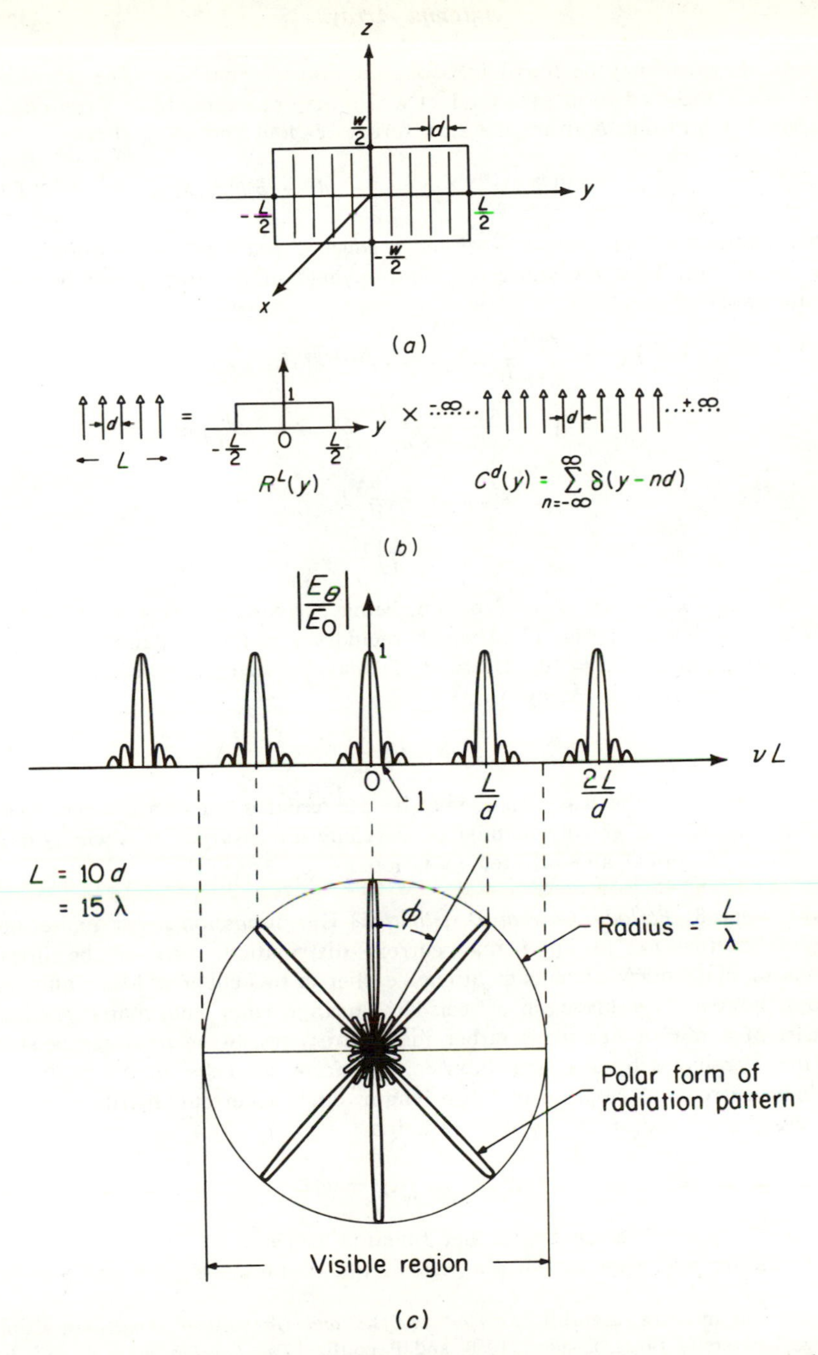

Figure 12-24. Uniform plane wave scattering by a thin-wire grating—normal incidence. The figure shows the grating (a), the induced current distribution (b), and the radiation pattern (c) for the scattered field in the x-y plane.

proofs of which may be found in books on Fourier analysis.* The transform of $J_a(y)$ is the transform of a product which may be expressed as a *convolution integral*; if $a(\nu)$ and $b(\nu)$ are the transforms of $A(y)$ and $B(y)$, then

$$\int_{-\infty}^{\infty} A(y)B(y)e^{j2\pi\nu y}\,dy = \int_{-\infty}^{\infty} a(\nu-\xi)b(\xi)\,d\xi \tag{12-24}$$

which states that *the transform of the product is equal to the convolution of the transforms.* Thus the scattered radiation field of the grating may be found as follows:

$$\begin{aligned} E(\nu) &= \int_{-\infty}^{\infty} \frac{1}{L} R^L(y) dC^d(y) e^{j2\pi\nu y}\,dy \\ &= \int_{-\infty}^{\infty} \frac{\sin\pi(\nu-\xi)L}{\pi(\nu-\xi)L} \sum_{n=-\infty}^{\infty} \delta\left(\xi - \frac{n}{d}\right) d\xi \\ &= \sum_{n=-\infty}^{\infty} \frac{\sin\pi\left(\nu-\frac{n}{d}\right)L}{\pi\left(\nu-\frac{n}{d}\right)L} \end{aligned} \tag{12-25}$$

in which $\nu = \lambda^{-1}\sin\phi$. This function, being periodic, is easy to sketch if $L/d \gg 1$ as shown in Fig. 12-24 which includes a radiation pattern in polar form for the case $L/d = 10$, $L/\lambda = 15$. It may be seen that the angles ϕ_n of the major lobes are given by

$$\sin\phi_n = n\frac{\lambda}{d} \tag{12-26}$$

which shows that the wavelength may be deduced from a measurement of ϕ_n if d is known. The grating method of wavelength measurement is widely used in studies of optical and infrared radiation.

EXAMPLE 8: *Periodic Current Distribution.* The discussion so far applies not only to gratings but to any planar current distribution such as the linear arrays of filamentary currents mentioned earlier in this chapter. Many antenna arrays, however, are made up of periodic arrangements and many gratings consist of a row of flat strips rather than thin wires. Some light can be shed on these problems by applying Fourier transform analysis to an arbitrary, periodic current distribution of finite length. Such a current distribution may be regarded as a uniform periodic function

$$F(y) = \sum_{n=-\infty}^{\infty} G(y - nd),$$

with an "envelope" given by another function $H(y)$.

There are two ways in which a periodic function may be given an envelope.

*See, for instance, Lighthill, *Fourier Analysis and Generalized Functions*, Cambridge University Press, London, 1960, and Papoulis, *The Fourier Integral and Its Applications*, McGraw-Hill Book Company, New York, 1962. An excellent summary may be found in *Reference Data for Radio Engineers* (4th ed.), International Telephone and Telegraph Corp., 1956.

The most obvious way is direct multiplication giving a y variation of surface current density of the form

$$J_a(y) = H(y)F(y) = H(y) \sum_{n=-\infty}^{\infty} G(y - nd) \tag{12-27}$$

which is illustrated in Fig. 12-25. Such a procedure is frequently useful but it is awkward to use when the current elements are identical in shape, differing only in amplitude. In these cases the current distribution has the form

$$J_a(y) = \sum_{n=-\infty}^{\infty} H(nd)G(y - nd) \tag{12-28}$$

which is also illustrated in Fig. 12-25. Because the current elements represented by (28) are identical in shape, one would expect the principle of multiplication of patterns to apply.

The radiation field of the current distribution in (27) may be calculated easily if one begins by expressing the periodic function in the form of a Fourier series. This is done by noting that $F(y)$ may be expressed as a convolution,

$$F(y) = \sum_{n=-\infty}^{\infty} G(y - nd) = \int_{-\infty}^{\infty} G(y - \zeta) \sum_{n=-\infty}^{\infty} \delta(\zeta - nd)\, d\zeta \tag{12-29}$$

from which it is clear that the transform of $F(y)$ is

$$f(v) = g(v)\frac{1}{d} \sum_{n=-\infty}^{\infty} \delta\left(v - \frac{n}{d}\right) \tag{12-30}$$

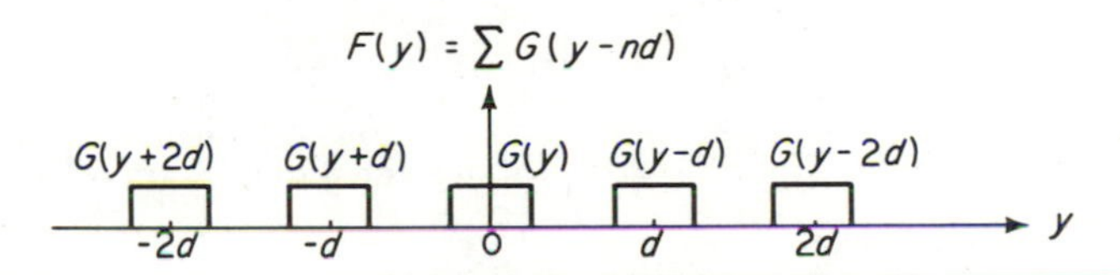

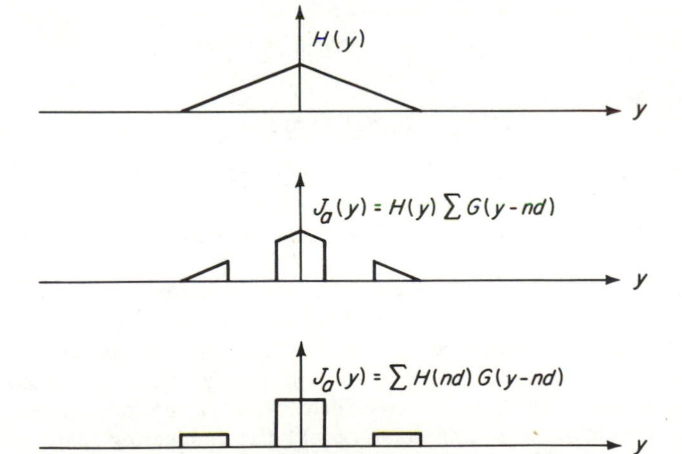

Figure 12-25. Two methods of applying an envelope function $H(y)$ to a periodic current distribution.

If it were needed, the Fourier series could be obtained by transforming (30). What is needed is the transform of (27) which may be expressed as the convolution,

$$E(\nu) = \int_{-\infty}^{\infty} h(\nu - \xi) f(\xi)\, d\xi \tag{12-31}$$

Substitution of (30) into (31) gives

$$\begin{aligned} E(\nu) &= \int_{-\infty}^{\infty} h(\nu - \xi) g(\xi) \frac{1}{d} \sum_{n=-\infty}^{\infty} \delta\left(\xi - \frac{n}{d}\right) d\xi \\ &= \frac{1}{d} \sum_{n=-\infty}^{\infty} g\left(\frac{n}{d}\right) h\left(\nu - \frac{n}{d}\right) \end{aligned} \tag{12-32}$$

which represents the radiation field. If the elements were isotropic (current filaments), g would be unity and (32) would have the form

$$E(\nu) = \frac{1}{d} \sum_{n=-\infty}^{\infty} h\left(\nu - \frac{n}{d}\right) \tag{12-33}$$

The radiation field of the current distribution in (28) may be obtained by noting that (28) can be expressed as a convolution:

$$\begin{aligned} J_a(y) &= \sum_{n=-\infty}^{\infty} H(nd) G(y - nd) \\ &= \sum_{n=-\infty}^{\infty} \int_{-\infty}^{\infty} H(\zeta) \delta(\zeta - nd) G(y - \zeta)\, d\zeta \\ &= \int_{-\infty}^{\infty} G(y - \zeta) M(\zeta)\, d\zeta \end{aligned} \tag{12-34}$$

in which $M(\zeta) = H(\zeta) \sum_{n=-\infty}^{\infty} \delta(\zeta - nd)$. Taking the Fourier transform of (34) gives the radiation field,

$$\begin{aligned} E(\nu) &= g(\nu) m(\nu) \\ &= g(\nu) \int_{-\infty}^{\infty} h(\nu - \xi) \frac{1}{d} \sum_{n=-\infty}^{\infty} \delta\left(\xi - \frac{n}{d}\right) d\xi \\ &= g(\nu) \frac{1}{d} \sum_{n=-\infty}^{\infty} h\left(\nu - \frac{n}{d}\right) \end{aligned} \tag{12-35}$$

Equation (35) expresses the idea of multiplication of patterns since it consists of the pattern function $g(\nu)$ of a single current element $G(y)$ multiplied by the pattern function of an array of isotropic sources as expressed in (33).

Problem 4. Obtain an expression for the radiation pattern of uniform (infinitely long) current strips of equal amplitude and phase. Take a to be the strip width and d to be the center-to-center spacing. Obtain expressions for the radiation field using both (32) and (35). This problem constitutes an approximate solution for the scattered field when a uniform plane wave is normally-incident on a grating composed of flat conducting strips.

12.06 Wave Polarization. The concept of wave polarization was introduced in chap. 5 where it was shown (for some special cases) that at a fixed point in space the endpoint of the $\tilde{\mathbf{E}}$ vector traces out an ellipse called the *polarization ellipse.* The polarization of the wave is therefore specified by the shape (axial ratio), orientation, and sense of rotation of the polarization ellipse. These three quantities are used frequently but there are also other important representations for the state of polarization. These are the polarization ratio (in both rectangular and rotating co-ordinates), the Stokes parameters, and the Poincare sphere. The purpose here is to introduce these representations and indicate how they are used.

Polarization Ratio Using Rectangular Co-ordinates. A wave of arbitrary polarization may be produced by combining the waves radiated by a pair of crossed-dipole antennas. If the dipoles are aligned parallel to the x and y axes, the waves radiated in the z direction will be of the form

$$E_x = A_x e^{j\delta_x} \tag{12-36}$$

$$E_y = A_y e^{j\delta_y} \tag{12-37}$$

in which A_x and A_y are positive real amplitude factors, and δ_x and δ_y are the phase angles associated with the field components. The relative amplitudes and phases of the field components may be adjusted to give any desired polarization. The state of polarization may be specified by a complex number called the *polarization ratio* P and defined by the relation

$$P = \frac{E_y}{E_x} = |P|\, e^{j\delta} \tag{12-38}$$

in which $|P| = A_y/A_x$ and $\delta = \delta_y - \delta_x$. The values of P may be represented as points in the complex P plane and it should be noted that there is a one-to-one correspondence between all the points in the P plane and all possible wave polarizations.

The path traced out by the endpoint of the time-varying electric vector $\tilde{\mathbf{E}}$ is given by (see problem 5 on page 459)

$$|P|^2\, \tilde{E}_x^2 - 2\,|P| \cos\delta\, \tilde{E}_x \tilde{E}_y + \tilde{E}_y^2 = A_x^2\, |P|^2 \sin^2\delta \tag{12-39}$$

which has the general form

$$A\tilde{E}_x^2 + B\tilde{E}_x\tilde{E}_y + C\tilde{E}_y^2 = D \tag{12-40}$$

and therefore represents a conic section. The discriminant is given by

$$B^2 - 4AC = -4\,|P|^2 \sin^2\delta \tag{12-41}$$

which is negative, indicating that the conic section is an ellipse. This ellipse is inscribed within a rectangle as shown in Fig. 12-26 and the ratio of the side dimensions of the rectangle is $|P|$. The orientation or tilt angle ψ is related to the polarization ratio by the equation

$$\tan 2\psi = \frac{2\,|P| \cos\delta}{1 - |P|^2} = \frac{B}{A - C} \tag{12-42}$$

in which customarily $0 \leq \psi < \pi$. Epuation (42) may be simplified by noting

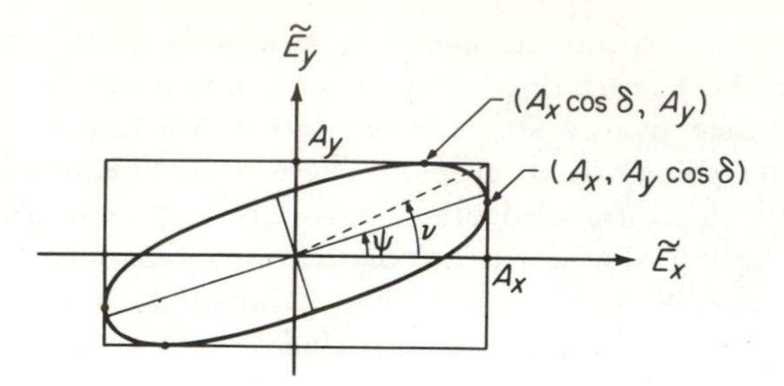

Figure 12-26. The polarization ellipse and its circumscribed rectangle.

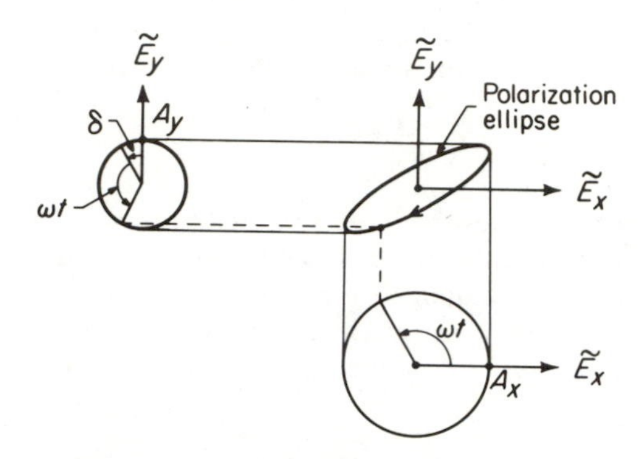

Figure 12-27. Sketching the polarization ellipse from the amplitude and phase of the rectangular field components (for the case $\delta_x = 0$, $\delta_y = \delta$).

that $|P| = \tan \nu$, the angle ν being indicated in Fig. 12-26. Thus (42) becomes

$$\tan 2\psi = \tan 2\nu \cos \delta \tag{12-43}$$

The points of contact with the rectangle are as shown in Fig. 12-26 and the sense of rotation may be determined by visualizing the time-varying vectors at some convenient point such as a point of contact.

Given the amplitude and phase of the rectangular field components, the discussion in the foregoing paragraph shows how to construct the polarization ellipse. Construction of the ellipse may be simplified even further by the use of the graphical technique illustrated in Fig. 12-27. Either component (x or y) of the field vector is a sinusoidally oscillating quantity and thus it may be represented as the projection of a line of fixed length rotating uniformly at the angular frequency ω. As time varies, the endpoint of the resultant vector traces out the polarization ellipse. Thus the polarization ellipse may be sketched easily for any point in the complex P plane as shown in Fig. 12-28. The upper half-plane represents the left-handed sense of polarization and the lower half-plane represents the right-handed sense of polarization, with reference to an imaginary screw advancing in the direction of propagation (assumed to be in the positive z direction). It should be noted that the real axis represents

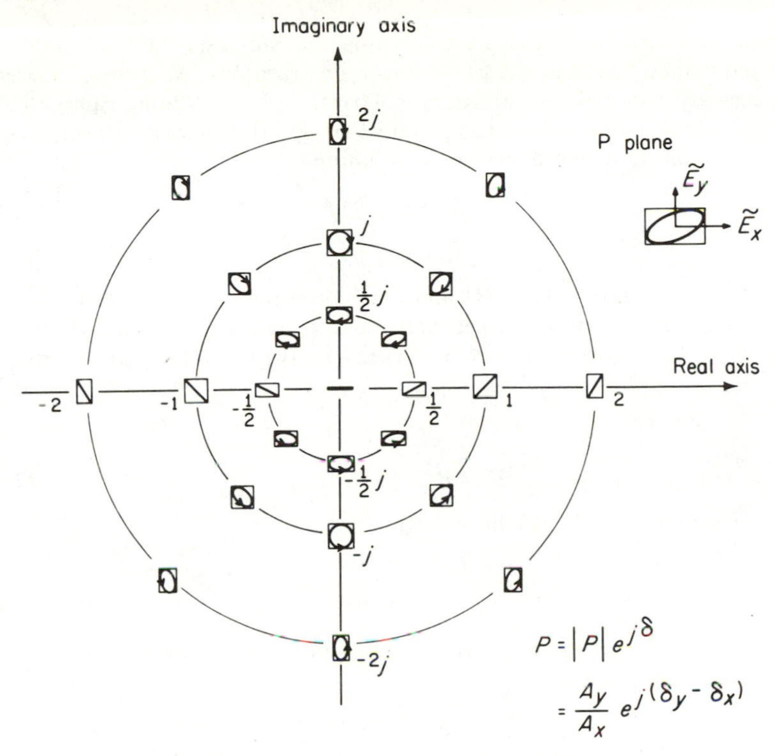

Figure 12-28. The complex P plane with polarization ellipses.

linear polarization and the points $+j$ and $-j$ represent left-circular and right-circular polarization, respectively. Linear polarization in the y direction is represented by the point at infinity, a difficulty which leads to frequent use of the *reciprocal* of the polarization ratio.

Problem 5. Using the time-varying fields corresponding to (36) and (37), show that (39) and (41) are satisfied.

Problem 6. Derive (42) by rotating the co-ordinates in Fig. 12-26 through the angle ψ and writing (39) in the rotated co-ordinate system, so that the cross-product term shall vanish.

Problem 7. Two identical straight dipole antennas lie in the x-y plane, antenna (1) parallel to the x axis and antenna (2) lying along the $\phi = \phi_0$ line. A transmission line from a transmitter is connected directly to antenna (1) and an attenuator phase-shifter with a factor $Ae^{j\theta}$ is connected between the two antennas. Find the values of A and θ required for the transmission of right- and left-circular polarization in the z direction, assuming each antenna radiates linear polarization oriented parallel to the wire. (Neglect any mutual impedance effects.) Work out numerical values for $\phi_0 = 30°$.

Polarization Ratio Using Rotating Co-ordinates. In the foregoing discussion, particular reference was made to combining the radiation fields of antennas which individually radiate waves of linear polarization. At times, however, it is necessary to obtain an arbitrary polarization by combining right-circular and left-circular polarizations from helical or spiral antennas. In such cases it is convenient to define a new pair of unit vectors,

$$\hat{\mathbf{R}} = \hat{\mathbf{x}} - j\hat{\mathbf{y}} \tag{12-44}$$

$$\hat{\mathbf{L}} = \hat{\mathbf{x}} + j\hat{\mathbf{y}} \tag{12-45}$$

It may be shown easily that (44) and (45) correspond to time-varying vectors of unit amplitude rotating, respectively, in the right-handed and left-handed senses. Thus any wave field may be written in the two alternative forms

$$\mathbf{E} = \hat{\mathbf{x}}E_x + \hat{\mathbf{y}}E_y \tag{12-46}$$

and

$$\mathbf{E} = \hat{\mathbf{R}}E_R + \hat{\mathbf{L}}E_L \tag{12-47}$$

The components of (47) may be written in the form*

$$E_R = A_R e^{j\xi_R} \tag{12-48}$$

$$E_L = A_L e^{j\xi_L} \tag{12-49}$$

This indicates that the state of polarization may be represented by a polarization ratio Q defined as

$$Q = \frac{E_L}{E_R} = |Q|e^{j\xi} \tag{12-50}$$

in which $|Q| = A_L/A_R$ and $\xi = \xi_L - \xi_R$. It may be shown easily that A_R and A_L are the fixed lengths of two time-varying, counter-rotating electric vectors. At $t = 0$, these vectors are in the directions indicated in Fig. 12-29. The resultant field is right-handed for $|Q| < 1$ and left-handed for $|Q| > 1$. If a and b are defined respectively as the semimajor and semiminor axes of the polarization ellipse, Fig. 12-29 shows that the *axial ratio* is

$$\frac{b}{a} = \left|\frac{1 - |Q|}{1 + |Q|}\right| \tag{12-51}$$

Also from Fig. 12-29 it may be shown that the orientation angle is given by

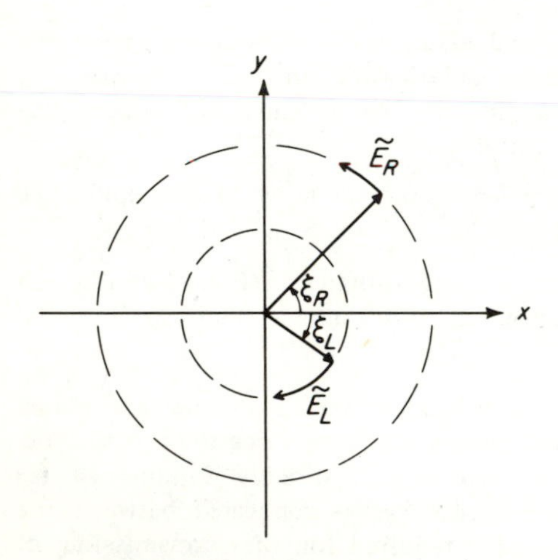

Figure 12-29. Right and left counter-rotating vectors.

*If an antenna radiates a circularly polarized wave, its phase factor ξ_R or ξ_L may be varied simply by a mechanical rotation of the antenna structure.

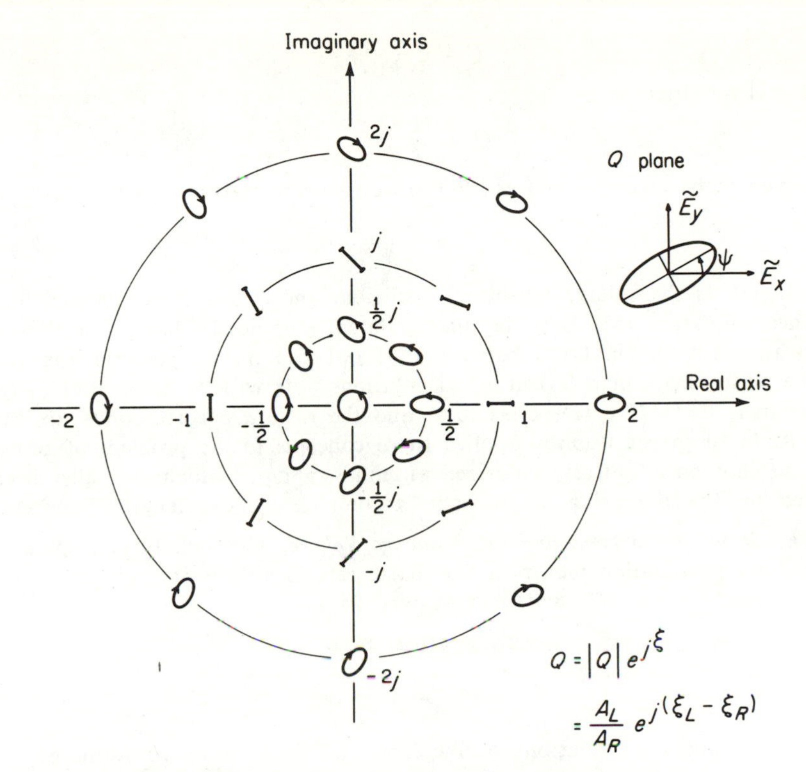

Figure 12-30. The complex Q plane with polarization ellipses.

$$\begin{aligned}\psi &= \xi_R - \left(\frac{\xi_R + \xi_L}{2}\right) + m\pi \\ &= \frac{\xi_R - \xi_L}{2} + m\pi \\ &= -\frac{\xi}{2} + m\pi\end{aligned} \tag{12-52}$$

Customarily, ψ is taken to be in the range $0 \leq \psi < \pi$ and thus the integer m may be chosen to satisfy this condition. This completes the information necessary to sketch the polarization ellipse at any point in the Q plane as illustrated in Fig. 12-30. It is frequently helpful to make use of the reciprocal of Q in order to map left-handed polarizations into the interior of the unit circle.

The relationship between the P and Q ratios may be found using (46) and (47). From these equations it may be shown easily that

$$E_R = \tfrac{1}{2}(E_x + jE_y) \tag{12-53}$$

and

$$E_L = \tfrac{1}{2}(E_x - jE_y) \tag{12-54}$$

The ratio of (54) to (53) gives the desired relationship,

$$Q = \frac{1 - jP}{1 + jP} \tag{12-55}$$

If (55) is rewritten as

$$-Q = \frac{jP - 1}{jP + 1},$$

it is seen to be identical in form to the well-known transmission-line relation

$$\Gamma = \frac{z - 1}{z + 1} \tag{12-56}$$

in which Γ is the voltage reflection coefficient and z is the normalized impedance. Equation (56) may be manipulated graphically using the Smith chart and thus the similarity between (55) and (56) makes possible the use of the Smith chart in polarization calculations.* It will be noted that $(-Q)$ corresponds to Γ, jP corresponds to z, and the ratio a/b corresponds to the VSWR. In his paper Rumsey applies these concepts to the problem of transmission between elliptically polarized antennas, a topic which has also been treated by Sinclair† using the concept of complex effective length.

The Stokes Parameters and the Poincaré Sphere. The remaining representations for polarization require a few basic relations. The first of these has already been derived (43) and may be written as

$$\tan 2\psi = \tan 2\nu \cos \delta \tag{12-43}$$

$$= \frac{2A_x A_y}{A_x^2 - A_y^2} \cos \delta \tag{12-57}$$

The second and third relations follow from the rotation of co-ordinates required to remove the cross-product term from the equation of the polarization ellipse (39). These may be worked out as an exercise or looked up in the literature.‡ The second relation is

$$a^2 + b^2 = A_x^2 + A_y^2 \tag{12-58}$$

which states that the corners of the circumscribed rectangles touch the same circle (see Fig. 12-31). In Fig. 12-31 the *ellipticity angle* χ is shown and normally it would be defined in the range $0 \leq \chi \leq \pi/2$; however, it is convenient to allow χ to take on negative values in order to distinguish between the two senses of polarization. That is,

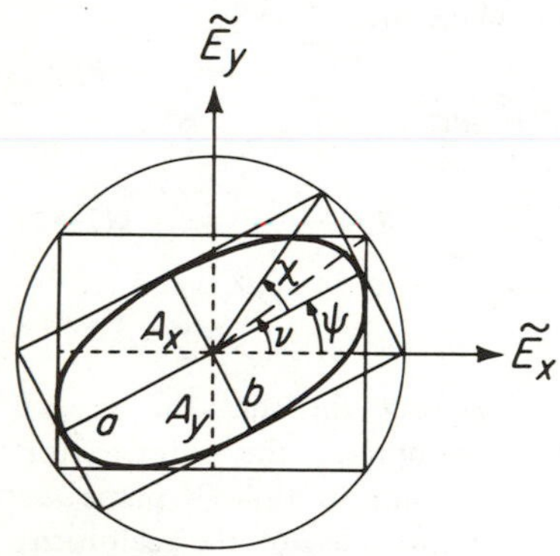

Figure 12-31. The relationship between the circumscribed rectangles.

*V. H. Rumsey, G. A. Deschamps, M. L. Kales, J. I. Bohnert, "Techniques for Handling Elliptically Polarized Waves with Special Reference to Antennas," *Proc. IRE*, **39**, No. 5, pp. 533–552, May, 1951. (See Part I by V. H. Rumsey.)

†G. Sinclair, "The Transmission and Reception of Elliptically Polarized Waves," *Proc. IRE*, **38**, pp. 148–151, February, 1950.

‡M. Born and E. Wolf, *Principles of Optics* [2nd (revised) ed.], Pergamon Press, Inc., New York, 1964.

$$\tan \chi = \pm \frac{b}{a} \tag{12-59}$$

in which $(-\pi/2) \leq \chi \leq (+\pi/2)$. The minus sign indicates the right-handed sense and the plus sign indicates the left-handed sense. The third relation is

$$\sin 2\chi = \sin 2\nu \sin \delta \tag{12-60}$$

$$= \frac{2A_x A_y}{A_x^2 + A_y^2} \sin \delta \tag{12-61}$$

$$= \pm \frac{2ab}{a^2 + b^2} \tag{12-62}$$

The Stokes parameters are usually used in connection with partially polarized radiation. This topic is beyond the scope of this work but it is possible to introduce the Stokes parameters for monochromatic (single-frequency) radiation. They are defined as follows:

$$s_0 = A_x^2 + A_y^2 \tag{12-63}$$

$$s_1 = A_x^2 - A_y^2 \tag{12-64}$$

$$s_2 = 2A_x A_y \cos \delta \tag{12-65}$$

$$s_3 = 2A_x A_y \sin \delta \tag{12-66}$$

The four parameters satisfy the relation

$$s_0^2 = s_1^2 + s_2^2 + s_3^2 \tag{12-67}$$

and thus any three of them are independent.

Equation (67) is the equation in (s_1, s_2, s_3) space of a sphere of radius s_0 known as the *Poincaré sphere*. The three basic relations discussed in the preceding paragraph may be used to prove that

$$s_1 = s_0 \cos 2\chi \cos 2\psi$$

$$s_2 = s_0 \cos 2\chi \sin 2\psi$$

$$s_3 = s_0 \sin 2\chi$$

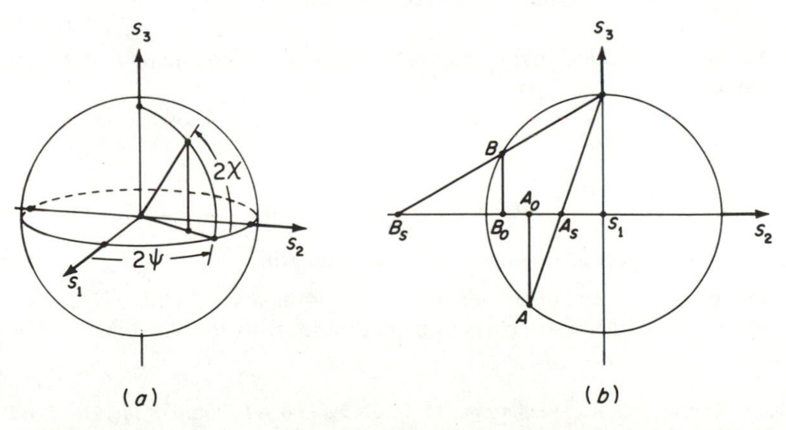

Figure 12-32. The Poincaré sphere, illustrating the Stokes parameters and stereographic and orthographic projections on the equatorial plane.

which indicate that 2χ and 2ψ are the angles shown in Fig. 12-32(*a*). A consideration of the parameters χ (ellipticity angle) and ψ (orientation angle) indicates that there is a one-to-one correspondence between all points on the surface of the sphere and all possible polarizations, the top and bottom hemispheres representing respectively the left-handed and right-handed senses of polarization. The poles represent circular polarization while points on the equator represent linear polarization.

Since representations on a spherical surface are awkward, it is desirable to map points on the sphere onto some plane, in particular the equatorial plane. Two such mappings, the orthographic projection and the stereographic projection, are illustrated in Fig. 12-32(*b*) for the points A and B. The orthographic projections, A_0 and B_0, are simple to carry out but they do not distinguish between the two senses of polarization. The stereographic projections A_s and B_s map right-handed polarizations inside the equator and left-handed polarizations outside the equator. It is interesting to note that the stereographic map is identical to the Q plane discussed previously, provided that the unit circle in the Q plane is made to coincide with the equator. Left-handed polarizations may be mapped stereographically inside the equator if the projection line is drawn to the "south" pole, instead of to the "north" pole as in Fig. 12-32.

Deschamps* has pointed out that distance measured on the surface of the Poincaré sphere is well suited for evaluation of the nearness of polarization states. If such an evaluation is being carried out using projections on the equatorial plane, it is clear that the orthographic projection is most accurate near the poles and that the stereographic projection is most accurate near the equator. Additional applications of the Poincaré sphere to problems of communication between elliptically polarized antennas, polarization measurement, and transformation of polarization can be found in the above referenced paper.

ADDITIONAL PROBLEMS

8. Design an end-fire array that will produce approximately the pattern described by

$$f(\phi) = 1 \qquad 0 < \phi < \frac{\pi}{3}$$

$$f(\phi) = 0 \qquad \frac{\pi}{3} < \phi < \pi$$

Use an element spacing of one-quarter wavelength.

9. Design a six-element broadside array having a spacing $d = \lambda/2$ between adjacent elements. The pattern is to be optimum, with the side-lobe level 20 db down.

*V. H. Rumsey, G. A. Deschamps, M. L. Kales and J. I. Bohnert, *loc. cit.* (See Part II by G. A. Deschamps)

BIBLIOGRAPHY

Articles

Booker, H. G. and P. C. Clemmow, "The Concept of an Angular Spectrum of Plane Waves, and its Relation to that of Polar Diagram and Aperture Distribution," *Jour. IEE*, **97**, Part III, pp. 11-17, 1949.

Chu, L. J., "Physical Limitations of Omnidirectional Antennas," *J. App. Phys.*, **19**, pp. 1163-1175, 1948.

Gilbert, E. N. and S. P. Morgan, "Optimum Design of Directive Antenna Arrays Subject to Random Variations," *Bell Sys. Tech. J.*, **34**, pp. 637-661, May, 1955.

Hansen, R. C., "Gain Limitations of Large Antennas," *IRE Trans. on Ant. and Prop.*, **8**, pp. 490-495, 1960.

Ko, H. C., "The Use of the Statistical Matrix and the Stokes Vector in Formulating the Effective Aperture of Antennas," *IRE Trans. Ant. and Prop.*, **9**, pp. 581-582, 1961.

Lo, Y, T. and S. W. Lee, "Affine Transformation and its Application to Antenna Arrays," *IEEE Trans. Ant. and Prop.*, Vol. AP-13, No. 6, pp. 890-896, Nov., 1965.

Ramsey, J. F., "Fourier Transforms in Aerial Theory," *Marconi Review*, **83**, p. 139, Oct.-Dec., 1946.

Rhodes, D. R., "The Optimum Line Source for the Best Mean-Square Approximation to a Given Radiation Pattern," *IEEE Trans. Ant. and Prop.*, **11**, pp. 440-446, 1963.

Rhodes, D. R., "On a Fundamental Principle in the Theory of Planar Antennas," *Proc. IEEE*, **52**, No. 9, pp. 1013-1020, Sept., 1964.

Schelkunoff. S. A., "A Mathematical Theory of Linear Arrays," *BSTJ*, **22**, No. 1, p. 80, 1943.

Taylor, T. T., "Design of Line-Source Antennas for Narrow Beam-Width and Low Sidelobes," *IRE Trans. Ant. and Prop.*, AP-3, p. 16, 1955.

Uzsoky, M. and L. Solymar, "Theory of Super-Directive Linear Arrays," *Acta Phys. Acad. Sci. Hung.*, **6**, p. 195, 1956.

Books

Hansen, R. C., "Aperture Antennas," in *Microwave Scanning Antennas*, Vol. I, R. C. Hansen, ed., Academic Press Inc., New York, 1964. This book contains a very complete list of references.

Ko, H. C., "Radio Telescope Antennas," in *Microwave Scanning Antennas*, Vol. I, R. C. Hansen, ed., Academic Press Inc., New York, 1964.

Mayes, P. E., *Electromagnetics for Engineers*, Edwards Brothers Inc., Ann Arbor, Michigan, 1965.

Schelkunoff, S. A., *Electromagnetic Waves*, D. Van Nostrand Co., Inc., Princeton, N. J., 1943.

——— and H. T. Friis, *Antennas, Theory and Practice*, John Wiley & Sons, Inc., New York, 1952.

Chapter 13

SECONDARY SOURCES AND APERTURE ANTENNAS

13.01 Magnetic Currents. In writing Maxwell's curl equations

$$\nabla \times \mathbf{H} = \dot{\mathbf{D}} + \mathbf{J} \qquad \nabla \times \mathbf{E} = -\dot{\mathbf{B}}$$

the quantities $\dot{\mathbf{D}}$, $\mathbf{J}$ and $\dot{\mathbf{B}}$ are interpreted as the densities of electric displacement current, electric conduction current, and magnetic displacement current, respectively. The absence of a magnetic quantity corresponding to $\mathbf{J}$, that is, to a magnetic conduction current, is explained by the fact that, so far as is yet known, there are no isolated magnetic charges. As a result, it has been found possible to derive the solution of electromagnetic problems in terms of electric currents and charges alone through the relations

$$\mu\mathbf{H} = \nabla \times \mathbf{A} \qquad \mathbf{E} = -\nabla V - \dot{\mathbf{A}} \tag{13-1}$$

where the electric potentials $\mathbf{A}$ and V are given by eqs. (11-20) and (10-21).

Although the above relations have proved adequate for the solution of problems considered up to the present, there are many other problems where the use of fictitious magnetic currents and charges is very helpful. In such problems the fields, which are actually produced by a certain distribution of electric current and charge, can be more easily computed from an "equivalent" distribution of fictitious magnetic currents and charges. An example of such "equivalent distribution" is the case of the electric current loop and the magnetic dipole. The electromagnetic field produced by a small horizontal electric current loop is identical with that produced by a vertical magnetic dipole. Conversely, the fields produced by a magnetic current loop and electric dipole are also identical. It will be found in many problems involving radiation from "aperture antennas," that the notion of magnetic currents and charges will prove an invaluable aid in arriving at solutions. Therefore expressions for the fields due to such magnetic charges and currents will now be developed. It should be emphasized that, although the magnetic charges and currents used in this procedure are fictitious, the fields calculated from them are physical fields that are actually produced by an equivalent distribution of electric charges and currents.

Written to include magnetic as well as electric conduction current, Maxwell's curl equations would be

$$\nabla \times \mathbf{H} = \dot{\mathbf{D}} + \mathbf{J} \qquad \nabla \times \mathbf{E} = -\dot{\mathbf{B}} - \mathbf{M} \tag{13-2}$$

or in the integral form

$$\oint \mathbf{H} \cdot d\mathbf{s} = \dot{\Psi} + I \qquad \oint \mathbf{E} \cdot d\mathbf{s} = -\dot{\Phi} - K \tag{13-3}$$

In these equations K is a magnetic conduction current and $\mathbf{M}$ is a magnetic conduction current density. K has the dimensions of volts and $\mathbf{M}$ has the dimensions of volts per square meter. For *surface magnetic current density* (corresponding to $\mathbf{J}_s$ for the electric case) the symbol $\mathbf{M}_s$ (V/m) will be used. It is apparent from (2) and (3) that (except for a matter of sign) complete symmetry now exists in Maxwell's equations. The positive sign in the first equations of (2) and (3) indicates that directions of magnetomotive force and electric current are related by the right-hand rule, whereas the negative sign in the second equation indicates that the directions of electromotive force and magnetic current are related through the left-hand rule.

In general it will be desired to solve problems having both electric and magnetic distributions. However, for the purpose of developing expressions due to magnetic currents and charges, consider, first the case where the fields are due to these alone. Equations (2), written for magnetic currents, and in the absence of electric currents, are

$$\nabla \times \mathbf{H}^m = \epsilon \dot{\mathbf{E}}^m \qquad \nabla \times \mathbf{E}^m = -\mu \dot{\mathbf{H}}^m - \mathbf{M} \tag{13-4}$$

These should be compared with the familiar relations written for electric currents alone (without magnetic currents)

$$\nabla \times \mathbf{E}^e = -\mu \dot{\mathbf{H}}^e \qquad \nabla \times \mathbf{H}^e = \epsilon \dot{\mathbf{E}}^e + \mathbf{J} \tag{13-5}$$

(The superscripts e refer to fields due to electric currents and superscripts m refer to fields due to magnetic currents.)

Comparison of (4) and (5) shows them to be identical sets (except for sign) if electric and magnetic quantities are interchanged, $\mathbf{H}^m, \mathbf{E}^m, \mathbf{M}, \mu$ and ϵ replacing $\mathbf{E}^e, \mathbf{H}^e, \mathbf{J}, \epsilon$ and μ respectively. Therefore, the procedures used in chap. 10 for electric currents can be followed to set up potentials due to magnetic currents, and the fields can be obtained from these potentials by differentiation. Corresponding to the magnetic vector potential $\mathbf{A}$ that yields the magnetic field strength through $\mu\mathbf{H}^e = \nabla \times \mathbf{A}$, there will be an electric vector potential $\mathbf{F}$ that will yield the electric field (due to magnetic currents) through $\epsilon\mathbf{E}^m = -\nabla \times \mathbf{F}$. Similarly, corresponding to the scalar electric potential V that is set up in terms of the electric charges, there will be a scalar magnetic potential $\mathscr{F}$ that is set up in terms of the magnetic charges. Rewriting eqs. (1), (10-20), and (10-21) for sinusoidal time variations and with suitable superscripts for fields due to electric currents and charges, we have

$$\mu \mathbf{H}^e = \nabla \times \mathbf{A} \qquad \mathbf{E}^e = -\nabla V - j\omega \mathbf{A} \tag{13-6}$$

$$\mathbf{A} = \frac{1}{4\pi}\int \frac{\mu \mathbf{J}\, e^{-j\beta R}}{R} dV \qquad V = \frac{1}{4\pi}\int \frac{\rho\, e^{-j\beta R}}{\epsilon R} dV \tag{13-7}$$

The corresponding relations for fields due to magnetic currents and charges are

$$\epsilon \mathbf{E}^m = -\nabla \times \mathbf{F} \qquad \mathbf{H}^m = -\nabla \mathscr{F} - j\omega \mathbf{F} \tag{13-8}$$

$$\mathbf{F} = \frac{1}{4\pi}\int \frac{\epsilon \mathbf{M}\, e^{-j\beta R}}{R} dV \qquad \mathscr{F} = \frac{1}{4\pi}\int \frac{\rho_m\, e^{-j\beta R}}{\mu R} dV \tag{13-9}$$

For problems where both electric and magnetic current and charge distributions are involved, the total electric and magnetic field strengths (indicated by no subscript) will be the sum of the field strengths produced by the distributions separately. Writing $\mathbf{E} = \mathbf{E}^e + \mathbf{E}^m$ and $\mathbf{H} = \mathbf{H}^e + \mathbf{H}^m$ the fields (in a region of no current densities **J** and **M**) will be given by*

$$\mathbf{E} = -\nabla V - j\omega \mathbf{A} - \frac{1}{\epsilon}\nabla \times \mathbf{F} \tag{13-10}$$

$$\mathbf{H} = -\nabla \mathscr{F} - j\omega \mathbf{F} + \frac{1}{\mu}\nabla \times \mathbf{A} \tag{13-11}$$

Alternatively, writing in terms of the vector potentials alone [as was done in (10-31)] gives

$$\begin{aligned} \mathbf{E} &= \frac{1}{j\omega\mu\epsilon}\nabla \times \nabla \times \mathbf{A} - \frac{1}{\epsilon}\nabla \times \mathbf{F} \\ \mathbf{H} &= \frac{1}{j\omega\mu\epsilon}\nabla \times \nabla \times \mathbf{F} + \frac{1}{\mu}\nabla \times \mathbf{A} \end{aligned} \tag{13-12}$$

There is one other relation connected with magnetic currents that must be considered. It was found that tangential H was discontinuous

*For dissipative media it is necessary to write the second term of (11) as $-(\sigma + j\omega\epsilon)\mathbf{F}$ instead of just $-j\omega\epsilon\,\mathbf{F}$ to include the effects of electric conduction currents due to magnetic sources in addition to electric displacement currents due to magnetic sources. There is, of course, no corresponding term in (10) for magnetic conduction currents due to electric sources. In this book there will be no occasion to deal with the effects of magnetic sources in dissipative media, and eqs. (10) and (11) will suffice.

across an electric-current sheet (though tangential E remained continuous). The discontinuity in tangential H is equal to the linear current density J_s as showh by the relation

$$H_{1(\tan)} - H_{2(\tan)} = J_s \qquad \text{(amp/m)}$$

This result was obtained directly from an application of Maxwell's mmf equation I; the continuity of tangential E followed from the emf equation II (sec. 4.03, page 103). The electric-current density J_s and tangential H are mutually perpendicular, and this fact is indicated by the vector relation

$$\mathbf{J}_s = \hat{\mathbf{n}} \times (\mathbf{H}_1 - \mathbf{H}_2) \qquad \text{(amp/m)} \quad (13\text{-}13)$$

where $\hat{\mathbf{n}}$, the unit vector normal to the current sheet, is regarded as positive when pointing to the side that contains H_1. In the same way it is found from equations (3) that tangential E is discontinuous across a magnetic-current sheet, whereas tangential H remains continuous. For a linear magnetic current density M_s (V/m) the relation corresponding to (13) is

$$\mathbf{M}_s = -\hat{\mathbf{n}} \times (\mathbf{E}_1 - \mathbf{E}_2) \qquad \text{(volt/m)} \quad (13\text{-}14)$$

The minus sign results from the minus sign in the second of eqs. (3). Equation (14) states that the tangential electric field strength is discontinuous across a magnetic-current sheet by an amount equal to the linear magnetic-current density.

Examples of the use of magnetic currents will appear in the sections that follow.

13.02 Duality. Eqs. (2), (3), (4) and (5) exhibit the property of *duality*. Duality means that it is possible to pass from one equation to another by suitable interchange of the *dual* quantities, which in this case are the electric and magnetic quantities. Written for sinusoidal time variations and regions with material media, and showing source currents explicitly (sec. 13.05), eqs. (2) may be written in the form

$$\nabla \times \mathbf{H} = Y\mathbf{E} + \mathbf{J} \qquad (13\text{-}15)$$

$$-\nabla \times \mathbf{E} = Z\mathbf{H} + \mathbf{M} \qquad (13\text{-}16)$$

where $Y = \sigma + j\omega\epsilon$ and $Z = j\omega\mu$. For complete duality it would be necessary to add, besides the magnetic conduction current density $\mathbf{M}$, a magnetic conductivity σ_m. However, because there will be no occasion to use this nonexistent quantity except in the perfect magnetic conductor case, it is generally omitted. The standard list of dual quantities is given in Table 13-1, where the first set is for electric current sources I or $\mathbf{J}$, and the second set is for magnetic current sources K or $\mathbf{M}$.

Table 13-1

(1)	I	$\mathbf{J}$	$\mathbf{E}$	$\mathbf{H}$	$\mathbf{A}$	ϵ	μ	$\eta = \sqrt{\frac{\mu}{\epsilon}}$
(2)	K	$\mathbf{M}$	$\mathbf{H}$	$-\mathbf{E}$	$\mathbf{F}$	μ	ϵ	$\eta_m = \sqrt{\frac{\epsilon}{\mu}} = \frac{1}{\eta}$

In problems involving electric conductors it is convenient to deal with *perfect* conductors ($\sigma = \infty$) for which the simplified boundary condition $\hat{\mathbf{n}} \times \mathbf{E} = 0$ or $E_{\tan} = 0$ applies. Similarly, in dealing with (nonexistent) magnetic conductors, the perfect magnetic conductor case ($\sigma_m = \infty$) is sometimes convenient; for this hypothetical case the boundary condition $\hat{\mathbf{n}} \times \mathbf{H} = 0$ or $H_{\tan} = 0$ applies. According to the principle of duality, any problem solution obtained using the quantities of set (1) will apply directly (numerically as well as formally) to the dual problem, with quantities of set (2) replacing the corresponding numbers of set (1).

13.03 Images of Electric and Magnetic Currents. The field of a magnetic current element or "magnetic dipole" $K\,dl$ can be derived using eqs. (8) and (9), as was done for the electric current element in chap. 10. However, because of the *duality* between electric and magnetic quantities displayed in eqs. (2) and (3) and also in eqs. (4) and (5), the solution for the magnetic dipole may be written down directly from eqs. (10-41), (10-42) and (10-43), with E_θ, E_r, H_ϕ, I and ϵ being replaced by H_θ, H_r, $-E_\phi$, K and μ respectively. It will be observed that for a magnetic current element the electric field is everywhere normal to any plane containing the element, and the magnetic field at the surface of every such plane is tangential to it. From this fact it follows that a magnetic current element may be split through the middle by a perfect (electric) conducting plane without in any way affecting the field. This result may be compared with that for an electric current element which can be bisected by an electric conducting plane *normal* to its axis, without in any way affecting the field distribution. For the distant or radiation field the directions of **E** and **H** produced by a magnetic current are as sketched in Fig. 13-1.

From the preceding paragraph it follows that if an electric current element is placed at the surface and normal to an electric conducting plane, or if a magnetic current element is placed at the surface and parallel to the electric conducting plane, a positive image results. The effect of the conducting plane is simply to double the electromagnetic field on the side of the current element, and consequently to reduce to zero the field on the shielded or shadow side. Conversely, it will be found that an electric dipole parallel to, or a magnetic dipole normal to an electric conducting plane results in a negative image.

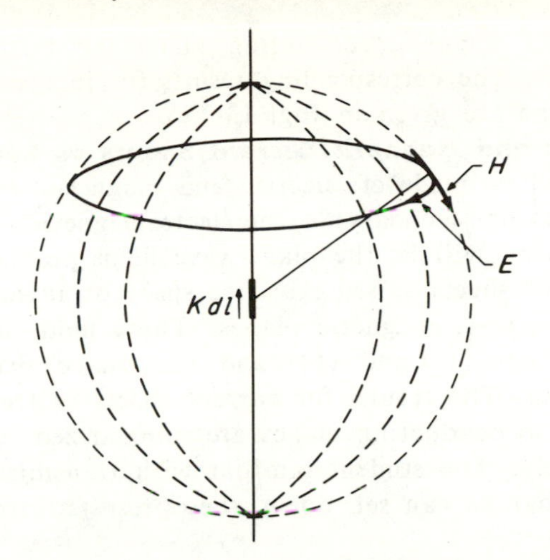

Figure 13-1. Lines of **E** (solid) and **H** (dashed) on a large spherical surface centered on a magnetic current element $K\,dl$.

In this case the effect of the electric currents induced on the conducting plane is to reduce the total fields, both electric and magnetic, to zero. These results on images in an electric conducting plane are summarized by Fig. 13-2(*a*).

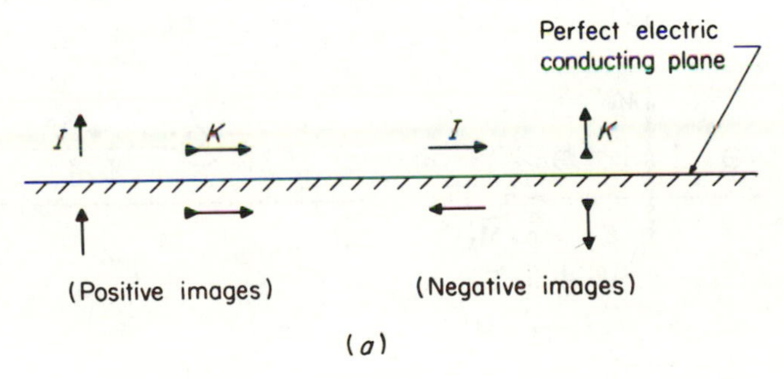

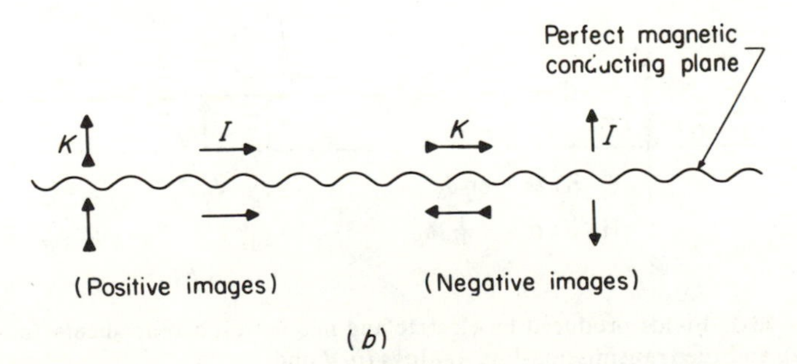

Figure 13-2.

For a perfect magnetic conducting plane the boundary condition $H_{\tan} = 0$ applies. The corresponding results for images in a magnetic conducting plane are given in Fig. 13-2(*b*).

13.04 Electric and Magnetic Current Sheets as Sources. In this chapter use will be made of electric and magnetic current *sheets* as sources (primary or secondary) of the electromagnetic field. Of interest in this connection will be the plane-wave fields produced by (infinitely) large current sheets existing in free space or adjacent to perfectly conducting electric or magnetic planes. These fields may be deduced through use of eqs. (13) and (14) and the image principles of the previous sections. The results for current sheets in free space and for those adjacent to conducting planes are summarized in Figs. 13-3 and 13-4, respectively. The student familiar with transmission-line theory will discover that he can set down these results directly from the

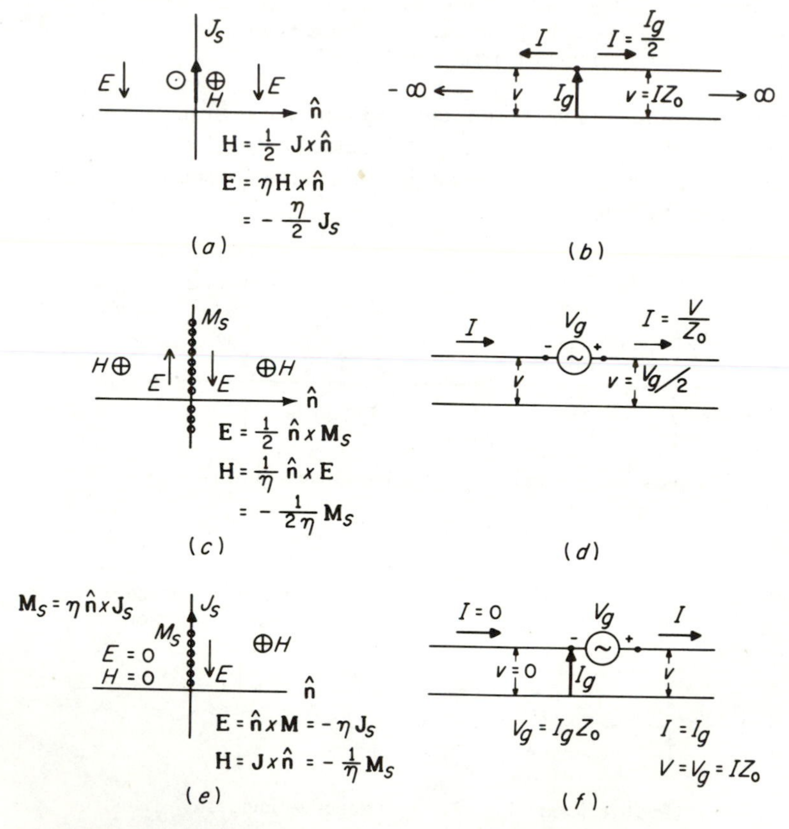

Figure 13-3. Fields produced by electric and magnetic current sheets (*a*, *c* and *e*), and the transmission-line analogs (*b*, *d* and *f*).

transmission-line analogy. Electric and magnetic current sheets correspond respectively to shunt current and series voltage generators, and electric and magnetic conducting planes are representable respectively by short and open circuits on the transmission line. Study of these figures reveals the following noteworthy points:

(1) An electric current sheet alone, or a magnetic current sheet alone, generates an electromagnetic field on both sides of the sheet [Fig. 13-3(*a*) and (*c*)].

(2) Crossed electric and magnetic current sheets of proper magnitude can be made to generate an electromagnetic field on one side only; perfect cancellation of fields occurs on the back side [Fig. 13-3(*e*)].

(3) The *same* fields produced in Fig. 13-3(*e*) by crossed electric and magnetic current sheets can be obtained from a magnetic current

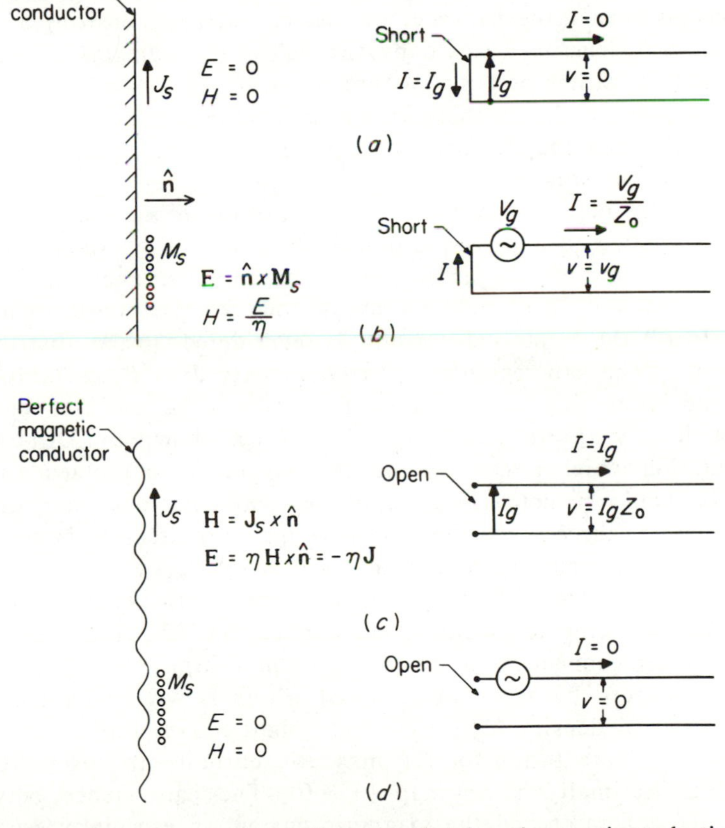

Figure 13-4. Current sheets adjacent to electric and magnetic conducting planes and the transmission-line analogs.

sheet alone in front of a perfect electric conductor [Fig. 13-4(*b*)] or from an electric current sheet alone in front of a perfect magnetic conductor [Fig. 13-4(*c*)]. These results are immediately evident from the transmission-line analogs.

Knowledge of the fields produced by these electric and magnetic current sheets proves useful in treating radiation from apertures.

13.05 Impressed and Induced Current Sources. In circuit theory it is convenient to regard the voltages and currents in a network as arising due to the presence of *impressed* voltages and currents supplied by hypothetical zero-impedance and infinite-impedance generators. Correspondingly in electromagnetic field theory it is often convenient to specify impressed magnetic currents (voltages) and electric currents as the sources of the field. These impressed currents are usually given in terms of the current densities $\mathbf{M}$ and $\mathbf{J}$, or in terms of the linear (surface) current densities $\mathbf{M}_s$ and $\mathbf{J}_s$.

In writing Maxwell's equations, as for example in eqs. (2), no distinction has been made between impressed current densities and induced current densities (caused by the field), the symbols $\mathbf{J}$ and $\mathbf{M}$ being used for either or both. Where necessary or desirable to make such a distinction the symbols $\mathbf{J}^0$, $\mathbf{M}^0$ will be used for *impressed* current densities and the symbols V^0 and I^0 will designate impressed voltages and currents.

Fig. 13-5(*a*) depicts the circuit representation of a constant current generator applied to a pair of antenna terminals. Fig. 13-5(*b*) shows an expanded view of the gap region of the antenna, and (*c*), (*d*) and (*e*) show commonly used field representations for the constant-current source. In (*c*) the impressed current is considered to be distributed uniformly in the gap region as a current density $J^0 = I^0/\pi a^2$ (amp/m^2). In (*d*) the impressed current takes the form of a circular current sheet of density $J_s^0 = I^0/2\pi a$ (amp/m). In (*e*) is shown the Dirac delta representation which frequently proves convenient in problem formulation. It should be noted in all of these representations, that turning off the current source leaves the terminals open-circuited, as is required for a constant-current (infinite-impedance) source.

Fig. 13-6 depicts circuit and field representations of a voltage source. In (*b*) a ring of magnetic current density, $\mathbf{M}^0$, circulates about a short-circuit connection between the antenna arms. In (*c*) the gap region is replaced by a conductor about which is wrapped a magnetic current sheet of density $\mathbf{M}_s^0$ (V/m) and height h. In Fig. 13-6(*d*) the Dirac delta representation for the magnetic current strip is indicated for an infinitesimally narrow gap ($h \to 0$). The equivalence between the voltage generator and the magnetic current is especially easy to demonstrate in this case. The voltage between B and A is

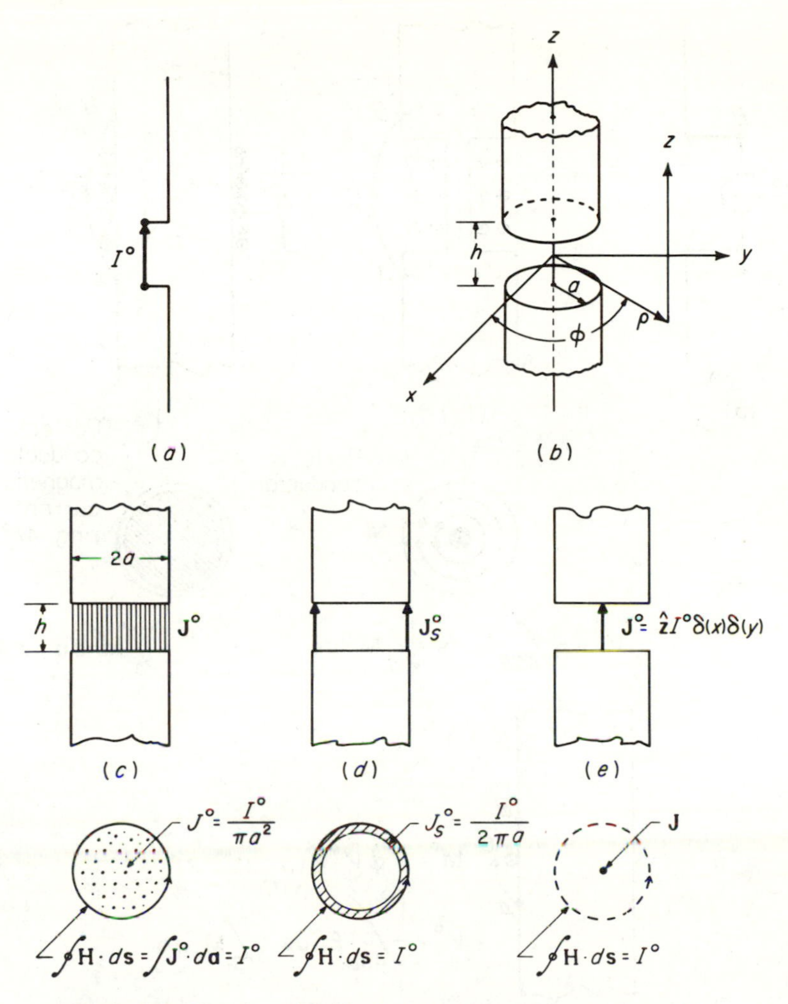

Figure 13-5. Circuit and field representations of a current source.

$$V_B - V_A = \int_{ACB} \nabla V \cdot d\mathbf{s} = -\int_{ACB} \mathbf{E} \cdot d\mathbf{s}$$

$$= -\oint \mathbf{E} \cdot d\mathbf{s} = \int_S \mathbf{M} \cdot d\mathbf{a}$$

$$= \int_S [-\hat{\boldsymbol{\phi}} V^0 \, \delta(\rho - a)\, \delta(z)] \cdot [-\hat{\boldsymbol{\phi}} \, da] = V^0 \qquad (13\text{-}17)$$

It should be noted that the line integral over the path ACB may be written as a closed-contour integral because the electric field within the conductor is zero.

For all of the above representations, when the magnetic current is

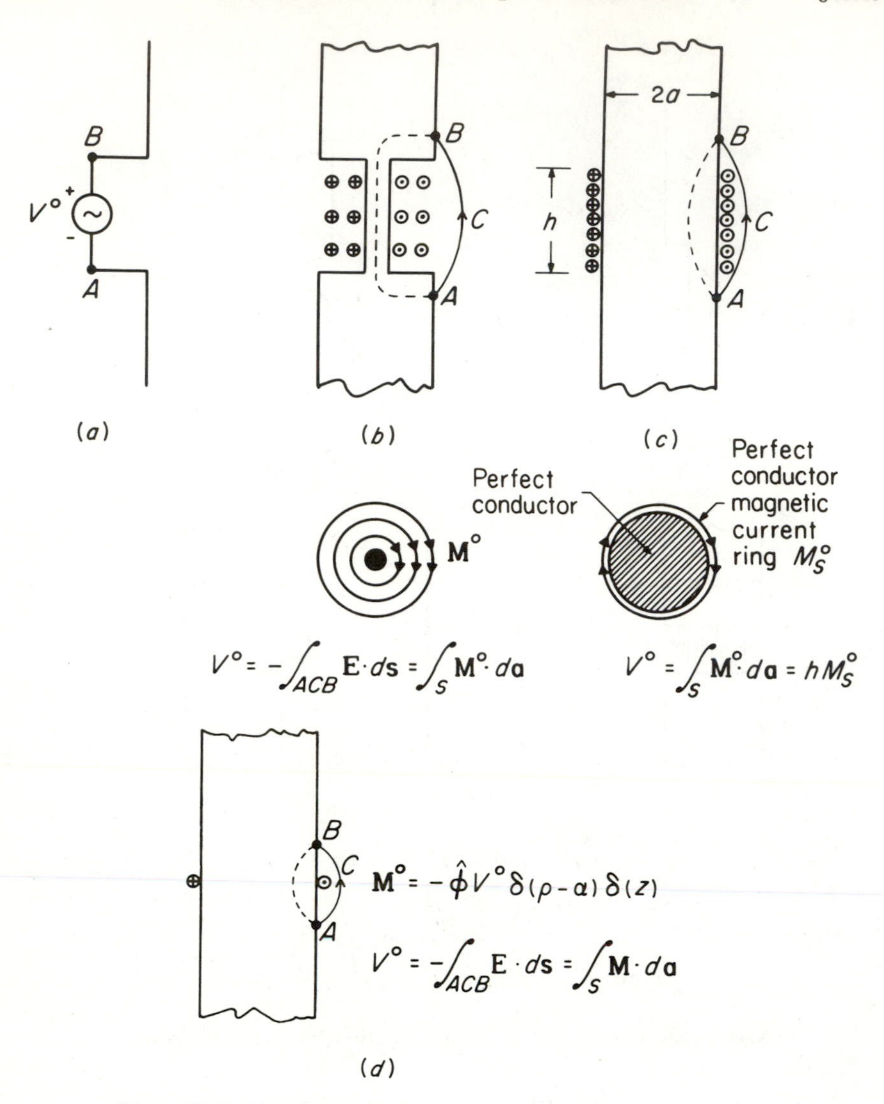

Figure 13-6. Circuit and field representations of a voltage source.

"turned off" a short circuit remains between the terminals, as is required for a constant-voltage (zero-impedance) generator. In both Figs. 13-5 and 13-6 the gap regions are considered to be very small with the contours of the integration paths of **H** and **E** wrapped closely about the electric and magnetic currents. This makes it possible to neglect without error the effects of electric and magnetic displacement currents.

Induced Currents. In the preceding paragraphs field equivalents were found for the circuit-theory notions of voltage and current generators. Another circuit-theory concept having a field-theory analog is the compensation theorem, which states that a passive circuit element may be replaced with an ideal generator. The analogous field-theory situation is shown in Fig. 13-7. An object with macroscopic properties $Y_1 = \sigma_1 + j\omega\epsilon_1$, $Z_1 = j\omega\mu_1$ occupies the volume V and is immersed in an infinite medium having properties $Y_0 = \sigma_0 + j\omega\epsilon_0$, $Z_0 = j\omega\mu_0$ and containing sources $\mathbf{J}^0$, $\mathbf{M}^0$.

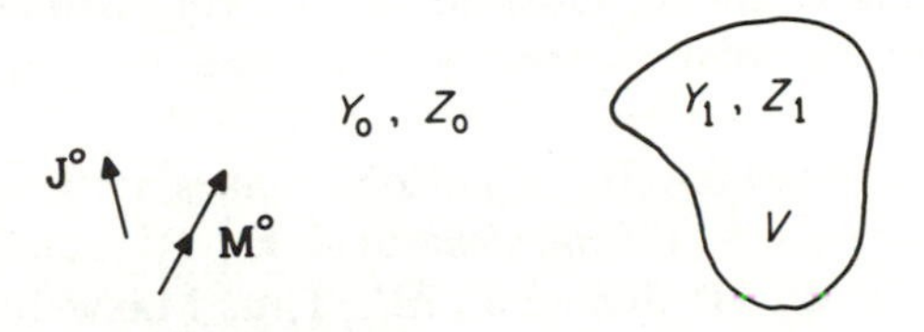

Figure 13-7. Scattering object in an electromagnetic field.

The object occupying the volume V may be regarded as disturbing or *scattering* the field that would exist if the object were not present, that is, if all of space had the properties Y_0, Z_0. *With the scattering object present* the total fields *outside* V satisfy Maxwell's curl equations in the form

$$\nabla \times \mathbf{H} = Y_0 \mathbf{E} + \mathbf{J}^0 \tag{13-18}$$

$$\nabla \times \mathbf{E} = - Z_0 \mathbf{H} - \mathbf{M}^0 \tag{13-19}$$

Inside V the curl equations are

$$\nabla \times \mathbf{H} = Y_1 \mathbf{E} \tag{13-20}$$

$$\nabla \times \mathbf{E} = - Z_1 \mathbf{H} \tag{13-21}$$

The latter two equations may be rewritten in the form

$$\nabla \times \mathbf{H} = Y_0 \mathbf{E} + \mathbf{J}^1 \tag{13-22}$$

$$\nabla \times \mathbf{E} = - Z_0 \mathbf{H} - \mathbf{M}^1 \tag{13-23}$$

in which

$$\mathbf{J}^1 = (Y_1 - Y_0)\mathbf{E} \tag{13-24}$$

and

$$\mathbf{M}^1 = (Z_1 - Z_0)\mathbf{H} \tag{13-25}$$

Relations (24) and (25) define the *induced sources* $\mathbf{J}^1$ and $\mathbf{M}^1$. These currents exist only within the volume V and are represented in (22) and (23) as flowing in a medium with properties Y_0, Z_0; they actually replace the scattering object insofar as electromagnetic effects are concerned. Because eqs. (18) and (22) and also (19) and (23) are of the same form, they may be written together as follows applying to all space both inside and outside the volume V:

$$\nabla \times \mathbf{H} = Y_0\mathbf{E} + \mathbf{J}^0 + \mathbf{J}^1 \tag{13-26}$$

$$\nabla \times \mathbf{E} = -\, Z_0\mathbf{H} - \mathbf{M}^0 - \mathbf{M}^1 \tag{13-27}$$

Thus the scattering problem in an inhomogeneous medium has been replaced with a problem involving sources in an infinite, uniform medium. This does not constitute a solution to the problem, however, because the fields $\mathbf{E}$ and $\mathbf{H}$ must be known (or estimated) inside the scattering object in order to specify the induced source currents $\mathbf{J}^1$ and $\mathbf{M}^1$.

The linearity of Maxwell's equations makes it possible to separate the field into two parts, the *incident* field $\mathbf{E}^i$, $\mathbf{H}^i$ due to $\mathbf{J}^0$, $\mathbf{M}^0$, and the *scattered* field $\mathbf{E}^s$, $\mathbf{H}^s$ due to $\mathbf{J}^1$, $\mathbf{M}^1$. Thus Maxwell's curl equations each may be written in two parts as follows:

$$\nabla \times \mathbf{H}^i = Y_0\mathbf{E}^i + \mathbf{J}^0 \tag{13-28}$$

$$\nabla \times \mathbf{E}^i = -\, Z_0\mathbf{H}^i - \mathbf{M}^0 \tag{13-29}$$

$$\nabla \times \mathbf{H}^s = Y_0\mathbf{E}^s + \mathbf{J}^1 \tag{13-30}$$

$$\nabla \times \mathbf{E}^s = -\, Z_0\mathbf{H}^s - \mathbf{M}^1 \tag{13-31}$$

Scattering problems usually involve a known incident field, for example a uniform plane wave or the field of a familiar type of antenna. The unknown is the scattered field and finding it can be an exceedingly difficult task. The induced current is the source of the scattered field (sometimes called the *induced* field) but specification of the induced current depends on accurate knowledge of the total field inside the scattering object. Fortunately in two important cases considered below the field inside the scattering object can be estimated with good accuracy.

Scattering when $Z_1 \approx Z_0$, $Y_1 \approx Y_0$. If the scattering object has properties which are but little different from those of the surrounding medium, it may be assumed that the field inside the scatterer is identical to the incident field; this is sometimes referred to as the *Born approximation*. For example, consider a cube of lossless dielectric material with $\epsilon_r \approx 1$ and measuring L meters on a side. Suppose that L is small compared to a given wavelength and that the cube is centered at the origin in free space. If an incident uniform plane wave has an electric field strength given by $\mathbf{E}^i = \hat{\mathbf{z}}E_0\, e^{-j\beta y}$, the induced current density $\mathbf{J}_1$ in the cube is approximately equal to $\hat{\mathbf{z}}\, j\omega\epsilon_v(\epsilon_r - 1)E_0$. This current flows for a distance L through a cross-sectional area L^2 and thus the induced current in the small cube comprises a current element whose moment $I\,dl$ is $j\omega\epsilon_v(\epsilon_r - 1)E_0L^3$. The scattered field may be obtained directly from the expression for the field of a current element (problem 4, chap. 10).

Scattering by a Conductor. If the scattering object is a conductor with conductivity σ and permittivity ϵ_v, eq. (24) shows that the induced current density is equal to the conduction current density. That is

$$\mathbf{J}_1 = \sigma \mathbf{E} \tag{13-32}$$

so that the scattered field may be calculated approximately from an estimate of the conduction current in the scatterer.

Problem 1. A long, thin dielectric rod with relative permittivity $\epsilon_r \approx 1$ has narrow dimensions equal to $W \ll \lambda$ and long dimension $L = \lambda$; the rod extends along the y axis from $(-L/2)$ origin to $(+L/2)$. An incident uniform plane wave has an electric field strength given by $\mathbf{E}^i = \hat{\mathbf{z}} E_0 e^{-j\beta y}$. Find an expression for the scattered electric field strength far from the rod and sketch the scattered-field radiation patterns in the three principal planes.

13.06 Reciprocity in Electromagnetic Field Theory. The principle of reciprocity is one of great generality, finding application in such diverse fields as international trade, human relations, science and engineering. Whereas in the political and economic fields a rather loose statement of the principle is usually sufficient, in scientific and engineering work a very precise statement is required. For example, it is not generally correct to say that according to reciprocity the response of a system is unchanged when transmitter and receiver, or generator and load, are interchanged, unless we first specify the generator and load impedances. In electromagnetic theory a suitably precise statement of the reciprocity principle can be derived, starting with Maxwell's equations.

Consider a region in space which is linear and isotropic though not necessarily homogeneous. Suppose that two sets of source currents $\mathbf{J}^a$, $\mathbf{M}^a$ and $\mathbf{J}^b$, $\mathbf{M}^b$ can exist in this region, producing the fields $\mathbf{E}^a$, $\mathbf{H}^a$ and $\mathbf{E}^b$, $\mathbf{H}^b$ respectively. These source-field pairs satisfy Maxwell's curl equations

$$\begin{aligned} \nabla \times \mathbf{H}^a &= Y\mathbf{E}^a + \mathbf{J}^a & \qquad \nabla \times \mathbf{H}^b &= Y\mathbf{E}^b + \mathbf{J}^b \\ \nabla \times \mathbf{E}^a &= -Z\mathbf{H}^a - \mathbf{M}^a & \qquad \nabla \times \mathbf{E}^b &= -Z\mathbf{H}^b - \mathbf{M}^b \end{aligned} \tag{13-33}$$

in which $Y = \sigma + j\omega\epsilon$ and $Z = j\omega\mu$. The two sets of sources are assumed to operate in the same medium and at the same frequency but whether or not they operate simultaneously is unimportant.

The principle of reciprocity is simply the statement of an interrelationship between the two source-field pairs. It is derived as follows:

$$\begin{aligned}\nabla\cdot(\mathbf{E}^b \times \mathbf{H}^a - \mathbf{E}^a \times \mathbf{H}^b) &= \mathbf{H}^a\cdot\nabla\times\mathbf{E}^b - \mathbf{E}^b\cdot\nabla\times\mathbf{H}^a - \mathbf{H}^b\cdot\nabla\times\mathbf{E}^a + \mathbf{E}^a\cdot\nabla\times\mathbf{H}^b \\ &= -\mathbf{H}^a\cdot(Z\mathbf{H}^b) - \mathbf{E}^b\cdot(Y\mathbf{E}^a) + \mathbf{H}^b\cdot(Z\mathbf{H}^a) + \mathbf{E}^a\cdot(Y\mathbf{E}^b) \\ &\quad - \mathbf{H}^a\cdot\mathbf{M}^b - \mathbf{E}^b\cdot\mathbf{J}^a + \mathbf{H}^b\cdot\mathbf{M}^a + \mathbf{E}^a\cdot\mathbf{J}^b\end{aligned} \tag{13-34}$$

For isotropic media* $\mathbf{H}^a\cdot(Z\mathbf{H}^b) = \mathbf{H}^b\cdot(Z\mathbf{H}^a)$ and $\mathbf{E}^b\cdot(Y\mathbf{E}^a) = \mathbf{E}^a\cdot(Y\mathbf{E}^b)$, permitting (34) to be simplified greatly. Integration over a volume V and application of the divergence theorem gives

$$\int_S (\mathbf{E}^b \times \mathbf{H}^a - \mathbf{E}^a \times \mathbf{H}^b)\cdot d\mathbf{a} = \int_V (\mathbf{E}^a\cdot\mathbf{J}^b - \mathbf{H}^a\cdot\mathbf{M}^b - \mathbf{E}^b\cdot\mathbf{J}^a + \mathbf{H}^b\cdot\mathbf{M}^a)dV \tag{13-35}$$

which constitutes a very general expression of the reciprocity concept known as the *Lorentz integral.* The requirement that cases (a) and (b) involve the same medium (or environment) now is seen to be necessary only on the surface S and throughout the enclosed volume V.

An especially useful statement of reciprocity may be derived by letting the surface S become a sphere of infinite radius. If all the sources are contained within a finite volume, the fields at infinite radius must be outgoing spherical waves satisfying the relations $E_\theta = \eta H_\phi$ and $E_\phi = -\eta H_\theta$. Under these conditions it is easy to show (the reader should do this as an exercise) that the surface integral in (35) is zero. Thus (35) becomes

$$\int (\mathbf{E}^a\cdot\mathbf{J}^b - \mathbf{H}^a\cdot\mathbf{M}^b)\,dV = \int (\mathbf{E}^b\cdot\mathbf{J}^a - \mathbf{H}^b\cdot\mathbf{M}^a)\,dV \tag{13-36}$$

Nominally these integrations are over all space, but of course the regions of integration reduce to the regions occupied by the source currents. Equation (36) is sometimes written in the form

$$\langle a, b\rangle = \langle b, a\rangle \tag{13-37}$$

in which $\langle a, b\rangle$ is referred to as the "reaction" of field a on source b.† The applications of the reciprocity concept are many and a few of them will be discussed in detail.

Derivation of Circuit Reciprocity from Field Reciprocity. The reciprocity theorem for fields reduces to the familiar circuit-theory statement of reciprocity when applied to the terminals of a pair of antennas. A pair of perfectly conducting antennas is shown in Fig. 13-8

*Although derived for scalar Y and Z, (35) also holds if Y and Z are symmetric tensors. If they are not symmetric tensors, the medium they represent is said to be *nonreciprocal* and (35) does not apply.

†V. H. Rumsey, "Reaction Concept in Electromagnetic Theory," *Phys. Rev.* **94**, No. 6, pp. 1483–1491, June 15, 1954.

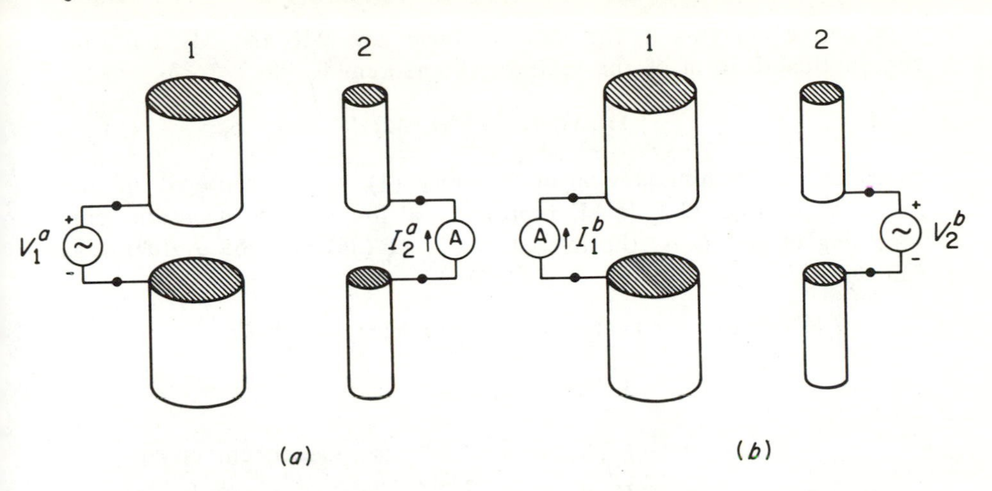

Figure 13-8. A pair of antennas connected to demonstrate reciprocity.

in which the circuit concepts of zero-impedance voltage generators and zero-impedance ammeters are used. Two situations are shown, one in which a generator and an ammeter are attached to the two antennas and another in which the ammeter and generator positions have been interchanged. The field formulation of reciprocity may be applied to this problem in several different ways, one making use of (36), which is valid in an infinite region. The two cases (*a*) and (*b*) involve similar environments, a similarity which is clearly evident if in each case the short circuit is represented as a continuation of the antenna structure and the voltage source is represented as a ring of magnetic current wrapped around a short circuit as depicted in Fig. 13-9. The

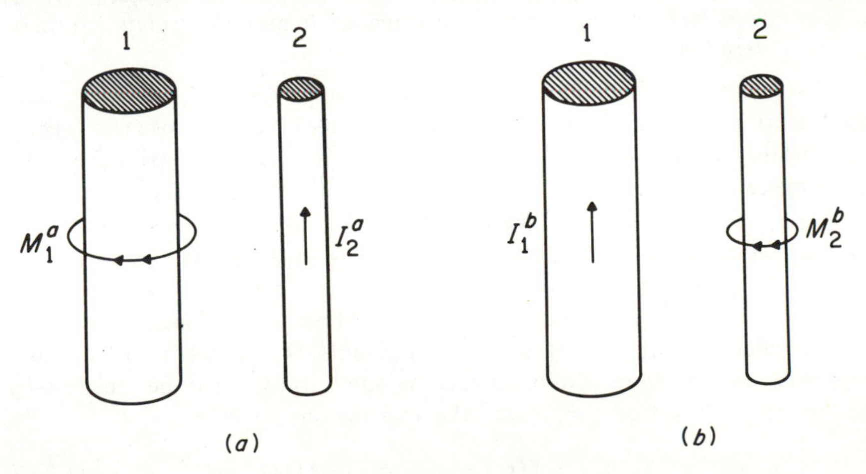

Figure 13-9. Field-theory representation of the reciprocity problem.

only source currents in the field problem are $\mathbf{M}_1^a$ and $\mathbf{M}_2^b$ and thus the applicable form of the reciprocity theorem is

$$\int \mathbf{H}_1^b \cdot \mathbf{M}_1^a \, dV = \int \mathbf{H}_2^a \cdot \mathbf{M}_2^b \, dV \tag{13-38}$$

Consider the terminal region of antenna (1) to be centered at the origin as in Fig. 13-5. If $\mathbf{M}_1^a$ represents a filamentary magnetic current ring of radius ρ_0, the left-hand side of (38) may be evaluated as follows:

$$\begin{aligned}
\int \mathbf{H}_1^b \cdot \mathbf{M}_1^a \, dV &= \int \mathbf{H}_1^b \cdot [-\hat{\boldsymbol{\phi}} V_1^a \delta(\rho - \rho_0)\delta(z)] \, \rho \, d\rho \, d\phi \, dz \\
&= -\int_0^{2\pi} H_{1\phi}^b \, V_1^a \, \rho_0 \, d\phi \\
&= -V_1^a H_{1\phi}^b \, 2\pi \rho_0 \, (\text{if } H_{1\phi}^b \text{ is independent of } \phi) \\
&= -V_1^a I_1^b
\end{aligned} \tag{13-39}$$

The right-hand side of (38) may be evaluated similarly, giving finally

$$\frac{V_1^a}{I_2^a} = \frac{V_2^b}{I_1^b} \tag{13-40}$$

which is the required circuit-theory statement of reciprocity (see sec. 11.02).

Problem 2. Obtain (40) from (38) for the voltage generator representations shown in Figs. 13-6(*b*) and (*c*).

Problem 3. Show how (40) may be derived if the conducting material in the antennas is replaced with the induced current flowing in free space, according to the "induced sources" concept.

Problem 4. Derive the equivalent of (40) using electric impressed currents and open-circuited terminals in place of magnetic impressed currents and short-circuited terminals.

Use of a Known Field to Find an Unknown Field. The electric field $\mathbf{E}$ due to sources $\mathbf{J}$, $\mathbf{M}$ may be calculated by the use of the vector potentials as outlined in sec. 13.01. The following alternative formulation makes use of reciprocity and the fact that the field of an electric current element is known. Referring to Fig. 13-10, suppose an electric current element at $\mathbf{r}_p$ with current density $\mathbf{J}_p(\mathbf{r}) = \hat{\mathbf{p}} \, \delta(\mathbf{r} - \mathbf{r}_p)$ produces the field $\mathbf{E}_p(\mathbf{r})$, $\mathbf{H}_p(\mathbf{r})$ in infinite free space. As shown, $\mathbf{J}$, $\mathbf{M}$ are source currents (impressed or induced) whose field is to be found. The current element with its field and the currents $\mathbf{J}$, $\mathbf{M}$ with their fields comprise two source-field pairs and as such they must be related by reciprocity. Equation (36) gives the expression

$$\hat{\mathbf{p}} \cdot \mathbf{E}(\mathbf{r}_p) = \int (\mathbf{J} \cdot \mathbf{E}_p - \mathbf{M} \cdot \mathbf{H}_p) dV \tag{13-41}$$

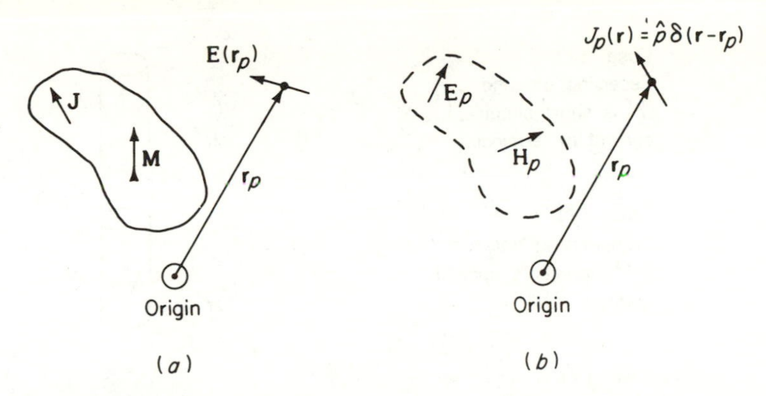

Figure 13-10. Use of the known field of an electric current element to find an unknown electric field.

for the component of electric field strength parallel to the current element.

At first glance, the evaluation of the electric field using (41) would appear to be quite different from the evaluation of the same field using the vector potential approach described in chap. 10 and in sec. 13.01. It turns out, however, that the mathematical operations involved in the two methods are essentially identical, the only difference between them being the manner in which the problem is set up initially. Choosing between the two methods must therefore be based on the ease with which each method can be visualized and used to set up the necessary equations.

Problem 5. Use a magnetic current element to derive an expression similar to (41) for the magnetic field strength due to the given currents **J, M.**

Problem 6. Use (41) to calculate the scattered field in problem 1.

Short-Circuit Current at the Terminals of a Receiving Antenna. In chap. 11, a formula was derived [eq. (11-5)] expressing the short-circuit current at the terminals of a receiving antenna in terms of the incident electric field in free space (no antenna present) and the current distribution on the antenna when it is used for transmitting. This formula may be derived from the field theory of reciprocity by noting that three distinctly different source-field situations are involved as sketched in Fig. 13-11. In the receiving case, the antenna responds to the field of some external source represented by the electric current density $\mathbf{J}^0$; the transmitting and incident field cases are self-explanatory.

The procedure to be followed is the application of the reciprocity relation (36) to the three situations in groups of two. Applying reci-

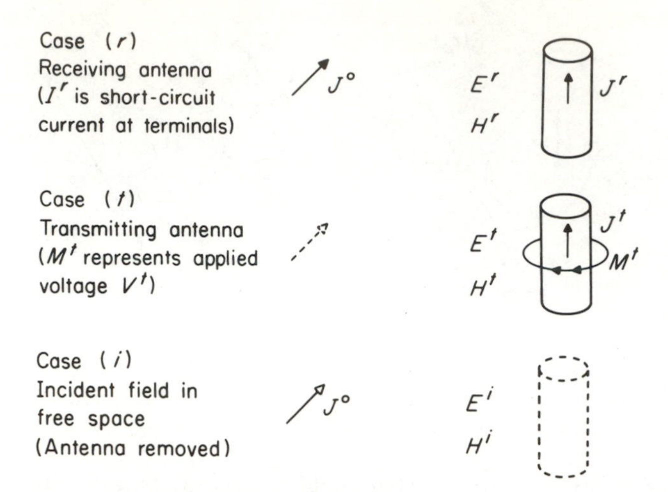

Figure 13-11.

procity to cases (r) and (t) gives

$$-\int \mathbf{H}^r \cdot \mathbf{M}^t \, dV = \int \mathbf{E}^t \cdot \mathbf{J}^0 \, dV \tag{13-42}$$

For cases (t) and (i) the restriction to identical environments requires that the metal antenna in the transmitting case be replaced with its induced current $\mathbf{J}^t$ flowing in free space. Applying reciprocity to cases (t) and (i) gives

$$\int \mathbf{E}^i \cdot \mathbf{J}^t \, dV - \int \mathbf{H}^i \cdot \mathbf{M}^t \, dV = \int \mathbf{E}^t \cdot \mathbf{J}^0 \, dV \tag{13-43}$$

Equations (42) and (43) may be combined to give

$$-\int \mathbf{H}^r \cdot \mathbf{M}^t \, dV = \int \mathbf{E}^i \cdot \mathbf{J}^t \, dV - \int \mathbf{H}^i \cdot \mathbf{M}^t \, dV \tag{13-44}$$

The left-hand side of (44) is just the product $I^r V^t$ as may be seen by inspection of eq. (39). The second integral on the right-hand side of (44) is a terminal-region contribution; it approaches zero as the size of the terminal region approaches zero and therefore may be neglected. Thus the short-circuit current is given by

$$I^r = \frac{1}{V^t} \int \mathbf{E}^i \cdot \mathbf{J}^t \, dV \tag{13-45}$$

which is a generalization of the formula (11-5) derived earlier using the circuit concept of reciprocity.

Problem 7. Evaluate the terminal-region integral in (44) for an incident uniform plane wave whose electric field is parallel to the antenna axis. Show that the integral approaches zero as the size of the terminal region approaches zero.

Problem 8. Derive an expression for the open-circuit voltage at the terminals of a receiving antenna in terms of the transmitting terminal current and the incident field in free space. Use an electric current source for the transmitting case and follow a procedure similar to the derivation of (45). Also show how the same result may be deduced from (45) using Thevenin's theorem.
Answer:

$$V^r = -\frac{1}{I^t}\int \mathbf{E}^i \cdot \mathbf{J}^t \, dV \tag{13-46}$$

13.07 The Induction and Equivalence Theorems. In sec. 13.05 the notion of induced or secondary sources was introduced to aid in the solution of problems involving scattering from an inhomogeneity. In that treatment the induced sources were current densities distributed throughout the volume occupied by the scatterer. It is also possible and frequently advantageous to consider the scattered field in terms of induced *surface* currents over the surface of the scattering volume; this treatment leads immediately to two important theorems in electromagnetic theory.

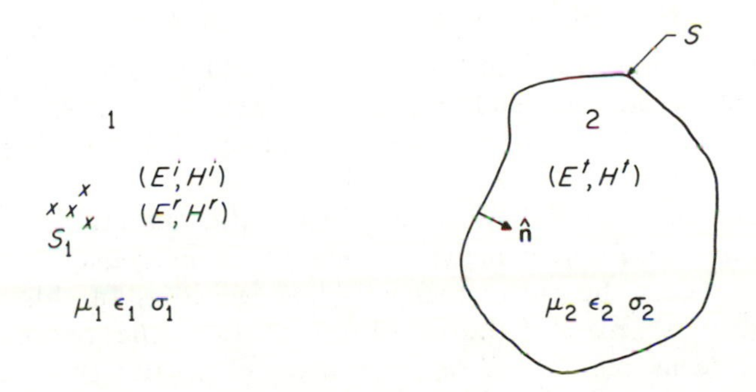

Figure 13-12. A closed surface S divides a region (1) containing sources from a source-free region (2).

In Fig. 13-12 the closed surface S separates two homogeneous media, one containing a system of sources s_1, and the other being source-free. In general the field in region (2) will be different from the value that it would have if media (1) and (2) were the same. The actual field in (2) can be determined by treating the problem as a reflection problem in which an incident field (E^i, H^i) sets up at the boundary surface S a reflected field (E^r, H^r) and a transmitted field (E^t, H^t). The incident field (E^i, H^i) is the field that would exist if there were no reflecting surface, that is, if the entire region were homogeneous. The actual field in region (1) is ($E^i + E^r$, $H^i + H^r$); the actual field in region (2) is (E^t, H^t). At any actual boundary surface S, the tangential components of these fields are continuous. That is,

$$E_t^i + E_t^r = E_t^t \qquad H_t^i + H_t^r = H_t^t \tag{13-47}$$

where the subscript t indicates the components tangential to the surface. Equation (47) can be rewritten in the form

$$(E_t^t - E_t^r) = E_t^i \qquad (H_t^t - H_t^r) = H_t^i \tag{13-47a}$$

Now let attention be concentrated on the "induced" or "scattered" fields (E^r, H^r) and (E^t, H^t). The reflected fields (E^r, H^r) satisfy Maxwell's equations in the homogeneous medium (1), and the transmitted or refracted fields satisfy Maxwell's equations in the homogeneous region (2). Together these fields constitute an electromagnetic field in the entire space. This field is source-free everywhere except on S, and the distribution of sources on S is calculable from the incident field, and therefore from the given sources s_1. In sec. 13.01 it was shown that the discontinuities in E and H across the surface S could be produced by current sheets on S of densities

$$\mathbf{J}_s = \hat{\mathbf{n}} \times (\mathbf{H}_t^t - \mathbf{H}_t^r) = \hat{\mathbf{n}} \times \mathbf{H}^i$$

$$\mathbf{M}_s = -\hat{\mathbf{n}} \times (\mathbf{E}_t^t - \mathbf{E}_t^r) = -\hat{\mathbf{n}} \times \mathbf{E}^i \tag{13-48}$$

Thus as far as the "induced" or scattered field is concerned, it could be produced by electric- and magnetic-current sheets over the surface S, the densities of these sheets being given by (48). This is the *induction theorem.*

A second theorem follows directly from the induction theorem for the particular case where region (2) has the *same* constants as region (1), that is where the entire region is homogeneous. In this case the reflected field is zero and the transmitted field is the actual field in the homogeneous region due to the sources of s_1. But this transmitted field can also be calculated from a suitable distribution of electric- and magnetic-current sheets over the surface S. The required surface current densities of these sheets will be

$$\mathbf{J}_s = \hat{\mathbf{n}} \times \mathbf{H}_t^t = \hat{\mathbf{n}} \times \mathbf{H}_t^i \qquad \mathbf{M}_s = -\hat{\mathbf{n}} \times \mathbf{E}_t^t = -\hat{\mathbf{n}} \times \mathbf{E}_t^i \tag{13-49}$$

Since the vector product of $\hat{\mathbf{n}}$ and the normal component of the field is zero, the t subscripts can be dropped and eqs. (49) written as

$$\mathbf{J}_s = \hat{\mathbf{n}} \times \mathbf{H}^i \qquad \mathbf{M}_s = -\hat{\mathbf{n}} \times \mathbf{E}^i \tag{13-50}$$

The vector $\hat{\mathbf{n}}$ is in the direction of the transmitted wave. Thus in a source-free region bounded by a surface S, in order to compute the electromagnetic field, the source distribution s_1 (outside of S) can be replaced by a distribution of electric and magnetic currents over the surface S, where the densities of this "equivalent" source distribution are given by (50). This is the *equivalence theorem.*

These theorems* prove to be powerful tools in solving electromagnetic problems that involve radiation from apertures.

The usefulness of the equivalence theorem is demonstrated by the two examples shown in Fig. 13-13. Although, as was pointed out in

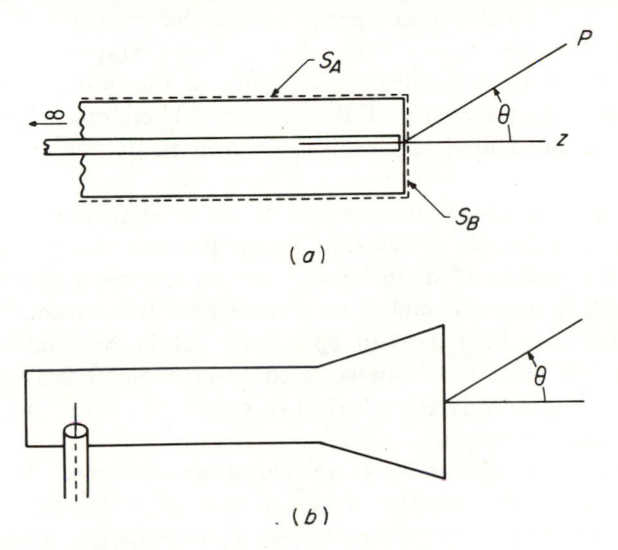

Figure 13-13. Radiation from (*a*) the open end of a coaxial line and (*b*) an electromagnetic horn.

sec. 13.01, it is always possible (in theory) to determine the electromagnetic field of a system entirely in terms of the electric currents and charges, this direct approach proves surprisingly difficult in these two cases. The first example concerns the radiation from the open end of a semi-infinite coaxial line. In this case, assuming that the transverse dimensions of the line are very small in wavelengths, the current distribution is known fairly accurately, but the problem is made difficult by the fact that all currents throughout the infinite length of the line must be considered in the integration to determine the radiated field. The second problem involves the radiation from the open end of a wave guide or from an electromagnetic horn. Here the currents are known only approximately, especially around the mouth of the horn. But even though only an approximate solution is required, the integration to obtain the fields from the known or guessed at currents is extremely difficult because all currents, including those on the probe antenna and in the coaxial feed line, must be included.

In both of these problems it is evident that a simpler way of obtain-

*There are several such theorems. Those given here are due to Schelkunoff. For reference to others see bibliography at end of chapter.

ing the radiated field must be found. In particular, although all currents of the system are involved in determining the radiation, it appears reasonable that for systems such as those of Fig. 13-13, the currents can affect the radiated field only through some change that they make in the fields that appear across the open end of the coaxial line or wave-guide horn. These latter fields may be known [Fig. 13-13(*a*)], or can be guessed at [as in (*b*)], to the same order of approximation as the currents of the system. Therefore a method for computing radiated fields in terms of known fields across an aperture will be sought.

The means for accomplishing the result is suggested by *Huygens' principle*. This principle states that "each particle in any wave front acts as a new source of disturbance, sending out secondary waves, and these secondary waves combine to form a new wave front." Huygens' principle has long been used in optics to obtain qualitative answers to diffraction problems. It can be used to give quantitative results for many aperture problems when suitably combined with the induction and equivalence theorems.

13.08 Field of a Secondary or Huygens' Source. It will be of interest to determine first the radiation field of a Huygens' source, or an element of area of an advancing wave front in free space. In Fig. 13-14 is shown an element of area $dx\,dy$ on the wave front of a uniform plane (TEM) wave, which is advancing in the z direction. By the theorems of the previous section this element of wave front having electric field strength E_x^0 and magnetic field strength $H_y^0 = E_x^0/\eta_v$ can be treated as a secondary source and can be replaced by electric and magnetic sheets. Using eqs. (50), the directions and densities of these current sheets will be

$$J_x = -H_y^0 = -\frac{E_x^0}{\eta_v} \qquad M_y = -E_x^0$$

The element of area $dx\,dy$ of electric surface current density J_x constitutes an electric current element $(J_x\,dy)\,dx$, and similarly the element of area of magnetic surface current density M_y constitutes a magnetic current element $(M_y\,dx)\,dy$. The problem is simply one of determining the radiation fields of these current elements. Because only the radiation fields are required, only the vector potentials need be considered.

At large distances, the electric field of the electric-current element will be

$$\mathbf{E}^e = -j\omega\mathbf{A} \qquad \mathbf{A} = \hat{\mathbf{x}}A_x = \hat{\mathbf{x}}\mu\left[\frac{(J_x\,dy)\,dx\,e^{-j\beta r}}{4\pi r}\right]$$

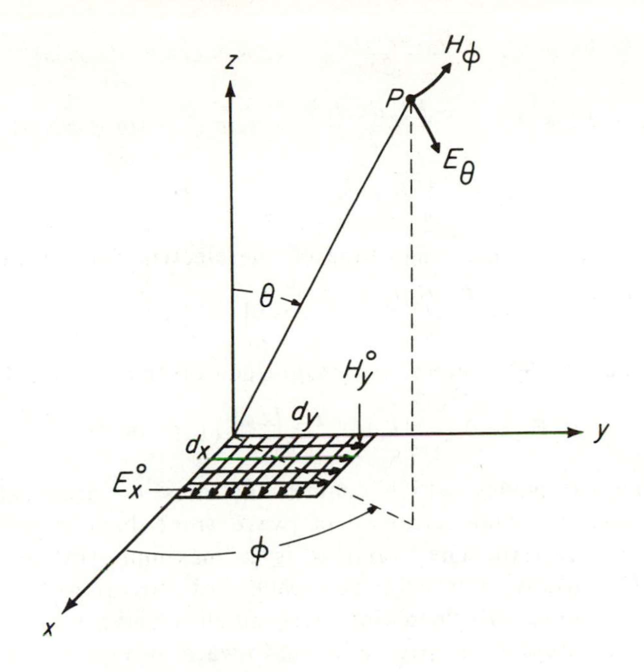

Figure 13-14. Radiation from a Huygens' source.

The θ and ϕ components of the radiation field of the electric-current element will be

$$E_\theta^e = -j\omega A_\theta \qquad H_\phi^e = \frac{E_\theta^e}{\eta_v}$$

$$E_\phi^e = -j\omega A_\phi \qquad H_\theta^e = -\frac{E_\phi^e}{\eta_v}$$

where

$$A_\theta = A_x \cos\phi \cos\theta \qquad \text{and} \qquad A_\phi = -A_x \sin\phi$$

Similarly for the magnetic-current element the radiation fields are

$$\mathbf{H}^m = -j\omega \mathbf{F} \qquad \mathbf{F} = \hat{\mathbf{y}}\mathbf{F}_y = \hat{\mathbf{y}}\epsilon\left[\frac{(M_y\,dx)\,dy}{4\pi r}e^{-j\beta r}\right]$$

$$H_\phi^m = -j\omega F_\phi \qquad E_\theta^m = \eta_v H_\phi^m$$

$$H_\theta^m = -j\omega F_\theta \qquad E_\phi^m = -\eta_v H_\theta^m$$

$$F_\phi = F_y \cos\phi \qquad F_\theta = F_y \sin\phi \cos\theta$$

Expressing all the fields in terms of E_x^0, the radiation field of the Huygens' source is found to be

$$E_\theta = E_\theta^e + E_\theta^m = \frac{jE_x^0\,dx\,dy\,e^{-j\beta r}}{2\lambda r}(\cos\phi\cos\theta + \cos\phi) \tag{13-51}$$

$$E_\phi = E_\phi^e + E_\phi^m = \frac{-jE_\phi^0\,dx\,dy\,e^{-j\beta r}}{2\lambda r}(\sin\phi + \sin\phi\cos\theta) \tag{13-52}$$

$$H_\phi = \frac{E_\theta}{\eta_v} \qquad H_\theta = \frac{-E_\phi}{\eta_v} \tag{13-53}$$

In the plane $\phi = 0$, the magnitude of the electric field strength is

$$|E_\theta| = \frac{E_y^0\,dx\,dy}{2\lambda r}[1 + \cos\theta] \qquad |E_\phi| = 0 \tag{13-54}$$

In the plane $\phi = 90$ degrees, the magnitude of the electric field is

$$|E_\theta| = 0 \qquad |E_\phi| = \frac{E_x^0\,dx\,dy}{2\lambda r}(1 + \cos\theta) \tag{13-55}$$

In the principal planes, which contain the axis of propagation, the radiation patterns of an element of wave front have a cardioid or unidirectional pattern. The radiation is a maximum in the forward direction ($\theta = 0$); it is zero in the backward direction ($\theta = 180$ degrees). In one sense, this "explains" why an electromagnetic wave, once launched, continues to propagate in the forward direction. An electric-current sheet alone radiates equally on both sides; similarly a magnetic-current sheet alone radiates equally on both sides; but crossed electric- and magnetic-current sheets of proper relative magnitude and phase can be made to radiate on one side only. A large square surface of a plane wave front constitutes a rectangular array of Huygens' sources, all fed in phase. The "radiation pattern" of the array is obtained by multiplying the unit pattern of the element (a cardioid pattern) by the group pattern or array factor, which in this case is a bidirectional pencil beam. The resultant pattern is a unidirectional pencil beam, the cone angle of which becomes very small as the area of the wave front becomes large.

13.09 Radiation from the Open End of a Coaxial Line. By application of the new approaches outlined in previous sections, the problem of radiation from the open end of a coaxial cable of small cross-sectional dimensions can now be solved quite easily. If the surface S, separating the source-free region from the region containing sources, is taken to be the surface shown dotted in Fig. 13-13(*a*), it is only necessary to specify the equivalent electric- and magnetic-current sheets over this surface. The surface can be divided into two parts: S_a is the cylindrical surface that encloses the outer wall of the coaxial line; S_b is the flat circular surface which caps the end of the line. Over S_a, tangential E is tangential to the metallic wall and has zero value. Therefore the equivalent magnetic-current sheet has zero density over S_a. Also the magnetic field strength at the outer surface of the outer conductor must be zero, since to the order of approximation used here it is assumed that no current flows on the outside of the cable.* With tangential H over S_a equal to zero, the equivalent electric-current sheet must also be zero over this part of the surface, and there remains only the contribution from S_b. Over S_b the electric field is radial and the magnetic field strength is circumferential; but whereas the electric field is relatively strong, to a first approximation the magnetic field strength is zero. Therefore the radiation from the open end can be computed by use of an equivalent magnetic-current sheet only, over the surface S_b. Having determined the radiation fields, and thence the power radiated, it is possible to compute from consideration of power flow through the open end, the small value of magnetic field strength that actually must exist there. If desired, an equivalent electric-current sheet could then be set up for this magnetic field strength, and the small radiation field of the electric-current sheet could be calculated. In practice this second approximation rarely needs to be made, because it produces only a very small correction to the radiation fields and power radiated.

Figure 13-15 shows the geometry appropriate for calculation of the radiation from the open end of a coaxial cable. Between the inner radius a and the outer radius b, the radial electric field E_ρ will have a value

$$E_\rho = \frac{k}{\rho} \qquad \text{where} \qquad k = \frac{V}{\ln b/a}$$

(V is the voltage between inner and outer conductors at the open end.) Using (50), E_ρ can be replaced by a magnetic-current sheet

$$M_\phi = -E_\rho = -\frac{k}{\rho}$$

The electric vector potential $\mathbf{F}$ will be in the ϕ direction and will have a value

$$F_\phi = \frac{\epsilon}{4\pi}\int_0^{2\pi}\int_a^b \frac{M_\phi\, e^{-j\beta r}}{r}\,\rho\, d\rho\, d\phi_1 \tag{13-56}$$

where the integration is over the area A between inner and outer con-

*It is assumed that the transverse dimensions of the cable are small so that, except very close to the open end, only the TEM mode exists. Close to the end of the cable the field structure becomes very complicated due to the presence of evanescent modes but to a first approximation these higher-order modes may be neglected. Such an approximation implies that the fringing fields also be disregarded so that as far as the transmission line is concerned the open end of the cable is an ideal "open circuit" at which the line current goes to zero. Thus, to this order of approximation, there are no currents flowing on the outer surface of the cable.

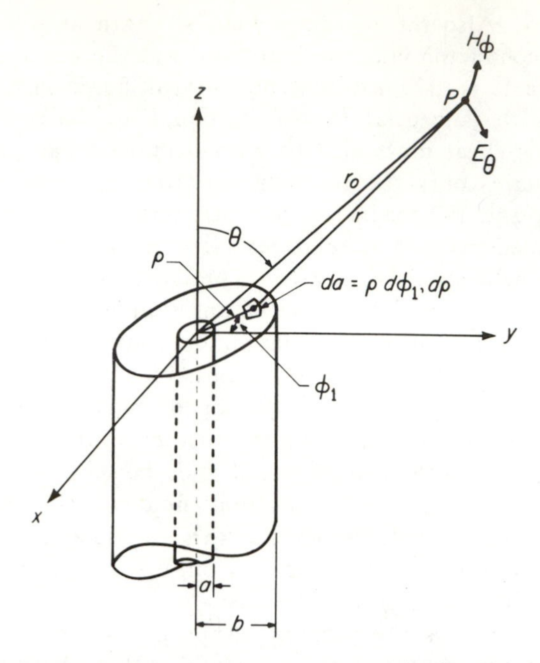

Figure 13-15. Geometry for calculation of power radiated by open end of a coaxial cable.

ductors. Without loss of generality the point P may be taken in the y-z plane, for which case only the x components, $M_\phi \cos \phi_1$, of magnetic current contribute to the potential, the y components cancelling out. Because only distant fields are being considered, the r in the denominator of (56) can be put equal to r_0, and the r in the phase factor in the numerator may be replaced by

$$r \approx r_0 - \rho \sin \theta \cos \phi_1$$

Making these substitutions, and remembering that for small values of δ the exponential $e^{j\delta}$ can be replaced by the first two terms of its power series expansion, viz.,

$$e^{j\delta} \approx 1 + j\delta$$

the integration indicated by (56) can be carried out. The result is

$$F_\phi = -\frac{j\beta\epsilon k \sin \theta \, e^{-j\beta r_0}}{8r_0}(b^2 - a^2) \tag{13-57}$$

The distant magnetic field is obtained from

$$H_\phi = -j\omega F_\phi$$

so that

$$H_\phi = \frac{-\beta\omega\epsilon k \sin\theta}{8r_0}(b^2 - a^2)\, e^{-j\beta r_0} \tag{13-58}$$

The strength of the distant electric field will be

$$E_\theta = \eta_v H_\phi$$

By integrating the Poynting vector over a large spherical surface, the radiated power is found to be

$$W = \frac{\pi^2 V^2}{360}\left(\frac{A}{\lambda^2 \ln b/a}\right)^2 \qquad \text{watts} \tag{13-59}$$

where $A = \pi(b^2 - a^2)$ is the area of the opening between inner and outer conductors.

13.10 Radiation through an Aperture in an Absorbing Screen. Another example of radiation through an aperture occurs in the problem of the transmission of electromagnetic energy through a rectangular aperture in a perfectly absorbing screen. Although admittedly not a very practical problem, because of the difficulties of obtaining a screen which is both infinitely thin and perfectly absorbing, the solution to this problem is required in obtaining answers to other, more practical problems.

In Fig. 13-16 the rectangle ab represents an aperture in a perfectly absorbing screen of infinite extent which occupies the $z = 0$ plane. A uniform plane electromagnetic wave traveling in the z direction is assumed to be incident upon the bottom side of the screen and aperture, and the problem is that of determining the radiation through the aperture in the positive z direction. Under the assumed conditions of the problem, the incident wave is completely absorbed at the sur-

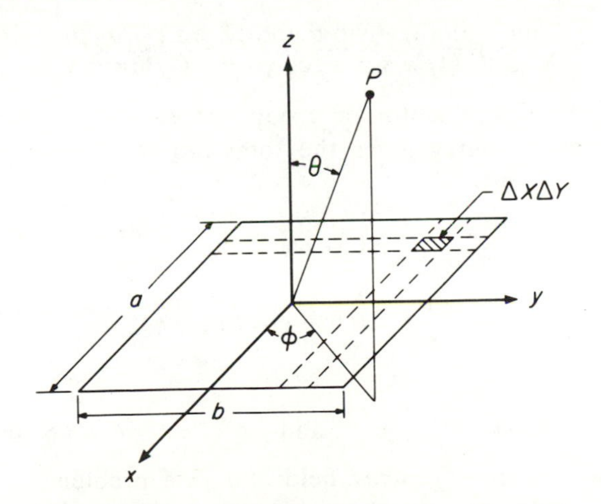

Figure 13-16. An element of area on an advancing wave front.

face of the screen. Over the aperture the field strength will be just that of the incident wave. By dividing up the aperture into a large number of Huygens' sources of area $da = \Delta x\, \Delta y$, the aperture may be treated as a rectangular array of such sources, all fed in phase.

For a line array of nondirective sources in the x direction, having a uniform spacing Δx, the radiation pattern or space factor is (from sec. 11.07)

$$S_x = |1 + e^{j\psi} + e^{2j\psi} + \ldots + e^{(m-1)j\psi}|$$

$$= \left|\sum_0^{m-1} e^{j\beta(\Delta x)\sin\theta\cos\phi}\right|$$

$$= \left|\frac{\sin\left[\frac{1}{2}m\beta\,(\Delta x)\sin\theta\cos\phi\right]}{\sin\left[\frac{1}{2}\beta\,(\Delta x)\sin\theta\cos\phi\right]}\right|$$

Similarly, for a line array in the y direction with a uniform spacing Δy, the space factor is

$$S_y = \left|\frac{\sin\left[\frac{1}{2}n\beta\,(\Delta y)\sin\theta\sin\phi\right]}{\sin\left[\frac{1}{2}\beta\,(\Delta y)\sin\theta\sin\phi\right]}\right|$$

The total space factor for the rectangular array of isotropic sources is then

$$S_{x,y} = S_x S_y$$

If Δx and Δy are now allowed to become small, but m and n are made large in such a manner that

$$(m-1)\,\Delta x = a \qquad (n-1)\,\Delta y = b$$

the space factor $S_{x,y}$ may be written

$$S_{x,y} = \frac{ab}{\Delta x\,\Delta y}\left|\frac{\sin\left(\frac{1}{2}\beta a\sin\theta\cos\phi\right)}{\left(\frac{1}{2}\beta a\sin\theta\cos\phi\right)}\cdot\frac{\sin\left(\frac{1}{2}\beta b\sin\theta\sin\phi\right)}{\left(\frac{1}{2}\beta b\sin\theta\sin\phi\right)}\right| \tag{13-60}$$

Multiplying this space factor or group pattern by the radiation from the unit Huygens' source gives the total radiation from the aperture. Then, using (51) and (52),

$$E_\theta = \frac{jE_x^0\,abe^{-j\beta r}}{2\lambda r}\left[(1+\cos\theta)\cos\phi\right]\left|\frac{\sin u_1}{u_1}\cdot\frac{\sin v_1}{v_1}\right| \tag{13-61}$$

$$E_\phi = \frac{-jE_x^0\,abe^{-j\beta r}}{2\lambda r}\left[(1+\cos\theta)\cos\phi\right]\left|\frac{\sin u_1}{u_1}\cdot\frac{\sin v_1}{v_1}\right| \tag{13-62}$$

where

$$u_1 = \tfrac{1}{2}\beta a\sin\theta\cos\phi \qquad \text{and} \qquad v_1 = \tfrac{1}{2}\beta b\sin\theta\sin\phi$$

This is known as the *diffracted* field and this problem is an example of *Fraunhofer diffraction.* In the principal x-z plane, $\phi = 0$ or π, $E_\phi = 0$, and the E_θ field is given by

$$E_\theta = \frac{jE_x^0 abe^{-j\beta r}}{2\lambda r}(1 + \cos\theta)\left|\frac{\sin u}{u}\right| \tag{13-63}$$

where

$$u = [\tfrac{1}{2}\beta a \sin\theta] = \frac{\pi a}{\lambda}\sin\theta$$

The expression $(\sin u)/u$, in (63), has already been encountered in sec. 11.07. It occurs in many radiation and diffraction problems and is plotted in Fig. 13-17 as a function of u.

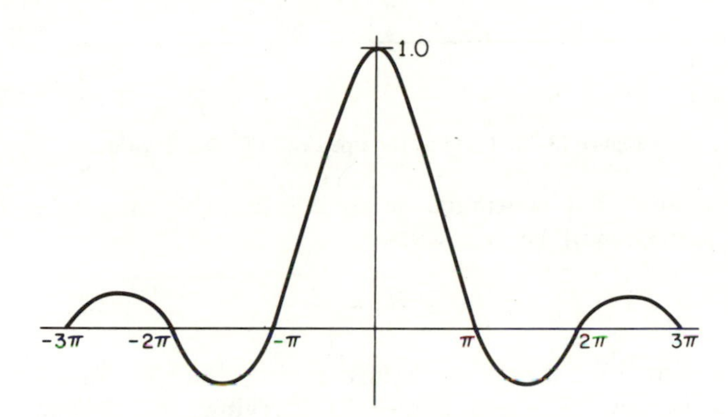

Figure 13-17. Plot of $(\sin u)/u$.

It will be noted that the first null in this general pattern occurs at $u = \pi$. Thus, for an aperture width $a = 1\lambda$, the first null of the pattern occurs at $\theta = 90$ degrees. For aperture widths smaller than 1λ, there is no null in the pattern. For very large apertures, the "beam" is quite narrow, and small angles of θ are of greatest interest. For small values of θ, $\sin\theta \approx \theta$, and

$$u \approx \frac{\pi a\theta}{\lambda}$$

For $a = 10\lambda$, the first null occurs at $\theta_0 = 0.1$ radian or 5.7 degrees.

Application to Open-ended Wave Guides. If it can be assumed that the currents on the outside walls of an open-ended wave guide have negligible effect on the radiation from the guide, the problem of diffraction through an aperture has direct application to this second, more practical, problem. Since experimentally measured radiation patterns are found to agree roughly with patterns computed by neglecting these outside currents, such calculations may be used if only approximate answers are sufficient.

Referring to Fig. 13-18, and assuming that the fields at the open end are approximately the same as they would be if the guide did not

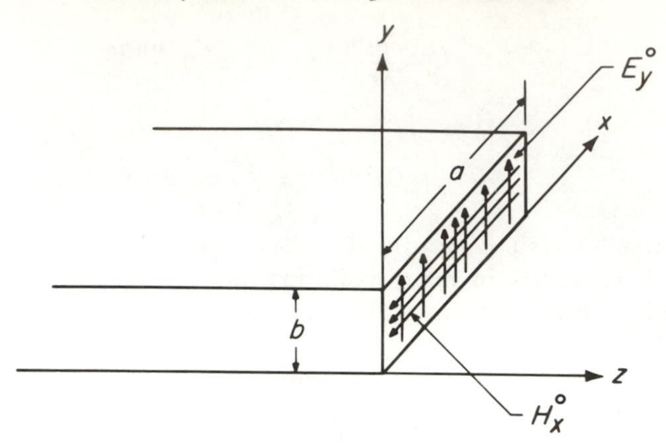

Figure 13-18. Field at the open end of a wave guide.

terminate there, but continued on to infinity, the tangential fields at the end surface will be

$$E_y^0 = E^0 \sin\frac{\pi x}{a} \qquad H_x^0 = -\frac{1}{\eta}\sqrt{1-\left(\frac{\lambda^2}{\lambda_c}\right)}\,E_y^0$$

The dominant TE_{10} mode has been assumed, and for this mode $\lambda_c = 2a$. For a very wide guide ($a \gg \lambda/2$) carrying the dominant mode, $|E_y^0/H_x^0| \approx \eta$, and the problem is the same as that of the rectangular aperture, except for the variation of E and H in the x direction across the mouth of the guide. For narrower guides, with operation closer to the cut-off frequency [i. e., as $\lambda \to 2a$, and $\sqrt{1-(\lambda/\lambda_c)^2} \to 0$], H_x^0 becomes very small and there are two important effects. First, the characteristic impedance of the guide becomes very great, so that there is now a large mismatch between the impedance of the guide and the effective terminating impedance. This means that more of the energy is reflected back from the open end, and less is radiated for a given value of E_y^0. Second, the radiation pattern approaches more closely that which would be calculated from a magnetic-current sheet alone, rather than from crossed electric- and magnetic-current sheets. Experimental and calculated radiation patterns of wave guides and horns may be found in the literature.*

13.11 Radiation through an Aperture in a Conducting Screen. When the open end of a wave guide or electromagnetic horn forms the aperture in a very large *conducting* plane, the electromagnetic field on the radiating side of the plane can be computed from a knowledge of the electric field alone across the aperture. To

*Radio Research Laboratory (Staff), *Very High Frequency Techniques*, Vol. 1, Sec. 6-4, McGraw-Hill Book Company, New York, 1947.

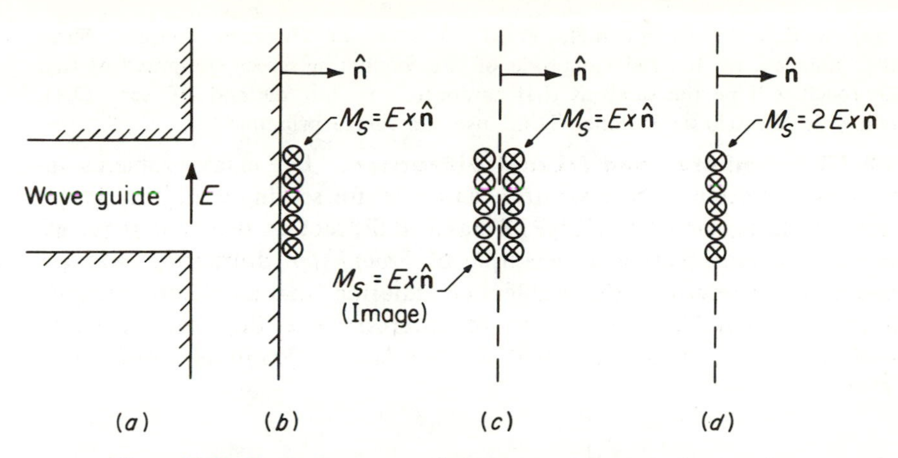

Figure 13-19. Equivalent source for an aperture in a conducting plane.

see that this is so, consider Fig. 13-19(*a*) which shows a wave guide feeding an aperture in a large conducting screen. In Fig. 13-19(*b*) the $\mathbf{E}$ field across the aperture is replaced by its equivalent magnetic-current sheet $\mathbf{M}_s = -\hat{\mathbf{n}} \times \mathbf{E}$ at the surface of a perfect conductor. The electromagnetic field to the right of the plane is the result of $\mathbf{M}_s$ *and* the electric currents flowing in the conducting plane, but by the image principle, as far as the field in the radiating region is concerned, the conducting plane can be removed and replaced by the image magnetic current. Because this image magnetic current sheet is positive and virtually coincident with $\mathbf{M}_s$, the double sheet of (*c*) becomes the single sheet of double strength shown in (*d*). By the uniqueness theorem (problem 23), because the tangential $\mathbf{E}$ fields are identical over the entire plane ($\hat{\mathbf{n}} \times \mathbf{E} = \mathbf{M}_s$ over the aperture, and $\hat{\mathbf{n}} \times \mathbf{E} = 0$ elsewhere), the electromagnetic fields to the right of the plane are the same in Figs. 13-19(*a*) and (*d*).

Problem 9. The open end of the coaxial cable of sec. 13.09 forms an aperture in a "ground screen" (a flat conducting screen, theoretically infinite in extent).

(a) Using the image principle show that the fields in the radiating region will be double those computed in sec. 13.09 for the coaxial opening in free space. Show that the power radiated and radiation conductance will also be doubled.

(b) Sketch the radiation pattern of the open end of the coaxial line, with and without the ground screen, and compare with those of a current element $I\,dl$ (or short antenna) in free space and at the surface of (and normal to) a conducting plane.

Problem 10. The open end of the wave guide in Fig. 13-18 forms the aperture in a large conducting ground screen. For a very wide guide E_y^0/H_x^0

$\approx \eta_v$ so that the aperture fields form an array of Huygens' sources. Show that the field on the radiating side of the conducting screen computed by that approach will be the same as that computed by the method of sec. 13.11, using the electric field alone. (Hint: use the image principle.)

13.12 Fraunhofer and Fresnel Diffraction. For many problems involving radiation from apertures, and also for solving certain propagation problems, some knowledge of classical diffraction theory is required. In the previous section an example of *Fraunhofer* diffraction was encountered, whereas in the problem of radiation from an electromagnetic horn, *Fresnel* diffraction will be of interest. The difference between these is illustrated in Fig. 13-20. In the case of Fraunhofer diffraction

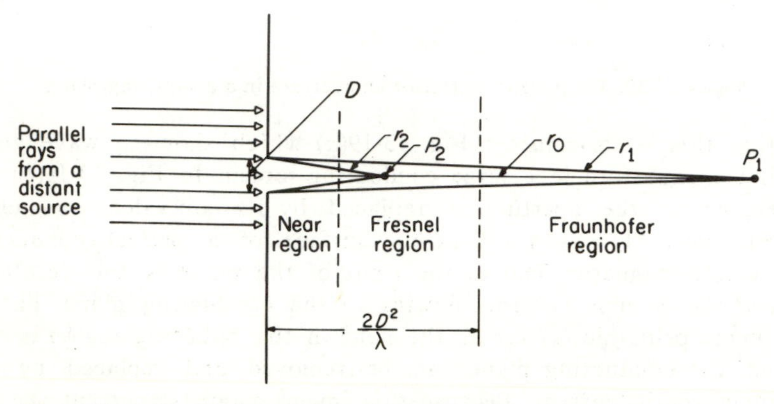

Figure 13-20. Illustration of Fresnel and Fraunhofer regions in diffraction theory.

both the source and receiving point are so remote from the aperture or screen that the rays may be considered as being essentially parallel. In Fig. 13-20, this means that rays arriving from the secondary source (the aperture D) may be considered to arrive in-phase at a point P_1, which is on a line drawn normal to the screen through the aperture. On the other hand, if the distance r to the receiving point P_2 is sufficiently large that the amplitude factor $1/r$ may still be considered constant, but is not so large that the phase difference of contributions from the various Huygens' sources over the aperture may be neglected, the point P_2 is in the region of Fresnel diffraction. The region so close to the aperture that both the amplitude and phase factors are variable with the position of the receiving point is sometimes called the *near region.* The dividing line between Fresnel and Fraunhofer diffraction depends upon the accuracy required; however, the distance to the dividing line is often taken* as $r = 2D^2/\lambda$. If the distance to the

**IRE Standards on Antennas,* 1948.

receiving point is very great, but the source is so close to the screen that the phase of the field varies over the aperture (with the source on the normal to the screen through the aperture), Fresnel diffraction theory is required.

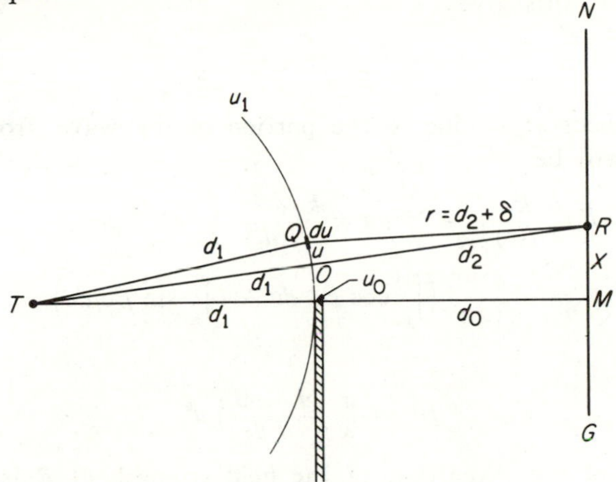

Figure 13-21. Diffraction at a straight edge.

Fresnel Diffraction at a Straightedge. Figure 13-21 illustrates a simple example of Fresnel diffraction. An obstacle, such as a straightedge (considered to be perfectly absorbing), is inserted between a transmitting source T and a receiving location R. To keep the problem two-dimensional, the source T is assumed to be a very long line source parallel to the long straightedge. The problem is to determine the field strength at the receiving point R, as R is moved along the line GMN. It is assumed that the distances d_0 and d_1 are sufficiently large that the approximations inherent in Fresnel diffraction theory are valid, but not large enough to permit the approximations used in Fraunhofer diffraction.

Assume that each elemental strip du of the wave front produces an effect at R given by

$$dE = \frac{k_1 \, du \, e^{-j\beta r}}{f(r)} \tag{13-64}$$

where $\beta = 2\pi/\lambda$, $f(r)$ is a function of r, and k_1 is a constant. For Fresnel diffraction, the r in the denominator of (64) can be considered constant but the variation of r in the phase-shift factor must be accounted for. By geometry,

$$(QR)^2 = r^2 = (d_1 + d_2)^2 + d_1^2 - 2d_1(d_1 + d_2)\cos\frac{u}{d_1}$$

$$\approx (d_1 + d_2)^2 + d_1^2 - 2d_1(d_1 + d_2)\left(1 - \frac{u^2}{2d_1^2}\right)$$

then

$$r^2 = (d_2 + \delta)^2 \approx d_2^2 + u^2 \frac{d_1 + d_2}{d_1}$$

Neglecting δ^2, this gives

$$\delta = u^2 \frac{d_1 + d_2}{2d_1 d_2} \approx u^2 \frac{d_1 + d_0}{2d_1 d_0}$$

The total effect at R, due to the portion of the wave front between u_0 and u_1, will be

$$E = \frac{k_1}{f(r)} \int_{u_0}^{u_1} e^{-j\beta r}\, du = \frac{k_1\, e^{-j\beta d_2}}{f(d_2)} \int_{u_0}^{u_1} e^{-j\beta\delta}\, du \qquad (13\text{-}65)$$

$$= \frac{k_1\, e^{-j\beta d_2}}{f(d_2)} \left(\int_{u_0}^{u_1} \cos \beta\delta\, du - j \int_{u_0}^{u_1} \sin \beta\delta\, du \right) \qquad (13\text{-}65a)$$

where

$$\beta\delta = \frac{\pi}{\lambda} \left(\frac{d_1 + d_2}{d_1 d_2} \right) u^2$$

The square of the magnitude of the field strength at R is given by

$$|E|^2 = \frac{k_1^2}{f^2(d_2)} \left[\left(\int_{u_0}^{u_1} \cos \beta\delta\, du \right)^2 + \left(\int_{u_0}^{u_1} \sin \beta\delta\, du \right)^2 \right] \qquad (13\text{-}66)$$

To evaluate and interpret this result consider the following integral

$$C(v) - jS(v) = \int_0^v e^{-j(\pi/2)v^2}\, dv \qquad (13\text{-}67)$$

which is a standard form of the *Fresnel integrals*. Plotting this integral in the complex plane, with C as the abscissa and S as the ordinate, results in a curve known as *Cornu's spiral* [Fig. 13-22(a)]. In this figure, positive values of v appear in the first quadrant and negative values of v in the third quadrant. The spiral has some interesting and important properties:

$$C = \int_0^v \cos \frac{\pi v^2}{2}\, dv \qquad S = \int_0^v \sin \frac{\pi v^2}{2}\, dv \qquad (13\text{-}68)$$

$$\delta s = \sqrt{(\delta C)^2 + (\delta S)^2} = \delta v \qquad v = s$$

$$\tan \phi = \frac{\delta S}{\delta C} = \tan \frac{\pi v^2}{2} \qquad \phi = \frac{\pi v^2}{2} = \frac{\pi s^2}{2}$$

$$\frac{d\phi}{ds} = \pi s; \qquad \text{radius of curvature} = \frac{ds}{d\phi} = \frac{1}{\pi s}$$

$$C(\pm \infty) = \pm \tfrac{1}{2}; \qquad S(\pm \infty) = \pm \tfrac{1}{2} \qquad (13\text{-}69)$$

The following properties follow from the above relations:

(1) A vector drawn from the origin to any point on the curve

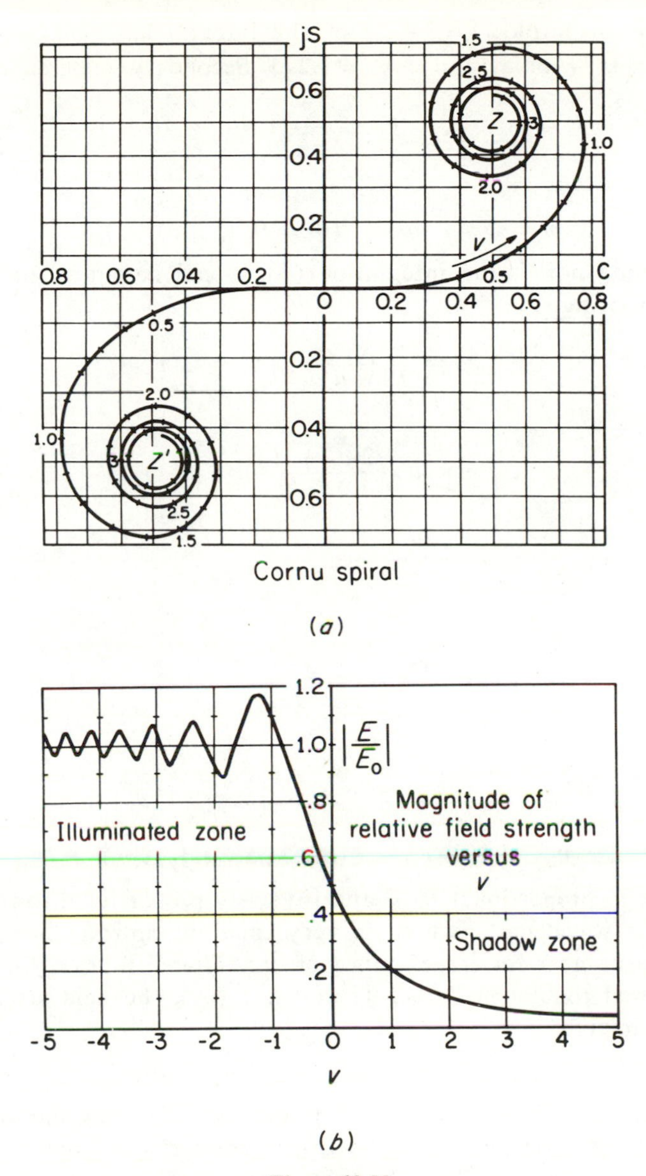

Figure 13-22.

represents in both magnitude and phase the value of the integral (67). (The phase of the vector is the negative of the phase of the integral.)

(2) The length s of arc along the spiral, measured from the origin, is equal to v. As v approaches plus or minus infinity, the spiral winds an infinity of times about the points $(\frac{1}{2}, \frac{1}{2})$ or $(-\frac{1}{2}, -\frac{1}{2})$.

(3) The magnitude $\sqrt{C^2 + S^2}$ of the integral has a maximum value when $\phi = 3\pi/4$, or at $v = \sqrt{\frac{3}{2}} = 1.225$. Secondary maxima occur at

$$\phi = \frac{3\pi}{4} + 2n\pi \qquad \text{or} \qquad v = \sqrt{\frac{3}{2} + 4n} \qquad (n = 1, 2, 3, \cdots)$$

Minima occur at

$$v = \sqrt{\tfrac{7}{2} + 4m} \qquad (m = 0, 1, 2, 3, \ldots)$$

Returning now to the integral of (65), it can be put in the standard form by writing

$$\beta\delta = \frac{\pi}{\lambda}\left(\frac{d_1 + d_2}{d_1 d_2}\right) u^2 = \frac{\pi}{2} v^2$$

or

$$v = u\sqrt{\frac{2(d_1 + d_2)}{\lambda d_1 d_2}} = k_2 u$$

Then

$$E = k_3 \int_{v_0}^{v_1} e^{-j(\pi/2)v^2}\, dv$$

where

$$k_3 = \frac{k_1 e^{-j\beta d_2}}{k_2 f(d_2)}$$

Using (68),

$$E = k_3\left(\int_0^{v_1} e^{-j(\pi/2)v^2}\, dv - \int_0^{v_0} e^{-j(\pi/2)v^2}\, dv\right)$$
$$= k_3[C(v_1) - C(v_0) - jS(v_1) + jS(v_0)]$$

Because v is proportional to u and inversely proportional to the square root of the wavelength (which is very small in optics), v_1 will be a very large number for large values of u_1. Therefore $C(v_1) \to C(\infty)$ as u_1 is allowed to become large. Then using (69), the field strength will be approximately

$$E = K\{[\tfrac{1}{2} - C(v_0)] - j[\tfrac{1}{2} - S(v_0)]\}$$

The quantities $(\frac{1}{2} - C)$ and $j(\frac{1}{2} - S)$ represent the real and imaginary parts of a vector drawn from the upper point of convergence $(\frac{1}{2}, \frac{1}{2})$ to a point on the spiral. Thus the magnitude of E is proportional to the length of the vector drawn from $(\frac{1}{2}, \frac{1}{2})$ to the appropriate point on the spiral. This makes it possible to visualize the strength variation as v_0 (and hence either u_0 or d_1 or d_2) is varied.

For u_0 equal to a large negative value, the free-space field strength E_0 results. Therefore

$$E_0 = K\{[\tfrac{1}{2} - (-\tfrac{1}{2})] - j[\tfrac{1}{2} - (-\tfrac{1}{2})]\} = K(1 - j)$$

and therefore

$$K = \frac{E_0}{1-j} = \frac{E_0}{2}(1+j)$$

The received field strength is given in terms of the free-space field by

$$E = \frac{E_0}{2}(1+j)\int_{v_0}^{\infty} e^{-j(\pi/2)v^2}\,dv \tag{13-70}$$

where

$$v_0 = u_0\sqrt{\frac{2(d_1+d_2)}{\lambda d_1 d_2}}$$

In order for this approximate treatment to be valid, the following inequalities must hold:

$$d_1, d_2 \gg u_0 \qquad d_1 d_2 \gg \lambda$$

In Fig. 13-22(*b*) is plotted the magnitude $|E/E_0|$ as taken off the spiral. The field strength in the shadow zone decreases smoothly to zero. Above the line of sight the field strength oscillates about its free-space value. On the line of sight the field strength is just one-half of its free-space value.

This approximate theory of diffraction was developed for use in optics, where the approximations and assumptions made are usually quite valid. However, it is found that even at radio frequencies, and especially at ultrahigh frequencies, there are many problems where the theory is applicable. An example occurs in computing the radiation from electromagnetic horns.

13.13 Radiation from Electromagnetic Horns. In order to secure greater directivity a wave guide can be flared out to form an electromagnetic horn. A rectangular guide flared out in one plane only constitutes a sectoral horn, whereas a guide flared in both planes forms a pyramidal horn. The sectoral horn flared out in the plane of the electric field is the easiest case to treat and will serve as an example of the method of attack.

Figure 13-23 shows a horn flared out in the electric plane with a flare angle 2ψ.

For best operation the angle ψ is usually sufficiently small that the area of the wave front is approximately equal to the area of the aperture. The total field at any distant point is obtained by summing the contributions from the Huygens' sources distributed over the wave front. It is permissible to assume that the field distribution over the aperture is approximately the same as it would be there if the horn did not terminate, but was infinitely long. For the case considered in

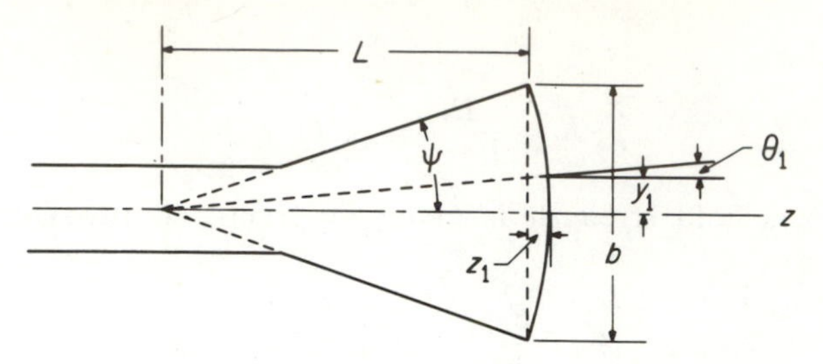

Figure 13-23. Electromagnetic horn.

Fig. 13-23, the field will be constant over the aperture in the y direction, but will vary in the x direction as $\cos (\pi x_1/a)$. The information of most interest will be the field strength, and hence the gain in the forward direction, that is, along the positive z axis.

At any distant point on the z axis the field due to a Huygens' source of strength E^0H^0 will be

$$\frac{E^0\, dx\, dy}{2\lambda r}(1 + \cos \theta_1) \approx \frac{E^0\, dx\, dy}{\lambda r}$$

since $\cos \theta_1 \approx 1$ for distant points in the forward direction. The strengths of the Huygens' sources over the aperture will be given by

$$E^0 = E_y^0 \cos \frac{\pi x_1}{a}$$

The total field at a distant point on the z axis is

$$E = \frac{E_y^0}{\lambda r}\int_{-b/2}^{+b/2}\int_{-a/2}^{+a/2} \cos \frac{\pi x_1}{a}\, e^{j\beta z_1}\, dx_1\, dy_1$$

where the reference phase has been taken as that due to a source in the plane of the aperture ($z_1 = 0$). From the geometry of Fig. 13-23,

$$\begin{aligned} z_1 &= L \cos \theta_1 - L \cos \psi \\ &\approx \frac{b^2}{8L} - \frac{y_1^2}{2L} \end{aligned}$$

Then

$$\begin{aligned} |E| &= \frac{E_y^0}{\lambda r}\int_{-a/2}^{+a/2} \cos \frac{\pi x_1}{a}\, dx_1 \left|\int_{-b/2}^{+b/2} e^{-j\beta\epsilon}\, dy_1\right| \\ &= \frac{4aE_y^0}{\pi\lambda r}\left|\int_0^{b/2} e^{-j\beta\epsilon}\, dy_1\right| \end{aligned} \qquad (13\text{-}71)$$

where

$$\epsilon = \frac{y_1^2}{2L}$$

Putting

$$\frac{\beta y_1^2}{2L} = \frac{\pi}{2} v^2$$

or

$$v = y_1 \sqrt{\frac{2}{\lambda L}}, \qquad dv = dy_1 \sqrt{\frac{2}{\lambda L}}$$

reduces (71) to the form of (67), and the expression for the square of the absolute magnitude of field strength will be

$$|E|^2 = \frac{2L}{\lambda} \left(\frac{2aE_y^0}{\pi r}\right)^2 \left[C^2 \left(\frac{b}{\sqrt{2\lambda L}}\right) + S^2 \left(\frac{b}{\sqrt{2\lambda L}}\right)\right] \tag{13-72}$$

The effect of changes in any of the horn dimensions is made evident by using (72) and Cornu's spiral. It is seen that, if b is increased, for a given L, the forward signal will first increase to a maximum and then decrease, increasing again to secondary maximum which is smaller than the first maximum. A similar variation results if b and L are increased together keeping the horn angle constant. The explanation for this result is that as b is increased, the contributions from some of the secondary sources on the wave fronts are out of phase with others, and so tend to decrease, instead of increase, the field strength in the forward direction.

13.14 Electromagnetic Theory, Geometrical Optics and Physical Optics. The treatment of diffraction in secs. 13.10 and 13.12 was an example of the use of *physical optics* or *wave optics*. Physical optics theory accounts for the wave nature of light, allowing for the phase differences due to different path lengths. In some cases (as in sec. 13.12) the field is treated as a scalar, but in other problems the vector and polarization characteristics of the field are accounted for. *Geometric optics* is the high-frequency limit of physical optics in which light is considered to travel along "rays" that can be treated by the methods of geometry. The methods of geometrical optics are by far the oldest and date back to the sixteenth and seventeenth centuries. Huygens formulated* his geometrical "wave" theory of light in 1690 but physical optics really began with Young's experiments on interference (1801) and Fresnel's mathematical formulation of Huygens' principle in 1815. In 1864 Maxwell presented his dynamical theory of the electromagnetic field in a paper which set forth the proposition that light was electromagnetic in character, and in 1887 Hertz confirmed Maxwell's theory by generating radio waves. With these waves ($\lambda \approx 60$

*C. Huygens, Traité de la Lumière, 1690. (English translation: University of Chicago Press, Chicago, 1945.)

cm) he was able to repeat many of the "optical" experiments on interference, diffraction, etc.

Inasmuch as Maxwell's theory has superseded the older theories of optics it might be supposed that these latter theories could be dispensed with. In principle every electromagnetic or light problem can be solved by an application of Maxwell's equations and the appropriate boundary conditions. In fact, there are very few problems which yield exact solutions, and we must resort to methods of approximation. The methods of geometrical optics and physical optics have long been used to give first-order solutions to problems involving large apertures and short wavelengths. With the successful generation of both submillimeter "radio" waves and coherent "light" waves, the domain of the radio engineer now extends without break through optical frequencies. Consequently the subject of "optics," both geometrical and physical, has become even more important to the radio engineer than in the past.

Since geometrical optics is the high-frequency limit ($\omega \to \infty$ or $\lambda \to 0$) of electromagnetic field theory it would be most satisfying to start with Maxwell's equations and the derived wave equations, and then demonstrate how physical optics and geometrical optics solutions represent approximations which improve as $\omega \to \infty$. Although several such approaches exist, the problem is more complex than it might at first appear. Indeed the relation between electromagnetic theory and geometrical optics is the subject of an entire book* which has as its objective the building of a mathematical bridge between the two fields. Such rigorous treatments provide the necessary background and understanding to use with confidence the newer intuitive approaches involved in extending ray theory to give a quantitative picture of the electromagnetic field. The recently developed *geometrical theory of diffraction* is a method† that improves upon physical optics approximations. It extends ray theory to include diffracted rays and takes into account phase differences and polarization. The method is based on the conclusion that the processes which give rise to refracted and diffracted rays are *local*; for example, the diffracted field near a point on an edge depends only upon the incident field and the properties of the medium in the neighborhood of the point. These local fields can be obtained as asymptotic expansions (as $\lambda \to 0$) of solutions of boundary value problems for Maxwell's equations.

Reflection from a Flat Rectangular Conductor. Fig. 13-24 illustrates

*Morris Kline and Irvin Kay, *Electromagnetic Theory and Geometrical Optics*, Interscience Publishers, New York, 1965.

†J. B. Keller, "Geometrical Theory of Diffraction," *J. Opt. Soc. Am.*, **52**, p. 116, Feb., 1962.

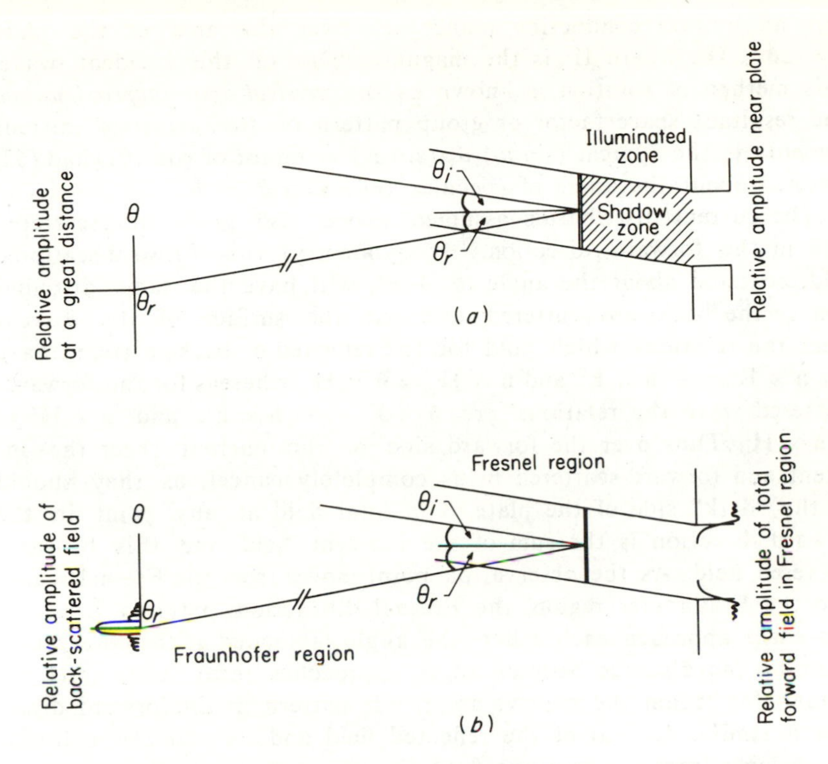

Figure 13-24. Scattering by a rectangular conducting plate: (*a*) geometrical optics approximation, (*b*) physical optics approximation.

geometrical optics and physical optics approximations applied to the problem of reflection or scattering from a flat conducting plate. No exact solution exists for this simple problem, but various approximate solutions, valid for various wavelength ranges, are available. In the geometrical optics formulation of Fig. 13-24(*a*) the reflection is considered to be specular with the angle of reflection equal to the angle of incidence. All of the reflected energy is concentrated at the angle θ_r and the transition between illuminated and shadow zones is abrupt. In the physical optics formulation of Fig. 13-24(*b*) the transition from illumination to shadow as it would be in the Fresnel region is depicted in the forward direction. This diffracted field can be obtained by the method of sec. 13.12, which yielded Fig. 13-22(*b*). The reflected or reradiated field is shown for the Fraunhofer or distant region. The approximation of this field can be obtained by the methods of secs. 13.08 and 13.10. To calculate the field approximately, the flat plate is considered to be replaced by a rectangular array of current elements $J_s\,\Delta x\,\Delta y$ where $\mathbf{J}_s$ is taken to be the same as it would be if the plate were an infinite conducting plane; i.e., over the area of the plate $\mathbf{J}_s = 2\hat{\mathbf{n}} \times \mathbf{H}^i$, where $\mathbf{H}^i$ is the magnetic field of the incident wave. This method of solution is known as the *physical optics approximation.* The resultant space factor or group pattern of the array of current elements is the typical $(\sin u_1)/u_1 \cdot (\sin v_1)/v_1$ factor of eqs. (61) and (62) centered about the angle of specular reflection $\theta_r = \theta_i$.

The current distribution assumed above also gives the scattered field in the forward direction. By symmetry, this forward-scattered field, centered about the angle $(\theta_i + \pi)$, will have the same distribution as the backward-scattered field. At the surface of the current sheet the relations which hold for the reflected or back-scattered wave are $\hat{\mathbf{n}} \times \mathbf{E}^r = -\hat{\mathbf{n}} \times \mathbf{E}^i$, and $\hat{\mathbf{n}} \times \mathbf{H}^r = \hat{\mathbf{n}} \times \mathbf{H}^i$, whereas for the forward-scattered wave the relations are $\hat{\mathbf{n}} \times \mathbf{E}^f = -\hat{\mathbf{n}} \times \mathbf{E}^i$, and $\hat{\mathbf{n}} \times \mathbf{H}^f = -\hat{\mathbf{n}} \times \mathbf{H}^i$. Thus over the forward side of the current sheet the incident and forward-scattered fields completely cancel, as they should on the "dark" side of the plate. The total field at any point in the forward direction is the sum of the incident field and this forward-scattered field. As the observation point moves from the Fresnel region into the Fraunhofer region, the Fresnel diffraction patterns from the two edges approach each other (the angle subtended at the observation point by the distance between edges approaches zero). Well into the Fraunhofer region the relative amplitude pattern in the forward direction is similar to that of the reflected field and its phase is such that it subtracts from the incident field.

No attempt has been made to illustrate in Fig. 13-24 the geometrical theory of diffraction which emphasizes the role of diffracted rays or waves from the edges. The flat conducting plate has been treated by this method in the literature* and the theoretical results compared with experiments.

In the long-wavelength or low-frequency region a different set of approximations becomes valid for the conducting-plate problem. For example, when its dimensions are small in terms of wavelengths, the flat plate can be treated as a short-circuited receiving antenna that reradiates the incident power.

Problem 11. Using the physical optics approximation derive expressions corresponding to eqs. (61) and (62) for the scattered field from a rectangular plate when a plane wave at frequency ω and with Poynting vector $\mathbf{P} = -\hat{\mathbf{z}} E_x H_y$ is incident normally on the plate (Fig. 13-16).

Problem 12. Expressing the power scattered per unit solid angle (radiation intensity Φ) in the z direction as

*R. A. Ross, "Radar Cross Section of Rectangular Flat Plates as a Function of Aspect Angle," *IEEE Trans.*, Vol. AP-14, **3**, p. 329, May, 1966.

$$\Phi = \frac{1}{4\pi}\sigma P$$

defines the *back-scattering cross section or radar cross section*,* σ, of the plate. Prove that under the approximation considered

$$\sigma = \frac{A^2}{\sigma_\lambda}$$

where

$$\sigma_\lambda = \lambda^2/4\pi$$

Problem 13. Assuming that a perfectly reflecting sphere which is large compared with a wavelength scatters approximately *isotropically*† all of the power incident upon it, show that for a large sphere of radius a, the scattering cross section is given by

$$\sigma = \pi a^2$$

13.15 Holography. An example of the merging of optical and radio techniques occurs in the intriguing new field of holography, or wave-front reconstruction, which was originally developed by Gabor‡ in 1947 and later improved by Leith and Upatnieks§ using coherent laser light. According to Huygens' principle (or the equivalence theorem) each point on a wave front acts as a secondary source, and the sum of the radiations from this surface sheet of secondary sources reproduces the wave at points beyond the surface. Accordingly, if the magnitude and phase of the field from some primary source is recorded at every point on some surface, e. g., a plane, and then each point on the surface is made to generate a signal having that magnitude and phase, the wave will be reconstructed beyond the surface, and appear to be coming from the original source (uniqueness theorem). At optical frequencies recording the magnitude of a wave is readily accomplished with a photographic plate, but recording phase is a more difficult task. For an electromagnetic field at radio frequencies, phase at a particular point can be measured by adding a signal of adjustable but known phase. A minimum in resulting amplitude indicates a 180-degree phase difference between known and unknown signals. Thus the addition of a reference signal converts phase variations into amplitude variations. When applied to light waves this principle suggests that the addition of a reference plane wave to a complex wave will result in the conversion of phase variations into amplitude variations which in turn may be recorded on a photographic plate.

*For a detailed discussion of these terms see *IEEE Test Procedure for Antennas*, No. 149, p. 28, Jan., 1965.

†L. N. Ridenour, "Radar System Engineering," McGraw-Hill M. I. T. Radiation Laboratory Series, **1**, p. 65, McGraw-Hill Book Company, New York, 1947.

‡D. Gabor, *Nature*, **161**, 777 (1948); *Proc. Roy. Soc.* (London), Vol. A-197, 454 (1949).

§E. Leith and J. Upatnieks, *J. Opt. Soc. Am.* **53**, 1377 (1963); **54**, 1295 (1964).

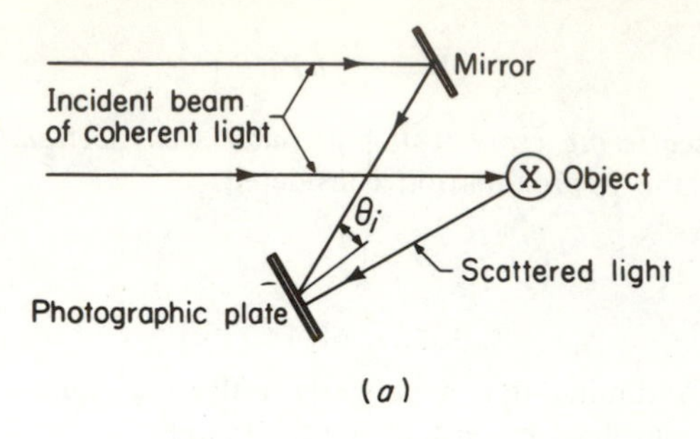

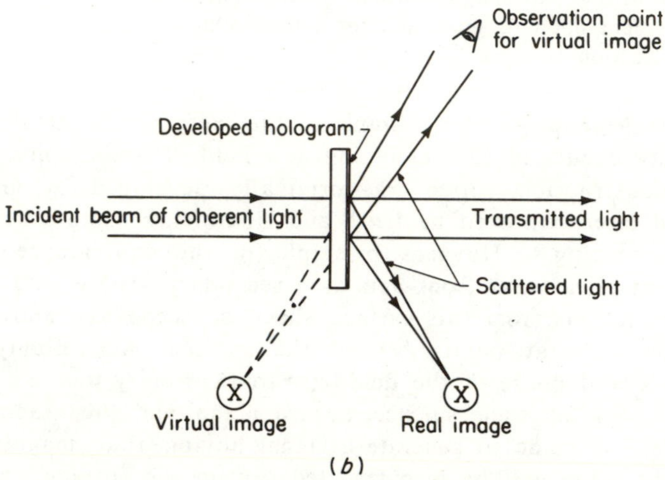

Figure 13-25. Basic holography experiment showing (*a*) preparation of the hologram and (*b*) reconstruction of the image.

An experimental arrangement for preparing a hologram is shown in Fig. 13-25(*a*). The reference light wave is obtained by using a mirror to divert part of the incident coherent light beam to the photographic plate. Consider for simplicity the situation in which the photographic plate lies in the *x-y* plane of a rectangular co-ordinate system. The time-varying scattered and reference wave fields on the plate may be written respectively in the form

$$W_1(x, y, t) = A_1(x, y) \cos [\omega t + \phi(x, y)] \tag{13-73}$$

and

$$W_0(x, y, t) = A_0(x, y) \cos [\omega t + kx] \tag{13-74}$$

in which $k = \beta \sin \theta_i$. The sum of (73) and (74) gives the total wave field $W(x, y, t)$, and it is this total wave field which acts on the photographic emulsion. The emulsion responds to energy rather than field strength and thus the effect upon the emulsion is proportional to the square of the total wave field,

$$\begin{aligned} W^2 &= (W_0 + W_1)^2 \\ &= A_0^2 \cos^2(\omega t + kx) + A_1^2 \cos(\omega t + \phi) \\ &\qquad + 2A_0A_1 \cos(\omega t + kx)\cos(\omega t + \phi) \qquad (13\text{-}75) \\ &= \tfrac{1}{2}A_0^2 + \tfrac{1}{2}A_1^2 + \tfrac{1}{2}A_0^2 \cos 2(\omega t + kx) + \tfrac{1}{2}A_1^2 \cos 2(\omega t + \phi) \\ &\qquad + A_0A_1 \cos(2\omega t + kx + \phi) + A_0A_1 \cos(kx - \phi) \qquad (13\text{-}76) \end{aligned}$$

The interaction with the emulsion involves a time-averaging process which results in a hologram having a light-transmission coefficient of the form

$$T = \tfrac{1}{2}A_0^2 + \tfrac{1}{2}A_1^2 + A_0A_1 \cos(kx - \phi) \qquad (13\text{-}77)$$

This transmission coefficient is illustrated in Fig. 13-26 for the case in which the scattered field is a uniform plane wave.

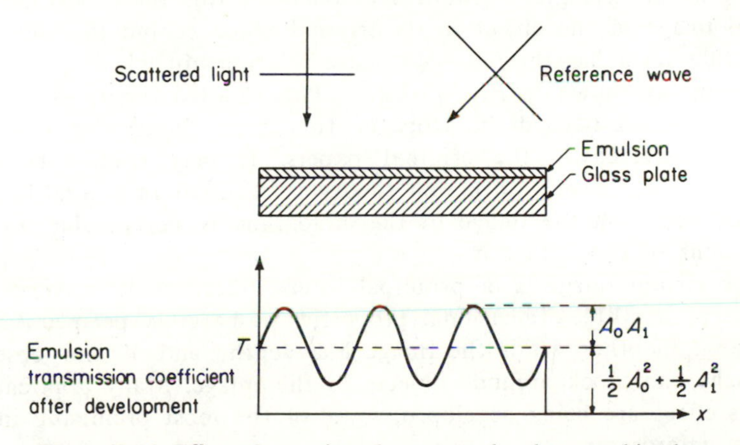

Figure 13-26. Effect of wave interference on the photographic plate.

It is clear that the interference between the scattered and reference waves turns the photographic plate into a kind of grating and that the structure of this grating contains complete information on both the amplitude and phase of the scattered light. The grating-like pattern on the photographic plate was given the name *hologram* by Gabor from the Greek word *holos* meaning *entire* or *complete*.

Reconstruction of the original scattered-light wavefront may be accomplished by shining a beam of coherent light normally on the hologram as illustrated in Fig. 13-25(*b*). According to Huygens' principle the light passing through the hologram may be regarded as setting up equivalent sources over the surface of the hologram, the strength of these sources being proportional to the transmission coefficient in (77). Thus the time-varying wave field on the surface of the hologram may be expressed in the form

$$W_S = A_2[\tfrac{1}{2} A_0^2 + \tfrac{1}{2} A_1^2 + A_0 A_1 \cos (kx + \phi)] \cos \omega t \qquad (13\text{-}78)$$

in which A_2 is the amplitude of the incident light wave falling on the hologram. Equation (78) also may be written in the form

$$W_S = \tfrac{1}{2} A_2[A_0^2 + A_1^2] \cos \omega t + \tfrac{1}{2} A_0 A_1 A_2[\cos (\omega t + \phi + kx) + \cos (\omega t - \phi - kx)] \qquad (13\text{-}79)$$

The $\cos \omega t$ term produces a wave which is transmitted with no change in direction. The other two terms contain $\pm kx$, indicating that they produce waves scattered to either side, the directions of the scattered rays making the angle θ_i with the normal to the photographic plate. One of these terms has the same $(\omega t + \phi)$ space-time phase relationship as the original scattered light wave and therefore an observer looking at the hologram toward the source of this light will see the virtual image of the object in its original place behind the hologram. The other term has the $(\omega t - \phi)$ phase relationship which produces a real image as shown in Fig. 13-25(*b*). The detailed theory of the imaging process is beyond the scope of this brief description and the reader is referred to the original papers. It may readily be seen, however, that Fresnel diffraction theory is required in general because the distance from the image to the hologram is comparable to the dimensions of the hologram.

The virtual image is of principal interest because it is very easy to observe its three-dimensional properties of parallax, perspective and focussing. In other words the image has depth, and if the observer moves, he can "look around" objects in the image. Many applications of this effect are being developed, some of the most promising in the field of microscopy.

An interesting property arises from the fact that any part of the photographic plate receives light from the entire object. As a result any part of the hologram is theoretically capable of reproducing the entire image. A very large hologram produces a clear image with little depth of field. As the hologram is made smaller the depth of field increases and the resolution decreases.

An important property of the holographic technique is the capability of recording more than one image on the same photographic plate. Two images may be recorded by the use of light beams at two different frequencies. Alternatively the same frequency may be used but with the reference wave striking the photographic plate at two different angles, say by rotating the plate through 90 degrees.

13.16 Complementary Screens and Slot Antennas. At frequencies above about 300 MHz slot antennas cut in a metallic surface, such as the skin of an aircraft or the wall of a wave guide, often prove to be convenient radiators. The slot may be fed by a generator or transmission line connected across it, or in the case of the wave guide, by a guided wave incident on the slot. To explore some of the characteristics of slot antennas we consider Fig. 13-27, which shows a slot

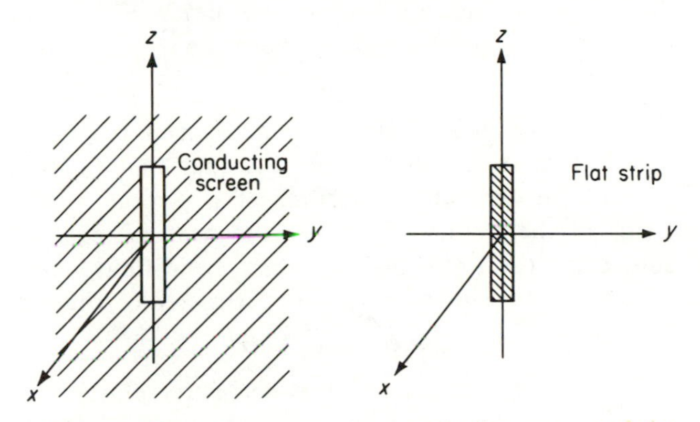

Figure 13-27. A slot antenna in a conducting screen, and the complementary flat dipole.

in a conducting plane and the complementary flat dipole antenna formed by the metal removed in making the slot. *Complementary screens* are defined in the following manner: An infinite-plane conducting screen is pierced with apertures of any shape or size and the resultant screen is called S_1. Consider then the screen which is obtained by interchanging the region of metal and aperture space in S_1 and call this second screen S_2. Then screens S_1 and S_2 are said to be complementary, because, added together they result in a complete infinite metal screen. In this sense the slot and flat dipole antennas described above are complementary.

For a first approach to the complementary slot and dipole problem consider these antennas as separate boundary-value problems. As such, the problem in each case is that of finding appropriate solutions to Maxwell's equations, or the derived wave equations, which will satisfy the boundary conditions. In both cases, for the free-space regions about the antennas, the wave equations to be solved are

$$\nabla^2 \mathbf{E} = \mu\epsilon \ddot{\mathbf{E}} \qquad \text{or} \qquad \nabla^2 \mathbf{H} = \mu\epsilon \ddot{\mathbf{H}} \tag{13-80}$$

Having obtained a solution for either **E** or **H** from one of the above equations, the other field strength can be obtained through one of the free-space relations

$$\nabla \times \mathbf{E} = -\mu\dot{\mathbf{H}} \qquad \nabla \times \mathbf{H} = \epsilon\dot{\mathbf{E}} \tag{13-81}$$

For the dipole, a solution to the second of eqs. (80) would be sought subject to the following boundary conditions:

(1) In the *y-z* plane, and outside the perimeter of the dipole, the tangential components of magnetic field are zero, i.e.,

$$H_y = 0 \qquad H_z = 0$$

(2) In the *y-z* plane, and within the perimeter of the dipole, the normal component of magnetic field is zero, i.e.,

$$H_x = 0$$

The first of these conditions follows from symmetry considerations. The second condition is a statement of the fact that the normal component of **H** must be zero at the surface of a perfect conductor.

For the slot in the conducting plane the first of eqs. (80) would be solved subject to the following boundary conditions:

(1) In the *y-z* plane, and outside the perimeter of the slot,

$$E_y = 0 \qquad E_z = 0$$

(2) In the *y-z* plane, and within the perimeter of the slot

$$E_x = 0$$

The first of these conditions results from the fact that the tangential component of **E** must be zero at the surface of a perfect conductor. The second condition results from symmetry considerations.

It is apparent that, mathematically, these two problems are identical. It is necessary only to interchange **E** and **H** to pass from one problem to the other. (Of course it will have to be shown that the driving forces are also similar.) Therefore, except for a constant, the solution obtained for **E** for the slot will be the same as the solution for **H** for the dipole, and it is possible to write for the fields at any corresponding points

$$\mathbf{E}_s = k_1 \mathbf{H}_d \tag{13-82}$$

where the subscripts *s* and *d* refer to the slot and dipole respectively. Similarly, the magnetic field of the slot and the electric field of the dipole will be related by

$$\mathbf{H}_s = k_2 \mathbf{E}_d \tag{13-83}$$

It follows that the field distribution of **E** for the slot is the same as the field distribution of **H** for the dipole, and vice versa. Also the impedance of the slot is proportional to the admittance of the dipole, and vice versa. The relations between the impedance properties can be studied by use of the integral expressions for the fields near the feed points.

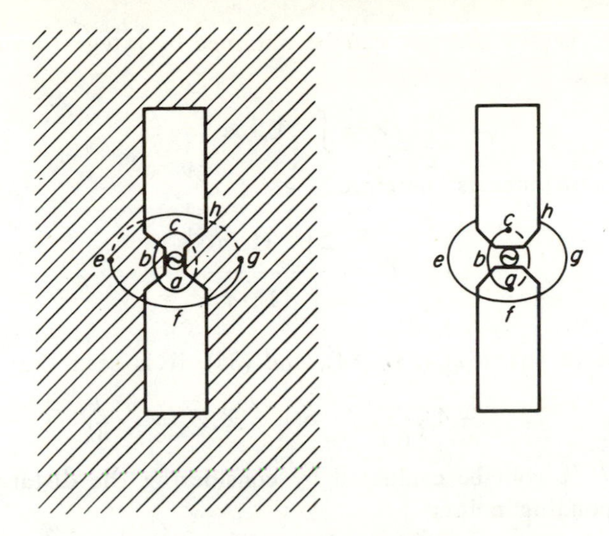

Figure 13-28. Slot and flat-strip dipole with small gaps.

Strictly, for impedance to have meaning, it is necessary to consider an infinitesimal gap in each case. However, the results will apply to a finite gap if this is kept small, as indicated in Fig. 13-28. For the dipole, the impedance is given by the voltage across the gap divided by the current through the generator and into one arm of the dipole. The voltage across the gap is obtained by integrating $\mathbf{E}_d \cdot d\mathbf{s}$ along a line of **E** between two closely spaced points a and c on opposite sides of the gap

$$V = -\int_{abc} \mathbf{E}_d \cdot d\mathbf{s}$$

The current into one arm of the dipole is equal to the magnetomotive force around the (small) closed loop *efghe*

$$I = \oint \mathbf{H}_d \cdot d\mathbf{s} = 2\int_{efg} \mathbf{H}_d \cdot d\mathbf{s}$$

The impedance of the dipole is therefore

$$Z_d = \frac{-\int_{abc} \mathbf{E}_d \cdot d\mathbf{s}}{2\int_{efg} \mathbf{H}_d \cdot d\mathbf{s}} \tag{13-84}$$

The admittance of the slot can be found by dividing the current into one edge by the voltage across the gap. The slot current is equal to the magnetomotive force around the closed path *abcda*.

$$I = \oint_{abcda} \mathbf{H}_s \cdot d\mathbf{s} = 2\int_{abc} \mathbf{H}_s \cdot d\mathbf{s}$$

The voltage across the slot can be obtained by integrating $\mathbf{E}_s \cdot d\mathbf{s}$ along the curve *efg*.

$$V = \int_{efg} \mathbf{E}_s \cdot d\mathbf{s}$$

The slot admittance is therefore

$$Y_s = \frac{2\int_{abc} \mathbf{H}_s \cdot d\mathbf{s}}{\int_{efg} \mathbf{E}_s \cdot d\mathbf{s}} \tag{13-85}$$

Making use of eqs. (82), (83), (84), and (85), it is seen that

$$Y_s = -4\frac{k_2}{k_1}Z_d \qquad \text{or} \qquad Z_s Z_d = -\frac{k_1}{4k_2}$$

The ratio k_1/k_2 can be evaluated by considering the distant fields. At any corresponding points

$$E_s = k_1 H_d \qquad H_s = k_2 E_d$$

At points sufficiently distant that the fields are essentially plane-wave fields

$$E_s = \eta_v H_s \qquad E_d = -\eta_v H_d$$

Combining these relations shows that

$$-\frac{k_1}{k_2} = \eta_v^2 = 377^2$$

Therefore

$$Z_s Z_d = \frac{\eta_v^2}{4} \tag{13-86}$$

For the theoretical half-wave dipole, $Z_d \approx 73 + j43$ ohms, so that for a theoretical half-wave slot,

$$Z_s \approx \frac{377^2}{4 \times (73 + j43)} \approx 418\underline{/-30.5^\circ}$$

For a resonant-length dipole, the input resistance depends upon the dipole thickness. It may be of the order of 65 ohms for practical

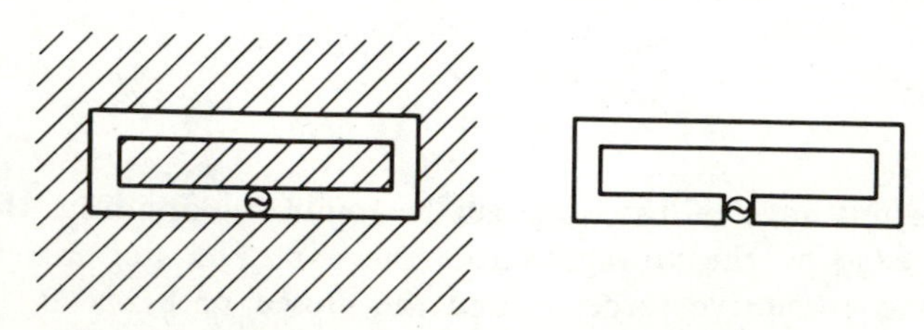

Figure 13-29. Folded slot and folded dipole.

dipoles. The corresponding impedance for a practical resonant-length slot would be of the order of $Z_s = 550$ ohms. For a folded half-wave dipole the input impedance is roughly four times that of an ordinary dipole, so that the input impedance of the *folded half-wave slot* of Fig. 13-29 is approximately $\frac{550}{4} = 138$ ohms.

13.17 Slot and Dipole Antennas as a Dual Problem. The slot and dipole antennas considered in the preceding section are complementary but *not* duals. However, it is easy to transform them into duals, and instructive to do so for later use. In Fig. 13-27, as far as the field in the region $x > 0$ is concerned, the slot in the conducting plane can be replaced by a distribution of magnetic current density $\mathbf{M}_s = -\hat{\mathbf{n}} \times \mathbf{E}_a$ at the surface of an infinite conducting screen, where $\mathbf{E}_a$ is the electric field that existed across the slot. This equivalence follows directly from the uniqueness theorem because of identical boundary conditions over the entire screen, viz., $\hat{\mathbf{n}} \times \mathbf{E} = 0$ outside the slot perimeter, and $\hat{\mathbf{n}} \times \mathbf{E} = \hat{\mathbf{n}} \times \mathbf{E}_a$ within the perimeter. Similarly for the region $x > 0$, the flat dipole with its surface current density J_a may be replaced by a similar distribution of electric current density $J_s = \frac{1}{2} J_a$ at the surface of an infinite perfect magnetic conducting screen. Again the boundary conditions over the plane remain unchanged, viz., $\hat{\mathbf{n}} \times \mathbf{H} = 0$ outside the perimeter of the dipole and $\hat{\mathbf{n}} \times \mathbf{H} = \hat{\mathbf{n}} \times \mathbf{H}_a = \mathbf{J}_s$ within the perimeter.

These equivalent problems are almost exact duals. To complete the duality it would be necessary to replace the free-space characteristic impedance $\eta = \sqrt{\mu/\epsilon}$ by its reciprocal $\eta_m = \sqrt{\epsilon/\mu} = 1/\eta$ as indicated in Fig. 13-30(b). If this were done, the solutions to the two situations depicted in Fig. 13-30(b) would be identical with H_d replacing E_s, etc., and H_d and E_s being numerically equal. Also the (electric) impedance Z_e of the electric dipole and the (magnetic) impedance Z_m of the magnetic dipole would be equal; that is $Z_e = Z_m$.

Now return to the original slot-dipole problem. In actuality the free-space medium is the same in both cases so we cannot have $\eta_m = 1/\eta$ (unless $\eta = 1$ which it does not). Accordingly the electric and magnetic dipoles must operate at different impedance levels, so that dual quantities are proportional rather than equal. The constant of proportionality can be determined as in the previous section; alternatively, one can note for example in the long-wire, long slit problems (sec. 14.10 and problem 2) that the external impedance for electric currents is proportional to η, whereas the external magnetic impedance (or electric admittance) for the slit is proportional to $1/\eta$. Hence for electric and magnetic dipoles immersed in the *same* free-space medium (of intrinsic impedance η) the ratio of electric impedance to magnetic

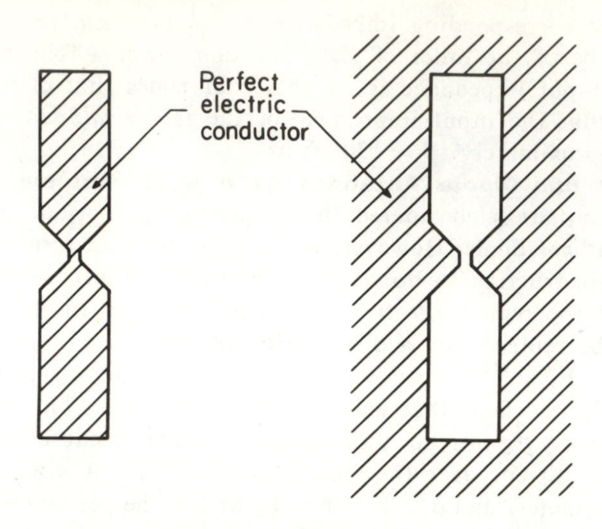

(*a*) Complementary

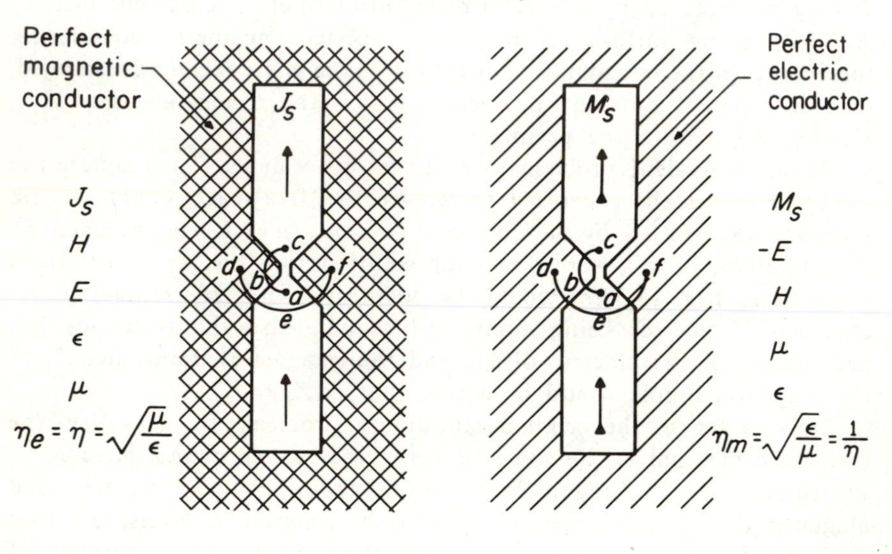

(*b*) Dual

$$Z_e = \frac{V}{I} = -\frac{\int_{abc} E \cdot ds}{\int_{def} H \cdot ds} \qquad Z_m = \frac{\mathcal{I}}{K} = -\frac{\int_{abc} H \cdot ds}{\int_{def} E \cdot ds}$$

$$\frac{Z_e}{Z_m} = 1$$

Figure 13-30. Complementary screens and dual problems.

impedance will be

$$\frac{Z'_e}{Z'_m} = \frac{\eta Z_e}{Z_m/\eta} = \eta^2 \frac{Z_e}{Z_m}$$

Finally, to complete the return to the complementary antennas of Fig. 13-30(*a*), note that the actual dipole radiates on both sides so it is equivalent in Fig. 13-30(*b*) to two electric dipoles, one on each side of the magnetic conducting plane, fed in parallel by a common voltage. Therefore $Z_d = \frac{1}{2} Z'_e$. The actual slot also radiates on both sides but in this case oppositely directed magnetic currents are fed by two magnetic voltages in series, with the result that the (electric) admittance of the slot will be

$$Y_s = 2Z'_m$$

The ratio of actual dipole impedance to actual slot admittance is therefore

$$Z_d/Y_s = \frac{\frac{1}{2} Z'_e}{2Z'_m} = \frac{\eta^2}{4} \frac{Z_e}{Z_m} = \frac{\eta^2}{4}$$

which is the same result that we obtained in sec. 13.16.

The ability to transform back and forth between complementary screens (which can be approximated physically) and hypothetical dual problems (whose solution is known) proves to be a powerful tool in the solution of certain electromagnetic problems. An example of this technique occurs in the extension of Babinet's principle in the next section.

13.18 Babinet's Principle. In optics, Babinet's principle relates to transmission through apertures in thin, perfectly absorbent complementary screens. Applied to such screens Babinet's principle for optics simply states that the sum of the fields, taken separately, beyond any two complementary absorbing screens will add to produce the field that would exist there without any screen. Thus if U_1 represents the ratio at every point of the field on the right of screen S_1 to the field without the screen, and U_2 similarly represents the ratio for screen S_2, then Babinet's principle for optics states that

$$U_1 + U_2 = 1$$

The correctness of this intuitively valid statement can be verified easily for the simple case of complementary screens consisting of semi-infinite absorbing planes (see problem 22).

In electromagnetics at radio frequencies, thin perfectly absorbing screens are not available, even approximately, and one is concerned with *conducting* screens and *vector fields* for which polarization plays an important role. Under such circumstances the simple statement of optics could not be expected to apply, but an extension of the principle, valid for conducting screens and polarized fields has been formulated.* This generalized formulation can be stated as follows: Let a source s_1 to the left of an infinite screen S_1 produce a field on the right of S_1, and let U_1 be the ratio of this field to the field strength that would exist there in the absence of the screen; then consider a *conjugate* source s_2 to the left of the *complementary* screen S_2, and let U_2 be the ratio of the field on the right of S_2 to the field that would exist there in the absence of the screen; then

$$U_1 + U_2 = 1 \tag{13-87}$$

In the above statement a *conjugate source* is one for which the distribution of electric and magnetic source currents is replaced by the corresponding distribution of magnetic and electric currents so that, for example, $I\,dl$ is replaced by $K\,dl$, etc. (Fig. 13-31). This replacement has the effect of interchanging E and H for the incident fields.

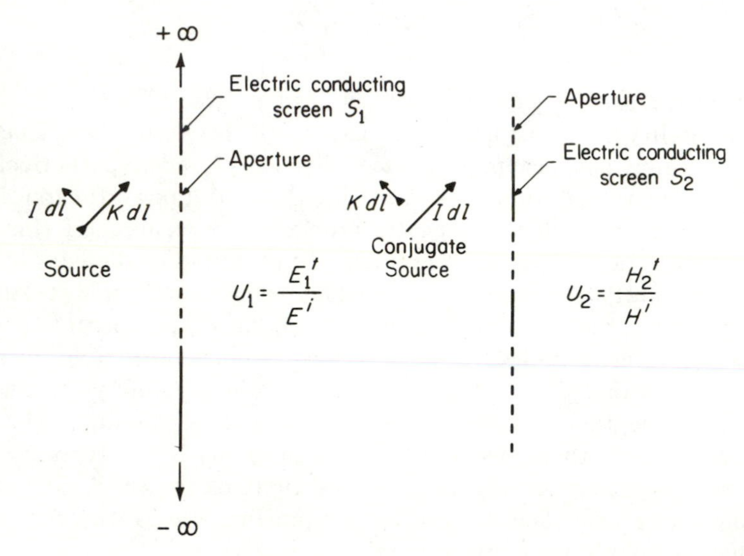

Figure 13-31. Illustrating Babinet's principle: $U_1 + U_2 = 1$.

The proof of this extension to Babinet's principle will be carried out by invoking the symmetry arguments applied earlier, and then transforming the problem to one of duals, for which the answers are known.

To prove the extended Babinet's principle, consider Fig. 13-32. In (*a*) screen S_1, irradiated by source s_1, has aperture area s_A and metal

*H. G. Booker, "Slot Aerials and Their Relation to Complementary Wire Aerials (Babinet's Principle)," *JIEE*, Vol. III A, 620–626 (1946).

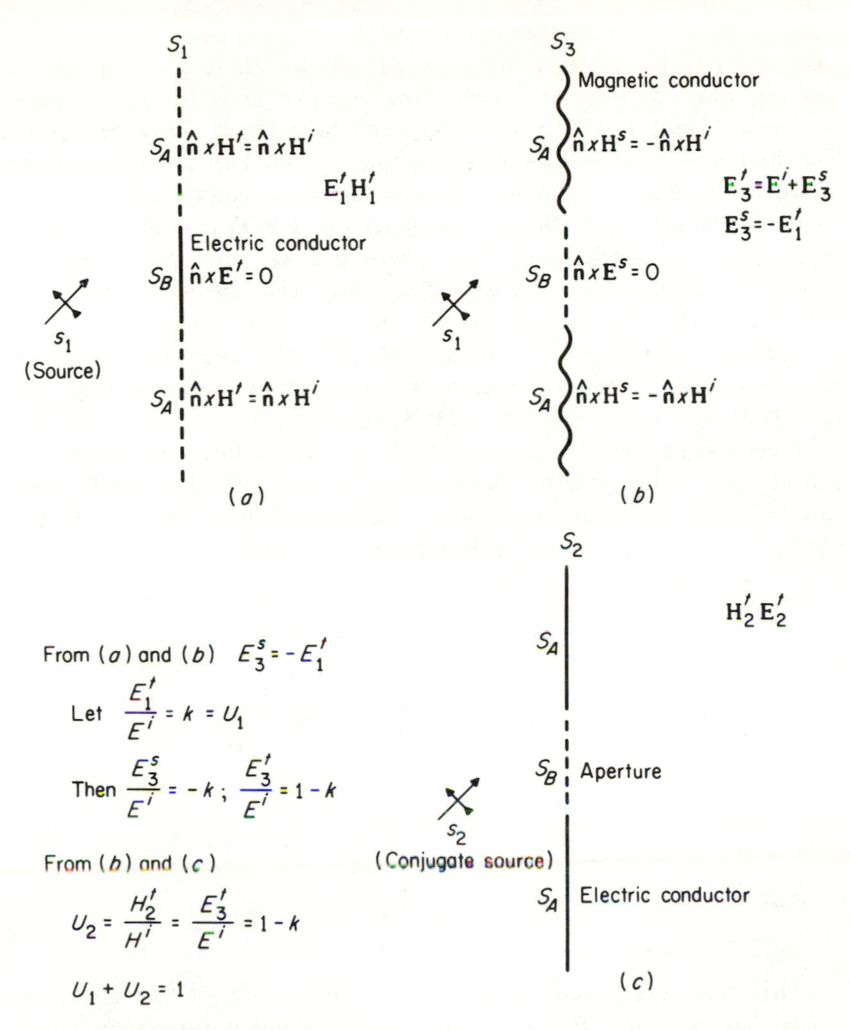

Figure 13-32. Duality (*b* and *c*) and the uniqueness theorem (*a* and *b*) are used to prove Babinet's principle (*a* and *c*).

area s_B. In (*c*) the complementary screen S_2 is irradiated by the conjugate source s_2. In (*b*) the source s_1 irradiates infinite screen s_3, obtained from S_1 by replacing aperture area s_A by a perfect *magnetic* conductor, and metal area s_B by an aperture.

In (*a*) the transmitted (total) field is determined uniquely by the specification of the tangential field over S_1, viz., $\hat{\mathbf{n}} \times \mathbf{E}^t = 0$ over s_B and $\hat{\mathbf{n}} \times \mathbf{H}^t = \hat{\mathbf{n}} \times \mathbf{H}^i$ over s_A. The first condition results from the requirement that tangential $\mathbf{E} = 0$ at the surface of a perfect electric conductor; the second condition results because $\hat{\mathbf{n}} \times \mathbf{H}^s = 0$ over s_A, which follows from the symmetry of the re-radiated or scattered field produced by electric currents induced on s_B. In Fig. 13-32(*b*), considering the scattered field alone, the specifications for the tangential field over s_3 are $\hat{\mathbf{n}} \times \mathbf{E}^s = 0$ over s_B and $\hat{\mathbf{n}} \times \mathbf{H}^s = -\hat{\mathbf{n}} \times \mathbf{H}^i$ over s_A. The first condition results from symmetry of the scattered E field produced by magnetic currents in the magnetic conductor; the second condition results from the requirement for $\hat{\mathbf{n}} \times \mathbf{H} = 0$ at the surface of a perfect magnetic conductor, where $\hat{\mathbf{n}} \times \mathbf{H} = \hat{\mathbf{n}} \times (\mathbf{H}^i + \mathbf{H}^s)$. Thus in (*b*) the scattered field is specified over the entire screen as the negative of the total or transmitted field in (*a*). Therefore from the uniqueness theorem $\mathbf{E}_3^s = -\mathbf{E}_1^t$ and $\mathbf{H}_3^s = -\mathbf{H}_1^t$. The total transmitted field in (*b*) is of course the sum of the incident and scattered fields, that is $\mathbf{E}_3^t = \mathbf{E}^i + \mathbf{E}_3^s$ and $\mathbf{H}_3^t = \mathbf{H}^i + \mathbf{H}_3^s$.

Now comparing (*b*) and (*c*) it is seen that they are exact duals except for the impedance level ($\eta_m = \eta$ instead of $\eta_m = 1/\eta$ as it should). Therefore the fields are proportional with E and H interchanged. Accordingly, for each rectangular component,

$$\frac{H_2^t}{H_2^i} = \frac{E_3^t}{E^i}$$

Finally, writing

$$U_1 = \frac{E_1^t}{E^i} = k \qquad \text{and} \qquad U_2 = \frac{H_2^t}{H_2^i}$$

it follows that

$$\frac{E_3^s}{E^i} = -k \qquad\qquad \frac{E_3^t}{E^i} = 1 - k$$

so that

$$U_1 + U_2 = 1 \tag{13-87}$$

This generalized statement of Babinet's principle for electromagnetics has the same form as the simple statement for optics, but the discussion leading to (87) will have pointed up the additional complexity of problems involving the polarized vector fields of electromagnetics.

In the statement of the generalized Babinet's principle leading to (87), conjugate sources were specified for use with the complementary screens. When the incident field is essentially a plane wave from a distant source this specification is met by simply rotating the plane of polarization through 90 degrees, and thus, in effect, interchanging E and H of the incident field.

Problem 14. Using Babinet's principle show that the diffracted field through a (resonant) half-wave length slot is given by

$$H_\theta = -\frac{j\,60\,\lambda\,H_i\,e^{-j\beta r}}{73\pi r}\cdot\frac{\cos(\pi/2\cos\theta)}{\sin\theta}$$

$$E_\phi = -\eta H_\theta$$

where H_i is oriented parallel to the length of the slot. For the complementary (resonant) half-wave dipole antenna, assume the antenna impedance to be $Z_a = 73$ ohms.

13.19 Slotted Cylinder Antennas. An antenna that has important applications at very high frequencies consists of a slot or slots cut in a conducting cylinder. For example, a longitudinal slot in a vertical cylinder produces a horizontally polarized signal suitable for FM or television. A method for obtaining the complete three-dimensional radiation pattern of a finite-length slot in a cylinder will be considered in the next section. The two-dimensional problem of an infinitely long slot in an infinite cylinder can be solved easily, and is an example that illustrates nicely a method of solution that is quite powerful for certain types of problems. The solution of this particular problem is useful because it gives the principal plane pattern (perpendicular to the axis) of a *finite length* slot in a cylinder.

Figure 13-33 shows a section of the infinitely long cylinder of radius a with a longitudinal slot of width $a\phi_0$. It is assumed that the slot is fed between its edges with a voltage $V = a\phi_0 E_0$, which is uniform in magnitude and phase along the length of the slot. This means that there will be no variations in the z direction. Also with this method of excitation there will be an E_ϕ and perhaps an E_ρ, but no E_z.

Writing Maxwell's equations in cylindrical co-ordinates for the free-space region external to the cylinder where $\sigma = 0$, and remembering that $E_z = 0$ and $\partial/\partial z = 0$, there results

$$\frac{\partial H_z}{\rho\,\partial\phi} = j\omega\epsilon E_\rho \qquad 0 = j\omega\mu H_\rho$$

$$\frac{\partial H_z}{\partial\rho} = -j\omega\epsilon E_\phi \qquad 0 = -j\omega\mu H_\phi$$

$$\frac{\partial(\rho H_\phi)}{\rho\,\partial\rho} - \frac{\partial H_\rho}{\rho\,\partial\phi} = 0 \qquad \frac{\partial(\rho E_\phi)}{\rho\,\partial\rho} - \frac{\partial E_\rho}{\rho\,\partial\phi} = -j\omega\mu H_z \tag{13-88}$$

It is seen that $H_\rho = H_\phi = 0$, and that for this problem all fields can be expressed in terms of the axial components of magnetic field strength H_z. Then

$$E_\rho = \frac{1}{j\omega\epsilon\rho}\frac{\partial H_z}{\partial\phi}$$

$$E_\phi = -\frac{1}{j\omega\epsilon}\frac{\partial H_z}{\partial\rho} \tag{13-89}$$

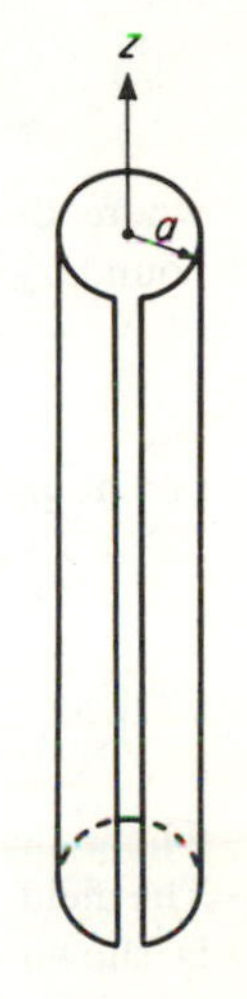

Figure 13-33. A section of an infinitely long slotted cylinder.

Substituting these expressions in the last expression of (88) gives a wave equation for H_z

$$\frac{1}{\rho}\frac{\partial}{\partial\rho}\left(\rho\frac{\partial H_z}{\partial\rho}\right)+\frac{1}{\rho^2}\frac{\partial^2 H_z}{\partial\phi^2}=-\beta^2 H_z \tag{13-90}$$

where

$$\beta^2=\omega^2\mu\epsilon$$

Solving in the usual manner by assuming a product solution results in*

$$H_z=[A_1 H_\nu^{(1)}(\beta\rho)+B_1 H_\nu^{(2)}(\beta\rho)]e^{j\nu\phi} \tag{13-91}$$

For this problem it is apparent that ν must be an integer n. Using only the outward traveling wave for this region outside the cylinder, the general solution for this region will be

$$H_z=\sum_{n=-\infty}^{n=+\infty} b_n H_n^{(2)}(\beta\rho)e^{jn\phi} \tag{13-92}$$

and

$$E_\phi=\frac{j\beta}{\omega\epsilon}\sum_{n=-\infty}^{n=+\infty} b_n H_n^{(2)\prime}(\beta\rho)e^{jn\phi} \tag{13-93}$$

where the b_n's are coefficients which are to be evaluated by applying the boundary conditions. At $\rho=a$, the expression for E_ϕ becomes

$$E_\phi|_{\rho=a}=\frac{j\beta}{\omega\epsilon}\sum_{n=-\infty}^{+\infty} b_n H_n^{(2)\prime}(\beta a)e^{jn\phi} \tag{13-94}$$

but at $\rho=a$, the boundary condition are

$$E_\phi=E_0 \qquad \left(-\frac{\phi_0}{2}<\phi<\frac{\phi_0}{2}\right) \tag{13-95}$$

$$E_\phi=0 \qquad |\phi|>\frac{\phi_0}{2}$$

where the electric field strength E_0 has been assumed uniform over the gap. The field distribution at $\rho=a$, as represented by these boundary conditions, is shown in Fig. 13-34.

This field distribution may be resolved into a Fourier series,

$$E_\phi|_{\rho=a}=\sum_{n=-\infty}^{+\infty} C_n e^{jn\phi} \tag{13-96}$$

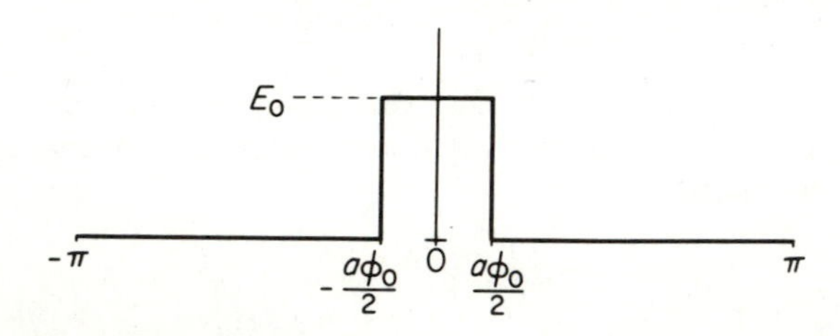

Figure 13-34. Distribution of field strength E_ϕ around the cylinder at $\rho=a$.

*Hankel function solutions are discussed in sec. 14.10 and in Appendix II.

where

$$C_n = \frac{1}{2\pi}\int_0^{2\pi} F(\alpha)e^{-jn\alpha}\,d\alpha$$

$$= \frac{E_0}{2\pi}\int_{-\phi_0/2}^{+\phi_0/2} e^{-jn\alpha}\,d\alpha$$

$$= -\frac{E_0}{2jn\pi}\left(e^{-jn\phi_0/2} - e^{+jn\phi_0/2}\right)$$

$$= \frac{E_0}{n\pi}\sin\frac{n\phi_0}{2}$$

The distribution represented by (96) is the *same* as that represented by (94), so

$$\frac{E_0}{n\pi}\sin\frac{n\phi_0}{2} = \frac{j\beta}{\omega\epsilon}b_n H_n^{(2)\prime}(\beta a)$$

or

$$b_n = \frac{\omega\epsilon}{j\beta}\frac{E_0}{n\pi}\sin\frac{n\phi_0}{2}\frac{1}{H_n^{(2)\prime}(\beta a)}$$

The external field at a distance ρ is then given by (93).

$$E_\phi = \frac{E_0}{\pi}\sum_{n=-\infty}^{+\infty}\frac{\sin(n\phi_0/2)H_n^{(2)\prime}(\beta\rho)}{nH_n^{(2)\prime}(\beta a)}e^{jn\phi}$$

It is possible to evaluate this expression at large distances from the cylinder where the asymptotic expressions* for the Hankel functions may be used. At large distance

$$H_n^{(2)\prime}(\beta\rho) \approx \frac{\partial}{\partial(\beta\rho)}\left[\sqrt{\frac{2}{\pi\beta\rho}}\,e^{-j(\beta\rho-(n\pi/2)-(\pi/4))}\right]$$

$$= \left[-\frac{1}{2}\sqrt{\frac{2}{\pi(\beta\rho)^3}} - j\sqrt{\frac{2}{\pi\beta\rho}}\right]e^{-j(\beta\rho-(n\pi/2)-(\pi/4))}$$

Neglecting the first term, at large distances

$$E_\phi \approx -\frac{jE_0}{\pi}\sqrt{\frac{2}{\pi\beta\rho}}\sum_{n=-\infty}^{+\infty}\frac{\sin(n\phi_0/2)\,e^{-j(\beta\rho-(n\pi/2)-(\pi/4)-n\phi)}}{nH_n^{(2)\prime}(\beta a)} \tag{13-97}$$

Using only the first few terms of this expansion will usually give results of sufficient accuracy. Using less than some fixed number, say N, it is possible to write, if ϕ_0 is sufficiently small,

$$\frac{1}{n}\sin\frac{n\phi_0}{2} \approx \frac{\phi_0}{2} \qquad \text{for } |n| < N$$

Therefore, approximately,

$$E_\phi = A\,e^{-j(\beta\rho-\pi/4)}\sum_{n=-N}^{n=+N}\frac{e^{jn(\phi+\pi/2)}}{H_n^{(2)\prime}(\beta a)} \tag{13-98}$$

where

$$A = \frac{-jE_0\phi_0}{2\pi}\sqrt{\frac{2}{\pi\beta\rho}}$$

*Appendix II, eq. (9).

Recalling that

$$H_{-n}^{(2)\prime}(\beta a) = (-1)^n H_n^{(2)\prime}(\beta a)$$

eq. (98) can be written

$$E_\phi = A\, e^{-j(\beta\rho - \pi/4)} \left[\frac{1}{H_0^{(2)\prime}(\beta a)} + 2 \sum_{n=1}^{n=N} \frac{(j)^n \cos n\phi}{H_n^{(2)\prime}(\beta a)} \right] \tag{13-99}$$

This expression gives the amplitude and phase of the electric field strength at any distant point (ρ, ϕ). The relative shape of the radiation pattern is given by the absolute value of the bracketed factor.

This factor has both real and imaginary parts and may be written as $C + jD$. The field strength pattern shape is then given by the absolute value

$$\sqrt{C^2 + D^2}$$

Figure 13-35 shows the radiation pattern calculated by this method for a $\lambda/20$-wide slot in a $5\lambda/4$-diameter cylinder. Shown for comparison is the measurement pattern of a $1\frac{1}{2}\lambda$-long slot of width $\lambda/20$ in a long, $5\lambda/4$-diameter cylinder. The radiation patterns of slots in cylinders of other diameters may be found in the literature.*

Transverse Slots. The same method can be used, although with less justification, to predict the radiation pattern perpendicular to the cylinder of a *transverse* slot in a cylinder. In this case the electric field strength applied across

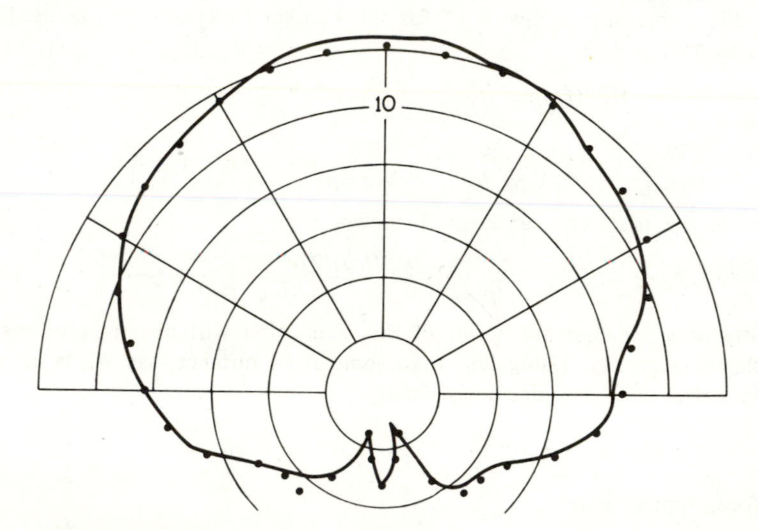

Figure 13-35. Experimental pattern of a long $\lambda/20$-wide axial slot in a long cylinder of diameter $5\lambda/4$. Points are calculated for a $\lambda/20$ slot that is 1.5λ long.

*G. Sinclair, E. C. Jordan, and E. W. Vaughan, "Measurement of Aircraft Antenna Patterns Using Models," *Proc. IRE,* **35**, 12, 1451–1462 (1947); E. C. Jordan and W. E. Miller, "Slotted-cylinder Antenna," *Electronics,* **20**, 2, 90 (1947).

the slot will be in the z direction. For a narrow transverse slot of length $L = a\phi_0$, the electric field across the slot will be assumed to have a distribution along the length of the slot similar to that which exists on a short-circuited lossless transmission line. That is, it will be assumed that at $\rho = a$

$$E_z = E_0 \sin \beta a \left(\frac{\phi_0}{2} - \phi\right) \qquad 0 < \phi < \frac{\phi_0}{2}$$

$$E_z = E_0 \sin \beta a \left(\frac{\phi_0}{2} + \phi\right) \qquad -\frac{\phi_0}{2} < \phi < 0$$

$$E_z = 0 \qquad |\phi| > \frac{\phi_0}{2}$$

Expressing this function by the appropriate Fourier series, and equating it to an expression similar to (94) for E_z leads to the following expression for the field at any point (ρ, ϕ)

$$E_z = \frac{\beta a}{\pi} E_0 \sum_{n=-\infty}^{+\infty} \frac{\cos\left(n\frac{\phi_0}{2}\right) - \cos\left(\beta a \frac{\phi_0}{2}\right)}{H_n^{(2)}(\beta a)[(\beta a)^2 - n^2]} H_n^{(2)}(\beta \rho)\, e^{jn\phi} \qquad (13\text{-}100)$$

In Fig. 13-36 expression (100) has been evaluated for a narrow $3\lambda/4$ transverse slot in a $5\lambda/4$-diameter cylinder. Although the agreement between calculated and experimental patterns is not as close as for longitudinal slots, it is sufficiently good to predict approximate radiation patterns.

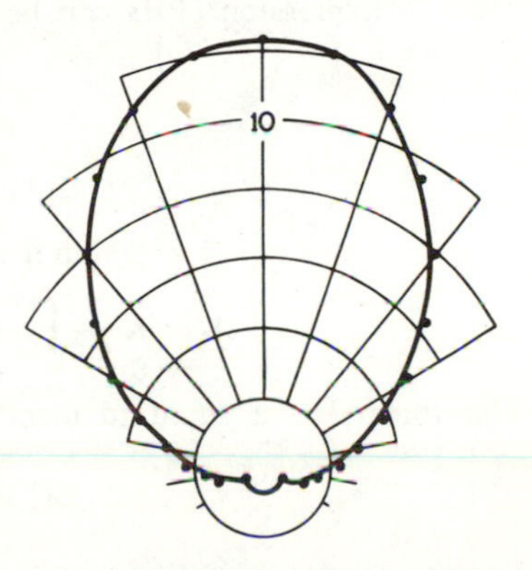

Figure 13-36. Experimental pattern of a $3\lambda/4$-long transverse slot in a $5\lambda/4$-diameter cylinder. Points are calculated.

13.20 Dipole and Slot Arrays around Cylinders.* A class of antenna arrays of practical interest in several different fields consists of an array of vertical or horizontal dipoles or slots about a vertical conducting cylinder. In practice, the "conducting cylinder" may be an existing structure such as the spire on a tall building, or part of the superstructure on a battleship, or it may be an actual cylinder constructed as part of the antenna system. The mathematical problem to be solved is the diffraction of waves of various polarizations by a conducting cylinder.

Line Source and Conducting Cylinder. Mathematically, the simplest problem of this type is the two-dimensional problem of an infinitely long line source parallel to an infinitely long conducting cylinder. Consider the problem of a very long wire carrying a uniform in-phase current I parallel to a very long conducting cylinder (Fig. 13-37). The field due to the current in the wire alone, without the cylinder, can first be obtained by integrating over the

*The approach used in this section is similar to that introduced by P. S. Carter (see Bibliography).

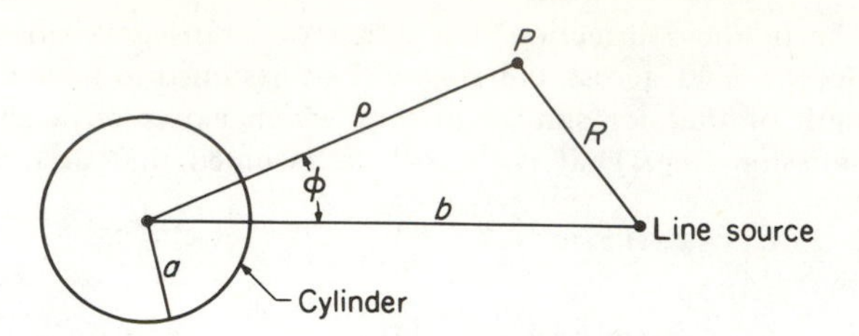

Figure 13-37. Line source parallel to a conducting cylinder.

length of the wire the expression for the vector potential due to a current element. For a point *P*, a distance ρ from the wire (located at the origin), the vector potential will be

$$A_z = \mu \frac{I}{4\pi} \int_{-\infty}^{+\infty} \frac{e^{-j\beta r}}{r}\, dz \tag{13-101}$$

where $r = \sqrt{\rho^2 + z^2}$ is the distance from the current element $I\,dz$ to the point *P*. Expression (101) can be integrated by changing the variable. Let

$$r = \rho \cosh \alpha$$

then

$$z^2 = \rho^2(\cosh^2 \alpha - 1) = \rho^2 \sinh^2 \alpha$$

$$z = \rho \sinh \alpha \qquad dz = \rho \cosh \alpha\, d\alpha = r\, d\alpha$$

$$A_z = \mu \frac{I}{4\pi} \int_{-\infty}^{+\infty} e^{-j\beta\rho \cosh \alpha}\, d\alpha \tag{13-102}$$

This integral is a standard form* and integrates to give

$$A_z = \frac{-j\mu I}{4} H_0^{(2)}(\beta\rho)$$

For the geometry of Fig. 13-37, the vector potential at *P* due to the line source alone will be

$$A_{z_1} = C H_0^{(2)}(\beta R)$$

where $\quad C = -\dfrac{j\mu I}{4} \quad$ and $\quad R = \sqrt{\rho^2 + b^2 - 2\rho b \cos\phi}$

The field due to the line source alone will be called the *primary wave.* The primary wave will induce currents in the conducting cylinder, and the field of these induced currents will be called the *secondary wave.* The total or resultant field at any point will be the sum of primary and secondary fields. The currents that are induced in the cylinder are of such magnitude and phase that the resultant electric field strength tangential to the (perfect) conducting cylinder is zero everywhere over the surface of the cylinder. In order

*E. Jahnke, F. Emde, and F. Lösch, *Tables of Higher Functions* (6th ed.), revised by F. Lösch, McGraw-Hill Book Company, New York, 1960.

to apply this boundary condition it is necessary to expand the primary wave in terms of a sum of cylindrical waves referred to the axis of the cylinder. Using the addition theorem* for Bessel functions, this expansion is given by

$$H_0^{(2)}(\beta R) = \sum_{n=0}^{\infty} \epsilon_n H_n^{(2)}(\beta b) J_n(\beta \rho) \cos n\phi \qquad \text{for } \rho < b$$

$$H_0^{(2)}(\beta R) = \sum_{n=0}^{\infty} \epsilon_n H_n^{(2)}(\beta \rho) J_n(\beta b) \cos n\phi \qquad \text{for } \rho > b$$

where ϵ_n is Neumann's number. ($\epsilon_n = 1$ for $n = 0$; $\epsilon_n = 2$ for $n \neq 0$)

The secondary waves that originate at the cylinder will be cylindrical waves, and the secondary field may be expressed as a sum of cylindrical waves originating at the axis of the cylinder. Thus for the secondary field

$$A_{z_2} = \sum_{n=0}^{\infty} \epsilon_n \mu b_n H_n^{(2)}(\beta \rho) \cos n\phi$$

The electric field strength parallel to the surface of the cylinder is given by

$$E_z = -j\omega A_z \tag{13-103}$$

since $\partial V/\partial z = 0$ for this case. At $\rho = a$, E_z (total) and therefore A_z (total) must be zero. Therefore

$$A_z\,(\text{total}) = A_{z_1} + A_{z_2} = \sum_{n=0}^{\infty} \epsilon_n \mu [H_n^{(2)}(\beta b) J_n(\beta a) + b_n H_n^{(2)}(\beta a)] \cos n\phi = 0$$

and therefore

$$b_n = -\frac{H_n^{(2)}(\beta b) J(\beta a)}{H_n^{(2)}(\beta a)}$$

Then at any point P, the total field will be given by

$$A_z\,(\text{total}) = \sum_{n=0}^{\infty} \epsilon_n \mu [H_n^{(2)}(\beta b) J_n(\beta \rho) + b_n H_n^{(2)}(\beta \rho)] \cos n\phi \quad (\rho < b) \tag{13-104}$$

or

$$A_z\,(\text{total}) = \sum_{n=0}^{\infty} \epsilon_n \mu \{H_n^{(2)}(\beta \rho)[J_n(\beta b) + b_n]\} \cos n\phi \qquad (\rho > b) \tag{13-105}$$

These expressions give the field due to the infinitely long line source and conducting cylinder. The electric field strength is obtained by using (103). The magnetic field strength is given by $\mu \mathbf{H} = \nabla \times \mathbf{A}$. The currents on the cylinder can be obtained by evaluating H at $\rho = a$.

Short Dipole near a Long Conducting Cylinder. A more practical problem than the one above is the case of a short dipole near a long conducting cylinder. If an attempt is made to solve this problem in the same manner as the preceding one, it is found that when the primary field of the dipole is expanded in cylindrical waves originating at the cylinder axis, the result is an infinite series in which each term of the series contains an infinite integral. The difficulties in evaluating such a series are formidable. Fortunately, the

*See. J. A. Stratton, *Electromagnetic Theory,* McGraw-Hill Book Company, New York, 1941, p. 372; S. A. Schelkunoff, *Electromagnetic Waves,* p. 300.

problem can be solved by another method, described by Carter,* which makes use of the reciprocity theorem.

In Carter's method the radiation pattern of the dipole at P near a conducting cylinder (Fig. 13-38) is obtained as a receiving antenna instead of as

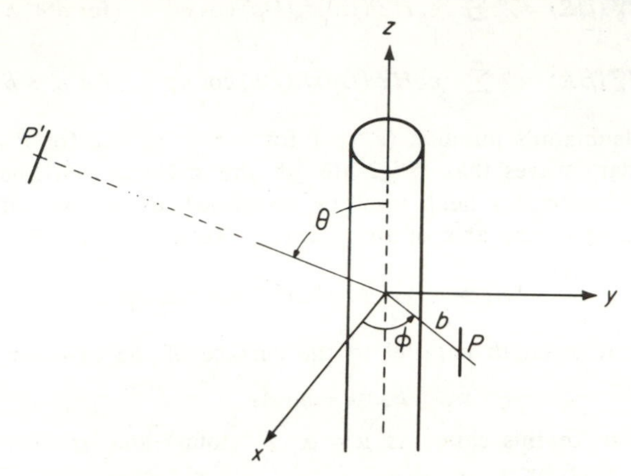

Figure 13-38. Dipole near a conducting cylinder.

a transmitting antenna. The wave received from a distant source will be essentially a *plane* wave that is easily expanded into a sum of standing cylindrical waves. Equating the sum of primary and secondary tangential electric fields to zero at the cylinder surface gives the magnitude of the secondary or re-radiated waves. The field at the dipole, and hence its open-circuit voltage, V_{oc}, can then be calculated. If it is assumed that the (essentially) plane wave is produced by a current I amperes flowing in a properly oriented distant dipole at P', application of the reciprocity principle shows that a current I amperes in the dipole at P will produce an open-circuit voltage, V_{oc}, at P'. In this manner the *distant* field of the transmitting dipole near the cylinder is determined. To obtain the complete radiation it is necessary to consider waves of both polarizations arriving at the cylinder.

Figure 13-38 shows a short vertical dipole near a very long conducting cylinder. Consider a wave, essentially plane, arriving at the cylinder from a distant dipole lying in the x-z plane, perpendicular to the radius vector, and at a polar angle θ. The magnetic field of such a wave will be horizontal and in the y direction, and can be represented by

$$\begin{aligned} H_y &= e^{j\beta(z \cos\theta + x \sin\theta)} \\ &= e^{j\beta(z \cos\theta + \rho \cos\phi \sin\theta)} \end{aligned} \tag{13-106}$$

where a wave of unit amplitude has been assumed. For this case the magnetic field will be entirely in the horizontal plane, with $H_z = 0$, so it will be possible to obtain $\mathbf{H}$ from a vector $\mathbf{A}' = \hat{\mathbf{z}}A_z$ which is everywhere parallel to the z axis. The appropriate relation for H_y is

*P. S. Carter, "Antenna Arrays Around Cylinders," *Proc. IRE,* **31**, 12, pp. 671–693 (1943).

$$\mu H_y = (\nabla \times \mathbf{A}')_y = -\frac{\partial A_z}{\partial x} \tag{13-107}$$

From (106) and (107)

$$A_{z_1} = -\mu \frac{e^{j\beta(x \sin\theta + z\cos\theta)}}{j\beta \sin\theta} = \frac{j\mu e^{j\beta(\rho \sin\theta \cos\phi + z\cos\theta)}}{\beta\sin\theta}$$

Using a standard Bessel function expansion

$$e^{j(\beta\rho\sin\theta\cos\phi)} = \sum_{n=0}^{\infty} \epsilon_n(j)^n J_n(\beta\rho\sin\theta)\cos n\phi$$

the expression for the vector potential due to the primary wave becomes

$$A_{z_1} = \frac{j\mu e^{j\beta z\cos\theta}}{\beta\sin\theta}\sum_{n=0}^{\infty} \epsilon_n(j)^n J_n(\beta\rho\sin\theta)\cos n\phi \tag{13-108}$$

In this expression the primary wave has been expanded in terms of standing waves (Bessel functions of the first kind) centered at the origin. The secondary waves, produced by induced currents on the cylinder, will have the same form, but will be outward traveling waves. For such waves the J_n's in (108) will be replaced by $H_n^{(2)}$'s, so that for the secondary waves

$$A_{z_2} = \frac{j\mu e^{j\beta z\cos\theta}}{\beta\sin\theta}\sum_{n=0}^{\infty} b_n\epsilon_n(j)^n H_n^{(2)}(\beta\rho\sin\theta)\cos n\phi$$

where the b_n's are arbitrary constants that must be evaluated from the boundary conditions. The total wave function is

$$A_z\,(\text{total}) = \frac{j\mu e^{j\beta z\cos\theta}}{\beta\sin\theta}\sum_{n=0}^{\infty} \epsilon_n(j)^n[J_n(\beta\rho\sin\theta) + b_n H_n^{(2)}(\beta\rho\sin\theta)]\cos n\phi \tag{13-109}$$

Recalling that

$$\nabla\cdot\mathbf{A} = -j\omega\mu\epsilon V$$

the electric field strength can be expressed in terms of the vector potential $\mathbf{A}$ by

$$\begin{aligned}\mathbf{E} &= -j\omega\mathbf{A} - \nabla V\\ &= -j\omega\mathbf{A} - \frac{j}{\omega\mu\epsilon}\nabla\nabla\cdot\mathbf{A}\end{aligned}$$

so that

$$\begin{aligned}E_z &= -j\omega A_z - \frac{j}{\omega\mu\epsilon}\frac{\partial^2 A_z}{\partial z^2}\\ &= \left[-j\omega - \frac{j}{\omega\mu\epsilon}(-\beta^2\cos^2\theta)\right]A_z = -j\omega\sin^2\theta\,A_z\end{aligned}$$

Since $E_z = 0$, and therefore $A_z = 0$ at $\rho = a$, it follows from (109) that

$$b_n = -\frac{J_n(\beta a\sin\theta)}{H_n^{(2)}(\beta a\sin\theta)}$$

Then the electric field strength at a vertical dipole located near the cylinder at $(b, \phi, 0)$ will be

$$E_z = \eta\sin\theta\sum_{n=0}^{\infty}\epsilon_n(j)^n\left[J_n(\beta b\sin\theta) - \frac{J_n(\beta a\sin\theta)}{H_n^{(2)}(\beta a\sin\theta)}H_n^{(2)}(\beta b\sin\theta)\right]\cos n\phi$$

By the reciprocity theorem this will also be the expression for the distant field of a vertical dipole, located a distance b from the cylinder. Therefore the relative radiation pattern for a short vertical dipole near a long vertical cylinder is

$$E_\theta = \sin\theta \sum_{n=0}^{\infty} \epsilon_n (j)^n \left[J_n(\beta b \sin\theta) - \frac{J_n(\beta a \sin\theta)}{H_n^{(2)}(\beta a \sin\theta)} H_n^{(2)}(\beta b \sin\theta) \right] \cos n\phi \quad (13\text{-}110)$$

Expression (110) gives the radiation pattern for all values of θ. For the special case of $\theta = 90$ degrees (the horizontal pattern) it should be observed that expression (110) gives the *same* pattern as was obtained with the infinitely long wire near the cylinder. This result, obtained here for a special case, is in fact quite general. For example, the horizontal pattern of a finite-length axial slot in the vertical cylinder is *independent* of the length of the slot, and is the same as the pattern for an infinitely long slot.

Expression (110) also gives (exactly) the vertical pattern of a short dipole near the cylinder. This result may be used as the approximate vertical pattern of a half-wave dipole at the same location. The vertical pattern of a vertical array of dipoles may be obtained by using this pattern as the unit pattern and by applying the principle of multiplication of patterns.

The radiation patterns of horizontal and radial dipoles also may be determined by this method. In general, it is necessary to consider both polarizations for the arriving plane wave.

Application to Slots in Cylinders. By replacing the electric field distribution across a slot by its equivalent magnetic-current sheet, it is evident that the above method should have direct application in obtaining the patterns of slots in cylinders. The problem of obtaining the radiation pattern of a slot in a cylinder is now just that of determining the field patterns produced by a magnetic dipole adjacent to the cylinder. The solution carries through, just as it did for the electric dipole, except for two differences. The distant dipole in this case will be a magnetic dipole, which results in an entirely horizontal electric field strength at the cylinder, so that the fields can be expressed in terms of an electric vector potential **F** which is in the z direction. When the boundary conditions are applied at the surface of the cylinder, they cannot be applied on H_z (corresponding to the application on E_z for the electric dipole), but must be applied to E_ϕ. When the problem is worked through, keeping these facts in mind, the expression obtained for H_θ due to a short axial slot in a long cylinder is

$$H_\theta = -\sin\theta \sum_{n=0}^{\infty} \epsilon_n (j)^n \left[J_n(\beta a \sin\theta) - J_n'(\beta a \sin\theta) \frac{H_n^{(2)}(\beta a \sin\theta)}{H_n^{(2)\prime}(\beta a \sin\theta)} \right] \cos n\phi \quad (13\text{-}111)$$

which should be compared with eq. (110). In expression (111) the magnetic dipole has been allowed to approach the surface of the cylinder so that $b = a$.

Remembering that $H_n^{(2)} = J_n - jN_n$ and using the following Bessel function relation*

$$J_n N_n' - J_n' N_n = J_{n+1} N_n - J_n N_{n+1} = \frac{2}{\pi x}$$

eq. (111) reduces to

$$H_\theta = K \sum_{n=0}^{\infty} \epsilon_n (j)^n \left[\frac{\cos n\phi}{H_n^{(2)\prime}(\beta a \sin \theta)}\right]$$

For $\theta = 90$ degrees

$$H_z = -H_\theta = -K \sum_{n=0}^{\infty} \epsilon_n (j)^n \frac{\cos n\phi}{H_n^{(2)\prime}(\beta a)} \tag{13-112}$$

Expression (112), which is for a short axial slot in a long cylinder, gives exactly the same pattern in the horizontal plane ($\theta = 90$ degrees) as was obtained for the infinitely long slot of eq. (99).

ADDITIONAL PROBLEMS

15. A coaxial line has an inner conductor of (outer) radius $a = \frac{1}{2}$ in., and an outer conductor of (inner) radius $b = 2$ in. Determine the power radiated at 100 MHz when the voltage across the open end is 1000 volts, and find the value of an equivalent resistance R that, when connected across the open end, would absorb the same amount of power as is radiated.

16. Derive the expressions (51, 52, and 53) for the electromagnetic field of a Huygens' source by direct use of eqs. (10) and (11).

17. Integrate the radiation fields due to all the Huygens' sources on an infinite plane to show that a plane wave results.

18. Using an "equivalent radius" $a = d/4$ for a flat-strip dipole of width d, calculate the approximate impedance of a slot in a large conducting plane at 300 MHz. The slot is 35 cm long and 1 cm wide.

19. Using Carter's method, derive eq. (111) for a short slot in a cylinder, following the procedure indicated in the text.

20. Verify that eq. (111) reduces to the form shown in (112).

21. From first principles prove that the horizontal pattern of an axial slot in an infinitely long vertical cylinder is independent of the length of the slot. Hint: Use the reciprocity theorem.

22. Verify Babinet's principle (the simple version used in optics) for the case of diffraction at a straightedge. That is, show that the vector sum of the two diffracted fields from two complementary straightedges is equal to the free-space field.

*S. A. Schelkunoff, *Electromagnetic Waves*, p. 56, eq. (7-13).

23. Prove that the specification of tangential **E** or tangential **H** at all points on a closed surface S (enclosing current sources) is sufficient to guarantee a unique field-strength solution everywhere inside S. This is one form of the *uniqueness theorem.* Hint: Assume that there exist two different solutions $\mathbf{E}^a$, $\mathbf{H}^a$ and $\mathbf{E}^b$, $\mathbf{H}^b$ and form the difference field. Then use the Poynting theorem, eq. (6-21), assuming that there is a small loss in the medium; the "lossless" situation can then be regarded as the limiting case in which the medium losses approach zero.

BIBLIOGRAPHY

Articles

Booker, H. G. and P. C. Clemmow, "The Concept of an Angular Spectrum of Plane Waves, and Its Relation to That of a Polar Diagram and Aperture Distribution," *JIEE*, **97**, Part III, pp. 11-17, 1949.

Carter, P. S., "Antenna Arrays Around Cylinders," *Proc. IRE*, **31**, No. 12, pp. 671-693, 1943.

Keller, J. B., "A Survey of Short-wavelength Diffraction Theory," in *Electromagnetic Theory and Antennas*, E. C. Jordan, ed., Part I, p. 3, Pergamon Press, Inc., New York, 1963.

Rumsey, V. H., "A Short Way of Solving Advanced Problems in Electromagnetic Fields and Other Linear Systems," *IEEE Trans.*, AP-11, No. 1, Jan., 1963.

Schelkunoff, S. A., "Some Equivalence Theorems of Electromagnetics and Their Application to Radiation Problems," *BSTJ*, **15**, No. 1, p. 92, 1936.

Silver, S., "Microwave Aperture Antennas and Diffraction Theory," *J. Opt. Soc. Am.*, **52**, No. 2, pp. 131-139, Feb., 1962.

Books

Andrews, C. L., *Optics of the Electromagnetic Spectrum*, Prentice-Hall, Inc., Englewood Cliffs, N. J., 1960.

Born, Max and E. Wolf, *Principles of Optics*, Second (revised) edition, Pergamon Press, Inc., New York, 1961.

Fradin, A. Z., *Microwave Optics*, R. C. Glass, trans., Pergamon Press, Inc., New York, 1961.

Hansen, R. C., *Microwave Scanning Antennas, Vol. 1, Apertures*, Academic Press Inc., New York, 1964.

Harrington, R. F., *Time-Harmonic Electromagnetic Fields*, McGraw-Hill Book Company, New York, 1961.

Jordan, E. C., ed., *Electromagnetic Theory and Antennas* (Proceedings of a Symposium held at Copenhagen, Denmark, June, 1962), Part I, Sec. A on Scattering and Diffraction Theory, Pergamon Press, Inc., New York, 1963.

Kline, Morris and I. W. Kay, *Electromagnetic Theory and Geometrical Optics*, Interscience Publishers, New York, 1965.

Schelkunoff, S. A., *Electromagnetic Waves*, D. Van Nostrand Co., Inc., Princeton, N. J., 1943.

Chapter 14

IMPEDANCE

14.01 Introduction. The notion of impedance is so basic to the work of the electrical engineer that sometimes he is inclined to forget its limitations (e.g., that it is defined for sinusoidal time variations, and, strictly speaking, applies only between closely spaced points). In electromagnetic wave propagation the notion of impedance has been extended to include the ratio of transverse components of electric and magnetic fields. In antenna theory the domains of fields and circuits come together. The antenna is the device that integrates the electric and magnetic fields to produce the voltages and currents required to actuate electrical devices. In this transition from fields to circuits it is necessary to re-examine critically the notion of impedance so that meaningful expressions involving both field and circuit quantities can be derived. From these expressions impedance can be calculated for a specified configuration of conductors and dielectrics.

In this chapter, first the impedance and then mutual impedance of antennas will be calculated by the induced-emf method. Applied to circuits this method consists of assuming a certain current flow, computing the voltages induced by the assumed current in the various parts of the circuit, and then summing these back-voltages around the circuit to determine the voltage that must be impressed at the input terminals to produce the assumed current. The application to antennas is similar but more complicated because all parts of the "circuit" interact with each other. Next the impedance-per-unit-length of a wire will be determined for the simple case of a long straight wire. Finally, the conventional relations of circuit theory will be developed from field theory applied to a circuit. This last development immediately exposes to view the power and the limitations of the circuit approach.

14.02 Induced-emf Method of Calculating Impedance. In chap. 10 the radiation resistance of an antenna was obtained by the Poynting vector method. It was also shown that only the inverse-distance or $1/r$ terms contribute to a *net* flow of power away from the antenna.

Consequently the integration of the Poynting vector to obtain radiated power can be carried out over any enclosing surface, either near to the antenna, or at great distance where the $1/r^2$ and $1/r^3$ terms are customarily discarded as being negligible compared with the $1/r$ terms. Because of the considerable simplification that results in the computation, a sphere of very large radius is ordinarily chosen for the surface of integration, so that the "Poynting vector method" is normally associated with the computation of radiation resistance using the *far* fields of the antenna. However, exactly the same result for radiation resistance is obtained if the surface of integration is made to coincide with the surface of the antenna. This result can be shown (sec. 14.07) to be identical with that obtained by an alternative approach known as the *induced-emf method*, described below.

In the induced-emf method the electric field E_z produced parallel to the antenna by a known (or assumed) current distribution is calculated and used to compute the voltage $E_z\,dz$ induced in each element dz of the antenna. Then the reciprocity theorem is invoked to determine the voltage at the antenna terminals. This induced-emf method was originally introduced* for the computation of radiation resistance, but because it deals with the near fields of the antenna, it can also be used to calculate mutual impedance between antennas and the reactance of a single antenna. In the interest of conceptual simplicity the induced-emf method will be used first in sec. 14.04 to compute the mutual impedance between antennas. The method will then be used in secs. 14.05 and 14.06 to determine approximately both the resistance and reactance of single thin antennas, using the sinusoidal current assumption.

14.03 Mutual Impedance between Antennas. When two or more antennas are used in an array, the driving-point impedance of each antenna depends upon the self-impedance of that antenna and in addition upon the mutual impedance between the given antenna and each of the others. For example, consider the two-element array of Fig. 14-1 in which base currents I_1 and I_2 flow. As far as voltages and currents at the terminals (1) and (2) are concerned, the two antennas of Fig. 14-1(*a*) can be represented by the general four-terminal network of Fig. 14-1(*b*).

Z_{11} is the impedance measured at the terminals (1) with terminals (2) open; that is, Z_{11} is the mesh impedance of mesh (1). Similarly Z_{22} is the mesh impedance of mesh (2), and is the impedance that would be measured at terminals (2) with (1) open. $Z_{12} = Z_{21}$ is the

*A. A. Pistolkors, The "Radiation Resistance of Beam Antennas," *Proc. IRE,* **17,** 562, March, 1929.

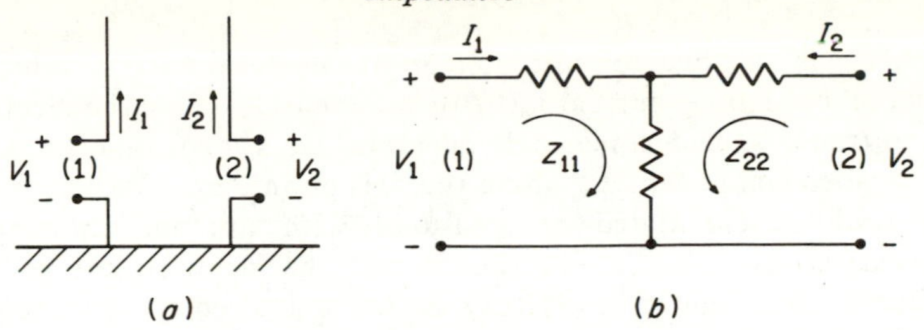

Figure 14-1.

mutual impedance between the antennas and is defined in both figures by

$$Z_{21} = \frac{V_{21}}{I_1} \qquad Z_{12} = \frac{V_{12}}{I_2}$$

where V_{21} is the open-circuit voltage induced across terminals (2) of antenna (2) owing to current I_1, flowing (at the base) in antenna (1). Similarly V_{12} is the open-circuit voltage across the teminals of antenna (1) owing to current I_2, flowing in antenna (2). Under most conditions the impedance Z_{11} is approximately equal to the *self-impedance* Z_{S_1} of antenna (1). The self-impedance of an antenna is its input impedance with all other antennas entirely removed. Except when antenna (2) is very near a resonant length (that is, $H \approx \lambda/2$) or when it is very close to antenna (1), the input impedance of antenna (1) will be nearly the same with antenna (2) open-circuited as it would be with antenna (2) entirely removed from the field of antenna (1).

The mesh equations for Fig. 14-1 are

$$\begin{aligned} V_1 &= I_1 Z_{11} + I_2 Z_{12} \\ V_2 &= I_1 Z_{21} + I_2 Z_{22} \end{aligned} \tag{14-1}$$

Let $r = I_1/I_2$, where in general r is a complex number. Then

$$\frac{V_1}{I_1} = Z_{11} + \frac{1}{r} Z_{12} \tag{14-2}$$

$$\frac{V_2}{I_2} = rZ_{21} + Z_{22} \tag{14-3}$$

It is seen that the input impedances, V_1/I_1 and V_2/I_2 are dependent upon the current ratio r. These are the impedances that any impedance-transforming networks must be designed to feed, and in order to calcuate them the mutual impedance must be known.

14.04 Computation of Mutual Impedance. The mutual impedance between the antennas of Fig. 14-1 is defined by

$$Z_{21} = \frac{V_{21}}{I_1(0)}$$

where V_{21} is the open-circuit voltage at the terminals of antenna (2) produced by a base current $I_1(0)$ in antenna (1). Now the electric field due to (the assumed sinusoidal) current I_1 is known, and in particular the parallel component $E_{z_{21}}$ along the axis of antenna (2) (which *incident* field would be calculated or measured with antenna (2) temporarily removed) is given by eq. (10-72). Hence the open-circuit voltage at the terminals of antenna (2) may be written directly from eq. (11-6) or eq. (13-46) as

$$\mathrm{V}_{21} = -\frac{1}{I_2(0)}\int_0^{H_2} E_{z_{21}} I_2(z)\, dz \tag{14-4}$$

so that the expression for mutual impedance becomes

$$Z_{21} = -\frac{1}{I_1(0)\, I_2(0)}\int_0^{H_2} E_{z_{21}} I_2(z)\, dz \tag{14-5}$$

In expression (5), $I_2(z)$ is the current distribution along antenna (2) for the transmitting case, and may be written approximately as

$$I_2(z) = I_{2m} \sin\beta\,(H_2 - z) = \frac{I_2(0)}{\sin\beta H_2} \sin\beta\,(H_2 - z) \tag{14-6}$$

Inserting eqs. (10-72) and (6) in (5), the expression (4) for mutual impedance becomes

$$Z_{21} = -\frac{30 I_{1m} I_{2m}}{I_1(0)\, I_2(0)}\int_0^{H_2} \left(\frac{-j\, e^{-j\beta r_1}}{r_1} - \frac{j\, e^{-j\beta r_2}}{r_2} + \frac{2j\cos\beta H_1 e^{-j\beta r_0}}{r_0}\right) \sin\beta(H_2 - z)\, dz \tag{14-7}$$

Equation (7) gives the mutual impedance *referred to the base*. The mutual impedance referred to the loop currents will be given by expression (7) multiplied by the ratio of the product of base currents to the product of loop currents, that is, by

$$\frac{I_1(0)\, I_2(0)}{I_{1m} I_{2m}}$$

Therefore the mutual impedance referred to loop currents will be

$$Z_{21} = -30\int_0^{H_2} \left(\frac{-j\, e^{-j\beta r_1}}{r_1} - \frac{j\, e^{-j\beta r_2}}{r_2} + \frac{2j\cos\beta H_1 e^{-j\beta r_0}}{r_0}\right) \sin\beta(H_2 - z)\, dz \tag{14-8}$$

It is usually this mutual impedance referred to the current loops that is plotted and shown in curves. When this impedance is known, the mutual impedance referred to the base or terminal currents can be easily calculated.

In Fig. 14-2 are two monopole antennas of height H mounted on a perfect reflecting plane and spaced a distance d apart. For this case

$$\left.\begin{aligned} r_0 &= \sqrt{d^2 + z^2} \\ r_1 &= \sqrt{d^2 + (H - z)^2} \\ r_2 &= \sqrt{d^2 + (H + z)^2} \end{aligned}\right\} \qquad (14\text{-}9)$$

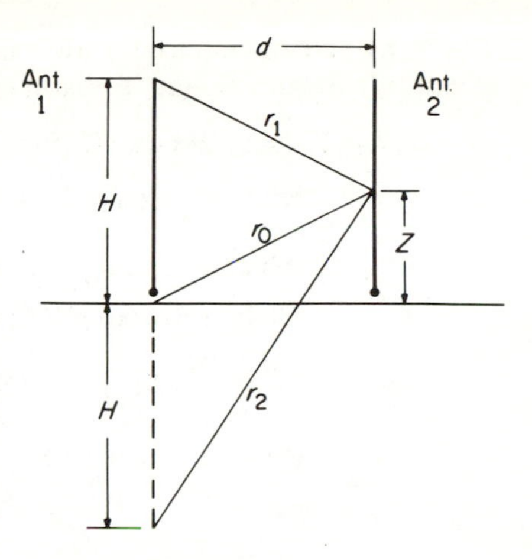

Figure 14-2.

Expression (8) for mutual impedance is complex. The real part gives the mutual resistance, and the imaginary part gives the mutual reactance. Substituting the relations (9) in the real part of (8) gives an expression for mutual resistance.

$$R_{21} = 30 \int_0^H \sin \beta(H - z) \left[\frac{\sin \beta \sqrt{d^2 + (H - z)^2}}{\sqrt{d^2 + (H - z)^2}} + \frac{\sin \beta \sqrt{d^2 + (H + z)^2}}{\sqrt{d^2 + (H + z)^2}} - 2 \cos \beta H \frac{\sin \beta \sqrt{d^2 + z^2}}{\sqrt{d^2 + z^2}} \right] dz \qquad (14\text{-}10)$$

Similarly the imaginary part of (8) yields the expression for mutual reactance

$$X_{21} = 30 \int_0^H \sin \beta(H - z) \left[\frac{\cos \beta \sqrt{d^2 + (H - z)^2}}{\sqrt{d^2 + (H - z)^2}} + \frac{\cos \beta \sqrt{d^2 + (H + z)^2}}{\sqrt{d^2 + (H + z)^2}} - \frac{2 \cos \beta H \cos \sqrt{d^2 + z^2}}{\sqrt{d^2 + z^2}} \right] dz \qquad (14\text{-}11)$$

The integrations indicated in (10) and (11) can be carried out in a manner that will be detailed in sec. 14.05. The case of mutual resistance between quarter-wave monopoles is not too difficult and is left as an exercise. The result for quarter-wave monopoles spaced at distance d is

$$R_{21} = 15 \left[2 \operatorname{Ci} \beta d - \operatorname{Ci} \left(\sqrt{(\beta d)^2 + \pi^2} - \pi\right) - \operatorname{Ci} \left(\sqrt{(\beta d)^2 + \pi^2} + \pi\right)\right] \qquad (14\text{-}12)$$

The expression for mutual reactance between quarter-wave monopoles is

$$X_{21} = 15 \left[\operatorname{Si} \left(\sqrt{(\beta d)^2 + \pi^2} - \pi\right) + \operatorname{Si} \left(\sqrt{(\beta d)^2 + \pi^2} + \pi\right) - 2 \operatorname{Si} (\beta d)\right] \qquad (14\text{-}13)$$

The general expressions for the mutual impedance between antennas of (equal) height H and a distance d apart are*

$$R_{21} = 30\,[\sin\beta H\cos\beta H\,(\mathrm{Si}\,u_2 - \mathrm{Si}\,v_2 - 2\,\mathrm{Si}\,v_1 + 2\,\mathrm{Si}\,u_1)$$
$$- \frac{\cos 2\beta H}{2}(2\,\mathrm{Ci}\,u_1 - 2\,\mathrm{Ci}\,u_0 + 2\,\mathrm{Ci}\,v_1 - \mathrm{Ci}\,u_2 - \mathrm{Ci}\,v_2)$$
$$- (\mathrm{Ci}\,u_1 - 2\,\mathrm{Ci}\,u_0 + \mathrm{Ci}\,v_1)] \tag{14-14}$$

$$X_{21} = -\,30\,[\sin\beta H\cos\beta H\,(2\,\mathrm{Ci}\,v_1 - 2\,\mathrm{Ci}\,u_1 + \mathrm{Ci}\,v_2 - \mathrm{Ci}\,u_2)$$
$$- \frac{\cos 2\beta H}{2}(2\,\mathrm{Si}\,u_1 - 2\,\mathrm{Si}\,u_0 + 2\,\mathrm{Si}\,v_1 - \mathrm{Si}\,u_2 - \mathrm{Si}\,v_2)$$
$$- (\mathrm{Si}\,u_1 - 2\,\mathrm{Si}\,u_0 + \mathrm{Si}\,v_1)] \tag{14-15}$$

where $u_0 = \beta d$

$$u_1 = \beta(\sqrt{d^2 + H^2} - H) \qquad u_2 = \beta(\sqrt{d^2 + (2H)^2} + 2H)$$
$$v_1 = \beta(\sqrt{d^2 + H^2} + H) \qquad v_2 = \beta(\sqrt{d^2 + (2H)^2} - 2H)$$

Because the evaluation of these expressions is a laborious task analog methods† and digital computers have been employed. Figure 14-3 shows the mutual impedance as a function of spacing between equal-height antennas. Curves for the mutual impedance between antennas of unequal height may be found in the last reference cited.

14.05 Radiation Resistance by Induced-emf Method. By definition the mutual impedance Z_{21} between two antennas is a measure of the open-circuit voltage V_{21} at terminals (2) due to a current $I_1(0)$ impressed at the terminals of antenna (1). If we consider two identical antennas and allow the distance d between them (Fig. 14-2) to approach zero, the voltage V_{21} becomes the voltage at the terminals of antenna (1) due to the impressed current $I_1(0)$ (supplied by an infinite-impedance generator). Hence the self-impedance Z_1 of an antenna of half-length H should be equal to the mutual impedance Z_{21} between two antennas for which $H_1 = H_2 = H$, and $d = 0$. That is, the expression for self-impedance becomes

$$Z_1 = Z_{21} \qquad (\text{for } d = 0, H_1 = H_2 = H)$$
$$= -\frac{1}{I^2(0)}\int_0^H E_z\,I(z)\,dz \tag{14-16}$$

*P. S. Carter, "Circuit Relations in Radiating Systems and Applications to Antenna Problems," *Proc. IRE*, **20,** 1004 (1932); J. Labus, "Mathematical Calcuation of the Impedance of Antennas," *Hochfrequenz. Technik*, **41,** 17 (1933); G. H. Brown and R. King, "High Frequency Models in Antenna Investigations," *Proc. IRE*, **22,** 457 (1934).

†E. C. Jordan and W. L. Everitt, "Acoustic Models of Radio Antennas," *Proc. IRE*, **29,** 186 (1941).

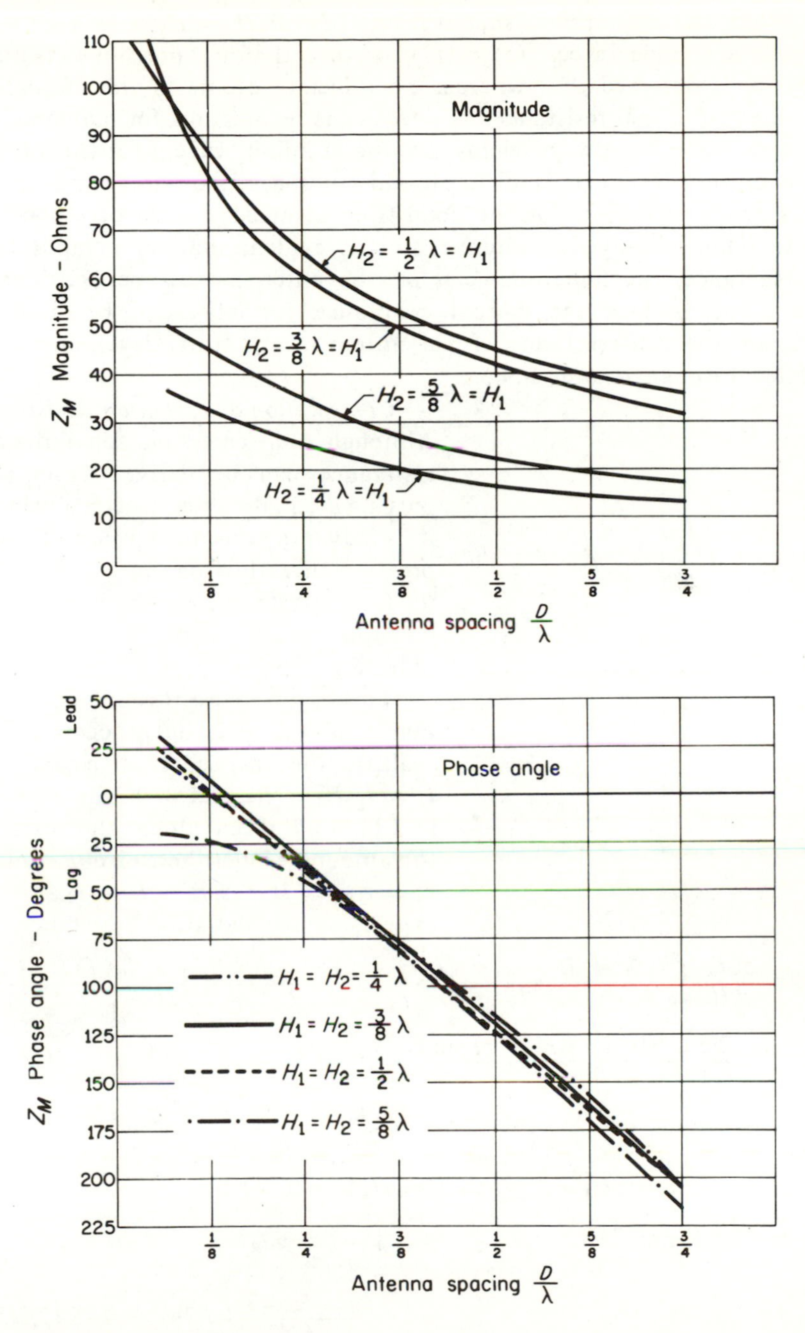

Figure 14-3. The mutual impedance (referred to the current loops) between monopole antennas of equal height.

When the appropriate substitutions ($d = 0, H_1 = H_2 = H$ and $I_1 = I_2 = I$) are made in eqs. (5) or (7), or directly in expressions (14) and (15), it is observed that whereas no difficulty occurs in the evaluation of the radiation resistance $R_r = R_{21}$, the expression for antenna reactance $X_1 = X_{21}$ always yields a value of infinity except for the special cases of $H_1 = H_2 = H$ equal to an odd multiple of a quarter-wavelength. This result is correct for the conditions assumed ($d = 0$, corresponding to an infinitesimally thin antenna) so it is evident that for computation of reactance, the finite diameter of the antenna must be considered. When this is done the mutual reactance formulas can be used to estimate the self-reactance of an antenna, and this development is carried through in sec. 14.06.

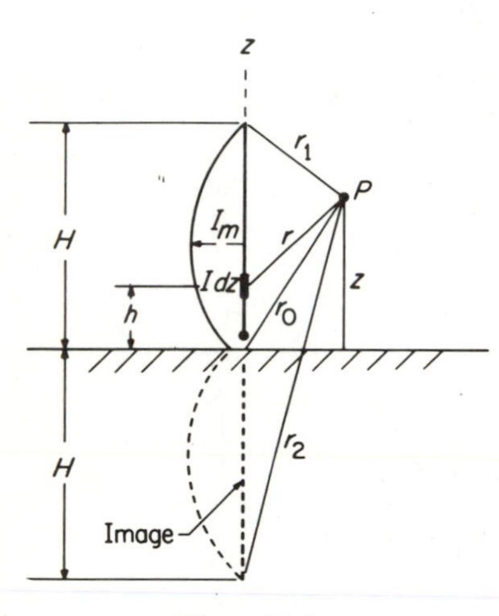

Figure 14-4.

Computation of Radiation Resistance. Although the expression for radiation resistance can be derived from (14) putting d equal to zero (see problem 8), it is advantageous to develop the expression directly from eq. (7) in order to demonstrate the evaluation of the integrals which occur frequently in radiation problems.

To obtain an expression for radiation resistance it is only necessary to evaluate the real part of eq. (7) for $I_1 = I_2, H_1 = H_2$, and with the point P in Fig. 14-4 taken along the axis of the antenna. Under these circumstances $r_0 = z, r_1 = H - z, r_2 = H + z$, and the expression for radiation resistance is

$$R_r = \frac{30I_m^2}{I^2(0)}\int_0^H \left[\frac{\sin\beta(H-z)\sin\beta(H-z)}{H-z} + \frac{\sin\beta(H-z)\sin\beta(H+z)}{H+z} - \frac{2\cos\beta H\sin\beta(H-z)\sin\beta z}{z}\right] dz \tag{14-17}$$

This expression can be integrated term by term. Consider the first term and let $u = \beta(H - z)$, $du = -\beta\, dz$, $du/u = -dz/(H - z)$ so that when $z = 0, u = \beta H$, and when $z = H, u = 0$. Then

$$\int_0^H \frac{\sin\beta(H-z)\sin\beta(H-z)}{H-z}\, dz = -\int_{\beta H}^0 \frac{\sin^2 u}{u}\, du$$

$$= \frac{1}{2}\int_0^{2\beta H} \frac{1-\cos(2u)}{(2u)}\, d(2u) = \frac{1}{2}S_1(2\beta H)$$

where the function $S_1(x)$ is defined by eq. (10-67).

To integrate the second term of eq. (3) let

$$w = 2\beta(H + z) \qquad dw = 2\beta dz \qquad 2\beta z = w - 2\beta H$$

Then $\displaystyle\int_0^H \frac{\sin\beta(H - z)\sin\beta(H + z)}{H + z}\,dz$

$$= \frac{1}{2}\int_0^H \frac{\cos 2\beta z - \cos 2\beta H}{(H + z)}\,dz$$

$$= -\frac{1}{2}\int_{2\beta H}^{4\beta H} \frac{\cos 2\beta H - \cos(w - 2\beta H)}{w}\,dw$$

$$= -\frac{1}{2}\int_{2\beta H}^{4\beta H} \frac{\cos 2\beta H(1 - \cos w) - \sin w \sin 2\beta H}{w}\,dw$$

$$= -\frac{1}{2}\cos 2\beta H\int_{2\beta H}^{4\beta H} \frac{1 - \cos w}{w}\,dw + \frac{1}{2}\sin 2\beta H\int_{2\beta H}^{4\beta H} \frac{\sin w}{w}\,dw$$

$$= -\frac{1}{2}\cos 2\beta H\left(\int_0^{4\beta H} \frac{1 - \cos w}{w}\,dw - \int_0^{2\beta H} \frac{1 - \cos w}{w}\,dw\right)$$

$$+ \frac{1}{2}\sin 2\beta H\left(\int_0^{4\beta H} \frac{\sin w}{w}\,dw - \int_0^{2\beta H} \frac{\sin w}{w}\,dw\right)$$

$$= -\tfrac{1}{2}\cos 2\beta H\,[S_1(4\beta H) - S_1(2\beta H)] + \tfrac{1}{2}\sin 2\beta H\,[\mathrm{Si}(4\beta H) - \mathrm{Si}(2\beta H)]$$

where Si (x) is the sine integral of x and is defined by eq. (10-66).

The third term of eq. (3) can be integrated almost directly. It is

$$-2\cos\beta H\int_0^H \frac{\sin\beta(H - z)\sin\beta z}{z}\,dz$$

$$= -\cos\beta H\int_0^H \frac{\cos\beta(H - 2z) - \cos\beta H}{z}\,dz$$

$$= \cos\beta H\int_0^{2\beta H} \frac{\cos\beta H(1 - \cos 2\beta z) - \sin\beta H\sin 2\beta z}{2\beta z}\,d(2\beta z)$$

$$= \cos^2\beta H S_1(2\beta H) - \cos\beta H\sin\beta H\,\mathrm{Si}\,(2\beta H)$$

Using the notation $b = 2\beta H$, the value A of the integral is given by

$$A = \tfrac{1}{2}\{S_1(b) - [S_1(2b) - S_1(b)]\cos b + [\mathrm{Si}(2b) - \mathrm{Si}(b)]\sin b + (1 + \cos b)\,S_1(b) - \sin b\,\mathrm{Si}\,(b)\} \tag{14-18}$$

Inserting (18) in eq. (17) the radiation resistance is

$$R_r = \frac{30 I_m^2}{I^2(0)}\,A \tag{14-19}$$

This is the radiation resistance referred to the terminal or base current $I(0)$. Frequently it is convenient to specify the radiation resistance *referred to the loop current* I_m. Because the radiated power W_r is the same in both cases,

$$W_r = \tfrac{1}{2} I_m^2 \, R_r\,(\text{loop}) = \tfrac{1}{2} I^2(0)\, R_r\,(\text{base}) \tag{14-20}$$

where I_m and $I(0)$ are both peak values in time. Using (18), (19) and (20) and with some rearrangement of terms the radiation resistance *referred to the loop current* is

$$R_r\,(\text{loop}) = 15\,[(2 + 2\cos b)\, S_1(b) - \cos b S_1(2b) - 2 \sin b \,\text{Si}\,(b) + \sin b\, \text{Si}\,(2b)] \tag{14-21}$$

This radiation resistance referred to the current loop is plotted in Fig. 14-5 for monopole antennas as a function of height H. For center-fed dipole antennas of half-length H, the values of resistance

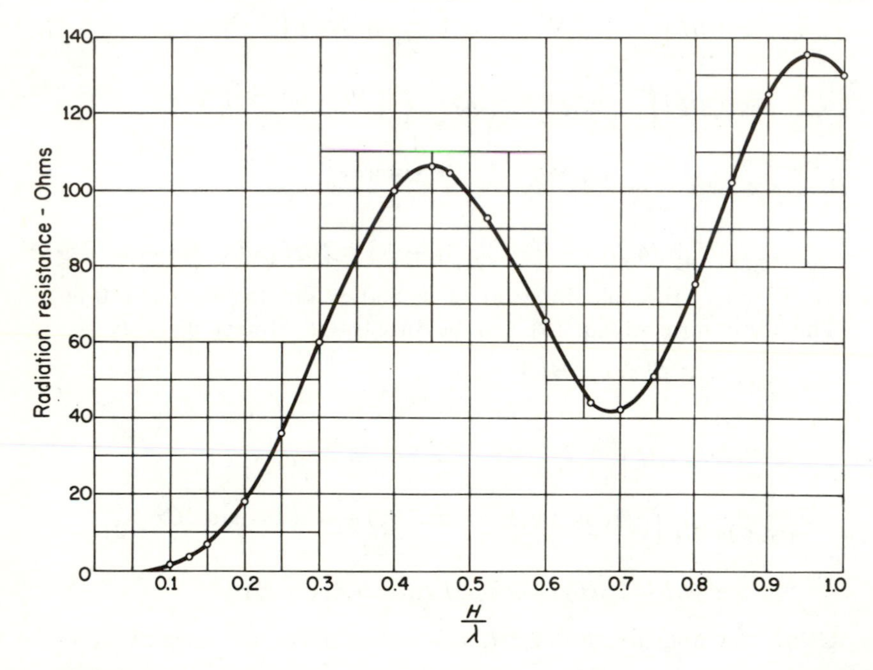

Figure 14-5. Radiation resistance (referred to the current loop) of a monopole antenna as a function of antenna height H/λ.

should be multiplied by 2. For example, for a quarter-wave monopole mounted on a reflecting plane the radiation resistance is shown as 36.5 ohms. For a half-wave dipole (in free space) it would be 73 ohms. For a half-wave monopole and a center-fed full-wave dipole the values of radiation resistance are 99.5 and 199 ohms, respectively.

Relation between R_r *(base) and* R_r *(loop).* For an antenna having the sinusoidal current distribution, the base or terminal current is related to the loop current by

$$I(0) = I_m \sin \beta H$$

Therefore, for this case,

$$R_r\,(\text{base}) = \frac{I_m^2}{I^2(0)} R_r\,(\text{loop})$$
$$= \frac{R_r\,(\text{loop})}{\sin^2 \beta H} \tag{14-22}$$

For antenna lengths for which H is a multiple of a half-wavelength, the assumed sinusoidal distribution gives a value of zero for the current at the feed point, and eq. (22) indicates that the input resistance will be infinite. For these lengths the actual input current will be small but not zero, and the input resistance will be large but not infinite. Although a value of infinity may be regarded as the first approximation to the actual resistance, it is a worthless approximation for practical purposes. For these cases it becomes necessary to use a method involving a better approximation than the sinusoidal for current distribution. Such methods are considered briefly in chap. 11, and in more depth later in this chapter. We see that the sinusoidal current distribution provides useful answers over a certain range of antenna lengths, but there are other ranges in which the approximation fails. In general this is true of every approximate method, and it is necessary for the engineer always to consider the limitations as well as the capabilities of any method he may employ.

14.06 Reactance of an Antenna. The reactance of an antenna having an assumed sinusoidal current distribution can be obtained in the same manner as that employed in sec. 14.05 for resistance, by using the imaginary part of eq. (7). As mentioned previously, when this computation is carried through with E_z evaluated along the axis of the antenna it is found that except for certain particular lengths $[H = (2n + 1)\lambda/4]$, a value of infinity is obtained for the reactance. This result is not incorrect, for in sec. 14.10 we shall see that the reactance per unit length of a very long wire approaches infinity as the radius of the wire approaches zero. We shall also show in sec. 14.10 that for a wire of finite radius the reactance is determined by evaluating E_z at the surface of the wire. It is evident that in order to obtain a more useful approximation than infinity for the reactance of an antenna, the finite diameter must be considered. This requires that E_z be evaluated at the radial distance a (where a is the radius of the antenna), rather that along the axis.

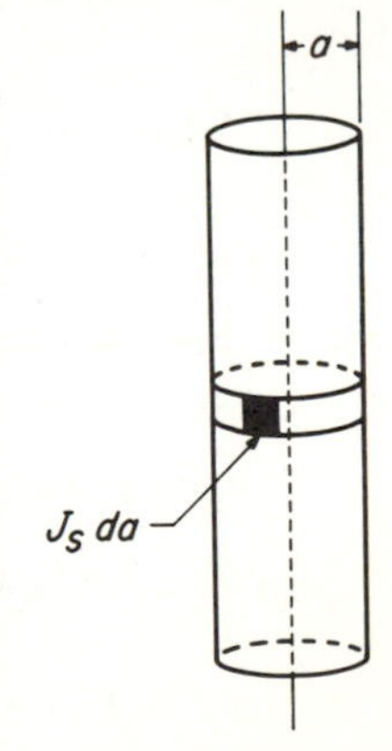

Figure 14-6.

Figure 14-6 shows a length of a cylindrical antenna of radius a, which is assumed to be carrying a sinusoidal distribution of current along its length. The current is uniformly distributed around the circumference of the cylinder, the major portion of it flowing in a very small thickness of conductor adjacent to the outer surface. For purposes of computation it may be assumed that the electric field at the surface, calculated from the actual current distributed around the cylinder, would be the same as that which can be calculated by considering the current to be concentrated along a filament at the axis of the cylinder. Then E_z will be evaluated along a line parallel to the axis and separated from it by a distance a. The expression for reactance will then be given by the imaginary part of eq. (7), with $H_1 = H_2 = H$, and $d = a$. That is, the reactance of an antenna of radius a, having an assumed sinusoidal current distribution, is equal to the mutual reactance between two filamentary antennas of the same height and with a spacing equal to a. Substituting in eq. (15) gives for the reactance

$$X = -30 \Big[\sin \beta H \cos \beta H (2 \,\mathrm{Ci}\, v_1 - 2 \,\mathrm{Ci}\, u_1 + \mathrm{Ci}\, v_2 - \mathrm{Ci}\, u_2) - \frac{\cos 2\beta H}{2} (2 \,\mathrm{Si}\, u_1 - 2 \,\mathrm{Si}\, u_0 + 2 \,\mathrm{Si}\, v_1 - \mathrm{Si}\, u_2 - \mathrm{Si}\, v_2) - (\mathrm{Si}\, u_1 - 2 \,\mathrm{Si}\, u_0 + \mathrm{Si}\, v_1)\Big] \tag{14-23}$$

where $u_0 = \beta a$

$$u_1 = \beta(\sqrt{H^2 + a^2} - H) \qquad u_2 = \beta(\sqrt{(2H)^2 + a^2} + 2H)$$

$$v_1 = \beta(\sqrt{H^2 + a^2} + H) \qquad v_2 = \beta(\sqrt{(2H)^2 + a^2} - 2H)$$

The radius of the antenna will normally be a very small fraction of a wavelength so that

$$\beta a \ll 1$$

and the following approximations may be used.

$$\mathrm{Si}\, (\beta a) = \beta a - \frac{(\beta a)^3}{3 \cdot 3!} + - \ldots \approx \beta a \approx 0$$

$$\mathrm{Ci}\, \beta(\sqrt{H^2 + a^2} - H) \approx \mathrm{Ci}\, \beta H \left(1 + \frac{a^2}{2H^2} - 2\right) \approx \mathrm{Ci} \left(\frac{\beta a^2}{2H}\right)$$

$$\mathrm{Ci}\, \beta(\sqrt{(2H^2 + a^2} - 2H) \approx \mathrm{Ci} \left(\frac{\beta a^2}{4H}\right)$$

$$\mathrm{Si}\, u_1 \approx 0 \qquad \mathrm{Si}\, v_1 \approx \mathrm{Si}\, 2\beta H \qquad \mathrm{Ci}\, v_1 \approx \mathrm{Ci}\, (2\beta H)$$

$$\mathrm{Si}\, v_2 \approx 0 \qquad \mathrm{Si}\, u_2 \approx \mathrm{Si}\, (4\beta H) \qquad \mathrm{Ci}\, u_2 \approx \mathrm{Ci}\, (4\beta H)$$

Using these approximations, expression (23) becomes

$$X = -30 \left\{\sin \beta H \cos \beta H \left[2 \,\mathrm{Ci}\, 2\beta H - 2 \,\mathrm{Ci} \left(\frac{\beta a^2}{2H}\right) + \mathrm{Ci} \left(\frac{\beta a^2}{4H}\right) - \mathrm{Ci}\, 4\beta H\right] - \frac{\cos 2\beta H}{2} [2 \,\mathrm{Si}\, 2\beta H - \mathrm{Si}\, 4\beta H] - \mathrm{Si}\, 2\beta H\right\}$$

Now when x is very small, $\mathrm{Ci}\, x \approx \gamma + \ln x$ where $\gamma = 0.5772\ldots$ is Euler's constant. Using this substitution, the second and third terms of the above expression may be combined

$$-2\,\mathrm{Ci}\left(\frac{\beta a^2}{2H}\right) + \mathrm{Ci}\left(\frac{\beta a^2}{4H}\right) = -\gamma + \ln\left(\frac{H}{\beta a^2}\right)$$

so that the final expression for the reactance of a monopole antenna of radius a and length H becomes

$$X = -15\left\{\sin 2\beta H\left[-\gamma + \ln\left(\frac{H}{\beta a^2}\right) + 2\,\mathrm{Ci}\,(2\beta H) - \mathrm{Ci}\,(4\beta H)\right]\right.$$
$$\left. - \cos 2\beta H\,[2\,\mathrm{Si}\,(2\beta H) - \mathrm{Si}\,(4\beta H)] - 2\,\mathrm{Si}\,(2\beta H)\right\} \qquad (14\text{-}24)$$

For the particular case of a quarter-wave antenna, $\sin 2\beta H = 0$, $\cos 2\beta H = -1$, and the expression for reactance reduces to

$$\begin{aligned} X &= 15\,\mathrm{Si}\,(4\beta H) \\ &= 15\,\mathrm{Si}\,(2\pi) = 21.25 \text{ ohms} \end{aligned}$$

Expression (24) gives the reactance (referred to the current loop) of a monopole antenna of length H and radius a as given by the induced-emf method, using the sinusoidal current distribution assumption. The reactance referred to the base can be obtained from (24) by dividing by $\sin^2 \beta H$. Figure 14-7 shows resistance and reactance values computed by this method for short monopole antennas of different thicknesses. The resistance or reactance of the corresponding dipole antennas of length $L = 2H$ is just double that of the monopole antenna of length H.

We see that under the assumed conditions of sinusoidal current distribution, a quarter-wavelength antenna has a positive reactance of 21.25 ohms, and this value of reactance is independent of antenna diameter as long as the latter is small in wavelengths. For lengths other than multiples of the quarter-wavelength, the reactance depends very greatly on the antenna diameter; it is very large for thin antennas. This fact indicates the desirability of using fat antennas for broadband applications such as television, where a low ratio of antenna reactance to resistance (low Q) is required.

It will also be seen that, as the antenna length is varied, the reactance goes through zero for some length shorter than a quarter-wavelength. This means that the "resonant" length is always somewhat less than a quarter wavelength, being shorter for fat antennas.

Problem 1. Verify that as the spacing d approaches zero, the expression (14-14) for mutual resistance between two antennas of equal height reduces to the expression for the radiation resistance of a single antenna.

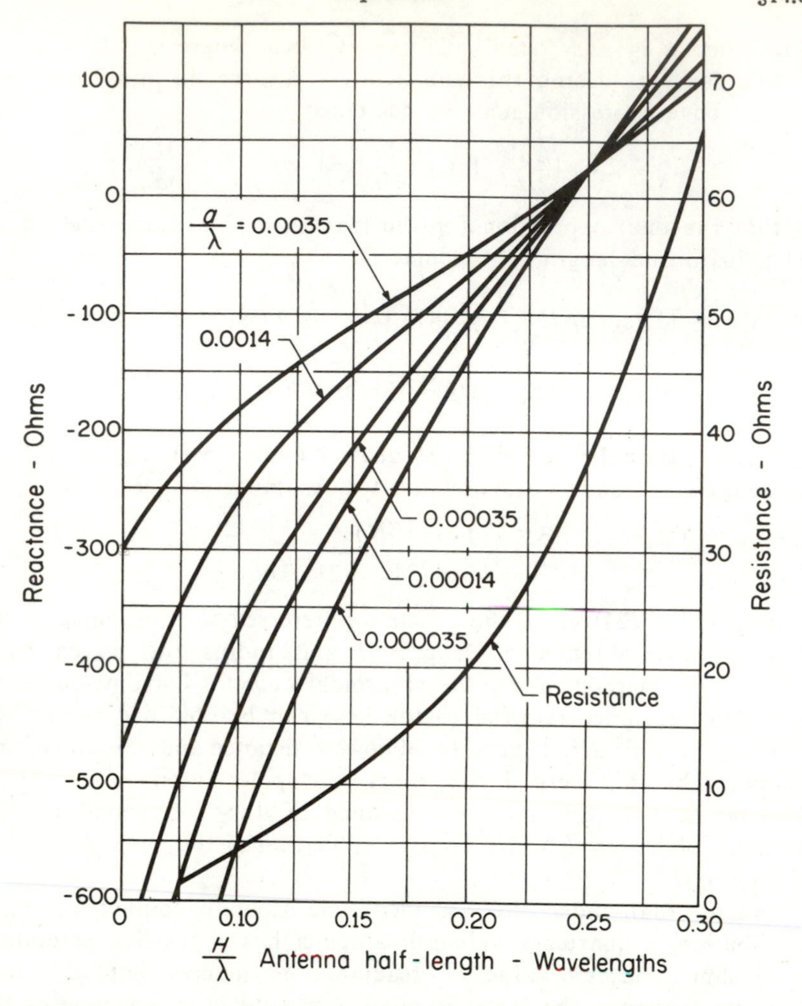

Figure 14-7. The resistance and reactance (referred to base) of short monopole antennas computed by the induced-emf method.

14.07 Equivalence of Induced-emf and Poynting Vector Methods.

It is easy to show that the induced-emf and Poynting vector methods for computing radiated power are one and the same method when the surface of integration coincides with the surface of the antenna. Consider an antenna of length L and radius R, which has some arbitrary current distribution I_z. The components of $\mathbf{E}$ and $\mathbf{H}$, tangential to the surface along the length of the antenna, are E_z and H_ϕ. At the top and bottom ends the tangential components are E_r and H_ϕ. If the real Poynting vector is integrated over the surface of the antenna, the following result is obtained:

$$\text{Re}\int_S \mathbf{E} \times \mathbf{H}^* \cdot d\mathbf{a} = \int_0^L |E_z||H_\phi| \cos \psi \, 2\pi R \, dz + 2\int_0^R |E_r||H_\phi| \cos \psi \, 2\pi r \, dr$$

where ψ is the time-phase angle between the tangential components of **E** and **H**.

The first integral covers the entire surface of the antenna except the end caps. The second integral covers these end caps. The quantity $2\pi RH_\phi$ is equal to the line integral of H_ϕ about the antenna and, by Maxwell's first equation, this is equal to the total current flowing through the closed path, so that

$$2\pi RH_\phi = \oint H \, dl = I_z$$

where I_z is the current along the antenna. Using this relation, the first integral becomes

$$\int_0^L |E_z||I_z| \cos \psi \, dz$$

The angle ψ is now the time-phase angle between E_z and I_z, because I_z and H_ϕ are in time phase.

In the end caps the current flows radially and must be zero at the center. Denoting by J_r the radial surface current density in the caps, the relation $\mathbf{H} = \mathbf{J} \times \hat{\mathbf{n}}$ becomes $H_\phi = J_r$ for the top and bottom caps. Then $2\pi rH_\phi = 2\pi rJ_r = I_r$ is the total radial current flowing across a circle of radius r on each of the caps. Using this relation the second integral becomes

$$2\int_0^R |E_r||I_r| \cos \psi \, dr$$

The total surface integral may then be written

$$\text{Re}\int_S \mathbf{E} \times \mathbf{H}^* \cdot d\mathbf{a} = \int_0^L |E_z||I_z| \cos \psi \, dz + 2\int_0^R |E_r||I_r| \cos \psi \, dr$$

It is evident that the contribution to the radiated power from the end surfaces of the antenna must be very small, since the current there is very small and in such directions that the various current elements produce radiation fields which cancel one another. Therefore, the second term of the above integral is usually dropped, and the power is obtained from

$$\int_0^L |E_z||I_z| \cos \psi \, dz$$

This expression, obtained by integrating the Poynting vector over the surface of the antenna, is seen to be identical with the real part of the induced-emf integral in eq. (5) from which radiation resistance and radiated power were computed.

14.08 Note on the Induced-emf Method. The radiation resistance and power radiated by an antenna have been calculated by two methods known as the Poynting vector and induced-emf methods. In the preceding section it has been shown that when the surface of integration in the Poynting vector method is taken to be the surface of the antenna these methods lead to identical results. Despite this demonstrated equivalence, certain questions inevitably are raised concerning the induced-emf method, and it is worthwhile to explore these questions in some depth by a detailed comparison of the two methods.

(*a*) *Poynting Vector Method.* A certain current distribution is assumed to exist along the antenna. The electric and magnetic field strengths due to the assumed current distributions are computed at a point P on some surface enclosing the antenna. The net outward flow of power through this surface is obtained by integrating the Poynting vector $\mathbf{E} \times \mathbf{H}$ over the entire surface and over a cycle in time. In practice, the enclosing surface is usually chosen to be a sphere of very large radius, so that the difference in distance to various points on the antenna affects only the phase, and not the magnitude, of the contributions to the total field, thus simplifying the computation. A sinusoidal current distribution is usually assumed, and the method is in error only by the amount that the *radiation* fields, produced by the actual current distribution, differ from the radiation fields calculated from this assumed sinusoidal distribution. Inasmuch as the actual current distribution is known to be very nearly sinusoidal for thin transmitting antennas, the answer obtained is a good approximation to the true power radiated.

The calculation is usually made assuming filamentary current, but the results hold for finite-diameter antennas as long as the diameter is very small compared with the length and compared with a wavelength.

(*b*) *Induced-emf Method.* In the second method a filamentary current distribution is assumed as before and the resulting electric and magnetic field strengths are computed. However, in this case the point P_1, at which the fields are computed, is taken right at the filament. Then, using reciprocity, the voltage at the antenna terminals produced by all the emf's induced along the length of the antenna is computed. The product of this terminal voltage and the in-phase terminal current represents power radiated, and this product is shown to be equal to $|E_z||I|\,dl \cos \psi$ integrated over the length of the antenna, where ψ is the time-phase angle between E_z and I at the point in question. This method gives exactly the same value for power radiated as the previous

Poynting vector method. This is as it should be because, as has been shown, this method can be derived directly from the Poynting vector method. The approximation involved in this method is the same as in the Poynting vector method and assumes a sinusoidal current distribution, whereas the actual current distribution is only approximately sinusoidal.

Although the induced-emf method is essentially the same as the Poynting vector method (when applied at the surface of the antenna) and gives exactly the same results, its validity is sometimes questioned when it is applied to an actual antenna having a finite diameter. The reason for this is as follows: If the antenna is assumed to be a perfect conductor (the usual assumption), the boundary conditions require that the total electric field strength **E** along the surface of the antenna be zero. In the case of a transmitting antenna, excited by a lumped voltage across a gap, the only electric field existing along the surface of the antenna is the field E_z, induced by the currents and charges along the antenna. The boundary conditions require that this electric field strength be zero everywhere on the surface, and therefore the product $|E_z||I|\,dl \cos\psi$ is zero at every point along the antenna. Then $|E_z||I|\,dl\cos\psi$ integrated along the antenna is zero and the computed power radiated from *the conducting part of the antenna* is zero. This also is as it should be, because the conductor contains in itself no source of electromagnetic energy, the energy coming from the generator. However, there are two questions raised that require clarification.

(1) Since the actual $|E_z||I|\,dl\cos\psi$ that exists along the surface is zero and, therefore, not even approximately the same as $|E_z||I|\,dl\cos\psi$, computed from the assumed sinusoidal distribution, is there any justification for expecting that the value given by the computed $\int|E_z||I|\,dl\cos\psi$ is even approximately correct?

(2) Since the actual $\int|E_z||I|\,dl\cos\psi$ over the surface of the conductor is zero, an incidental question is "from where is the power radiated?"

The answer to the first question regarding the validity of the method can be readily obtained by considering initially a receiving antenna of resonant length that has the load terminals *a-b* short-circuited and which, therefore, re-radiates all the received energy. Assume first that the current flowing in the antenna owing to the received electric field has a true sinusoidal distribution. The self-induced electric field strength or "back-voltage" due to this current flow (and the corresponding charge distribution) can be calculated in the usual manner and will be designated by E_s. (The subscript s indicates that this is the electric field strength computed from the assumed sinusoidal distribution.) Then, *if* the received or applied tangential field—which will be designated by E'—were exactly equal and opposite to E_s at all points along the surface of the antenna, the assumed current would flow in the antenna. The boundary conditions at the surface of the antenna would be satisfied because the total electric field strength parallel to it would be $E = E' + E_s = 0$. The power re-radiated by the antenna is obtained by integrating the Poynting vector over the surface of the antenna. As pointed out above, this is equal to $|E_s||I|\, dl \cos\psi$ integrated along the length of the antenna. Similarly, the power per unit length flowing into the antenna from the received field is $-|E'||I|\, dl \cos\psi$ (outward flow of energy is assumed positive). The net flow of power out of the antenna, which is the difference between these two, is equal to zero.

Next consider same short-circuited receiving antenna under conditions where the received field E' does not have the particular configuration required in the above case, but instead has some arbitrary value along the length of the antenna. In particular, consider the case where E' is *uniform*, as it would be for reception of a plane wave at $\theta = 90$ degrees. Then the current distribution will *not* be sinusoidal, and the actual current distribution will be such as to produce a self-induced field E'' along the antenna, such that $E' + E'' = 0$. That is, E'' will be uniform or constant along the antenna and will have a value $E'' = -E'$. Now, although the actual current distribution cannot be sinusoidal, it is known to be very closely sinusoidal for the resonant length. Evidently then, it requires but a very small change in current distribution from the sinusoidal to change the self-induced parallel component of E from that calulated for the sinusoidal current cases, E_s, to the value $E'' = -E'$ that must exist in the actual case. Since the current distribution is but little changed from the sinusoidal, the power radiated for a given loop current must be very nearly equal to the case for the true sinusoidal distribution. (Small changes in current produce only small changes in the *radiation* terms of the electric field.) The actual power re-radiated in this case is $\int |E''||I_a|\, dl \cos\psi$, where I_a is the actual current and E'' is the self-induced parallel component of electric field due to it. But from the previous statement this must be very nearly equal to the power re-radiated in the sinusoidal case, which is $\int |E_s||I_s|\, dl \cos\psi$. That is,

$$\int |E''||I_a|\, dl \cos\psi = \int -|E'||I_a|\, dl \cos\psi \approx \int |E_s||I_s|\, dl \cos\psi$$

This means that, although the actual current is not sinusoidal and the actual self-induced voltage $E'' = -E'$ differs *greatly* from that calculated from a sinusoidal distribution, nevertheless, the radiated power computed from an assumed sinusoidal distribution with its resulting E_s gives an answer that is very close to that which would be obtained from $\int -|E'||I_a|\,dl\cos\psi$ if the actual current I_a were known. However, it should be noted that this is true only because the actual current distribution is nearly sinusoidal.

Finally, consider the case of a transmitting antenna in which the applied electric field is concentrated over a short section at the center.

If V is the applied voltage and S is the separation of the terminals *a-b* (Fig. 14-8), then the field across *a-b* can be considered to be $E' = V/S$. The applied field is zero everywhere else along the antenna. The actual current that flows in the antenna as a result of the applied voltage V must be such as to produce a self-induced electric field opposite to the applied field everywhere along the antenna. That is, the self-induced field must be zero everywhere along the antenna except between a and b, where it has a value of $-V/S$. It is an experimental fact that the actual antenna current that flows and necessarily produces the above electric field distribution, is *very closely sinusoidal* for thin transmitting antennas. Therefore, as in the discussion of receiving antennas, the radiated power computed from $\int |E_s||I_s|\,dl\cos\psi$ must be very nearly equal to the actual power radiated. In this case, the actual power radiated is

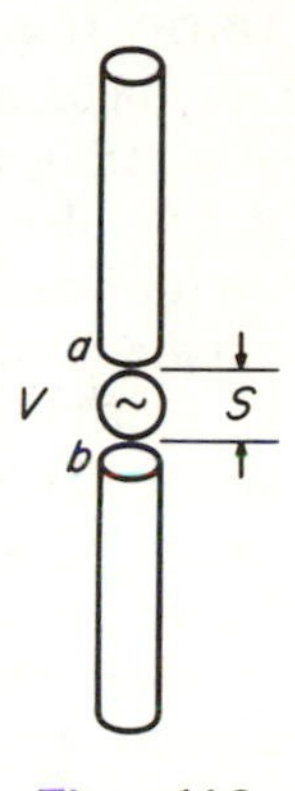

Figure 14-8.

$$\int_0^L -|E'||I_a|\,dl\cos\psi = \int_a^b -|E'||I_a|\,dl\cos\psi = |V||I_0|\cos\theta$$

where I_0 is the current at the feed point and θ is the angle between V and I. Therefore, actual power radiated $= |V||I_0|\cos\theta \approx \int_0^L |E_s||I_s|dl\cos\psi$.

It should be noted in passing that this latter integration should be performed over the whole of the antenna including the section between a and b. However, since E_s between a and b is of the same order of magnitude as E_s at adjacent points on the antenna, the error incurred in neglecting the section *a-b* becomes very small when the gap length is small compared with the length of the antenna. However, the situation is very different in the case of the actual current distribution with the resulting actual distribution of the self-induced field. In this latter case, the integral is zero everywhere, except at the gap or generator. As the gap is made very small, the actual E' across it becomes very large for a given applied voltage V and the gap generator section cannot be neglected. Indeed, it may be said that all the power flows out from this generator section, being guided into space by the antenna conductors.

The above discussion has attempted to give a physical explanation* of an interesting and somewhat puzzling result, viz., that the impedance of an antenna calculated from the induced-emf integral expression (16) and using the sinusoidal current distribution yields a useful approximation, even though E_z calculated from the assumed current distribution differs markedly from the true electric field tangential to a finite-diameter antenna. The mathematical explanation for this result is that the input impedance as calculated from eq. (16) is stationary with respect to variations in $I(z)$, subject to the condition that the input current is constant.†

14.09 The Self-Impedance Formula: Reciprocity Derivation. In secs. 14.05 and 14.06 the input resistance and reactance of a perfectly conducting antenna were obtained by evaluating the mutual impedance of two identical filamentary antennas and then allowing the two antennas to come together. Both the resistance and the reactance may be calculated by this method if the two current filaments remain separated by a distance equal to the radius of the actual antenna wire. It was shown that the resistance calculation (but not the reactance) may be simplified considerably with negligible loss in accuracy by allowing the distance between the two filaments to go to zero.

The expression for the self-impedance of a perfectly conducting antenna also may be derived directly from the field statement of the reciprocity theorem without introducing mutual impedance. Reciprocity involves two source-field situations which are chosen so as to permit the derivation of a useful result. There is a great deal of freedom in the choice of the two situations and for the impedance problem one possible choice is shown in Fig. 14-9. Case (*a*) shows a perfectly conducting antenna located in free space and fed by a voltage source (magnetic current $\mathbf{M}^a$). Case (*b*) shows an auxiliary current density $\mathbf{J}^b$ flowing in free space and occupying the same volume (inside surface S) as the antenna of case (*a*). The reciprocity theorem for an infinite region may be applied to the two cases provided they exist in identical environments. The environments may be made identical simply by replacing the conducting antenna material of case (*a*) with its induced current density $\mathbf{J}^a$ flowing in free space. It is important to remember that the use of induced current has no effect on the total fields; for

*The clarification of the induced-emf method is due to R. E. Burgess, "Aerial Characteristics," *Wireless Engineer*, **21,** No. 247, p. 154, 1944.

†For a discussion of this point see S. A. Schelkunoff, *Advanced Antenna Theory*, John Wiley & Sons, Inc., New York, 1952, pp. 137–138.

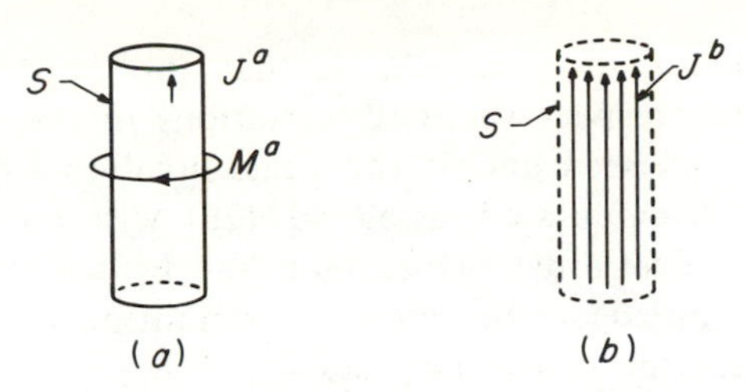

Figure 14-9. Calculation of self-impedance using (*a*) a transmitting antenna fed by a voltage source (magnetic current) and (*b*) an auxiliary current distribution.

instance, in case (*a*) the total field in free space is the sum of the fields due to the sources $\mathbf{M}^a$ and $\mathbf{J}^a$, and thus the total tangential electric field on the surface S must be zero just as it would be in the presence of the perfectly conducting antenna wire. In most cases of interest the terminal region is very small and thus the fields of the magnetic current source are highly localized.

The general statement of the reciprocity theorem gives the relation

$$\int \mathbf{E}^a \cdot \mathbf{J}^b \, dV = \int \mathbf{E}^b \cdot \mathbf{J}^a \, dV - \int \mathbf{H}^b \cdot \mathbf{M}^a \, dV \tag{13-36}$$

The left-hand side of eq. (13-36) is zero because $\mathbf{E}^a$ is zero within S and the tangential component of $\mathbf{E}^a$ is zero on S. The second integral on the right-hand side may be expressed in terms of V^a, the applied voltage and I^b, the total auxiliary current passing through the terminal region (see sec. 13.06). Thus eq. (13-36) becomes

$$-I^b V^a = \int \mathbf{E}^b \cdot \mathbf{J}^a \, dV \tag{13-36a}$$

The applied voltage may be expressed as $V^a = I^a Z^a$ in which I^a is the terminal current in the transmitting case and Z^a is the antenna impedance. Substitution of this last expression into eq. (13-36a) gives a very general expression for the antenna impedance,

$$Z^a = -\frac{1}{I^a I^b} \int \mathbf{E}^b \cdot \mathbf{J}^a \, dV \tag{14-25}$$

An important special case of eq. (25) arises when the current distributions in cases (*a*) and (*b*) are identical, that is when $\mathbf{J}^b = \mathbf{J}^a = \mathbf{J}$ (and consequently $I^b = I^a = I$). If $\mathbf{E}$ is defined as the field of the induced current $\mathbf{J}$ (flowing in free space), then eq. (25) becomes the widely used impedance formula

$$Z^a = -\frac{1}{I^2} \int \mathbf{E} \cdot \mathbf{J} \, dV \tag{14-26}$$

The particular advantage of eq. (26) is that the impedance given by it is *stationary* with respect to small variations in the current density **J**. That is, if **J** is guessed and if the guess involves a first-order error, then the impedance calculated using eq. (26) will contain a second-order error and no first-order error. Thus the procedure of calculating impedance from a guess at the current distribution can lead to results sufficiently accurate for many purposes.*

The Filamentary Current Approximation. For a thin wire antenna fed at the center it is known that the longitudinal current distribution is approximately sinusoidal and therefore suitable for use in eq. (26). Because the current flows on the surface of a good conductor, **J** is in the form of a hollow cylindrical sheet of the same radius as the wire. This approximation for **J** may be inserted easily into eq. (26); much more difficult, however, is the calculation of **E** due to this cylindrical current distribution.†

The field of a current *filament* is much easier to calculate than the field of a hollow cylindrical "pipe" of current. The longitudinal electric field of a filamentary, sinusoidal current distribution has a particularly simple analytical form, given by eq. (10-72). In order to take advantage of this simplicity it is necessary to return to the general impedance expression, eq. (25). In this equation one may use the cylindrical sinusoidal current sheet for $\mathbf{J}^a$ and the filamentary, sinusoidal current for $\mathbf{J}^b$ from which $\mathbf{E}^b$ may be obtained readily using eq. (10-72). The approximate expression for impedance then may be found by integrating $\mathbf{E}^b \cdot \mathbf{J}^a$ as indicated in eq. (25). Due to the cylindrical symmetry of the problem it is not necessary to regard $\mathbf{J}^a$ as a hollow pipe of current; rather it may be regarded as a current filament spaced the radius of the wire from the current filament $\mathbf{J}^b$. This "picture" of two closely spaced current filaments is the same as that described in the induced-emf method and the calculated impedances are identical.

It should be pointed out that when $\mathbf{J}^a$ and $\mathbf{J}^b$ are different, eq. (25) is not stationary with respect to small changes in $\mathbf{J}^a$. However, if $\mathbf{J}^a$ and $\mathbf{J}^b$ are chosen to have the same total current so that they differ only in their cross-sections, then for thin wire antennas eq. (25) is very nearly stationary and thus is insensitive to small errors in the assumed longitudinal current distribution. It is interesting to note that the use of the filamentary current for $\mathbf{J}^b$ would result in the correct impedance if the antenna current $\mathbf{J}^a$ were known exactly.

*The stationary property of the impedance formula is discussed in the books by Schelkunoff, Harrington, Collin and King, all listed at the end of the chapter.

†Some notion of the difficulties involved may be obtained by reading the paper by O. Zinke, "Fundamentals of Voltage and Current Distributions along Antennas," *Arch. Elektrotech.*, **35,** pp. 67–84 (1941).

14.10 Uniform Cylindrical Waves and the Infinitely Long Wire. In foregoing sections the impedances of finite-length antennas have been computed by the induced-emf method, using an assumed sinusoidal current distribution. A simpler problem is that of determining the impedance per unit length of an infinitely long wire, which is assumed to carry a uniform, in-phase current I. Although this may appear to be a rather unreal situation, it can be approximated in practice by a very long wire that is excited by a parallel electric field of constant value. This particular problem has the definite advantage that its solution is simple enough to permit easy interpretation. Before solving it a brief discussion of uniform cylindrical waves will be in order.

For a homogeneous medium having the constants μ, ϵ, and σ, Maxwell's equations in cylindrical co-ordinates are

$$\left.\begin{aligned} \frac{\partial H_z}{\rho\,\partial\phi} - \frac{\partial H_\phi}{\partial z} &= (\sigma + j\omega\epsilon)\,E_\rho & \frac{\partial E_z}{\rho\,\partial\phi} - \frac{\partial E_\phi}{\partial z} &= -j\omega\mu H_\rho \\ \frac{\partial H_\rho}{\partial z} - \frac{\partial H_z}{\partial \rho} &= (\sigma + j\omega\epsilon)\,E_\phi & \frac{\partial E_\rho}{\partial z} - \frac{\partial E_z}{\partial \rho} &= -j\omega\mu H_\phi \\ \frac{\partial(\rho H_\phi)}{\rho\,\partial\rho} - \frac{\partial H_\rho}{\rho\,\partial\phi} &= (\sigma + j\omega\epsilon)\,E_z & \frac{\partial(\rho E_\phi)}{\rho\,\partial\rho} - \frac{\partial E_\rho}{\rho\,\partial\phi} &= -j\omega\mu H_z \end{aligned}\right\} \qquad (14\text{-}27)$$

For fields that have no variation with ϕ or z, such as would be generated by an infinitely long wire carrying a uniform current I, eqs. (27) reduce to

$$\begin{aligned} \frac{\partial H_z}{\partial\rho} &= -\,(\sigma + j\omega\epsilon)\,E_\phi & \frac{\partial E_z}{\partial\rho} &= j\omega\mu H_\phi \\ \frac{\partial(\rho H_\phi)}{\rho\,\partial\rho} &= (\sigma + j\omega\epsilon)E_z & \frac{\partial(\rho E_\phi)}{\rho\,\partial\rho} &= -j\omega\mu H_z \end{aligned} \qquad (14\text{-}28)$$

The waves obtained with these fields are *uniform cylindrical* waves, having no variation of amplitude or phase over any cylindrical surface represented by $\rho = \rho_0$. It is evident that uniform cylindrical waves are transverse electromagnetic, and that they may be divided into two types, viz., (a) those having E_z and H_ϕ components, and (b) those having E_ϕ and H_z components. The former would be generated by the infinitely long wire mentioned, whereas the latter would be produced by an infinitely long line of closely spaced coaxial loops carrying equal and uniform currents that are everywhere in phase.

Considering the first of these types, the two relations

$$\frac{\partial(\rho H_\phi)}{\rho\,\partial\rho} = (\sigma + j\omega\epsilon)E_z \qquad \frac{\partial E_z}{\partial\rho} = j\omega\mu H_\phi \qquad (14\text{-}29)$$

can be combined to yield a wave equation in cylindrical co-ordinates.

$$\frac{\partial^2 E_z}{\partial\rho^2} + \frac{1}{\rho}\frac{\partial E_z}{\partial\rho} - \gamma^2 E_z = 0 \tag{14-30}$$

where as usual $\gamma^2 = j\omega\mu(\sigma + j\omega\epsilon)$

For the special case of wave propagation in a nonconducting medium, $\sigma = 0$ and $\gamma^2 = -\omega^2\mu\epsilon$, so that the wave equation becomes

$$\frac{\partial^2 E_z}{\partial\rho^2} + \frac{1}{\rho}\frac{\partial E_z}{\partial\rho} + \beta^2 E_z = 0 \tag{14-31}$$

where $\beta = \omega\sqrt{\mu\epsilon}$ is a real number. Dividing through by β^2 in (31) shows it to be an ordinary Bessel equation of order zero with the independent variable $(\beta\rho)$:

$$\frac{\partial^2 E_z}{\partial(\beta\rho)^2} + \frac{1}{(\beta\rho)}\frac{\partial E_z}{\partial(\beta\rho)} + E_z = 0 \tag{14-32}$$

As in sec. 8.05 the general solution may be written in terms of zero-order Bessel functions of the first and second kinds.

$$E_z = AJ_0(\beta\rho) + BN_0(\beta\rho) \tag{14-33}$$

In this form the solution represents standing waves. An alternative solution may be written in terms of linear combinations of J_0 and N_0.

$$E_z = A_1 H_0^{(1)}(\beta\rho) + B_1 H_0^{(2)}(\beta\rho) \tag{14-34}$$

where

$$\left.\begin{aligned} H_0^{(1)}(\beta\rho) &= J_0(\beta\rho) + jN_0(\beta\rho) \\ H_0^{(2)}(\beta\rho) &= J_0(\beta\rho) - jN_0(\beta\rho) \end{aligned}\right\} \tag{14-35}$$

$H_0^{(1)}$ and $H_0^{(2)}$ are called *Hankel functions* of zero order, first and second kinds, respectively. When appropriately combined with the time factor $e^{j\omega t}$, these functions represent inward- and outward-*traveling* waves respectively. That this is so, is evident from the asymptotic expressions for large values of $(\beta\rho)$. These expressions are:

$$\left.\begin{aligned} H_0^{(1)}(\beta\rho) &\to \sqrt{\frac{2}{\pi\beta\rho}}\, e^{j(\beta\rho - \pi/4)} \\ H_0^{(2)}(\beta\rho) &\to \sqrt{\frac{2}{\pi\beta\rho}}\, e^{-j(\beta\rho - \pi/4)} \end{aligned}\right\} \quad \text{for } \beta\rho \to \infty \tag{14-36}$$

which should be compared with the corresponding asymptotic expressions for J_0 and N_0.

$$\left.\begin{aligned} J_0(\beta\rho) &\to \sqrt{\frac{2}{\pi\beta\rho}} \cos\left(\beta\rho - \frac{\pi}{4}\right) \\ N_0(\beta\rho) &\to \sqrt{\frac{2}{\pi\beta\rho}} \sin\left(\beta\rho - \frac{\pi}{4}\right) \end{aligned}\right\} \quad \text{for } \beta\rho \to \infty \tag{14-37}$$

It is also apparent from (35) that the Hankel functions bear a relation to the Bessel functions similar to the relation between the exponential functions (with imaginary exponents) and the trigonometric (sine and cosine) functions.

For propagation in a conducting medium solutions to eq. (30) will be required. Dividing through by γ^2 in (30) shows it to be a *modified Bessel equation* of order zero in the variable $(\gamma\rho)$

$$\frac{\partial^2 E_z}{\partial(\gamma\rho)^2} + \frac{1}{\gamma\rho}\frac{\partial E_z}{\partial(\gamma\rho)} - E_z = 0 \tag{14-38}$$

Solutions to this modified Bessel equation are called *modified Bessel functions* and are denoted by $I_0(\gamma\rho)$ and $K_0(\gamma\rho)$ (for the zero order.) Expressions for the I and K functions are given in Appendix II. For *small* values of $(\gamma\rho)$,

$$\left.\begin{aligned} I_0(\gamma\rho) &\to 1 \\ K_0(\gamma\rho) &\to -[\ln(\gamma\rho) + C - \ln 2] \end{aligned}\right\} \quad \text{for } \gamma\rho \to 0 \tag{14-39}$$

Since eq. (30) reduces to (31) when γ is a pure imaginary, it is not surprising to find that the modified and ordinary Bessel functions are related to each other. The relations are

$$\begin{aligned} I_0(jz) &= J_0(z) \\ K_0(jz) &= \frac{\pi}{2j}[J_0(z) - jN_0(z)] \\ &= -\frac{\pi}{2}[N_0(z) + jJ_0(z)] \end{aligned} \tag{14-40}$$

The modified functions I and K are most suitable for propagation in a dissipative medium. For a lossless medium, for which γ is a pure imaginary, the corresponding Bessel or Hankel functions are usually more convenient.

Field about an Infinitely Long Wire. Consider now the electromagnetic field about a long wire carrying a current I. In the region external to the wire the Hankel function solutions of eq. (31) will be appropriate, and the expression for $\mathbf{E}$ can be written as

$$E_z = A_1 H_0^{(1)}(\beta\rho) + B_1 H_0^{(2)}(\beta\rho) \tag{14-41}$$

Only the zero-order functions appear, because there is no variation of the field in the ϕ direction. If the region is assumed to extend to infinity, there is no reason for retaining the first term of (41) which represents an inward-traveling wave, and so the solution is given by

$$E_z = B_1 H_0^{(2)}(\beta\rho)$$

which represents an outward-traveling wave. Using eq. (29), the expression for magnetic field strength will be

$$H_\phi = \frac{1}{j\omega\mu}\frac{\partial E_z}{\partial\rho} = -\frac{\beta B_1}{j\omega\mu} H_1^{(2)}(\beta\rho) = -\frac{B_1}{j\eta} H_1^{(2)}(\beta\rho)$$

since
$$\frac{d}{du}[H_0^{(2)}(u)] = -H_1^{(2)}(u)$$

At
$$\rho = a \qquad H_\phi = \frac{I}{2\pi a}$$

Therefore,
$$B_1 = -\frac{j\eta I}{2\pi a H_1^{(2)}(\beta a)} \approx -\frac{\beta\eta I}{4} \qquad (\text{for } \beta a \ll 1)$$

Then

$$E_z = -\frac{\beta\eta I}{4} H_0^{(2)}(\beta\rho) \xrightarrow[\rho\to\infty]{} -\frac{\eta I}{2\sqrt{\rho\lambda}} e^{-j(\beta\rho-\pi/4)}$$
$$H_\phi = -\frac{j\beta I}{4} H_1^{(2)}(\beta\rho) \xrightarrow[\rho\to\infty]{} \frac{I}{2\sqrt{\rho\lambda}} e^{-j(\beta\rho-\pi/4)} \tag{14-42}$$

At large distances from the wire the fields decrease in amplitude as $1/\sqrt{\rho}$. Also at large distances the fields are periodic in 2π radians (this is not true close to the source) and appear to have originated at an "effective" source, which is one-eighth of a wavelength out from the center of the wire.

The outward radial impedance is

$$Z_\rho^+ = -\frac{E_z}{H_\phi} = j\eta\frac{H_0^{(2)}(\beta\rho)}{H_1^{(2)}(\beta\rho)} \tag{14-43}$$

At large distances, where the asymptotic expressions for the Hankel functions can be used, the radial impedance becomes a pure resistance

$$Z_\rho^+ \approx \eta = 377 \text{ ohms} \qquad (\text{for } \beta\rho \gg 1)$$

The *impedance of the wire* can be obtained from a consideration of the field strengths at its surface. Assuming first a perfectly conducting wire of radius a, the total tangential electric field strength at its surface, $E(a)$, must be zero. Then

$$E(a) = E_a + E_z(a) = 0$$

or
$$E_a = -E_z(a)$$

where E_a is the applied electric field strength that causes the current I to flow, and $E_z(a)$ is the self-induced electric field strength (due to the current I) evaluated at the surface of the wire, $\rho = a$. (In this problem the "applied" field E_a might be the incident field from a distant transmitter.) Then the *external impedance* of the wire per unit length will be

$$Z_{\text{ext}} = \frac{E_a}{I} = -\frac{E_z(a)}{2\pi a H_\phi(a)}$$

Therefore,

$$Z_{\text{ext}} = \frac{j\eta}{2\pi a}\frac{H_0^{(2)}(\beta a)}{H_1^{(2)}(\beta a)} \tag{14-44}$$

For $\beta a \ll 1$, as would normally be the case, (44) reduces to

$$Z_{\text{ext}} = \frac{j\eta}{2\pi a}\left[\frac{J_0(\beta a) - jN_0(\beta a)}{J_1(\beta a) - jN_1(\beta a)}\right]$$

$$\approx \frac{j\eta}{2\pi a}\left[\frac{1 - j\dfrac{2}{\pi}(\ln \beta a + C - \ln 2)}{\dfrac{\beta a}{2} + j\dfrac{2}{\pi\beta a}}\right]$$

$$\approx \frac{60\pi^2}{\lambda} + \frac{j\eta}{\lambda}\ln\frac{\lambda}{2\pi a} \tag{14-45}$$

$$= \pi\omega \times 10^{-7} + j\omega\left(\frac{\mu}{2\pi}\ln\frac{\lambda}{2\pi a}\right) \tag{14-45a}$$

The real part of this external impedance is the radiation resistance per unit length, and the imaginary part is the external inductive reactance per unit length of the wire. The former is independent of wire diameter, whereas the latter becomes logarithmically infinite as the wire diameter approaches zero. The quantity $(\mu/2\pi)\ln \lambda/2\pi a$ is the *high-frequency external inductance* of the wire (per meter length).

If the assumption of perfect conductivity is not made, the total tangential field strength $E(a)$ at the surface will not be zero, but it will have the (small) value required to drive the current I against the *internal impedance* of the wire. This total or resultant field at the surface of the wire is as before

$$E(a) = E_a + E_z(a)$$

so that

$$E_a = E(a) - E_z(a) \tag{14-46}$$

Dividing by the current I gives the impedance per unit length of the wire.

$$Z = \frac{E_a}{I} = \frac{E(a)}{I} - \frac{E_z(a)}{I} = Z_{\text{int}} + Z_{\text{ext}} \tag{14-47}$$

For the fields *within* the wire it is the appropriate solutions of (38), which must be used. Therefore within the wire

$$E_z\,(\text{int}) = AI_0(\gamma\rho) + BK_0(\gamma\rho)$$

The second of these functions becomes infinite at $\rho = 0$. Since E_z must always remain finite this requires that $B = 0$, so

$$E_z\,(\text{int}) = AI_0(\gamma\rho)$$

and from (29), remembering* that $I_0' = I_1$

$$H_\phi = \frac{\gamma A}{j\omega\mu} I_1(\gamma\rho)$$

*Recurrence formulas for the I and K functions differ from the other Bessel functions. These formulas are listed in the Appendix.

At the surface of the wire, $E_z\,(\text{int})$ must equal the total or resultant electric field strength $E(a)$, and $2\pi a H_\phi = I$.

Therefore

$$E(a) = AI_0(\gamma a) \qquad I = 2\pi a \frac{\gamma A}{j\omega\mu} I_1(\gamma a)$$

and

$$Z_{\text{int}} = \frac{E(a)}{I} = \frac{j\omega\mu}{2\pi a\gamma}\frac{I_0(\gamma a)}{I_1(\gamma a)}$$

$$= \frac{\eta_m}{2\pi a}\frac{I_0(\gamma a)}{I_1(\gamma a)} \tag{14-48}$$

where $\eta_m = \sqrt{j\omega\mu/(\sigma + j\omega\epsilon)}$ is the intrinsic impedance of the metal.

Equation (48) gives the exact expression for the internal impedance of the wire. The evaluation of this expression is simplified by recalling that for all metallic conductors at frequencies less than optical, $\sigma \gg \omega\epsilon$ and $\gamma \approx \sqrt{j\omega\mu\sigma} = \sqrt{\omega\mu\sigma}\sqrt{j} = \sqrt{\omega\mu\sigma}\,\underline{/45^\circ}$. To assist in obtaining numerical values for expressions such as (48), the following auxiliary functions have been defined and tabulated

$$I_0(v\sqrt{j}) = \text{ber}\; v + j\;\text{bei}\; v$$

Tables of these "farmyard functions" may be found in the reference noted.* Curves showing the internal impedance of wires as given by (48) may be found in several texts.† Two special cases of this general expression are of importance practically and will be considered.

The first of these special cases occurs for thin wires at low (power) frequencies, where the wire radius is small compared with the depth of penetration. For this case, $|\gamma a| \ll 1$, and only the first two terms of the power series expansion for $I_0(\gamma a)$ and $I_1(\gamma a)$ need be used. Then

$$I_0(\gamma a) \approx 1 + \frac{(\gamma a)^2}{4} \qquad I_1(\gamma a) \approx \frac{\gamma a}{2} + \frac{(\gamma a)^3}{16}$$

$$Z_{\text{int}}\,(\text{low freq}) \approx \frac{1}{\pi a^2(\sigma + j\omega\epsilon)} + \frac{j\omega\mu}{8\pi}$$

$$\approx \frac{1}{\pi a^2\sigma} + j\omega\frac{\mu}{8\pi} \tag{14-49}$$

The terms represent respectively the *low-frequency resistance* and *internal inductive reactance* of the wire, per unit length. The low-frequency internal inductance of the wire is $\mu/8\pi$ henry/m.

The second special case of practical importance occurs for frequencies sufficiently high that the depth of penetration is small compared with the radius of the wire. This makes $|\gamma a| \gg 1$. Except for quite

*McLachlan, *Bessel Functions for Engineers*, Oxford Press, London, 1934; Dwight, *Tables of Integrals*, The Macmillan Company, New York, 1961.

†For example, Ramo, Whinnery and Van Duzer, *Fields and Waves in Communication Electronics*, John Wiley & Sons, Inc., New York, 1965.

thin wires, this case covers all radio frequencies. Using the asymptotic expansions for I_0 and I_1, the internal impedance becomes

$$Z_{\text{int}}\,(\text{high freq}) \approx \frac{\eta_m}{2\pi a} = \frac{Z_s}{2\pi a} = \frac{R_s}{2\pi a} + \frac{jX_s}{2\pi a} \tag{14-50}$$

As would be expected, when the depth of penetration is small compared with the radius, the internal impedance per unit length of the wire is equal to the surface impedance of a thick plane sheet of the metal 1 meter long and $2\pi a$ meters wide. Evaluating (50) in terms of the constants of the metal shows that

$$Z_{\text{int}}\,(\text{high freq}) = \frac{1}{2\pi a}\sqrt{\frac{\omega\mu}{2\sigma}} + \frac{j}{2\pi a}\sqrt{\frac{\omega\mu}{2\sigma}} \tag{14-50a}$$

The first term is high-frequency ohmic resistance of the wire per unit length, and the second term is the high-frequency internal inductive reactance of the wire per unit length.

The Infinitely-long Slit. The narrow infinitely long slit in a perfectly conducting plane is the "complementary screen" of the infinitely long, perfectly conducting wire.* Accordingly the solution to the slit problem can be deduced from the long-wire problem by the methods described in secs. 13.16 to 13.18. Alternatively, the infinitely long slit can be solved directly as a boundary-value problem, applying Maxwell's equations to a "wedge" transmission line consisting of two semi-infinite conducting planes (Fig. 14-10). When the included angle ϕ_0 between the planes is 180 degrees the slit-in-plane problem results. The solutions for this type of problem will not be carried through here but will be left as special problems [2(*a*) and 2(*b*)] at the end of this section. In this second method it is evident that uniform cylindrical waves will be excited, so expressions for H_z similar to eqs. (41) and (42) will be expected. The input voltage between the edges at $\rho = a$ will be $V = -\phi_0\, aE_\phi(a)$, and then it must be remembered that for the slit-in-plane problem the admittances of regions (1) and (2) will be in parallel. For the slit in the conducting plane the voltage V can be evaluated in terms of the incident fields in the following way: If there were no slit, the current per meter width in the conducting planes would be $J_s = 2H^0$ where H^0 is the magnetic field strength of the (normally) incident wave. With the slit in the screen the *total* conduction current at the edge of the slit is zero, so the induced voltage (due to charge concentrations at the edge of the slit) must be just sufficient to produce a current per meter, J_ρ, which is equal and opposite to $J_s = 2H^0$. Therefore $VY_{\text{slit}} = -2H^0$.

*For true complementarity the wire should be a flat strip, but for thin wires, and using the appropriate equivalent radius ($a = d/4$, where d is the strip width), the difference between flat strip and round wire is inconsequential.

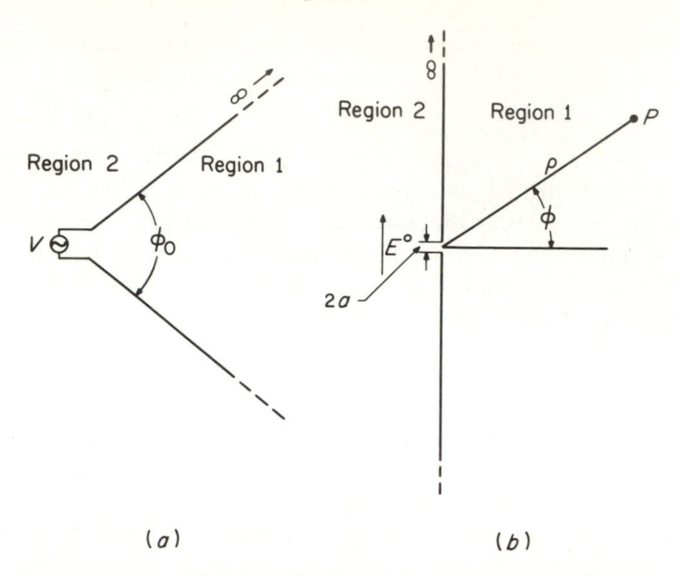

Figure 14-10. (*a*) A "wedge" transmission line that supports uniform cylindrical waves. (*b*) Diffraction through a narrow slit in an infinite conducting plane.

Problem 2(a) Using the methods of secs. 13.16 to 13.18, use the results of sec. 14.10 to write the expressions for the fields and admittance of an infinitely long thin slot of width d in a conducting plane when a plane wave field E^0, H^0 is incident normally on the plane.

Problem 2(b) Solve the infinitely long slit directly as a boundary value problem and compare the result with that for the infinitely long wire to verify (the extended) Babinet's principle in this case.

14.11 The Cylindrical Antenna Problem. The methods considered earlier in this chapter for the calculation of antenna impedance involved making an estimate of the current distribution, and substituting this estimate into an integral formula. Except for very thin antennas there is considerable uncertainty as to the accuracy of the assumed current distribution; for this reason it is desirable to look for more precise methods of calculating antenna impedance. The purpose of this sections to describe three important methods of solution that have been applied to the antenna problem. In a later section one of these methods will be considered in detail.

Any solution of the antenna problem will of course have Maxwell's equations as a starting point. In fact, the problem is essentially one of solving Maxwell's equations subject to the boundary conditions imposed by the antenna and the source. For the simple center-fed cylindrical antenna this turns out to be a surprisingly difficult problem.

Three general methods of attack have been used. The first of these methods (historically) treats the problem as a boundary-value problem. The second method sets up the problem as that of finding the solution of an integral equation for the current. The third method treats the antenna as an open-ended wave guide or electromagnetic horn.

(*a*) *As a Boundary-value Problem.* For certain symmetrical antenna shapes (e.g., the ellipsoid or prolate spheroid) it is possible to solve for the *free* oscillations or natural modes, and so determine the proper frequencies and corresponding damping factors. This problem was worked out many years ago by Abraham* for very thin ellipsoids and later by Brillouin† for prolate spheroids of any eccentricity. When the antenna is excited or fed, the solution is given in terms of an infinite series of the free-oscillation modes with coefficients chosen so as to satisfy the force function. Page and Adams,‡ Ryder,§ Stratton‖ and Chu° are among those who have worked on this problem. This method has the advantage of yielding very reliable results, but is restricted to a relatively few shapes, among which, unfortunately, the cylinder is not included. There are two main disadvantages of the method. First, although the method is useful near resonance, for lengths considerably different from the resonant length the series converge very slowly so that an excessive amount of labor is involved in obtaining numerical answers. Second, actual antennas generally are not prolate spheroids, but have various shapes, the circular cylinder being most common. About the best that can be done in obtaining a solution for the actual antenna by this method is to assume that the solution for an "equivalent" thin prolate spheroid will hold approximately for the cylindrical antenna. The troublesome question of just what size of prolate spheroid is "equivalent" to a cylinder prevents this method from being as useful as it might otherwise be. An excellent summary and comparison of the work of different writers using this and other methods is contained in an article by Brillouin./

(*b*) *Integral Equation Solution.* Hallén# has used a different approach

*Max Abraham, *Ann. Physik*, **66,** 435 (1898): *Math. Ann.*, **52,** 81 (1899).

†L. Brillouin, *Propagation de l'Electricite*, Hermann, Paris, 1904, Vol. 1.

‡L. Page and N. I. Adams, *Phys. Rev.*, **53,** 819 (1938).

§Robert M. Ryder, *J. Applied Phys.*, **13,** 327 (1942).

‖J. A. Stratton, *Proc Nat. Acad. Science*, **21,** 51 (317) 1935.

°J. A. Stratton and L. J. Chu, *J. App. Phys.*, **19,** 236 (1941).

/L. Brillouin, "Antennae for Ultrahigh Frequencies," *Elec. Comm.*, **21,** No. 4, p. 257 (1944); and **22,** No. 1, p. 11 (1944).

#Erik Hallén," Theoretical Investigations into the Transmitting and Receiving Qualities of Antennae, "*Nova Acta Upsal* **11,** 4, (1938); "Further Investigations into the Receiving Qualities of Antennae; The Absorption of Transient Unperiodic Radiation," *Årsskrift*, Upsala, No. 4, (1939).

to the antenna problem. Starting with an arbitrary current distribution, general expressions for the field are obtained by the use of retarded potentials. Application of the boundary conditions at the surface of the antenna then leads to an integral equation for the current. Thus, instead of a set of partial differential equations, it is now an integral

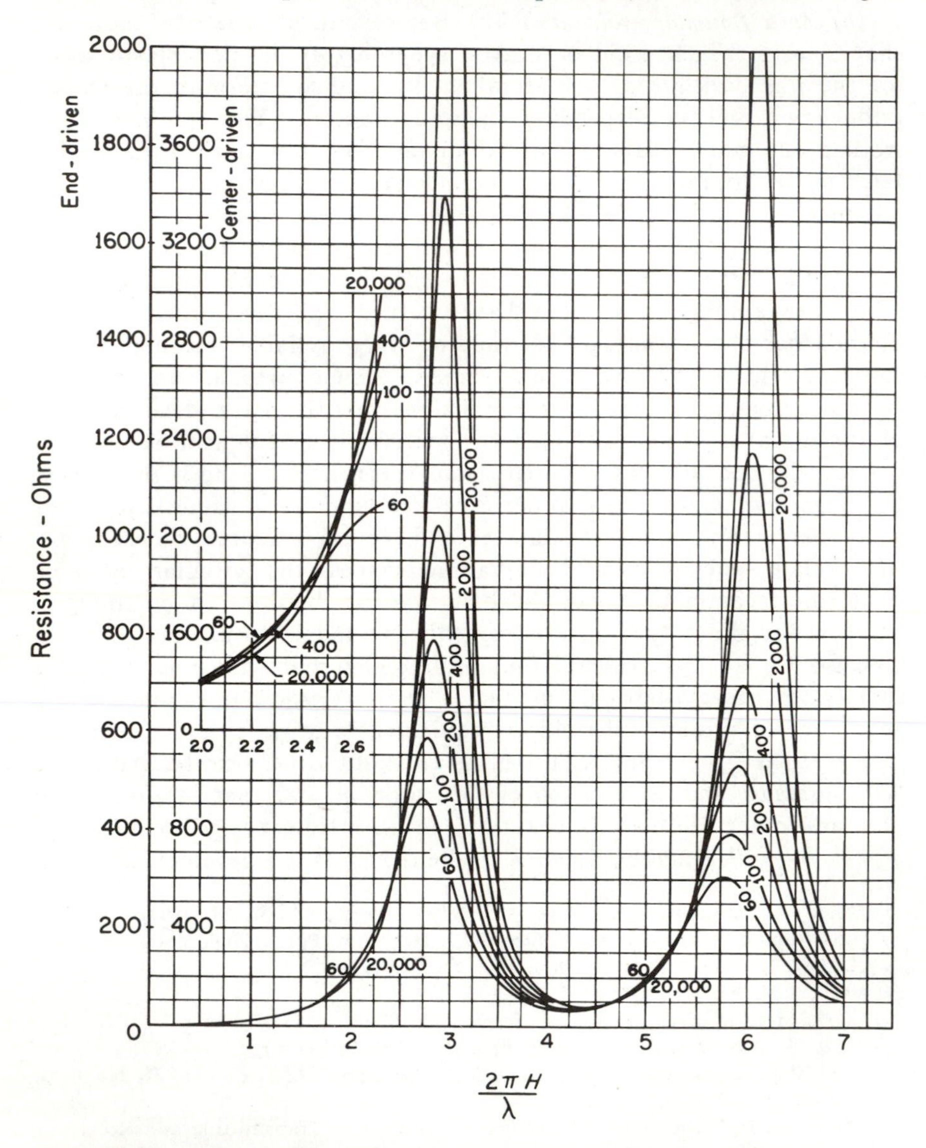

Figure 14-11. Antenna resistance according to Hallén. Resistance is shown as a function of $2\pi H/\lambda$ for various ratios of H/a, half length to radius, for monopoles (end-driven) and dipoles (center-driven).

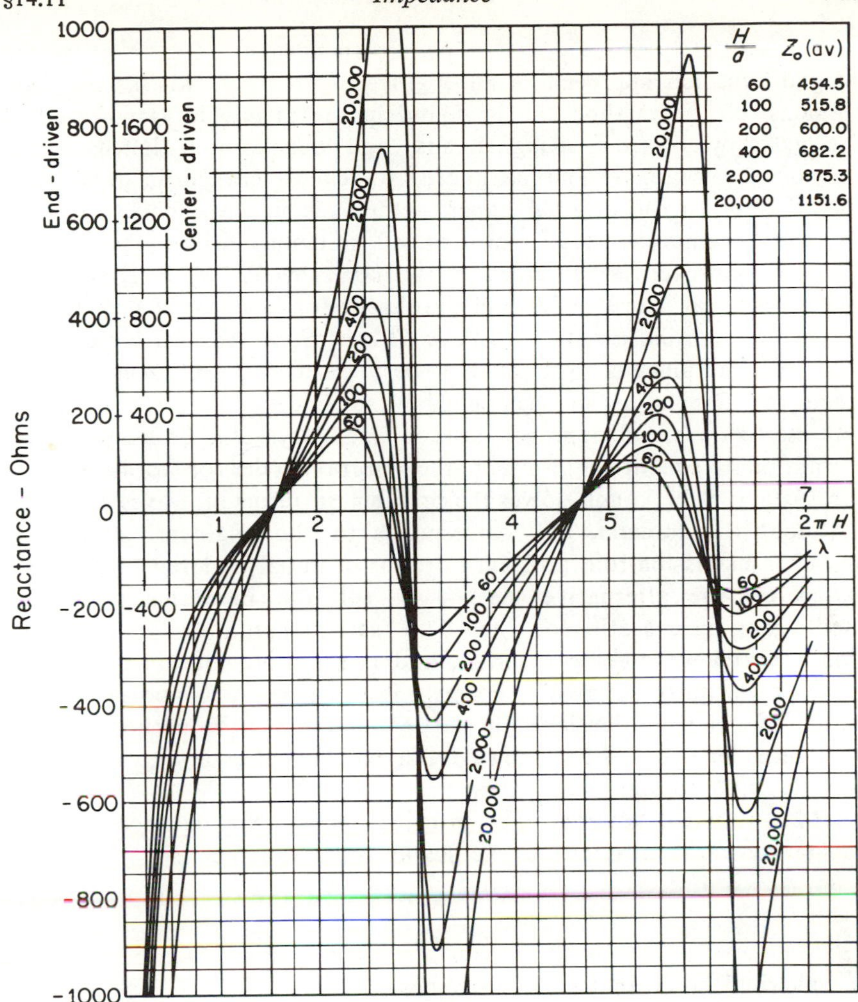

Figure 14-12. Antenna reactance according to Hallén (see legend for Fig. 14-11).

equation that must be solved. The method is general and applicable to antennas of different shapes, but the accurate evaluation of the resulting expressions is very difficult. Hallén succeeded in reducing the integrals to ordinary sine and cosine integrals, and iterated sine and cosine integrals, whose values were then tabulated for arguments from 0 to 7. From these tables admittance and impedance diagrams were constructed for cylindrical antennas over a wide range of antenna dimensions. The impedance diagrams showing antenna resistance and reactance separately, are reproduced in Figs. 14-11 and 14-12. In these curves, resistance and reactance are shown as a function of antenna length (in radians or wavelengths) for various ratios of half-length to radius.

Unfortunately space does not permit an adequate treatment of the integral equation approach in an engineering text of this scope. Discussions of the method can be found in the article by Bowkamp and the books by Aharoni, King, Kraus, and Schelkunoff, listed at the end of the chapter. The advanced student interested in this topic should refer to the original papers referenced in these books. By far the most comprehensive treatment of antennas using the integral equation approach is contained in the book by R. W. P. King which includes several score tables and graphs useful in antenna design.

(*c*) *The Antenna as a Wave Guide or Electric Horn.* An entirely different attack on the antenna problem has been made by Schelkunoff, who treats the antenna as an open-ended wave guide or electric horn. In contrast to the usual approach used in boundary-value problems, where a solution is sought in terms of the natural modes or oscillations of the system, Schelkunoff solves the problem in terms of waves transmitted along the antenna. This corresponds to the engineering solution of the transmission-line problem in terms of initial and reflected waves, as against the alternative method of solution in terms of natural oscillations on a section of line. The method uses familiar transmission-line and wave-guide theories, and is an approach which the engineer finds quite satisfying. Because it represents an important application of concepts developed in earlier chapters, this method will be considered in detail. First, however, it will be necessary to give some consideration to spherical waves.

14.12 Spherical Waves. For propagation in a homogeneous medium having constants μ, ϵ, and σ the scalar wave equation is

$$\nabla^2 V = \gamma^2 V \tag{14-51}$$

where

$$\gamma = \sqrt{j\omega\mu(\sigma + j\omega\epsilon)}$$

In spherical co-ordinates (51) becomes

$$\frac{1}{r^2}\frac{\partial}{\partial r}\left(r^2\frac{\partial V}{\partial r}\right) + \frac{1}{r^2 \sin\theta}\frac{\partial}{\partial\theta}\left(\sin\theta\frac{\partial V}{\partial\theta}\right) + \frac{1}{r^2\sin^2\theta}\frac{\partial^2 V}{\partial\phi^2} = \gamma^2 V \tag{14-51a}$$

Although in general the propagation constant γ may be complex, for dissipationless media $\sigma = 0$ and γ^2 reduces to $-\omega^2\mu\epsilon$. Separating eq. (51a) by letting

$$V = R(r)P(\theta)\Phi(\phi)$$

results in the three equations

$$\frac{d}{dr}\left(r^2\frac{dR}{dr}\right) = (b^2 + \gamma^2 r^2)\,R \tag{14-52}$$

$$\frac{d^2\Phi}{d\phi^2} = -m^2\Phi \tag{14-53}$$

$$\frac{d^2P}{d\theta^2} + \cot\theta\,\frac{dP}{d\theta} + \left(b^2 - \frac{m^2}{\sin^2\theta}\right)P = 0 \qquad (14\text{-}54)$$

where the constants b^2 and $-m^2$ may be real or complex. When m is an integer, Φ is periodic with a period 2π.

Equation (54) is the *associated Legendre equation.* When b^2 is real and has the form $b^2 = n(n+1)$, eq. (54) may be written

$$(1 - x^2)\frac{d^2P}{dx^2} - 2x\frac{dP}{dx} + \left[n(n+1) - \frac{m^2}{1-x^2}\right]P = 0 \qquad (14\text{-}55)$$

where

$$x = \cos\theta \qquad 1 - x^2 = \sin^2\theta \qquad \frac{d}{d\theta} = -\sin\theta\,\frac{d}{dx}$$

For those problems in which there is no variation with ϕ, the constant m in eq. (53) is zero, and for these cases eq. (55) becomes

$$(1 - x^2)\frac{d^2P}{dx^2} - 2x\frac{dP}{dx} + n(n+1)P = 0 \qquad (14\text{-}56)$$

which is the *ordinary Legendre equation.*

For nonintegral values of n the solutions of (56) are given by the functions $P_n(\cos\theta)$ and $P_n(-\cos\theta)$, where

$$P_n(\cos\theta) = \sum_{q=0}^{\infty}\frac{(-1)^q(n+q)!}{(n-q)!\,(q!)^2}\sin^{2q}\left(\frac{\theta}{2}\right) \qquad (14\text{-}57)$$

For integral values of n, expression (57) reduces to the Legendre polynomials and $P_n(\cos\theta)$ and $P_n(-\cos\theta)$ are no longer linearly independent. Under these circumstances it is convenient to use $P_n(\cos\theta)$ (where n is a positive integer) for one solution and $Q_n(\cos\theta)$ for the other, where the Q_n functions are defined by

$$Q_n(\cos\theta) = P_n(\cos\theta)\log\cot\frac{\theta}{2} - \sum_{s=1}^{s=n}\frac{P_{n-s}P_{s-1}}{s} \qquad (14\text{-}58)$$

The Q functions become infinite at $\theta = 0$ and $\theta = \pi$, and so can be used to represent physically realizable fields only when the $0 - \pi$ axis is excluded from the region being considered.

When m is not equal to zero, the associated Legendre equation must be considered. For integral values of n its solutions are $P_n^m(\cos\theta)$ and $Q_n^m(\cos\theta)$ where

$$P_n^m(\cos\theta) = (-1)^m\sin^m\theta\,\frac{d^m[P_n(\cos\theta)]}{d(\cos\theta)^m} \qquad (14\text{-}59)$$

$$Q_n^m(\cos\theta) = (-1)^m\sin^m\theta\,\frac{d^m[Q_n(\cos\theta)]}{d(\cos\theta)^m} \qquad (14\text{-}60)$$

For the first few values of n, the *Legendre* and *associated Legendre polynomials* represented by (7) and (9) are,

$$
\begin{aligned}
P_0(\cos\theta) &= 1\\
P_1(\cos\theta) &= \cos\theta\\
P_2(\cos\theta) &= \tfrac{1}{2}(3\cos^2\theta - 1)\\
P_3(\cos\theta) &= \tfrac{1}{2}(5\cos^3\theta - 3\cos\theta)\\
P_4(\cos\theta) &= \tfrac{1}{8}(35\cos^4\theta - 30\cos^2\theta + 3)\\
P_1^1(\cos\theta) &= -\sin\theta\\
P_2^1(\cos\theta) &= -3\sin\theta\cos\theta\\
P_2^2(\cos\theta) &= 3\sin^2\theta\\
P_3^1(\cos\theta) &= -\tfrac{3}{2}\sin\theta(5\cos^2\theta - 1)\\
P_3^2(\cos\theta) &= 15\sin^2\theta\cos\theta
\end{aligned}
$$

Considering now the solutions to eq. (52) for R, this equation may be written as

$$r^2\frac{d^2R}{dr^2} + 2r\frac{dR}{dr} - (\gamma^2r^2 + b^2)R = 0 \tag{14-61}$$

This equation is slightly different from the ordinary Bessel equation (62) or the modified Bessel equation (63) with which it should be compared.

Ordinary Bessel $$z^2\frac{d^2w}{dz^2} + z\frac{dw}{dz} + (z^2 - \nu^2)w = 0 \tag{14-62}$$

Modified Bessel $$z^2\frac{d^2w}{dz^2} + z\frac{dw}{dz} - (z^2 + \nu^2)w = 0 \tag{14-63}$$

Equation (61) can be reduced to a standard form by suitable change of variable. Let

$$w = rR \qquad \text{or} \qquad R = \frac{w}{r}$$

Then $$\frac{dR}{dr} = \frac{1}{r}\frac{dw}{dr} - \frac{w}{r^2} \qquad \frac{d^2R}{dr^2} = \frac{1}{r}\frac{d^2w}{dr^2} - \frac{2}{r^2}\frac{dw}{dr} + \frac{2w}{r^3}$$

and eq. (61) becomes

$$\frac{d^2w}{dr^2} - \left(\gamma^2 + \frac{b^2}{r^2}\right)w = 0 \tag{14-64}$$

Now put $b^2 = n(n+1)$ as in the Legendre equations and let $z = \gamma r$. Then (64) becomes

$$\frac{d^2w}{dz^2} - \left[1 + \frac{n(n+1)}{z^2}\right]w = 0 \tag{14-65}$$

Solutions to this equation are denoted by $\hat{K}_n(z)$ and $\hat{I}_n(z)$ where

$$\hat{K}_n(z) = e^{-z}\sum_{p=0}^{n}\frac{(n+p)!}{p!(n-p)!(2z)^p} \tag{14-66}$$

$$\hat{I}_n(z) = \frac{1}{2}\left[e^z \sum_{p=0}^{n} \frac{(-1)^p(n+p)!}{p!(n-p)!(2z)^p} + (-1)^{n+1} e^{-z} \sum_{p=0}^{n} \frac{(n+p)!}{p!(n-p)!(2z)^p}\right] \tag{14-67}$$

For the first few values of n these are

$$\hat{K}_0(z) = e^{-z} \qquad \hat{I}_0(z) = \sinh z$$

$$\hat{K}_1(z) = e^{-z}\left(1 + \frac{1}{z}\right) \qquad \hat{I}_1(z) = \cosh z - \frac{\sinh z}{z}$$

$$\hat{K}_2(z) = e^{-z}\left(1 + \frac{3}{z} + \frac{3}{z^2}\right) \qquad \hat{I}_2(z) = \left(1 + \frac{3}{z^2}\right)\sinh z - \frac{3}{z}\cosh z$$

In general the propagation constant γ of eq. (61) is complex. For the particular, but important, practical case where the attenuation factor α is zero, γ is a pure imaginary equal to $j\beta$, and eq. (61) becomes

$$r^2 \frac{d^2 R}{dr^2} + 2r \frac{dR}{dr} + (\beta^2 r^2 - b^2) R = 0 \tag{14-68}$$

which should be compared with the ordinary Bessel equation (62). Reducing (68) in the same manner that (61) was reduced, but letting $z = \beta r$, there results

$$\frac{d^2 w}{dz^2} + \left[1 - \frac{n(n+1)}{z^2}\right] w = 0 \tag{14-69}$$

instead of (65).

Solutions of this equation are denoted by $\hat{J}_n(z)$ and $\hat{N}_n(z)$, where for the first few values of n, these functions* are

$$\hat{J}_0(z) = \sin z \qquad \hat{N}_0(z) = -\cos z$$

$$\hat{J}_1(z) = \frac{\sin z}{z} - \cos z \qquad \hat{N}_1(z) = -\sin z - \frac{\cos z}{z}$$

$$\hat{J}_2(z) = \left(\frac{3}{z^2} - 1\right)\sin z - \frac{3}{z}\cos z \qquad \hat{N}_2(z) = \left(1 - \frac{3}{z^2}\right)\cos z - \frac{3}{z}\sin z$$

The $\hat{I}$ and $\hat{K}$ functions are simply related to the $\hat{J}$ and $\hat{N}$ functions by

$$\hat{I}_n(jz) = j^{n+1}\hat{J}_n(z)$$
$$\hat{K}_n(jz) = j^{-n-1}[\hat{J}_n(z) - j\hat{N}_n(z)] \tag{14-70}$$

In addition the functions $\hat{J}$, $\hat{N}$, $\hat{I}$, $\hat{K}$, are related to the ordinary and modified Bessel functions, J, N, I, K. Indeed they are just the

*The $\hat{J}$ and $\hat{N}$ functions used are as defined by Schelkunoff, *Electromagnetic Waves*, D. Van Nostrand Co., Inc., Princeton, N. J., 1943, p. 51. They are just z times the *Spherical Bessel Functions* as defined by Morse, *Vibration and Sound*, McGraw-Hill Book Company, New York, 1936, p. 246.

half-integral orders of the corresponding ordinary and modified Bessel functions. The relations are:

$$\left.\begin{aligned} \hat{J}_n(z) &= \sqrt{\frac{\pi z}{2}}\, J_{(n+\frac{1}{2})}(z) \qquad & \hat{N}_n(z) &= \sqrt{\frac{\pi z}{2}}\, N_{(n+\frac{1}{2})}(z) \\ \hat{I}_n(z) &= \sqrt{\frac{\pi z}{2}}\, I_{(n+\frac{1}{2})}(z) \qquad & \hat{K}_n(z) &= \sqrt{\frac{2z}{\pi}}\, K_{(n+\frac{1}{2})}(z) \end{aligned}\right\} \tag{14-71}$$

For propagation in a lossless medium, the propagation constant γ will be a pure imaginary equal to $j\beta$. Under these conditions (imaginary arguments), the $\hat{I}$ and $\hat{K}$ functions reduce to half-order Bessel and Hankel functions as follows:

$$\begin{aligned} \hat{I}_n(j\beta r) &= j^{n+1}\, \hat{J}_n(\beta r) \\ &= j^{n+1} \sqrt{\frac{\pi\beta r}{2}}\, J_{(n+\frac{1}{2})}(\beta r) \\ \hat{K}_n(j\beta r) &= j^{-n-1}\, [\hat{J}_n(\beta r) - j\hat{N}_n(\beta r)] \\ &= j^{-n-1} \sqrt{\frac{\pi\beta r}{2}}\, [J_{(n+\frac{1}{2})}(\beta r) - jN_{(n+\frac{1}{2})}(\beta r)] \\ &= j^{-n-1} \sqrt{\frac{\pi\beta r}{2}}\, H^{(2)}{}_{(n+\frac{1}{2})}(\beta r) \end{aligned} \tag{14-72}$$

The first of these functions represents a spherical standing wave and is suitable for regions that include the origin. The second function represents an outward-traveling spherical wave, and is appropriate for regions that may extend to infinity, but do not include the origin. Applications of these functions* will be made in the following sections.

14.13 Spherical Waves and the Biconical Antenna. Schelkunoff has obtained a solution to the antenna problem by treating the antenna as an open-ended wave guide, or electromagnetic horn. To accomplish this, he has started with a biconical antenna as a prototype for which a solution can be obtained from Maxwell's equations. In the process, it is demonstrated that for the biconical antenna the input impedance depends only on the principal wave. Therefore, for biconical antennas, the input impedance can be represented *exactly* as the input impedance of a uniform transmission line, terminated in an appropriate terminal impedance. Two methods for calculating the terminal impedance are given. Then, using the solution for the biconical antenna as a guide, the solution for cylindrical antennas is obtained by analogy. Whereas

*Of necessity, discussion of these functions in this section has been very brief. For a more thorough treatment, reference should be made to a mathematics text. An excellent treatment is given in S. A. Schelkunoff, *Applied Mathematics for Engineers and Scientists*, D. Van Nostrand Co., Inc., Princeton, N. J., 1948.

this approach necessarily involves making some approximations, the approximations are justifiable on the basis of the physical picture gained from the biconical antenna theory.

Before investigating radiation from biconical antennas, it is desirable to give consideration to some of the general properties of spherical waves. It will be recalled that for *plane* waves it was found possible to divide the waves into transverse magnetic (TM), transverse electric (TE), and transverse electromagnetic (TEM) waves. For TM waves traveling in the z direction, $H_z = 0$, and the divergence equation for **H** is

$$\frac{\partial H_x}{\partial x} + \frac{\partial H_y}{\partial y} = 0 \tag{14-73}$$

It follows that it should be possible to derive **H** from a *stream function* Π_m through the relations

$$H_x = \frac{\partial \Pi_m}{\partial y} \qquad H_y = -\frac{\partial \Pi_m}{\partial x} \tag{14-74}$$

which satisfy (73). Since $H_z = 0, \Pi_m$ may be regarded as the magnitude of a vector $\mathbf{A}'$ that is parallel to the z axis. Then eqs. (74) are given by

$$\mathbf{H} = \nabla \times \mathbf{A}'$$

where $\quad \mathbf{A}' = \hat{\mathbf{z}} A_z \qquad A_z = \Pi_m \qquad A_x = A_y = 0$

Similarly for TE waves traveling in the z direction, $E_z = 0$, and the divergence equation for **E** (in a charge-free region) is

$$\frac{\partial E_x}{\partial x} + \frac{\partial E_y}{\partial y} = 0$$

so that, in this case, it is possible to obtain **E** from a stream function Π_e through the relations

$$E_x = \frac{\partial \Pi_e}{\partial y} \qquad E_y = -\frac{\partial \Pi_e}{\partial x}$$

Since $E_z = 0, \Pi_e$ may be regarded as the scalar magnitude of a vector $\mathbf{F}'$ that is parallel to the z axis. Then

$$\mathbf{E} = \nabla \times \mathbf{F}'$$

where $\quad \mathbf{F}' = \hat{\mathbf{z}} F_z \qquad F_z = \Pi_e \qquad F_x = F_y = 0$

Since TEM waves may be considered a special case of either TM or TE waves, it follows that the most general plane-wave field traveling in the z direction can be expressed in terms of two scalar stream functions A_z and F_z.

The theory of spherical waves is similar to that of plane waves. There are TM spherical waves for which $H_r = 0$, TE spherical waves for which $E_r = 0$, and TEM spherical waves for which both E_r and H_r are zero. For TM spherical waves, the divergence equation for **H** reduces to

$$\frac{\partial}{\partial\theta}(\sin\theta H_\theta) + \frac{\partial H_\phi}{\partial\phi} = 0$$

Therefore it should be possible to obtain **H** from a stream function Π_m through the relations

$$H_\theta = \frac{1}{r\sin\theta}\frac{\partial\Pi_m}{\partial\phi} \qquad H_\phi = -\frac{1}{r}\frac{\partial\Pi_m}{\partial\theta}$$

Since $H_r = 0, \Pi_m$ may be regarded as the magnitude of a vector $\mathbf{A}'$ which* at every point is in the direction of the r co-ordinate. Then **H** is obtained from

$$\mathbf{H} = \nabla \times \mathbf{A}'$$

where $\quad \mathbf{A}' = \hat{\mathbf{r}}A_r \qquad A_r = \Pi_m \qquad A_\theta = A_\phi = 0$

In a similar manner the electric field of a spherical TE wave is found to be expressible in terms of a stream function F_r. However, in dealing with biconical and cylindrical antennas, only TM (and TEM) spherical waves are encountered.

Figure 14-13 shows a biconical antenna that is assumed to be excited by a voltage applied across an infinitesimal gap at the apices.

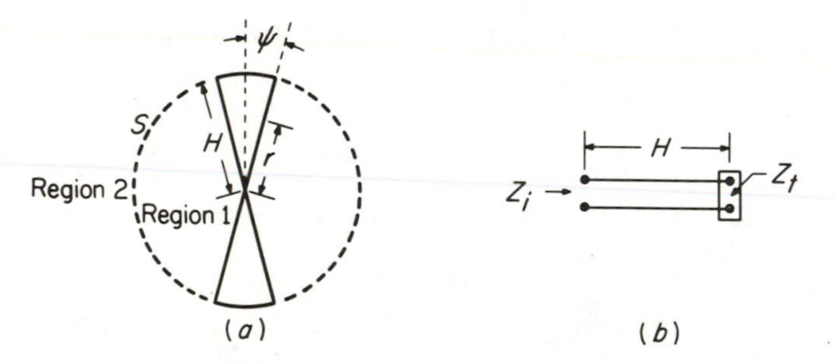

Figure 14-13. (*a*) Biconical antenna and (*b*) its equivalent circuit (insofar as impedance is concerned).

The spherical surface S divides the space about the antenna into two regions: region (1) is the antenna region and region (2) is the outside or free-space region. The conducting cones and dielectric in region (1) can be considered as a wave guide that is "terminated" in a second wave guide, consisting of region (2). Because of circular symmetry, currents along the cones will be radial (except at the end surfaces)

*In general, for spherical waves, the vector $\mathbf{A}' = \hat{\mathbf{r}}A_r$ will not be the same as the magnetic vector potential **A**.

and magnetic lines will be circular about the axis of the cones. That is, only transverse magnetic waves will be present.

With $H_r = 0$, the electromagnetic field about the cones can be completely specified in terms of a radial vector $\mathbf{A}' = \hat{\mathbf{r}}A_r$. Taking the curl of $\mathbf{A}_r'$ and remembering that $\partial/\partial\phi = 0$, the magnetic field strength is given by

$$H_\phi = -\frac{1}{r}\frac{\partial A_r}{\partial\theta} \tag{14-75}$$

Then, assuming a nondissipative dielectric ($\sigma = 0$), Maxwell's equations become

$$\left.\begin{aligned} \frac{\partial}{\partial\theta}(\sin\theta H_\phi) &= j\omega\epsilon r\sin\theta E_r \\ -\frac{\partial}{\partial r}(rH_\phi) &= j\omega\epsilon r E_\theta \\ \frac{\partial}{\partial r}(rE_\theta) - \frac{\partial E_r}{\partial\theta} &= -j\omega\mu r H_\phi \end{aligned}\right\} \tag{14-76}$$

with $$H_r = H_\theta = E_\phi = 0$$

Since $E_\phi = 0$, the lines of electric field strength lie in axial planes. Also, since there is no radial magnetic current ($H_r = 0$), $(\nabla \times \mathbf{E})_r = 0$ and the transverse electric field strength can be expressed as the gradient of a scalar potential V. That is

$$E_\theta = -\frac{1}{r}\frac{\partial V}{\partial\theta} \tag{14-77}$$

The stream function or potential A_r, from which the fields are to be obtained through (75) and (76), can be determined from the following considerations. For this problem, where there is no variation with ϕ, the separation of the wave equation in spherical co-ordinates leads to the Legendre equation (56), sec. 14.12. In region (2), the free-space region, where the axis ($\theta = 0, \pi$) is included, n will be integral and the θ function solution will be given in terms of the Legendre polynomials $P_n(\cos\theta)$. The Q functions cannot be used in this region because they become infinite at the axis. Because this region extends to infinity, the appropriate radial functions will be those that represent outward-traveling waves, that is, the K functions or the spherical Hankel functions. Therefore, the expression for vector potential in this region must be of the form

$$A_r(r, \theta) = \hat{K}_n(j\beta r)\, P_n(\cos\theta) \tag{14-78}$$

From (78), (75), and (76) it follows that

$$rH_\phi = -\hat{K}_n(j\beta r)\, P_n^1(\cos\theta) \tag{14-79}$$

$$rE_\theta = \eta\hat{K}_n'(j\beta r)\, P_n^1(\cos\theta) \tag{14-80}$$

Expressions for E_r can be obtained from the three parts of eq. (76). Equating these expressions and using (75) the following equation in A_r is obtained:

$$r^2 \frac{\partial^2 A_r}{\partial r^2} + \omega^2 \mu \epsilon r^2 A_r + \frac{1}{\sin\theta}\frac{\partial}{\partial\theta}\left(\sin\theta \frac{\partial A_r}{\partial\theta}\right) = 0 \tag{14-81}$$

This equation is sometimes called a wave equation, although it is different from eq. (51a), which is the wave equation in spherical coordinates. Separating, and letting the separation constant be $n(n+1)$ as before, results in

$$\frac{\partial^2 A_r}{\partial r^2} = \left[\frac{n(n+1)}{r^2} - \omega^2 \mu\epsilon\right] A_r \tag{14-82}$$

$$\frac{1}{\sin\theta}\frac{\partial}{\partial\theta}\left(\sin\theta\frac{\partial A_r}{\partial\theta}\right) = -n(n+1)\, A_r \tag{14-83}$$

Combining (83), (76), and (75) gives an expression for E_r directly in terms of A_r:

$$\begin{aligned} j\omega\epsilon r^2 E_r &= n(n+1)\, A_r \\ &= n(n+1)\, \hat{K}_n(j\beta r)\, P_n(\cos\theta) \end{aligned} \tag{14-84}$$

In this region [region (2)], when $n = 0$ all the fields vanish, so the lowest-order or principal wave is given by $n = 1$, for which

$$\left.\begin{aligned} rH_\phi &= e^{-j\beta r}\left(1 + \frac{1}{j\beta r}\right)\sin\theta \\ rE_\theta &= \eta\, e^{-j\beta r}\left(1 + \frac{1}{j\beta r} - \frac{1}{\beta^2 r^2}\right)\sin\theta \\ r^2 E_r &= 2\eta\, e^{-j\beta r}\left(\frac{1}{j\beta} - \frac{1}{\beta^2 r}\right)\cos\theta \end{aligned}\right\} \tag{14-85}$$

When multiplied by the factor $j\beta I\, dl/4\pi$, expressions (85) are exactly the expressions obtained in chap. 10 for the fields due to a current element. Thus it is seen that the simple current element generates fields that are representable by the lowest-order transverse magnetic spherical waves.

A sketch of the first- and second-order transverse magnetic spherical waves is shown in Fig. 14-14. These are for the waves in free space.

In the presence of two coaxial conductors [that is, in the antenna region (2)], the first- and second-order TM waves appear as shown in Figs. 14-15(*b*) and 14-15(*c*). However, in this latter region the zero-order TM wave, that is the TEM wave, can and does exist. Its electric-field lines are shown in Fig. 14-15(*a*).

The radial impedance for outgoing waves is defined by

$$Z_r^+ = \frac{E_\theta}{H_\phi}$$

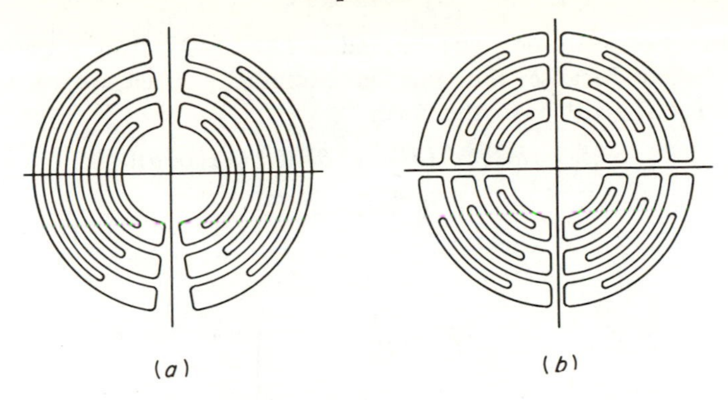

Figure 14-14. Electric field lines for first- and second-order transverse magnetic spherical waves in free space.

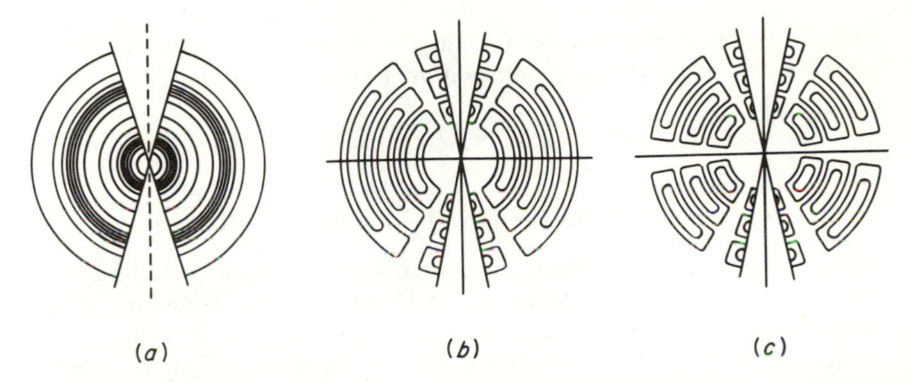

Figure 14-15. Electric-field lines for zero-, first-, and second-order TM spherical waves between coaxial cones. The zero-order wave is the TEM wave.

For the first-order TM wave of eqs. (85), it is

$$Z_r^+ = \frac{\eta\beta^2 r^2}{1 + \beta^2 r^2} - \frac{j\eta}{\beta r(1 + \beta^2 r^2)} \qquad \text{(14-86)}$$

We note, in passing, that at large distances this impedance approaches η, the intrinsic impedance. On the other hand, for small values of r, the radial impedance becomes a small resistance in series with a large capacitive reactance of value $-j/\omega\epsilon r$.

Within the antenna region there will exist a TEM wave as well as the higher-order waves. In general, to meet the boundary conditions, n will be nonintegral in this region and the θ function solution will be given in terms of P functions in the form

$$AP_n(-\cos\theta) + BP_n(\cos\theta)$$

An exception to this occurs for $n = 0$, which gives the TEM wave. For $n = 0$, $P_n(-\cos\theta)$ and $P_n(\cos\theta)$ are not independent solutions, and the Q function solution must be added. (The Q function is permissible in this region because the axis is excluded.) Then, for $n = 0$, the solution for A_r will have the form

$$\begin{aligned} A_r &= \hat{K}_0(j\beta r)\,[aP_0(\cos\theta) + bQ_0(\cos\theta)] \\ &= e^{-j\beta r}\left[a + b \ln \cot \frac{\theta}{2}\right] \end{aligned} \tag{14-87}$$

Then

$$\left.\begin{aligned} rE_\theta &= \frac{b\eta}{\sin\theta}\, e^{-j\beta r} \\ rH_\phi &= \frac{b}{\sin\theta}\, e^{-j\beta r} \\ E_r &= 0 \end{aligned}\right\} \tag{14-88}$$

The electric-field distribution for the TEM wave is seen to be the same as that obtained as a solution to Laplace's equation in the static case.

For the higher-order waves between the cones, the solution will be of the form

$$A_r(r, \theta) = [a\hat{J}_n(\beta r) + b\hat{N}_n(\beta r)][a_2 P_n(-\cos\theta) + b_2 P_n(\cos\theta)] \tag{14-89}$$

where in general n will have nonintegral values. Now recalling from (84) that E_r is proportional to A_r, and applying the boundary conditions $E_r = 0$ at $\theta = \psi$ and at $\theta = \pi - \psi$, it follows that the second bracketed term of (89), containing the θ function, must be zero at $\theta = \psi$ and at $\theta = \pi - \psi$. Applying these conditions, it is found that

$$b_2 = -a_2$$

and

$$P_n(\cos\psi) = P_n(-\cos\psi) \tag{14-90}$$

Equation (90) may be solved for n. When this is done there results,* for small-cone angles, ψ,

$$\begin{aligned} n &\approx (2m + 1) + \frac{1}{\ln \dfrac{2}{\psi}} \\ &\approx (2m + 1) + \frac{120}{Z_0} = (2m + 1) + \Delta \end{aligned} \tag{14-91}$$

where m is an integer, $\Delta = 120/Z_0$, and $Z_0 \approx 120 \ln 2/\psi$ is the characteristic impedance of the biconical antenna. As Z_0 approaches infinity

*S. A. Schelkunoff, *Electromagnetic Waves*, p. 446.

(that is, as the cone angle approaches zero) n approaches an integral value, and the transmission modes approach the corresponding free-space modes.

The appropriate radial functions in the antenna region are the spherical Bessel function $\hat{J}_n$ and $\hat{N}_n$, which represent standing waves. However, the $\hat{N}_n$ cannot be used as they become infinite at the origin, which is not excluded. Except for the zero order, the $\hat{K}_n$ functions are ruled out for the same reason. It follows that the higher-order waves ($n > 0$) in the antenna region will be given by

$$\left.\begin{aligned} A_r &= a\hat{J}_n(\beta r)\, T(\theta) \\ rH_\phi &= -a\hat{J}_n(\beta r)\frac{dT(\theta)}{d\theta} \\ rE_\theta &= -ja\eta\hat{J}'_n(\beta r)\frac{dT(\theta)}{d\theta} \\ j\omega\epsilon r^2 E_r &= n(n+1)\, a\hat{J}_n(\beta r)\, T(\theta) \end{aligned}\right\} \qquad (14\text{-}92)$$

where

$$T(\theta) = [P_n(\cos\theta) - P_n(-\cos\theta)] \qquad (14\text{-}93)$$

The current in the cones is proportional to rH_ϕ, evaluated at the surface, so the current associated with the higher-order waves can be obtained from the second of eqs. (92). Now for $n > 0$, and for $r \to 0$, $J_n(r)$ varies as r^n which, of course, goes to zero as $r \to 0$. Therefore the current *at the input* $I_n(0)$, associated with the higher-order waves is zero. Also the voltage caused by the higher-order waves and taken along *any* meridian between the cones, can also be shown to be zero, for

$$\begin{aligned} V_n(r) &= \int_\psi^{\pi-\psi} rE_\theta\, d\theta \\ &= -j\eta a\hat{J}'_n(\beta r)\,[T(\pi-\psi) - T(\psi)] = 0 \end{aligned} \qquad (14\text{-}94)$$

for $n > 0$. Therefore the input voltage and current, and hence the input impedance, *depend only on the principal or TEM wave*. This is a very important result, because it makes it possible, without approximation, to treat the input impedance of the biconical antenna as the input impedance of a transmission line that is terminated in an appropriate impedance.

14.14 Equivalent Transmission-line and Terminal Impedance. Considering the biconical antenna as a transmission line, the voltage and current at a distance r from the origin or input terminals will be

$$\begin{aligned} V(r) &= V_0(r) \\ I(r) &= I_0(r) + \bar{I}(r) \end{aligned} \qquad (14\text{-}95)$$

where V_0 and I_0 are the principal mode ($n = 0$) values, and $\bar{I}$ is the "complementary" current due to all the higher-order waves. As has already been noted, $\bar{I}(0) = 0$. Then, in terms of principal mode values, the lossless transmission-line equations may be written (Fig. 14-13)

$$V_0(r) = V_0(H) \cos \beta(H - r) + jZ_0 I_0(H) \sin \beta(H - r)$$

$$I_0(r) = I_0(H) \cos \beta(H - r) + j\frac{V_0(H)}{Z_0} \sin \beta(H - r)$$

where $Z_0 = 120 \ln \cot \psi/2$ is the characteristic impedance of the coaxial cones. The input impedance will be

$$Z_i = \frac{V_0(0)}{I_0(0)} = Z_0 \left[\frac{V_0(H) \cos \beta H + jZ_0 I_0(H) \sin \beta H}{Z_0 I_0(H) \cos \beta H + jV_0(H) \sin \beta H}\right] \tag{14-96}$$

The equivalent terminal impedance Z_t will be

$$Z_t = \frac{V_0(H)}{I_0(H)} = \frac{V(H)}{I(H) - \bar{I}(H)} \tag{14-97}$$

The equivalent terminal admittance is

$$Y_t = \frac{1}{Z_t} = \frac{I(H)}{V(H)} - \frac{\bar{I}(H)}{V(H)} = Y_{\text{caps}} + Y' \tag{14-98}$$

The current $I(H)$ is the total current on the antenna at $r = H$ and is, therefore, just the current flow out of and into the spherical caps that are assumed to close the ends of the antenna. That is, $I(H)$ is the current flow through the capacitance between the caps, and $I(H)/V(H)$ is the admittance between the two caps. For thin antennas, the capacitance between caps is very small and $I(H)$ is approximately zero. Then

$$I_0(H) + \bar{I}(H) = I(H) \approx 0$$

or

$$I_0(H) \approx -\bar{I}(H) \tag{14-99}$$

and

$$Y_t = \frac{I_0(H)}{V(H)} \approx -\frac{\bar{I}(H)}{V(H)}$$

also

$$Z_t = \frac{V(H)}{I_0(H)} \approx -\frac{V(H)}{\bar{I}(H)}$$

In general, the terminal admittance consists of two admittances in parallel as diagrammed in Fig. 14-16. The admittance, Y_{caps}, between caps of small radius is approximately just the capacitive susceptance $j\omega C$, where C is the electrostatic capacitance between the caps. This may be obtained by calculating the capacitance between the outside surfaces of two thin disks having radii very much smaller than their separation. The capacitance of an isolated thin circular disk treated as a very flat spheroid is found* to be $20a/9\pi$ pF where a is the

*J. H. Jeans, *Electricity and Magnetism*, Cambridge Press, London, 1946, p. 249.

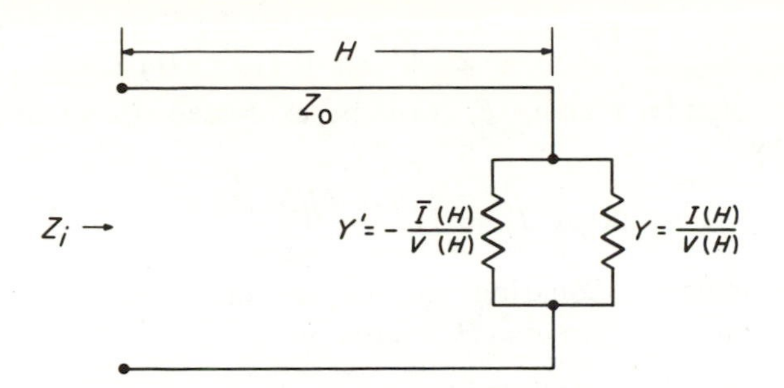

Figure 14-16. The terminal admittance $Y_t (Y_t = Y_1 - Y$ and $Z_t = 1/Y_t)$.

radius in centimeters, so the admittance between caps will be

$$Y_{\text{caps}} = \frac{ja}{30\lambda} \qquad \text{mhos}$$

where a is now the radius of the circular disks in meters.

The other part of the terminal admittance, $Y' = -\bar{I}(H)/V(H)$, is calculated from the higher-order current waves at $r = H$. When the terminal impedance Z_t has been determined the input impedance will be given by

$$Z_i = Z_0 \left(\frac{Z_t \cos \beta H + jZ_0 \sin \beta H}{Z_0 \cos \beta H + jZ_t \sin \beta H} \right) \tag{14-100}$$

Schelkunoff has carried out the evaluation of Y', and hence Z_t, in the following manner. Since the detailed calculations are lengthy, only an outline of the method is given here.

First, expressions for E_r in the antenna region and in the outside or free-space region are written and compared. For the antenna region, the resultant field due to the higher-order waves can be expressed in the form

$$2\pi j\omega\epsilon r^2 E_r = \sum_n a_n \frac{\hat{J}_n(\beta r)}{\hat{J}_n(\beta H)} T_n(\theta) \tag{14-101}$$

where $T_n(\theta)$ is defined by (93) and where n is nonintegral, being defined by (91). Making use of (84) and (75), the corresponding expression for the complementary current will be

$$\begin{aligned} \bar{I}(r) &= 2\pi r \sin \psi H_\phi |_{\theta=\psi} \\ &= -\sum_n \frac{a_n \hat{J}_n(\beta r)}{n(n+1)\hat{J}_n(\beta H)} \sin \psi \frac{dT_n(\psi)}{d\psi} \end{aligned} \tag{14-102}$$

As $\psi \to 0, Z_0 \to \infty$, and $n \to 2m + 1 + \Delta$,

so that
$$\frac{dT_n(\psi)}{d\psi} \approx \frac{\Delta}{\psi} = \frac{120}{Z_0 \psi}$$

Then, for thin antennas,

$$\bar{I}(r) = -\frac{120}{Z_0}\sum_n \frac{a_n \hat{J}_n(\beta r)}{n(n+1)\hat{J}_n(\beta H)} \tag{14-103}$$

In the outside region, E_r can be expressed in spherical Hankel functions by

$$2\pi j\omega r^2 E_r = \sum_{\bar{n}=1}^{\infty} \frac{b_{\bar{n}} \hat{K}_{\bar{n}}(j\beta r)\, P_{\bar{n}}(\cos\theta)}{\hat{K}_{\bar{n}}(j\beta H)} \tag{14-104}$$

where $\bar{n}$ is integral. Equating the expressions (104) and (101) for E_r at the boundary surface $r = H$ results in

$$\sum_n a_n T_n(\theta) = \sum_{\bar{n}=1}^{\infty} b_{\bar{n}} P_{\bar{n}}(\cos\theta)$$

Now, as $\psi \to 0$ and $Z_c \to \infty$, then $n \to 2m + 1$, and $T_n(\theta) \to P_{2m+1}(\cos\theta)$. Therefore, in the limit, for infinitely thin antennas, $a_n = b_{2m+1} = \bar{n}$. Then for thin antennas, it is permissible to use the b_{2m+1} terms as first approximations for the a_n terms. The expression for the complementary current on thin antennas is then given by

$$\bar{I}(r) \approx -\frac{60}{Z_c}\sum_{m=0}^{\infty} \frac{b_{2m+1}\hat{J}_{2m+1}(\beta r)}{(2m+1)(m+1)\,\hat{J}_{2m+1}(\beta H)} \tag{14-105}$$

The b_{2m+1} terms can be evaluated by again considering the limiting case as $\psi \to 0$ and $Z_0 \to \infty$. For very thin antennas the current distribution approaches the sinusoidal distribution of the principal wave

$$I(r) = I_0 \sin\beta(H - r)$$

with

$$I_0 = \frac{jV_0(H)}{Z_0} \tag{14-106}$$

For this distribution the fields have been calculated in chap. 10. By expanding in terms of Legendre polynomials the distant field expression for E_r obtained from chap. 10 and comparing it with eq. (104), the b coefficients can be evaluated.

The result is:

$$b_{2m+1} = -jI_0(4m+3)\,\hat{J}_{2m+1}(\beta H)[\hat{J}_{2m+1}(\beta H) - j\hat{N}_{2m+1}(\beta H)]$$

Inserting this in (105) and combining with (104) gives for the complementary current

$$\bar{I}(r) = -\frac{60V_0(H)}{Z_0^2}\sum_{m=0}^{\infty}\frac{4m+3}{(m+1)(2m+1)}$$

$$[\hat{J}_{2m+1}(\beta H) - j\hat{N}_{2m+1}(\beta H)]\,\hat{J}_{2m+1}(\beta r) \tag{14-107}$$

Then the terminal admittance is

$$Y_t \approx -\frac{\bar{I}(H)}{V(H)} = \frac{Z_a(\beta H)}{Z_0^2} = \frac{R_a(\beta H) + jX_a(\beta H)}{Z_0^2} \tag{14-108}$$

where $$R_a(\beta H) = 60 \sum_{m=0}^{\infty} \frac{4m+3}{(m+1)(2m+1)} \hat{J}^2_{2m+1}(\beta H) \tag{14-109}$$

$$X_a(\beta H) = -60 \sum_{m=0}^{\infty} \frac{4m+3}{(m+1)(2m+1)} \hat{J}_{2m+1}(\beta H)\, \hat{N}_{2m+1}(\beta H)$$

The terminal impedance Z_t is

$$Z_t = \frac{Z_0^2}{Z_a(\beta H)} \tag{14-108a}$$

$Z_a(\beta H)$ is the *inverse* of the terminal impedance. The input impedance of a quarter-wave section of lossless line having a characteristic impedance Z_0 and terminated in $Z_a(\beta H)$ is Z_t.

Although it is possible to calculate Z_a directly from expressions (109), the series converge slowly and are not useful for computations except when βH is small. Schelkunoff has circumvented this difficulty by providing an ingenious alternative method for calculating Z_a. Using (100), the input-impedance can be expressed in terms of Z_a by

$$Z_i = Z_0 \frac{Z_a \sin \beta H - jZ_0 \cos \beta H}{Z_0 \sin \beta H - jZ_a \cos \beta H} \tag{14-110}$$

$$= \frac{Z_a - jZ_0 \cot \beta H}{1 - jZ_a/Z_0 \cot \beta H}$$

As $Z_0 \rightarrow \infty$,

$$Z_i \approx (Z_a - jZ_0 \cot \beta H)\left(1 + j\frac{Z_a}{Z_0} \cot \beta H\right)$$

$$Z_i \rightarrow \frac{Z_a}{\sin^2 \beta H} - jZ_0 \cot \beta H \tag{14-110a}$$

and, since the input current approaches $I_0 \sin \beta H$, the input power (complex) becomes

$$\tfrac{1}{2} Z_i I_0^2 \sin^2 \beta H = \tfrac{1}{2} [Z_a - jZ_0 \sin \beta H \cos \beta H]\, I_0^2$$

However, as $Z_0 \rightarrow \infty$, the current distribution approaches the sinusoid, and the complex input power can be obtained by the induced-emf method of secs. 14.05 and 14.06. The real part, which gives the radiation resistance, will be *independent* of the antenna shape (for thin antennas) and will, therefore, be the same as that already calculated for the infinitely thin cylindrical antenna. However, the reactive part, which determines the reactance, will be a function of shape even for thin antennas, and so must be calculated for conical antennas (with $\psi \rightarrow 0$). Using this approach, Schelkunoff obtains the following results

$$R_a(\beta H) = 60(\gamma + \ln 2\beta H - \mathrm{Ci}\, 2\beta H) + 30(\gamma + \ln \beta H - 2\,\mathrm{Ci}\, 2\beta H + \mathrm{Ci}\, 4\beta H)\cos 2\beta H + 30(\mathrm{Si}\, 4\beta H - 2\,\mathrm{Si}\, 2\beta H)\sin 2\beta H$$

$$X_a(\beta H) = 60\,\mathrm{Si}\, 2\beta H + 30(\mathrm{Ci}\, 4\beta H - \ln \beta H - \gamma)\sin 2\beta H - 30\,\mathrm{Si}\, 4\beta H \cos 2\beta H$$

where $\gamma = 0.5772$ (Euler's constant)

These expressions are plotted in Fig. 14-17. The input impedance is then obtained from

$$Z_i = Z_0 \frac{Z_a \sin \beta H - jZ_0 \cos \beta H}{Z_0 \sin \beta H - jZ_a \cos \beta H} \tag{14-110}$$

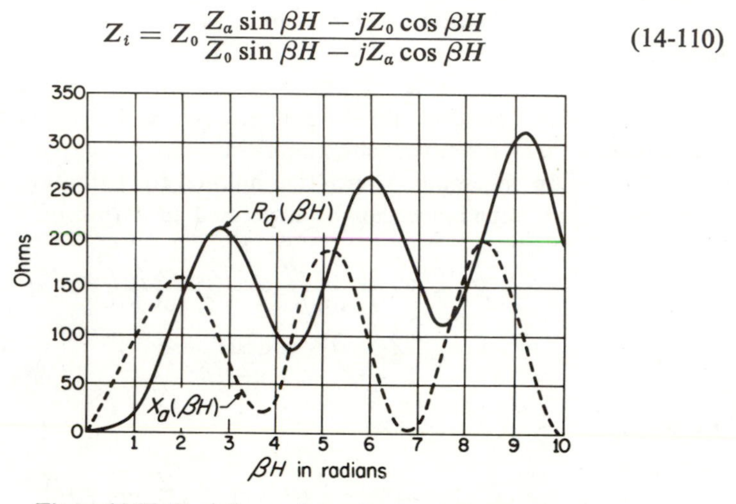

Figure 14-17. Resistive and reactive components of the inverse terminal impedance $Z_a = R_a + jX_a$.

It is important to note that, although the approximate relation (110a) was used in calculating Z_a, it is necessary to use the exact expression (110) for calculating Z_i. Use of the approximate expression here would lead to the same answer as is given by the induced-emf method.

The final result of this attack on the problem is seen to be a surprisingly simple one. The input impedance of the conical antenna is calculated as the input impedance of a lossless transmission line which is terminated by an impedance Z_t. This terminal impedance is just the inverse of an impedance Z_a, which can be calculated by the induced-emf method.

The input impedance of hollow conical antennas is shown in Fig. 14-18 for various values of Z_0. The term *hollow* refers to the fact that the cap capacitance has not been taken into account. For thin antennas, the cap capacitance has negligible effect, but for thicker

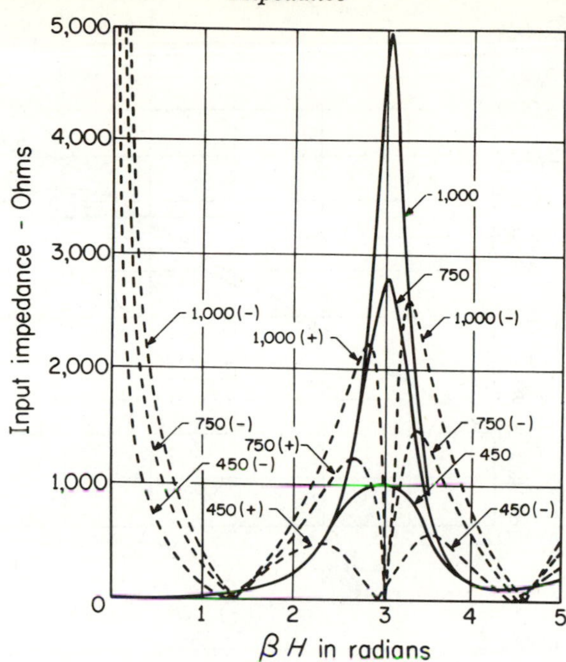

Figure 14-18. The input impedance of hollow conical antennas for various values of Z_0 as given by Schelkunoff.

antennas it must be accounted for. This can be done by adding the admittance Y_{caps} to the calculated value of $Y_t \approx Y'$. With this correction (108) becomes

$$Y_t = \frac{R_a(\beta H)}{Z_0^2} + j\left[\frac{X_a(\beta H)}{Z_0^2} + \omega C\right] \tag{14-111}$$

Two effects of considerable practical importance can be observed in the curves of Fig. 14-18. The first of these is that the fatter antennas (lower Z_0) have very much smaller impedance variations with frequency, so that a fat cone is inherently a wide-band antenna. The second effect is the shortening of the resonant length for the thicker antennas. For very thin antennas resonance occurs for lengths just slightly shorter than multiples of $\lambda/4$, but for thicker antennas the shortening effect becomes quite large, especially for first resonance. The cap capacitance acts to decrease the resonant lengths still further.

14.15 Impedance of Cylindrical Antennas. The analysis for conical antennas can be extended to cover antennas of other shapes in the following manner. If the transverse dimensions of the antenna are small, the waves along it will be nearly spherical, whatever its shape. Then such antennas can be treated as nonuniform transmission lines

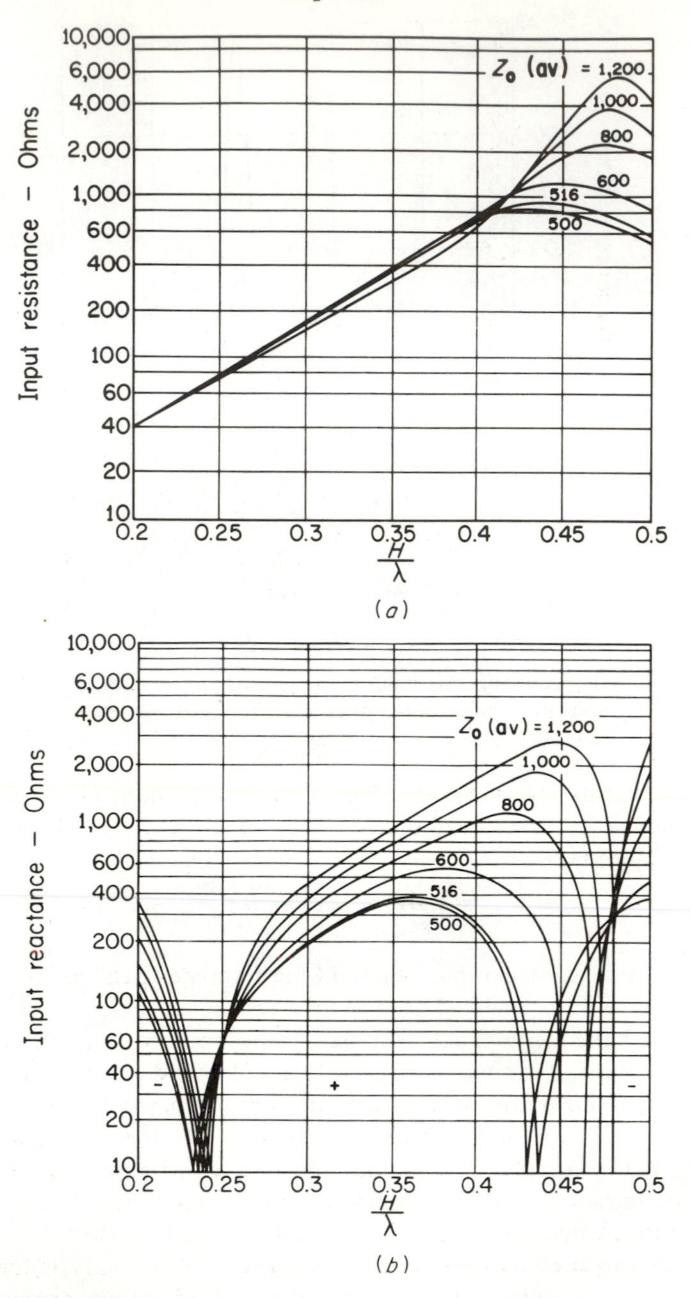

Figure 14-19. Input resistance (*a*) and input reactance (*b*) of hollow cylindrical center-fed dipole antennas as given by Schelkunoff. For monopole antennas of height H, divide the ordinates and the value shown for Z_0 (av) by two.

whose inductances and capacitances per unit length and characteristic impedance vary along the line. The terminating impedance will be as calculated from (108a), except that an "average" characteristic impedance must be used for Z_0. From the theory of nonuniform transmission lines, Schelkunoff has obtained for the input impedance of antennas

$$Z_i = Z_0(\text{av}) \left[\frac{R_a \sin \beta H + j[(X_a - N) \sin \beta H - (Z_0(\text{av}) - M) \cos \beta H]}{[(Z_0(\text{av}) + M) \sin \beta H + (X_a + N) \cos \beta H] - jR_a \cos \beta H} \right] \tag{14-112}$$

where, for cylindrical* dipoles of radius a and half-length H,

$$M = 60(\ln 2\beta H - \text{Ci } 2\beta H + \gamma - 1 + \cos 2\beta H)$$

$$N = 60(\text{Si } 2\beta H - \sin 2\beta H)$$

$$Z_0\,(\text{av}) = 120 \left(\ln \frac{2H}{a} - 1 \right)$$

In Fig. 14-19 are shown curves for the input resistance and reactance of hollow cylindrical antennas for various values of Z_0 (av). Z_0 (av) for cylindrical antennas of half-length H is plotted in Fig. 14-20 as

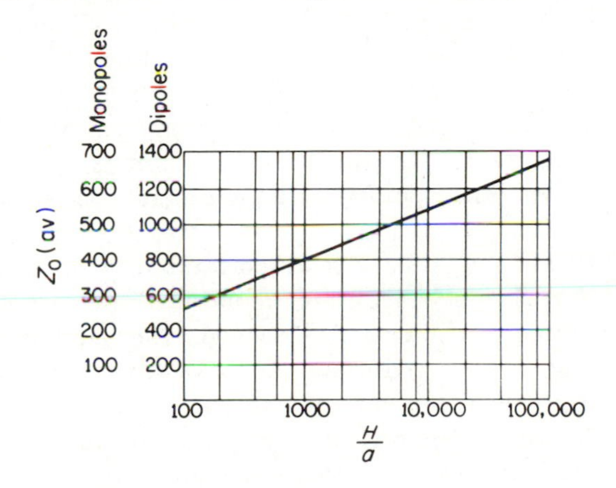

Figure 14-20. Average characteristic impedance, Z_0 (av), for cylindrical antennas.

a function of the ratio H/a. For a monopole antenna Z_0 (av) has just one-half the value it has for the corresponding dipole. The input resistance and reactance of a monopole antenna are just one-half those of the corresponding dipole antenna that has the same H/a. Therefore, the input impedance of monopole antennas can be obtained from Fig.

*For the M and N functions for antennas of other shapes the reader should refer to S. A. Schelkunoff, *Electromagnetic Waves,* D. Van Nostrand Co., Inc., Princeton, N. J., 1943, p. 463.

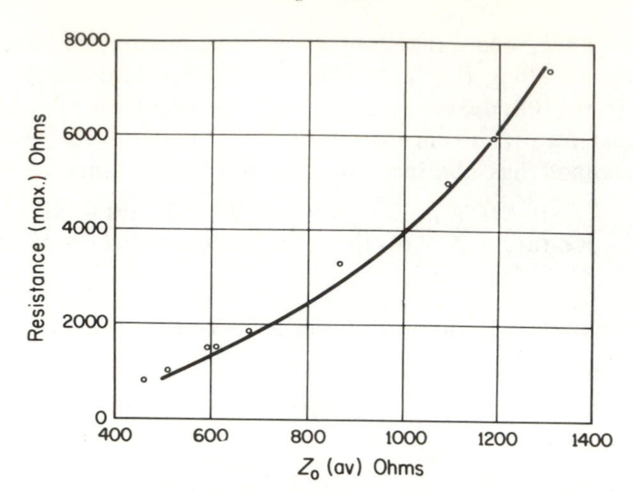

Figure 14-21. Resonant impedance of hollow cylindrical antennas as a function of Z_0 (av) at second resonance (antiresonance).

14-19 by dividing the ordinates and Z_0 (av) by 2. Figure 14-21 shows the resonant impedance of hollow cylindrical dipole antennas at the second-resonance points. Also shown are some experimental values (circles) obtained from various sources. The agreement between theory and experimental results at this critical point is seen to be quite good. Agreement at other points will, in general, be found to be even closer.

Schelkunoff's antenna theory is important for two reasons. First, it has provided reasonably accurate numerical answers over a fairly wide range of antenna dimensions. Second, the method itself is an excellent example of how and when to make approximations. In engineering, most problems are not amenable to exact solutions. Therefore the ability to make approximations can spell the difference between success and failure in the solution of the problem.

Problem 3. Using Schelkunoff's method, calculate the input impedance of a uniform cross section tower antenna at 1300 kHz. The tower is 400 ft high and $6\frac{1}{2}$ ft square. The base-insulator capacitance is 30 pf.

Problem 4. The antenna for a portable test transmitter consists of a tubular steel mast 2 in. in diameter and 50 ft high. The base insulator has an effective shunting capacitance of 15 pf. (a) If a test survey is to be made at 650 kHz, determine (1) the radiation resistance, (2) the antenna reactance, (3) the input impedance, including the effect of the base capacitance. (b) For 1 amp through an ammeter in the lead to the antenna, what is the current in the mast near the base? (c) For an ammeter reading of 1 amp, what is the field strength at 1 mile and how much power is radiated?

14.16 Circuit Relations and Field Theory. In the first part of this chapter the relations in field theory have been used to develop expression for the *impedance* of a straight wire in two rather special cases. In one case the wire was of finite length and was assumed to carry a sinusoidally distributed current. The impedance was calculated at the terminals. In the other case the wire was assumed infinitely long with a uniform current distribution, and the *impedance per unit length* was calculated. It is apparent that it should be possible, in a somewhat similar manner, to derive an expression for the impedance at the terminals of a wire circuit of any configuration. This is indeed the case, and it will be shown that the so-called *circuit relations*, by means of which the engineer solves for the current in a circuit in terms of the applied voltage and the circuit impedances, are derivable from field theory as special and approximate cases. Before carrying through such a derivation, it is desirable to re-examine circuit concepts for a simple closed circuit, to see how these concepts follow directly from the integral statement of Maxwell's equations.

Circuit Relations and Maxwell's Equations in the Integral Form. Consider the Maxwell emf equation (Faraday's law) applied to the simple circuit of Fig. 14-22 consisting of a loop of wire with terminals *a-b*.

$$\int_{(bca)} E_s\,ds + \int_{(ab)} E_s\,ds = -j\omega\Phi \tag{14-113}$$

where E_s is the component of $\mathbf{E}$ parallel to the wire and Φ is the magnetic flux through the loop.

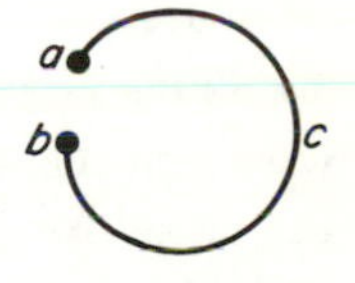

Figure 14-22.

The first integral is taken along the wire, and the second integral is along a straight line joining the terminals. The E_s along the path of integration is the self-induced electric field strength, produced by the charges and current in the circuit. The voltage V^0, which must be applied or impressed at the terminals of the circuit to transfer the charges against this self-induced field, will be equal and opposite to the second integral. That is,

$$V^0 = -\int_{(ab)} E_s\,ds$$

Then eq. (113) can be rewritten as

$$V^0 = -\int_{(ab)} E_s\,ds = \int_{(bca)} E_s\,ds + j\omega\Phi \tag{14-114}$$

Dividing through by the current I, an impedance equation is obtained.

$$Z = \frac{V^0}{I} = \frac{\int_{(bca)} E_s\,ds}{I} + \frac{j\omega\Phi}{I} \tag{14-115}$$

The first term on the right-hand side is the internal impedance Z_i of the wire, and the second term represents the external reactance. If the external inductance L_e is defined by Φ/I, then eq. (115) becomes

$$Z = Z_i + j\omega L_e \tag{14-116}$$

In the d-c case ($\omega = 0$), the external reactance is zero and the internal impedance is the resistance of the wire. In the alternating case, the internal impedance Z_i is complex, and consists of a resistance R_i and an internal reactance $X_i = j\omega L_i$. If a perfectly conducting wire is assumed, the first integral on the right-hand side of (114) is zero, and the applied voltage is equal to the external reactive voltage drop $j\omega\Phi = j\omega L_e I$. For an actual conductor the total inductance L is the sum of the external and internal inductances, that is $L = L_e + L_i$. In practice $L_i \ll L_e$, so that the total inductance L is very nearly equal to the external inductance L_e. The inductance L can be increased by winding the wire in the form of a coil. In this manner the magnetic flux per ampere is increased, and the same magnetic flux is caused to link several turns. If the inductive reactance of the coil is large enough so that the inductive reactances of the rest of the circuit may be neglected, the inductance is said to be "lumped."

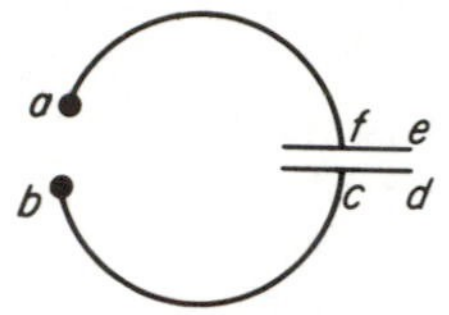

Figure 14-23.

If a capacitor is connected in series with the loop as in Fig. 14-23, the emf equation becomes

$$\int_{(bcd)} E_s\,ds + \int_{(de)} E_s\,ds + \int_{(efa)} E_s\,ds + \int_{(ab)} E_s\,ds = -j\omega\Phi \tag{14-117}$$

The first and third terms are due to the internal impedance of the wire and capacitor plates. The second term is the voltage between the capacitor plates. This is proportional to the charge Q on the plates, and the ratio

$$\frac{Q}{\int_{(de)} E_s\,ds}$$

is the capacitance C of the capacitor. Q is related to the current I by

$$Q = \int I\,dt = \frac{I}{j\omega}$$

so the second term of eq. (117) may be written

$$\int_{(de)} E_s\,ds = \frac{1}{j\omega C} I$$

The applied voltage V^0 is equal to the negative of the fourth integral, so eq. (117) becomes

$$V^0 = Z_i I + j\omega\Phi + \frac{1}{j\omega C} I \tag{14-118}$$

If the small internal reactance of the wire is lumped with the external inductive reactance, (118) may be written as

$$V^0 = I\left(R + j\omega L + \frac{1}{j\omega C}\right) \tag{14-119}$$

where L is now the total or effective inductance of the circuit. This is the usual form of the circuit equation.

There are several approximations and assumptions involved in writing eq. (119). Some of these will be evident from the manner of its derivation from eq. (113), but others are more obscure. They will be listed here and discussed in greater detail later in this section.

(1) The current I has been assumed to have the same magnitude in all parts of the circuit. This means that "distributed capacitance" effects, or displacement currents from one conductor to another have been neglected.

(2) An inductance L has been defined for low frequencies (actually at $\omega = 0$) and has then used in (119) as though it were independent of frequency.

(3) Retardation effects, e.g., radiation, have been neglected.

At power frequencies, the approximations are excellent and the neglected quantities are indeed negligible. However, at radio frequencies and more especially at ultrahigh frequencies, some of the neglected factors become important, and the circuit approach breaks down unless appropriate steps are taken to make circuit concepts carry over, for example, by generalizing definitions. Generalized definitions for circuit constants can be obtained by considering the circuit as a problem in field theory. The direct derivation of **E** and **H** from Maxwell's equations in the integral form was easily done for the *closed* or *quasi-closed* circuits of Figs. 14-22 and 14-23. However, for open circuits such as antennas, where radiation is important, it is generally simpler to obtain **E** and **H** indirectly through the retarded potentials **A** and V. It is instructive to use this more general field method to derive the simple circuit relations already considered. Such a derivation points up the approximations involved in the latter relations and indicates the extent of the errors incurred when ordinary circuit theory is used at high frequencies. In addition, generalized definitions can be obtained for the circuit "constants," by means of which it becomes possible to extend the use of the circuit approach to the ultrahigh frequencies.

14.17 Derivation of Circuit Relations from Field Theory. The electric circuit laws of Ohm, Faraday, and Kirchhoff were based on experimental observations and antedated the electromagnetic theory of Maxwell and Lorentz. Indeed, the theory was developed as a generalization from these simpler and more restricted laws. It is interesting, but not surprising, then, to find that the circuit relations are just special cases of the more general field relations, and that they may be developed from the latter when suitable approximations are made. Nevertheless, the importance of the simple (and approximate) circuit relations should not be underestimated. With these beautifully simple relations the electrical engineer has been enabled to design and construct electrical systems and circuits of amazing intricacy. Without the simplifying assumptions of circuit theory the vast power and communication networks of today would not have been possible, for the exact field solutions to many of the problems would have been of overwhelming complexity.

In this section the circuit relations dealing with voltages and currents will be derived as special cases of electromagnetic field theory, which deals with charge and current densities and their associated fields.

Consider again the simple series circuit of Fig. 14-24 for which can be written the circuit equation

$$V^0 = IR + \frac{I}{j\omega C} + j\omega LI \tag{14-120}$$

The applied or impressed voltage V^0 is assumed to be independent of the resultant current I. Equation (120) can be rewritten in the form

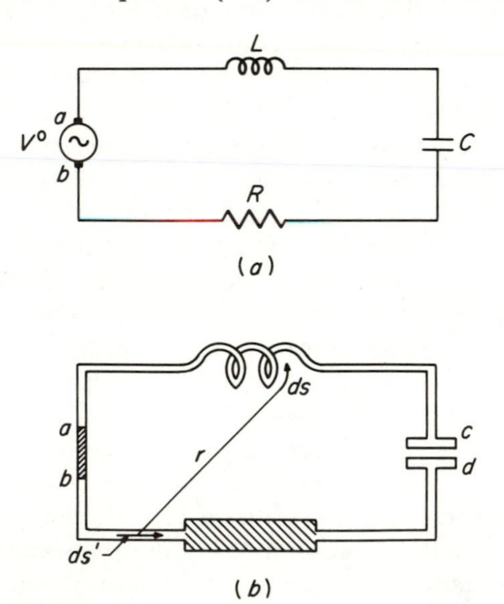

Figure 14-24. (*a*) A simple *RLC* circuit. (*b*) A representation suitable for the application of field theory.

$$V^0 + V' = V_R \tag{14-121}$$

where
$$V' = -j\omega LI - \frac{1}{j\omega C} I \tag{14-122}$$

is the sum of the reactive voltages across the circuit elements and

$$V_R = IR \tag{14-123}$$

is the net voltage left to drive the current I through the resistance R after the reactive voltage drops have been subtracted from the applied voltage.

In field theory relations similar to (121), (122), and (123) may be written for electric fields and conduction current densities. Thus at the surface of, or within, a conductor, the conduction current density is given by Ohm's law.

$$\frac{\mathbf{J}}{\sigma} = \mathbf{E} \tag{14-124}$$

where σ is the conductivity of the conductor and $\mathbf{E}$ is the *total* electric field strength tangential to the surface. In general, this total $\mathbf{E}$ is the sum of an *applied* or *impressed* electric field strength $\mathbf{E}^0$ and a "self-induced" or back electric field strength $\mathbf{E}'$ that is due to the charges and currents in the system. In mathematical terms,

$$\mathbf{E} = \mathbf{E}^0 + \mathbf{E}' \tag{14-125}$$

The impressed field strength $\mathbf{E}^0$ is assumed to be independent of the charges and currents in the system under consideration. This would be the case for example if $\mathbf{E}^0$ were the electric field of a *distant* antenna. In this circumstance, to use circuit terminology, the coupling between the two systems is sufficiently loose that the charges and currents in the second system do not affect (to any significant extent) the current flowing in the distant antenna.

The "self-induced" electric field strength $\mathbf{E}'$ that is due to the charges and currents in the system under consideration may be determined from Maxwell's equations, either directly or through the scalar and vector potentials. In terms of the potentials,

$$\mathbf{E}' = -\nabla V - j\omega \mathbf{A} \tag{14-126}$$

where V and $\mathbf{A}$ are related to the charge and current densities of the system through

$$V = \frac{1}{4\pi\epsilon}\int_V \frac{\rho\left(t - \dfrac{r}{c}\right)}{r}\,dV \qquad \mathbf{A} = \frac{\mu}{4\pi}\int_V \frac{\mathbf{J}\left(t - \dfrac{r}{c}\right)}{r}\,dV \tag{14-127}$$

Then, rewriting (125) and using (124) and (126), the field relations at the surface of a conductor may be written as

$$\mathbf{E}^0 = \frac{\mathbf{J}}{\sigma} + \nabla V + j\omega\mathbf{A} \tag{14-128}$$

Integrating along the *conducting* portion of a circuit (Fig. 14-23), the general circuital relation is obtained, viz.:

$$\int_{(cad)} E_s^0\, ds = \int_{(cad)} \frac{J_s}{\sigma}\, ds + \int_{(cad)} \frac{\partial V}{\partial s}\, ds + \int_{(cad)} j\omega A_s\, ds \tag{14-129}$$

Making suitable assumptions and approximations, the general relation (129) can be reduced to the simple circuit equation (120). The steps required to derive (120) from (129) show clearly the approximations involved in the simple circuit equation.

In eq. (129) the path of integration is taken along the surface of the conductor parallel to its axis. The expression (128) cannot be integrated across the capacitor gap, because there both $\mathbf{J}$ and σ are zero and the first expression on the right-hand side is indeterminate. If there is no series capacitor in the circuit, the points c and d are coincident, and the integration is performed around a *completely closed* conducting path. For this case of a completely closed path the circuital relation corresponding to (129) would be

$$\oint E_s^0\, ds = \oint \frac{J_s}{\sigma}\, ds + \oint j\omega A_s\, ds \tag{14-130}$$

The second term on the right-hand side of (129) has dropped out because the gradient of a scalar potential integrated around a *closed* path is always zero. That is

$$\oint \nabla V \cdot d\mathbf{s} = \oint \frac{\partial V}{\partial s}\, ds \equiv 0$$

The various terms of eq. (129) will now be considered one at a time. The term on the left-hand side of (129) evidently corresponds to the applied voltage V^0. In circuit work V^0 is supplied by an electric generator, which is usually a complicated circuit in itself. However, for purposes of solving for voltages and currents in the circuit under consideration (the driven circuit), V^0 is assumed to be supplied across a pair of terminals by a zero-impedance generator, or by a zero-impedance generator connected in series with a lumped impedance equal to the generator impedance. Similarly, in considering the field relations, the impressed or applied field $\mathbf{E}^0$ usually exists along a complicated configuration of conductors (in the generator winding) and may extend over an appreciable portion of the circuit under consideration. However, for purposes of analysis the impressed field $\mathbf{E}^0$ is often assumed to exist only along a section of conductor of very short length; that is, a "point" or "slice" generator is assumed to exist between the points a and b [Fig. 14-24(a)], so that the applied voltage is

$$\int_c^d E_s^0\, ds = \int_a^b E_s^0\, ds = V^0 \tag{14-131}$$

Now consider the first term on the right-hand side of eq. (129). For the direct-current case, the interpretation of this term would be very simple. The current density $\mathbf{J}_s$ would be uniform throughout the conductor cross section and would be given by

$$J_s = \frac{I}{A} \tag{14-132}$$

where A is the cross-sectional area of the conductor. The conductivity is the reciprocal of the resistivity ρ, and so

$$\frac{1}{A\sigma} = \frac{\rho}{A} = R'$$

where R' is the resistance per unit length of the conductor. Then

$$\frac{J_s}{\sigma} = \frac{J_s}{A} \cdot \frac{A}{\sigma} = IR' \tag{14-133}$$

is just the voltage drop (due to resistance) per unit length, and the first term on the right-hand side of (129) becomes

$$\int_c^d \frac{J_s}{\sigma}\, ds = \int_c^d IR'\, ds = IR \tag{14-134}$$

which is the total IR drop around the circuit.

From (133) and (124) it is seen that for the direct-current case, the ratio of total tangential electric field strength E to total current I is the resistance per unit length, that is

$$\frac{E}{I} = R'$$

For alternating currents, especially at high frequencies, the current density is no longer uniform throughout the cross section of the conductor. Instead it varies—both in magnitude and phase through the cross section—so that the total current I, in general, differs in phase from the current density at the surface of the conductor. The ratio E/I is now complex and defines z_i, the "internal impedance" per unit length of the conductor. Now the first term on the right-hand side of eq. (129) may be written

$$\int_c^d \frac{J_s}{\sigma}\, ds = \int_c^d E_s\, ds = \int_c^d Iz_i\, ds \tag{14-135}$$

When the current I is uniform around the circuit, as it is in the low-frequency case, (135) becomes

$$\int_c^d \frac{J_s}{\sigma}\, ds = I \int_c^d z_i\, ds = IZ_i \tag{14-136}$$

$Z_i = R_i + jX_i$ is the so-called internal impedance of the conductors of the circuit. For direct-current it reduces to the circuit resistance. Even in the case of high-frequency alternating currents, the internal reactance X_i is very small compared with the "external reactance" of the circuit obtained from the second and third terms on the right-hand side of (129) and may usually be neglected. In any event, in circuit work the internal reactance of the conductors is usually lumped with the external reactance to give the total circuit reactance, and the resistive or in-phase component of the first term of (129) is shown explicitly as IR.

Consider next the second term on the right-hand side of eq. (129). When integrated around a closed path, as in the case of a circuit containing only resistance and inductance, this term is zero. However, for a circuit with a capacitor [Fig. 14-24(*b*)], where the integration is carried from one plate c to the other plate d, there results

$$\int_c^d \frac{\partial V}{\partial s}\, ds = V_d - V_c \tag{14-137}$$

This is the potential difference between the plates of the capacitor. If these plates are considered to be very close together, and if the charge distributed along the wire is small compared with the charge concentrated on the capacitor plates (that is, if stray capacitance is negligible compared with the capacitance of the capacitor), the potential difference $(V_d - V_c)$ will be proportional to the charge on the capacitor plates. That is

$$V_d - V_c = \frac{Q}{C} = \frac{I}{j\omega C} \tag{14-138}$$

The proportionality factor C is just the capacitance of the capacitor as defined for the static case. Thus, for the second term of eq. (129), it is possible to write

$$\int_c^d \frac{\partial V}{\partial s}\, ds = \frac{I}{j\omega C} \tag{14-139}$$

Finally, consider the third term on the right-hand side of eq. (129). This could be written

$$\int_c^d j\omega A_s\, ds = j\omega L' I' \tag{14-140}$$

where

$$L' = \frac{1}{I'} \int_c^d A_s\, ds \tag{14-141}$$

is a "generalized inductance" of the circuit. This generalized inductance depends both on the circuit geometry and on the current distribution, and, as defined by (141), it also depends upon where in the circuit the current I' is measured. For low frequencies, where the current amplitude is constant around the circuit, this generalized definition reduces to a well-known formula for low-frequency inductance (148). To see how the "inductance" of the circuit changes as the frequency increases, it is necessary to examine more closely the integral expression of equation (141).

For a current flowing in a thin wire, the expression for the vector potential at any point in space, due to an elemental length, is

$$d\mathbf{A} = \frac{\mu}{4\pi} \frac{I\, d\mathbf{s}'\, e^{-j\beta r}}{r} \tag{14-142}$$

where I is the integrated value of current density over the cross section of the wire, and r is the distance from an element of length ds' along the center of the wire to the point at which $\mathbf{A}$ is evaluated. The total vector potential due to current flow in the entire circuit will be

$$\mathbf{A} = \frac{\mu}{4\pi} \oint \frac{I\, e^{-j\beta r}}{r}\, d\mathbf{s}' \tag{14-143}$$

In the third term of eq. (129) the component of $\mathbf{A}$ parallel to the axis of the wire is evaluated at the surface of the wire, and integrated around the conducting part of the circuit from c to d. This term can then be written

$$j\omega \int_c^d A_s\, ds = j\omega \int_c^d \oint \frac{I\, e^{-j\beta r}}{4\pi r}\, d\mathbf{s}' \cdot d\mathbf{s} \tag{14-144}$$

In the general case I varies with the position of $d\mathbf{s}'$ and must be retained under the integral sign. The usual low-frequency approximations are to assume that I is constant around the circuit with no change of phase, and also to neglect the phase-shift factor $e^{-j\beta r}$. At very high frequencies where the circuit dimensions become appreciable fractions of a wavelength, both of these approximations lead to error. However, the effect of neglecting the phase-shift factor is much the more important because it is responsible for radiation from the circuit. For circuits that are not too large in terms of wavelengths—say less than one-tenth wavelength around—the current distribution usually departs a surprisingly small amount from the low-frequency, constant-amplitude, constant-phase condition. (The reason for this can be seen by considering the current amplitude and phase variations along a short-circuited low-loss transmission line, the length of which is less than one-twentieth of a wavelength.) Because of these facts it is often permissible to neglect variations of current amplitude and phase around the circuit while still accounting for the phase-shift factor $e^{-j\beta r}$. Under such conditions (144) becomes

$$j\omega I \int_c^d \oint \frac{\mu\, e^{-j\beta r}}{4\pi r}\, d\mathbf{s}' \cdot d\mathbf{s} \tag{14-145}$$

Comparison with $j\omega LI$ shows that

$$\int_c^d \oint \frac{\mu\, e^{-j\beta r}}{4\pi r}\, ds' \cdot ds \tag{14-146}$$

is the factor that, at low frequencies, is identified as the inductance of the circuit. At frequencies sufficiently low that the phase shift is negligible (that is, for which $\beta r <<< 1$)

$$e^{-j\beta r} \approx 1$$

and the low-frequency inductance is

$$L_{\mathrm{LF}} = \int_c^d \oint \frac{\mu\, ds' \cdot ds}{4\pi r} \tag{14-147}$$

If the circuit is *closed*, c and d coincide and

$$L_{\mathrm{LF}} = \oint \oint \frac{\mu\, ds' \cdot ds}{4\pi r} \tag{14-148}$$

which is known as Neumann's formula for the external inductance of a circuit.

At higher frequencies, where it is no longer permissible to neglect the factor $e^{-j\beta r}$, its effect can be determined by expanding it in series form and using the first few terms·

$$e^{-j\beta r} = 1 - j\beta r - \frac{\beta^2 r^2}{2!} + \frac{j\beta^3 r^3}{3!} + \frac{\beta^4 r^4}{4!} - \dots$$

$$= \left(1 - \frac{\beta^2 r^2}{2!} + \dots\right) - j\left(\beta r - \frac{\beta^3 r^3}{3!} + \dots\right)$$

It is seen that the expression (146) for "inductance" now has both real and imaginary parts. The real part represents the high-frequency external inductance of the circuit. The imaginary part is the so-called radiation resistance of the circuit. From expression (145) it is evident that this imaginary part combines with the factor $j\omega I$ to yield a voltage *in-phase* with I. The power required to drive I against this in-phase component of voltage is radiated from the circuit.

The value of the radiated power is given by

$$W_{\mathrm{rad}} = I^2 \int_c^d \oint \frac{j\omega\mu}{4\pi r}\left(- j\beta r + j\frac{\beta^3 r^3}{3!} - \dots\right) ds' \cdot ds = I^2 R_{\mathrm{rad}} \tag{14-149}$$

where R_{rad} is the radiation resistance of the circuit and is given by

$$R_{\mathrm{rad}} = \int_c^d \oint \mu \left(\frac{\omega^2}{4\pi c} - \frac{\omega^4 r^2}{24\pi c^3} - \dots\right) ds' \cdot ds \tag{14-150}$$

When integrated around a closed path, the first term drops out, leaving

$$R_{\text{rad}} = 10^{-7} \int_c^d \oint \left(-\frac{\omega^4 r^2}{3!\,c^3} + \frac{\omega^6 r^4}{5!\,c^5} - \dots \right) d\mathbf{s}' \cdot d\mathbf{s} \qquad (14\text{-}151)$$

Consideration of the real part of expression (146) shows how the inductance depends upon the phase factor $e^{-j\beta r}$.

$$L = \int_c^d \oint \frac{\mu}{4\pi r} \left(1 - \frac{\beta^2 r^2}{2!} + \dots \right) d\mathbf{s}' \cdot d\mathbf{s} \qquad (14\text{-}152)$$

$$= 10^{-7} \int_c^d \oint \frac{1}{r} \left(1 - \frac{\omega^2 r^2}{2c^2} + \dots \right) d\mathbf{s}' \cdot d\mathbf{s}$$

Using eqs. (136), (137), (139), and (140), it is seen that at low frequencies eq. (129) reduces directly to eq. (120). Comparison of eqs. (129) and (120) shows clearly the approximations involved in the simple circuit relations, and makes it possible to determine the magnitude of the neglected factors. With this knowledge circuit concepts may be extended to much higher frequencies.

The extension of circuit concepts to higher frequencies is accomplished in practice by the addition of appropriately located lumped-circuit constants. For example, "distributed" inductance and capacitance effects are accounted for by suitably located series inductors and shunt capacitors, and radiation effects by the inclusion of a "radiation resistance." An outstanding example in electrical engineering of the extension of circuit concepts to systems not necessarily small in wavelengths is the ordinary transmission line. Here, by suitably representing the distributed constants of the line by lumped constants, a circuit results that can be solved by ordinary circuit methods. Although the circuit is complicated, the solution is relatively simple in the important practical case of a uniform transmission line. In this manner it is possible in some problems to extend circuit concepts even to the microwave range.

BIBLIOGRAPHY

Articles

Bouwkamp, C. J., "Hallén's Theory for a Straight Perfectly Conducting Wire, Used as a Transmitting or Receiving Aerial, *Physica*, **9**, July, 1942, pp. 609–631.

Brillouin, L., "The Antenna Problem," *Quart. Appl. Math.*, **1**, October, 1943, pp. 201–214.

Burgess, R. E., "Aerial Characteristics," *Wireless Engineer*, **21**, No. 247, April, 1944 p. 154.

Carter, P. S., "Circuit Relations in Radiating Systems," *Proc. IRE*, **20**, June, 1932 p. 1004.

Hallén, E., "Theoretical Investigations into the Transmitting and Receiving Qualities of Antennae," *Nova Acta*, Upsala, **11**, No. 4 (1938).

———, "Further Investigations into the Receiving Qualities of Antennae: The Absorbing of Transient, Unperiodic Radiation, *Upsala Universitets Årsskrift*, 1939, No. 4.

———, "Admittance Diagrams for Antennas and the Relation between Antenna Theories," Cruft Laboratory Report No. 46, Harvard University, Cambridge, Mass., June 1, 1948.

King, R. and D. Middleton, "The Cylindrical Antenna: Current and Impedance," *Quart. Appl. Math.*, **3**, January, 1946, pp. 302–335; Corrections, *ibid.*, **4**, July, 1946, pp. 199–200; Addit. corrections, *ibid.*, **6**, July, 1948, p. 192.

Middleton, D. and R. King, "The Thin Cylindrical Antenna: A Comparison of Theories," *J. Appl. Phys.*, **17**, 273 (1946).

Pistolkors, A. A., "The Radiation Resistance of Beam Antennas," *Proc. IRE*, **17**, 562, March, 1929.

Rhodes, D. R., "On a Fundamental Principle in the Theory of Planar Antennas," *Proc. IEEE*, **52**, 1013 (Sept., 1964).

Schelkunoff, S. A., "Transmission Theory of Spherical Waves," *Trans. AIEE*, **57**, 744–750 (1938).

———, "Theory of Antennas of Arbitrary Size and Shape," *Proc. IRE*, **29**, 439 (1941).

———, "Antenna Theory and Experiment," *J. Appl. Phys.*, **15**, January, 1944, pp. 54–60.

———, "On the Antenna Problem," *Quart. Appl. Math.*, **1**, January, 1944, pp. 354–355.

Books

Aharoni, J., *Antennae*, Clarendon Press, Oxford, 1946.

Collin, R. E., *Field Theory of Guided Waves*, McGraw-Hill Book Company, New York, 1961.

Harrington, Roger F., *Time-Harmonic Electromagnetic Fields*, New York: McGraw-Hill Book Company, New York, 1961.

King, Ronald W. P., *Theory of Linear Antennas*, Harvard University Press, Cambridge, Mass., 1956.

Kraus, John D., *Antennas*, McGraw-Hill Book Company, New York, 1950.

Ramo, S., John R. Whinnery and T. Van Duzer, *Fields and Waves in Communication Electronics*, John Wiley & Sons, Inc., New York, 1965.

Schelkunoff, S. A., *Electromagnetic Waves*, D. Van Nostrand Co., Inc., Princeton, N. J., 1943.

———, *Advanced Antenna Theory*, John Wiley & Sons, Inc., New York, 1952.

——— and Harold T. Friis, *Antennas, Theory and Practice*, John Wiley & Sons, New York, 1952.

Chapter 15

PRINCIPLES OF BROADBAND ANTENNA DESIGN

15.01 Introduction. Depending upon application and frequency range involved, an antenna may be a wire, rod, tower, slot, an array of any of these elements, or (at the higher frequencies) an "aperture" consisting of a horn or shaped reflector. The practical design of antennas is now well covered in various handbooks,* so this chapter will be restricted to a brief mention of the principles of design as they apply in the various frequency ranges, with a more extended discussion of extremely broad-band antennas.

At the lower frequencies, ELF (< 3 kHz), VLF (3–30 kHz) and LF (30–300 kHz), efficiency is usually the most important factor. Although antennas in these frequency ranges are customarily very large physically, they normally fall in the class of electrically small antennas because of the large dimensions of a wavelength. At MF (300 kHz–3 MHz) which includes the "broadcast" band, tower antennas are mostly used for transmitting, where efficiency is important, and arrays of towers are employed to produce the desired directional characteristics. For reception in this band efficiency is unimportant and relatively short wire antennas suffice. At HF (3–30 MHz), VHF (30–300 MHz) and UHF (300 MHz–3 GHz), elevated wires and rods are used, and arrays of such antennas are designed to provide more directivity. Because operation over a large band width is often required (the HF band covers a 10-to-1 band width and the American television band extends from 54–890 MHz), a major problem is often that of designing an antenna or array to maintain desired impedance and pattern characteristics over a wide band of frequencies. At SHF (3–30 GHz) where aperture antennas are employed, the problems are those of producing a narrow beam with low side-lobe level, and (sometimes) maintaining the desired radiation pattern characteristics over a wide range of frequencies or over an appreciable range of scan

*See bibliography at end of chapter.

angles. The theory of radiation from apertures has been treated in chap. 13 along with its application to some simple cases. An extensive treatment of design considerations for aperture antennas is available in a number of books.*

15.02 Antenna Band Width. The useful band width of an antenna is that range of frequencies over which the antenna maintains certain required impedance, pattern or polarization characteristics. Because these requirements vary so much with function there is no unique definition of antenna band width; instead, the specifications are set in each case to meet the needs of the particular application. For electrically small antennas having linear dimensions less than about a half-wavelength, the pattern is usually that of an electric dipole or magnetic dipole (or some combination of these), and is relatively insensitive to frequency. Under these circumstances the limiting factor on antenna performance is usually impedance variation and the problem can be formulated in terms of the Q of the antenna as discussed in sec. 11.13. For antennas or arrays that are *large* in wavelengths the antenna Q as defined in chap. 12 is near unity (excluding superdirective designs) and the design statement is normally formulated in terms of beam-width and side-lobe level requirements, and the avoidance of "pattern breakup" at certain frequencies. For intermediate cases where the antenna has dimensions of the order of a wavelength, the useful band width may be limited by either impedance or pattern considerations depending upon the particular application. In these instances a usable band width of two-to-one is considered good, although the rhombic and discone antennas perform satisfactorily over a four-to-one frequency range.

In recent years it has become possible to design antennas having *unlimited* band width, in the sense that the upper and lower frequency limits of useful performance may be independently specified by the designer. Such antennas are known as frequency-independent or log-periodic structures, and are the subjects of the following sections.

15.03 Frequency-independent Antennas. The concept of frequency-independent antennas was introduced by Rumsey† who proposed that a

*See R. C. Hansen, ed., *Microwave Scanning Antennas*, Vol. 1, "Apertures," Academic Press Inc., New York, 1964; Henry Jasik, ed., *Antenna Engineering Handbook*, McGraw-Hill Book Company, New York, 1961; Samuel Silver, *Microwave Antenna Theory and Design*, McGraw-Hill Book Company, New York, 1949.

†V. H. Rumsey, "Frequency Independent Antennas," *IRE National Convention Record*, Part I, p. 114, 1957. Also see V. H. Rumsey, *Frequency-Independent Antennas* (Academic Press Inc., New York, 1966).

For a survey of frequency-independent and log-periodic antennas including a history of their development, see E. C. Jordan, G. A. Deschamps, J. D. Dyson and P. E. Mayes, "Developments in Broadband Antennas," *IEEE Spectrum*, pages 58–71, April, 1964. Much of the material of this and subsequent sections is based on this article.

structure which could be defined entirely by angles (without any characteristic length dimension) should have characteristics that are *independent* of frequency. Because all such structures extend to infinity the design problem includes determination of which (if any) of these structures retain the frequency-independent properties when truncated to a finite length. For example, the well-known biconical structure satisfies the angle requirement when it is infinitely long, but its performance is far from frequency-independent when it is truncated to form a biconical antenna. In contrast, the infinite equiangular spiral structure illustrated in Fig. 15-1 is an angle structure which does retain its frequency-independent properties (over a certain range of frequencies) when truncated to form the antenna shown in Fig. 15-2.

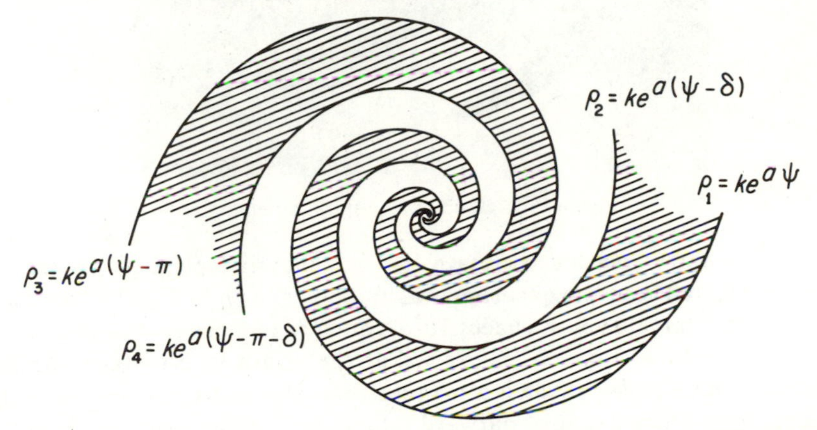

Figure 15-1. An (infinite) equiangular spiral structure.

The Equiangular Spiral Antenna. The geometry of the equiangular spiral antenna is shown in Fig. 15-1. The equiangular or logarithmic spiral is defined by

$$\rho = e^{a(\phi-\delta)} \qquad \text{or} \qquad (\phi - \delta) = \frac{1}{a}\ln \rho$$

where ρ and ϕ are conventional polar co-ordinates, and a and δ are constants. In Fig. 15-1 the edges of the metallic arms are defined by $\rho_1 = k\,e^{a\phi}$, $\rho_2 = k\,e^{a(\phi-\delta)}$ for one arm, and $\rho_3 = k\,e^{a(\phi-\pi)}$, $\rho_4 = k\,e^{a(\phi-\pi-\delta)}$ for the other arm, where the constants a, k, and δ determine the rate of spiral, size of the terminal region, and arm width, respectively. This particular spiral has the property that the angle between the spiral and the radius vector remains the same for all points on the curve; hence "equiangular" spiral. Experimental investigation by Dyson* established that this par-

*See the following articles by J. D. Dyson: "The Equiangular Spiral Antenna," TR No. 21, Contract 33 (616)–3220, Dept. of Electrical Engineering, University of Illinois, Urbana, Illinois, Sept. 15, 1957; "The Equiangular Spiral," *IRE Trans.*, AP-7, p. 181, April, 1959.

Figure 15-2. Equiangular spiral antenna.

ticular geometry does indeed retain its frequency-independent properties when truncated, and the experimental antenna shown in Fig. 15-2 became the first of a large class of successful frequency-independent antennas.

The antenna of Fig. 15-2 is excited at the origin by a voltage applied between the two arms in such a manner that the arms are balanced to ground. The currents flow outward along the spiral arms with small attenuation until a region of certain size in wavelengths is reached. In this region, known as the active or radiating region, essentially all of the power guided along the spiral arms is radiated. Beyond this region the presence or absence of the arms is of little consequence. Because the radiating region has fixed dimensions in wavelengths it moves inward or outward as the frequency is raised or lowered. Consequently the size of the effective radiating aperture automatically adjusts or "scales" with frequency to produce an antenna that has the same pattern and impedance at all frequencies. (Because of the spiraling of the arms the radiation pattern actually rotates about the axis of the antenna as frequency is changed.) This remarkable property of scaling, or automatic adjustment of the size of the radiating region to suit the frequency of operation, is characteristic of all successful frequency-independent antennas.

At this point it is necessary to define the term "frequency-independent" when used with a practical finite-sized structure. The antenna of Fig. 15-2, excited by a voltage applied between the two arms at the origin, has an impedance and radiation pattern which are *essentially constant** (that is, independent of frequency) for all frequencies from that for which the outer diameter of the truncated structure is approximately one-half wavelength to the frequency at which the diameter of the feed region (as determined by the transmission-line feed) is comparable with a half-wavelength. Since these two dimensions can be specified independently, the design band width can be made arbitrarily large; in actuality it is limited only by practical considerations of construction: how large the outer diameter is made, and how finely the geometry at the feed region can be modeled.

The planar equiangular spiral antenna is bidirectional, radiating a broad-lobed beam on each side of the plane. As long as the expansion rate of the spiral is not too rapid, the field is circularly polarized and the pattern is approximately $\cos\theta$, where θ is the angle measured from the axis.

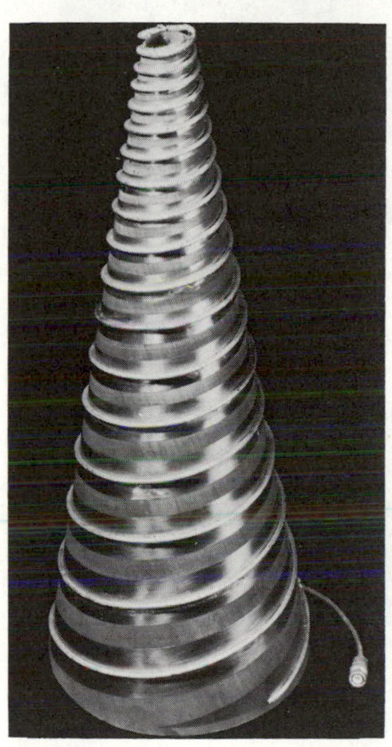

Figure 15-3. Conical equiangular spiral with coaxial cable feed.

Conical Equiangular Spiral. The usefulness of the equiangular spiral design was greatly enhanced when it was discovered† that a unidirectional pattern could be obtained when the balanced spiral arms were wrapped on the surface of a cone. For conical angles less than 45 degrees and appropriately chosen rates of spiral, this conical equiangular spiral produces a single broad-lobed beam in the backward direction (off the apex of the cone), and the beam shape remains essentially independent of frequency.

The conical equiangular spiral antenna is a balanced structure which may be fed (at the apex) by means of a balanced transmission line carried

*As was noted above, the pattern actually rotates with frequency about an axis perpendicular to the plane of the spiral. If the pattern-measuring co-ordinate system is allowed to rotate at the same rate, the measured pattern remains constant; otherwise there will be a (generally small) periodic variation of magnitude proportional to the rotational asymmetry of the pattern.

†See J. D. Dyson, "The Unidirectional Equiangular Antenna," *IRE Trans.*, AP-7, pp. 330–334, Oct., 1959.

up inside and along the axis of the cone. Alternatively it may be fed as illustrated in Fig. 15-3 by a coaxial cable carried along and soldered in contact with one of the arms. Because the amplitude of antenna current on the arms, and also on the outside of the coaxial cable, falls off quite rapidly beyond the active region, the end of the arm where the cable enters is essentially a field-free region. This type of feed automatically provides a frequency-independent balun, permitting the balanced antenna to be fed by means of an unbalanced coaxial line. To maintain physical symmetry a dummy cable is usually soldered to the other arm. Conical equiangular or log-spiral antennas have been constructed to operate over band widths with a frequency range of more than 40 to 1. The band width obtained is at the discretion of the designer; the upper usable frequency is determined by the truncated region at the apex which must remain small in terms of wavelengths; the lowest usable frequency is set by the base diameter of the cone which must be at least $\frac{3}{8}$ wavelength at the lowest frequency of operation for fairly tightly wrapped spirals.

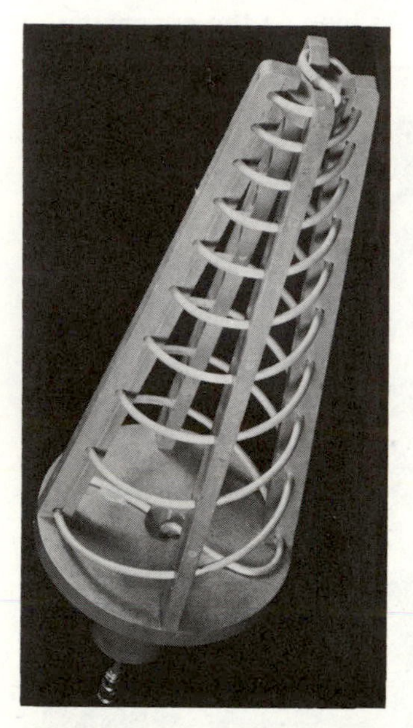

Figure 15-4. A simple and practical conical spiral antenna.

A further modification of the conical equiangular spiral results in a very practical, simple-to-construct antenna. By narrowing the width of the expanding arms and allowing them to degenerate to constant-width structures, the cables alone can form the arms. For fairly tightly spiraled antennas there is little change in the characteristics from those obtained from an antenna with narrow expanding arms. Fig. 15-4 shows a model of this version of the conical spiral antenna.

15.04 Log-periodic Antennas. A development which closely paralleled the frequency-independent concept was that of log-periodic antennas. Starting with the angle concept, DuHamel* reasoned that it should be possible to force radiation from otherwise "angle structures" by means of appropriately located discontinuities. One of the first successful appli-

*See R. H. DuHamel and D. E. Isbell, "Broadband Logarithmically Periodic Antenna Structures," *IRE 1957 Nat'l Convention Record*, Part I, pp. 119–128; R. H. DuHamel and F. R. Ore, "Logarithmically Periodic Antenna Designs," *IRE 1958 Nat'l Convention Record*, Part I, pp. 139–151.

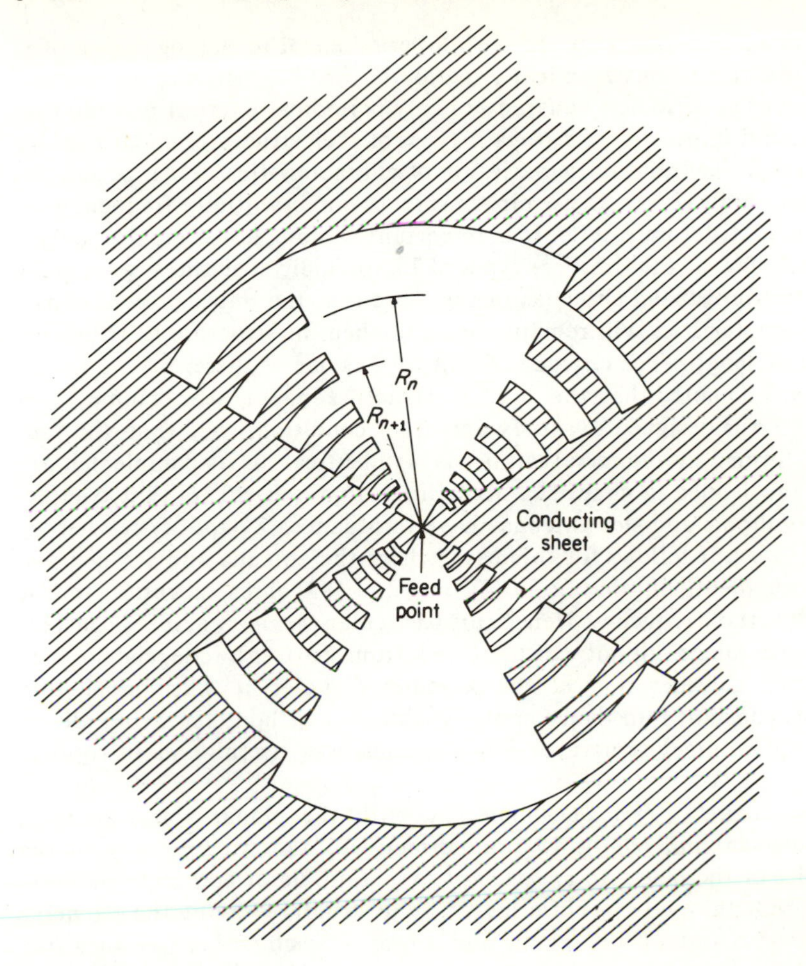

Figure 15-5. A planar log-periodic structure. (The infinite structure satisfies complementarity conditions.)

cations of this idea was an antenna formed using the geometry of Fig. 15-5.

Here two wedge-shaped metallic angle structures have teeth cut into them along circular arcs. The radii of the arms which define the location of successive teeth are chosen to have a constant ratio $\tau = R_{n+1}/R_n$. This same ratio τ defines the lengths and the widths of successive teeth. From the principle of modeling it is evident for this structure extending from zero to infinity, and energized at the vertex, that whatever properties it may have at a frequency, f, will be repeated at all frequencies given by $\tau^n f$, where n is an integer. When plotted on a logarithmic scale these frequencies are equally spaced with a period equal to the logarithm of τ; hence the name "logarithmically periodic" or "log-periodic" structure. Although log-periodicity guarantees only periodically repeating radiation pattern and impedance, for certain types of such structures and for values of τ not too far from unity, variation of characteristics over a period can be quite small. Under these circumstances an essentially frequency-independent structure results. It is important to note, however, that only a relatively few of the limitless types of log-periodic structures will make successful broadband antennas in the sense that the impedance and pattern characteristics will remain constant when the structure is truncated to a finite length. The antenna of Fig. 15-5 is one of these.

It will be noted that the infinite structure having the geometry of Fig. 15-5 was designed to satisfy one other condition, viz., that the "antenna" formed by the metal cut away is identical to its complementary screen (as defined in chap. 13). Recalling from sec. 13.16 that complementary dipole and slot antennas have impedances Z_d and Z_s related by $Z_d Z_s = \eta^2/4$, it follows for this case, where the antenna and its complement are identical, that $Z_d = Z_s = \eta/2 \approx 189$ ohms, a result that is independent of frequency. Hence this particular geometry assured constant impedance (although not constant radiation pattern) independently of the other consideration of log-periodicity. Later it was found that (almost) constant impedance could be obtained without requiring identical complementary structures but the idea was useful and intriguing, nevertheless.

Nonplanar Log-Periodic Antenna. As with the planar spiral antenna, the planar antenna of Fig. 15-5 is bidirectional, radiating equally on the two sides of the plane. In an attempt to obtain unidirectional radiation Isbell* bent the two arms of a log-periodic structure having the geometry of Fig. 15-5 out of the plane and towards each other to form the V-shaped antenna of Fig. 15-6. Two important and surprising results were observed. As the angle between the arms was decreased from 180 degrees the radiation changed gradually from bidirectional to unidirectional as had been hoped for. However, instead of radiating out the front end as would have been expected from this open-sided horn, the antenna radiated in the *backward* direction with the major radiation off the apex of the antenna. The second result was equally important. Although no longer the planar structure required to satisfy Babinet's principle, the impedance of this non-planar version continued to remain nearly constant with frequency, but at a different value which depended on the angle between the arms.

*See D. E. Isbell, "Non-planar Logarithmically Periodic Antenna Structure," Rept. No. 30, University of Illinois Antenna Lab., Feb., 1958.

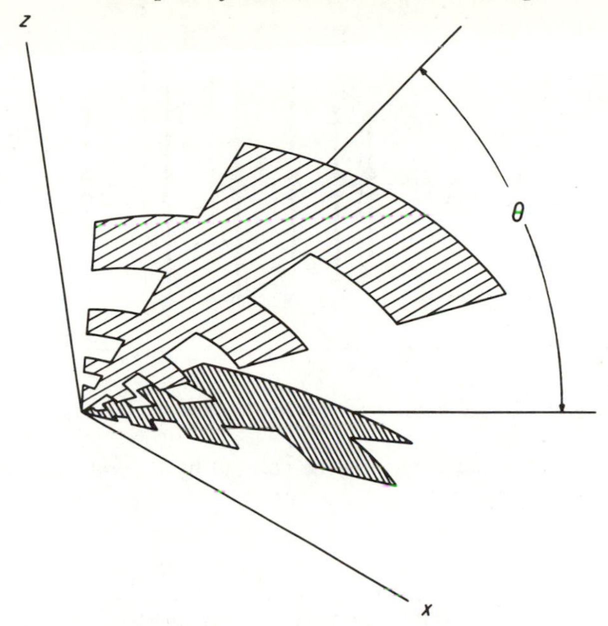

Figure 15-6. Nonplanar log-periodic antenna.

Sheet metal structures of the type of Figs. 15-5 or 15-6 are limited by mechanical considerations to microwave frequencies, but it was discovered later* that wire versions of log-periodic antennas could be built successfully. These antennas which permit operation down into the important HF band are described in sec. 15.05.

Log-Periodic Dipole Array.† Since the early days of radio many attempts had been made to increase the operating band width of dipole and monopole arrays. Many such attempts included the basic idea of using elements of varying lengths which would resonate at different frequencies. Most of these experiments will not be found described in the literature because, in the main, they were unsuccessful. A partially successful design was the early fishbone or comb antennas which used dipoles or monopoles capacitively coupled to a transmission line in such a manner that fewer elements were excited at higher frequencies, thus resulting in an array whose effective length decreased with decreasing wavelength. With the advent of the log-periodic concept it was natural to try to apply it to dipole arrays; Fig. 15-7 shows the required geometry. As with all log-periodic geometries all dimensions increase in proportion to the distance from the origin. Following this rule, the lengths and

*R. H. DuHamel and F. R. Ore, *op. cit.*

†D. E. Isbell, "Log-Periodic Dipole Arrays," *IRE Trans.* AP-8, No. 3, p. 260, 1960.

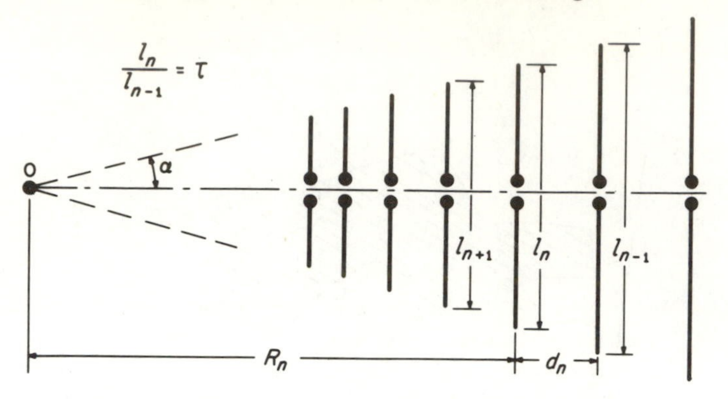

Figure 15-7. A dipole array having log-periodic geometry.

spacings of adjacent elements must be related by a constant scale factor τ, so that

$$\frac{L_n}{L_{n-1}} = \frac{d_n}{d_{n-1}} = \tau$$

When applied in straightforward fashion to a practical array, fed as in Fig. 15-8(*a*), an unsuccessful design usually results. Recognition of the cause of failure, and devising means for overcoming the difficulty, ranks as a major step in understanding the principles of successful log-periodic design. It was reasoned that with elements closely spaced in wavelengths the phase progression along the array of Fig. 15-8(*a*) was such as to produce a beam in the forward direction (to the right in the figure). In this case the larger elements, to the right of the active region, are in the beam and will produce interference effects. The solution arrived at in this instance was to reverse the phasing of alternate elements [Fig. 15-8(*b*)] to produce a beam in the backward direction (to the left). In this case the energy radiated in the active region is beamed through the short end of the array, where the elements are short and closely spaced in wavelengths, have alternate phasings, and consequently produce negligible interference. The magnitude and phasing of the element currents are considered in detail in sec. 15.05.

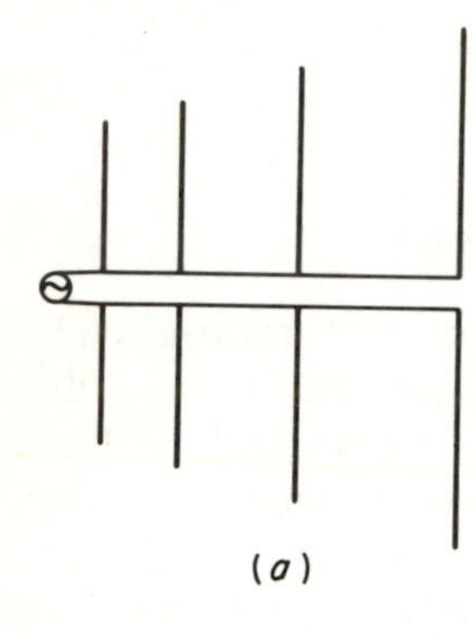

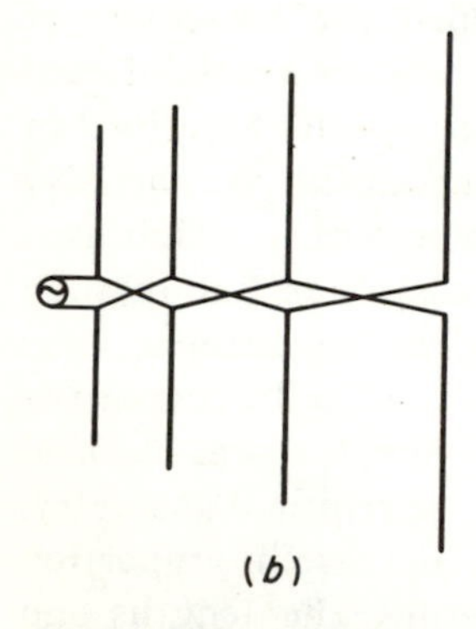

Figure 15-8. (*a*) Unsuccessful and (*b*) successful method of exciting the log-periodic dipole array.

15.05 Array Theory for LP and FI Structures. The formulation of an array theory for log-periodic or frequency-independent structures could be expected to be most difficult because all of the usual parameters, such as element length in wavelengths, element spacing in wavelengths, magnitude and phasing of element currents, vary along the array, and of course are also functions of frequency. Fortunately a method exists for arriving at a first approximation of array behavior. This method* consists of viewing the log-periodic structure as a locally periodic structure whose period varies slowly, increasing linearly with the distance to the apex. From this viewpoint, analysis of a uniform periodic structure as a function of frequency corresponds approximately to an examination of the log-periodic structure as a function of distance from the apex. Using this approach, the standard uniform array theory of chap. 12 can be used directly to give a first approximation to the performance of log-periodic and frequency-independent structures.

From eq. (12-1), for an n-element array of equispaced isotropic radiators the magnitude of the relative field strength is given by

$$|E| = |1 + e^{j\psi} + e^{j2\psi} + \cdots + e^{j(n-1)\psi}| \tag{15-1}$$

where $\psi = kd\cos\phi + \alpha$. The angle α is the progressive phase shift from left to right, and ϕ and d are defined by Fig. 12-1 on page 422. The letter symbol $k = 2\pi/\lambda_0 = \omega/c$ is the *free-space* phase-shift constant. (It is used in this and the next section to distinguish it from $\beta = 2\pi/\lambda = \omega/v$, the phase-shift constant along the array.) Expression (1) has a maximum for $\psi = 0$, so for element spacings less than one-half wavelength the angle ϕ_m for maximum radiation is given by

$$\phi_m = \cos^{-1}\frac{-\alpha}{kd} \tag{15-2}$$

If the elements are fed in phase, $\alpha = 0$, and $\phi_m = 90°$, so the maximum radiation is broadside. If successive elements are fed with a lagging phase of value $\alpha = -kd$, then $\phi_m = 0$, so the maximum radiation is end-fire in the forward direction. If successive elements are fed with a leading phase of value $\alpha = +kd$, then ϕ_m will equal 180 degrees, and the maximum radiation will be end-fire in the backward direction. For values of α between $-kd$ and $+kd$, the angle of maximum radiation is at an angle between 0 and 180 degrees as given by eq. (2). [Of course, by symmetry about the axis of the array, there is another maximum at an angle between 0 and -180 degrees, which is also given by eq. (2)]. When $|\alpha| > kd$, eq. (2) cannot be satisfied for any real value of ϕ. That is, there is no value of ϕ in "visible" range between 0 and 180 degrees (and, therefore,

*P. E. Mayes, G. A. Deschamps, W. T. Patton, "Backward-wave Radiation from Periodic Structures and Application to the Design of Frequency-Independent Antennas," *Proc. IRE*, **49**, No, 5, p. 962, May, 1961.

also between 180 and 360 degrees) that will produce a maximum (in the sense that all the radiations add in phase). However, if $|\alpha|$ is only slightly greater than kd, so that ψ is not much larger than zero, the total field can still be quite strong in the forward direction ($\phi = 0$) for negative α, or in the backfire direction ($\phi = 180°$) for positive α. This case is illustrated by the sketch of Fig. 15-9(*a*). On the other hand, if $|\alpha|$ is considerably greater than kd (that is, the phase shift between elements is large) the phase diagram might be as illustrated in Fig. 15-9(*b*), with a resulting small total E_t for *all* values of ϕ. For these cases of large phase shift [Fig. 15-9(*b*)] there is no major lobe anywhere, and the array radiates only feebly, scattering its small radiated energy in various directions.

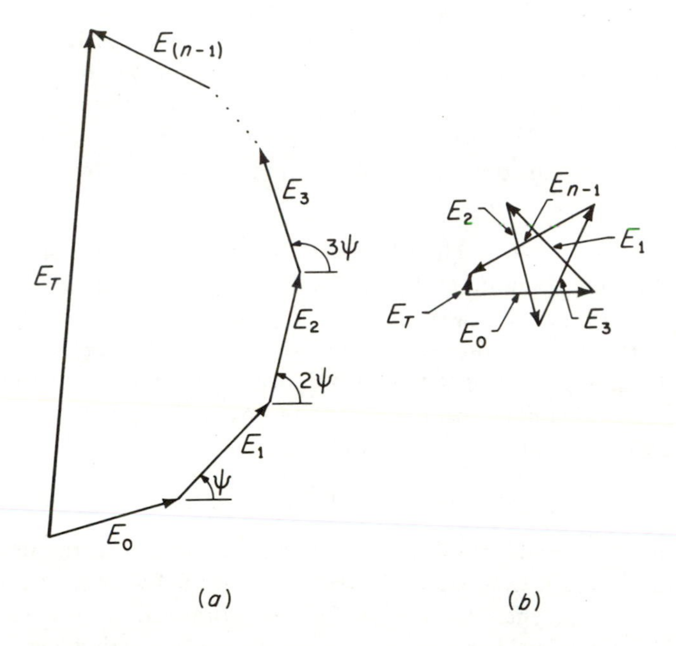

Figure 15-9. Phasor addition: (*a*) for small ψ; (*b*) for large ψ.

The elementary notions just discussed can be applied with some slight modification to an analysis of the log-periodic dipole array sketched in Fig. 15-10(*a*). For this purpose, it is helpful to consider separately three main regions of the array designated in Fig. 15-10(*b*).

(1) *Transmission-line Region.* In this region the antenna elements are short compared with the resonant length (i.e., $l \ll \lambda/2$) so the element presents a relatively high capacitive impedance. The element current is small and leads the base voltage supplied by the transmission line by approximately 90 degrees. The element spacing is small in wavelengths and the phase reversal introduced by transposition of the transmission

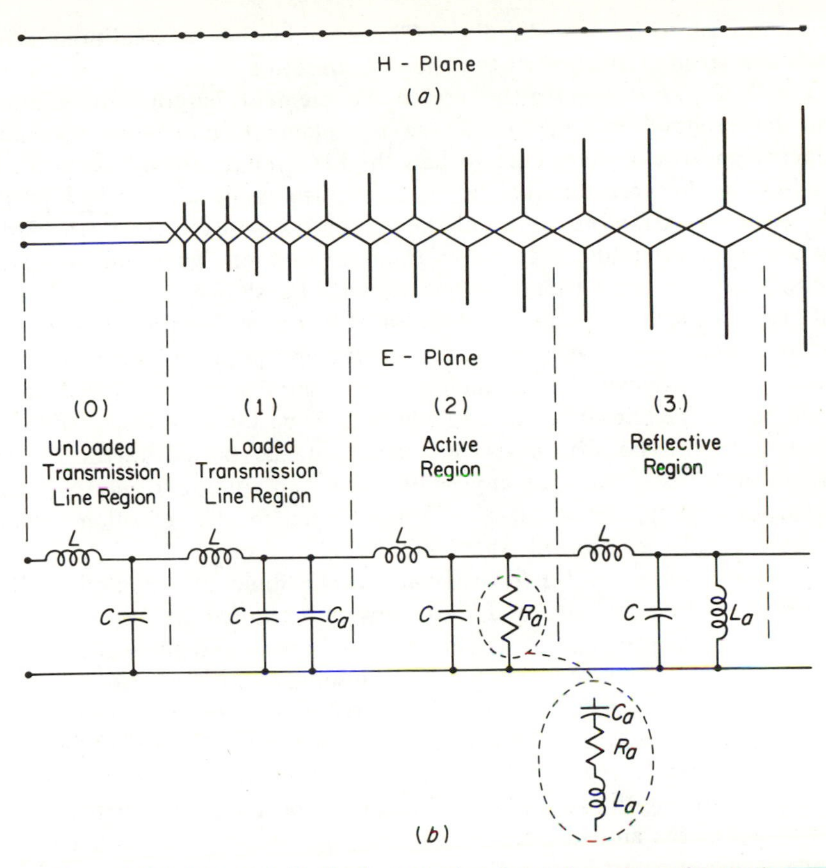

Figure 15-10. Log-periodic dipole array showing main regions of operation.

line means that adjacent elements are nearly 180 degrees out of phase. More precisely, each element current leads the preceding element current approximately by $\alpha = (\pi - \beta d)$, where d is the element separation and $\beta = 2\pi/\lambda = \omega/v$ is the phase-shift constant along the line. In general, β, λ, and v will differ from their free-space values owing to the loading effect of the elements on the transmission line. Because of the phasing and close spacing of the elements, radiation from this region will be very small and in the backfire direction.

(2) *Active Region.* In this region the element lengths approach the resonant length (l slightly less than $\lambda/2$), so the element impedance has an appreciable resistive component. The element current is large and more nearly in phase with the base voltage; the current is slightly leading just below resonance and slightly lagging just above resonance. The element spacing in now sufficiently large that the phase of current in a given element leads that in the preceding element by an angle $\alpha = (\pi - \beta d)$

which may approximate $\pi/2$ radians. This combination of conditions will produce a strong radiation in the backfire direction.

(3) *Reflection Region.* In this region the element lengths are greater than the resonant length ($l \geq \lambda/2$), so the element impedance becomes inductive and the element current lags the base voltage. The base voltage provided by the transmission line is now quite small because in a properly designed array, nearly all of the energy transmitted down the line has been abstracted and radiated by the active region. The element spacing may now be larger than $\lambda/4$, but as will be shown later, the phase shift per unit length along the line in this region is small so that the resulting phasing between elements (including the phase reversal introduced by the transposition) is such that any small amount of radiation is still in the backfire direction. In addition, it will be demonstrated later that the transmission-line characteristic impedance becomes reactive in this region. This means that any small amount of incident energy transmitted through the active region is not accepted in the reflection region but is reflected back towards the source.

The Array as a Loaded Transmission Line. Some of the remarkable properties of log-periodic and frequency-independent antennas are attributable to the propagation characteristics of the equivalent loaded transmission line which conveys energy from the source to the radiating portion of the antenna. These effects are particularly easy to see in the case of the log-periodic dipole array, and are indicated in Fig. 15-10(*b*). On the feed line to the antenna, region (0), the series inductance and shunt capacitance per unit length are shown as L and C, respectively. In the transmission region of the antenna, region (1), the transmission line is loaded by a capacitance per unit length, C_a, which represents the loading effect of the short dipoles which have a capacitance reactance. It is noted that to the first approximation C_a is nearly constant throughout this region because at the beginning of the region the capacitance per element is small, but the elements are closely spaced, whereas near the end of the region the capacitance per element is larger, but so is the spacing. The effect of the augmented shunt capacitance of the line ($C + C_a$) is to increase the phase delay per unit length, and since $\beta = 2\pi/\lambda = \omega/v$, a decrease of wavelength λ and a decrease of phase velocity v along the line below the free-space values result. This is said to be a "slow-wave" region of the transmission line. Note, however, that because of the transposition of the feed line between elements, successive elements are fed with a *leading* phase shift of $(\pi - \alpha)$ per section. This rapid phase shift in the reverse direction corresponds to a slow wave in the *backward* direction along the antenna elements.

In the active region (2), the element lengths approach the resonant length and the transmission line loading becomes resistive, designated by the shunt resistance R_a in series with the antenna capacitance C_a and antenna inductance L_a. The phase shift per unit length, the wavelength and phase velocity all approach their free-space values. Because of the transposition between elements, and accounting for the fact that the element current leads the base voltage by lesser amounts in successive elements as the resonant length is approached, it turns out that phasing of currents in the elements corresponds to a backward traveling wave having a velocity v somewhat less than c, the velocity of light.

In the inactive or reflection region (3), the element lengths become longer than the resonant length, the antenna inductive reactance predominates, and the loading effect on the line is represented by the shunt inductance L_a. When (and if) the parallel combination of L_a and C is inductive, we have the equivalent of the attenuation region of a filter. The phase shift per unit length is then zero (for the lossless case) and the phase velocity is infinite (that is, there is no wave motion). The incident energy propagating down the line is no longer accepted but is reflected back toward the source. (These results are strictly true only in

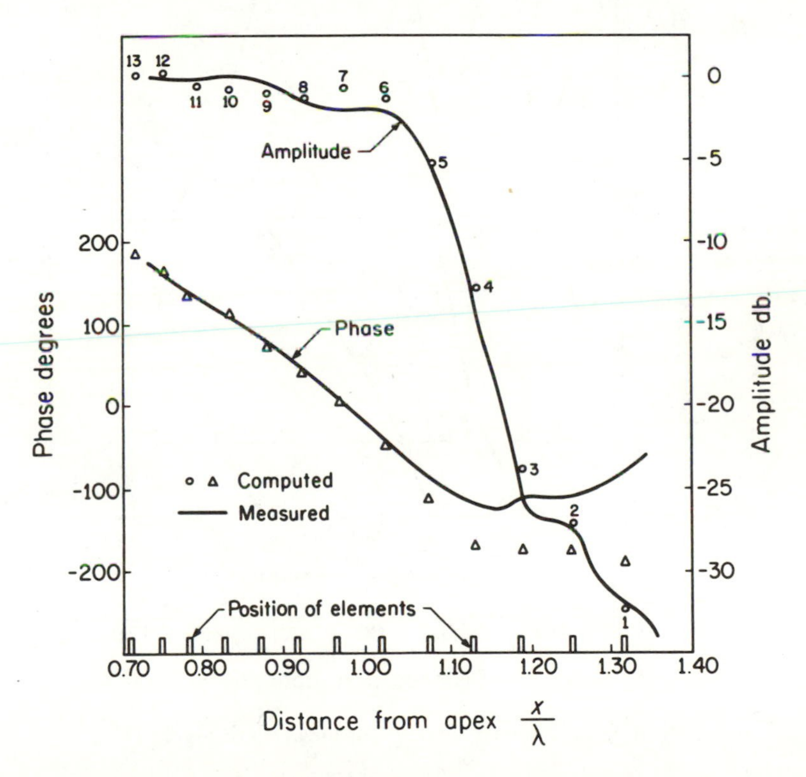

Figure 15-11. Amplitude and phase of transmission-line voltages along a particular LP dipole array.

the case of a lossless filter, but they form the first approximation for this lossy-filter case.)

The general features outlined above will be illustrated for a particular log-periodic dipole array which has been analyzed in considerable detail.* Fig. 15-11 shows the amplitude and phase of the transmission-line voltage along a particular 13-element LP dipole array. Distance is shown measured from the apex of the array, and the elements are numbered starting with the largest element as number 1. This set of data is for a frequency f for which element number 4 is one-half wavelength long. Several interesting aspects of the data are immediately apparent. In the transmission region (elements 13 to 7), the amplitude of voltage along the line is approximately constant and the phase shift between element positions

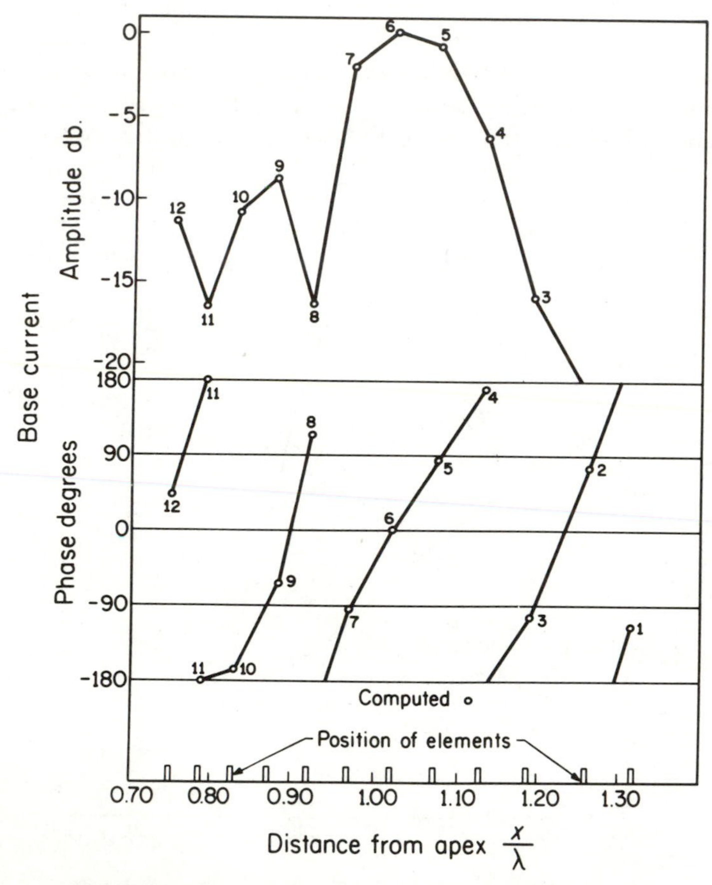

Figure 15-12. Element currents for the array of Fig. 15-11.

*R. L. Carrel, "Analysis and Design of the Log-Periodic Dipole Antenna," Antenna Laboratory Report No. 52, University of Illinois. (Contract AF33(616)–6079.)

increases gradually from about 20 degrees to 30 degrees. (Because of the transposition between elements, this means that adjacent elements are fed with a progressive phase lead 160 degrees to 150 degrees.) In the active region (elements 7 to 4) the amplitude drops sharply because of power absorbed by the strongly radiating elements, and the phase shift averages about 90 degrees between adjacent elements. Finally, in the unexcited or reflection region (elements 3 to 1), the amplitude drops to very low values and the phase shift between element positions is nearly zero (corresponding to the zero phase shift or infinite phase velocity in the attenuation region of a low pass filter).

The resulting *element currents* for the LP dipole array of Fig. 15-11 are shown in Fig. 15-12, both in amplitude and phase. From the current amplitudes (and noting that small contributions from elements 12 through 8 tend to cancel one another because of the nearly 180-degree phase shift between them), it is evident that the only elements which will contribute appreciably to the radiation are elements 7, 6, 5 and 4. For these elements, the phase difference between adjacent members is approximately 90 degrees leading, so a backfire radiation will be expected. The phasor diagrams for $\phi = 0°$, 90°, and 180° are shown in Fig. 15-13 and the resulting radiation patterns are shown in Fig. 15-14. (The *E*-plane pattern is the *H*-plane pattern modified by the directivity of the individual elements in this plane.)

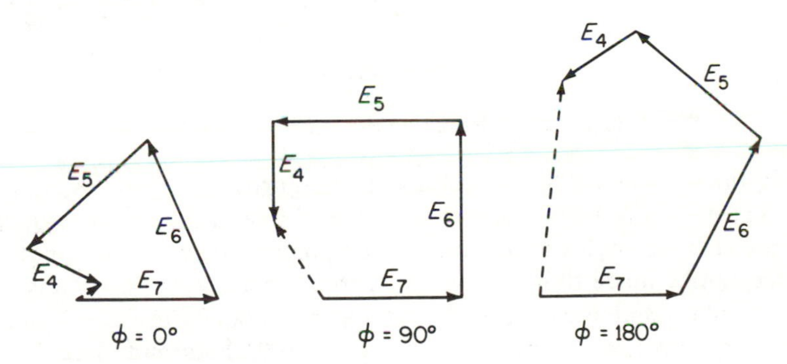

Figure 15-13. Phasor diagrams for addition of fields produced by 4 elements in the active region.

As the operating frequency is decreased or increased the active region moves up or down the array, but the radiation pattern and input impedance remain almost constant.

Directivity and Impedance. The radiation characteristics of the log-periodic dipole array are functions of the array geometry as expressed by the factor τ and the angle α (Fig. 15-7); the input impedance depends mainly on the characteristic impedance, Z_0, of the transmission line

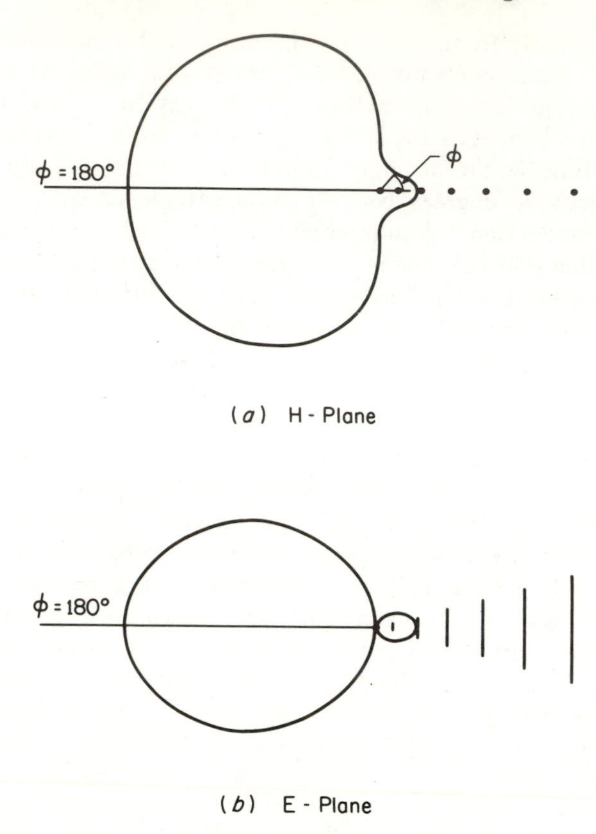

Figure 15-14. Resulting radiation patterns.

feeding the elements. The gain of a well-designed array ranges from 7.5 to 12 db (over an isotropic radiator) with the larger-gain values associated with the smaller angles of α, that is, with longer arrays. The input impedance can be made to have any value from about 50 to 200 ohms. For the particular 13-element array for which the data of Figs. 15-11 and 15-12 were obtained, $\tau = 0.95, \alpha = 12.5^\circ$, $Z_0 = 100$ ohms, and these parameters produced a directive gain of $9\frac{1}{2}$ db and an input impedance of 65 ohms. The pattern remained constant and the VSWR was less than 1.17 over the operating band width (in this case about 2 to 1). This directive gain is only slightly better (≈ 1 db) than an optimum 3-element Yagi-Uda antenna, but in contrast to the quite narrow band width of the latter antenna, the pattern-and-impedance band width of the log-periodic dipole array can be increased to any desired value by adding more elements.

General Properties of Log-Periodic and Frequency-Independent Antennas. The manner of operation of the LP dipole array has been described in some detail because of the insight it gives into what are believed to be general requirements for successful frequency-independent operation. These requirements appear to be as follows:

(1) An excitation of the antenna or array from the high-frequency or small end of the antenna.

(2) A backfire radiation (in the case of unidirectional radiators) so that the antenna fires through the small part of the antenna, with the radiation in the forward direction being zero, or at least very small. For bidirectional antennas the backfire requirement is replaced by a requirement for broadside radiation. In any case the radiation in the forward direction along the surface of the antenna (which theoretically extends to infinity) must be zero, or very small.

(3) A transmission region formed by the inactive portion of the antenna between the feed point and the active region. This transmission-line region should have the proper characteristic impedance and negligible radiation.

(4) An active region from which the antenna radiates strongly because of a proper combination of current magnitudes and phasings. The position and phasing of these radiating currents are such as to produce a very small radiation field along the surface of the antenna or array in the forward direction, and a maximum radiation field in the backward (broadside for bidirectional antennas) direction. For successful backfire antennas these requirements are frequently met with separations less than quarter-wavelength and phasings near 90 degrees leading, for adjacent elements in the active region. For broadside radiation the phasings must, of course, be zero.

(5) An inactive or reflection region beyond the active region. The *essential characteristic* of all successful frequency-independent antennas is a rapid decay of current within and beyond the active region, so that the structure can be truncated without affecting its operation. A major cause of the rapid current decay is, of course, the large radiation of energy from the active region. An additional cause, in at least some types of frequency-independent and log-periodic antennas, is the attenuation resulting from the rejection of incident energy by the reflection region (the filter stop-band effect mentioned above).

15.06 Other Types of Log-periodic Antennas. Although the number and type of log-periodic antennas that can be conceived is unlimited, very few of these types will produce successful designs having acceptable patterns, small impedance variations, and insensitivity to small changes in parameters. In this section several typical types will be described briefly to indicate the extent and variety of designs available.

Fig. 15-15 shows a *uniform* monopole array (of equal-length, equi-

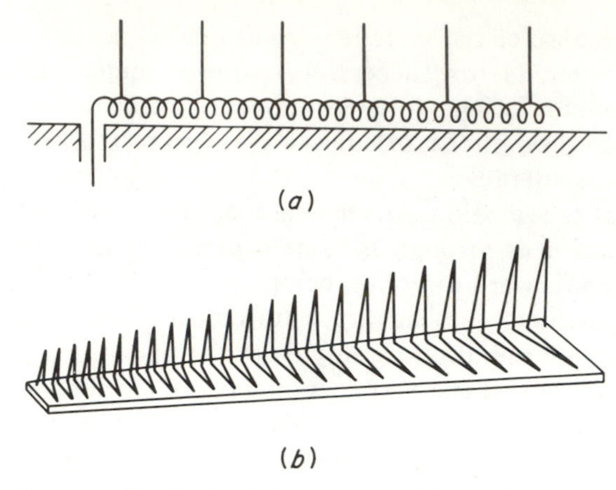

Figure 15-15. (*a*) A uniform monopole array fed by a slow-wave structure. (*b*) The bent log-periodic zigzag antenna.

spaced elements) fed by a helical transmission line.* Although not a particularly practical antenna, this simple structure has the distinct advantage that the radiating portion (the monopole elements) and the helical transmission line can be adjusted independently over a wide range of dimensions to explore the radiating properties theoretically and experimentally. To the first approximation, and assuming the velocity of propagation along the wire to be c, the phase velocity along the axis of the helix will be $v = ac/w$, where a and w are the corresponding lengths measured along the axis and along the wire. Hence the electromagnetic energy traveling from left to right along the slow-wave structure is slowed by the adjustable ratio a/w. This slow wave is then sampled periodically by the monopoles, and the rate of sampling determines the phase shift between elements and consequently the radiation characteristics of the array. As an example, for a slowness factor of 0.2, there is a phase shift of $5 \times 360°$ per wavelength in the axial direction. With four elements per wavelength ($\lambda/4$ spacing), the phase shift (lead) between successive elements will be $-5 \times 360°/4 = -450° = -90°$, so the array is phased for end-fire radiation. With five elements per wavelength, the successive phasing is $-360°$ so that all elements are fed in phase with a resulting broadside radiation of energy. With six elements per wavelength, the successive phasing is $-300°$ which is equivalent to $+60°$, and the resultant radiation will be backfire. It is seen that by varying the element spacing the beam can be swept from end-fire through broadside

*The theory and experimental results for this structure are given by John W. Greiser and Paul E. Mayes, "Vertically Polarized Log-Periodic Zig-Zag Antennas, *Proc. N. E. C.*, **17**, pp. 193–204, 1961.

to backfire. By decreasing the element spacing slightly beyond the backfire condition, the effects of operating just within the "invisible" region can be explored. This region is of interest because it is here that best operation of log-periodic structures usually occurs.

For a given *periodic* structure of the type shown in Fig. 15-15, and assuming that the velocity of propagation along the wire is approximately independent of frequency, it is evident that the beam angle will scan with frequency because the number of elements per wavelength changes with frequency. In a *log-periodic structure* the element spacing increases along the structure, so that at a given frequency the regions encountered will be successively: backfire (invisible region), backfire, broadside, end-fire, and end-fire (invisible region). For proper operation it is essential that nearly all of the energy be radiated before reaching the broadside and end-fire conditions in the visible region.

Log-Periodic Zig-Zag Antenna. The principles deduced from an analysis of the structure of Fig. 15-15(a) have been applied* to the bent log-periodic zig-zag antenna sketched in Fig. 15-15(b). In this antenna the vertical portion of the structure provides the radiating elements and the nonradiating horizontal portion provides the additional phase-shift required for proper operation. This simple structure fed against ground provides an effective vertically polarized log-periodic antenna.

Figure 15-16. Log-periodic wire trapezoid antenna. (Courtesy Collins Radio Co.)

*Greiser and Mayes *op. cit.*

LP Trapezoidal Antenna. The log-periodic wire trapezoid (Fig. 15-16) was the first of the log-periodic wire antennas. Although evolved* directly from the trapezoidal sheet structure, its operation is analogous to that of the simple zig-zag.

The LP Resonant V Array. A major shortcoming of an ordinary log-periodic dipole array is the excessive length of the array when designed to cover a very wide frequency band. The log-periodic resonant V array† shown in Fig. 15-17 overcomes this difficulty in an ingenious manner by

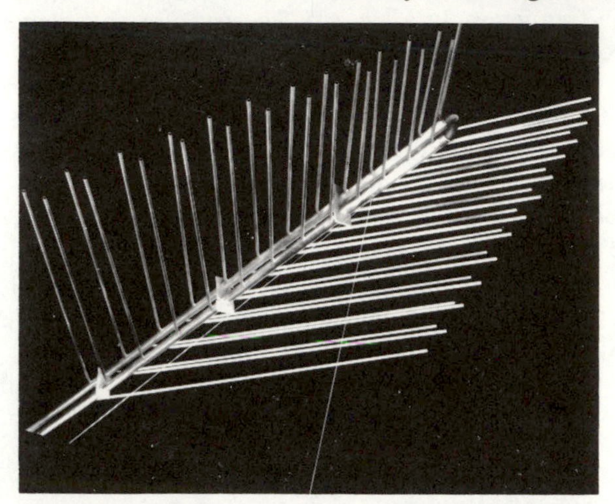

Figure 15-17. Log-periodic resonant-vee array.

providing operation in any of several modes. In the lowest-order $\lambda/2$ mode, the operation is similar to that of the LP dipole array because the forward tilt of the elements has small effect for this mode. However, as the frequency of operation is increased beyond that at which the shortest elements are resonant, that is, when the active region runs off the front end of the array, the largest elements at the rear become active in the $3\lambda/2$ resonance mode. In this mode the forward tilting of the elements ensures a good unidirectional pattern of high directivity. As the frequency is further increased the active region moves forward through the array in the $3\lambda/2$ mode until once again it runs off the front end, to return to the rear in the $5\lambda/2$ mode. This scheme makes it possible to obtain large band widths of the order of 20 to 1 with a relatively compact array. The pattern and impedance characteristics remain good over the entire frequency spectrum except for intervals about the mode-transition frequencies. Using these principles, arrays have been designed to cover

*R. H. DuHamel and F. R. Ore, *op. cit*

†P. E. Mayes and R. L. Carrel, "Log-Periodic Resonant-V Arrays," *IRE Wescon Conv. Rec.* 1961. Also P. E. Mayes, "Broadband Backward Wave Antennas," *Microwave Journal*, **6**, No. 1, p. 2, Jan., 1963.

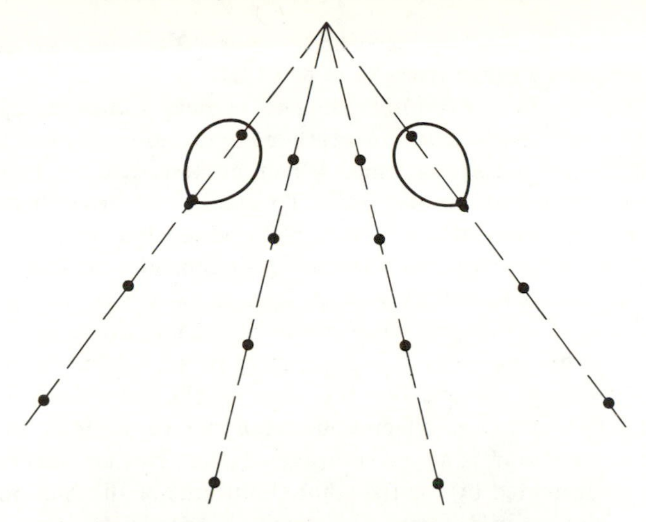

Figure 15-18. Directive array of LP antennas.

Figure 15-19. Two-element array of LP dipole antennas.

the entire American television band (channels 2 through 83) corresponding to a frequency range from 54 to 890 MHz.

Directive Arrays of LP Antennas. The typical radiation pattern of a log-periodic antenna or array is a rather broad unidirectional beam quite similar to that of a Yagi antenna. When greater directivity is desired, arrays of LP antennas can be used. To retain the broad band width of the individual log-periodic structure in the directive array, it is necessary to array the elements in a frequency-independent manner.* This can be done, for example, as indicated in Fig. 15-18 where the locations of the elements of the array with respect to each other are defined by angles. With this geometry it is evident that as frequency is decreased and the active region moves to the rear of the individual log-periodic structures, the effective spacing between active regions increases to maintain a constant spacing in wavelengths. Design theory for such arrays is complicated by the fact that the beams of the individual structures are oriented in different directions. Figure 15-19 is a picture of a 2-element array of log-periodic dipole arrays designed to operate as a broadband feed for a parabolic reflector. The 100-ohm impedances of the individual log-periodic structures are effectively connected in parallel to produce a 50-ohm input.

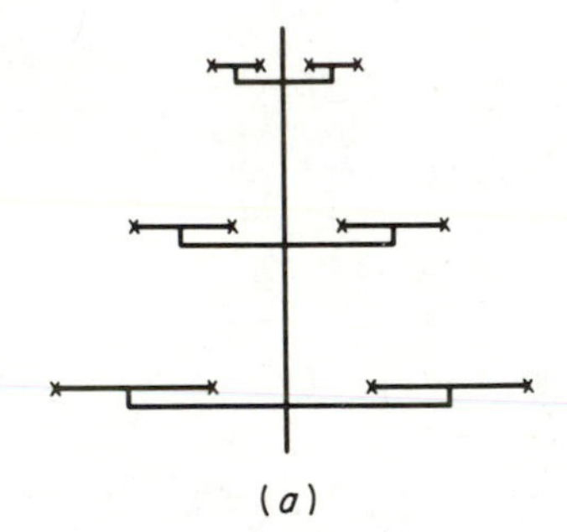

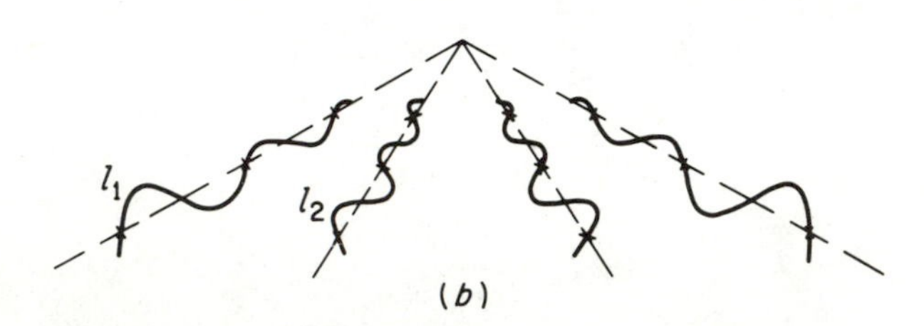

Figure 15-20. Alternative methods for arraying LP structures.

*R. H. DuHamel and D. G. Berry, "Logarithmically Periodic Antenna Arrays," *Wescon Convention Record,* Part I, pp. 161–174, 1958.

Alternative methods* for arraying log-periodic structures are illustrated in Fig. 15-20 (*a* and *b*). In these arrangements the active regions lie in a straight line normal to the direction of the beam, and the principle of operation requires that the corresponding lengths l_1 and l_2 of the log-periodic feeding structure be equal. This requirement can be satisfied by either of the arrangements (*a*) or (*b*), where the crosses represent individual dipoles. In a typical successful design a slowness factor of 0.53 was found satisfactory for the snaking transmission line in (*b*).

15.07 General Observations. In the preceding sections we have discussed the basic principles of operation of frequency-independent and log-periodic structures using elementary array theory. Although this simple approach explains many of the observed characteristics of these antennas, there are additional factors that need to be considered in any detailed analysis. Among these factors are the effects of mutual impedance between elements in a log-periodic array and the effect of curvature on the radiation properties of the equiangular structures. Analyses of these factors can be found in the extensive literature on the subject, and particularly in the book by Rumsey, listed in the bibliography.

Finally, there are two general observations that can be made about these unlimited band-width structures. Some of them (e.g., those of Figs. 15-2 through 15-6 and 15-16) have been referred to as antennas whereas others (e.g., Figs. 15-7 and 15-17) have been called arrays. In fact, it appears that all frequency-independent and log-periodic antennas may be considered as *antenna arrays*, with the array factor playing the major role in the formation of a proper end-fire or broadside pattern. The localization of the individual radiating elements is easier to see for the case of the log-periodic dipole array but the array action can also be observed in other cases; it is particularly evident in the case of the fairly tightly wrapped conical log-spiral.

The second observation relates to the similarity between antennas derived from the "angle" and "log-periodic" concepts.† Both lead to a solution of the unlimited band-width problem and for this reason both have come to be known as *frequency-independent*. An example of the similarity between these two antenna types can be demonstrated in the case of the log-periodic wire antenna of Fig. 15-16, which produces a linearly polarized beam off the apex with the electric vector parallel to the transverse elements. If two such antennas are arranged in space quad-

*K. K. Mei, M. W. Moberg, V. H. Rumsey and Y. S. Yeh, "Directive Frequency Independent Arrays," *IEEE Trans. on Ant. and Prop.*, AP-13, No. 5, pp. 807–809, Sept., 1965.

†For a detailed treatment of the relations between the two types of antennas, see chapter 18, "Frequency Independent Antennas," by George A. Deschamps, in the *Antenna Engineering Handbook*, ed. Henry Jasik. McGraw-Hill Book Company, New York, 1960.

rature along a common axis, and with a common origin but with one structure scaled a quarter period from the other, the resultant combination produces a circularly polarized beam with a pattern which rotates about the axis with frequency exactly as in the case of the conical equiangular spiral antenna. Conversely, of course, if the pattern of a conical equiangular spiral is probed with a linear receiving antenna of fixed plane of polarization, the measured pattern will vary log-periodically with frequency as does the pattern of the antenna of Fig. 15-16. Also, it is worth noting that if a narrow-armed conical equiangular spiral (an angle structure) is flattened sideways (along the axis) it becomes a log-periodic zig-zag antenna.

BIBLIOGRAPHY

Articles

DuHamel, R. H., and D. E. Isbell, "Broadband Logarithmically Periodic Antenna Structures," *IRE 1957 Nat'l Convention Record*, Part I, pp. 119-128.

DuHamel, R. H., and F. R. Ore, "Logarithmically Periodic Antenna Designs," *IRE 1958 Nat'l Convention Record*, Part I, pp. 139-151.

Dyson, J. D., "The Equiangular Spiral Antenna," TR No. 21, Contract 33 (616)-3220, Dept. of Electrical Engineering, University of Illinois, Sept. 15, 1957.

______, "The Equiangular Spiral Antenna," *IRE Trans.*, AP-7, p. 181, April, 1959.

______, "The Unidirectional Equiangular Spiral Antenna," *IRE Trans.*, AP-7, pp. 330-334, Oct., 1959.

Isbell, D. E., "Non-planar Logarithmically Periodic Antenna Structure," Rept. No. 30, University of Illinois Antenna Lab., Feb., 1958.

______, "Log-Periodic Dipole Arrays," *IRE Trans. on Antennas and Propagation*, AP-8, May, 1960, pp. 260-267.

Kraus, J. D., "Helical Beam Antenna for Wide-Band Applications," *Proc. IRE*, **36**, Oct., 1948, pp. 1236-1242.

Rumsey, V. H., "Frequency-Independent Antennas," *IRE Nat'l Convention Record*, Part I, 1957, pp. 114-118; also Rept. TR-20, University of Illinois Antenna Lab., Contract AF33(616)-3220, Oct., 1957.

Springer, P. S., "End-Loaded and Expanding Helices as Broadband Circularly Polarized Radiators," *Proc. Nat'l Electronics Conf.*, **5**, 1949, pp. 161-171.

Turner, E. M., "Spiral Slot Antenna," Tech. Note WCLR 55-8, Wright Air Development Center, Dayton, Ohio, June, 1955.

Wickersham, A. F., R. F. Franks, and R. L. Bell, "Further Developments in Tapered Ladder Antennas," *Proc. IRE*, **49**, Jan. 1961, p. 378.

Books

Hansen, R. C., ed., *Microwave Scanning Antennas*, Vol. 1, Apertures, Academic Press Inc., New York, 1964.

Jasik, Henry, ed, *Antenna Engineering Handbook*, McGraw-Hill Book Company, New York, 1961.

Log-Periodic Handbook (prepared for Bureau of Ships, U. S. Navy), Smith Electronics, Inc., Cleveland, Ohio, 1961.

Rumsey, V. H., *Frequency Independent Antennas*, Academic Press Inc., New York, 1966.

Silver, Samuel, *Microwave Antenna Theory and Design*, McGraw-Hill Book Company, New York, 1949.

Chapter 16

GROUND-WAVE PROPAGATION

The energy radiated from a transmitting antenna may reach the receiving antenna over any of many possible propagation paths, some of which are illustrated in Fig. 16-1. Waves that arrive at the receiver

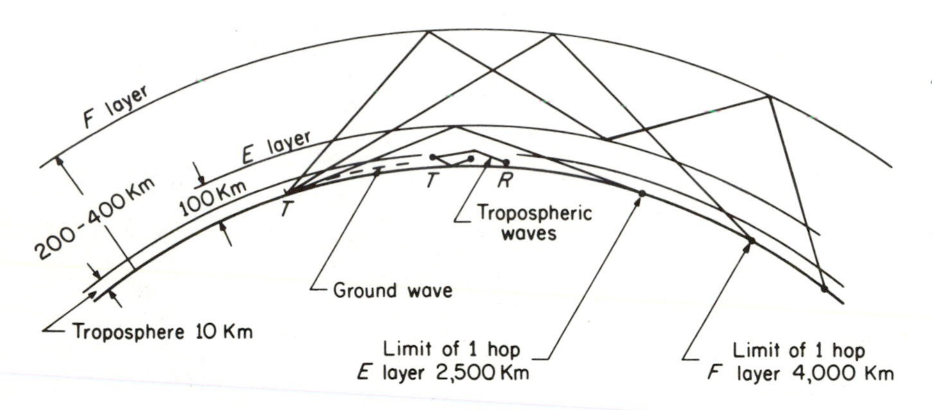

Figure 16-1. Some possible propagation paths.

after reflection or scattering in the ionosphere are known as *sky waves*, or alternatively, as *ionospherically reflected* and *ionospherically scattered* waves. Waves that are reflected or scattered in the troposphere (that region of the atmosphere within 10 kilometers of the earth's surface) are termed *tropospheric* waves. Energy propagated over other paths near the earth's surface is considered to be *ground-wave*. It is convenient to divide the ground-wave signal into the *space wave* and *surface wave*. The space wave is made up of the *direct wave*, the signal that travels the direct path from transmitter to receiver, and the *ground-reflected* wave, which is the signal arriving at the receiver after being reflected from the surface of the earth. The space wave also includes that portion of the energy received as a result of diffraction around the earth's surface and refraction in the upper atmosphere.

The surface wave* is a wave that is guided along the earth's surface, much as an electromagnetic wave is guided by a transmission line. Energy is abstracted from the surface wave to supply the losses in the ground; so the attenuation of this wave is directly affected by the constants of the earth along which it travels. When both antennas are located right at the earth's surface, the direct and ground-reflected terms in the space wave cancel each other, and transmission is entirely by means of this surface wave (assuming no sky wave or tropospheric wave). The surface wave is not shown in Fig. 16-1.

The factors that affect propagation over each of these various paths will be considered in detail. This chapter will deal with ground-wave and tropospheric propagation. Ionospheric reflection and scattering will be treated in chap. 17.

16.01 Plane-earth Reflection. For elevated transmitting and receiving antennas within line-of-sight of each other, the direct and ground-reflected wave combine to produce the resultant signal. For a smooth-plane, finitely conducting earth the magnitude and phase of the reflected wave can be calculated by an extension of the analysis of sec. 5.09 for reflection at the surface of a perfect dielectric. When the earth is rough the reflected wave tends to be scattered and may be much reduced in amplitude compared with smooth-earth reflection. A measure of "roughness" is given by the Rayleigh criterion which is

$$R = \frac{4\pi\sigma \sin \phi}{\lambda}$$

where σ is the standard deviation of the surface irregularities relative to the mean surface height, ϕ is the angle of incidence measured from the grazing angle, and λ is the wavelength. For $R < 0.1$, there is a well-defined specular reflection and the reflecting surface may be considered as being "smooth," so that the reflection factors may be computed by the methods of this section. For $R > 10$, the surface is "rough" and the reflected wave has a small magnitude. It will be noted that a surface which might be considered rough for waves incident at high angles (that is, large ϕ), may approach being a smooth surface as the angle of incidence approaches grazing. It will be found that when the incident wave is near grazing over a smooth earth the reflection coefficient approaches -1.0 for both polarizations.

The problem of reflection at the surface of a perfect (nonconducting) dielectric has already been solved in chap. 5 and the reflection

*This surface wave is sometimes called the "Norton surface wave." For a comprehensive discussion of other connotations which have been given to the term "surface wave," see the section on Surface Waves, *IRE Transactions on Antennas and Propagation*, AP-7, pp. S132-S231, Dec., 1959.

factors obtained for both perpendicular (horizontal) and parallel (vertical) polarizations. The earth, although not a good conductor in the sense that copper and silver are good conductors, is by no means a perfect dielectric, and its finite conductivity must be taken into account.

For a medium which has a dielectric constant ϵ and a conductivity σ, Maxwell's equation (I) is

$$\nabla \times \mathbf{H} = \epsilon \dot{\mathbf{E}} + \sigma \mathbf{E} \tag{16-1}$$

If the variation of $\mathbf{E}$ with time is sinusoidal, that is, if the expression for $\mathbf{E}$ at any point may be written

$$\mathbf{E} = \mathbf{E}_0 e^{j\omega t}$$

then

$$\begin{aligned} \dot{\mathbf{E}} &= j\omega \mathbf{E}_0 e^{j\omega t} \\ &= j\omega \mathbf{E} \end{aligned} \tag{16-2}$$

Putting this in eq. (1), there results

$$\begin{aligned} \nabla \times \mathbf{H} &= \left(\epsilon + \frac{\sigma}{j\omega}\right) \dot{\mathbf{E}} \\ &= \epsilon' \dot{\mathbf{E}} \end{aligned} \tag{16-3}$$

From eq. (3) it is apparent that a partially conducting dielectric can be considered as a dielectric that has a complex dielectric constant ϵ', where

$$\epsilon' = \epsilon \left(1 + \frac{\sigma}{j\omega\epsilon}\right)$$

The wave equations and reflection coefficients derived for perfect dielectrics will apply directly to dielectrics having loss or conductance, if the dielectric constant ϵ is replaced by an equivalent complex dielectric constant

$$\epsilon' = \left(\epsilon + \frac{\sigma}{j\omega}\right) \tag{16-4}$$

Reflection Factor for Perpendicular (Horizontal) Polarization. The reflection factor R_h for a plane wave having horizontal or perpendicular polarization is obtained directly from eq. (5-90) on page 146. It is

$$R_h = \frac{E_r}{E_i} = \frac{\sqrt{\epsilon_v} \cos\theta - \sqrt{\left(\epsilon + \dfrac{\sigma}{j\omega}\right) - \epsilon_v \sin^2\theta}}{\sqrt{\epsilon_v} \cos\theta + \sqrt{\left(\epsilon + \dfrac{\sigma}{j\omega}\right) - \epsilon_v \sin^2\theta}} \tag{16-5}$$

For the case of a wave incident at the surface of the earth, medium

(1) is air and so ϵ_1 has been replaced by ϵ_v, the dielectric constant of free space. Also the dielectric constant ϵ_2 of the second medium has been replaced by the complex dielectric constant $[\epsilon + (\sigma/j\omega)]$. θ is the angle of incidence measured from the normal. In dealing with reflection by the earth, it is usual to express the direction of the incident wave in terms of the angle ψ which is measured from the earth's surface. That is

$$\psi = 90^\circ - \theta$$

so that

$$\cos\theta = \sin\psi \qquad \sin\theta = \cos\psi$$

Equation (5) may then be written

$$R_h = \frac{\sin\psi - \sqrt{\left(\frac{\epsilon}{\epsilon_v} - \frac{j\sigma}{\omega\epsilon_v}\right) - \cos^2\psi}}{\sin\psi + \sqrt{\left(\frac{\epsilon}{\epsilon_v} - \frac{j\sigma}{\omega\epsilon_v}\right) - \cos^2\psi}} \tag{16-6}$$

where (5) has been divided through by $\sqrt{\epsilon_v}$. It is also customary to state the earth's dielectric constant relative to that of free space by means of a relative dielectric constant ϵ_r, where

$$\epsilon_r = \frac{\epsilon}{\epsilon_v}$$

(This is the familiar dielectric constant of electrostatic units where $\epsilon_v = 1$.) The final form of the expression for the reflection factor for horizontal polarization is

$$R_h = \frac{\sin\psi - \sqrt{(\epsilon_r - jx) - \cos^2\psi}}{\sin\psi + \sqrt{(\epsilon_r - jx) - \cos^2\psi}} \tag{16-7}$$

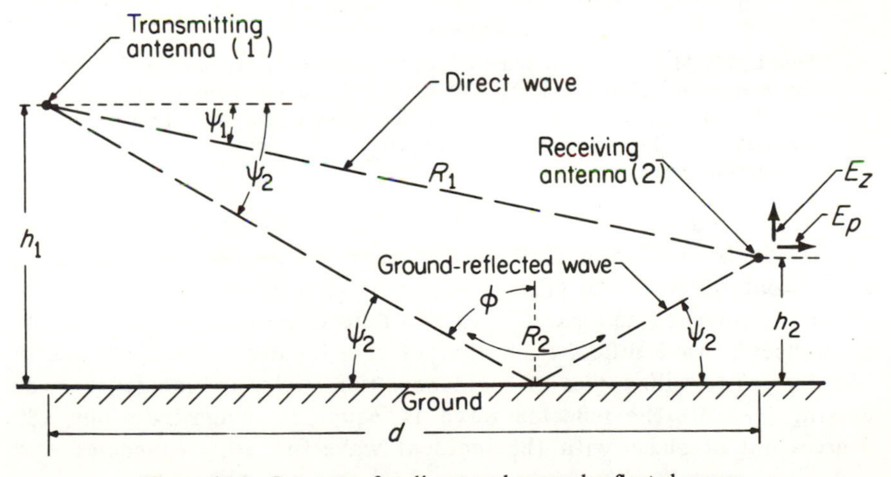

Figure 16-2. Geometry for direct and ground-reflected waves.

where

$$x = \frac{\sigma}{\omega\epsilon_v} = \frac{18 \times 10^9\sigma}{f} = \frac{18 \times 10^3\sigma}{f_{\text{MHz}}}$$

Reflection Factor for Parallel (Vertical) Polarization. In a manner similar to the above, the reflection factor for parallel or vertical polarization is obtained from eq. (5-91). It is

$$R_v = \frac{(\epsilon_r - jx)\sin\psi - \sqrt{(\epsilon_r - jx) - \cos^2\psi}}{(\epsilon_r - jx)\sin\psi + \sqrt{(\epsilon_r - jx) - \cos^2\psi}} \tag{16-8}$$

It is evident from eqs. (7) and (8) that the reflection factors are complex and that the reflected wave will differ both in magnitude and phase from the incident wave. The manner in which the reflection factors vary with angle of incidence is shown in Figs. 16-3 and 16-4.

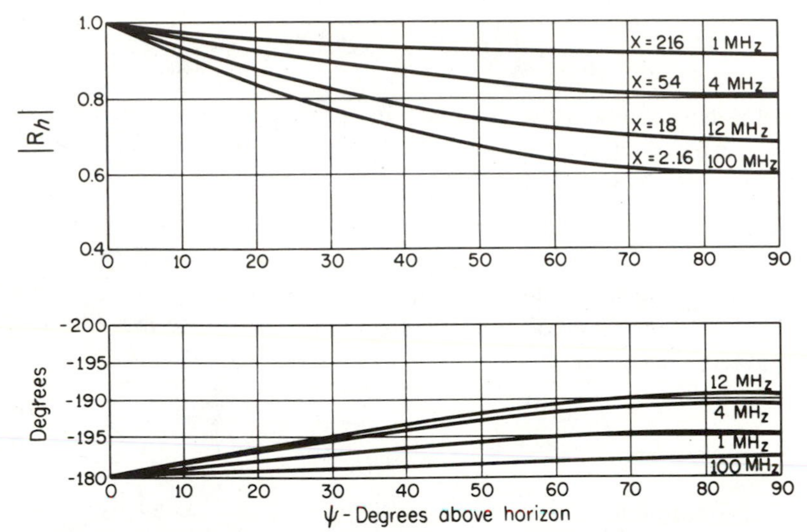

Figure 16-3. Magnitude and phase of the plane-wave reflection coefficient for *horizontal* polarization. The curves are for a relatively good earth ($\sigma = 12 \times 10^{-3}$, $\epsilon_1 = 15$) but can be used to give approximate results for other earth conductivities and other frequencies by calculating the appropriate value of $x = 18 \times 10^3\sigma/f_{\text{MHz}}$.

The various curves are for different frequencies. A study of these figures yields some interesting information. When the incident wave is horizontally polarized (Fig. 16-3), so that **E** is perpendicular to the plane of incidence and parallel to the reflecting surface, the phase of the reflected wave differs from that of the incident wave by nearly 180 degrees for all angles of incidence. For angles of incidence near grazing ($\psi = 0$), the reflected wave is equal in magnitude but 180 degrees out of phase with the incident wave for all frequencies and

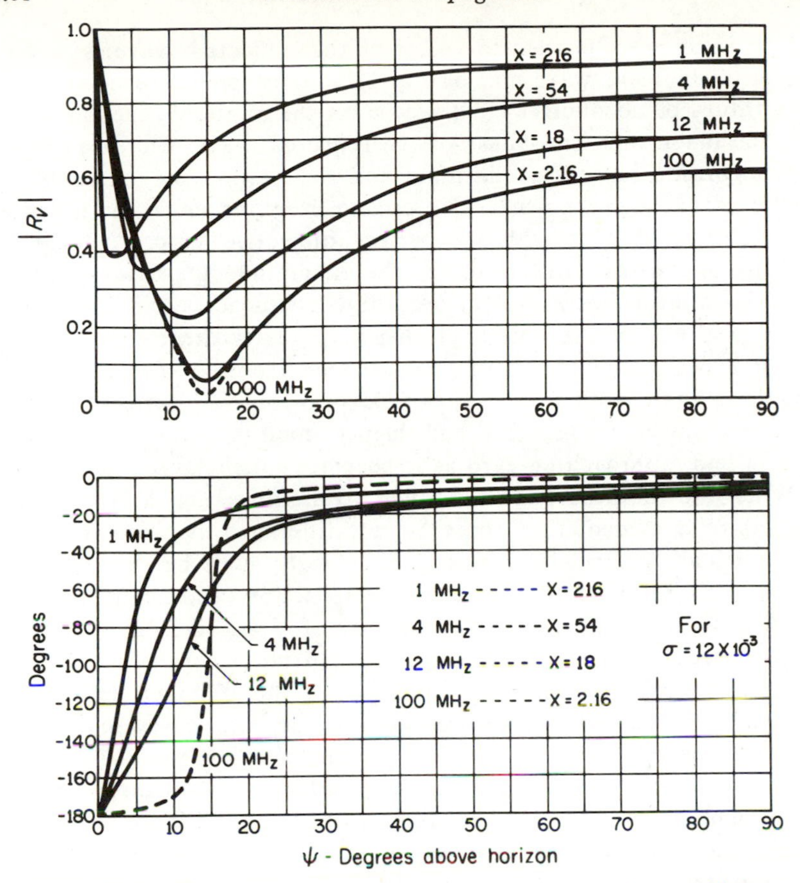

Figure 16-4. Magnitude and phase of the plane-wave reflection coefficient for vertical polarization. (See legend for Fig. 16-3.)

all ground conductivities. As the angle of incidence is increased, both the magnitude and phase of the reflection factor change, but not to any large extent. The change is greater for the higher frequencies and lower ground conductivities. The curves of Fig. 16-3 are drawn for an earth having a "good" conductivity and can be used for a range of frequencies from 0.5 to 1000 MHz. The relative dielectric constant ϵ_r varies from about 7 for a "poor" (low conductivity) earth to about 30 for a "good" (high conductivity) earth, so an average value of $\epsilon_r = 15$ has been used.

Figure 16-4 shows the manner in which the reflection factor R_v for vertical polarization varies with angle of incidence. In this case the electric vector **E** is parallel to the plane of incidence and the magnetic vector **H** is parallel to the boundary surface. The results are quite different from those obtained for horizontal polarization. As before, at grazing incidence the electric vector of the reflected wave is equal to that of the incident wave and has a 180-degree phase reversal for all finite values of conductivity. However, as the angle ψ increases from zero, the magnitude and phase of the reflected wave decrease rapidly. The magnitude reaches a minimum and the phase goes through -90 degrees at an angle known as the pseudo-Brewster angle (or just Brewster angle) by analogy with the perfect dielectric case. At angles of incidence above this critical angle, the magnitude increases again and the phase approaches zero. For very high frequencies and low conductivities ($x \ll \epsilon_r$), the Brewster angle has very nearly the same value as it has for a perfect dielectric. This can be seen from eq. (8). (For $\epsilon_r = 15$, Brewster's angle occurs at $\psi = 14.5$ degrees for the perfect dielectric case.) For lower frequencies and higher conductivities the Brewster angle is less, approaching zero as x becomes much larger than ϵ_r.

When the incident wave is normal to the reflecting surface ($\psi = 90$ degrees), it is evident that there is no difference between horizontal and "vertical" polarization. The electric vector will be parallel to the reflecting surface in both cases and the reflection coefficients R_v and R_h should have the same values. Comparison of Figs. 16-3 and 16-4 shows that, whereas they do have the same magnitude, there is a 180-degree difference in phase. This comes about from the different definitions of positive direction for the reflected wave in the two cases and requires some explanation. For the case of reflection of a horizontally polarized wave from the surface of a perfect conductor, if the electric vector of the incident wave is in the positive x direction [Figure 5-8(*a*)], the electric vector of the reflected wave will also be in the positive x direction, but will be 180 degrees out of phase with the incident wave. This could also be interpreted as a wave in phase with the incident wave, but having its electric vector in the opposite direction. In the vertical polarization case, the positive directions for incident and reflected electric field strengths are usually assumed to be as shown in Fig. 5-8(*b*), that is, both in the positive z direction when $\psi = 0$. As ψ increases from zero, the horizontal components of both electric fields increase, but one horizontal component is positive and the other negative. At $\psi = 90$ degrees the electric field strengths are wholly horizontal, but oppositely directed, one being in the positive y direction and the other in the negative y direction. From Fig. 16-4(*b*) the phase angle between these fields at the surface of the reflector is zero degrees (for a perfect conductor). But two vectors oppositely directed in space and having the same time phase give the same result as two vectors having the same direction and opposite phases; so this result is identical with that obtained from Fig. 16-3(*b*) at $\psi = 90$ degrees.

For angles of incidence near grazing (ψ nearly equal to zero), a more accurate plot of reflection coefficients than those given by Figs. 16-3 and 16-4 is often required. In Fig. 16-5, the magnitudes and phases of the reflection coefficients are shown* on a logarithmic scale for a relative dielectric constant $\epsilon_r = 10$.

Figure 16-6 shows how the earth's conductivity varies throughout the United States. In general, hilly or mountainous regions have low conductivity (from 10^{-3} to 5×10^{-3} mho/m) whereas the flat prairies are regions of relatively high conductivities (from 10×10^{-3} to 30×10^{-3} mho/m). The curves of Figs. 16-3 and 16-4 may be used with other conductivities than those shown on the figures by computing the appropriate values of x and interpolating between curves. For example, the curve labeled $x = 18$ corresponds to a frequency of 12 MHz and a fairly good ground conductivity ($\sigma = 12 \times 10^{-3}$ mho/m). Since $x = (18 \times 10^3 \times \sigma)/f_{\text{MHz}}$, it is seen that this same curve would also apply for 1 MHz over an earth having a conductivity $\sigma = 1 \times 10^{-3}$ mho/m (that is a very poor earth). The curves of Fig. 16-5 are labeled directly in terms of x.

16.02 Space Wave and Surface Wave. The general problem of radiation from a vertical antenna above a plane earth having finite conductivity was originally solved by Sommerfeld† in 1909. Similar solutions have since been obtained by other writers using different attacks. All of these leave the solution in complicated forms that are difficult to evaluate.

Norton‡ has reduced the complex expressions of the Sommerfeld theory to a form suitable for use in engineering work. In his original discussion, Sommerfeld stated that it was possible to divide the ground-wave field strength into two parts, a space wave and a surface wave. The space wave predominates at large distances above the earth, whereas the surface wave is the larger near the earth's surface. As given by Norton, the expressions for the electric field of an electric dipole above the surface of a finitely conducting plane earth are in a form that clearly shows this separation into space and surface waves.

*C. R. Burrows, "Radio Propagation Over a Plane Earth," *Bell System Tech. J.*, **16**, 45 (1937). In the curves of Fig. 16-5 as well as those of Figs. 16-3 and 16-4 the phase angle shown for the reflection coefficient is the angle by which the reflected wave leads the incident wave.

†A. Sommerfeld, "The Propagation of Waves in Wireless Telegraphy," *Ann. Physik*, **28**, 665 (1909).

‡K. A. Norton, "The Propagation of Radio Waves over the Surface of the Earth and in the Upper Atmosphere," *Proc. IRE*, **24**, 1367 (1936); *Proc. IRE*, **25**, 1203 (1937); *Proc. IRE*, **25**, 1192 (1937).

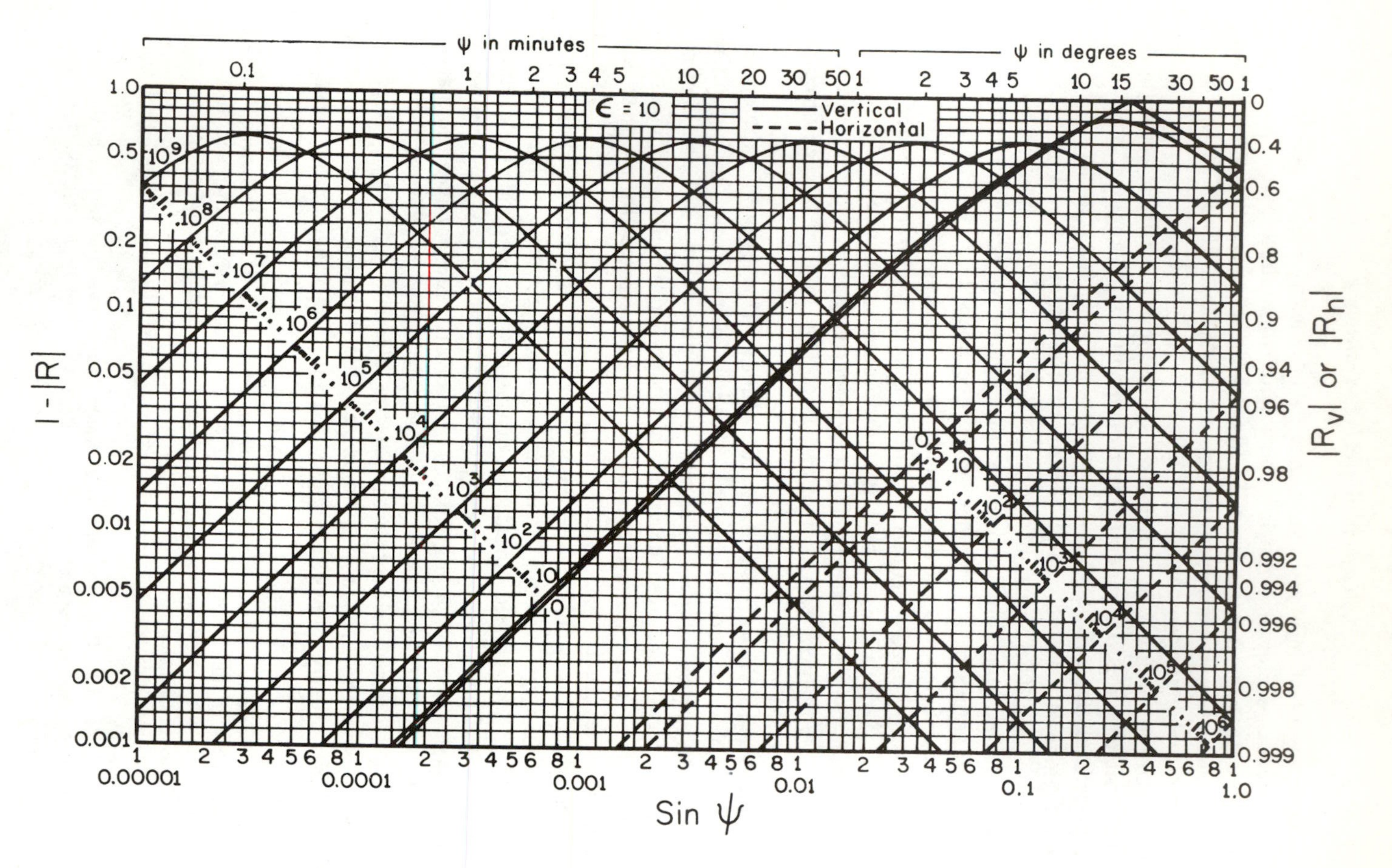
ψ in minutes
ψ in degrees
ε = 10
Vertical
Horizontal
1 - |R|
|Rv| or |Rh|
Sin ψ
0.00001
0.0001
0.001
0.01
0.1
1.0
1.0
0.5
0.2
0.1
0.05
0.02
0.01
0.005
0.002
0.001
0
0.4
0.6
0.8
0.9
0.94
0.96
0.98
0.992
0.994
0.996
0.998
0.999

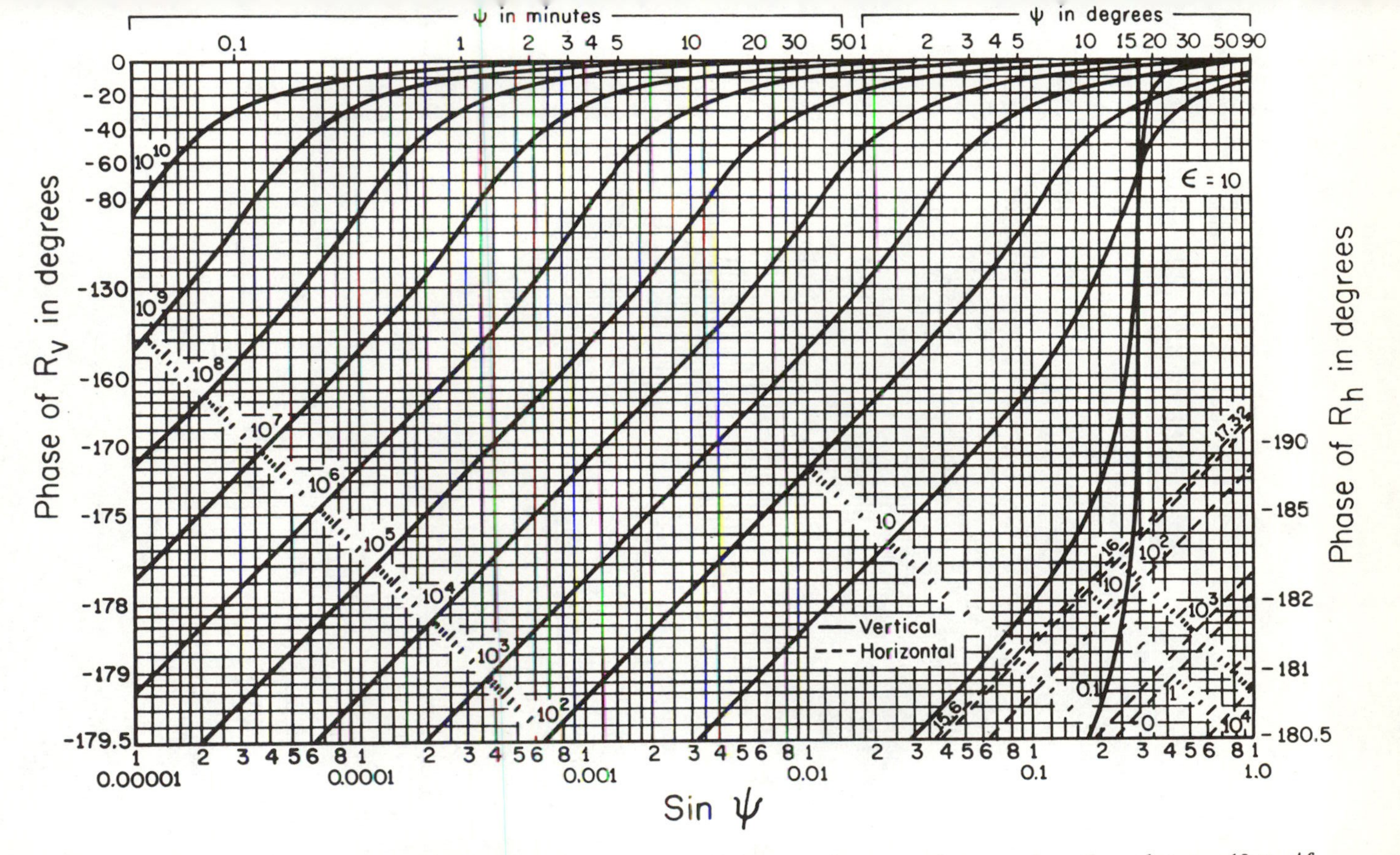

Figure 16-5. Magnitude and phase of the reflection coefficient for $\epsilon_r = 10$. The number on each curve gives the value $x = 18 \times \sigma / f_{\mathrm{MHz}}$. Vertical polarization is shown by solid lines, horizontal polarization by dashed lines. (Courtesy *Bell System Technical Journal*)

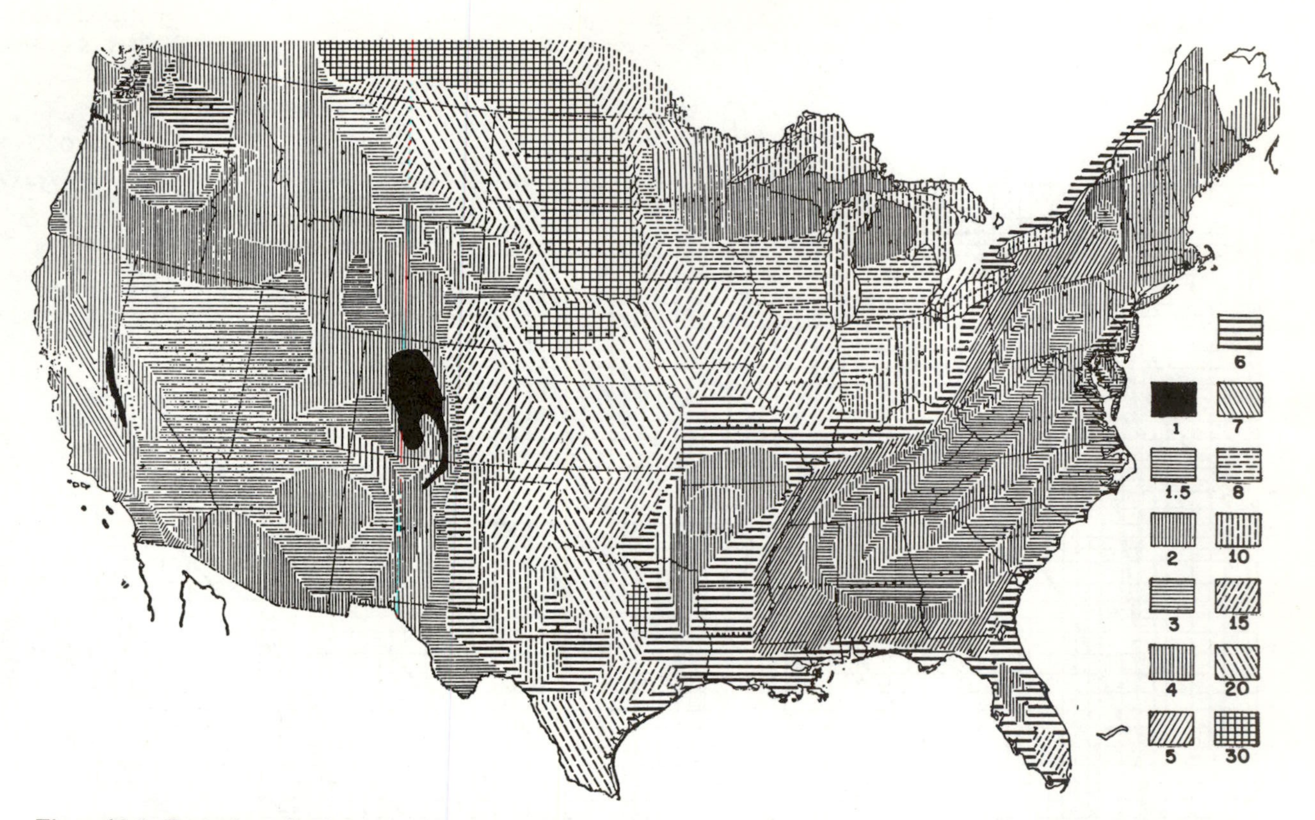

Figure 16-6. Ground conductivity map of the United States. Numbers on the legend, when multiplied by 10^{-3}, indicate ground conductivity in mhos/meter. (To obtain ground conductivity in e.m.u. multiply the numbers by 10^{-14}.) (Map by FCC)

At large distances from the dipole, such that the terms containing the higher orders of $1/R_1$ and $1/R_2$ may be neglected, the expressions for the vertical dipole above a finitely conducting plane earth reduce to

$$E_z = j30\beta I\,dl\left[\cos^2\psi\left(\frac{e^{-j\beta R_1}}{R_1} + R_v\frac{e^{-j\beta R_2}}{R_2}\right) + (1 - R_v)(1 - u^2 + u^4\cos^2\psi)F\frac{e^{-j\beta R_2}}{R_2}\right] \tag{16-9}$$

$$E_\rho = -j30\beta I\,dl\left[\sin\psi\cos\psi\left(\frac{e^{-j\beta R_1}}{R_1} + R_v\frac{e^{-j\beta R_2}}{R_2}\right) - \cos\psi(1 - R_v)u\sqrt{1 - u^2\cos^2\psi}\,F\frac{e^{-j\beta R_2}}{R_2}\left(1 + \frac{\sin^2\psi}{2}\right)\right] \tag{16-10}$$

In these expressions, E_z is the z component of electric field and E_ρ is the radial component (cylindrical co-ordinates, see Fig. 16-2); R_1 and R_2 are the distances from the dipole and its image, respectively, to the field point P. R_v is the plane-wave reflection coefficient, the expression for which has already been developed. F is an attenuation function that depends upon the earth's constants and upon the distance to the receiving point. It will be discussed under "surface wave." Also

$$u^2 = \frac{1}{\epsilon_r - jx}$$

where

$$x = \frac{1.8 \times 10^4\sigma \text{ mho/m}}{f_{\text{MHz}}}$$

σ = conductivity of the earth, mho/m

$\epsilon_r = \epsilon/\epsilon_v$ = relative dielectric constant of the earth

$\beta = 2\pi/\lambda$

Inspection of eqs. (9) and (10) shows that the total field may be divided into two parts, a "space wave," given by the inverse-distance terms, and a "surface wave"* that contains the additional attenuation function F. Combining (9) and (10) and separating into these two types of waves, there results

$$E_{\text{total space}} = E_\psi\,(\text{space}) = \sqrt{E_z^2\,(\text{space}) + E_\rho^2\,(\text{space})} = j30\beta I\,dl\cos\psi\left(\frac{e^{-j\beta R_1}}{R_1} + R_v\frac{e^{-j\beta R_2}}{R_2}\right) \tag{16-11}$$

$$E_{\text{total surface}} = j30\beta I\,dl(1 - R_v)F\frac{e^{-j\beta R_2}}{R_2}\sqrt{1 - 2u^2 + (\cos^2\psi)u^2\left(1 + \frac{\sin^2\psi}{2}\right)^2} \tag{16-12}$$

*This is the Norton Surface Wave—see footnote on page 629.

In equations (11) and (12), terms involving the factor u^4 have been discarded.

The Space Wave. The expression for the space wave of a vertical dipole over a plane earth as given by eq. (11) consists of two terms. The first term $e^{-j\beta R_1}/R_1$ represents a spherical wave originating at the position of the dipole. $e^{-j\beta R_1}$ is the phase factor (the time factor $e^{j\omega t}$ has been dropped) and $1/R_1$ is the inverse-distance factor. Similarly the second term represents a spherical wave originating at the position of the image of the dipole, but in this case the magnitude and phase of the wave have been modified by the *plane*-wave reflection factor R_v. Thus the space-wave part of the field consists of a direct wave and a reflected wave, and the expression for the reflected wave contains the reflection factor R_v that would apply if the incident wave were plane. When the dipole is located far from the earth, the incident wave is essentially a plane wave, and, in this case, the space-wave field is the total (ground-wave) field. On the other hand, when the dipole is located close to the earth, the incident wave will not be plane, and the expression for the total reflected field must contain terms in addition to those given by the space-wave field. These additional terms are just those which account for the surface wave.

Space-wave Patterns of a Vertical Dipole. In order to determine the effect of a finitely conducting earth upon the radiation pattern of an actual antenna, it is desirable first to investigate the radiation pattern of an elementary dipole above the earth. Expression (11) gives the space-wave field of a vertical dipole located at any height above a finitely conducting earth having the reflection coefficient R_v. The expression has been evaluated and plotted as a function of frequency for a range of ground conductivities and several dipole heights (Figs. 16-7 to 16-9).

Figure 16-7 shows the vertical radiation pattern of a vertical dipole located at the surface of a finitely conducting earth. The parameter $n = x/\epsilon_r$, where as before

$$x = \frac{\sigma}{\omega\epsilon_v} = \frac{18 \times 10^3\sigma}{f_{\mathrm{MHz}}}$$

and where σ is the earth conductivity in mhos per meter and f_{MHz} is the frequency in megahertz. An average value of 15 has been used for ϵ_r, the relative dielectric constant of the earth. The curve $n = \infty$ represents the case of a perfectly conducting earth. $n = 100$ represents conditions at low broadcast frequencies over a good (high conductivity) earth. $n = 10$ corresponds to high broadcast frequencies over an earth of average conductivity. The curve $n = 1$ represents conditions at the

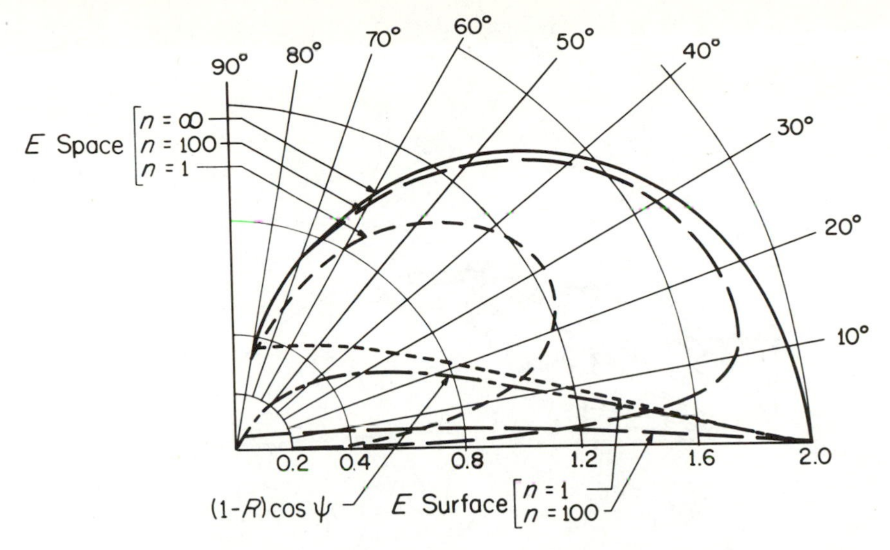

Figure 16-7. Vertical radiation pattern of a vertical dipole at the surface of an earth having finite conductivity. The parameter $n = x/\epsilon_r$ and an average value $\epsilon_r = 15$ has been used. Both space wave and unattenuated surface wave are shown.

medium-high frequencies. The top three curves are the space-wave patterns. Shown dotted is the unattenuated surface-wave curve, which will be discussed later. Figures 16-8 and 16-9 show the vertical radiation patterns that result when the dipole is elevated one-quarter wavelength and one-half wavelength above the earth.

From these figures it is apparent that the chief effects of the finite conductivity of the earth on the vertical radiation patterns occur at the low angles where the space wave is much reduced from its value over a perfectly conducting earth. This is because of the phase of the reflection factor R_v, which changes rapidly for angles of incidence near the pseudo-Brewster angle. Above this angle the phase of R_v is nearly zero, whereas below this angle near grazing incidence the phase of R_v approaches -180 degrees. The phase of R_v is always -90 degrees at the pseudo-Brewster angle. This rapid change of phase of the reflection coefficient near the critical pseudo-Brewster angle is responsible for many of the propagation characteristics peculiar to vertical polarization.

The patterns shown in Figs. 16-7 to 16-11 have been plotted for equal currents in the dipoles. A small radiated field, as for example in the case of $n = 1$, indicates small power radiated for a given current and, therefore, a low radiation resistance. For a given *power radiated* the dipole currents would be larger for this case ($n = 1$) and the resul-

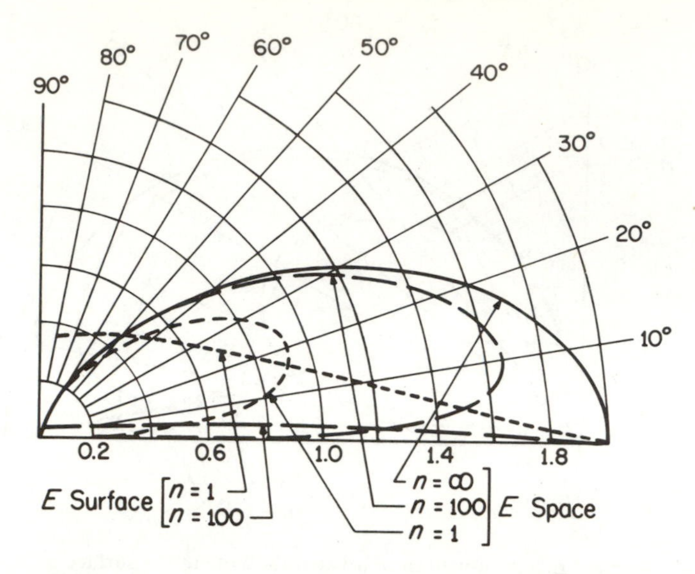

Figure 16-8. Vertical radiation pattern of a vertical dipole located a quarter wavelength above an earth of finite conductivity. $n = x/\epsilon_r$ and $\epsilon_r = 15$.

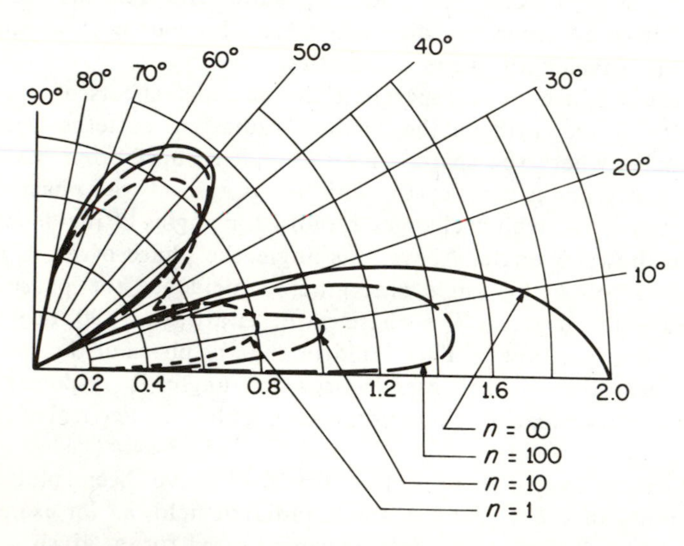

Figure 16-9. Vertical radiation pattern of a vertical dipole one-half wavelength above an earth of finite conductivity. $n = x/\epsilon_r$ and $\epsilon_r = 15$.

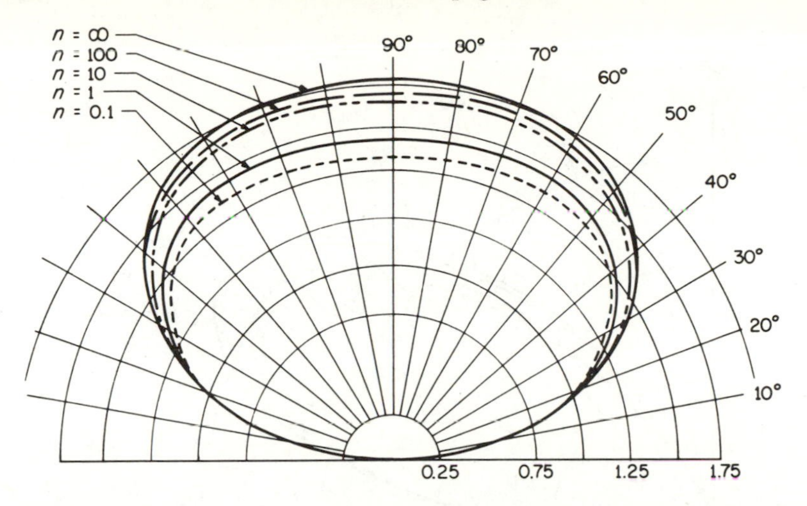

Figure 16-10. Vertical radiation (in the plane perpendicular to the axis of the dipole) of a horizontal dipole a quarter wavelength above an earth having finite conductivity. $n = x/\epsilon_r$ and $\epsilon_r = 15$.

tant field would also be larger than shown. The relative shape of the patterns shown is the important thing; their relative size has less significance.

Space-wave Patterns for the Horizontal Dipole. The expression for the space-wave field of a horizontal dipole in the plane perpendicular to the axis of the dipole is similar to that for the vertical dipole, except that R_v is replaced by R_h and the cos ψ factor is absent. It is

$$E^h_{\text{space}} = j30\beta I\,dl\left(\frac{e^{-j\beta R_1}}{R_1} + R_h\frac{e^{-j\beta R_2}}{R_2}\right)$$

The absence of the cos ψ factor is due to the fact that the horizontal dipole by itself is a uniform radiator in the plane perpendicular to its axis.

Figs. 16-10 and 16-11 show the space-wave patterns of a horizontal dipole at heights of one-quarter wavelength and one-half wavelength above a finitely conducting earth. These are the patterns in the plane perpendicular to the axis of the dipole. The effects of finite conductivity are much less marked than in the vertical dipole case because the reflection factor R_h never deviates much from the value -1, which it has for the perfect conductor case. In the plane *parallel* to the axis of the dipole, the electric field is given by the expression

$$E^h_{\text{space}} = j30\beta I\,dl\sin\psi\left(\frac{e^{-j\beta R_1}}{R_1} - R_v\frac{e^{-j\beta R_2}}{R_2}\right)$$

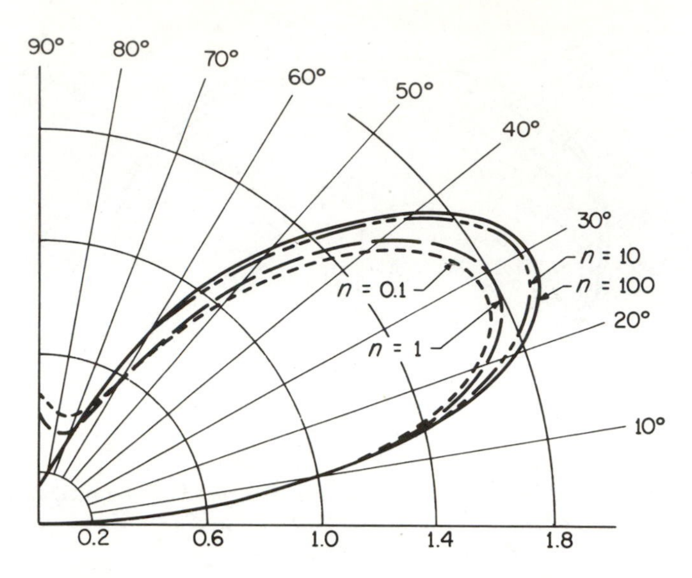

Figure 16-11. Vertical radiation pattern (in the plane perpendicular to the axis of the dipole) of a horizontal dipole one-half wavelength above an earth having finite conductivity. $n = x/\epsilon_r$ and $\epsilon_r = 15$.

In this case the incident wave is polarized parallel to the plane of incidence, and the reflection factor R_v for "vertical" polarization is required. The minus sign comes about from the assumed positive directions of electric fields for the incident and reflected waves, as explained earlier in the chapter. Note that in this plane, parallel to the dipole axis, the electric field of a horizontal dipole is "vertically" polarized.

16.03 The Surface Wave. The expressions for the electric field of a vertical dipole above a finitely conducting plane earth were given in eqs. (9) and (10). When the dipole is at the surface of the earth, the expression for the surface-wave part of this field reduces to

$$\mathbf{E}_{\text{surface}} = j30\beta I\,dl(1 - R_v)F\left(\frac{e^{-j\beta R}}{R}\right)\cdot \left[\hat{\mathbf{k}}(1 - u^2) + \hat{\mathbf{r}}\cos\psi\left(1 + \frac{\sin^2\psi}{2}\right)u\sqrt{1 - u^2\cos^2\psi}\right] \qquad (16\text{-}13)$$

In this expression R is the distance from the dipole to the point at which the field is being considered ($R \gg \lambda$). $\hat{\mathbf{k}}$ and $\hat{\mathbf{r}}$ are unit vectors respectively parallel to and perpendicular to the vertical dipole. Also

$$F = \{1 - j\sqrt{\pi\omega}\,e^{-\omega}[\text{erfc}\,(j\sqrt{\omega})]\}$$

$$\omega = \frac{-j\beta R u^2(1 - u^2 \cos^2 \psi)}{2}\left[1 + \frac{\sin \psi}{u\sqrt{1 - u^2 \cos^2 \psi}}\right]^2$$

$$u^2 = \frac{1}{\epsilon_r - jx}$$

$$x = \frac{18 \times 10^3 \sigma}{f_{\text{MHz}}}$$

$$\operatorname{erfc}(j\sqrt{\omega}) = \frac{2}{\sqrt{\omega}}\int_{j\sqrt{\omega}}^{\infty} e^{-v^2}\, dv$$

The function F introduces an attenuation that is dependent upon distance, frequency, and on the constants of the earth along which the wave is traveling. For distances within a few wavelengths of the dipole, F has a value of very nearly unity, and it approaches unity as the distance R approaches zero. Putting $F = 1$ in eq. (13), it is possible to evaluate and plot what is called "unattenuated surface wave." This is shown in Fig. 16-7 for two values of the parameter n. For low frequencies and good ground conductivity ($n = 100$), the unattenuated surface wave is very small, except for angles near grazing ($\psi = 0$). At $\psi = 0$, it has the value 2. At this same angle the space wave is always zero because the direct and ground-reflected waves cancel. For higher frequencies and poorer conductivity ($n = 1$), the unattenuated surface wave still has a value of 2 at $\psi = 0$, but it also has appreciable value at high angles as well. However, this wave attenuates very rapidly with distance because of the factor F.

At the surface of the earth ($\psi = 0$), the absolute value of F has been evaluated and is called the "ground-wave attenuation factor." It is designated by the symbol A. That is, at $\psi = 0$

$$\begin{aligned} A &= |F| \\ &= |1 - j\sqrt{\pi\omega}\, e^{-\omega} \operatorname{erfc}(j\sqrt{\omega})|_{\psi=0} \\ &= |1 - j\sqrt{\pi p_1}\, e^{-p_1} \operatorname{erfc}(j\sqrt{p_1})| \end{aligned} \tag{16-14}$$

p_1 is the value of ω at the angle $\psi = 0$. In general, it is a complex quantity and may be written

$$\omega|_{\psi=0} = p_1 = pe^{jb}$$

where p is known as the *numerical distance* and b as the *phase constant.*

Evaluating ω at $\psi = 0$ shows that

$$p = \frac{\pi R}{\lambda x}\frac{\cos^2 b''}{\cos b'} \cong \frac{\pi R}{\lambda x}\cos b$$

$$b = (2b'' - b') \cong \tan^{-1} \frac{\epsilon_r + 1}{x}$$

where

$$b'' = \tan^{-1} \frac{\epsilon_r}{x}$$

$$b' = \tan^{-1} \frac{\epsilon_r - \cos^2 \psi}{x} \cong \tan^{-1} \frac{\epsilon_r - 1}{x}$$

$$x = \frac{18 \times 10^3 \sigma}{f_{\mathrm{MHz}}}$$

The Ground-wave Attenuation Factor A. A plot of the ground-wave attenuation factor A, as given by eq. (14), is shown in Fig. 16-12 in terms of p and b. The numerical distance p depends upon the frequency and the ground constants, as well as upon the actual distance to the transmitter. It is proportional to the distance and the square of the frequency and varies almost inversely with the ground conductivity. The phase constant b is a measure of the power-factor angle of the earth (the actual power-factor angle is b''). When the earth constants and the frequency are such that $x \gg \epsilon_r$, the power-factor angle will be nearly zero, and the earth will be mainly resistive. This is the case for average or better-than-average earth at broadcast frequencies. At very high frequencies and over poor earths the condition $\epsilon_r \gg x$ may be obtained, and the earth impedance will then be reactive. It will be noticed that the same earth which acts as a conductor at very low frequencies will act as a dielectric that has a small loss at very high frequencies.

The attenuation factor A can also be represented approximately by the following empirical formulas:

For $b < 5$ degrees,

$$A_1 \cong \frac{2 + 0.3p}{2 + p + 0.6p^2} \tag{16-15}$$

For all values of b,

$$A \cong A_1 - \sin b \sqrt{\frac{p}{2}}\, e^{-(5/8)p} \tag{16-16}$$

For $b < 5$ degrees and $p < 4.5$ (that is, for short numerical distances),

$$A \cong e^{-0.43p + 0.01p^2} \tag{16-17}$$

This relation shows that A varies almost exponentially with p for short numerical distance.

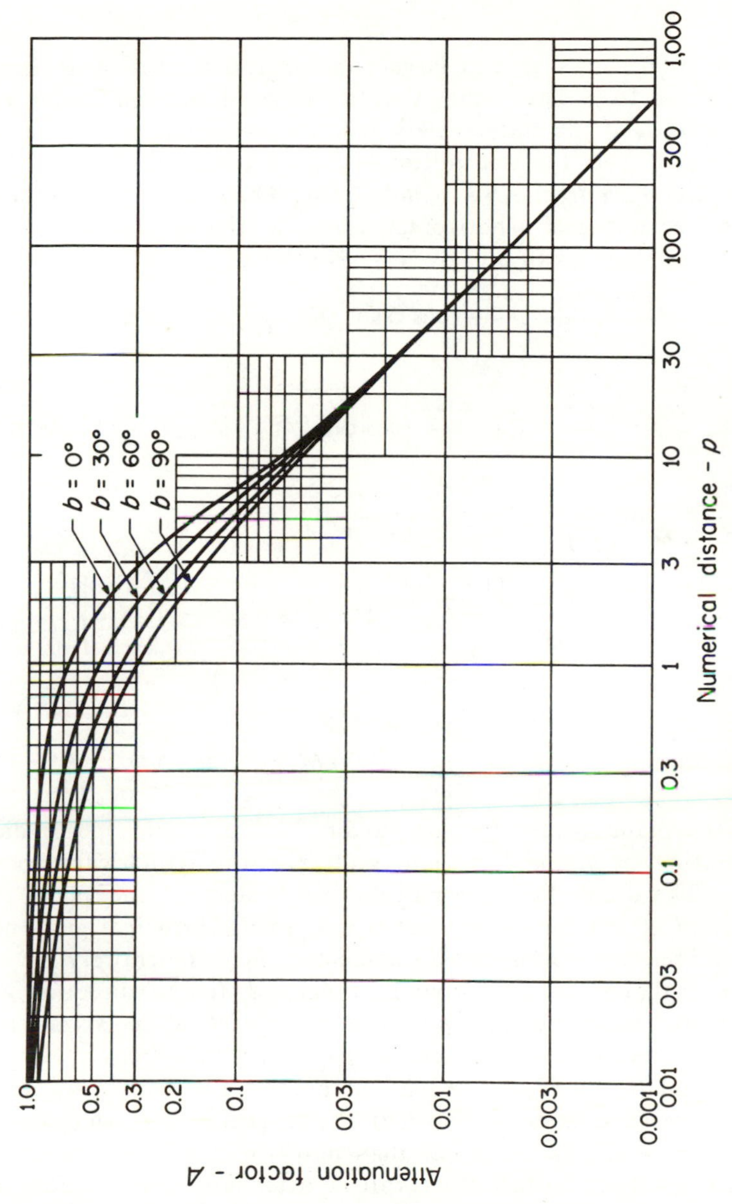

Figure 16-12. Ground-wave attenuation factor A.

For $b < 5$ degrees and $p \geqq 4.5$,

$$A \cong \frac{1}{2p - 3.7} \tag{16-18}$$

This relation shows that at large numerical distances A is inversely proportional to p. This means that at large numerical distances the field strength of the surface wave will vary inversely as the *square* of the distance from the transmitter.

Surface Wave from a Horizontal Dipole. The expressions for the space and surface waves of a horizontal dipole at the surface of a finitely conducting plane earth are given by Norton as

$$\mathbf{E}_{\text{space}} = \frac{j30\beta I\, dl\, e^{-j(\beta R - \omega t)}}{R} [\cos\phi \sin\psi(1 - R_v)\hat{\mathbf{k}} + \sin\phi(1 + R_h)\hat{\boldsymbol{\phi}}] \tag{16-19}$$

$$\mathbf{E}_{\text{surface}} = \frac{j30\beta I\, dl\, e^{-j(\beta R - \omega t)}}{R} \Bigg\{ \cos\phi(u\sqrt{1 - u^2\cos^2\psi})(1 - R_v)F \left[\cos\psi\left(1 + \frac{\sin^2\psi}{2}\right)\hat{\mathbf{k}} + u\sqrt{1 - u^2\cos^2\psi} \left(\frac{1 - \sin^2\psi - \dfrac{(1 - R_h)G}{(1 - R_v)u^2F}}{1 - u^2\cos^2\psi}\right)\hat{\boldsymbol{\rho}}\right] + \sin\phi(1 - R_h)G\hat{\boldsymbol{\phi}}\Bigg\} \tag{16-20}$$

where

$$G = [1 - j\sqrt{\pi v}\, e^{-v} \operatorname{erfc}(j\sqrt{v})]$$

$$v = -\frac{j\beta R(1 - u^2\cos^2\psi)}{2u^2}\left(1 + \frac{u\sin\psi}{\sqrt{1 - u^2\cos^2\psi}}\right)^2$$

R_h is the plane-wave reflection factor for horizontal (perpendicular) polarization. $\hat{\mathbf{k}}$, $\hat{\boldsymbol{\rho}}$, and $\hat{\boldsymbol{\phi}}$ are unit vectors in the cylindrical co-ordinate system. The dipole lies perpendicular to $\hat{\mathbf{k}}$ and in the plane $\phi = 0$. Inspection of the expressions shows that in the principal plane normal to the dipole ($\phi = 90$ degrees) the electric field is entirely in the ϕ direction, that is, it is horizontally polarized. In the direction $\phi = 0$, the electric vector lies in the plane $\phi = 0$ ("vertical" polarization). For intermediate directions the field is elliptically polarized.

The function G is an attenuation function for horizontal polarization. At large numerical distances G approaches u^4F, and since $u^2 = 1/(\epsilon_r - jx)$ is always much less than unity, it is evident that the horizontally polarized surface wave will be attenuated more rapidly than a vertically polarized wave of the same frequency.

In practical computations the attenuation of a horizontally polarized wave along the surface of the earth is determined by using the same ground-wave attenuation factor A as is used for vertical polarization. However, now the numerical distance p and phase factor b are given by

$$p = \frac{\pi R}{\lambda} \frac{x}{\cos b'}$$

$$b = 180^\circ - b'$$

where, as before,

$$x = \frac{18 \times 10^3 \sigma}{f_{\text{MHz}}}$$

$$b' = \tan^{-1} \frac{\epsilon_r - 1}{x}$$

For a given actual distance R, the numerical distance p will be greater for horizontal polarization than for vertical polarization. This means greater attenuation for the horizontally polarized surface wave than for the vertically polarized wave. At low and medium frequencies, where x is large, this difference in attenuation is very great and only vertically polarized surface waves need be considered. In this frequency range the antennas used will be designed to radiate and receive vertically polarized signals. At high and very high frequencies the attenuation of the surface wave is very large for both polarizations, with the result that surface wave propagation is limited to very short distances. However, in this frequency range elevated antennas are used, and propagation paths are provided by the space wave. For this wave either vertical or horizontal polarization may be used.

16.04 Elevated Dipole Antennas above a Plane Earth. When both transmitting and receiving antennas are located at the surface of the earth, the angle ψ of the ground-wave propagation path between the antennas is zero. Under these conditions the earth's reflection coefficient is -1, so the direct and ground-reflected waves cancel. Propagation is then entirely by means of the surface wave. This is the case, for example, in the daytime reception of ordinary broadcast program signals. At high and very high frequencies, however, where a wavelength becomes sufficiently short, it is possible to elevate the antennas a quarter wavelength or more above the ground. When the antennas are elevated, the space wave is no longer zero, and the resultant signal at the receiving antenna is the vector sum of space and surface wave.

Consider the case of two vertical antennas elevated at heights h_1 and h_2 above the surface of the earth (Fig. 16-2). From eq. (9) the vertical component of the electric field at the receiving antenna (2) due to a vertical dipole at (1) will be

$$E_z = j30\beta I\,dl\left\{\cos^2\psi\left[\frac{e^{-j\beta R_1}}{R_1} + R_v\frac{e^{-j\beta R_2}}{R_2} + (1 - R_v)F\frac{e^{-j\beta R_2}}{R_2}\right]\right\} \quad (16\text{-}21)$$

In expression (21) u^2 and u^4 have been neglected as being small compared with unity. The first two terms of the expression constitute the space wave, and the third term is the surface wave. This expression is accurate at distances from the antenna larger than a few wavelengths. However, as it stands, it is rather involved for actual computations. Fortunately, the case of interest in practice is usually that in which the distance between antennas is very large compared with their heights above the ground, that is, for which

$$r \gg (h_1 + h_2)$$

Under these circumstances considerable simplification of the expression (21) results. The following relations will then hold approximately.

$$\cos \psi = 1$$

$$R_1 = R_2 = d \qquad \text{(for the magnitude factor in the denominator)}$$

Also for large numerical distances, the asymptotic expansion for the error function *erfc* can be used so that

$$F = 1 - j\sqrt{\pi\omega}\, e^{-\omega} \operatorname{erfc}(j\sqrt{\omega})$$

$$= 1 - \frac{1 + \dfrac{1}{2\omega} + \dfrac{(1)(3)}{(2\omega)^2} + \cdots}{1 + \dfrac{h_1 + h_2}{uR_2}}$$

$$\cong -\frac{1}{2\omega}$$

where $\quad uR_2 \gg (h_1 + h_2) \quad$ and $\quad |\omega| > 20.$

Introducing these approximations into (21) gives

$$E_z = \frac{j30\beta I\, dl}{d}\left\{e^{-j\beta R_1} + e^{-j\beta R_2}\left[R_v - \frac{(1 - R_v)}{2\omega}\right]\right\} \tag{16-22}$$

The expression for the numerical distance ω for this case will be

$$\omega = p_1\left(1 + \frac{h_1 + h_2}{R_2 u\sqrt{1 - u^2}}\right)^2 \tag{16-23}$$

When the distance between antennas is very large compared with their height above the ground, it is evident from an inspection of expression (23) that the numerical distance ω is very nearly equal to p_1, the numerical distance along the surface of the earth. Also under the same conditions the attenuation factor F, which is approximately equal to $1/2\omega$, will not change much with height of either transmitting or receiving antenna. It will have a value approximately equal to A, the surface-wave attenuation factor of Fig. 16-12. Thus, for elevated antennas, the magnitude of the surface wave will be given approximately by

$$E_{z\,\text{surface}} \approx \frac{j30\beta I\,dl}{d}\left(\frac{1-R_v}{2p}\right) = \frac{j30\beta I\,dl}{d}(1-R_v)A \qquad (16\text{-}24)$$

as long as the distance between antennas is very much greater than their heights above the ground.

Expression (22) has been obtained for short vertical dipoles, but it also holds for elevated half-wave dipoles under the same conditions if dl is replaced by λ/π, the effective length of the half-wave dipole.

The corresponding expression for horizontal half-wave dipoles would be

$$E_\phi = \frac{j60\sin\phi}{d}[e^{-j\beta R_1} + R_h\,e^{-j\beta R_2} + (1-R_h)G\,e^{-j\beta R_2}] \qquad (16\text{-}25)$$

At large numerical distances ($|\omega| > 20$), the attenuation factor G approaches u^4F, and so G is a very small quantity. The surface-wave attenuation for horizontal polarization is so large that the surface wave becomes negligibly small at very short distances, and ordinarily only the space wave needs to be considered. At large numerical distances the factor G in expression (25) can be replaced by $A \approx 1/2p$, where p has already been defined for horizontal polarization.

EXAMPLE 1: A half-wave dipole radiator is elevated 100 ft above the ground. A receiving dipole 3 miles distant is elevated 30 ft. Determine the space- and surface-wave field strengths at the receiving antenna when the transmitting antenna carries a current of 1 ampere at a frequency of 50 MHz. Assume an average earth having $\epsilon_r = 15$ and $\sigma = 5 \times 10^{-3}$, (a) for vertical half-wave dipoles and (b) for horizontal half-wave dipoles.

CASE (a)—*Vertical half-wave dipoles.*

$$E_{sp} = \frac{j30\beta I l_{\text{eff}}}{d}(e^{-j\beta R_1} + e^{-j\beta R_2}R_v)$$

$$= \frac{j60}{d}e^{-j\beta R_1}[1 + R_v\,e^{-j\beta(R_2-R_1)}]$$

$$\psi = \tan^{-1}\frac{h_1+h_2}{r} = \tan^{-1}\frac{130}{3\times 5280} = 0.47^\circ$$

$$x = \frac{5\times 10^{-3}\times 10^3\times 18}{50} = 1.8$$

From Fig. 16-5

$$R_v = 0.94\angle{-180^\circ}$$

Referring to Fig. 16-2,

$$R_1 = \sqrt{d^2 + (h_1 - h_2)^2} = d\sqrt{1 + \left(\frac{h_1 - h_2}{d}\right)^2}$$

$$= d\sqrt{1.0000196} = d(1.0000098)$$

$$R_2 = \sqrt{d^2 + (h_1 + h_2)^2} = d\sqrt{1 + \left(\frac{h_1 + h_2}{d}\right)^2}$$

$$= d\sqrt{1.0000677} = d(1.0000339)$$

$$R_2 - R_1 = 3 \times 0.304 \times 5280(1.0000339 - 1.0000098)$$

$$= 0.116 \text{ meters}$$

$$\frac{360}{\lambda}(R_2 - R_1) = \frac{360}{6} \times 0.116 = 7.0^\circ$$

$$|E_{\text{sp}}| = \frac{60}{d}\,|1 + 0.94\underline{/-180^\circ - 7^\circ}|$$

$$= \frac{60}{3 \times 1609}\,|1 - 0.935 + j0.113|$$

$$= \frac{60 \times 0.13}{3 \times 1609} = 1.62 \text{ mv/m}$$

$$|E_{\text{su}}| = \frac{60}{d}\left|\frac{(1 - R_v)}{2p}\right|$$

$$b \cong \tan^{-1}\frac{\epsilon_r + 1}{x} = \tan^{-1}\frac{16}{1.8} = 83.6^\circ$$

$$p \cong \frac{\pi R}{\lambda x}\cos b = \frac{\pi \times 3 \times 1609 \times 0.112}{6 \times 1.8} = 157$$

$$|E_{\text{su}}| = \frac{60}{3 \times 1609} \times \frac{1.94}{2 \times 157} = 0.077 \text{ mv/m}$$

CASE (b)—*Horizontal half-wave dipoles.* From Fig. 16-5, $R_h = 0.995\underline{/-180^\circ}$.

$$E_{\text{sp}} = \frac{60}{d}e^{-j\beta R_1}[1 + R_h\,e^{-j\beta(R_2 - R_1)}]$$

$$|E_{\text{sp}}| = \frac{60}{3 \times 1609}\,|1 + 0.995\underline{/-180^\circ - 7^\circ}|$$

$$= \frac{60 \times 0.122}{3 \times 1609} = 1.52 \text{ mv/m}$$

$$|E_{\text{su}}| = \frac{60}{d}\left|\frac{(1 - R_h)}{2p}\right|$$

$$b' = \tan^{-1}\frac{\epsilon_r - 1}{x} = \tan^{-1}\frac{14}{1.8} = 82.6^\circ$$

$$p = \frac{\pi R}{\lambda}\frac{x}{\cos b'} = \frac{\pi \times 3 \times 1609}{6} \times \frac{1.8}{0.128} = 35{,}600$$

$$|E_{\text{su}}| = \frac{60}{3 \times 1609}\,\frac{1.995}{2 \times 35{,}600} = 0.000349 \text{ mv/m}$$

Approximate Formula for VHF Propagation. The preceding example indicates that certain simplifying assumptions can be made when the elevated transmitting and receiving antennas are far apart. When these approximations are used, a quite simple formula for VHF propagation between elevated antennas results. These approximations are:

(1) The surface wave can be neglected in comparison with the space wave.

(2) The angle ψ is very small so that the reflection factor R_v or $R_h \cong -1$.

Then the field at the receiving antenna due to a current I amperes in a half-wave transmitting antenna is given by

$$|E| = \frac{60I}{d}\,|1 + R_{\substack{v\\h}}\underline{/-\alpha}|$$

$$= \frac{60I}{d}\,|1 - 1\underline{/-\alpha}| \tag{16-26}$$

where α is the difference in path length between direct and reflected waves expressed in degrees. That is,

$$\alpha = \frac{2\pi}{\lambda}(R_2 - R_1) \tag{16-27}$$

Referring to Fig. 16-2,

$$R_2 \approx d\sqrt{1 + \left(\frac{h_1 + h_2}{d}\right)^2} \qquad R_1 \approx d\sqrt{1 + \left(\frac{h_1 - h_2}{d}\right)^2}$$

Using the binomial expansion, when $x \ll 1$,

$$(1 + x)^{1/2} \approx 1 + \tfrac{1}{2}x$$

Then

$$R_2 - R_1 \approx d\left[1 + \frac{1}{2}\left(\frac{h_1 + h_2}{d}\right)^2\right] - d\left[1 + \frac{1}{2}\left(\frac{h_1 - h_2}{d}\right)^2\right]$$

$$\approx \frac{2h_1h_2}{d} \tag{16-28}$$

It should be observed that, in actual computations, this *approximate* expression (28) for $R_2 - R_1$, obtained by using the first two terms of a series, will give a more accurate numerical answer than the "exact" computation, using a reasonable number of significant figures. This is because when two large and nearly equal numbers are subtracted one from the other, significant figures are lost, so that it is necessary to start with a very large number of significant figures in order to end up with only fair accuracy. In the "approximate" method one works directly on the difference between the numbers and no significant figures are lost. Then, from (27) and (28)

$$\alpha = \frac{4\pi}{\lambda}\frac{h_1 h_2}{d}$$

From (26)

$$|E| = \frac{60I}{d}\,|1 - \cos\alpha + j\sin\alpha|$$

When the angle α is small, so that $\cos\alpha \cong 1$,

$$|E| \approx \frac{60I}{d}\sin\alpha \approx \frac{60I}{d}\sin\frac{4\pi h_1 h_2}{\lambda d} \tag{16-29}$$

If α is sufficiently small so that $\sin\alpha \approx \alpha$, this reduces to

$$|E| = \frac{60I\alpha}{d} = \frac{240\pi I h_1 h_2}{\lambda d^2} \qquad \text{V/m} \tag{16-30}$$

Where the approximations used are valid, the received field strength is proportional to the height of the transmitting antenna, the height of the receiving antenna, and inversely proportional to the *square* of the distance between them. In most propagation problems met with in frequency-modulation and television applications, the above approximations will hold so that the simple expression of eq. (30) may be used in these important practical cases.

16.05 Wave Tilt of the Surface Wave. A vertically polarized wave at the surface of the earth will have a forward tilt, the magnitude of which depends upon the conductivity and permittivity of the earth. The slight tilt forward of the electric field strength is responsible for a small vertically downward component of the Poynting vector, sufficient to furnish the power dissipated in the earth over which the wave is passing. In general, the component of electric field strength parallel to the earth will not be in phase with the component perpendicular to it, so that the electric field just above the surface of the earth will be elliptically polarized.

In chap. 7 the problem of a wave guided along the surface of a good conductor was solved. The results obtained will apply directly to this case of a radio wave along the surface of the earth, as long as the same assumption (depth of penetration not too large a fraction of the wavelength) is valid. This will be true over most of the range of frequencies and conductivities that are of interest in surface-wave propagation. Then the surface impedance of the earth is given approximately by

$$Z_s \approx \sqrt{\frac{\omega\mu}{\sqrt{\sigma^2 + \omega^2\epsilon^2}}}\,\Big/\,\tfrac{1}{2}\tan^{-1}\frac{\sigma}{\omega\epsilon}$$

where σ, μ, and ϵ are respectively the conductivity, permeability, and permittivity of the earth. The horizontal component of electric field strength will be $E_h = J_s Z_s$ and the vertical field strength will be approximately $E_v = H\eta_v$, so that the ratio of horizontal to vertical field will be

$$\frac{E_h}{E_v} = \frac{J_s Z_s}{H\eta_v} = \frac{Z_s}{\eta_v}$$

$$= \frac{1}{377}\sqrt{\frac{\omega\mu}{\sqrt{\sigma^2 + \omega^2\epsilon^2}}} \angle \tfrac{1}{2}\tan^{-1}\frac{\sigma}{\omega\epsilon} \qquad (16\text{-}31)$$

As an example consider a one-MHz radio wave at the surface of an average earth having $\sigma = 5 \times 10^{-3}$ and $\epsilon_r = 10$. For this case

$$\omega\epsilon = 5.55 \times 10^{-4}$$

$$\frac{E_h}{E_v} = 0.105\angle 41.8^\circ$$

The horizontal component of **E** is about one-tenth of the vertical component and leads it by an angle 41.8°. If the electric vector wave were plotted at various instants of time, the locus of the endpoint would trace out an ellipse. This elliptical polarization of the field at the earth's surface is shown in Fig. 16-13 for $\epsilon_r = 5$ and various values of x, where

$$x = \frac{\sigma}{\omega\epsilon_v} = \frac{18 \times 10^3\sigma}{f_{\text{MHz}}}$$

In the example above $x = 90$.

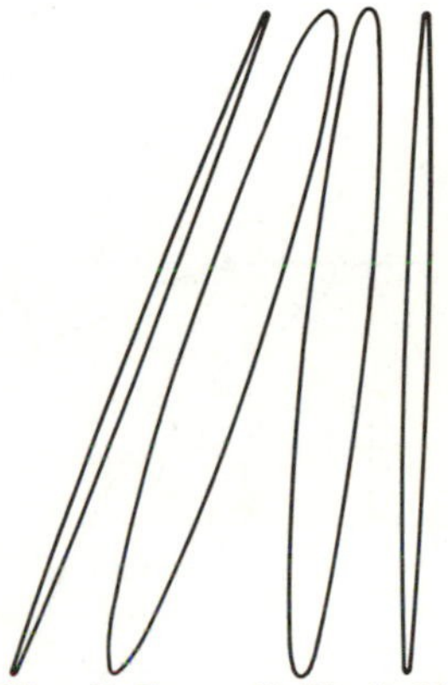

Figure 16-13. Elliptical polarization of the electric vector at the surface of an earth for which $\epsilon_r = 5$ and for various values of $x = 18 \times 10^3\sigma/f_{\text{MHz}}$.

16.06 Spherical-earth Propagation. The formulas for ground-wave propagation developed in the preceding sections were obtained on the assumption of a flat or plane earth. Whereas such an assumption gives answers that are approximately correct at short distances, it cannot be expected to yield correct results at large distances. The distance up to which the curves of Fig. 16-12 can be used without serious error is given by the relation $d = 50/f_{\text{MHz}}^{1/3}$ miles. Beyond this distance the actual field strength starts to deviate from that computed on the plane earth assumption. The curvature of the earth affects the propagation of the ground-wave signal in several ways. First, the bulge of the earth prevents the surface wave from reaching the receiving point by a straight-line path. The surface wave, which does arrive at the receiver, reaches it by diffraction around the earth and refraction in the lower atmosphere above the earth. Secondly, for elevated antennas the space wave is affected in two different ways. The ground-reflected wave is now reflected from a curved surface, and its energy is diverged more than in the case when it is reflected from a flat surface. This means that the ground-reflected wave reaching the receiver will be weaker than for a flat earth by the divergence factor D, which is less than unity. Finally, for a spherical earth, the heights h_1' and h_2' of the transmitting and receiving antennas above the plane tangent to the surface of the earth at the point of reflection of the ground-reflected wave are less than the antenna heights h_1 and h_2 above the surface of the earth (Fig. 16-14).

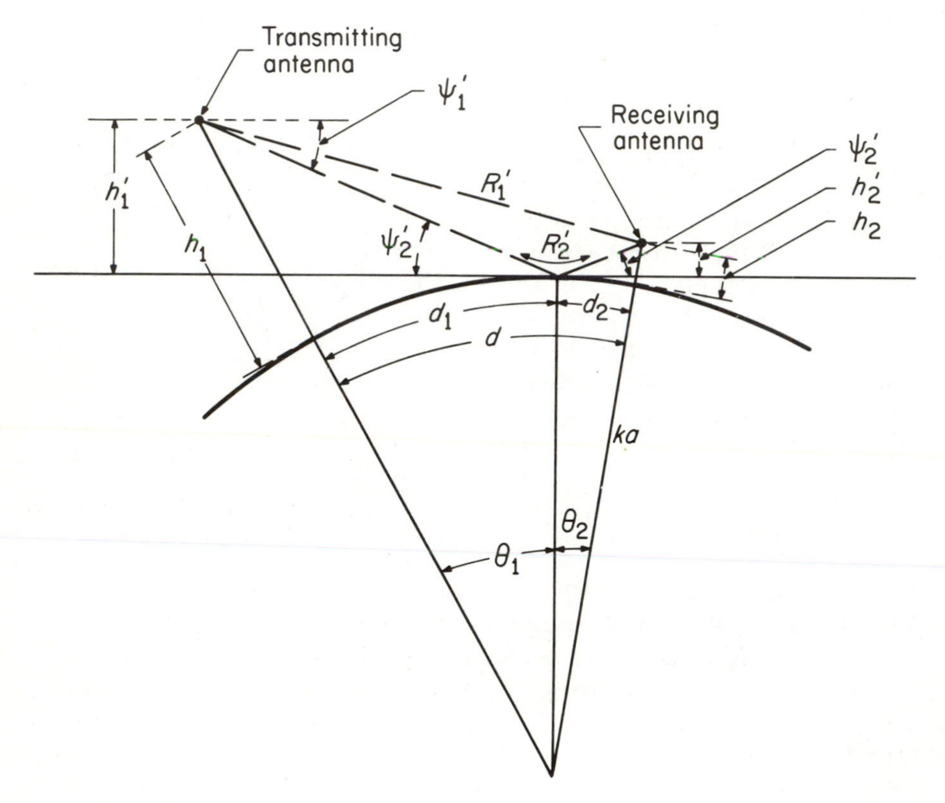

Figure 16-14. Geometry for a spherical earth.

It would seem that it should be possible to obtain an exact solution to the problem of an antenna above a spherical finitely conducting earth by solving Maxwell's equations subject to the appropriate boundary conditions. Although formal solutions to this problem have been set up, these solutions are much more involved than even the rigorous plane-earth solution. For example, one such solution is in the form of an infinite series of spherical harmonics with coefficients containing twelve Bessel functions. The convergence of the series is extremely slow, the main contribution being given by those terms for which n is of the order of the ratio $2\pi R/\lambda$, where R/λ is the radius of the earth in wavelengths. For commonly used radio frequencies, this ratio is of the order of 10^3 to 10^8! It is thus apparent that a different approach must be used if numerical answers are desired. Answers of engineering accuracy can be obtained by considering separately various particular cases. The detailed analysis is complex, and in general the expressions that result are complicated. However, the results may be put in a graphical form suitable for engineering use, and such results may be found in various articles and books.*

16.07 Tropospheric Waves. The troposphere is considered to be the region of the atmosphere adjacent to the earth and extending up to about 10 kilometers. It is in this region that clouds are formed. The temperature in the troposphere decreases with height at the rate of about 6.5°C per kilometer to a value of about −50°C at its upper boundary. Above the troposphere lies the stratosphere. Wave propagation beyond line-of-sight within the troposphere can result from several mechanisms which may be classified as diffraction, normal refraction, abnormal reflection and refraction, and tropospheric scatter. Diffraction has already been treated briefly, and the other mechanisms will be considered in this section.

Normal Refraction. A radio wave traveling horizontally in the earth's atmosphere follows a path which has a slight downward curvature due to refraction of the wave in the atmosphere. This curvature of the path tends to overcome partially the loss of signal due to curvature of the earth and permits the direct ray to reach points slightly beyond the horizon as determined by the straight-line path. In making computations the effect of refraction is accounted for by using an effective radius of curvature for the earth that is somewhat larger than the actual radius, and then assuming straight-line paths (that is, no refraction) in the atmosphere.

The refraction of a radio wave in the atmosphere occurs because the dielectric constant, and hence the refractive index of the atmosphere, varies with height above the earth. The dielectric constant of dry air is slightly greater than the value of unity that applies for a

*K. A. Norton, "The Calculation of Ground Wave Field Intensity over a Finitely Conducting Spherical Earth," *Proc. IRE*, **29**, No. 11, 632 (1941).

H. Bremmer, *Terrestrial Radio Waves,* American Elsevier Publishing Co., Inc., New York, 1949.

Ya. L. Alpert, *Radio Wave Propagation and the Ionosphere*, Authorized Translation from the Russian by Consultants Bureau, New York.

vacuum, and the presence of water vapor increases the dielectric constant still further. For this reason, the dielectric constant of the atmosphere is greater than unity near the earth's surface, but decreases to unity at great heights where the air density approaches zero.

Normal refraction results from this gradual decrease of the effective dielectric constant of the atmosphere with height. The refractive index $n = \sqrt{\epsilon}$ of the atmosphere has a value of approximately 1.0003 at the earth's surface. Because we are dealing with very small changes in n it is usual to use a scaled-up value of n called the *refractivity* and defined by $N = (n - 1)\,10^6$. The refractivity can be calculated through the relation

$$N = \frac{77.6}{T}\,p + 4810\,\frac{e}{T}$$

where p is the total pressure in millibars, e is the partial pressure of water vapor in millibars and T is the absolute temperature in degrees Kelvin. Of chief significance as far as bending of the radio waves is concerned is the *gradient* of the refractive index, that is, the rate of change of N with height. If a linear N profile (that is, uniform gradient) is assumed, the downward bending of the rays may be accounted for as described below by using straight rays and an "effective" earth radius greater than the actual radius. For a "standard atmosphere" the appropriate factor for the effective earth radius is $k = \frac{4}{3}$, so that the effective radius is $\frac{4}{3}$ times the actual radius. For other refractive index gradients the effective radius can be adjusted accordingly. It has been found that there is a high correlation between the monthly median transmission loss and monthly median values of ∇N. The quantity ∇N is the difference in values, N_1, at the height of one kilometer above the surface and N_s, the value at the surface. It is a quantity which can be obtained directly from radiosonde observations or can be predicted with good accuracy from N_s. Maps

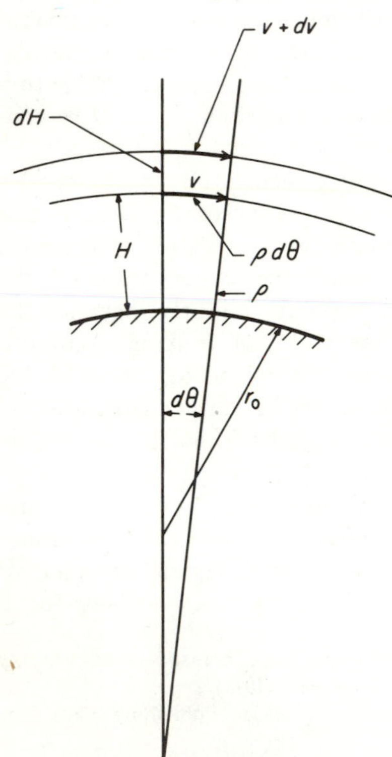

Figure 16-15.

of ∇N have been prepared for the United States for several months of the year.

The relation between the radius of curvature of the path and the change of refractive index with height can be derived as follows. Let ρ be the radius of curvature of the path and v the velocity of propagation at a height H above the earth. Then from Fig. 16-15

$$\rho\, d\theta = v\, dt$$

or

$$\frac{d\theta}{dt} = \frac{v}{\rho} \tag{16-32}$$

Also

$$v = \frac{1}{\sqrt{\mu_v \epsilon_r \epsilon_v}} = k_1 \epsilon_r^{-1/2} \tag{16-33}$$

At a height $H + dH = H + d\rho$, the velocity

$$(v + dv) = \frac{(\rho + dH)\, d\theta}{dt}$$

Therefore

$$\frac{dv}{dH} = \frac{d\theta}{dt} = \frac{v}{\rho}$$

or

$$\rho = \frac{v}{dv/dH} \frac{k_1 \epsilon_r^{-1/2}}{-\frac{1}{2} k_1 \epsilon_r^{-3/2} \dfrac{d\epsilon_r}{dH}} = -\frac{2\epsilon_r}{d\epsilon_r/dH} \tag{16-34}$$

Since $\epsilon_r \approx 1$,

$$\rho = -\frac{2}{d\epsilon_r/dH} \tag{16-35}$$

The radius of curvature of the path, being a function of the rate of change of the dielectric constant with height, varies from hour to hour, day to day, and season to season. However, in practice an average value of four times the radius of the earth is used for the purposes of calculations.

In working propagation problems it is often convenient to consider the ray paths as straight lines instead of being curved as they actually are, and to compensate for the curvature by using a larger value for the "effective" radius of the earth. The relations involved are shown in Fig. 16-16(a) and (b). In Fig. 16-16(a), the actual path is shown above an earth of radius a. In order for the straight-line path of Fig. 16-16(b) to be the equivalent of that shown in Fig. 16-16(a), it is necessary that the change in height dH be the same in the two cases for the same horizontal distance D. In Fig. 16-16(b),

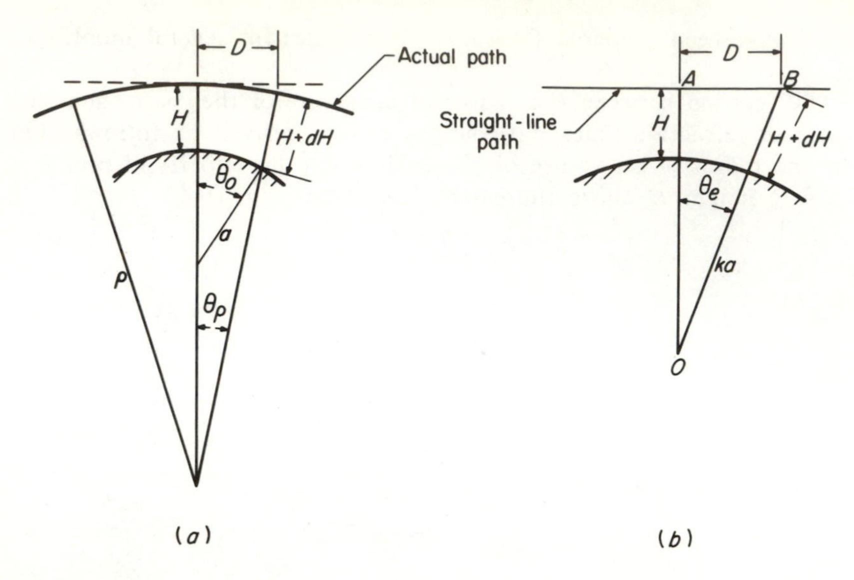

Figure 16-16. Curved paths become straight lines when an *effective* radius ka is used for the earth.

$$dH = BO - AO = (ka + H)\left(\frac{1}{\cos\theta_e} - 1\right)$$

For small angles,

$$\frac{1}{\cos\theta_e} \approx \frac{1}{1 - \frac{\theta_e^2}{2}} \approx 1 + \frac{\theta_e^2}{2}$$

$$dH \approx \frac{ka\theta_e^2}{2}$$

when H is small compared to ka. But

$$\theta_e \approx \sin\theta_e = \frac{D}{(ka + H)} \approx \frac{D}{ka}$$

Therefore

$$dH \approx \frac{D^2}{2ka} \tag{16-36}$$

On the other hand, in Fig. 16-16(a)

$$dH = \frac{D^2}{2a} - \frac{D^2}{2\rho} \tag{16-37}$$

therefore

$$\frac{1}{ka} = \frac{1}{a} - \frac{1}{\rho} \tag{16-38}$$

The effective radius of the earth required is therefore

$$ka = a\left(\frac{1}{1 - \dfrac{a}{\rho}}\right)$$

so that

$$k = \frac{1}{1 - \dfrac{a}{\rho}} \tag{16-39}$$

For a radius of curvature ρ, equal to four times the radius a of the earth, the effective radius of the earth is $\frac{4}{3}$ times the actual radius. By using this effective radius instead of the actual radius in making ground-wave path computations, the systematic bending of the waves in the atmosphere is accounted for, and straight-line paths may be drawn.

Abnormal Refraction and Reflection. In addition to the systematic refraction of waves that occurs in the troposphere under normal conditions there are also the possibilities of abnormal refraction and of reflections occurring at places of abrupt change in the refractive index or its gradient. The case of reflection at abrupt changes in the dielectric constant is easily treated using the reflection factors developed in chap. 5. For a wave propagating in a dielectric medium of permittivity ϵ_1, and incident upon a second medium of permittivity $\epsilon_2 = \epsilon_1 + \Delta\epsilon$, where $\Delta\epsilon$ is the change in permittivity at the layer in the troposphere, it is easy to show (problem 1, page 665) that eq. (5-91) for vertically polarized waves reduces to

$$R_v \approx \frac{\Delta\epsilon}{2} - \frac{\Delta\epsilon}{4\cos^2\theta_1} \tag{16-40}$$

Similarly for horizontally polarized waves the reflection coefficient of eq. (5-90) can be reduced to

$$R_h \approx -\frac{\Delta\epsilon}{4\cos^2\theta_1} \tag{16-41}$$

Using these reflection coefficients and various assumed conditions for $\Delta\epsilon$ and reflecting layer height, the field strengths of tropospheric waves can be calculated. These calculations show that when abrupt changes of permittivity occur in the troposphere, the resulting reflections can produce usable signals at distances considerably beyond those that result when only ground-wave propagation paths are considered. Propagation beyond normal ground-wave range also results from abnormal refraction in the troposphere. These effects are best handled through use of modified index curves considered next.

Modified Index Curves and Duct Propagation. The atmospheric condition that gives rise to the tropospheric reflection just considered is a *nonstandard* condition. Thare are many different types of such nonstandard atmospheres, each of which affects wave propagation in a different way. The standard dry atmosphere has already been defined as one for which the temperature decreases at the rate of 6.5°C per kilometer. When the temperature increases with height over a certain range of heights, it is known as *temperature inversion.* Actually the water content of the atmosphere has much more effect than temperature on its dielectric constant and on the manner in which it affects radio waves. The *moist standard atmosphere* is specified as one which has a water-vapor pressure of 10 millibars at sea level, decreasing with altitude at the rate of 1 millibar per thousand feet, up to 10,000 ft. If the temperature or water content differs from these standard conditions, *nonstandard* propagation will result. The effects to be expected can be estimated most readily by transforming the meteorological data, temperature, water content, and so on, into *M* curves. *M curves* are curves that show the variation of the modified index of refraction with height. (The term "modified" refers to the fact that the actual index has been modified to account for the curvature of the earth. When this is done, straight rays above a curved earth come out as curved rays (with an upward curvature) above a flat earth. This procedure, which simplifies computations when rays of different curvatures must be considered, is just the reverse of that used previously when curved rays over a curved earth were transformed to straight rays over an earth of lesser curvature.)

When *M* curves are available, it is possible to predict, at least roughly, the type of transmission path that can be expected. Standard propagation occurs when the modified index of refraction increases linearly with height. In this case, the *M* curve is a straight line with a positive slope. If the slope of the *M* curve decreases near the surface of the earth, *substandard* propagation results, with the rays curving upward (over the *flat* earth) more than for normal conditions. If the slope of the *M* curve increases near the surface of the earth, the upward curvature of the rays is less, so that greater coverage is achieved and *superstandard* conditions result. If the *M* curve becomes vertical (no change of modified index with height), the rays over the flat earth are straight and very great coverage can be obtained. (In this condition the actual rays have the same curvature as the curvature of the earth.)

If the modified index decreases with height (*M* curve slopes to left) over a portion of the range of height, the rays will be curved downward (over the flat earth) and a condition known as *trapping* or *duct* propagation can occur. Under such conditions the wave tends to be trapped or guided along the duct, much as a wave is guided by a leaky wave guide. If the lower side of the duct is at the surface of the earth, it is known as a surface duct. Sometimes when the inverted portion of the M curve is elevated above the surface of the earth, the lower side of the duct is also elevated, and the duct is called an *elevated duct*. If the receiving antenna is elevated to within the duct, the signal may be very large. However, if the receiving antenna is outside the duct, either below or above it, the received signal will be very small. Elevated ducts are due to a subsidence of large air masses and are common in Southern California and certain areas of the Pacific. They are found at elevations of 1000 to 5000 ft and may vary in thickness from a few feet to a thousand feet. In the trade wind belt over sea areas there appears to be a continuous surface duct about 5 ft thick. Over land areas surface ducts are produced by radiation cooling of the earth. As with ordinary wave-guide propagation, there is a certain critical frequency (which depends on the thickness of the duct) below which duct propagation will not occur. Since these nonstandard refraction effects appear to be restricted to waves that make a very small angle with the horizontal, it is evident that the required thickness of the duct would have to be large in wavelengths. For this reason trapping is more likely to occur at the ultrahigh frequencies than at very high frequencies.

Although trapping or ducting is prevalent over ocean areas in the trade wind belt it is estimated that ducting over land areas, such as the continental United States, occurs for less than 25% of the time. In contrast, tropospheric forward scatter, considered next, can provide a consistent and reliable beyond-the-horizon signal for distances up to 300 or 400 miles.

Tropospheric Scattering. Tropospheric scatter is the name given to still another kind of extended-range tropospheric propagation. The mechanisms involved appear to be scattering and reflections from inhomogeneities in the refractive index of the atmosphere within the common volume of the troposphere occupied by transmitting and receiv-

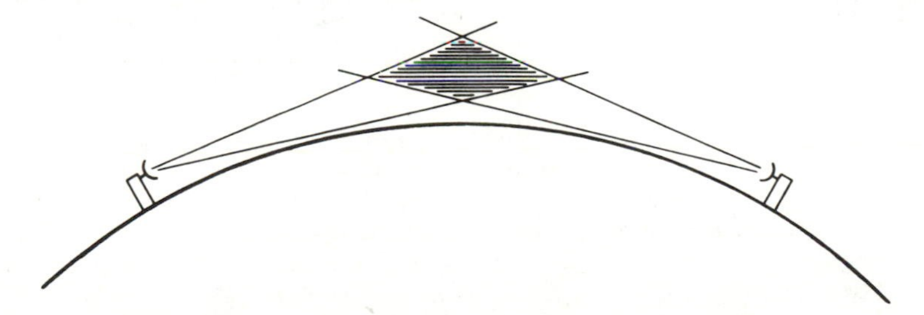

Figure 16-17. Scattering in the common volume of transmitting and receiving antenna beams.

ing beams (Figure 16-17). In various scattering theories that have been developed the received signal is ascribed to scattering from eddies or "blobs" in the atmosphere due to turbulence, or alternatively, to uncorrelated reflections from many layers of limited extent and arbitrary aspect. Measurements have shown that relatively sharp variations in the effective dielectric constant do indeed occur in both horizontal and vertical planes.

Experimental results obtained for beyond-the-horizon transmission may be summarized as follows.* Beyond the horizon the received power decreases at about the 7th or 8th power of the distance. The signal level has seasonal variations of ± 10 db which seem to be proportional to variations in k, the effective earth-radius factor. In addition there is fast fading which is essentially random and follows a Rayleigh distribution. Although the transmission loss increases slightly with frequency, the range of frequencies from 100 to 10,000 MHz appears to be useful. High-gain antennas are required, but when the antenna diameter is greater than about 50λ, the antennas fail to realize their full gain because of phase incoherence across the aperture. Diversity reception is almost essential and requires antennas spaced at least 100λ apart. Available band width can be sufficient to provide television transmission over a 200-mile

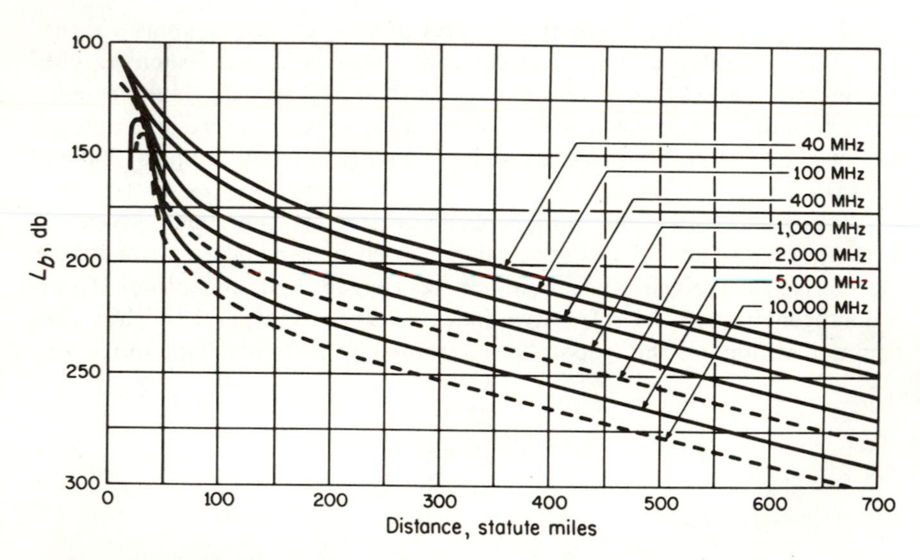

Figure 16-18. Median basic transmission loss for ground-wave and tropospheric scatter modes over smooth earth in CRPL reference atmosphere with $N_s = 301$; $H_1 = 30'$, $H_2 = 500'$. (K. A. Norton, Courtesy *NBS Journal of Research*)

*Kenneth Bullington, "Characteristics of Beyond-the-Horizon Radio Transmission," *Proc. IRE*, **43**, No. 10, pp. 1175–1180, October, 1955.

link. Estimated median basic transmission loss is shown as a function of distance in Figure 16-18.

A useful rule of thumb for tropospheric forward-scatter propagation has been given* as follows: At 100 miles range the median field strength is approximately 57 db below free-space value, and a further loss of about 0.12 db per mile occurs at greater distances. Depending on the degree of reliability required, an allowance of 8 to 16 db should be made for slow fading, and a further allowance of 3 to 8 db for Rayleigh type fading. This propagation mode is most effective for ranges up to about 350 miles, but useful narrow-band signals have been received up to 600 miles. An indication of the distances that can be achieved for wide-band, high-quality transmission is given by Fig. 16-19 for various antenna sizes. Using Figs. 16-18 and 16-19 together with the conclusions from

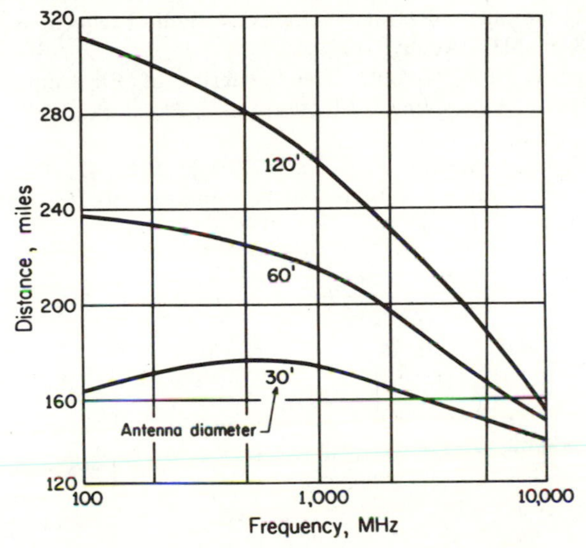

Figure 16-19. Estimated range for beyond-the-horizon transmission. (Bullington, Courtesy *IEEE*, formerly *IRE*)

this section an estimate can be made of the power and antenna requirements to establish an extended-range tropospheric-scatter communications link.

Problem 1. It is known that the abrupt changes in dielectric constant at the surface of a layer in the troposphere must be very small, say of the order of 10^{-6} to 10^{-4}. On this basis, and remembering that the relative dielectric constant is approximately unity, derive expressions (16-40) and (16-41) from expressions (5-91) and (5-90).

*I. H. Gerks, "Factors Affecting Spacing of Radio Terminals in a UHF Link," *Proc. IRE*, **43**, No. 10, pp. 1290–1297, October, 1955.

BIBLIOGRAPHY

Bean, B. R., and F. M. Meaney, "Some Applications of the Monthly Median Refractivity Gradient in Tropospheric Propagation," *Proc. IRE*, **43**, 1419-1431, October, 1955.

Booker, H. G., and W. E. Gordon, "Radio Scattering in the Troposphere," *Proc. IRE*, **38**, No. 4, pp. 401-412, April, 1950.

Bremmer, H., *Terrestrial Radio Waves*. American Elsevier Publishing Co., Inc., New York, 1949.

Friis, H. T., A. B. Crawford, and D. C. Hogg, "A Reflection Theory for Propagation Beyond the Horizon," *BSTJ*, **36**, No. 3, p. 627, May, 1957.

Gordon, W. E., "Radio Scattering in the Troposphere," *Proc. IRE*, **43**, pp. 23-28, Jan., 1955.

Smith, Ernest K., Jr., and Stanley Weintraub, "The Constants in the Equation for Atmospheric Refractive Index at Radio Frequencies," *Proc. IRE*, **41**, No. 8, p. 1035, August, 1953.

Villars, F., and V. F. Weisskopf, "The Scattering of Electromagnetic Waves by Turbulent Atmospheric Fluctuations," *Phys. Rev.*, **94**, pp. 232-240, April 15, 1954.

IRE Special Issue on Scatter Propagation, *Proc. IRE*, **43**, Oct., 1955. This special issue contains a comprehensive summary of tropospheric scatter propagation.

Chapter 17

IONOSPHERIC PROPAGATION

17.01 Introduction. The ground-wave propagation paths considered in the previous chapter are not the only paths along which a transmitted wave may travel to reach the receiver. This was demonstrated to a surprised scientific world in 1901 by Marconi's successful transmission of radio signals across the Atlantic. Calculations had already been made to show that diffraction effects would be insufficient to permit such long-distance transmission around the curvature of the earth, and immediately other explanations were sought. The existence of a reflecting region in the earth's upper atmosphere was proposed (independently) by A.E. Kennelly and Oliver Heaviside, and the Kennelly-Heaviside layer, or *ionosphere* as it is now known, became a much discussed part of radio propagation phenomena.

Knowledge of the characteristics of the ionosphere is based largely upon its effect on radio waves, either those that penetrate it in the case of signals from rockets and satellites, or those that are reflected from it in the case of signals from earth-based transmitters. Experimentally it is found that at night signals from earth-based transmitters in the broadcast frequency range are reflected back, but in the daytime the reflected signal is very weak or entirely absent. As the frequency is raised, however, these daytime reflected waves become stronger, and for frequencies between 10 and 30 MHz, they may provide strong signals over distances of several thousand miles. As the frequency is increased still higher, a point is reached where the waves cease to be reflected back, but instead, penetrate the ionosphere to be lost in outer space. Thus there is a range of frequencies roughly between 3 and 30 MHz where, although the surface wave is greatly attenuated, long-distance transmission between points on the earth's surface may still occur because of reflections from the ionosphere. In general, these "sky-wave" signals are less stable than ground-wave signals, their strength depending upon the frequency, and upon the condition of the ionosphere. The state of the ionosphere is found to vary from hour to hour, day to day, and season to season in much the same way as does the weather. Also, as with the weather, there may be periods of sudden storms, but many of the variations are fairly regular and may be predicted several days or even weeks ahead of time. Indeed, the art of ionosphere prediction is in much the same state as, and is very similar to, that of weather forecasting. Ionosphere stations set up in various parts of the world continuously gather and record information about the ionosphere in those regions. This information is assembled, correlated, interpreted, and issued in the form of charts that show past conditions and also make predictions for the future. Using these charts it is possible to determine in advance the optimum frequency to use for communication between any two points on the earth's surface at any given time. Thus, although long-distance ionospheric propagation does not have the stable characteristics of short-distance ground-wave propagation, it does, in general, provide a predictable, and therefore usable, means of radio communication. A knowledge of some of the more important characteristics of the ionosphere will aid the engineer in an intelligent overall design of a communications system.

17.02 The Ionosphere. The ionosphere is that region of the earth's atmosphere in which the constituent gases are ionized by radiations from outer space (chiefly solar radiation). This region extends from about 50 km above the earth to several earth radii (mean earth radius = 6371 km) with the maximum in ionization density at about 300 km as illustrated in Fig. 17-1.

During the day, the bulk of the ionization is at altitudes between 90 km and 1000 km, where the electron density is approximately 10^4 electrons/cc. Throughout the ionosphere there are in fact several "layers," or regions in which the ionization density either reaches a maximum or remains roughly constant. These regions are designated *D*, *E*, and *F* in order of height. During the daytime the *F* layer splits into separate layers called F_1 and F_2. Also at times a peak in electron density has been observed in the lower *D* region, suggesting that the designation "*C* region" might be appropriate for the range 50–70 km. It should be emphasized that Fig. 17-1 represents a long-term average and that the actual electron-density distribution is a function of many variables, including time of day, season, latitude, and year (in relation to the eleven-year sunspot cycle).

The existence of the ionosphere in the form of a layer is explained on the following basis: At great heights the ionizing radiations are very intense, but the atmosphere is rare and there are few molecules present to be ionized. Therefore in this region the ionization density (number of ions or electrons per unit volume) is very low. As height is decreased, the atmospheric pressure and ionization density increase until a height is reached where the ionization density is a maximum. Below this height the atmospheric pressure continues to increase, but

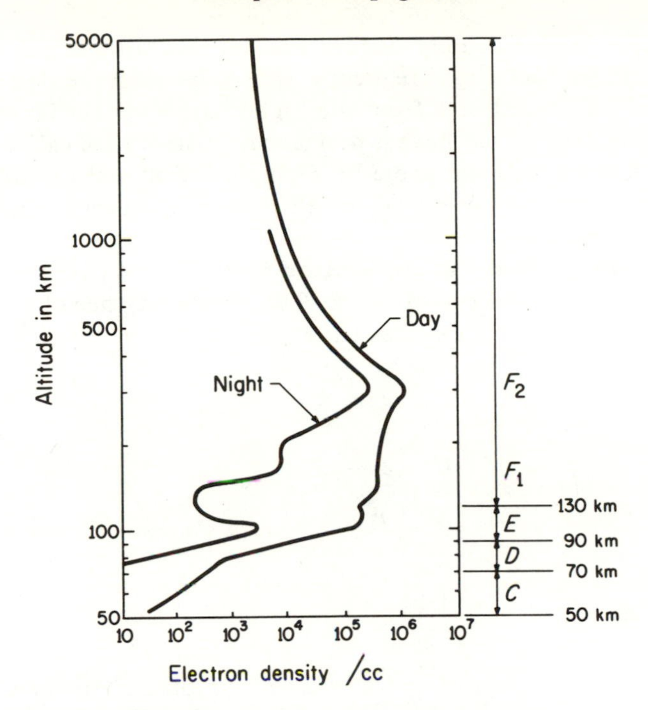

Figure 17-1. Electron density profiles (typical).

the ionization density decreases because the ionizing radiation has been absorbed or used up in the process of ionization. This explains in general why there should be a layer. The existence of layers within a region is accounted for by the fact that the atmosphere is a mixture of several gases that differ in their susceptibility to the ionizing radiations, and so produce maximum ionization at different altitudes.

Although the number and heights of the layers vary with time, the *E* and *F* layers have a permanent existence and make possible long-distance radio communication by reflecting radio waves (Fig. 17-2).

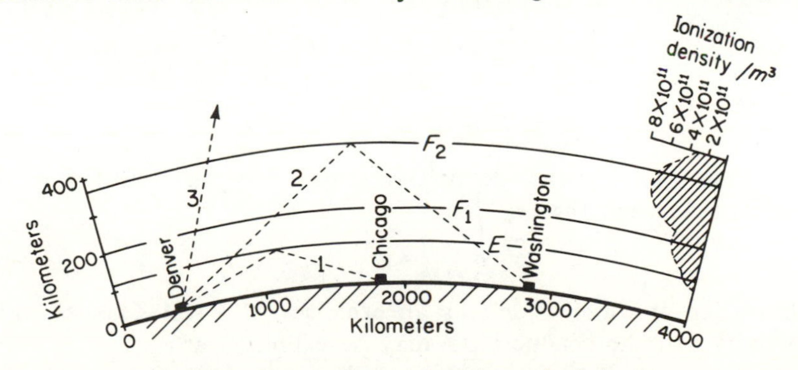

Figure 17.2. Reflection from ionospheric layers.

The D layer is present only during the day and although it does not normally reflect back high-frequency waves, its presence decreases the intensity of signals reflected from the higher layers. Other layers within the E region that do not have a permanent existence are called *sporadic* E layers. Reflections from sporadic E patches often make possible long-distance reception of waves of much higher frequency than would normally be possible.

17.03 Effective ϵ and σ of an Ionized Gas. The relative permittivity of an ionized gas was derived in sec. 9.06. In the absence of an external magnetic field it may be expressed as

$$K_0 = 1 - \frac{X}{U} \tag{17-1}$$

in which

$$U = 1 - jZ$$

$$Z = \frac{\nu}{\omega} \qquad (\nu = \text{collision frequency})$$

$$X = \frac{\omega_p^2}{\omega^2}$$

$$\omega_p^2 = \frac{Ne^2}{m\epsilon_v} \qquad (\omega_p = \text{plasma frequency})$$

The collision frequency ν represents the combined effects of collisions with all species of particles present. In the ionosphere, collisions with molecules and ions predominate and approximate values for ν are 10^6 sec^{-1} at 90 km and $10^3\ \text{sec}^{-1}$ at 300 km. The variation in ν with height arises because of its dependence on gas pressure, electron thermal velocity and ion density.

The relative permittivity may be separated into real and imaginary parts as follows:

$$K_0 = 1 - \frac{X}{1 - jZ}$$

$$= 1 - \frac{X}{1 + Z^2} - j\frac{XZ}{1 + Z^2} \tag{17-2}$$

$$= \epsilon_r' - j\epsilon_r'' \tag{17-3}$$

The imaginary part of the permittivity also may be regarded as a conductivity through the relation

$$\sigma = \omega\epsilon_v\epsilon_r'' \tag{17-4}$$

the derivation of which is left as an exercise. The real part of the permittivity and the conductivity may be expressed as

$$\epsilon_r' = 1 - \frac{Ne^2}{m\epsilon_v(\nu^2 + \omega^2)} \qquad \text{and} \qquad \sigma = \frac{Ne^2\nu}{m(\nu^2 + \omega^2)} \tag{17-5}$$

It will be observed that for a given frequency the effective conductivity in a region is a maximum when the collision frequency ν is equal to ω. At great heights where ν is small and $\omega \gg \nu$, the conductivity will become vanishingly small and the effective dielectric constant will be given by $\epsilon_v[1 - (Ne^2/\epsilon_v m\omega^2)]$. On the other hand, at low heights such that $\nu \gg \omega$, the conductivity again becomes small and the reduction of the dielectric constant approaches zero. These effects that occur with decreasing height are augmented by the fact that the electron density N also decreases rapidly below about 80 km. The result is that the region of high conductivity (and therefore high absorption when the wave penetrates it) is confined to a relatively thin layer at the lower edge of the E region and to the upper part of the D region.

17.04 Reflection and Refraction of Waves by the Ionosphere. The mechanism of reflection and refraction of radio waves by the ionosphere is very much a function of frequency. At low frequencies, say below 100 kHz, the change in electron and ion density within the distance of a wavelength is so great that the layer presents virtually an abrupt discontinuity in the medium. Under these circumstances, the reflection may be treated in the same manner as the reflection of waves at the surface of a dielectric that may or may not have loss. On the other hand, at the high end of the high-frequency band, the length of a wavelength is sufficiently short that the ionization density changes only slightly in the course of a wavelength. Under such conditions the ionosphere may be treated (by methods well-known in optics) as a dielectric with a continuously variable refractive index. For in-between frequencies, not covered by these two cases, it is possible to treat the reflection region as though it consisted of several thin but discrete layers, each layer having a constant ionization density that differs from that of the adjacent layer. It follows that the incident wave will be partially refracted. The refracted wave penetrates to the second layer where it is partially reflected and partially refracted, and so on. In this case the resultant reflected signal may be considered as the sum of reflections from various parts of the ionized layer. Because they suffer greater attenuation, these in-between frequencies are of less practical interest than the others, and only the first two cases will be treated.

CASE I: *Reflection at Low Frequencies* In this case the wavelength is considered to be sufficiently long that there is a great change in the ionization density in the course of a wavelength. The layer then may be considered a reflecting surface, for which the following reflection coefficients may be written:

$$R_1 = \frac{\cos\theta - \sqrt{\left(\epsilon_r' + \frac{\sigma}{j\omega\epsilon_v}\right) - \sin^2\theta}}{\cos\theta + \sqrt{\left(\epsilon_r' + \frac{\sigma}{j\omega\epsilon_v}\right) - \sin^2\theta}} \tag{17-6}$$

$$R_v = \frac{\left(\epsilon_r' + \frac{\sigma}{j\omega\epsilon_v}\right)\cos\theta - \sqrt{\left(\epsilon_r' + \frac{\sigma}{j\omega\epsilon_v}\right) - \sin^2\theta}}{\left(\epsilon_r' + \frac{\sigma}{j\omega\epsilon_v}\right)\cos\theta + \sqrt{\left(\epsilon_r' + \frac{\sigma}{j\omega\epsilon_v}\right) - \sin^2\theta}} \tag{17-7}$$

where the effective values of σ and ϵ_r' are given by eqs. (5).

It is apparent that for this type of reflection, the reflection coefficient of the medium will depend upon the frequency, polarization, and angle of incidence of the wave. When the effective conductivity can be neglected the reflection curves will be those for reflection from a perfect dielectric that has a refractive index of less than unity. For angles of incidence greater than a certain critical angle (which depends upon the refractive index), there will be complete reflection of the signal for both polarizations of the wave. For angles less than the critical (that is, closer to the normal), the reflection coefficient will be less than unity and will depend on the angle of incidence.

Case II: *Reflection or Refraction at High Frequencies* (This is an important case, practically.) When the change in phase velocity within the course of a wavelength is small, the well-known methods of ray optics may be used to obtain a solution. The requirement of small change in phase velocity means, in this case, a small change in electron density, as can be seen from the following considerations: The phase velocity of a wave in a medium having negligible loss is given by

$$v_p = \frac{1}{\sqrt{\mu\epsilon}} = \frac{c}{\sqrt{\mu_r\epsilon_r}} \tag{17-8}$$

where, as usual, $c = 1/\sqrt{\mu_v\epsilon_v}$ is the velocity of light in a vacuum. Assuming the permeability of the ionosphere to be unchanged by the presence of electrons so that $\mu_r = 1$, the phase velocity will be

$$v_p = \frac{c}{\sqrt{\epsilon_r}} \tag{17-9}$$

where ϵ_r depends upon the electron density N as indicated in eqs. (5). If the change in electron density in the distance of a wavelength is small the change in phase velocity will also be small.

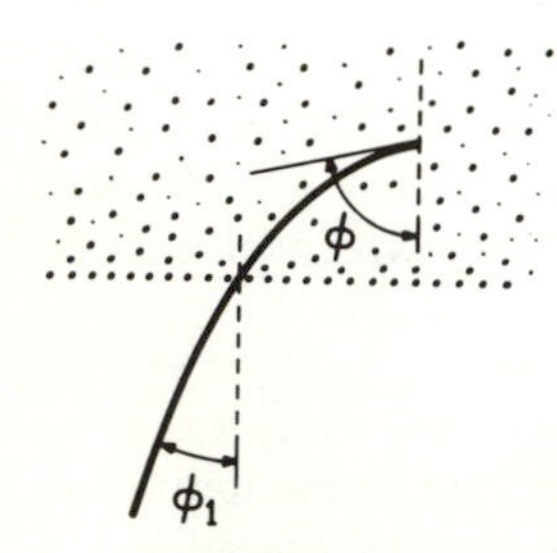

Figure 17-3.

Under the conditions for this case the wave penetrates the lower edge of the ionosphere without reflection, but within the ionosphere travels a path that is curved away from the region of greater electron density (smaller index of refraction). At any point along the path, the angle ϕ between the path and the normal (Fig. 17-3) is given by Snell's law of refraction

$$\sin \phi_i = n \sin \phi$$

or

$$\sin \phi = \frac{\sin \phi_i}{n} \tag{17-10}$$

n is the index of refraction at the point where ϕ is observed, and ϕ_i is the angle of incidence (measured from the normal to the ionosphere layer). The refractive index for any medium is given by

$$n = \frac{c}{v_p} = \frac{\text{velocity of light in vacua}}{\text{phase velocity in the medium}} \tag{17-11}$$

For the lossless case, where (9) is true, eq. (11) gives for the refractive index

$$n = \sqrt{\epsilon_r} \tag{17-12}$$

In general the effective conductivity of the ionosphere cannot be neglected, but at the higher frequencies where this present analysis is applicable, reflection takes place in the F layers where the collision frequency is very small and the conductivity is correspondingly low. Therefore, for a first approximation at least, it is permissible to neglect the effects of conductivity and use the simple expression given in (12).

For $\omega^2 \gg \nu^2$, the expression for ϵ_r [from eq. (5)] is

$$\epsilon_r = \left(1 - \frac{Ne^2}{\epsilon_v m \omega^2}\right) \tag{17-13}$$

For an electron, $e = 1.59 \times 10^{-19}$ coulombs, $m = 9 \times 10^{-31}$ kg, so that (13) becomes

$$\epsilon_r = \left(1 - \frac{81N}{f^2}\right) \tag{17-14}$$

N is the number of electrons per cubic meter and f is the frequency in Hertz. (However, if N is expressed as the number of electrons per cubic centimeter and the frequency is expressed in kHz, relation (14) is still true.)

From (12) the refractive index is

$$n = \sqrt{1 - \frac{81N}{f^2}} \tag{17-15}$$

The refractive index decreases as the wave penetrates into regions of greater electron density and the angle of refraction increases correspondingly. When n has decreased to the point where $n = \sin \phi_i$, the angle of refraction ϕ will be 90 degrees and the wave will be traveling horizontally. The highest point reached by the wave is therefore that point at which the electron density N satisfies the relation

$$\sqrt{1 - \frac{81N'}{f^2}} = \sin \phi_i$$

or

$$N' = \frac{f^2 \cos^2 \phi_i}{81} \tag{17-16}$$

If the electron density at some level in a layer is sufficiently great to satisfy relation (16), the wave will be returned to earth from that level. If the maximum electron density in a layer is less than that required by (16), the wave will penetrate the layer (though it may be reflected back from a higher layer for which N is greater).

The largest electron density required for reflection occurs when the angle of incidence ϕ_i is zero, that is, for vertical incidence. For any given layer the highest frequency that will be reflected back for vertical incidence will be

$$f_{cr} = \sqrt{81 N_{max}} \tag{17-16a}$$

where N_{max} is the maximum ionization density (electrons per cubic meter) and f_{cr} is the *critical frequency* for the layer.

Experimental Determination of Critical Frequencies and Virtual Heights. Ionosphere characteristics are determined experimentally by measuring the amplitude and time delay of reflected signals as functions of frequency. The commonest method is that in which the transmitted signal consists of pulses of rf energy of short duration. The receiver, which is located close to the transmitter, picks up both the direct and the reflected signal. The spacing between these signals on the time axis of a cathode ray oscilloscope gives a measurement of the height of the layer. The height so measured is the *virtual height* of the layer and is higher than the true height of the lower edge of the layer as indicated in Fig. 17-4.

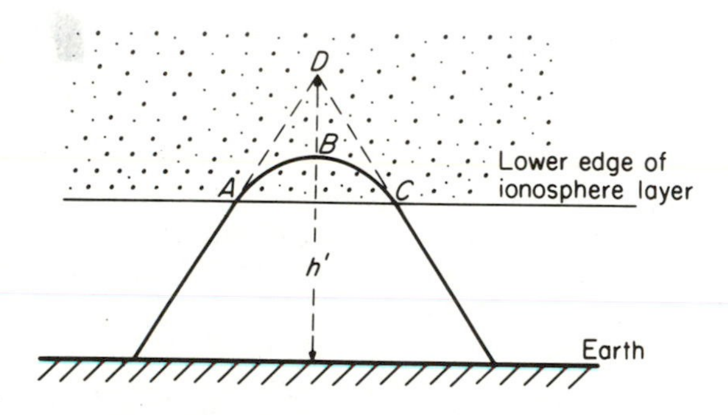

Figure 17-4. Virtual height h' of a layer.

The virtual height h' is that height from which a wave sent up at an angle appears to be reflected. It is also the height obtained by pulse measurements, because the time delay for the actual curved path ABC is approximately the same as it would be for a wave to travel the path ADC if the ionosphere were replaced by a mirrorlike reflecting surface at the level of D. Although the path length ADC is greater than ABC, the group velocity in the ionosphere is less than in free space by just the amount required to make the time delays of the two paths equal.

As the frequency of the transmitted signal is increased, starting say at 2 MHz, the measured virtual height increases slightly, indicating that for the higher frequencies the wave is returned back from higher levels within the layer. This continues until a critical frequency is approached

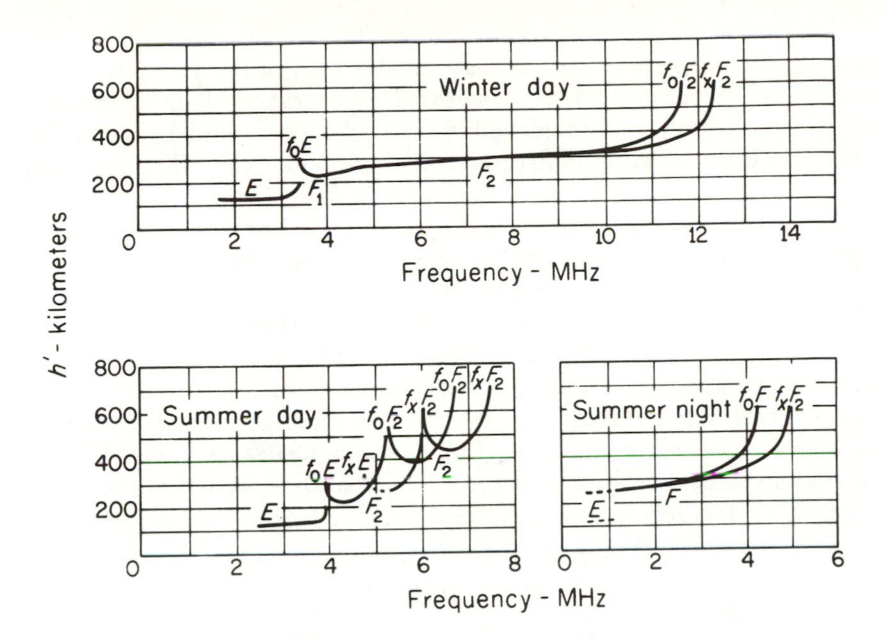

Figure 17-5. Virtual height-frequency curves.

near which the virtual height increases suddenly to quite high values as shown in the virtual height-frequency curves (ionograms) of Fig. 17-5. As the critical frequency is passed, the measured virtual height drops back to a second more or less steady value which is higher than was obtained for frequencies well below the critical frequency. The wave is now penetrating the first layer and is reflected from a second, higher layer. The critical frequency of a layer is that frequency for which a vertically incident wave just fails to be reflected back from the layer. As the frequency is increased above the critical frequency for the lower layer, a second and sometimes a third critical frequency may be reached corresponding to the critical frequencies for the higher layers. The apparent increase in the measured height of the layer in the neighborhood of the critical frequency is due to a large time delay in the ionized medium, occurring as a result of a much reduced group velocity near this frequency. At the higher frequencies the earth's magnetic field causes the virtual height trace to split into two parts corresponding to the two characteristic waves (ordinary and extraordinary) which have different phase and group velocities. Traces of a similar type are obtained by the "topside sounder" technique in which signals are transmitted from a satellite at about 1000 km and reflected from the part of the ionosphere above the F region maximum in electron density.

A particular type of radio wave echo called *spread F* is observed frequently in ionograms, especially in the equatorial and polar regions. In this phenomenon, the duration of a pulse reflected from the *F* region may be as much as ten times greater than the duration of the incident pulse, causing the ionogram to have a "smeared" appearance. This is believed to be due to scattering from irregularities in the ionosphere, irregularities which are distributed both directly over the sounder and also away from the zenith.

Maximum Usable and Optimum Frequencies. Although the critical frequency for any layer represents the highest frequency that will be reflected back from that layer at vertical incidence, it is not the highest frequency that can be reflected from the layer. The highest frequency that can be reflected depends also upon the angle of incidence, and hence, for a given layer height, upon the distance between the transmitting and receiving points. The maximum frequency that can be reflected back for a given distance of transmission is called the *maximum usable frequency* (*MUF*) for that distance. From eq. (16) and using (16a), it is seen that the maximum usable frequency is related to the critical frequency and the angle of incidence by the simple expression

$$\text{MUF} = f_{cr} \sec \phi_i \tag{17-17}$$

The maximum usable frequency for a layer is greater than the critical frequency by the factor $\sec \phi_i$. Because of curvature of the earth and the ionospheric layer, the largest angle of incidence ϕ_i that can be obtained in *F*-layer reflection is of the order of 74 degrees. This occurs for a ray that leaves the earth at the grazing angle. The geometry for this case is shown by Fig. 17-6, where $\phi_i\,(\text{max}) = \sin^{-1}(r/r + h)$. The maximum usable frequency at this limiting angle is related to the critical frequency of the layer by

$$\text{MUF}\,(\text{max}) \approx \frac{f_{cr}}{\cos 74^\circ} \approx 3.6 f_{cr} \tag{17-18}$$

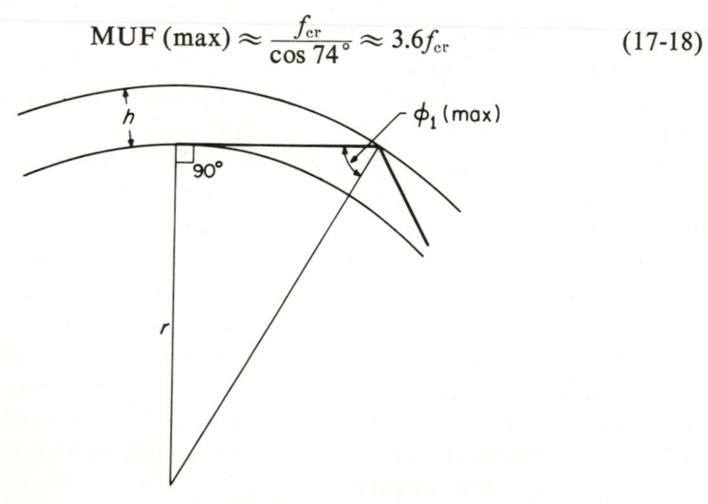

Figure 17-6.

When the critical frequency is known, the maximum usable frequency can be calculated for any given distance through use of eq. (17). Figure 17-7 shows a set of maximum usable frequency curves for the latitude of Washington, D. C., for a winter month during a period of maximum sunspot activity. It is evident that, whereas a given frequency, say 28 MHz,

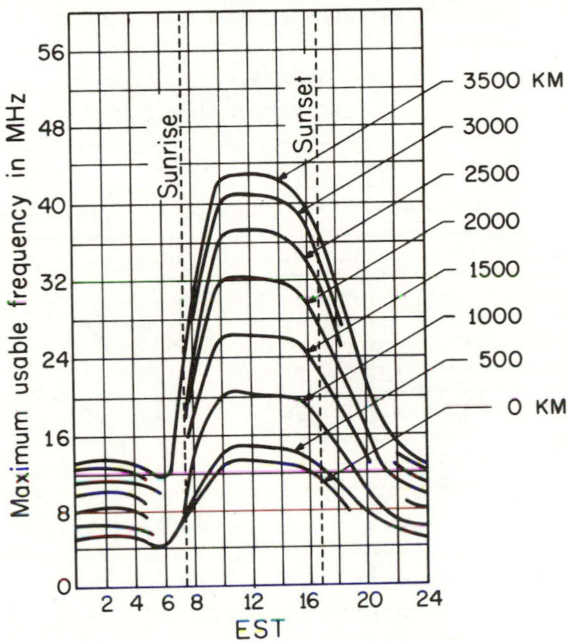

Figure 17-7. Typical set of maximum usable frequency curves for a winter month.

would have been satisfactory for transmitting over distances of 2000 kilometers or more near midday, the same signal would have failed to be reflected back at points less than 1500 kilometers from the transmitter at the same time. The distance within which a signal of given frequency fails to be reflected back is the *skip distance* for that frequency. The higher the frequency the greater is the skip distance.

Because the maximum usable frequency may show small daily variations about the monthly average of up to 15 per cent, it is customary to use a frequency somewhat lower than the predicted maximum usable frequency. Also because it is desirable to restrict the number of different frequencies required to a reasonable number, some latitude must be permitted in the choice of frequency actually used. The *optimum* frequency for transmitting between any two points is therefore selected as some frequency lying between about 50 and 85 per cent of the predicted maximum usable frequency between those points.

17.05 Regular and Irregular Variation of the Ionosphere. Conditions in the ionosphere vary throughout the day with the angle of the sun's rays, and in addition vary quite regularly with the season of year. A plot of critical frequencies and virtual heights as a function of time of day gives a reasonably good picture of ionospheric variations. It is customary to show monthly averages of these quantities because the day-to-day variations are usually quite small, except during periods of ionosphere storminess. Figure 17-8 shows some typical virtual-height and critical-frequency curves for summer and winter. The F layer, which has virtual

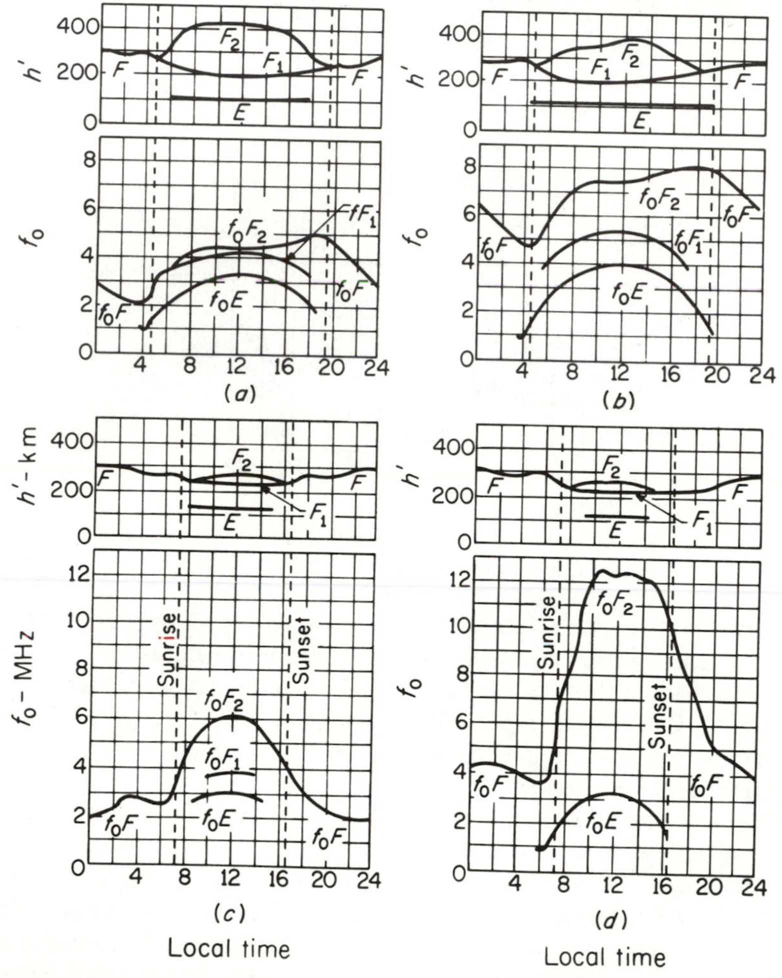

Figure 17-8. Diurnal variation of critical frequency and virtual height of the regular ionospheric layers: (*a*) summer at period of sunspot minimum; (*b*) summer at sunspot maximum; (*c*) winter at sunspot minimum; (*d*) winter at sunspot maximum.

height of about 300 kilometers during the night, splits into two separate layers during the day. The lower of these is designated F_1 and the upper is designated F_2. The E layer exists as an effective radio-wave reflector only during the day. At night its critical frequency falls below 1 MHz. Its virtual height remains almost constant at 110–120 kilometers from season to season and year to year.

Besides the diurnal and seasonal variation of virtual height and critical frequency, these quantities also vary in synchronism with the 11-year sunspot cycle, as shown in Fig. 17-8. The critical frequencies are considerably higher during a year near a sunspot maximum than during a period near a sunspot minimum.

In curves such as those of Figs. 17-5 and 17-8, the critical frequencies for the different layers are designated by appropriate letter subscripts denoting the layer. Because of the presence of the earth's magnetic field, there are actually two different critical frequencies for each layer, one for the so-called *ordinary* wave and a higher one for the *extraordinary* wave. The respective critical frequency is therefore denoted by the subscript o or x; for example f_oF_1 is the critical frequency of the F_1 layer for the ordinary wave. For the E layer the ordinary wave predominates and the extraordinary is usually not considered because it has negligible effect on radio reception. Ordinary and extraordinary waves will be discussed further in a later section on the effect of the earth's magnetic field.

The critical frequency of the E layer has a regular diurnal and seasonal variation. It increases with the altitude of the sun and is a maximum at noon on a summer day. For the E layer it has been found that the critical frequency is given approximately by the simple relation

$$f_E = K\sqrt[4]{\cos \psi}$$

where ψ is the zenith angle of the sun and K is a factor that depends upon the intensity of the radiation from the sun. The critical frequencies of the F layers do not obey any such simple law. For the F_2 layer the diurnal maximum lags behind the altitude of the sun and the daytime critical frequencies are higher in winter than in summer (for the northern hemisphere). Figure 17-9 shows typical curves of the distribution of the ionization density N with height. The curves shown are for day and night conditions in both summer and winter for a mid-latitude region. The values of N are obtained from sweep-frequency virtual-height measurements through the relation

$$N = \frac{\omega^2 m\epsilon_v}{e^2} \tag{17-19}$$

The maximum value of N for any layer is given by

$$N_{\text{max}} = \frac{\omega_{\text{cr}}^2 m\epsilon_v}{e^2} \tag{17-20}$$

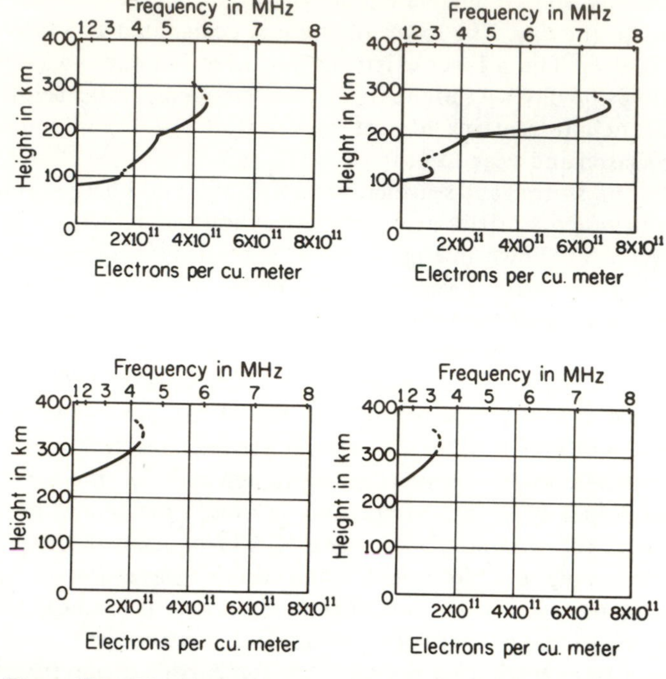

Figure 17-9. Distribution of ionization density with height for quiet conditions: (*a*) summer noon; (*b*) winter noon; (*c*) summer midnight; (*d*) winter midnight.

where N is the number of electrons per cubic meter and f_{cr} is the critical frequency in Hz for that layer. It will be observed from these curves that for daytime conditions the ionization density falls off very rapidly below about 100 kilometers.

Irregular Variation of the Ionosphere. In addition to the regular or normal variation of ionospheric characteristics indicated by Fig. 17-8, there are also irregular variations that are often unpredictable and that sometimes have a marked influence on radio-wave propagation. One of these irregular variations is a sudden ionospheric disturbance, known as an S. I. D. (or sometimes as the *Dellinger effect*), which produces a complete radio "fade out" lasting from a few minutes to an hour or more. The phenomenon is caused by sudden bright eruptions on the sun that produce a large increase in the ionizing radiations that reach the D layer. The resulting increase of ionization density in this layer results in a complete absorption of all sky-wave signals having a frequency greater than about 1 MHz. However, for the very low frequencies that are normally reflected from this layer, the sky-wave signals will increase in intensity. This sudden ionospheric disturbance is often accompanied by disturbances in terrestrial magnetism and earth currents. The effect never occurs at night.

A second type of irregular variation is somewhat similar in origin and effect to the sudden disturbance mentioned, but its beginning and ending are more gradual and it may last for several hours. Usually the absorption of radio signals is not as complete and communication may be carried on at higher frequencies.

In a third type of irregularity, known as *ionospheric storms*, the ionosphere is turbulent and loses its normal stratification. The result is that radio-wave propagation becomes very erratic, and it is often necessary to lower the working frequency in order to maintain communication. The cause of the storm is thought to be the emission of a burst of electrified particles from the sun, and the fact that the storms tend to recur at 27-day intervals, the period of rotation of the sun, seems to indicate that there are active areas on the sun which produce the phenomenon. The effects of ionosphere storms may last for several days.

A fourth type of irregularity occurring only in polar regions during a period of sunspot maximum is known as Polar Cap Absorption (P. C. A.).

17.06 Attenuation Factor for Ionospheric Propagation. In sec. 17.03 the equivalent conductivity σ and dielectric constant ϵ of the ionosphere were obtained in terms of the ionization density N and the collision frequency ν. The attenuation factor α for wave propagation through this region will be given directly by eq. (5-54). It is

$$\alpha = \omega\sqrt{\frac{\mu\epsilon}{2}\left(\sqrt{1+\frac{\sigma^2}{\omega^2\epsilon^2}}-1\right)} \tag{5-54}$$

where

$$\epsilon = \epsilon_r\epsilon_v = \epsilon_v\left(1 - \frac{Ne^2}{\epsilon_v m(\nu^2+\omega^2)}\right) \tag{17-21a}$$

$$\sigma = \frac{Ne^2\nu}{m(\nu^2+\omega^2)} \tag{17-21b}$$

and

$$\mu = \mu_v.$$

Substituting (21a) and (21b) in (5-54), the expression for α may be written

$$\alpha = \frac{\omega}{c}\sqrt{\frac{\epsilon_r}{2}\left(\sqrt{1+\frac{\sigma^2}{\omega^2\epsilon^2}}-1\right)}$$

$$= \frac{\omega}{c}\sqrt{\frac{1}{2}\left(\sqrt{\epsilon_r^2+\frac{\sigma^2}{\omega^2\epsilon_v^2}}-\epsilon_r\right)}$$

$$= \frac{\omega}{c}\sqrt{\frac{1}{2}\sqrt{\epsilon_r^2+\left[(1-\epsilon_r)\frac{\nu}{\omega}\right]^2}-\frac{\epsilon_r}{2}} \tag{17-22}$$

For frequencies not too near the maximum usable frequency, and for the important practical case of a section of the ionosphere where the relation $\sigma/\omega\epsilon \ll 1$ holds, eq. (5-54) for α reduces by use of the binomial expansion to

$$\alpha = \frac{\sigma}{2}\sqrt{\frac{\mu}{\epsilon}} = \frac{60\pi\sigma}{\sqrt{\epsilon_r}} = \frac{60\pi Ne^2 v}{\sqrt{\epsilon_r}\, m(v^2 + \omega^2)} \tag{17-23}$$

For the frequency range where $\omega \gg v$, eq. (23) shows that the attenuation varies approximately as the inverse square of the frequency. Therefore it is desirable to use as high a frequency as possible without approaching too close to the maximum usable frequency. If the ionization density and collision frequency are known, the attenuation per unit length can be calculated by means of (21a) and (22). The total attenuation of the wave in the ionosphere would then be obtained by integrating α along the whole length of path through the ionosphere. In general, it is found that attenuation is negligibly small, except in the region near the lower edge of the ionosphere (the D region) and at the top of the path where the ray is being bent. The absorption that occurs in the region where the wave is bent is called *deviative absorption*, whereas that which occurs in the D region is known as *nondeviative absorption*. For high frequencies, where reflection takes place from the F layer, deviative absorption is usually small because the collision frequency in this layer is low. Exceptions to this occur for frequencies near the maximum usable frequency where the wave is abnormally retarded and appreciable absorption of energy may take place.

Since it is known from theoretical considerations that the collision frequency v is high near the surface of the earth but decreases very rapidly with increasing height, it can be deduced through the use of eq. (23) that the main region of nondeviative absorption will be confined to a relatively narrow range of heights lying somewhere between 60 and 100 km.

17.07 Sky-wave Transmission Calculations. *Maximum Usable Frequency.* For ionospheric transmission to be possible, the frequency used must lie between the maximum usable frequency (MUF) and the lowest useful high frequency (LUHF). The MUF was defined in sec. 17.04, and the LUHF will be discussed later in this section.

The MUF for any given path at any time of day is calculated quite simply through use of convenient world contour charts obtainable from the Environmental Science Services Administration in Washington. These charts, an example of which appears in Fig. 17-10, show the world-wide variation of MUF with local time for all latitudes. The MUF figures predicted on these charts are the monthly *median* values of maximum usable frequency. Thus, communication at the MUF calculated from these charts should be effective approximately 50 per cent of the time during undisturbed periods.

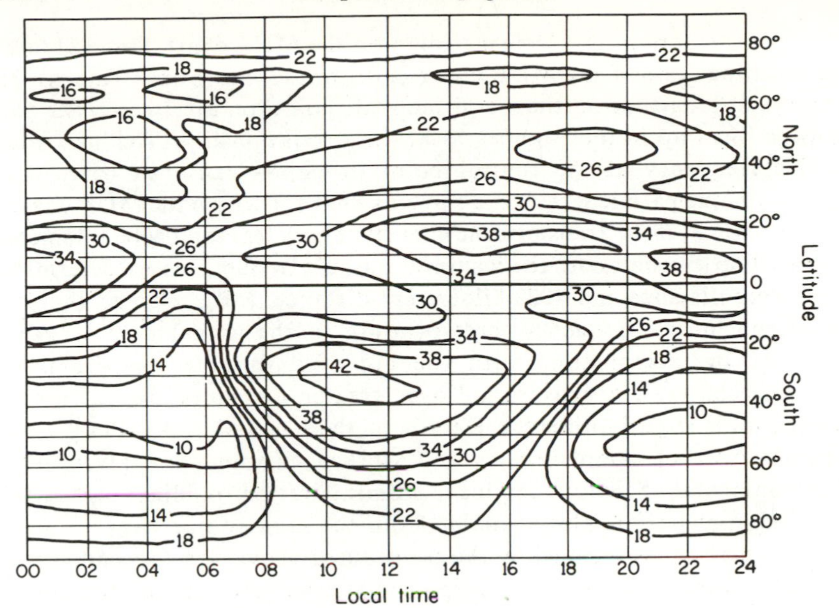

Figure 17-10. Typical world contour MUF chart for the F_2 layer. Chart is for distances of 4000 km and summertime conditions.

To calculate the MUF it is necessary to know the length of the transmission path and the location of certain "control" points on the path. The control points of a transmission path are points along the path, the ionospheric conditions of which seem to control transmission along the path. For paths shorter than 4000 kilometers (that is, single-hop paths) the control point is midway along the path, as would be expected. For longer paths the control points are taken as 2000 kilometers from each end for F_2-layer reflection and 1000 kilometers from each end for *E*-layer reflection. Although the choice is empirical, it can be justified to some extent from a consideration of the probable paths in multi-hop transmission.

To calculate the maximum usable frequency the appropriate MUF chart is used in conjunction with a world map and a great-circle chart of the same size. A sheet of transparent paper is placed over the world map, and on it are marked the location of the transmitting and receiving points and the equatorial line. The transparent paper is then placed on the great-circle chart. Keeping the equatorial lines on chart and paper lined up, the transparency is moved sideways until the transmitting and receiving points both lie on the same great-circle line, which is then sketched in. The transparency is then placed over the MUF chart, and the meridian whose local time is to be used for the calculation is lined up with the appropriate time meridian on the MUF chart. Since 24 hours on the time scale of the MUF chart is drawn to the same scale as 360 degrees of longitude on the world map, all points on the great-circle path will be lined up in their proper local time relationship. The maximum usable frequency at the control point or points can then be read off directly if the path length is the same as that for which the MUF chart is drawn. For F_2-layer transmission over distances less than 4000 kilometers (single-hop transmission) the maximum usable frequencies are determined from zero-distance and 4000-kilometers-distance MUF charts; then the maximum usable frequency corresponding to the actual path length is obtained by interpolation with the aid of a suitable nomogram. For transmission via other layers and over greater distances the calculation procedure is slightly different. Details of these calculations, along with a complete set of sample charts and worked examples are given in an Environmental Science Services Administration publication.* MUF prediction charts are also available from the same source.

The frequencies selected by the procedures above are for undisturbed periods. During periods of "ionospheric storms" the critical frequencies are lower than usual, and it may be necessary to lower the working frequency in order to insure communication. As the frequency is lowered the absorption of the wave increases. If it is necessary to lower the operating frequency below the lowest useful high frequency, communication becomes impossible. Since the "ionospheric storm" type of disturbance is most severe in the polar regions, with less severity toward the equator, communication can often be maintained during "storm" periods by relaying through points closer to the equator. This last statement does not apply in the case of "sudden ionospheric disturbances" or "radio fadeouts" (Dellinger effect). During radio fadeouts high-frequency communication becomes impossible on all paths in the daylight side of the world. Fortunately this latter type of disturbance, which is unpredictable, rarely lasts more than two hours.

Sky-wave Absorption and Lowest Useful High Frequency. As has already been pointed out, when the operating frequency is reduced from the maximum usable frequency the absorption of the wave in the ionosphere increases, and the received signal strength becomes less. The *lowest useful high frequency* (LUHF) for a given distance and given transmitter power is defined as the lowest frequency (in the high-frequency band) that will give satisfactory reception for that distance and power. Unlike the MUF, which depends only upon the state of the ionosphere and the

**Ionospheric Radio Propagation*, National Bureau of Standards Monograph 80, issued April 1, 1965.

distance between transmitting and receiving points, the LUHF depends upon the following factors:

(*a*) The effective radiated power.

(*b*) The absorption characteristics of the ionosphere for the paths between transmitter and receiver.

(*c*) The *required* field strength, which in turn depends upon radio noise at the receiving location and the type of service involved.

These factors will be considered in turn.

(*a*) *Effective Radiated Power and Unabsorbed Field Strength.* The *effective radiated power* is the power actually radiated by the antenna, multiplied by the antenna gain in the direction of propagation. This second factor requires a knowledge of the vertical angle of radiation that is effective in producing a signal at the receiver. This angle depends upon the layer involved, the distance to the receiver, and the number of hops. The *unabsorbed field strength* of a sky-wave signal at a given distance for a transmitter is defined as the median incident field strength that would be observed by use of an antenna of fixed linear polarization if no absorption were introduced by the ionosphere. The unabsorbed field strength is less than that which would result from inverse-distance attenuation alone because of (1) interference and polarization fading and (2) loss of energy upon reflection at the ground between hops. In practice the unabsorbed field strength for any distance is obtained from a graph that takes the above factors into account.

(*b*) *Absorption Characteristics of the Ionosphere.* Absorption in the ionosphere can be classified as deviative or nondeviative absorption. *Deviative Absorption* occurs in that region of the ionosphere where the wave is bent back to earth. Except for frequencies near the critical frequency for the reflecting layer this type of absorption is small. During daylight hours a much greater absorption of the wave occurs in the *D* region, where the collision frequency is high. This latter absorption is called *nondeviative* because it is not associated with a bending of the wave. In the *D* region recombination is rapid and the ionization density, and hence the absorption, varies almost in synchronism with the elevation of the sun. *D*-layer absorption is a maximum at noon and decreases to negligibly small values within two hours after sunset. As is pointed out in the following sections, the absorption has a broad maximum in the neighborhood of the gyromagnetic (or cyclotron) frequency (1.2–1.4 MHz) for the electron in the earth's magnetic field. As the frequency is increased above the resonance frequency, the absorption decreases steadily except for frequencies close to the critical frequency of each layer.

In addition to frequency and time of day, sky-wave absorption is also dependent upon the season of year and sunspot activity. When all of these factors have been taken into account, the absorption of the wave in the ionosphere, and hence the expected incident field strength, can be calculated. As in the case of the MUF, such calculations are greatly facilitated by the use of numerous charts and nomograms which have been prepared, and which are available.*

(*c*) *Required Field Strength.* The field strength required for satisfactory reception for a given type of service depends among other things on the receiver sensitivity, the receiving-set noise, the radio noise level prevailing at the receiving location, and the type of modulation. Radio "noise" can be divided into *man-made noise*, that is, local electrical disturbances produced by electrical machinery, and *atmospheric noise*, or static. The latter noise depends upon the frequency, the time of day, the season of year, and location with respect to the sources of thunderstorms. Because most atmospheric noise has its origin in thunderstorms, those areas in which thunderstorms are most prevalent will have the highest atmospheric-noise level. The world can be divided into noise zones that correspond roughly to the zones showing the incidence of thunderstoms. The principal noise centers or active thunderstorm areas are located in Central Africa, Central America, and the East and West Indies. Areas of very low thunderstorm incidence are the north and south frigid zones. In general the temperate zones are areas of moderate thunderstorm incidence. The actual atmospheric noise at any location depends upon the local noise sources (thunderstorms) and also upon the sky-wave propagation characteristics between that location and the principal noise centers. It should be noted that the same factors that make for good transmission of radio signals from distant transmitters also provide good transmission of noise signals from distant noise sources.

The distribution of noise intensities throughout the world has been plotted on world maps for each month of the year. Using these noise-grade charts in conjunction with curves that show required field strength *vs.* frequency for different times of day and different noise grades, it is a simple matter to figure the required field strength for any given set of conditions.† For any frequency less than the MUF, if the incident field strength calculated under (*b*) is greater than the required field strength calculated under (*c*), communication can be established. The lowest frequency for which this occurs is the lowest useful high frequency.

*Such charts are available in Bureau of Standards Monograph 80.

†The required field strength curves and noise-grade charts are available in the references already mentioned.

It is seen that the calculation of the probable received field strength at any point is an engineering problem that can be solved when sufficient data are given. The data required are the time of day, the season of year, the transmission path, the frequency, and the effective radiated power. Although the ionosphere is a complex natural phenomenon not under the control of man, by familiarizing himself with its characteristics man has learned to predict its behavior and so has been enabled to use it to serve his needs.

17.08 Effect of the Earth's Magnetic Field. Electrons and ions in the ionosphere are influenced not only by the fields of a passing electromagnetic wave but also by the earth's magnetic field, which causes the charged particles to move in circular or spiral paths. It will be shown that the effects of the earth's magnetic field may be taken into account by a suitable modification in the permittivity, in much the same way as for the case in which no external magnetic field is present (sec. 9.07). That is, Maxwell's equations may be combined with the equation of motion in such a way as to give a new set of Maxwell's equations with a permittivity different from that in free space.

Assuming that the ions are stationary and that there is no steady drift velocity, the time-varying electron number density and electron current are given by

$$\tilde{n} = N + \mathrm{Re}\,\{ne^{j\omega t}\} \tag{17-25}$$

$$\begin{aligned}\tilde{\mathbf{J}} &= Ne\tilde{\mathbf{v}} \\ &= Ne\,\mathrm{Re}\,\{\mathbf{v}e^{j\omega t}\}\end{aligned} \tag{17-26}$$

Maxwell's equations in phasor form for electrons in a vacuum are

$$\nabla \times \mathbf{H} = j\omega\epsilon_v \mathbf{E} + \mathbf{J} \tag{17-27}$$

$$\nabla \times \mathbf{E} = -j\omega\mu_v \mathbf{H} \tag{17-28}$$

$$\nabla \cdot \mathbf{E} = \frac{ne}{\epsilon_v} \tag{17-29}$$

$$\nabla \cdot \mathbf{H} = 0 \tag{17-30}$$

The equation of motion in time-varying form is

$$m\dot{\tilde{\mathbf{v}}} = e(\tilde{\mathbf{E}} + \tilde{\mathbf{v}} \times \tilde{\mathbf{B}}) - m\nu\tilde{\mathbf{v}} \tag{17-31}$$

in which

$$\begin{aligned}\tilde{\mathbf{E}} &= \mathrm{Re}\,\{\mathbf{E}e^{j\omega t}\} \\ \tilde{\mathbf{B}} &= \mathbf{B}_0 + \mathrm{Re}\,\{\mathbf{B}e^{j\omega t}\} \\ &= \mu_v\tilde{\mathbf{H}}\end{aligned}$$

In linearized phasor form, (31) becomes

$$j\omega m\mathbf{v} = e(\mathbf{E} + \mathbf{v} \times \mathbf{B}_0) - m\nu\mathbf{v} \tag{17-32}$$

The objective now is to express $\mathbf{v}$ in terms of $\mathbf{E}$ and to substitute the expression for $\mathbf{v}$ into the current density term in (27). Rearranging (32) gives

$$e\mathbf{E} = j\omega m\left(1 - j\frac{\nu}{\omega}\right)\mathbf{v} + e\mathbf{B}_0 \times \mathbf{v} \tag{17-33}$$

which expresses $\mathbf{E}$ in terms of $\mathbf{v}$. At this point the introduction of the following notation simplifies the appearance of the equation of motion.

$$\begin{array}{lll} Z = \dfrac{\nu}{\omega} & \omega_p^2 = \dfrac{Ne^2}{m\epsilon_v} & \boldsymbol{\omega}_c = \dfrac{-e\mathbf{B}_0}{m} \\ U = 1 - jZ & X = \dfrac{\omega_p^2}{\omega^2} & \mathbf{Y} = \dfrac{\boldsymbol{\omega}_c}{\omega} \\ & & Y = \dfrac{\omega_c}{\omega} = \dfrac{-eB_0}{m\omega} \end{array} \tag{17-34}$$

ν = collision frequency
ω_p = plasma frequency
ω_c = cyclotron frequency

Thus the equation of motion may be written in the form

$$-j\omega\epsilon_v X\mathbf{E} = U\mathbf{J} + j\mathbf{Y} \times \mathbf{J} \tag{17-35}$$

Equation (35) now must be inverted (or in other words, $\mathbf{J}$ must be expressed in terms of $\mathbf{E}$) but this procedure is complicated by the presence of the cross-product. If a vector is represented by a column matrix, the cross-product may be expressed as

$$\mathbf{Y} \times \mathbf{J} = \begin{bmatrix} 0 & -Y_z & Y_y \\ Y_z & 0 & -Y_x \\ -Y_y & Y_x & 0 \end{bmatrix} \begin{bmatrix} J_x \\ J_y \\ J_z \end{bmatrix} \tag{17-36}$$

For the case in which the magnetic field is in the z direction, that is $\mathbf{Y} = \hat{\mathbf{z}}Y$, the equation of motion becomes

$$-j\omega\epsilon_v X \begin{bmatrix} E_x \\ E_y \\ E_z \end{bmatrix} = \begin{bmatrix} U & -jY & 0 \\ jY & U & 0 \\ 0 & 0 & U \end{bmatrix} \begin{bmatrix} J_x \\ J_y \\ J_z \end{bmatrix} \tag{17-37}$$

Expression of $\mathbf{J}$ in terms of $\mathbf{E}$ requires inversion of the coefficient matrix, which gives

$$\begin{bmatrix} J_x \\ J_y \\ J_z \end{bmatrix} = \frac{-j\omega\epsilon_v X}{U(U^2 - Y^2)} \begin{bmatrix} U^2 & jUY & 0 \\ -jUY & U^2 & 0 \\ 0 & 0 & U^2 - Y^2 \end{bmatrix} \begin{bmatrix} E_x \\ E_y \\ E_z \end{bmatrix} \tag{17-38}$$

The quantity multiplying the $\mathbf{E}$ vector must be the conductivity matrix, since (38) is of the form

$$\mathbf{J} = \bar{\boldsymbol{\sigma}}\cdot\mathbf{E} \tag{17-39}$$

in which $\bar{\boldsymbol{\sigma}}$ is the dyadic whose elements are taken from (38). The permittivity of the plasma may be obtained by substituting (38) into (27). This procedure gives, in matrix form,

$$\begin{bmatrix} 0 & -\dfrac{\partial}{\partial z} & \dfrac{\partial}{\partial y} \\ \dfrac{\partial}{\partial z} & 0 & -\dfrac{\partial}{\partial x} \\ -\dfrac{\partial}{\partial y} & \dfrac{\partial}{\partial x} & 0 \end{bmatrix}\begin{bmatrix} H_x \\ H_y \\ H_z \end{bmatrix} = j\omega\epsilon_v \begin{bmatrix} K' & jK'' & 0 \\ -jK'' & K' & 0 \\ 0 & 0 & K_0 \end{bmatrix}\begin{bmatrix} E_x \\ E_y \\ E_z \end{bmatrix} \tag{17-40}$$

in which

$$K' = 1 - \frac{UX}{U^2 - Y^2} \qquad K_0 = 1 - \frac{X}{U} \qquad K'' = \frac{-XY}{U^2 - Y^2}$$

Equation (40) also may be written in dyadic form as

$$\nabla \times \mathbf{H} = j\omega\epsilon_v \bar{\mathbf{K}}\cdot\mathbf{E} \tag{17-41}$$

in which

$$\bar{\mathbf{K}} = K'\hat{\mathbf{x}}\hat{\mathbf{x}} + jK''\hat{\mathbf{x}}\hat{\mathbf{y}} - jK''\hat{\mathbf{y}}\hat{\mathbf{x}} + K'\hat{\mathbf{y}}\hat{\mathbf{y}} + K_0\hat{\mathbf{z}}\hat{\mathbf{z}}$$

Gauss's law (29) also may be rewritten using the relative permittivity dyadic $\bar{\mathbf{K}}$. The equation of continuity

$$\nabla\cdot\mathbf{J} = -j\omega ne \tag{17-42}$$

makes it possible to write (29) in the form

$$\nabla\cdot\bar{\mathbf{K}}\cdot\mathbf{E} = 0 \tag{17-43}$$

Equations (43) and (41) show that the displacement density (electric flux density) $\mathbf{D}$ given by

$$\mathbf{D} = \epsilon_v \bar{\mathbf{K}}\cdot\mathbf{E} \tag{17-44}$$

is not parallel to $\mathbf{E}$, indicating that the magnetized plasma is an *anisotropic* medium.

17.09 Wave Propagation in the Ionosphere. Many types of wave motion can occur in an anisotropic plasma but for ionospheric propagation by far the most important type is the familiar uniform plane wave having the form

$$\begin{aligned} \mathbf{E}(\mathbf{r}) &= \mathbf{E}_0 e^{-\gamma\hat{\mathbf{n}}\cdot\mathbf{r}} \\ \mathbf{H}(\mathbf{r}) &= \mathbf{H}_0 e^{-\gamma\hat{\mathbf{n}}\cdot\mathbf{r}} \end{aligned} \tag{17-45}$$

The principal properties of uniform plane waves in a plasma may be deduced by considering two special cases, propagation in a direction perpendicular to the magnetic field ($\hat{\mathbf{n}} = \hat{\mathbf{x}}$) and propagation parallel to the magnetic field ($\hat{\mathbf{n}} = \hat{\mathbf{z}}$).

Perpendicular Propagation. A uniform plane wave propagating in the x direction has a spatial variation of the form $e^{-\gamma x}$. With this spatial variation, substitution of (45) into (40) gives

$$\begin{aligned} 0 &= j\omega\epsilon_v(K' E_x + jK'' E_y) \\ \gamma H_z &= j\omega\epsilon_v(-jK'' E_x + K' E_y) \\ -\gamma H_y &= j\omega\epsilon_v K_0 E_z \\ 0 &= -j\omega\mu_v H_x \\ \gamma E_z &= -j\omega\mu_v H_y \\ -\gamma E_y &= -j\omega\mu_v H_z \end{aligned} \tag{17-46}$$

The above are six homogeneous equations in six unknowns and thus the existence of a solution requires that the determinant of the coefficients be zero. This condition gives an equation for the propagation constant γ. Writing a six-by-six determinant is unnecessary, however, because the equations in (46) are very simple. To begin with, it is clear that $H_x = 0$. H_y and H_z may be eliminated easily, giving

$$0 = j\omega\epsilon_v(K' E_x + jK'' E_y) \tag{17-47}$$

$$\gamma^2 E_y = -\omega^2 \mu_v\epsilon_v(-jK'' E_x + K' E_y) \tag{17-48}$$

$$\gamma^2 E_z = -\omega^2 \mu_v\epsilon_v K_0 E_z \tag{17-49}$$

The above equations may be simplified somewhat by introducing the refractive index n defined by

$$\gamma = j\omega\sqrt{\mu_v\epsilon_v}\, n \tag{17-50}$$

Thus (47), (48) and (49) may be written

$$\begin{bmatrix} K' & jK'' \\ jK'' & n^2 - K' \end{bmatrix} \begin{bmatrix} E_x \\ E_y \end{bmatrix} = 0 \tag{17-51}$$

$$(n^2 - K_0)E_z = 0 \tag{17-52}$$

If $E_z \neq 0$, (52) indicates that one possible value for the refractive index is given by

$$n_1^2 = K_0 \tag{17-53}$$

This value of the refractive index does not in general make the determinant of the coefficient matrix in (51) equal to zero so that $E_x = E_y = 0$ and thus $H_z = 0$. What remains is the set of field quantities E_z, H_y. From (46) it may be shown that these are related by

$$-\frac{E_z}{H_y} = \frac{1}{n_1}\sqrt{\frac{\mu_v}{\epsilon_v}} \tag{17-54}$$

Under lossless, propagating conditions (n_1 positive real) the wave is unaffected by the steady magnetic field, a situation which is not surprising when one realizes that the electric field is in the z direction and thus electrons accelerated by it are not deflected by $\mathbf{B}_0$.

If E_x and E_y are not zero, then setting the determinant of the coefficient matrix in (51) equal to zero gives

$$n_2^2 = \frac{(K' - K'')(K' + K'')}{K'} \tag{17-55}$$

Since in general $n_2^2 \neq K_0$, the anisotropic plasma is *doubly refracting* or *birefringent*. Furthermore (52) indicates that $E_z = 0$ and from (46), $H_y = 0$. The remaining field quantities are E_y, H_z, E_x, related by

$$\frac{E_y}{H_z} = \frac{1}{n_2}\sqrt{\frac{\mu_v}{\epsilon_v}} \tag{17-56}$$

$$\frac{E_x}{E_y} = -j\frac{K''}{K'} \tag{17-57}$$

As one would expect, (56) indicates power flow in the direction of propagation (the x direction), provided n_2^2 is positive and real. Equation (57) indicates the existence of a longitudinal electric field 90 degrees out of phase with the transverse electric field (under lossless conditions).

Summarizing, for wave propagation perpendicular to the earth's magnetic field $\mathbf{B}_0$, two distinctly different cases result for $\mathbf{E}$ parallel to or perpendicular to $\mathbf{B}_0$. In the former case the wave is unaffected by $\mathbf{B}_0$, and has the same phase velocity as without the earth's magnetic field. This wave is called the *ordinary wave*, or *ordinary ray*. When $\mathbf{E}$ is perpendicular to $\mathbf{B}_0$, the effective refractive index and phase velocity are affected by the presence of the magnetic field; this latter wave is called the *extraordinary wave*, or *extraordinary ray*. In general a wave propagating perpendicular to $\mathbf{B}_0$ will have components of $\mathbf{E}$ both parallel and perpendicular to $\mathbf{B}_0$. Such a wave will split into ordinary and extraordinary rays which travel different paths with different time delays. The extraordinary ray suffers greater absorption (at high frequencies), and has a slightly higher critical frequency than the ordinary ray.

Parallel Propagation. A uniform plane wave propagating in the z direction has a spatial variation of the form $e^{-\gamma z}$. Substitution into (40) gives

$$\begin{aligned}
\gamma H_y &= j\omega\epsilon_v(K' E_x + jK'' E_y)\\
-\gamma H_x &= j\omega\epsilon_v(-jK'' E_x + K' E_y)\\
0 &= j\omega\epsilon_v K_0 E_z\\
\gamma E_y &= -j\omega\mu_v H_x\\
-\gamma E_x &= -j\omega\mu_v H_y\\
0 &= -j\omega\mu_v H_z
\end{aligned} \tag{17-58}$$

As in the case of perpendicular propagation, the above are six homogeneous equations in six unknowns and setting the determinant of the coefficient matrix equal to zero would give the required equation for the unknown propagation constant γ. It is simpler in this case, however, to carry out first some direct reduction of the size of the determinant. It is evident from (58) that E_z and H_z are both zero, that is, there are no field components in the direction of propagation. Elimination of H_x and H_y reduces (58) to

$$\begin{aligned}\gamma^2 E_x &= -\omega^2 \mu_v \epsilon_v (K' E_x + jK'' E_y) \\ \gamma^2 E_y &= -\omega^2 \mu_v \epsilon_v (-jK'' E_x + K' E_y)\end{aligned} \tag{17-59}$$

With the refractive index as defined in (50), (59) reduces to

$$\begin{bmatrix} n^2 - K' & -jK'' \\ jK'' & n^2 - K' \end{bmatrix} \begin{bmatrix} E_x \\ E_y \end{bmatrix} = 0 \tag{17-60}$$

Setting the determinant of the coefficient matrix equal to zero gives

$$(n^2 - K')^2 - K''^2 = 0 \tag{17-61}$$

which gives the two refractive indices

$$n_1^2 = K' + K'' \tag{17-62}$$

$$n_2^2 = K' - K'' \tag{17-63}$$

The polarization ratios are obtained from (60) and are

$$P_1 = -j$$

$$P_2 = +j$$

where the subscripts 1 and 2 denote right and left circular polarization, respectively.

Problem 1. For the wave whose refractive index is given by (55) calculate the real part of the complex Poynting vector and the time average of the instantaneous Poynting vector (consider the collisionless case only).

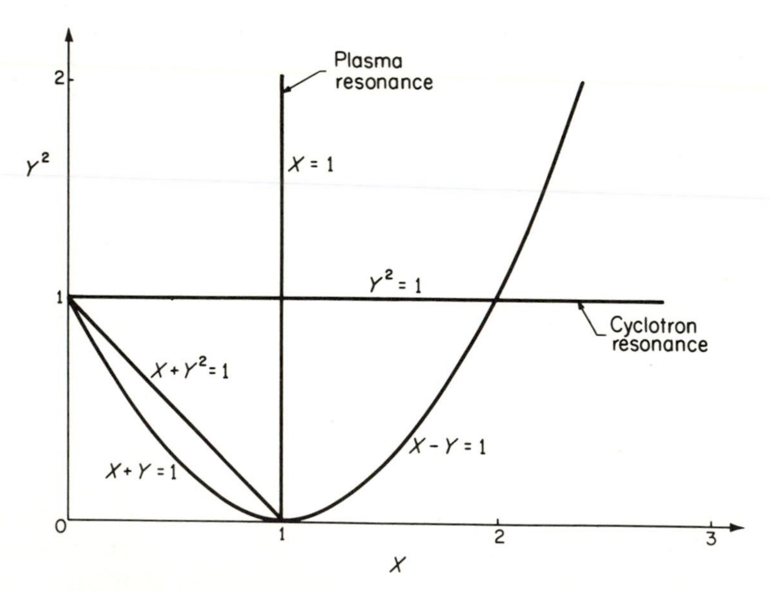

Figure 17-11. The parameters for wave propagation in a collisionless plasma. The axes are $X = \omega_p^2/\omega^2 \propto N/\omega^2$ and $Y^2 = \omega_c^2/\omega^2 \propto B_0^2/\omega^2$. Ion motion is not taken into account.

Problem 2. Fig. 17-11 shows several regions in the X-Y^2 plane. Classify each region as being a "cutoff" or a "propagating" region for a collisionless plasma. Refer to the refractive indices n_1 and n_2 in the classification and consider both perpendicular and parallel propagation.

17.10 Faraday Rotation and the Measurement of Total Electron Content. Any linearly polarized wave may be regarded as the vector sum of two counter-rotating, circularly polarized waves. If such a wave propagates in the direction of the magnetic field in a lossless plasma, the two circularly polarized components will travel at different phase velocities and thus the plane of polarization will rotate with distance, a phenomenon known as *Faraday rotation.*

The propagation of two equal-magnitude, circularly polarized waves in a lossless plasma may be represented by writing the electric field strength in the form

$$\mathbf{E} = \hat{\mathbf{R}}e^{-j\beta_R z} + \hat{\mathbf{L}}e^{-j\beta_L z} \tag{17-64}$$

in which $\hat{\mathbf{R}}$ and $\hat{\mathbf{L}}$ are right-handed and left-handed rotating unit vectors and β_R and β_L are real. The complex polarization ratio, as defined in eq. (12-50) on page 460, gives

$$Q = \frac{E_L}{E_R} = e^{-j(\beta_L - \beta_R)z} \tag{17-65}$$

Because $|Q| = 1$, the polarization is linear and from eq. (12-52) the tilt angle of the plane of polarization is given by

$$\psi = \frac{1}{2}(\beta_L - \beta_R)z = \frac{\omega}{2c}(n_L - n_R)z \tag{17-66}$$

measured from the x axis. Under lossless conditions (62) and (63) show that the refractive indices are given by

$$n_L^2 = 1 - \frac{X}{1 + Y} \tag{17-67}$$

$$n_R^2 = 1 - \frac{X}{1 - Y} \tag{17-68}$$

If the signal frequency is high compared to the plasma and cyclotron frequencies, the binomial approximation gives

$$n_L = 1 - \tfrac{1}{2}X(1 - Y) \tag{17-69}$$

$$n_R = 1 - \tfrac{1}{2}X(1 + Y) \tag{17-70}$$

and thus the tilt angle of the plane of polarization is

$$\psi = \frac{\omega}{2c}XYz = \frac{|e^3|\,B_0}{2cm^2\epsilon_v\omega^2}Nz \tag{17-71}$$

That is, the rotation of the plane of polarization is proportional to the distance z and also to electron density N. If the electron density varies with distance and if B_0 is approximately constant, the change in tilt angle is given by

$$\psi_{\text{transmitted}} - \psi_{\text{incident}} = \frac{|e^3|\, B_0}{2cm^2 \epsilon_v \omega^2} \int N(z)\, dz \tag{17-72}$$

Equation (72) shows that the amount of Faraday rotation on a high-frequency signal passing through the ionosphere is proportional to the total electron content in a one-square-meter column extending through the most highly ionized part of the ionosphere. For signals at several hundred megahertz most of the Faraday rotation occurs in the 90–1000-km alitude range. Measurements of total electron content are made regularly using satellite-to-earth transmissions and signals reflected from the moon.*

17.11 Other Ionospheric Phenomena. *Whistlers.* During World War I sensitive audio amplifiers were being used to eavesdrop on distant telephone conversations. In addition to the usual atmospheric noises the listeners frequently heard whistles of descending tone, and over the years it has been determined that these "atmospheric whistlers" are associated with electromagnetic pulses of *audio* frequency radiation propagating along the lines of the earth's magnetic field between conjugate points in the northern and southern hemispheres.† Usually lightning discharges generate the pulses which may "bounce" back and forth between the northern and southern hemispheres several times before disappearing. This behavior is illustrated in Fig. 17-12 which shows idealized graphs of frequency as a function of time. Illustrated are the *long whistlers* originating at the receiving point, the *short whistlers* originating at the conjugate point, and the *nose whistlers* which exhibit a minimum in propagation time (maximum in group velocity).

The characteristics of a whistler are determined mainly by the properties of the ionosphere at the top of the pulse "trajectory," at an alitude of two to ten earth radii. For this reason whistlers are used in the study of the composition of the upper ionosphere.

*An excellent review of Faraday rotation theory and experiments is contained in R. S. Lawrence, C. G. Little and H. J. A. Chivers, "A Survey of Ionospheric Effects Upon Earth-Space Radio Propagation," *Proc. IEEE*, **52**, No. 1, pp. 4–27, January, 1964. Plots of ionospheric Faraday rotation and electron content versus time are given by H. D. Webb, *Atlas of Lunar Data*, Vols. I to IV, 1961–1965, University of Illinois report prepared under Contract DA 36-039-AMC-0373(E). (Available through O. T. S.)

†For a detailed discussion, the reader is referred to the review by R. A. Helliwell and M. G. Morgan, "Atmospheric Whistlers," *Proc. IRE*, **47**, No. 2, pp. 200–208, February, 1959. See also, R. A. Helliwell, *Whistlers and Related Ionospheric Phenomena*, Stanford University Press, Stanford, Calif., 1965.

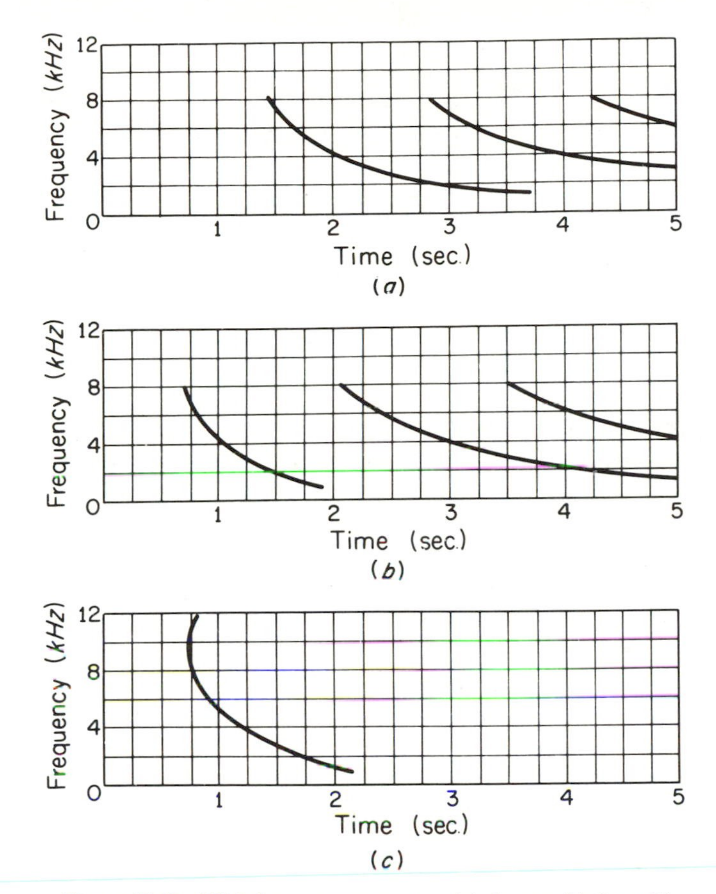

Figure 17-12. Whistler spectrograms: (*a*) long whistler with echoes; (*b*) short whistler with echoes; (*c*) nose whistler.

The refractive indices in (67) and (68) are appropriate for propagation along the magnetic field lines under the near collisionless conditions prevailing far from the earth's surface. At very low frequencies, X and Y are both large and thus the only propagating wave is the right-handed (extraordinary) wave whose refractive index (67) may be written in the form

$$n^2 = 1 + \frac{\omega_p^2}{\omega(\omega_c - \omega)} \tag{17-73}$$

The above relation may be used to calculate the group velocity v_g which gives a useful measure of the time taken by an electromagnetic pulse in passing from one point to another through a lossless medium. The reciprocal of the group velocity is given by

$$\frac{1}{v_g} = \frac{d\beta}{d\omega} = \frac{d}{d\omega}\left(\frac{\omega}{c}n\right)$$

$$= \frac{1}{c}\left(n + \omega\frac{dn}{d\omega}\right) \tag{17-74}$$

$$= \frac{n'}{c} \tag{17-75}$$

which defines n' as the *group refractive index*. From (74) it may be shown (problem 3) that the group velocity is

$$v_g = \frac{cn}{1 + \dfrac{\omega_p^2\omega_c}{2\omega(\omega_c - \omega)^2}} \tag{17-76}$$

In the upper ionosphere for low-freqency propagation it is reasonable to make the approximation $\omega\omega_c \ll \omega_p^2$ which allows (76) to be expressed as

$$v_g = 2c\,\frac{\omega^{1/2}(\omega_c - \omega)^{3/2}}{\omega_p\omega_c} \tag{17-77}$$

At very low frequencies the group velocity decreases with decreasing frequency, thus corresponding to the decreasing tone observed on most whistlers. In addition it may be shown easily that (76) exhibits a maximum at $\omega = \frac{1}{4}\omega_c$, corresponding to minimum time delay observed with nose whistlers.

Problem 3. Verify eq. (76).

Problem 4. Using Fig. 17-1, calculate approximately the altitudes at which the plasma frequency is equal to the cyclotron frequency. Consider both day and night conditions and assume that the earth's magnetic field has a flux density equal to 0.5 gauss (5×10^{-5} webers/m^2).

Problem 5. An incident electromagnetic wave causes an electron to oscillate, thus producing an electric dipole radiator. The ratio of the power radiated (or scattered) per unit solid angle, Φ, to the incident power density $P = E^2/\eta$, defines the "scattering cross section" σ_E of the electron. Using relations developed in chaps. 9 and 11, show that

$$\sigma_E = \left[\frac{\mu_v e^2}{4\pi m}\sin\theta\right]^2$$

where e and m are the charge and mass, respectively, of the electron. The scattered signal, known as *Thomson scatter*, is the subject of the remainder of this section.

Incoherent Scatter (*Thomson Scatter*). VHF radar backscatter from the ionsosphere can be used to deduce charged-particle densities and temperatures, even above the F-region maximum in electron density.* An individual electron scatters an incident electromagnetic wave weakly and furthermore the thermal motion of the electron gives the scattered wave a Doppler frequency shift Δf. Since the scattered wave frequency for each electron is not the same as the incident wave frequency and since the medium contains many electrons in random motion, the scattering is is said to be *incoherent*. To put it another way, the phases of the scattered waves vary randomly from electron to electron. Incoherent scattering predominates over coherent scattering if the wave frequency is greater than the plasma frequency and the spatial extent of the scattering electrons is large compared with the wavelength. Scattering may be regarded as incoherent at altitudes above 100 km if $f > 200$ MHz ($\lambda <$ 1.5 m). When the scattering from a cloud of electrons is incoherent, the total power in the scattered wave is the sum of the powers scattered from the individual electrons, making the total scattered power proportional to the electron density. Thus if the incident wave is a short pulse, the power in the received signal as a function of time gives directly the electron density as a function of altitude.

The electron temperature may be deduced from the frequency spread of the received signal spectrum, a spread which would be of the order of 100 kHz for ionospheric backscatter of an incident wave at 200 MHz. At this frequency it turns out that the frequency spread of the backscatter from the electrons can be associated with the ion temperature rather than the electron temperature, an effect due to the Coulomb interactions between the the charged particles. The electrostatic field of an ion in free space is essentially that of a point charge, the electric field strength varying as r^{-2} at all radii. In a plasma the ion is surrounded by electrons which tend to shield the plasma from the influence of the ion. This shielding is effective up to a distance from the ion of the order of a *Debye length* λ_D, given by

$$\lambda_D = \sqrt{\frac{\epsilon_v k T_e}{N e^2}}$$

in which T_e is the electron temperature, N is the electron density, and k is Boltzmann's constant. Within a distance λ_D from the ion the potential varies sharply and the electric field strength is stronger than for an ion

*W. E. Gordon, "Incoherent Scattering of Radio Waves by Free Electrons with Applications to Space Exploration by Radar," *Proc. IRE*, **46**, pp. 1824–1829, November, 1958.

W. E. Gordon, "Radar Backscatter from the Earth's Ionosphere," *IEEE Transactions on Military Electronics*, Vol. MIL-8, Nos. 3 and 4, pp. 206–210, July–October, 1964.

in free space; beyond λ_D the potential is very nearly constant and the electric field strength is much smaller than it is in the free-space case. Within a distance λ_D from the ion the electron motion is strongly influenced by the ion's field so that the scattering may be thought of as occurring from fluctuations in electron density having dimensions of the order of λ_D. These fluctuations move with the ion velocity and thus when scattering from them predominates (i.e., when the wavelength is large compared to λ_D), the frequency spread in the scattered wave is characteristic of the ion temperature. When the wavelength is small compared to λ_D the frequency spread in the scattered wave is characteristic of the electron temperature. In the ionosphere λ_D varies from about 2 cm at 90 km altitude to about 3 mm at the *F*-region peak in electron density.

The deduction of electron temperature (and electron density) from radar backscatter is further complicated by the fact that the electrons and ions are not in thermal equilibrium, that is, they have different temperatures (here *temperature* refers to kinetic temperature which is associated with the mean thermal velocity of the particles).

VLF Propagation. The efficacy of the very low frequencies (3–30 kHz) for long-distance communication has been known since the early days of radio. The advent of continuous-wave navigation systems such as OMEGA, the increasing need for accurate frequency comparison and transmission of timing signals, and the problems of (submerged) submarine communications and nuclear explosion detection have all caused a resurgence of interest in this frequency band. The great size required for reasonably efficient transmitting antennas tends to restrict the usable part of the band to the upper end (10–30 kHz) where the wavelengths range from 30,000 to 10,000 meters. Very long-distance propagation at these frequencies may be viewed (to a first approximation) as transmission within the earth-ionosphere wave guide. In this approach it is convenient to use the mode theory of VLF propagation which has been treated in detail by several authors.* Summaries of measured field-strength data indicate an average attenuation rate of 2 to 3 decibels per 1000 kilometers for very long-distance transoceanic propagation in the frequency range of 16–20 kHz. Attenuation over a land path may be as much as 3 db higher compared with propagation over sea water and nonreciprocity (due to the earth's magnetic field) causes about 1 db more attenuation for east-west propagation than for west-east propagation.

*See references in the bibliography.

BIBLIOGRAPHY

Articles

Bowhill, S. A., ed., "VLF Ionospheric Radio Propagation," *Radio Science* **1** (new series), 1356–1357 (1966).

Crombie, D. D. and A. G. Jean, "The Guided Propagation of ELF and VLF Radio Waves between the Earth and the Ionosphere," *Radio Science* **68D**, 584–588 (1964).

Wait, J. R., "A Diffraction Theory for LF Skywave Propagation," *Jour. Geophy. Res.*, **66**, 1713 (1961).

______, "Terrestrial Propagation of VLF Radio Waves," *J. Res. NBS*, **64D**, 153 (1960).

Books

Budden, K. G., *Radio Waves in the Ionosphere,* Cambridge University Press, New York, 1961.

Davies, Kenneth, *Ionospheric Radio Propagation*, National Bureau of Standards Monograph 80, U. S. Govt. Printing Office, Washington, D. C., 1965.

Kelso, J. M., *Radio Ray Propagation in the Ionosphere,* McGraw-Hill Book Company, New York, 1964.

Ratcliffe, J. A., *The Magneto-Ionic Theory and Its Application to the Ionosphere,* Cambridge University Press, New York, 1959.

Wait, J. R., *Electromagnetic Waves in Stratified Media,* Pergamon Press, Inc., New York, 1962.

Chapter 18

ELECTROMAGNETIC THEORY AND SPECIAL RELATIVITY

18.01 Introduction. The electromagnetic theory of Maxwell, formulated during the period 1855 to 1865, was based upon the experimental laws of Coulomb, Ampere, and Faraday. The special theory of relativity was enunciated in 1904–1905. As long ago as 1912, L. Page pointed out that "if the principle of (special) relativity had been enunciated before the date of Oersted's discovery (of the magnetic effects of electric currents), the fundamental relations of Maxwellian electrodynamics could have been predicted on theoretical grounds as a direct consequence of the fundamental laws of electrostatics, extended so as to apply to charges in relative motion as well as to charges at rest."*

It is the purpose of this chapter to demonstrate that the laws of Ampere and Faraday, and their generalization into Maxwell's equations, do indeed follow directly from the fundamental law of electrostatics (Coulomb's law) and the transformations of the special theory of relativity. The advantages of this approach are several:

(1) What were originally thought to be unrelated experimental laws are shown to be different facets of a single comprehensive theory.

(2) Correct solutions are indicated for certain problems which appear indeterminate or ambiguous using the conventional flux-cutting rules of electrodynamics.

(3) The groundwork is laid for more advanced study of electromagnetic effects in moving media.

(4) By no means least, a deeper understanding of all aspects of electromagnetic theory is provided by viewing established concepts in an entirely new light.

It is not our intent to study relativity theory in detail. A few of the many books which cover this subject at various levels are listed in the references. Nor shall we attempt to treat the electromagnetics of moving media. Admittedly, recently renewed interest in this topic

*Leigh Page, "A Derivation of the Fundamental Relations of Electrodynamics from those of Electrostatics," *Amer. J. of Sci.*, **34,** p. 57, 1912.

provides a major impetus for the study of relativity in relation to electromagnetic theory. However, an adequate treatment in anything less than a complete book seems to be hardly possible. Fortunately an excellent engineering treatment of this important subject is available.*

18.02 Galileian Relativity. The notions of simple relativity as used in classical mechanics are so commonplace as to be employed almost without thinking. If a boat on a river moves downstream with a velocity v relative to the water, and the river has a velocity u, then the velocity of the boat relative to a stationary observer on the bank is $v + u$. To an observer in the boat the observer on the bank appears to be moving by with a velocity $-(v + u)$. The position of the boat at any instant can be specified with respect to a fixed point on the bank, or with respect to a freely drifting buoy moving with the water. Mentally attaching a frame of reference or co-ordinate system S to a fixed point on the bank, and a second co-ordinate system S' to the drifting buoy (Fig. 18-1), we write

$$x' = x - ut \tag{18-1a}$$

where x' is the position of the boat in S' and x is the position in S. Equation (1a) expresses the position of the boat relative to the buoy in terms of position and time as measured by an observer on the bank. We could equally well have written

$$x = x' + ut', \qquad (t' = t) \tag{18-1b}$$

where now the position relative to the point on the bank is expressed

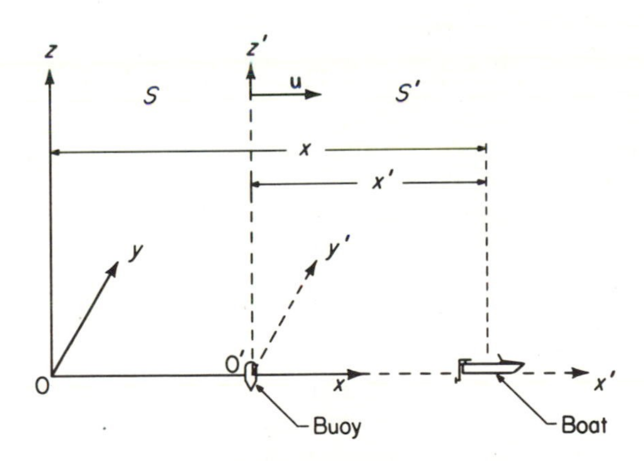

Figure 18-1. Co-ordinate systems S and S'.

*E. G. Cullwick, *Electromagnetism and Relativity, With Particular Reference to Moving Media and Electromagnetic Induction* (2nd ed.), John Wiley & Sons, Inc., New York, 1959.

in terms of position and time as measured by an observer moving with the buoy.

The transformation from one co-ordinate system to the other as given by eq. (1a) or (1b) is known as the *Galileian transformation* (after Galileo Galilei). In classical Newtonian mechanics it is a basic hypothesis that laws of motion hold equally in all rigid co-ordinate frames moving with uniform rectilinear velocity with respect to each other; in Fig. 18-1 the equations of motion will remain unchanged in the two systems by substitution of the appropriate Galileian transformation relations (eqs. 1a and b). This invariance of the laws of mechanics under a Galileian transformation is called the principle of *Galileian relativity*.

Associated with this simple relativity principle and the "common-sense" transformation of eqs. (1a and b) are certain other common-sense ideas. We expect the mass and dimensions of the anchor-weight carried in the moving boat to remain the same as they were when measured at rest on land, and we assume that time marches on at the same rate for both the "moving" and "fixed" observers.

Most readers will be aware that these common-sense notions associated with the Galileian relativity principles of classical mechanics are, in fact, only first approximations, valid for velocities, $v \ll c$, where c is the velocity of light. For velocities attainable with *macroscopic* bodies they are extremely good approximations; so good that for such bodies no *earthbound* experiment of sufficient sensitivity has been yet performed to detect the error. The situation is different of course for the velocities associated with the high-speed microscopic particles of physics where "relativistic" effects can become important.

18.03 Galileian Relativity and Electromagnetic Theory. In electromagnetism some of the effects of relative motion are of first-order importance, so it was natural to incorporate into electromagnetic theory, as it developed, the principle of Galileian relativity which had proven so successful in mechanics. However, as will be seen, a conflict arose between the requirements of this simple relativity principle and those of classical electromagnetic theory. Out of the attempts to resolve this conflict came a bold new principle known as Special Relativity. Before pursuing this interesting subject it is in order to recall some of the well-known effects of relative motion in classical electromagnetic theory. To do this we shall examine electromagnetic relations in a particular region from the two frames of reference shown in Fig. 18-1, where the primed system S' moves with a constant velocity u in the x direction with respect to the unprimed system S.

Consider the region between two layers of electric charge parallel to the x-y plane as in Fig. 18-2. The electric charge is stationary in

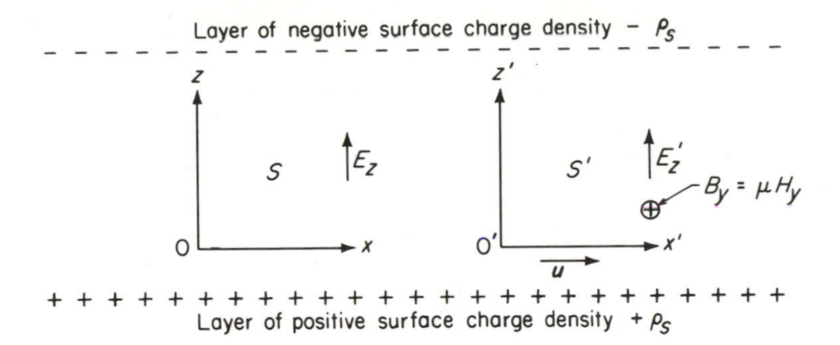

Figure 18-2. Parallel layers of electric charge.

S, so observer O at the origin of co-ordinate system S would measure an electric field given by $E_z = D_z/\epsilon = \rho_s/\epsilon$, where ρ_s is the surface charge density in coulombs per square meter. To observer O', stationary at the origin of S', the layers of charge are gliding by with a velocity $-u$. Hence O' sees two parallel current sheets of surface current density $J'_{sx} = \mp\rho_s u$ amperes per meter, and so would measure a magnetic field $H'_y = -J'_{sx} = \rho_s u$ and a magnetic flux density $B'_y = \mu H'_y = \mu\epsilon E_z u = (u/c^2)E_z$. In addition, of course, observer O' would measure an electric field $E'_z = E_z$ produced by the layers of charge.

Conversely, consider the case (Fig. 18-3) where observer O, stationary in S, detects a magnetic flux density $B_y = \mu H_y$ which might be produced, for example, by equal and opposite currents $\pm J_{sx}$ flowing in a pair of *conducting* planes parallel to the *x-y* axis and stationary in O. (Whether observer O also sees an electric field E_z depends on whether or not there is a voltage V_1 between the conducting planes. In this event $E_z = V_1/d$ and there is a layer of unneutralized charge density of magnitude $\rho_s = \epsilon E_z$ on each of the conducting planes. For the moment assume $E_z = 0$.) Now according to the familiar flux-cutting rules,

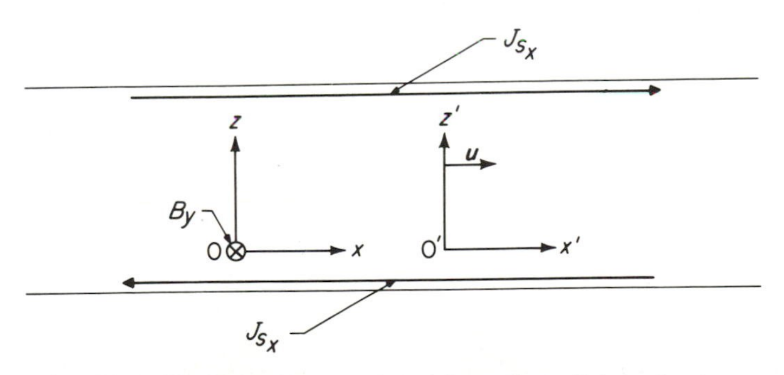

Figure 18-3. Current sheets at the surface of parallel conducting planes.

if we consider a wire parallel to the z axis and fixed in S' (hence moving with a velocity u relative to the current-carrying conducting planes and the associated magnetic field), there will be induced in the wire of length l, a voltage $V_i = luB_y$. Because the wire is stationary with respect to his laboratory, observer O' will infer that this voltage is induced by an electric field $E'_{zi} = uB_y$. In addition he can measure, for example with a rotating loop, a magnetic field $B'_y = B_y$. Of course if there already exists an electric field E_z in S, due to voltage between the conducting planes, O' will detect this field also and will therefore measure a total electric field $E'_z = E_z + E'_{zi} = E_z + uB_y$. Also in this latter case (where there exists an E_z in S) when he comes to measure the magnetic field, O' will detect the presence of an additional small magnetic flux density equal to $(u/c^2)E_z$, and so will measure a total magnetic field $B'_y = B_y + (u/c^2)E_z$. Summarizing, for the fields between parallel planes normal to the z axis,

$$E'_z = E_z + uB_y \qquad B'_y = B_y + \frac{u}{c^2}E_z$$

By following the same reasoning for a pair of planes normal to the y axis it is easily demonstrated that the corresponding relations for this axis will be

$$E'_y = E_y - uB_z \qquad B'_z = B_z - \frac{u}{c^2}E_y$$

For field components in the x direction, that is, parallel to the direction of motion of S' relative to S, there are no transverse field effects so that

$$E'_x = E_x \qquad B'_y = B_y$$

Using superposition, the relations between field components in the primed and unprimed systems are

$$\begin{aligned} E'_x &= E_x \qquad E'_y = E_y - uB_z \qquad E'_z = E_z + uB_y \\ B'_x &= B_x \qquad B'_y = B_y + \frac{u}{c^2}E_z \qquad B'_z = B_z - \frac{u}{c^2}E_y \end{aligned} \tag{18-2}$$

It will be recognized at once that these relations can be combined in vector form as

$$\mathbf{E}' = \mathbf{E} + \mathbf{u} \times \mathbf{B} \tag{18-3}$$

$$\mathbf{B}' = \mathbf{B} - \frac{1}{c^2}\mathbf{u} \times \mathbf{E} \tag{18-4}$$

where $\mathbf{u} = \hat{\mathbf{x}}u_x, \quad u_y = u_z = 0.$

Finally, in the light of eq. (3), consider the force on a charge q moving with a velocity $v = u$ in the co-ordinate system S in which exist an electric field E and a magnetic field B. Because the charge is moving with respect to an observer O in S, he will expect to measure both an electric force and a magnetic force on the charge. From the point of view of an observer riding along in co-ordinate system S' in which q is stationary, the stationary charge will experience only an electric force of value qE'. Of course, the force on q should be the same regardless of the point of view. From eq. (3) it is seen that the force on q will be

$$\mathbf{F} = q\mathbf{E}' = q(\mathbf{E} + \mathbf{v} \times \mathbf{B}) \tag{18-5}$$

which shows that an observer O stationary in frame S would measure an electric force $q\mathbf{E}$ and a magnetic force $q(\mathbf{v} \times \mathbf{B})$ on the moving charge. The force $\mathbf{F}$ in eq. (5) is the well-known *Lorentz force*.

18.04 Transformation of Electric and Magnetic Fields. Relations (2) between electric and magnetic fields in systems having relative motion were developed using the principles of Newtonian mechanics, Galileian or "common-sense" relativity, and classical electromagnetics. According to these same principles the physical results obtained should be independent of whether system S or S' is regarded as the "fixed" system. That is, it is equally valid to regard S' as fixed, and S as moving in the x' direction with a velocity $-u$. Under these circumstances the unprimed fields should be obtainable from the primed fields by a set of relations corresponding exactly to eqs. (2) but with primed and unprimed quantities interchanged, and u replaced by $-u$. Hence we should expect to find that eqs. (2) transform to

$$\begin{aligned} E_x &= E'_x & E_y &= E'_y + uB'_z & E_z &= E'_z - uB'_y \\ B_x &= B'_x & B_y &= B'_y - \frac{u}{c^2}E'_z & B_z &= B'_z + \frac{u}{c^2}E'_y \end{aligned} \tag{18-6}$$

Now the relations between unprimed and primed quantities are quite easily obtained by solving for the former in eqs. (2). When this is done it is discovered that instead of eqs. (6) the following set of relations results

$$\begin{aligned} E_x &= E'_x & E_y &= \beta^2(E'_y + uB'_z) & E_z &= \beta^2(E'_z - uB'_y) \\ B_x &= B'_x & B_y &= \beta^2\left(B'_y - \frac{u}{c^2}E'_z\right) & B_z &= \beta^2\left(B'_z + \frac{u}{c^2}B'_y\right) \end{aligned} \tag{18-7}$$

where $\quad \beta^2 = (1 - u^2/c^2)^{-1}$.

Problem 1. Solve eqs. (2) to obtain eqs. (7).

It is evident that for conditions under which they were obtained eqs. (2) are not self-consistent in that they do not transform properly into a similar set when solved (with primed and unprimed quantities interchanged and u replaced by $-u$). It is interesting to inquire at this point what would constitute a self-consistent set, and this query is left as an exercise for the student to answer.

Problem 2(a). Prove that $\beta^2 u^2/c^2 = \beta^2 - 1$.

Problem 2(b). Verify that a self-consistent set corresponding to (2) would be

$$\begin{aligned} E'_x &= E_x \qquad E'_y = \beta(E_y - uB_z) \qquad E'_z = \beta(E_z + uB_y) \\ B'_x &= B_x \qquad B'_y = \beta\left(B_y + \frac{u}{c^2} E_z\right) \qquad B'_z = \beta\left(B_z - \frac{u}{c^2} E_y\right) \end{aligned} \tag{18-8a}$$

which transforms to

$$\begin{aligned} E_x &= E'_x \qquad E_y = \beta(E'_y + uB'_z) \qquad E_z = \beta(E'_z - uB'_y) \\ B_x &= B'_x \qquad B_y = \beta\left(B'_y - \frac{u}{c^2} E'_z\right) \qquad B_z = \beta\left(B'_z + \frac{u}{c^2} E'_y\right) \end{aligned} \tag{18-8b}$$

where $\quad \beta = (1 - u^2/c^2)^{-1/2}$.

The modifications set forth in eqs. (8), which make the electromagnetic relations self-consistent, were first proposed* by H. A. Lorentz in a study of electromagnetic phenomena in a moving system.

The fact that eqs. (2) are not self-consistent suggests that there might be something wrong with the underlying principles, that is, with Newtonian mechanics and simple Galileian relativity, or classical electromagnetics or both. This inconsistency is symptomatic of a more general basic difficulty that confronted physicists toward the end of the nineteenth century. This general difficulty was pointed up by the uncertainty and controversy associated with the correct formulation of the velocity of light. According to Newtonian-Galileian principles, all inertial frames are equivalent and as far as the laws of mechanics are concerned uniform (unaccelerated) motion is indistinguishable from no motion. If these principles are assumed to apply also to the laws of physics and electromagnetics then it might be concluded that the velocity of light measured by observer O between two stationary points in S (Fig. 18-1) should be the same as that measured by O' between two points stationary in S'. However, in the Maxwellian theory, and assuming light to be an electromagnetic phenomenon, the velocity of light given by $c = 1/\sqrt{\mu\epsilon}$ is characteristic of the medium, and presumably this velocity is measured *relative to the medium*. Under these circumstances, if frame S is assumed to be stationary in the medium, frame S' will be moving through the medium with a velocity u. Then a light wave traveling through the medium in the x direction with a velocity c would have a velocity $(c - u)$ relative to observer O' in S' using

*H. A. Lorentz, "Electromagnetic Phenomena in a System Moving with any Velocity Less than that of Light," *Proc. Acad. of Sciences of Amsterdam*, **6**, 1904.

ordinary Galileian addition of velocities. This result conflicts with the above-mentioned conclusion.

At the root of this conflict was the question of the existence of an "aether." Because light waves can propagate through many kinds of media, including a vacuum, an all-pervasive aetherial substance called the *aether* was hypothesized to exist, and the velocity of light was considered to be relative to this aether. Maxwell himself was a strong advocate of the aether theory and made it "a constant aim to construct a mental representation of all the details of its action."*

18.05 Michelson-Morley Experiment. The question of the existence of an aether, and more importantly the problem of the correct statement for the propagation velocity of light in moving systems, was the subject of many experiments, of which the Michelson-Morley experiment is perhaps the most famous. It was Maxwell who pointed out that under the aether hypothesis the round-trip time for propagation of light between two points stationary with respect to each other, but having a common velocity through the aether, would depend upon this velocity. The Michelson-Morley experiment was designed to use the earth's velocity through space (the aether) to confirm the aether hypothesis.

Using an extremely sensitive interferometer designed by Michelson, the experiment consisted of comparing the round-trip times for light to traverse two separate paths, one parallel and the other perpendicular to the earth's velocity v. In terms of the geometry of Fig. 18-4 and using the arguments of classical physics and Galileian relativity the round-trip time for the path PM_1P which is assumed to be parallel to v is

$$t_1 = l\left(\frac{1}{c - v} + \frac{1}{c + v}\right) = \frac{2l}{c(1 - v^2/c^2)}$$

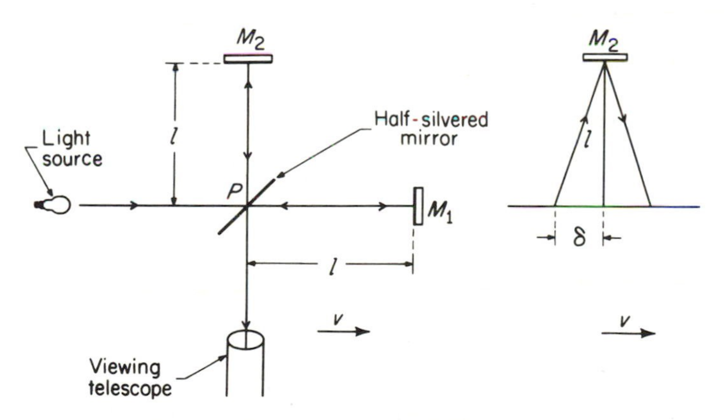

Figure 18-4. Michelson-Morley experiment.

*J. Clerk Maxwell, *Electricity and Magnetism,* Vol. II, Sec. 866

For the path perpendicular to v the distance traversed by the light beam will be $2\sqrt{l^2 + \delta^2}$, where δ is the distance traversed by P while the light travels from P to M_2. Hence $\delta/\sqrt{l^2 + \delta^2} = v/c$. Therefore the round-trip time for path PM_2P will be

$$t_2 = \frac{2\sqrt{l^2 + \delta^2}}{c} = \frac{2l}{c}\frac{1}{\sqrt{1 - v^2/c^2}}$$

This time is shorter by the factor $\sqrt{1 - v^2/c^2}$. By rotating the apparatus to interchange the paths and measuring the shift in the interference pattern a determination of v can be made.

As is well known the results of this experiment were negative. Within the accuracy of the quite sensitive equipment no shift was observed and it was concluded that if there were a nonzero value for v it would have to be much less than the known velocity of the earth in its orbit about the sun. Several attempts followed to explain the results in a manner which would allow retention of the aether as a preferred frame. G. F. Fitzgerald pointed out that if lengths in the direction of motion were contracted by the factor $\sqrt{1 - v^2/c^2}$, this would account for the null results of the experiment. Lorentz extended this idea, showing that with a suitable modification in the measure of time for a system moving through the aether, Maxwell's equations take the same form in the moving system as they do in a system stationary in the aether. From this it follows that the velocity of light (in a vacuum) would be the same in the two systems and hence the Michelson-Morley experiment could not be expected to yield a positive result.

18.06 The Lorentz Transformation. The modifications proposed by Lorentz for the measures of length and time in a system moving with a velocity u in the x direction are given by eqs. (9).

$$x' = \beta(x - ut) \qquad y' = y \qquad z' = z \qquad t' = \beta\left(t - \frac{ux}{c^2}\right) \tag{18-9a}$$

and the transformed set

$$x = \beta(x' + ut') \qquad y = y' \qquad z = z' \qquad t = \beta\left(t' + \frac{ux'}{c^2}\right) \tag{18-9b}$$

where $\qquad \beta = (1 - u^2/c^2)^{-1/2}$

Equations (9) are known as the *Lorentz transformation.*

Problem 3. By solving for x, y, z and t in terms of x', y', z', t' in eqs. (9a), show that the transformed set (9b) results.

That the Lorentz transformation does indeed satisfy the requirement that the velocity of light be the same in two systems having uniform motion relative to one another is easily demonstrated. In Fig. (18-1) let the origins of the primed and unprimed systems coincide at $t = t' = 0$. Assume that at this instant a light pulse is emitted from the coincident origins. An observer in the unprimed reference system will see a spherical wave front diverging from 0 and given by

$$x^2 + y^2 + z^2 - c^2t^2 = 0 \tag{18-10}$$

An observer moving with the primed system should also see a spherical wave front given by

$$x'^2 + y'^2 + z'^2 - c^2t'^2 = 0 \tag{18-11}$$

Because of symmetry we may assume that $y = y', z = z'$ so that eqs. (10) and (11) require that

$$x^2 - c^2t^2 = x'^2 - c^2t'^2 \tag{18-12}$$

Problem 4. Demonstrate that relation (12) will be satisfied under the Lorentz transformation (9).

18.07 Theory of Special Relativity. Subsequent to the null results of the Michelson-Morley experiments (1881, 1887), various hypotheses were advanced and numerous experiments performed to try to resolve the impasse at which classical physics seemed to have arrived. However, in all cases the explanations proposed were contradicted by one or more experiments. This presented a most challenging problem which called forth the efforts of many of the best thinkers of the time. Among these Henri Poincaré extended Lorentz's work to derive a complete set of consistent relations for electromagnetic theory for systems in relative motion. In 1904 he proposed as a basic relativity principle that:*

1. Physical laws, or the mathematical equations describing them, take the same general form in all reference systems which are in uniform rectilinear motion relative to each other.

In 1905, working independently and from a fundamentally different viewpoint, Albert Einstein proposed that:†

2(a). The laws of electrodynamics (including the velocity of propagation of light in a vacuum), as well as the laws of mechanics, are the same in all reference systems having uniform rectilinear motion with respect to one another.

2(b). It is impossible to devise an experiment defining a state of absolute motion, or to determine for any physical phenomenon a preferred reference frame.

*H Poincaré, *Bull. des Sc. Math.* (2) xxviii, p. 302, 1904. See also Sir Edmund Whittaker, *History of Theories of Aether and Electricity: 1900–1929*, Thomas Nelson & Sons, Camden, N. J., 1953.

†A. Einstein, "Zur Elektrodynamik bewegter Korper," *Ann. d. Phys.* **17** p. 891, Sept., 1905.

The statements (1) and (2) are essentially equivalent. However, Einstein arrived at his postulates after a thorough study of the meaning of the measurements of time, length, and "simultaneity." The transformations developed in the course of his analysis were just those of Lorentz, but his theory provided a rationale for their use.

18.08 Einstein's Definition of Simultaneity. Einstein's approach to the new relativity principle was through a fundamental discussion and precise definition of simultaneity. An understanding of this definition (and of the need for it) leads at once to an understanding of Special Relativity. In this section simultaneity will be discussed briefly, but the reader surely can do no better than to read Einstein's own presentation of the principle written for the layman.*

Using the same simple example given in that presentation, let it be supposed that lightning strikes simultaneously at two widely separated points on the tracks of a straight railroad. The question arises as to how an observer should prove that two such widely separated events occurring at A and B are indeed simultaneous. A simple and reasonable method of doing this would be to measure the distance between A and B, and then to locate the observer at a point M exactly midway between A and B. By providing the observer with two mirrors inclined at an angle of 90°, so that he can look towards A and B at the same time, he is able to judge the relative time of events at A and B if these events are indicated, for example, by light flashes. If, stationed at M, he sees the light flashes at the same instant, he judges that the events at A and B occurred simultaneously. This definition of simultaneity *stipulates* that the velocity of light in the direction A to M is the same as the velocity from B to M. Having thus provided a means for determining when two separated events are simultaneous we are led at once to a meaningful definition of "time." For if now two clocks of identical construction (and assumed to run at exactly the same rate) are placed at points A and B, they can be set so the positions of their hands are simultaneously the same (where simultaneity is judged as defined above). Taking clocks in pairs, this same procedure of setting can be followed for any number of identical clocks located at specified positions. Hence the "time" of an event is the reading on that clock (assumed to be identical and set as above) in the immediate vicinity of the event.

Having arrived at a definition of simultaneity and time for separated events, it is now in order to consider the effect of relative motion of observers on their measurements of time. For this purpose suppose a long train to be traveling along the straight track. When the train is

*Albert Einstein, *Relativity, The Special and General Theory,* Henry Holt & Company, New York, 1920.

stationary relative to the track, corresponding to points A, M and B along the track there are points A', M' and B' immediately above on the train. Now let the train be moving relative to the track with a velocity v. At the instant when an observer on the train occupying the position M' is exactly above M let two flashes of light be emitted simultaneously (as judged by observer at M) from points A and B. The question arises as to whether the two flashes appear to be simultaneous to the observer on the train; the answer obviously is no. For if he remained fixed above the point M he would observe the flashes simultaneously, but because he is moving away from A and towards B with a velocity v, he will see the flash from B before that from A. Therefore events which are simultaneous to an observer on the embankment are not simultaneous to an observer moving relative to the embankment, and conversely. Hence two clocks at A and B, set to the same time as determined by an observer at M on the embankment appear to an observer on the moving train to be set to different times. Each observer has his own measure of time for an event depending upon the velocity of his co-ordinate system. It is noted at this point that this conclusion differs from the statement of Galileian relativity [eqs. (1a and b)] that $t' = t$, which says that time is independent of the state of motion of the reference body or co-ordinate system.

In addition to the dependence of time on an observer's velocity, his measure of length is also a function of relative velocity. For suppose an observer on the train is set the task of determining the distance between points A and B on the track. The observer on the embankment can do this quite simply by measuring the distance with a measuring rod or tape. For the observer on the train the task is more involved. He can make his measurement by arranging to have two marks made simultaneously on the train at points directly above points A and B, and then measuring off along the train the distance between these markings. Of course the result depends upon the markings having been made at the same instant of time, but as has been seen already, his judgment of simultaneity differs from that of an observer at M on the embankment. Hence his measurement of the length between points A and B can be expected to differ from that made by an observer on the embankment. It seems reasonable to conclude that the measurement made of the length of a body having motion relative to the observer will depend upon the relative velocity between observer and body, and this proves to be the case.

From the above considerations it is evident that the measures of length and time between two events, made by two observers moving with a velocity v relative to each other, may be expected to depend upon v. With this as a starting point, and hypothesizing that physical laws (and in particular the velocity of light) should be the same for both observers,

Einstein derived the mathematical relationships required to attain the desired result. The resulting transformations proved to be just those already derived by Lorentz for moving systems. However, Lorentz considered one system to be stationary in the aether and ascribed the apparent change in length of a moving body to a real contraction of the body. In Einstein's view absolute motion is meaningless and only relative motion has significance. Observers having relative velocities v will obtain different measures for the length of a body, but there is no actual contraction or expansion of the body as one changes his point of view from that of one observer to the other.

18.09 The Special and General Theories. The Poincaré-Einstein hypotheses (1, 2 on page 709) form the basis of the *Special Theory of Relativity*. The term "Special" or "Restricted" theory is used to differentiate between this theory and the "General Theory" which deals with accelerated motion. An alternative statement of the Special Theory is:

"There is no meaning in absolute velocity."

The corresponding statement for the General Theory is:

"There is no meaning in absolute acceleration."

Surprisingly, the Special Theory is not just a special case of, and deducible from, the General Theory (except from a certain restricted mathematical viewpoint). The General Theory is not required for our present purposes and will not be considered further in this book.

18.10 Transformation Relations for Systems in Relative Motion. Starting with the Special Theory of Relativity as a hypothesis it was demonstrated* by Einstein that the Lorentz transformation for length and time follows automatically. In addition, corresponding transformations can be derived for mass, velocity, force, etc. These transformations are developed in detail in most of the references cited and will only be summarized here.

A. *Contraction of length.* Suppose it is desired to measure the length of a rod lying parallel to the x axis and at rest in S' (Fig. 18-1 on page 701). Observer O' who is at rest with respect to the rod measures a "rest length" $L_0 = L' = x_1' - x_2'$. Observer O, who sees the rod moving with a velocity u can determine the length L (as he sees it) by making a simultaneous observation (at $t_1 = t_2$) of the positions x_1 and x_2 of the ends of the rod. Using the Lorentz transformation (9),

$$L_0 = x_1' - x_2' = \beta(x_1 - ut_1) - \beta(x_2 - ut_2) = \beta(x_1 - x_2)$$

*A. Einstein, "Zur Elektrodynamik bewegter Korper," *Ann. der Phys.* **17,** 1905. This article has been translated by W. Perrett and G. B. Jeffery in *The Principle of Relativity*, Dover Publications, Inc., New York.

Therefore

$$L = (x_1 - x_2) = \frac{1}{\beta}(x_1' - x_2')$$

or

$$L = \sqrt{1 - u^2/c^2}\, L_0 \tag{18-13}$$

Thus to observer O, the rod moving with a velocity u appears to be contracted from its "rest length" by the factor $\sqrt{1 - u^2/c^2}$. Similarly a rod at rest in S (if parallel to the x axis) appears to observer O' to be contracted by the same factor. Since $y' = y$ and $z' = z$, no contraction is observed for rods lying perpendicular to the direction of motion.

In expressing relations between systems having relative uniform rectilinear motion it is the "generalized length," $L_0\sqrt{1 - u^2/c^2}$, that is involved in translating measurements from one reference system to the other. This generalized length is a function of the relative velocity between systems, but as pointed out earlier there is no physical contraction of the rod when we change our mental standard of rest from one system to the other.

B. *Time Dilation.* Let observer O' send two light signals from a fixed point in S' at times t_1' and t_2' to observer O stationary in S. The time interval in O' will be $t' = t_1' - t_2'$. The corresponding time interval as determined by O will be

$$\begin{aligned}\Delta t &= t_2 - t_1 \\ &= \beta\left(t' + \frac{u}{c^2}\, x_1'\right) - \beta\left(t_2' + \frac{u}{c^2}\, x_2'\right) \\ &= \beta(t_1' - t_2') = \beta\Delta t'\end{aligned} \tag{18-14}$$

since $\quad x_1' = x_2'$.

Thus observer O measures a time interval Δt that is greater than $\Delta t'$ by the factor $1/\sqrt{1 - u^2/c^2}$.

C. *Transformation of Velocity.* If a body moves with a velocity v' relative to O' it can be shown that the velocity as determined by O will be

$$v_x = \frac{v_x' + u}{1 + uv_x'/c^2} \qquad v_y = \frac{v_y'}{\beta(1 + uv_x'/c^2)} \qquad v_z = \frac{v_z'}{\beta(1 + uv_x'/c^2)} \tag{18-15}$$

D. *Variation of Mass with Velocity.* If m_0 is the "rest mass" of a body as measured by an observer relative to whom the mass is stationary, its "apparent mass" when moving with velocity v relative to an observer is

$$m = \frac{m_0}{\sqrt{1 - v^2/c^2}} \tag{18-16}$$

E. *Transformation of Force.* If a particle moving with a velocity v' in S' experiences a force F' the relation $F = d(mv)/dt$ can be used to calculate the components of force as determined by O in S. These components are

$$
\begin{aligned}
F_x &= F'_x + \frac{v'_y u}{c^2 + uv'_x} F'_y + \frac{v'_z u}{c^2 + uv'_x} F'_z \\
F_y &= \frac{c^2}{\beta(c^2 + uv'_x)} F'_y \\
F_z &= \frac{c^2}{\beta(c^2 + uv'_x)} F'_z
\end{aligned}
\tag{18-17}
$$

The corresponding transformed relations are

$$
\begin{aligned}
F'_x &= F_x - \frac{v_y u}{c^2 - uv_x} F_y - \frac{v_z u}{c^2 - uv_x} F_z \\
F'_y &= \frac{c^2}{\beta(c^2 - uv_x)} F_y \\
F'_z &= \frac{c^2}{\beta(c^2 - uv_x)} F_z
\end{aligned}
\tag{18-18}
$$

where v_x, v_y, v_z are the velocity components referred to S.

18.11 Derivation of Electromagnetic Relations from Theory of Special Relativity. As was stated earlier, the purpose of this chapter is to demonstrate that if we assume the validity of Coulomb's law and the viewpoint of Special Relativity as expressed through the Lorentz transformation, many of the relations of classical electromagnetic theory, including the experimental laws of Biot-Savart, Ampere, and Faraday, as well as Maxwell's equations, can be shown to follow.* To illustrate how "magnetic" forces on moving charges can be derived from "electrostatic" forces (in a different reference frame) and the appropriate Lorentz transformations, consider the following elementary, and rather special, example shown in Fig. 18-5.

An infinitely long line-charge ρ_L coulombs per meter moving in the x direction with a velocity u constitutes a current $I = \rho_L u$ in the laboratory frame S. A test-charge q_t, stationary in this frame at the position shown [Fig. 18-5(a)] experiences an electric force given by

$$F_0 = q_t E_z = \frac{q_t D_z}{\epsilon} = \frac{q_t \rho_L}{2\pi\epsilon z}$$

It is desired to determine the additional velocity-dependent force on q_t when the test-charge has the velocity v_x, where for simplicity v_x is made equal to u in this example [Fig. 18-5(b)]. This determination is readily made by considering the same problem in the reference frame S', where

*This elementary treatment is based on *uniform* relative motion. When acceleration of charges is being considered, at least one other postulate (that of "causality") is required.

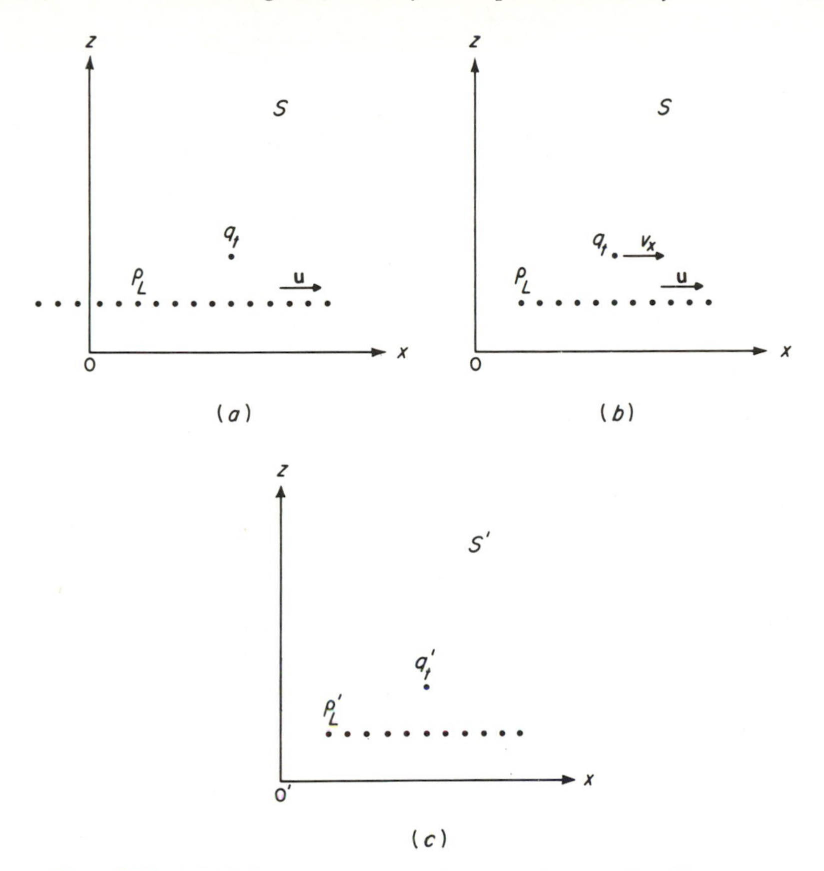

Figure 18-5. A test charge q_t experiences a force due to a line charge ρ_L.

S' has the x directed velocity u with respect to S. In S' [Fig. 18-5(c)] both the line-charge and test-charge are stationary, so the force on q_t is given by

$$F' = \frac{q'_t \rho'_L}{2\pi\epsilon z'}$$

Now the assumption is made (and found to be justified by consistent experimental results) that the magnitude of charge Q, and the total displacement $\Psi = Q$, are *invariant* under the Lorentz transformation.* Therefore we may put $q'_t = q_t$, and of course the transverse length $z' = z$. However, a length x' in S' contracts to $x = x'\sqrt{1 - u^2/c^2} = x'/\beta$ in S, so the line-charge density ρ_L is greater than ρ'_L by the same factor, β. That is, $\rho_L = \beta\rho'_L$ or $\rho'_L = \rho_L/\beta$. Then we have

*For a discussion of this point, see Edward M. Purcell, *Electricity and Magnetism*, Berkeley Physics Course, Vol. 2, p. 153, McGraw-Hill Book Company, New York, 1963.

$$F' = \frac{q_t \rho_L}{\beta 2\pi\epsilon z}$$

This is the force measured in S'. Using the transformation relation in eq. (17) for $v'_x = 0$ shows that the force on the particle as measured by observer O in S would be

$$F_z = \frac{1}{\beta} F'_z = \frac{q_t \rho_L}{\beta^2 2\pi\epsilon z}$$
$$= \frac{F_0}{\beta^2} = F_0\left(1 - \frac{u^2}{c^2}\right) = F_0 - F_m$$

It is observed that for a test-charge moving with velocity $v_x = u$ the force on the test-charge has been modified by an amount

$$F_m = \frac{u^2}{c^2} \frac{q_t \rho_L}{2\pi\epsilon z}$$

which may be written as

$$F_m = \frac{(q_t v_x)(\rho_L u)}{2\pi\epsilon z c^2} = \frac{\mu I(q_t v_x)}{2\pi z}$$

Putting $B = \mu H = \mu I/2\pi z$ shows that the "magnetic" force on a test-charge moving with velocity $v_x = u$ should be $F_m = q_t v_x B$. In the next section it is demonstrated that this relation continues to hold for $v_x \neq u$, and that for an arbitrary particle velocity the expression for force becomes $\mathbf{F}_m = q_t \mathbf{v} \times \mathbf{B}$.

18.12 Coulomb's Law and the Lorentz Force. Coulomb's law, expressing the force between two charges q'_0 and q'_1 in S' may be written as

$$\mathbf{F}' = \frac{q'_0 q'_1 \hat{\mathbf{r}}'}{4\pi\epsilon r'^2} \tag{18-19}$$

where q'_0 is at rest in S', and q'_1 has an arbitrary velocity v' with respect to S'. If the primed system is moving with a velocity u with respect to S, the force as observed by O, at rest in S, will be given by eqs. (17). It is possible* to start with (19) and (17), and derive the Lorentz force, the law of Biot-Savart, and Ampere's law. However, it is also possible to achieve this same end more simply and directly, by considering the force on q'_1 due to the particular distribution of stationary charge consisting of two parallel planes of surface charge density $\pm\rho'_s$, as shown in Fig. 18-2 on page 703, except that now the charge density ρ'_s is stationary in S'. Recalling that the flux density of a point charge q'_0 is given by

$$\mathbf{D}' = \frac{q'_0 \hat{\mathbf{r}}'}{4\pi r'^2}$$

*Leigh Page, "A Derivation of the Fundamental Relations of Electrodynamics from those of Electrostatics," *Amer. J. of Sci.*, **34,** p. 57, 1912. Also see R. S. Elliott, *Electromagnetics*, McGraw-Hill Book Company, New York, 1966.

Equation (19) may be written

$$\mathbf{F}' = \frac{q_1'\mathbf{D}'}{\epsilon} \tag{18-20}$$

Now the displacement density between the parallel layers of charge is $\mathbf{D}' = \hat{\mathbf{z}}D_z'$ where $D_z' = \rho_s'$. Then $F_z' = q_1'\rho_s'/\epsilon$. From eqs. (18) it is found that in the general case the force observed by O will be

$$\begin{aligned} F_x &= F_x' + \frac{\beta u v_y}{c^2} F_y' + \frac{\beta u v_z}{c^2} F_z' \\ F_y &= \frac{\beta}{c^2}(c^2 - uv_x)F_y' \\ F_z &= \frac{\beta}{c^2}(c^2 - uv_x)F_z' \end{aligned} \tag{18-21}$$

For the charge distribution of (revised) Fig. 18-2 the force on q_1' as observed by O reduces to

$$\begin{aligned} F_x &= \frac{\beta u v_z}{c^2} F_z' = \frac{\beta q_1' u v_z D_z'}{\epsilon c^2} \\ F_y &= 0 \\ F_z &= \frac{\beta q_1'(c^2 - uv_x)D_z'}{\epsilon c^2} \end{aligned} \tag{18-22}$$

In eqs. (22) all quantities are referred to the unprimed system S except the charge magnitude q_1' and the displacement density D'. Using invariance of charge, we may write directly $q_1' = q_1$. For the surface charge density $\rho_s' = \Delta Q'/\Delta x'\,\Delta y'$ it is noted that $\Delta Q = \Delta Q'$, $\Delta y = \Delta y'$, but $\Delta x = \Delta x'/\beta = \Delta x\sqrt{1 - u^2/c^2}$, so that

$$\rho_s = \frac{\Delta Q}{\Delta x\,\Delta y} = \beta\rho_s'$$

and consequently

$$D_z = \beta D_z' \tag{18-23}$$

Then relations (22) become

$$\begin{aligned} F_x &= q_1 \frac{u}{c^2\epsilon} v_z D_z \\ F_y &= 0 \\ F_z &= q_1\left(\frac{D_z}{\epsilon} - \frac{uv_x}{c^2\epsilon} D_z\right) \end{aligned} \tag{18-24}$$

For a geometry similar to that of Fig. 18-2 but with the two layers of charge density parallel to the x-z plane the corresponding relations would be found to be

$$F_x = q_1 \frac{u}{c^2 \epsilon} v_y D_y$$

$$F_y = q_1\left(\frac{D_y}{\epsilon} - \frac{u v_x}{c^2 \epsilon} D_y\right) \tag{18-25}$$

$$F_z = 0$$

and for two layers of charge density parallel to the y-z plane

$$F_x = q_1 \frac{D_x}{\epsilon}$$

$$F_y = F_z = 0 \tag{18-26}$$

All of these cases may be combined in the vector statement

$$\mathbf{F} = q_1\left(\frac{\mathbf{D}}{\epsilon} + \frac{\mathbf{v}}{c^2 \epsilon} \times \mathbf{v}_0 \times \mathbf{D}\right) \tag{18-27}$$

where $\mathbf{v}_0 = \hat{\mathbf{x}}u$ is the x-directed velocity of S' with respect to S.

Recall that

$$\mathbf{D} = \epsilon \mathbf{E} \qquad \text{and} \qquad \frac{1}{c^2 \epsilon} = \mu,$$

and *define*

$$\mathbf{B} = \mu \mathbf{H} = \mu \mathbf{v}_0 \times \mathbf{D} \tag{18-28}$$

Then (27) becomes

$$\mathbf{F} = q_1(\mathbf{E} + \mathbf{v} \times \mathbf{B}) \tag{18-29}$$

It is interesting to note that this expression for the *Lorentz force* follows automatically from Coulomb's law, the Lorentz transformation and the definition of B without need for an additional postulate as required in the classical development of electromagnetic theory.

The Lorentz force (29) is a fundamental relation of major significance. Using it as the starting point it is possible to develop a complete theory of electromagnetics and electrodynamics. From the development in the foregoing sections, and the vector expression (27), we see that the magnetic field vectors are definable in terms of the electric field (E, D) and the force on a moving (with respect to the observer) charge in the electric field of other charges which are also in motion with respect to the observer. Thus it should be possible to develop a complete "electromagnetic" theory in terms of forces between moving sets of charges without reference to a magnetic field; however, such a procedure, involving expressions of the form of (27), is much too unwieldy to be useful. There is, of course, no justification for believing the magnetic field to be any less fundamental or important than the electric field. The former expresses the force on a moving charge whereas the latter expresses the force on a stationary charge. But the terms "moving" and "stationary" depend upon one's frame of reference, and what is an electric field to one observer can be a magnetic field to another, and vice versa, as was pointed out in secs. 18.03 and 18.04.

18.13 Biot-Savart Law. The law of Biot-Savart (or Ampere's law for the magnetic field of a current element) can be shown to follow directly from the definition for B and H used in the Lorentz force law (29). The Biot-Savart law expresses the magnetic field in terms of electric *current*, which is related to the flow of charge by the relation

$$\mathbf{J} = \rho \mathbf{v} \qquad \text{amp/m}^2 \tag{18-30}$$

At this point it is desirable to digress briefly to discuss current flow. In eq. (30) charge density ρ coulombs/cubic meter flowing with velocity v meters/second gives rise to a current density $\mathbf{J}$ amperes/square meter. When this current flows as a convection current of charges of a single sign, as for example in an electron beam, both electric and magnetic forces exist on a nearby moving charge q_1 as indicated by the Lorentz force (29). When the velocities of q_1 and of the electron beam are small compared with the velocity of light, as is usually the case, the electric force on q_1 will be much greater than the magnetic force and the latter can often be neglected. However, when the current flows in a *conductor* the negative charge of the electrons is compensated by an equal positive charge (the positive atomic lattice of the metal through which the electrons drift) so that, in the absence of a surface charge on the conductor, the electric force is zero, and there is left only a magnetic force due to the velocities of the charges. With this cancellation of electric force resulting from the coexistence of equal numbers of positive and negative charges, the charge densities can be increased to very high values (for example $\rho_e \approx 10^{29}$ electrons/m^3 $\approx 1.6 \times 10^{10}$ coulombs/m^3) in a conductor. Such high charge densities would be quite impossible in an electron beam owing to the tremendous electrostatic forces that would tend to blow the beam apart. The velocities associated with current flow in a metal are very low (ordinarily less than 1 cm per second) but with the high charge densities possible the resultant *magnetic* forces are far from negligible as evidenced by the fact that they account for the operation of nearly all practical electric motors.

Returning to discussion of the Biot-Savart law, recall the definition of B and H in the Lorentz force relation, viz., $\mathbf{B} = \mu\mathbf{H} = \mu\mathbf{v}_0 \times \mathbf{D}$. Now consider the magnetic field strength H produced by an infinitely long line-charge ρ_L coulombs per meter moving along a conducting wire with a velocity v_0 meters per second. For this case $I = v_0\rho_L$, and $D = \rho_L/2\pi R$ in the radial direction, so that (28) gives

$$B = \mu H = \frac{\mu I}{2\pi R} \tag{18-31}$$

which is the Biot-Savart law expressed for a current in a very long wire.

On integrating (31) around any closed path circling the current there results the usual form of Ampere's work law

$$\oint \mathbf{H} \cdot d\mathbf{s} = I \tag{4-1}$$

The corresponding vector point relation is

$$\nabla \times \mathbf{H} = \mathbf{J} \tag{4-2}$$

18.14 Ampere's Law for a Current Element. When applied to a *current element* $I\,d\mathbf{s} = \mathbf{J}\,dV = \rho\mathbf{v}_0\,dV$, where

$$\mathbf{D} = \frac{dQ\,\hat{\mathbf{r}}}{4\pi R^2} = \frac{\rho\,dV\,\hat{\mathbf{r}}}{4\pi R^2}$$

the definition of B and H results in

$$\mathbf{B} = \mu\mathbf{H} = \mu\mathbf{v}_0 \times \mathbf{D} = \frac{\mu\rho\,dV\,\mathbf{v}_0 \times \hat{\mathbf{r}}}{4\pi R^2}$$

$$= \frac{\mu I\,d\mathbf{s} \times \hat{\mathbf{r}}}{4\pi R^2} \tag{18-32}$$

which is Ampere's law for a current element.

18.15 Ampere's Force Law. The force between two current elements $I\,d\mathbf{s}_1$, and $I\,d\mathbf{s}_2$ follows directly from (29) and (32). Writing

$$I\,d\mathbf{s}_1 = q_1\mathbf{v} \qquad \text{and} \qquad \mathbf{B} = \frac{\mu I\,d\mathbf{s}_2 \times \hat{\mathbf{r}}}{4\pi R^2}$$

there results

$$d\mathbf{F} = \frac{\mu I\,d\mathbf{s}_1 \times (I\,d\mathbf{s}_2 \times \hat{\mathbf{r}})}{4\pi R^2} \tag{18-33}$$

which is Ampere's law of force between current elements.

18.16 Faraday's Law. In sec. 18.03 it was noted that a *uniform* electrostatic field (for which $\partial/\partial x' = 0$) in system S' gives rise to an additional uniform and constant (non-time-varying) magnetic field to observer O in S (and vice versa). A little throught suggests that a non-uniform electrostatic field in S' should appear as a nonuniform *time-varying* magnetic field in S. Indeed, from the transformation relations (9) it is at once evident that a function of x' in S' becomes a function of $\beta(x - ut)$ in S, so that a static distribution in S' becomes a traveling wave in S.

Before deriving the general relationships between fields in S and S' it is instructive to work through a simple particular case for which $f(x') = \sin bx'$. Assume that we have in S' a charge-free region in which exists a two-dimensional electrostatic field having no variation in the y' direction. For an *electrostatic* field, $\nabla' \times \mathbf{E}' = 0$, so $\mathbf{E}$ must be the gradient of a scalar potential and we may write $\mathbf{E}' = -\nabla' V'$. For a charge-free region $\nabla' \cdot \mathbf{E}' = 0$ so that we have

$$\nabla' \cdot (\nabla' V') = \nabla'^2 V' = 0$$

which says simply that the field must satisfy Laplace's equation. The first step, then, must be to determine a two-dimensional field $f(x', z')$ satisfying Laplace's equation, for which the variation in the x' direction is given by $\sin bx'$.

Recalling that for a two-dimensional problem any function of a complex variable $w = u + jv$ satisfies Laplace's equation, let us choose to represent the potential function V' by the imaginary part of $f(w)$ where we let

$$f(w) = \sinh b(z' + jx')$$

so that there results

$$V' = \sin bx' \cosh bz'$$

This result produces the desired sinusoidal variation with x', and from $\mathbf{E}' = -\nabla' V'$ we obtain

$$\begin{aligned} E'_x &= -b \cos bx' \cosh bz' \\ E'_z &= -b \sin bx' \sinh bz' \end{aligned} \tag{18-34}$$

Using the transformation relations (8b and 9a) the fields in S as seen by observer O are

$$\begin{aligned} E_x &= -b \cos b\beta(x - ut) \cosh bz & B_x &= 0 \\ E_y &= 0 & B_y &= \frac{\beta u}{c^2} b \sin b\beta(x - ut) \sinh bz \\ E_z &= -\beta b \sin b\beta(x - ut) \sinh bz & B_z &= 0 \end{aligned} \tag{18-35}$$

The fields appear to observer O in S as traveling waves with a phase velocity equal to u.

Problem 5. Verify that the relations (35) satisfy Faraday's law or Maxwell's equation in the form $\nabla \times \mathbf{E} = -\partial \mathbf{B}/\partial t$.

In the above example it was seen that a particular electrostatic field distribution in S' which satisfied Laplace's equation gave rise to time-varying electric and magnetic fields in S which satisfy Faraday's law. It will now be shown that this follows for the general case. To do this start with the Lorentz transformation (9) and note that

$$\frac{\partial x}{\partial x'} = \beta \qquad \frac{\partial t}{\partial x'} = \frac{\beta u}{c^2} \qquad \frac{\partial t}{\partial t'} = \beta \qquad \frac{\partial x}{\partial t'} = \beta u$$

so that

$$\frac{\partial}{\partial x'} = \frac{\partial}{\partial x}\frac{\partial x}{\partial x'} + \frac{\partial}{\partial t}\frac{\partial t}{\partial x'} = \beta\left(\frac{\partial}{\partial x} + \frac{u}{c^2}\frac{\partial}{\partial t}\right)$$

$$\frac{\partial}{\partial t'} = \frac{\partial}{\partial x}\frac{\partial x}{\partial t'} + \frac{\partial}{\partial t}\frac{\partial t}{\partial t'} = \beta\left(\frac{\partial}{\partial t} + u\frac{\partial}{\partial x}\right) \tag{18-36}$$

$$\frac{\partial}{\partial y'} = \frac{\partial}{\partial y} \qquad \frac{\partial}{\partial z'} = \frac{\partial}{\partial z}$$

Applying these results to

$$\nabla' \times \mathbf{E}' = 0 \tag{18-37}$$

yields

$$\frac{\partial E_z'}{\partial y} - \frac{\partial E_y'}{\partial z} = 0$$

$$\frac{\partial E_x'}{\partial z} - \beta\left(\frac{\partial E_z'}{\partial x} + \frac{u}{c^2}\frac{\partial E_z'}{\partial t}\right) = 0$$

$$\beta\left(\frac{\partial E_y'}{\partial x} + \frac{u}{c^2}\frac{\partial E_y'}{\partial t}\right) - \frac{\partial E_x'}{\partial y} = 0$$

Introducing relations (8a) into the above results in

$$\left[\beta\frac{\partial E_z}{\partial y} + u\beta\frac{\partial B_y}{\partial y} - \beta\frac{\partial E_y}{\partial z} + \beta u\frac{\partial B_z}{\partial z}\right] =$$

$$\left[\beta\left(\frac{\partial E_z}{\partial y} - \frac{\partial E_y}{\partial z}\right) + u\beta\,\nabla\cdot\mathbf{B} - u\beta\frac{\partial B_x}{\partial x}\right] = 0$$

$$\left[\frac{\partial E_x}{\partial z} - \beta^2\frac{\partial E_z}{\partial x} - \beta^2 u\frac{\partial B_y}{\partial x} - \frac{\beta^2 u}{c^2}\frac{\partial E_z}{\partial t} - \frac{\beta^2 u^2}{c^2}\frac{\partial B_y}{\partial t}\right] =$$

$$\left[\frac{\partial E_x}{\partial z} - \frac{\partial E_z}{\partial x} + (1-\beta^2)\frac{\partial E_z}{\partial x} - \beta^2 u\frac{\partial E_y}{\partial x} - \frac{\beta^2 u}{c^2}\frac{\partial E_z}{\partial t} - \frac{\beta^2 u^2}{c^2}\frac{\partial B_y}{\partial t}\right] = 0$$

$$\left[\beta^2\frac{\partial E_y}{\partial x} - \beta^2 u\frac{\partial B_z}{\partial x} + \frac{\beta^2 u}{c^2}\frac{\partial E_y}{\partial t} - \frac{\beta^2 u^2}{c^2}\frac{\partial B_z}{\partial t} - \frac{\partial E_x}{\partial y}\right] =$$

$$\left[\frac{\partial E_y}{\partial x} - \frac{\partial E_x}{\partial y} - (1-\beta^2)\frac{\partial E_y}{\partial x} - \beta^2 u\frac{\partial B_z}{\partial x} + \frac{\beta^2 u}{c^2}\frac{\partial E_y}{\partial t} - \frac{\beta^2 u^2}{c^2}\frac{\partial B_z}{\partial t}\right] = 0$$

Now use the fact that for *static* fields in S'

$$\frac{\partial}{\partial t'} = 0$$

so that from (36) there results

$$\frac{\partial}{\partial t} = -\frac{u\partial}{\partial x} \tag{18-38}$$

and the above equations simplify to

$$\left[\frac{\partial E_z}{\partial y} - \frac{\partial E_y}{\partial z} + \frac{\partial B_x}{\partial t} + u\,\nabla\cdot\mathbf{B}\right] = 0$$

$$\left[\frac{\partial E_x}{\partial z} - \frac{\partial E_z}{\partial x} + \frac{\partial B_y}{\partial t}\right] = 0$$

$$\left[\frac{\partial E_y}{\partial x} - \frac{\partial E_x}{\partial y} + \frac{\partial B_z}{\partial t}\right] = 0$$

Assuming that we may put $\nabla\cdot\mathbf{B} = 0$ (see problem 6 at the end of this section), these equations become

$$\nabla \times \mathbf{E} = -\frac{\partial \mathbf{B}}{\partial t} \tag{18-39}$$

which is the statement of Faraday's law or Maxwell's emf equation in point vector form. Integration over a surface and application of Stokes' theorem leads to the integral form

$$\oint \mathbf{E}\cdot d\mathbf{s} = -\frac{\partial \Phi}{\partial t} \tag{18-40}$$

Problem 6. Using the Lorentz transformation and the fact that $\partial/\partial t' \equiv 0$ for static fields, show that $\nabla\cdot\mathbf{B} = 0$ if $\nabla'\cdot\mathbf{B}' = 0$.

18.17 Maxwell's Assumption and the Generalized Mmf Law. If, instead of an electrostatic field in S', we start with a static, source-free ($J' = 0$) magnetic field in a given region in S' then instead of eq. (37) we have

$$\nabla' \times \mathbf{H}' = 0 \tag{18-41}$$

Carrying through the derivation as in the preceding section one obtains for the fields in S the relation

$$\nabla \times \mathbf{H} = \frac{\partial \mathbf{D}}{\partial t} \tag{18-42}$$

Eq. (42) will be recognized as Maxwell's assumption (p. 102) that a time changing displacement density $\partial D/\partial t$ is equivalent to an electric current density in producing a magnetic field. Combining with Ampere's work law [eq. (3-6), page 84] or the Biot-Savart law already derived, we have Maxwell's first equation

$$\nabla \times \mathbf{H} = \mathbf{J} + \frac{\partial \mathbf{D}}{\partial t} \tag{18-43}$$

or in the integral form

$$\oint \mathbf{H}\cdot d\mathbf{s} = \int_S (\dot{\mathbf{D}} + \mathbf{J})\cdot d\mathbf{a} \tag{18-44}$$

In the preceding sections it has been demonstrated under rather special conditions that electrostatic and magnetostatic fields in one reference frame give rise to or are, in another frame, time-changing fields which satisfy Maxwell's equations. The elementary analysis was purposely restricted to source-free regions in which $\rho = 0$, and $J = 0$. The more general analysis for regions containing charge and current densities requires consideration of how these quantities transform from one reference frame to another.

The above analysis was also restricted in that one started with a very special set of fields, namely those fields which are static in a given reference frame. However, it is easily shown that this restriction is not severe. For the static fields in S' can be transformed to yet another reference frame S'' moving relative to S' with a velocity $u_1 \neq u$. Now if the resultant time-varying fields in S'', satisfying Maxwell's equations, are transformed to S, they should produce the same set of time-varying fields in S as was obtained by direct transformation from S'. Hence a set of time-varying fields satisfying Maxwell's equations in one frame result in a different set of time-varying fields which also satisfy Maxwell's equations in another frame.* The general analysis, including transformation of current and charge densities as well as Maxwell's equations, is required for any consideration of electromagnetic effects in moving media. Because it is not our intent to treat this subject here, the more general analysis has not been included. The interested student is referred to the several references listed on this topic. The treatments of Cullwick and Elliott are particularly suitable.

18.18 Summary. In this chapter the interrelationships between electromagnetic theory and Special Relativity have been explained. We saw that the original conflict between Maxwellian electrodynamics and simple Galileian relativity was resolved by the introduction of the Special Theory of Poincaré and Einstein. Using the Special Theory and the Lorentz transformation the effects of relative motion in electromagnetic theory were examined, and it was seen that stationary charges and the accompanying electrostatic field in one reference frame appeared as currents and a magnetic field in another. By assuming the validity of Coulomb's law and applying the Lorentz transformation to these relatively moving frames, we were led directly to the hypotheses and the seemingly unrelated experimental laws of electromagnetics, including the Lorentz force and the laws of Biot-Savart, Ampere and Faraday, as well as Maxwell's assumption; that is, the foundation was laid for Maxwellian electromagnetic theory.

Finally, it was shown that, in the words of Einstein, "—the electrodynamics of Maxwell-Lorentz in no way opposes the (Special) theory of

*See R. S. Elliott, *Electromagnetics,* Appendix F, McGraw-Hill Book Company, New York, 1966.

relativity. Rather, the latter has been developed from electrodynamics as an astoundingly simple combination and generalization of the hypotheses, formerly independent of each other, on which electrodynamics was built."*

BIBLIOGRAPHY

ARTICLE

Elliott, R. S., "Relativity and Electricity," *IEEE Spectrum*, 3, p. 140, March, 1966.

BOOKS

Bergman, P. G., *An Introduction to the Theory of Relativity*, Prentice-Hall, Inc., Englewood Cliffs, N. J., 1942.

Carter, G. W., *The Electromagnetic Field in its Engineering Aspects*, Longmans, Green and Co., London, 1954.

Cullwick, E. G., *Electromagnetism and Relativity* (2nd ed.), John Wiley & Sons, Inc., New York, 1959.

Einstein, A., H. A. Lorentz, H. Minkowski, and H. Weyl, *The Principle of Relativity*, "A Collection of Original Memoirs of the Special and General Theory of Relativity, together with notes by A. Sommerfeld," translated by W. Perrett and G. B. Jeffery, 1923. (Reprinted by Dover Publications, Inc., New York.)

Elliott, R. S., *Electromagnetics*, McGraw-Hill Book Company, New York, 1966.

Fano, R. M., L. J. Chu, and R. B. Adler, *Electromagnetic Fields, Energy and Forces*, Chapters 9, 10, and Appendix 1, John Wiley & Sons, Inc., New York, 1960.

Panofsky, W. K. H., and M. Phillips, *Classical Electricity and Magnetism*, Addison-Wesley Publishing Company, Inc., Reading, Massachusetts, 1955.

Whittaker, Sir Edmund, *A History of the Theories of Aether and Electricity: Volume I, Classical Theories*; *Volume II, The Modern Theories 1900–1926*, Thomas Nelson & Sons, Camden, N. J., 1951 and 1953.

Whittaker, Sir Edmund, *From Euclid to Eddington*, Cambridge University Press, London, 1949.

*A. Einstein, *Relativity*, p. 49, 1920.

Appendix I

VELOCITIES OF PROPAGATION

Group Velocity and Phase Velocity. In the discussion of wave propagation between parallel planes and in wave guides two different velocities were encountered. The first of these was the phase velocity v, which represented the velocity of propagation of equiphase surfaces along the guide. The second velocity was the group velocity v_g, which in those particular cases, could be considered as the velocity of energy propagation in the direction of the axis of the guide. For wave-guide propagation the phase velocity is always greater than $v_0 = 1/\sqrt{\mu\epsilon}$, whereas the group velocity is always less than v_0. The term group velocity has a more general significance than was indicated in that discussion.

In order to convey intelligence it is always necessary to modulate by some means or other the carrier frequency being transmitted. When this is done, there is a group of frequencies, usually centered about the carrier, that must be propagated along the guide or transmission line. If the phase velocity is a function of frequency, the waves of different frequencies in the group will be transmitted with slightly different velocities. The component waves combine to form a "modulation envelope," which is propagated as a wave having the group velocity v_g defined by

$$v_g = \frac{d\omega}{d\beta}$$

The frequency spread of the group is assumed to be small compared with the mean frequency of the group, and the derivative is evaluated at this mean frequency. The significance of the definition is made clear by consideration of a simple and well-known example.

Consider the case of a carrier $E_0 \cos \omega t$, amplitude-modulated by a modulation frequency $\Delta f = \Delta\omega/2\pi$. Such a signal would be represented by

$$E = E_0(1 + m \cos \Delta\omega t) \cos \omega t$$

where m is the modulation factor. This expression can be expanded in the usual manner to show the presence of the carrier and side-band frequencies.

$$E = E_0 \cos \omega t + \frac{mE_0}{2} [\cos (\omega + \Delta\omega)t + \cos (\omega - \Delta\omega)t]$$

If now such a signal is propagated in the z direction under conditions where the phase velocity varies with frequency, the resultant wave would be written as

$$E = E_0 \cos (\omega t - \beta z) + \frac{mE_0}{2} \{\cos [(\omega + \Delta\omega)t - (\beta + \Delta\beta)z]$$

$$+ [\cos (\omega - \Delta\omega)t - (\beta - \Delta\beta)z]\}$$

This expression can be recombined to show an amplitude-modulated wave progressing in the z direction

$$E = E_0[1 + m \cos (\Delta\omega t - \Delta\beta z)] \cos (\omega t - \beta z)$$

The bracketed part of this expression represents the envelope of the wave. It is seen that the envelope progresses in the z direction with a velocity

$$v_g = \frac{\Delta\omega}{\Delta\beta}$$

If the frequency spread of the group is small enough that $\Delta\omega/\Delta\beta$ may be considered constant throughout the group, this may be written as the limit

$$v_g = \frac{d\omega}{d\beta}$$

To simplify the evaluation of v_g, this may also be written as

$$v_g = \frac{1}{d\beta/d\omega}$$

The phenomenon of phase and group velocities can be illustrated by sketching the addition of the two side-band frequencies waves at a certain instant of time (the carrier frequency is omitted to simplify the discussion and the sketch). Figure A-1 shows two waves of slightly different

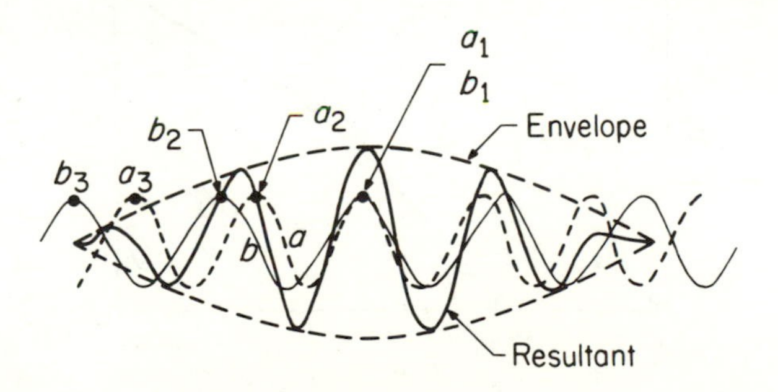

Figure A-1.

frequencies combining to form a single amplitude-modulated wave. If the component waves have the same velocity, the two crests a_1 and b_1 will move along together and the maximum of the modulation envelope will move along with them at the same velocity. Under these circumstances, phase and group velocity are the same. If it is assumed that the lower frequency wave b with the longer wavelength has a velocity slightly greater than that of a, the crests a_1 and b_1 will move apart and the crests a_2 and b_2 will come together. Therefore, at some later instant of time the maximum of the envelope will occur at the point where a_2 and b_2 are coincident, and at a still later instant where a_3 and b_3 are coincident. It is evident that the envelope is slipping backward with respect to the component waves. In other words, it is moving forward with the group velocity v_g, which is less than the phase velocity of either of the component waves. Visually, as, for example, in the case of water waves, it appears as though the envelope were slipping behind the component waves, or, on the other hand, as though the component waves were slipping forward through the envelope. Under those conditions, where the shorter wavelength (high frequency) wave has the greater phase velocity, the situation is reversed and the modulation envelope slips forward. The group velocity is then greater than the phase velocity of the component waves. If β is plotted as a function of ω, the phase velocity and group velocity may be determined directly from the graph. Figure A-2 shows such a plot for wave-guide propagation. It is observed that the slope of $\beta/\omega = 1/\bar{v}$ is always less than that of $d\beta/d\omega = 1/v_g$, so that $\bar{v}$ is always greater than v_g, but both approach v_0 as ω approaches infinity.

Figure A-3 shows a typical plot of β *vs.* ω for a two-conductor transmission line having loss. For this case the slope of $1/\bar{v}$ is always greater than that of $d\beta/d\omega = 1/v_g$, so that the phase velocity is always less than

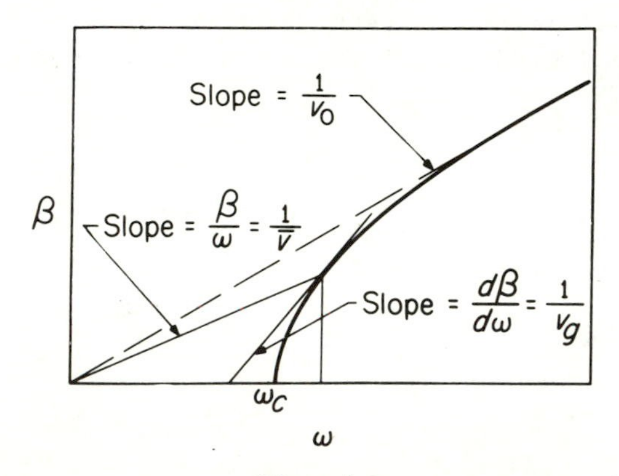

Figure A-2.

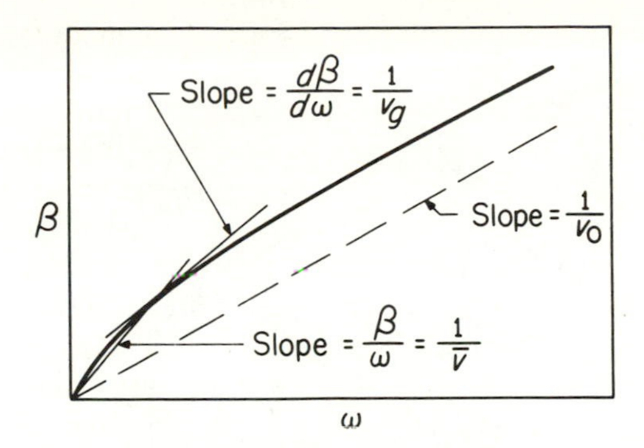

Figure A-3.

the group velocity, but as the frequency is increased both velocities approach the velocity $v_0 = 1/\sqrt{LC}$, which applies in the lossless case.

Signal Velocity. It is to be noted that both phase velocity and group velocity are terms that apply only under *steady-state* conditions. If a signal be impressed suddenly at one end of a transmission line or wave guide, the time required for the disturbance to reach the other end is a measure of what is sometimes called the *signal velocity*. However, it is difficult to state just what is the value of this signal velocity, because the signal at the other end builds up more or less gradually to a steady state as the initial transient condition dies out. The first impulse always reaches the receiving end with a velocity equal to the velocity of light, with other impulses arriving at later times. However, the amplitude of the first impulse is zero and the build-up to the steady-state condition is gradual, so the time required for the signal to reach (and be indicated by) the detector is dependent on the sensitivity of the detector. A thorough discussion of this rather complex phenomenon has been given by Brillouin.*

*L. Brillouin, Congrès international d'électricité, Vol II, Paris, 1932. See also, Leon Brillouin, *Wave Propagation and Group Velocity*. Academic Press Inc., New York, 1960.

Appendix II

BESSEL FUNCTIONS

Bessel and Hankel functions have been discussed briefly in various parts of the text. In working electromagnetic problems the series expansions of the functions, and a knowledge of the differentiation and recurrence formulas are often required. For convenience a few of these series and formulas are assembled together here. For a more detailed treatment of these functions, reference should be made to one of the standard texts on Bessel functions.*

Bessel Functions and Hankel Functions. Bessel functions are solutions of the following differential equation, which is known as Bessel's equation

$$z^2 \frac{d^2 w}{dz^2} + z \frac{dw}{dz} + (z^2 - \nu^2)w = 0 \tag{1}$$

The *order* of the equation is given by the value of ν. In general ν will be nonintegral. For integral values of ν the symbol n is usually used.

For nonintegral values of ν two linearly independent solutions of (1) are

$$\begin{aligned} J_\nu(z) &= \sum_{m=0}^{\infty} \frac{(-1)^m z^{\nu+2m}}{m!\Gamma(m+\nu+1)2^{\nu+2m}} \\ J_{-\nu}(z) &= \sum_{m=0}^{\infty} \frac{(-1)^m z^{-\nu+2m}}{m!\Gamma(m-\nu+1)2^{-\nu+2m}} \end{aligned} \tag{2}$$

where $J_\nu(z)$ is Bessel function of the first kind, of order ν. The function $\Gamma(m + \nu + 1) = \Gamma(p)$ is the generalized factorial function known as the *Gamma function.* It is defined by

$$\Gamma(p) = \int_0^\infty x^{p-1} e^{-x}\, dx$$

When $\nu = n$ (an integer), the Gamma function becomes the factorial

*In addition, an excellent summary treatment of Bessel functions is given by S. A. Schelkunoff, *Electromagnetic Waves,* Chapter III: *also Applied Mathematics for Engineers and Scientists,* Chapter XX. Where differences in definitions and notation exist among various texts, those used by Schelkunoff have been followed here.

$\Gamma(m + n + 1) = (m + n)!$ The two solutions given by (2) are then no longer independent, but instead are related by

$$J_{-n}(z) = (-1)^n J_n(z)$$

where now

$$J_n(z) = \sum_{m=0}^{\infty} \frac{(-1)^m z^{n+2m}}{2^{n+2m} m!(m + n)!} \tag{3}$$

A second independent solution of (1) is defined by

$$N_\nu(z) = \frac{J_\nu(z) \cos \nu\pi - J_{-\nu}(z)}{\sin \nu\pi} \tag{4}$$

where $N_\nu(z)$ is known as a Neumann function or, more commonly, as a Bessel function of *second kind*, of order ν. When ν is integral, $N_n(z)$ continues to be a solution of Bessel's equation, and is still defined* by (4). For integral values of ν, a complete solution of (1) is

$$w = AJ_n(z) + BN_n(z) \tag{5}$$

Bessel functions of the second kind become infinite at $z = 0$, and so cannot be used to represent physical fields except in those problems in which the region $z = 0$ is excluded.

Solutions to eq. (1) may also be written in terms of *Hankel functions*. Hankel functions are linear combinations of Bessel functions defined by

$$\begin{aligned} H_\nu^{(1)}(z) &= J_\nu(z) + jN_\nu(z) \\ H_\nu^{(2)}(z) &= J_\nu(z) - jN_\nu(z) \end{aligned} \tag{6}$$

Like the N functions, Hankel functions become infinite at $z = 0$, and so in physical problems are restricted to cases where $z = 0$ is excluded.

Bessel Functions for Small and Large Arguments. For $z \ll 1$, the J and N functions are given approximately by the expressions

$$J_\nu(z) \approx \frac{z^\nu}{2^\nu \nu!} \qquad N_\nu(z) \approx -\frac{2^\nu(\nu - 1)!}{\pi z^\nu} \tag{7}$$

and in particular

$$J_0(z) \approx 1 \qquad N_0(z) \approx \frac{2}{\pi}(\ln z + C - \ln 2) \tag{8}$$

On the other hand for *large* values of z the asymptotic expressions are

*When ν is an integer, n, $N_n(z)$ as defined by (4) is indeterminate. However, it can be evaluated by usual methods.

$$\left.\begin{aligned}
J_\nu(z) &\approx \sqrt{\frac{2}{\pi z}}\cos\left(z - \frac{\nu\pi}{2} - \frac{\pi}{4}\right)\\
N_\nu(z) &\approx \sqrt{\frac{2}{\pi z}}\sin\left(z - \frac{\nu\pi}{2} - \frac{\pi}{4}\right)\\
H_\nu^{(1)}(z) &\approx \sqrt{\frac{2}{\pi z}}\, e^{j(z-(\nu\pi/2)-(\pi/4))}\\
H_\nu^{(2)}(z) &\approx \sqrt{\frac{2}{\pi z}}\, e^{-j(z-(\nu\pi/2)-(\pi/4))}
\end{aligned}\right\} \tag{9}$$

From these expansions it is apparent that the J and N functions correspond to cosine and sine functions and as such represent standing waves when used with the time factor $e^{j\omega t}$. On the other hand, the $H^{(1)}$ and $H^{(2)}$ functions correspond to exponential functions with imaginary exponents and, when used with the time factor e^{jwt}, represent inward- and outward-traveling cylindrical waves.

Differentiation and Integration of Bessel Functions. Using the series definition for $J_0(z)$, and differentiating term by term, shows that

$$\frac{d}{dz}[J_0(z)] = -J_1(z)$$

Similarly it can be shown that the following relations are true: [In the expressions listed here $Z_\nu(z)$ may denote any of the functions $J_\nu(z)$, $N_\nu(z)$, $H_\nu^{(1)}(z)$, or $H_\nu^{(2)}(z)$; also $Z_\nu'(z)$ means $(d/dz)Z_\nu(z)$].

$$\left.\begin{aligned}
Z_0'(z) &= -Z_1(z)\\
Z_1'(z) &= Z_0(z) - \frac{1}{z}Z_1(z)\\
zZ_\nu'(z) &= \nu Z_\nu(z) - zZ_{\nu+1}(z)\\
&= zZ_{\nu-1}(z) - \nu Z_\nu(z)\\
\frac{d}{dz}[z^\nu Z_\nu(z)] &= z^\nu Z_{\nu-1}(z)\\
\frac{d}{dz}[z^{-\nu} Z_\nu(z)] &= -z^{-\nu} Z_{\nu+1}(z)
\end{aligned}\right\} \tag{10}$$

A useful recurrence formula is

$$2\nu Z_\nu(z) = z[Z_{\nu-1}(z) + Z_{\nu+1}(z)] \tag{11}$$

Some integration formulas are

$$\left.\begin{aligned}
\int Z_1(z)\,dz &= -Z_0(z)\\
\int z^\nu Z_{\nu-1}(z)\,dz &= z^\nu Z_\nu(z)\\
\int z^{-\nu} Z_{\nu+1}(z)\,dz &= -z^{-\nu} Z_\nu(z)
\end{aligned}\right\} \tag{12}$$

Modified Bessel Functions. Modified Bessel functions of zero order were encountered in sec. 14.10. The modified Bessel equation of order ν is

$$z^2 \frac{d^2 w}{dz^2} + z \frac{dw}{dz} - (z^2 + \nu^2)w = 0 \tag{13}$$

For nonintegral values of ν two independent solutions are

$$I_\nu(z) \qquad \text{and} \qquad I_{-\nu}(z) \tag{14}$$

where

$$I_\nu(z) = \sum_{m=0}^{\infty} \frac{z^{\nu+2m}}{2^{\nu+2m} m! \Gamma(\nu + m + 1)} \tag{15}$$

As in the case of the J functions, when ν is integral, these two solutions are related by

$$I_{-n}(z) = I_n(z)$$

and it becomes necessary to seek another solution. Another solution is given by

$$K_\nu(z) = \frac{\pi}{2 \sin \nu\pi} [I_{-\nu}(z) - I_\nu(z)] \tag{16}$$

For integral values of n this reduces to

$$K_n(z) = \frac{2}{\cos n\pi} \left(\frac{\partial I_{-n}}{\partial n} - \frac{\partial I_n}{\partial n} \right) \tag{17}$$

and gives a second independent solution.

For $z \ll 1$, the I and K functions are given approximately by

$$I_\nu(z) \approx \frac{z^\nu}{2^\nu \nu!} \qquad K_\nu(z) \approx \frac{2^{\nu-1}(\nu - 1)!}{z^\nu} \tag{18}$$

and in particular

$$I_0(z) \approx 1 \qquad K_0(z) \approx -[\ln z + C - \ln 2] \tag{19}$$

Series expansions for the I and K functions are given by Schelkunoff.* As noted in sec. 14.10, the I and K functions are appropriate for dissipative media. For propagation in a lossless dielectric, the arguments of these functions would be pure imaginaries, and the functions would reduce to ordinary Bessel functions. For imaginary arguments the relations between modified and ordinary Bessel functions are

$$\left.\begin{aligned} I_\nu(jz) &= e^{j\nu\pi/2} J_\nu(z) \\ K_\nu(jz) &= \frac{\pi}{2} e^{-j(\nu+1)\pi/2} [J_\nu(z) - jN_\nu(z)] \\ &= \frac{\pi}{2} e^{-j(\nu+1)\pi/2} H_\nu^{(2)}(z) \end{aligned}\right\} \tag{20}$$

**Electromagnetic Waves*, pp. 50–51: *Applied Mathematics for Engineers and Scientists*, p. 396.

It is apparent that for imaginary arguments the I functions represent standing waves and the K functions represent outward traveling waves.

Recurrence Formulas for I and K Functions. In general the differentiation and recurrence formulas for the modified Bessel functions are different from those for Bessel and Hankel functions. For the modified functions the recurrence formulas corresponding to those of eqs. (10) are

$$\left.\begin{aligned}
zI_\nu'(z) &= \nu I_\nu(z) + zI_{\nu+1}(z),\\
zK_\nu'(z) &= \nu K_\nu(z) - zK_{\nu+1}(z)\\
zI_\nu'(z) &= zI_{\nu-1} - \nu I_\nu(z),\\
zK_\nu'(z) &= -zK_{\nu-1}(z) - \nu K_\nu(z)\\
\frac{d}{dz}[z^\nu I_\nu(z)] &= z^\nu I_{\nu-1}(z),\\
\frac{d}{dz}[z^\nu K_\nu(z)] &= -z^\nu K_{\nu-1}(z)\\
\frac{d}{dz}[z^{-\nu} I_\nu(z)] &= z^{-\nu} I_{\nu+1}(z),\\
\frac{d}{dz}[z^{-\nu} K_\nu(z)] &= -z^{-\nu} K_{\nu+1}(z)\\
2\nu I_\nu(z) &= z[I_{\nu-1}(z) - I_{\nu+1}(z)]\\
2\nu K_\nu(z) &= -z[K_{\nu-1}(z) - K_{\nu+1}(z)]
\end{aligned}\right\} \qquad (21)$$

LIST OF SYMBOLS

(*See also pages 23–27*)

The following list contains those symbols that have been used consistently throughout the text. The symbols shown on pages 23–27 are not repeated here. Because of the large number of quantities to be represented and the undesirability of using alphabets other than English and Greek, it has been necessary to use some symbols to reperesent different quantities at different times and places. In every instance the symbol has been defined where introduced to avoid misinterpretation of its meaning.

Two common symbols are the tilde (~), used to denote a time-varying quantity in those instances where it must be distinguished from a phasor quantity, and the circumflex (^), used to denote a unit vector.

Symbol	*Quantity*	*Page*
$\mathbf{A}$	Magnetic vector potential	92
A	Effective area	377
C	Euler's constant (.5772157)	332
$\mathrm{Ci}(x)$	Cosine integral	332
$C^d(y)$	Comb function	452
D	Directivity	375
e	Charge on electron or on singly-charged ion (negative for electron)	282
erfc	Error function	645
$\mathbf{F}$	Electric vector potential	468
F_N	Receiver noise figure	419
$\mathscr{F}$	Scalar magnetic potential	468
f_c	Cut-off frequency	186
G	Green's function	39
G_d	Directive gain (*db*)	375
g_d	Directive gain	375
g_p	Power gain	375
$H_\nu^{(1)}(x)$	Hankel function, first kind, order ν	558
$H_\nu^{(2)}(x)$	Hankel function, second kind, order ν	558

Symbol	Quantity	Page
$I_\nu(x)$	Modified Bessel function, first kind, order ν	571
$\hat{I}_\nu(x)$	Modified Bessel function, half-order	571
$\mathbf{J}_s$	Surface current density	174
$J_\nu(x)$	Bessel function, first kind, order ν	571
$\hat{J}_\nu(x)$	Bessel function, half-order	571
K_0	Relative permittivity	298
K	Magnetic current	467
$\overline{\mathbf{K}}$	Relative permittivity dyadic	689
K_ν	Modified Bessel function, second kind, order ν	559
$\hat{K}_\nu$	Modified Bessel function, half-order	571
k	Boltzmann's constant	697
$\hat{\mathbf{L}}$	Unit left-hand rotating vector	460
L'	Generalized inductance	596
$\mathbf{M}_s$	Magnetic surface current density	467
$\mathscr{M}$	Magnetization	303
$\mathbf{m}$	Magnetic dipole moment	303
N	Electron number density	673
$N_\nu(x)$	Bessel function, second kind, order ν	255
$\hat{N}_\nu(x)$	Bessel function, half-order	571
n	Refractive index	658
n'	Group refractive index	696
$\tilde{n}$	Time-varying electron density	687
$\mathbf{P}$	Poynting vector	164
P_n^m	Legendre polynomial	569
P	Polarization ratio	457
$\mathscr{P}$	Polarization	299
$\mathbf{p}$	Electric dipole moment	298
Q	Quality factor	272
Q	Polarization ratio	480
Q_n^m	Associated Legendre polynomial	569
$\hat{\mathbf{R}}$	Unit right-hand rotating vector	460
$R^L(y)$	Gating function, unit amplitude, width L	449
$\mathrm{Si}(x)$	Sine integral	332
$S_1(x)$	Integral related to cosine integral	332
$T_n(x)$	Tchebyscheff polynomial, order n	438
v_0	Phase velocity	239
v_g	Group velocity	192
$\tilde{\mathbf{v}}$	Time-varying particle velocity	687
X	Reactance	154
Z_0	Characteristic impedance	231
Z_s	Surface impedance	654

Symbol	*Quantity*	*Page*
α	Attenuation factor	128
α_e	Electronic polarizability	299
β	Phase factor	128
γ	Propagation constant	154
Γ	Voltage reflection coefficient	332
$\Gamma(x)$	Gamma function	730
δ	Penetration depth	130
$\delta(x)$	Dirac delta	18
η	Intrinsic impedance	128
λ	Wavelength	135
λ_D	Debye length	697
ν	Collision frequency	670
Π	Stream function	573
ρ	Standing-wave ratio	224
$\bar{\boldsymbol{\sigma}}$	Conductivity dyadic	689
σ	Scattering cross section	420
Φ	Radiation intensity	375
χ	Susceptibility	299
χ	Ellipticity angle	462
ω_c	Cyclotron (gyro) frequency	285
ω_p	Plasma frequency	671
Ω	Solid angle	375

INDEX

B

C

D

E

F

I

J

S

W

Y